P9-EKY-897

Fluid Mechanics for Chemical Engineers

Second Edition

with Microfluidics and CFD

FLUID MECHANICS FOR CHEMICAL ENGINEERS

Second Edition

with Microfluidics and CFD

JAMES O. WILKES

Department of Chemical Engineering
The University of Michigan, Ann Arbor, MI

with contributions by

STACY G. BIRMINGHAM: *Non-Newtonian Flow*
Mechanical Engineering Department
Grove City College, PA

BRIAN J. KIRBY: *Microfluidics*
Sibley School of Mechanical and Aerospace Engineering
Cornell University, Ithaca, NY

COMSOL (FEMLAB): *Multiphysics Modeling*
COMSOL, Inc., Burlington, MA

CHI-YANG CHENG: *Computational Fluid Dynamics and FlowLab*
Fluent, Inc., Lebanon, NH

Prentice Hall Professional Technical Reference
Upper Saddle River, NJ • Boston • Indianapolis • San Francisco
New York • Toronto • Montreal • London • Munich • Paris • Madrid
Capetown • Sydney • Tokyo • Singapore • Mexico City

Many of the designations used by manufacturers and sellers to distinguish their products are claimed as trademarks. Where those designations appear in this book, and the publisher was aware of a trademark claim, the designations have been printed with initial capital letters or in all capitals.

The author and publisher have taken care in the preparation of this book, but make no expressed or implied warranty of any kind and assume no responsibility for errors or omissions. No liability is assumed for incidental or consequential damages in connection with or arising out of the use of the information or programs contained herein.

The publisher offers excellent discounts on this book when ordered in quantity for bulk purchases or special sales, which may include electronic versions and/or custom covers and content particular to your business, training goals, marketing focus, and branding interests.

For more information, please contact:

> U.S. Corporate and Government Sales
> (800) 382–3419
> corpsales@pearsontechgroup.com

For sales outside the U.S., please contact:

> International Sales
> international@pearsoned.com

Visit us on the Web: www.phptr.com

Library of Congress Cataloging-in-Publication Data
Wilkes, James O.
  Fluid mechanics for chemical engineers, 2nd ed., with microfluidics
      and CFD/James O. Wilkes.
      p. cm.
  Includes bibliographical references and index.
  ISBN 0–13–148212–2 (alk. paper)
  1. Chemical processes. 2. Fluid dynamics. I. Title.
  TP155.7.W55 2006
  660'.29–dc22   2005017816

ISBN 0–13–148212–2
Text printed in the United States on recycled paper at Courier in Westford, Massachusetts.
First printing, September 2005

CONTENTS

PART II—MICROSCOPIC FLUID MECHANICS

CHAPTER 9—TURBULENT FLOW

CHAPTER 10—BUBBLE MOTION, TWO-PHASE FLOW, AND
            FLUIDIZATION

CHAPTER 11—NON-NEWTONIAN FLUIDS

CHAPTER 12—MICROFLUIDICS AND ELECTROKINETIC
            FLOW EFFECTS

PREFACE

THIS text has evolved from a need for a single volume that embraces a wide range of topics in fluid mechanics. The material consists of two parts—four chapters on *macroscopic* or relatively large-scale phenomena, followed by ten chapters on *microscopic* or relatively small-scale phenomena. Throughout, I have tried to keep in mind topics of industrial importance to the chemical engineer. The scheme is summarized in the following list of chapters.

Part I—Macroscopic Fluid Mechanics

1. Introduction to Fluid Mechanics
2. Mass, Energy, and Momentum Balances
3. Fluid Friction in Pipes
4. Flow in Chemical Engineering Equipment

Part II—Microscopic Fluid Mechanics

5. Differential Equations of Fluid Mechanics
6. Solution of Viscous-Flow Problems
7. Laplace's Equation, Irrotational and Porous-Media Flows
8. Boundary-Layer and Other Nearly Unidirectional Flows
9. Turbulent Flow
10. Bubble Motion, Two-Phase Flow, and Fluidization
11. Non-Newtonian Fluids
12. Microfluidics and Electrokinetic Flow Effects
13. An Introduction to Computational Fluid Dynamics and FlowLab
14. COMSOL (FEMLAB) Multiphysics for Solving Fluid Mechanics Problems

In our experience, an undergraduate fluid mechanics course can be based on Part I plus selected parts of Part II, and a graduate course can be based on much of Part II, supplemented perhaps by additional material on topics such as approximate methods and stability.

Second edition. I have attempted to bring the book up to date by the major addition of Chapters 12, 13, and 14—one on microfluidics and two on CFD (computational fluid dynamics). The choice of software for the CFD presented a difficulty; for various reasons, I selected FlowLab and COMSOL Multiphysics, but there was no intention of "promoting" these in favor of other excellent CFD programs.[1] The use of CFD examples in the classroom really makes the subject

---

[1] The software name "FEMLAB" was changed to "COMSOL Multiphysics" in September 2005, the first release under the new name being COMSOL 3.2.

come "alive," because the previous restrictive necessities of "nice" geometries and constant physical properties, etc., can now be lifted. Chapter 9, on turbulence, has also been extensively rewritten; here again, CFD allows us to venture beyond the usual flow in a pipe or between parallel plates and to investigate further practical situations such as turbulent mixing and recirculating flows.

Example problems.   There is an average of about six completely worked examples in each chapter, including several involving COMSOL (dispersed throughout Part II) and FlowLab (all in Chapter 13). The end of each example is marked by a small, hollow square: □. All the COMSOL examples have been run on a Macintosh G4 computer using FEMLAB 3.1, but have also been checked on a PC; those using a PC or other releases of COMSOL/FEMLAB may encounter slightly different windows than those reproduced here. The format for each COMSOL example is: (a) problem statement, (b) details of COMSOL implementation, and (c) results and discussion (however, item (b) can easily be skipped for those interested only in the results).

The numerous end-of-chapter problems have been classified roughly as easy (E), moderate (M), or difficult/lengthy (D). The University of Cambridge has given permission—kindly endorsed by Professor J.F. Davidson, F.R.S.—for several of their chemical engineering examination problems to be reproduced in original or modified form, and these have been given the additional designation of "(C)".

Acknowledgments.   I gratefully acknowledge the written contributions of my former Michigan colleague Stacy Birmingham (non-Newtonian fluids), Brian Kirby of Cornell University (microfluidics), and Chi-Yang Cheng of Fluent, Inc. (FlowLab). Although I wrote most of the COMSOL examples, I have had great help and cooperation from COMSOL Inc. and the following personnel in particular— Philip Byrne, Bjorn Sjodin, Ed Fontes, Peter Georen, Olof Hernell, Johan Linde, and Rémi Magnard. At Fluent, Inc., Shane Moeykens was instrumental in identifying Chi-Yang Cheng as the person best suited to write the FlowLab chapter. Courtney Esposito and Jordan Schmidt of The MathWorks kindly helped me with MATLAB, needed for the earlier 2.3 version of FEMLAB.

I appreciate the assistance of several other friends and colleagues, including Nitin Anturkar, Stuart Churchill, John Ellis, Kevin Ellwood, Scott Fogler, Leenaporn Jongpaiboonkit, Lisa Keyser, Kartic Khilar, Ronald Larson, Susan Montgomery, Donald Nicklin, the late Margaret Sansom, Michael Solomon, Sandra Swisher, Rasin Tek, Robert Ziff, and my wife Mary Ann Gibson Wilkes. Also very helpful were Bernard Goodwin, Elizabeth Ryan, and Michelle Housley at Prentice Hall PTR, and my many students and friends at the University of Michigan and Chulalongkorn University in Bangkok. Others are acknowledged in specific literature citations.

Further information.   The website `http://www.engin.umich.edu/~fmche` is maintained as a "bulletin board" for giving additional information about the book—hints for problem solutions, *errata*, how to contact the authors, etc.—as proves desirable. My own Internet address is `wilkes@umich.edu`. The text was composed on a Power Macintosh G4 computer using the TEXtures "typesetting" program. Eleven-point type was used for the majority of the text. Most of the figures were constructed using MacDraw Pro, Excel, and KaleidaGraph.

Professor Terence Fox, to whom this book is dedicated, was a Cambridge engineering graduate who worked from 1933 to 1937 at Imperial Chemical Industries Ltd., Billingham, Yorkshire. Returning to Cambridge, he taught engineering from 1937 to 1946 before being selected to lead the Department of Chemical Engineering at the University of Cambridge during its formative years after the end of World War II. As a scholar and a gentleman, Fox was a shy but exceptionally brilliant person who had great insight into what was important and who quickly brought the department to a preeminent position. He succeeded in combining an industrial perspective with intellectual rigor. Fox relinquished the leadership of the department in 1959, after he had secured a permanent new building for it (carefully designed in part by himself).

T.R.C. Fox

Fox was instrumental in bringing Peter Danckwerts, Kenneth Denbigh, John Davidson, and others into the department. He also accepted me in 1956 as a junior faculty member, and I spent four good years in the Cambridge University Department of Chemical Engineering. Danckwerts subsequently wrote an appreciation[2] of Fox's talents, saying, with almost complete accuracy: "Fox instigated no research and published nothing." How times have changed—today, unless he were known personally, his résumé would probably be cast aside and he would stand little chance of being hired, let alone of receiving tenure! However, his lectures, meticulously written handouts, enthusiasm, genius, and friendship were a great inspiration to me, and I have much pleasure in acknowledging his positive impact on my career.

James O. Wilkes
August 18, 2005

---

[2] P.V. Danckwerts, "Chemical engineering comes to Cambridge," *The Cambridge Review*, pp. 53–55, February 28, 1983.

PART I

MACROSCOPIC

FLUID MECHANICS

Some Greek Letters

α	alpha	ν	nu
β	beta	ξ, Ξ	xi
γ, Γ	gamma	o	omicron
δ, Δ	delta	π, ϖ, Π	pi
ϵ, ε	epsilon	ρ, ϱ	rho
ζ	zeta	$\sigma, \varsigma, \Sigma$	sigma
η	eta	τ	tau
$\theta, \vartheta, \Theta$	theta	υ, Υ	upsilon
ι	iota	ϕ, φ, Φ	phi
κ	kappa	χ	chi
λ, Λ	lambda	ψ, Ψ	psi
μ	mu	ω, Ω	omega

Chapter 1

INTRODUCTION TO FLUID MECHANICS

1.1 Fluid Mechanics in Chemical Engineering

A knowledge of fluid mechanics is essential for the chemical engineer because the majority of chemical-processing operations are conducted either partly or totally in the fluid phase. Examples of such operations abound in the biochemical, chemical, energy, fermentation, materials, mining, petroleum, pharmaceuticals, polymer, and waste-processing industries.

There are two principal reasons for placing such an emphasis on fluids. First, at typical operating conditions, an enormous number of materials normally exist as gases or liquids, or can be transformed into such phases. Second, it is usually more efficient and cost-effective to work with fluids in contrast to solids. Even some operations with solids can be conducted in a quasi-fluidlike manner; examples are the fluidized-bed catalytic refining of hydrocarbons, and the long-distance pipelining of coal particles using water as the agitating and transporting medium.

Although there is inevitably a significant amount of theoretical development, almost *all* the material in this book has some application to chemical processing and other important practical situations. Throughout, we shall endeavor to present an understanding of the *physical* behavior involved; only then is it really possible to comprehend the accompanying theory and equations.

1.2 General Concepts of a Fluid

We must begin by responding to the question, "What *is* a fluid?" Broadly speaking, a fluid is a substance that will deform *continuously* when it is subjected to a tangential or *shear* force, much as a similar type of force is exerted when a water-skier skims over the surface of a lake or butter is spread on a slice of bread. The rate at which the fluid deforms continuously depends not only on the magnitude of the applied force but also on a property of the fluid called its *viscosity* or resistance to deformation and flow. Solids will also deform when sheared, but a position of equilibrium is soon reached in which elastic forces induced by the deformation of the solid exactly counterbalance the applied shear force, and further deformation ceases.

3

A simple apparatus for shearing a fluid is shown in Fig. 1.1. The fluid is contained between two concentric cylinders; the outer cylinder is stationary, and the inner one (of radius R) is rotated steadily with an angular velocity ω. This shearing motion of a fluid can continue indefinitely, provided that a source of energy—supplied by means of a torque here—is available for rotating the inner cylinder. The diagram also shows the resulting *velocity profile*; note that the velocity in the direction of rotation varies from the peripheral velocity $R\omega$ of the inner cylinder down to zero at the outer stationary cylinder, these representing typical *no-slip* conditions at both locations. However, if the intervening space is filled with a solid—even one with obvious elasticity, such as rubber—only a limited rotation will be possible before a position of equilibrium is reached, unless, of course, the torque is so high that *slip* occurs between the rubber and the cylinder.

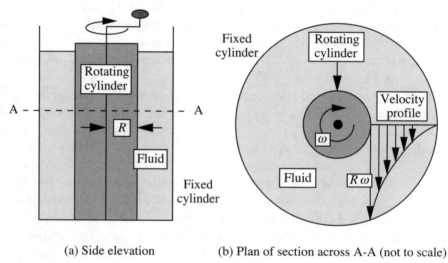

(a) Side elevation   (b) Plan of section across A-A (not to scale)

Fig. 1.1   Shearing of a fluid.

There are various classes of fluids. Those that behave according to nice and obvious simple laws, such as water, oil, and air, are generally called *Newtonian* fluids. These fluids exhibit constant viscosity but, under typical processing conditions, virtually no elasticity. Fortunately, a very large number of fluids of interest to the chemical engineer exhibit Newtonian behavior, which will be assumed throughout the book, except in Chapter 11, which is devoted to the study of non-Newtonian fluids.

A fluid whose viscosity is not constant (but depends, for example, on the intensity to which it is being sheared), or which exhibits significant elasticity, is termed *non-Newtonian*. For example, several polymeric materials subject to deformation can "remember" their recent molecular configurations, and in attempting to recover their recent states, they will exhibit *elasticity* in addition to viscosity. Other fluids, such as drilling mud and toothpaste, behave essentially as solids and will *not* flow when subject to *small* shear forces, but *will* flow readily under the influence of *high* shear forces.

Fluids can also be broadly classified into two main categories—liquids and gases. *Liquids* are characterized by relatively high densities and viscosities, with molecules close together; their volumes tend to remain constant, roughly independent of pressure, temperature, or the size of the vessels containing them. *Gases*, on the other hand, have relatively low densities and viscosities, with molecules far apart; generally, they will rapidly tend to fill the container in which they are placed. However, these two states—liquid and gaseous—represent but the two extreme ends of a continuous *spectrum* of possibilities.

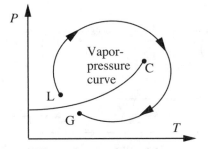

Fig. 1.2   When does a liquid become a gas?

The situation is readily illustrated by considering a fluid that is initially a gas at point G on the pressure/temperature diagram shown in Fig. 1.2. By increasing the pressure, and perhaps lowering the temperature, the vapor-pressure curve is soon reached and crossed, and the fluid condenses and apparently becomes a liquid at point L. By continuously adjusting the pressure and temperature so that the clockwise path is followed, and circumnavigating the critical point C in the process, the fluid is returned to G, where it is presumably once more a gas. But where does the transition from liquid at L to gas at G occur? The answer is at no single point, but rather that the change is a continuous and gradual one, through a whole spectrum of intermediate states.

1.3   Stresses, Pressure, Velocity, and the Basic Laws

Stresses. The concept of a *force* should be readily apparent. In fluid mechanics, a force per unit area, called a *stress*, is usually found to be a more convenient and versatile quantity than the force itself. Further, when considering a specific surface, there are two types of stresses that are particularly important.

1. The first type of stress, shown in Fig. 1.3(a), acts *perpendicularly* to the surface and is therefore called a *normal* stress; it will be tensile or compressive, depending on whether it tends to stretch or to compress the fluid on which it acts. The normal stress equals F/A, where F is the normal force and A is the area of the surface on which it acts. The dotted outlines show the volume changes caused

by deformation. In fluid mechanics, *pressure* is usually the most important type of *compressive* stress, and will shortly be discussed in more detail.

2. The second type of stress, shown in Fig. 1.3(b), acts *tangentially* to the surface; it is called a *shear* stress τ, and equals F/A, where F is the tangential force and A is the area on which it acts. Shear stress is transmitted through a fluid by interaction of the molecules with one another. A knowledge of the shear stress is very important when studying the flow of viscous Newtonian fluids. For a given rate of deformation, measured by the time derivative $d\gamma/dt$ of the small angle of deformation γ, the shear stress τ is directly proportional to the *viscosity* of the fluid (see Fig. 1.3(b)).

Fig. 1.3(a)   Tensile and compressive normal stresses F/A, acting on a cylinder, causing elongation and shrinkage, respectively.

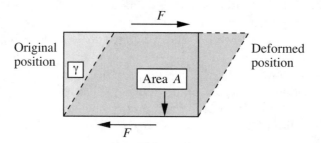

Fig. 1.3(b)   Shear stress $\tau = F/A$, acting on a rectangular parallelepiped, shown in cross section, causing a deformation measured by the angle γ (whose magnitude is exaggerated here).

Pressure.   In virtually all *hydrostatic* situations—those involving fluids at rest—the fluid molecules are in a state of *compression*. For example, for the swimming pool whose cross section is depicted in Fig. 1.4, this compression at a typical point P is caused by the downwards gravitational weight of the water *above* point P. The degree of compression is measured by a scalar, p—the *pressure*.

A small inflated spherical balloon pulled down from the surface and tethered at the bottom by a weight will still retain its spherical shape (apart from a small distortion at the point of the tether), but will be diminished in size, as in Fig. 1.4(a). It is apparent that there must be forces acting *normally inward* on the surface of the balloon, and that these must essentially be uniform for the shape to remain spherical, as in Fig. 1.4(b).

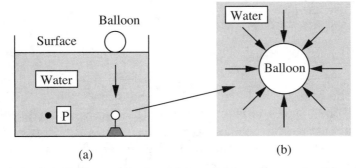

(a)   (b)

Fig. 1.4   (a) Balloon submerged in a swimming pool; (b) enlarged view of the compressed balloon, with pressure forces acting on it.

Although the pressure p is a scalar, it typically appears in tandem with an area A (assumed small enough so that the pressure is uniform over it). By definition of pressure, the surface experiences a normal compressive force $F = pA$. Thus, pressure has units of a force per unit area—the same as a stress.

The value of the pressure at a point is *independent* of the orientation of any area associated with it, as can be deduced with reference to a differentially small wedge-shaped element of the fluid, shown in Fig. 1.5.

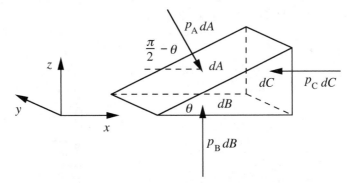

Fig. 1.5   Equilibrium of a wedge of fluid.

Due to the pressure there are three forces, $p_A dA$, $p_B dB$, and $p_C dC$, that act on the three rectangular faces of areas dA, dB, and dC. Since the wedge is not moving, equate the two forces acting on it in the horizontal or x direction, noting that $p_A dA$ must be resolved through an angle $(\pi/2 - \theta)$ by multiplying it by $\cos(\pi/2 - \theta) = \sin\theta$:

$$p_A dA \sin\theta = p_C dC. \tag{1.1}$$

The vertical force $p_B dB$ acting on the bottom surface is omitted from Eqn. (1.1) because it has no component in the x direction. The horizontal pressure forces acting in the y direction on the two triangular faces of the wedge are also omitted, since again these forces have no effect in the x direction. From geometrical considerations, areas dA and dC are related by:

$$dC = dA \sin\theta. \tag{1.2}$$

These last two equations yield:

$$p_A = p_C, \tag{1.3}$$

verifying that the pressure is independent of the orientation of the surface being considered. A force balance in the z direction leads to a similar result, $p_A = p_B$.[1]

For *moving* fluids, the normal stresses include both a pressure *and* extra stresses caused by the motion of the fluid, as discussed in detail in Section 5.6.

The amount by which a certain pressure exceeds that of the atmosphere is termed the *gauge* pressure, the reason being that many common pressure gauges are really *differential* instruments, reading the difference between a required pressure and that of the surrounding atmosphere. *Absolute* pressure equals the gauge pressure plus the atmospheric pressure.

Velocity. Many problems in fluid mechanics deal with the *velocity* of the fluid at a point, equal to the rate of change of the position of a fluid particle with time, thus having both a magnitude and a direction. In some situations, particularly those treated from the *macroscopic* viewpoint, as in Chapters 2, 3, and 4, it sometimes suffices to ignore variations of the velocity with position. In other cases—particularly those treated from the *microscopic* viewpoint, as in Chapter 6 and later—it is invariably essential to consider variations of velocity with position.

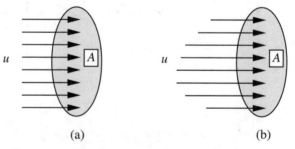

(a)   (b)

Fig. 1.6 Fluid passing through an area A:
(a) Uniform velocity, (b) varying velocity.

Velocity is not only important in its own right, but leads immediately to three *fluxes* or flow rates. Specifically, if u denotes a *uniform* velocity (not varying with position):

---

[1] Actually, a force balance in the z direction demands that the gravitational weight of the wedge be considered, which is proportional to the *volume* of the wedge. However, the pressure forces are proportional to the *areas* of the faces. It can readily be shown that the volume-to-area effect becomes vanishingly small as the wedge becomes infinitesimally small, so that the gravitational weight is inconsequential.

1. If the fluid passes through a plane of area A normal to the direction of the velocity, as shown in Fig. 1.6, the corresponding *volumetric* flow rate of fluid through the plane is $Q = uA$.
2. The corresponding *mass* flow rate is $m = \rho Q = \rho u A$, where ρ is the (constant) fluid density. The alternative notation with an overdot, $\dot{m}$, is also used.
3. When velocity is multiplied by mass it gives *momentum*, a quantity of prime importance in fluid mechanics. The corresponding *momentum* flow rate passing through the area A is $\dot{\mathcal{M}} = mu = \rho u^2 A$.

If u and/or ρ should vary with position, as in Fig. 1.6(b), the corresponding expressions will be seen later to involve *integrals* over the area A: $Q = \int_A u\,dA$, $m = \int_A \rho u\,dA$, $\dot{\mathcal{M}} = \int_A \rho u^2\,dA$.

Basic laws.   In principle, the *laws* of fluid mechanics can be stated simply, and—in the absence of relativistic effects—amount to conservation of mass, energy, and momentum. When applying these laws, the procedure is first to identify a system, its boundary, and its surroundings; and second, to identify how the system interacts with its surroundings. Refer to Fig. 1.7 and let the quantity X represent either mass, energy, or momentum. Also recognize that X may be *added* from the surroundings and transported into the system by an amount X_{in} across the boundary, and may likewise be *removed* or transported out of the system to the surroundings by an amount X_{out}.

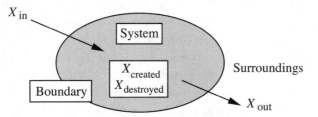

Fig. 1.7   *A system and transports to and from it.*

The general conservation law gives the increase ΔX_{system} in the X-content of the system as:

$$X_{in} - X_{out} = \Delta X_{system}. \tag{1.4a}$$

Although this basic law may appear intuitively obvious, it applies *only* to a very restricted selection of properties X. For example, it is not *generally* true if X is another extensive property such as volume, and is quite meaningless if X is an intensive property such as pressure or temperature.

In certain cases, where X^i is the mass of a definite chemical species i, we may also have an amount of creation $X^i_{created}$ or destruction $X^i_{destroyed}$ due to chemical reaction, in which case the general law becomes:

$$X^i_{in} - X^i_{out} + X^i_{created} - X^i_{destroyed} = \Delta X^i_{system}. \tag{1.4b}$$

The conservation law will be discussed further in Section 2.1, and is of such fundamental importance that in various guises it will find numerous applications throughout all of this text.

To solve a physical problem, the following information concerning the fluid is also usually needed:

1. The physical properties of the fluid involved, as discussed in Section 1.4.
2. For situations involving fluid *flow*, a *constitutive equation* for the fluid, which relates the various stresses to the flow pattern.

1.4 Physical Properties—Density, Viscosity, and Surface Tension

There are three physical properties of fluids that are particularly important: density, viscosity, and surface tension. Each of these will be defined and viewed briefly in terms of molecular concepts, and their dimensions will be examined in terms of mass, length, and time (M, L, and T). The physical properties depend primarily on the particular fluid. For liquids, viscosity also depends strongly on the temperature; for gases, viscosity is *approximately* proportional to the square root of the absolute temperature. The density of gases depends almost directly on the absolute pressure; for most other cases, the effect of pressure on physical properties can be disregarded.

Typical processes often run almost isothermally, and in these cases the effect of temperature can be ignored. Except in certain special cases, such as the flow of a compressible gas (in which the density is not constant) or a liquid under a very high shear rate (in which viscous dissipation can cause significant internal heating), or situations involving exothermic or endothermic reactions, we shall ignore any variation of physical properties with pressure and temperature.

Densities of liquids. Density depends on the mass of an individual molecule and the number of such molecules that occupy a unit of volume. For liquids, density depends primarily on the particular liquid and, to a much smaller extent, on its temperature. Representative densities of liquids are given in Table 1.1.[2] (See Eqns. (1.9)–(1.11) for an explanation of the specific gravity and coefficient of thermal expansion columns.) The accuracy of the values given in Tables 1.1–1.6 is adequate for the calculations needed in this text. However, if highly accurate values are needed, particularly at extreme conditions, then specialized information should be sought elsewhere.

Density.   The *density* ρ of a fluid is defined as its mass per unit volume, and indicates its inertia or resistance to an accelerating force. Thus:

$$\rho = \frac{\text{mass}}{\text{volume}} \ [=] \ \frac{\text{M}}{\text{L}^3}, \tag{1.5}$$

---

[2] The values given in Tables 1.1, 1.3, 1.4, 1.5, and 1.6 are based on information given in J.H. Perry, ed., *Chemical Engineers' Handbook,* 3rd ed., McGraw-Hill, New York, 1950.

in which the notation "[=]" is consistently used to indicate the *dimensions* of a quantity.[3] It is usually understood in Eqn. (1.5) that the volume is chosen so that it is neither so small that it has no chance of containing a representative selection of molecules nor so large that (in the case of gases) changes of pressure cause significant changes of density throughout the volume. A medium characterized by a density is called a *continuum*, and follows the classical laws of mechanics— including Newton's law of motion, as described in this book.

*Table 1.1 Specific Gravities, Densities, and
Thermal Expansion Coefficients of Liquids at 20 °C*

Liquid	Sp. Gr. s	Density, ρ kg/m^3	Density, ρ lb$_m$/ft^3	α °C^{-1}
Acetone	0.792	792	49.4	0.00149
Benzene	0.879	879	54.9	0.00124
Crude oil, 35°API	0.851	851	53.1	0.00074
Ethanol	0.789	789	49.3	0.00112
Glycerol	1.26 (50 °C)	1,260	78.7	—
Kerosene	0.819	819	51.1	0.00093
Mercury	13.55	13,550	845.9	0.000182
Methanol	0.792	792	49.4	0.00120
n-Octane	0.703	703	43.9	—
n-Pentane	0.630	630	39.3	0.00161
Water	0.998	998	62.3	0.000207

Degrees A.P.I. (American Petroleum Institute) are related to specific gravity s by the formula:

$$°\text{A.P.I.} = \frac{141.5}{s} - 131.5. \tag{1.6}$$

Note that for water, °A.P.I. = 10, with correspondingly *higher* values for liquids that are *less* dense. Thus, for the crude oil listed in Table 1.1, Eqn. (1.6) indeed gives $141.5/0.851 - 131.5 \doteq 35\,°\text{A.P.I.}$

Densities of gases. For *ideal* gases, $pV = nRT$, where p is the *absolute* pressure, V is the volume of the gas, n is the number of moles (abbreviated as "mol" when used as a unit), R is the gas constant, and T is the *absolute* temperature. If M_w is the molecular weight of the gas, it follows that:

$$\rho = \frac{nM_w}{V} = \frac{M_w p}{RT}. \tag{1.7}$$

---

[3] An early appearance of the notation "[=]" is in R.B. Bird, W.E. Stewart, and E.N. Lightfoot, *Transport Phenomena*, Wiley, New York, 1960.

Thus, the density of an ideal gas depends on the molecular weight, absolute pressure, and absolute temperature. Values of the gas constant R are given in Table 1.2 for various systems of units. Note that degrees Kelvin, formerly represented by "°K," is now more simply denoted as "K."

Table 1.2   Values of the Gas Constant, R

Value	Units
8.314	J/g-mol K
0.08314	liter bar/g-mol K
0.08206	liter atm/g-mol K
1.987	cal/g-mol K
10.73	psia ft^3/lb-mol °R
0.7302	ft^3 atm/lb-mol °R
1,545	ft lb$_f$/lb-mol °R

For a *nonideal* gas, the compressibility factor Z (a function of p and T) is introduced into the denominator of Eqn. (1.7), giving:

$$\rho = \frac{nM_{\mathrm{w}}}{V} = \frac{M_{\mathrm{w}}p}{ZRT}. \tag{1.8}$$

Thus, the extent to which Z deviates from unity gives a measure of the nonideality of the gas.

The *isothermal compressibility* of a gas is defined as:

$$\beta = -\frac{1}{V}\left(\frac{\partial V}{\partial p}\right)_T,$$

and equals—at constant temperature—the fractional *decrease* in volume caused by a unit *increase* in the pressure. For an ideal gas, $\beta = 1/p$, the reciprocal of the absolute pressure.

The *coefficient of thermal expansion* α of a material is its isobaric (constant pressure) fractional increase in volume per unit rise in temperature:

$$\alpha = \frac{1}{V}\left(\frac{\partial V}{\partial T}\right)_p. \tag{1.9}$$

Since, for a given mass, density is inversely proportional to volume, it follows that for moderate temperature ranges (over which α is essentially constant) the density of most liquids is approximately a linear function of temperature:

$$\rho \doteq \rho_0[1 - \alpha(T - T_0)], \tag{1.10}$$

where ρ_0 is the density at a reference temperature T_0. For an ideal gas, $\alpha = 1/T$, the reciprocal of the absolute temperature.

The *specific gravity s* of a fluid is the ratio of the density ρ to the density ρ_{SC} of a reference fluid at some *standard condition*:

$$s = \frac{\rho}{\rho_{SC}}. \tag{1.11}$$

For *liquids*, ρ_{SC} is usually the density of water at $4\,°C$, which equals 1.000 g/ml or 1,000 kg/m^3. For *gases*, ρ_{SC} is sometimes taken as the density of air at $60\,°F$ and 14.7 psia, which is approximately 0.0759 lb$_m$/ft^3, and sometimes at $0\,°C$ and one atmosphere absolute; since there is no *single* standard for gases, care must obviously be taken when interpreting published values. For *natural gas*, consisting primarily of methane and other hydrocarbons, the *gas gravity* is defined as the ratio of the molecular weight of the gas to that of air (28.8 lb$_m$/lb-mol).

Values of the molecular weight M_w are listed in Table 1.3 for several commonly occurring gases, together with their densities at standard conditions of atmospheric pressure and $0\,°C$.

Table 1.3   Gas Molecular Weights and Densities
(the Latter at Atmospheric Pressure and $0\,°C$)

Gas	M_w	Standard Density kg/m^3	lb$_m$/ft^3
Air	28.8	1.29	0.0802
Carbon dioxide	44.0	1.96	0.1225
Ethylene	28.0	1.25	0.0780
Hydrogen	2.0	0.089	0.0056
Methane	16.0	0.714	0.0446
Nitrogen	28.0	1.25	0.0780
Oxygen	32.0	1.43	0.0891

Viscosity.   The viscosity of a fluid measures its resistance to flow under an applied shear stress, as shown in Fig. 1.8(a). There, the fluid is ideally supposed to be confined in a relatively small gap of thickness h between one plate that is stationary and another plate that is moving steadily at a velocity V relative to the first plate.

In practice, the situation would essentially be realized by a fluid occupying the space between two concentric cylinders of large radii rotating relative to each other, as in Fig. 1.1. A steady force F to the right is applied to the upper plate (and, to preserve equilibrium, to the left on the lower plate) in order to maintain a constant motion and to overcome the viscous friction caused by layers of molecules sliding over one another.

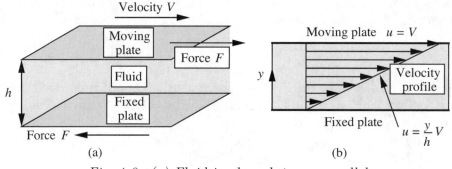

(a)   (b)

Fig. 1.8   (a) Fluid in shear between parallel plates; (b) the ensuing linear velocity profile.

Under these circumstances, the velocity u of the fluid to the right is found experimentally to vary linearly from zero at the lower plate ($y = 0$) to V itself at the upper plate, as in Fig. 1.8(b), corresponding to no-slip conditions at each plate. At any intermediate distance y from the lower plate, the velocity is simply:

$$u = \frac{y}{h}V. \tag{1.12}$$

Recall that the *shear stress* τ is the tangential applied force F per unit area:

$$\tau = \frac{F}{A}, \tag{1.13}$$

in which A is the area of each plate. Experimentally, for a large class of materials, called *Newtonian* fluids, the shear stress is directly proportional to the velocity *gradient*:

$$\tau = \mu\frac{du}{dy} = \mu\frac{V}{h}. \tag{1.14}$$

The proportionality constant μ is called the *viscosity* of the fluid; its dimensions can be found by substituting those for F (ML/T^2), A (L^2), and du/dy (T^{-1}), giving:

$$\mu \ [=] \ \frac{M}{LT}. \tag{1.15}$$

Representative units for viscosity are g/cm s (also known as *poise*, designated by P), kg/m s, and lb$_m$/ft hr. The *centipoise* (cP), one hundredth of a poise, is also a convenient unit, since the viscosity of water at room temperature is approximately 0.01 P or 1.0 cP. Table 1.11 gives viscosity conversion factors.

The viscosity of a fluid may be determined by observing the pressure drop when it flows at a known rate in a tube, as analyzed in Section 3.2. More sophisticated methods for determining the *rheological* or flow properties of fluids—including viscosity—are also discussed in Chapter 11; such methods often involve containing the fluid in a small gap between two surfaces, moving one of the surfaces, and measuring the force needed to maintain the other surface stationary.

Table 1.4   Viscosity Parameters for Liquids

Liquid	a	b	a	b
	(T in K)		(T in °R)	
Acetone	14.64	−2.77	16.29	−2.77
Benzene	21.99	−3.95	24.34	−3.95
Crude oil, 35° API	53.73	−9.01	59.09	−9.01
Ethanol	31.63	−5.53	34.93	−5.53
Glycerol	106.76	−17.60	117.22	−17.60
Kerosene	33.41	−5.72	36.82	−5.72
Methanol	22.18	−3.99	24.56	−3.99
Octane	17.86	−3.25	19.80	−3.25
Pentane	13.46	−2.62	15.02	−2.62
Water	29.76	−5.24	32.88	−5.24

The *kinematic viscosity* ν is the ratio of the viscosity to the density:

$$\nu = \frac{\mu}{\rho}, \tag{1.16}$$

and is important in cases in which significant viscous *and* gravitational forces coexist. The reader can check that the dimensions of ν are L^2/T, which are identical to those for the diffusion coefficient $\mathcal{D}$ in mass transfer and for the thermal diffusivity $\alpha = k/\rho c_p$ in heat transfer. There is a definite analogy among the three quantities—indeed, as seen later, the value of the kinematic viscosity governs the rate of "diffusion" of momentum in the laminar and turbulent flow of fluids.

Viscosities of liquids. The viscosities μ of liquids generally vary approximately with absolute temperature T according to:

$$\ln \mu \doteq a + b \ln T \quad \text{or} \quad \mu \doteq e^{a + b \ln T}, \tag{1.17}$$

and—to a good approximation—are independent of pressure. Assuming that μ is measured in centipoise and that T is either in degrees Kelvin or Rankine, appropriate parameters a and b are given in Table 1.4 for several representative liquids. The resulting values for viscosity are approximate, suitable for a first design only.

Viscosities of gases. The viscosity μ of many gases is approximated by the formula:

$$\mu \doteq \mu_0 \left(\frac{T}{T_0} \right)^n, \qquad (1.18)$$

in which T is the absolute temperature (Kelvin or Rankine), μ_0 is the viscosity at an absolute reference temperature T_0, and n is an empirical exponent that best fits the experimental data. The values of the parameters μ_0 and n for atmospheric pressure are given in Table 1.5; recall that to a *first* approximation, the viscosity of a gas is independent of pressure. The values μ_0 are given in centipoise and correspond to a reference temperature of $T_0 \doteq 273$ K $\doteq 492\,°$R.

Table 1.5   Viscosity Parameters for Gases

Gas	μ_0, cP	n
Air	0.0171	0.768
Carbon dioxide	0.0137	0.935
Ethylene	0.0096	0.812
Hydrogen	0.0084	0.695
Methane	0.0120	0.873
Nitrogen	0.0166	0.756
Oxygen	0.0187	0.814

Surface tension.[4] Surface tension is the tendency of the surface of a liquid to behave like a stretched elastic membrane. There is a natural tendency for liquids to minimize their surface area. The obvious case is that of a liquid droplet on a horizontal surface that is not wetted by the liquid—mercury on glass, or water on a surface that also has a thin oil film on it. For small droplets, such as those on the left of Fig. 1.9, the droplet adopts a shape that is almost perfectly spherical, because in this configuration there is the least surface area for a given volume.

Fig. 1.9   The larger droplets are flatter because gravity is becoming more important than surface tension.

---

[4] We recommend that this subsection be omitted at a first reading, because the concept of surface tension is somewhat involved and is relevant only to a small part of this book.

For larger droplets, the shape becomes somewhat flatter because of the increasingly important gravitational effect, which is roughly proportional to a^3, where a is the approximate droplet radius, whereas the surface area is proportional only to a^2. Thus, the ratio of gravitational to surface tension effects depends roughly on the value of $a^3/a^2 = a$, and is therefore increasingly important for the larger droplets, as shown to the right in Fig. 1.9. Overall, the situation is very similar to that of a water-filled balloon, in which the water accounts for the gravitational effect and the balloon acts like the surface tension.

A fundamental property is the *surface energy*, which is defined with reference to Fig. 1.10(a). A molecule I, situated in the *interior* of the liquid, is attracted equally in all directions by its neighbors. However, a molecule S, situated in the *surface*, experiences a net attractive force into the bulk of the liquid. (The vapor above the surface, being comparatively rarefied, exerts a negligible force on molecule S.) Therefore, work has to be done against such a force in bringing an interior molecule to the surface. Hence, an energy σ, called the *surface energy*, can be attributed to a unit area of the surface.

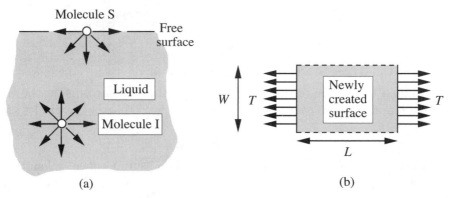

Fig. 1.10   (a) Molecules in the interior and surface of a liquid; (b) newly created surface caused by moving the tension T through a distance L.

An equivalent viewpoint is to consider the surface tension T existing per unit distance of a line drawn in the surface, as shown in Fig. 1.10(b). Suppose that such a tension has moved a distance L, thereby creating an area WL of fresh surface. The work done is the product of the force, TW, and the distance L through which it moves, namely TWL, and this must equal the newly acquired surface energy σWL. Therefore, $T = \sigma$; both quantities have units of force per unit distance, such as N/m, which is equivalent to energy per unit area, such as J/m^2.

We next find the amount $p_1 - p_2$ by which the pressure p_1 inside a liquid droplet of radius r, shown in Fig. 1.11(a), exceeds the pressure p_2 of the surrounding vapor. Fig. 1.11(b) illustrates the equilibrium of the upper hemisphere of the droplet, which is also surrounded by an imaginary cylindrical "control surface" ABCD, on which forces in the vertical direction will soon be equated. Observe that the internal pressure p_1 is trying to blow apart the two hemispheres (the lower one is not shown), whereas the surface tension σ is trying to pull them together.

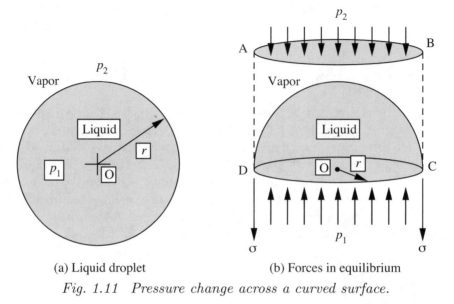

(a) Liquid droplet   (b) Forces in equilibrium

Fig. 1.11 *Pressure change across a curved surface.*

In more detail, there are two different types of forces to be considered:

1. That due to the *pressure difference* between the pressure inside the droplet and the vapor outside, each acting on an area πr^2 (that of the circles CD and AB):

$$(p_1 - p_2)\pi r^2. \qquad (1.19)$$

2. That due to surface tension, which acts on the circumference of length $2\pi r$:

$$2\pi r\sigma. \qquad (1.20)$$

At equilibrium, these two forces are equated, giving:

$$\Delta p = p_1 - p_2 = \frac{2\sigma}{r}. \qquad (1.21)$$

That is, there is a *higher* pressure on the concave or droplet side of the interface. What would the pressure change be for a *bubble* instead of a droplet? Why?

More generally, if an interface has principal radii of curvature r_1 and r_2, the increase in pressure can be shown to be:

$$p_1 - p_2 = \sigma \left(\frac{1}{r_1} + \frac{1}{r_2} \right). \qquad (1.22)$$

For a sphere of radius r, as in Fig. 1.11, both radii are equal, so that $r_1 = r_2 = r$, and $p_1 - p_2 = 2\sigma/r$. Problem 1.31 involves a situation in which $r_1 \neq r_2$. The radii r_1 and r_2 will have the same sign if the corresponding centers of curvature are on the same side of the interface; if not, they will be of opposite sign. Appendix A contains further information about the curvature of a surface.

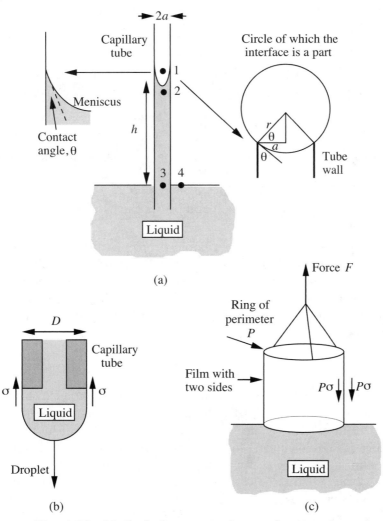

Fig. 1.12   *Methods for measuring surface tension.*

A brief description of simple experiments for measuring the surface tension σ of a liquid, shown in Fig. 1.12, now follows:

(a) In the *capillary-rise* method, a narrow tube of internal radius a is dipped vertically into a pool of liquid, which then rises to a height h inside the tube; if the *contact angle* (the angle between the free surface and the wall) is θ, the meniscus will be approximated by part of the surface of a sphere; from the geometry shown in the enlargement on the right-hand side of Fig. 1.12(a) the radius of the sphere is seen to be $r = a/\cos\theta$. Since the surface is now concave on the air side, the reverse of Eqn. (1.21) occurs, and $p_2 = p_1 - 2\sigma/r$, so that p_2 is *below* atmospheric pressure p_1. Now follow the path 1–2–3–4, and observe that $p_4 = p_3$ because points

3 and 4 are at the same elevation in the same liquid. Thus, the pressure at point 4 is:

$$p_4 = p_1 - \frac{2\sigma}{r} + \rho g h.$$

However, $p_4 = p_1$ since both of these are at atmospheric pressure. Hence, the surface tension is given by the relation:

$$\sigma = \frac{1}{2}\rho g h r = \frac{\rho g h a}{2\cos\theta}. \tag{1.23}$$

In many cases—for complete wetting of the surface—θ is essentially zero and $\cos\theta = 1$. However, for liquids such as mercury in glass, there may be a complete *non*-wetting of the surface, in which case $\theta = \pi$, so that $\cos\theta = -1$; the result is that the liquid level in the capillary is then *depressed* below that in the surrounding pool.

(b) In the *drop-weight* method, a liquid droplet is allowed to form very slowly at the tip of a capillary tube of outer diameter D. The droplet will eventually grow to a size where its weight just overcomes the surface-tension force $\pi D\sigma$ holding it up. At this stage, it will detach from the tube, and its weight $w = Mg$ can be determined by catching it in a small pan and weighing it. By equating the two forces, the surface tension is then calculated from:

$$\sigma = \frac{w}{\pi D}. \tag{1.24}$$

(c) In the *ring tensiometer,* a thin wire ring, suspended from the arm of a sensitive balance, is dipped into the liquid and gently raised, so that it brings a thin liquid film up with it. The force F needed to support the film is measured by the balance. The downward force exerted on a unit length of the ring by one side of the film is the surface tension; since there are two sides to the film, the total force is $2P\sigma$, where P is the circumference of the ring. The surface tension is therefore determined as:

$$\sigma = \frac{F}{2P}. \tag{1.25}$$

In common with most experimental techniques, all three methods described above require slight modifications to the results expressed in Eqns. (1.23)–(1.25) because of imperfections in the simple theories.

Surface tension generally appears only in situations involving either free surfaces (liquid/gas or liquid/solid boundaries) or interfaces (liquid/liquid boundaries); in the latter case, it is usually called the *interfacial* tension.

Representative values for the surface tensions of liquids at 20 °C, in contact either with air or their vapor (there is usually little difference between the two), are given in Table 1.6.[5]

---

[5] The values for surface tension have been obtained from the *CRC Handbook of Chemistry and Physics,* 48th ed., The Chemical Rubber Co., Cleveland, OH, 1967.

Table 1.6   Surface Tensions

Liquid	σ dynes/cm
Acetone	23.70
Benzene	28.85
Ethanol	22.75
Glycerol	63.40
Mercury	435.5
Methanol	22.61
n-Octane	21.80
Water	72.75

1.5   Units and Systems of Units

Mass, weight, and force.   The *mass M* of an object is a measure of the amount of matter it contains and will be constant, since it depends on the number of constituent molecules and their masses. On the other hand, the *weight w* of the object is the gravitational force on it, and is equal to Mg, where g is the local gravitational acceleration. Mostly, we shall be discussing phenomena occurring at the surface of the earth, where g is approximately 32.174 ft/s^2 = 9.807 m/s^2 = 980.7 cm/s^2. For much of this book, these values are simply taken as 32.2, 9.81, and 981, respectively.

Table 1.7   Representative Units of Force

System	Units of Force	Customary Name
SI	kg m/s^2	newton
CGS	g cm/s^2	dyne
FPS	lb$_m$ ft/s^2	poundal

Newton's second law of motion states that a force F applied to a mass M will give it an acceleration a:

$$F = Ma, \tag{1.26}$$

from which is apparent that force has *dimensions* ML/T^2. Table 1.7 gives the corresponding *units* of force in the SI (meter/kilogram/second), CGS (centimeter/gram/second), and FPS (foot/pound/second) systems.

The poundal is now an archaic unit, hardly ever used. Instead, the *pound force*, lb_f, is much more common in the English system; it is defined as the gravitational force on 1 lb_m, which, if left to fall freely, will do so with an acceleration of 32.2 ft/s^2. Hence:

$$1 \ lb_f = 32.2 \ lb_m \ \frac{ft}{s^2} = 32.2 \ \text{poundals}. \qquad (1.27)$$

Table 1.8   SI Units

Physical Quantity	Name of Unit	Symbol for Unit	Definition of Unit
Basic Units			
Length	meter	m	—
Mass	kilogram	kg	—
Time	second	s	—
Temperature	degree Kelvin	K	—
Supplementary Unit			
Plane angle	radian	rad	—
Derived Units			
Acceleration			m/s^2
Angular velocity			rad/s
Density			kg/m^3
Energy	joule	J	$kg \ m^2/s^2$
Force	newton	N	$kg \ m/s^2$
Kinematic viscosity			m^2/s
Power	watt	W	$kg \ m^2/s^3 \ (J/s)$
Pressure	pascal	Pa	$kg/m \ s^2 \ (N/m^2)$
Velocity			m/s
Viscosity			$kg/m \ s$

When using lb_f in the ft, lb_m, s (FPS) system, the following conversion factor, commonly called "g_c," will almost invariably be needed:

$$g_c = 32.2 \ \frac{lb_m \ ft/s^2}{lb_f} = 32.2 \ \frac{lb_m \ ft}{lb_f \ s^2}. \qquad (1.28)$$

Some writers incorporate g_c into their equations, but this approach may be confusing since it virtually implies that one particular set of units is being used, and hence tends to rob the equations of their generality. Why not, for example, also incorporate the conversion factor of 144 in^2/ft^2 into equations where pressure is expressed in lb$_f$/in^2? We prefer to omit all conversion factors in equations, and introduce them only as needed in evaluating expressions numerically. If the reader is in any doubt, units should *always* be checked when performing calculations.

SI Units. The most systematically developed and universally accepted set of units occurs in the *SI* units or *Système International d'Unités*[6]; the subset we mainly need is shown in Table 1.8.

The basic units are again the meter, kilogram, and second (m, kg, and s); from these, certain *derived* units can also be obtained. Force (kg m/s^2) has already been discussed; energy is the product of force and length; power amounts to energy per unit time; surface tension is energy per unit area or force per unit length, and so on. Some of the units have names, and these, together with their abbreviations, are also given in Table 1.8.

Table 1.9   Auxiliary Units Allowed in Conjunction with SI Units

Physical Quantity	Name of Unit	Symbol for Unit	Definition of Unit
Area	hectare	ha	10^4 m^2
Kinematic viscosity	stokes	St	10^{-4} m^2/s
Length	micron	μm	10^{-6} m
Mass	tonne	t	10^3 kg = Mg
	gram	g	10^{-3} kg = g
Pressure	bar	bar	10^5 N/m^2
Viscosity	poise	P	10^{-1} kg/m s
Volume	liter	l	10^{-3} m^3

Tradition dies hard, and certain other "metric" units are so well established that they may be used as *auxiliary* units; these are shown in Table 1.9. The *gram* is the classic example. Note that the basic SI unit of mass (kg) is even represented in terms of the gram, and has not yet been given a name of its own!

Table 1.10 shows some of the acceptable prefixes that can be used for accommodating both small and large quantities. For example, to avoid an excessive number of decimal places, 0.000001 s is normally better expressed as 1 μs (one microsecond). Note also, for example, that 1 μkg should be written as 1 mg—one prefix being better than two.

---

[6] For an excellent discussion, on which Tables 1.8 and 1.9 are based, see *Metrication in Scientific Journals*, published by The Royal Society, London, 1968.

Table 1.10   Prefixes for Fractions and Multiples

Factor	Name	Symbol	Factor	Name	Symbol
10^{-12}	pico	p	10^{3}	kilo	k
10^{-9}	nano	n	10^{6}	mega	M
10^{-6}	micro	μ	10^{9}	giga	G
10^{-3}	milli	m	10^{12}	tera	T

Some of the more frequently used conversion factors are given in Table 1.11.

Example 1.1—Units Conversion

Part 1. Express 65 mph in (a) ft/s, and (b) m/s.

Solution

The solution is obtained by employing conversion factors taken from Table 1.11:

$$(a) \quad 65 \ \frac{\text{mile}}{\text{hr}} \ \times \ \frac{1}{3{,}600} \ \frac{\text{hr}}{\text{s}} \ \times \ 5{,}280 \ \frac{\text{ft}}{\text{mile}} \ = \ 95.33 \ \frac{\text{ft}}{\text{s}} \ .$$

$$(b) \quad 95.33 \ \frac{\text{ft}}{\text{s}} \ \times \ 0.3048 \ \frac{\text{m}}{\text{ft}} \ = \ 29.06 \ \frac{\text{m}}{\text{s}} \ .$$

Part 2. The density of 35 °API crude oil is 53.1 $\text{lb}_\text{m}/\text{ft}^3$ at 68 °F and its viscosity is 32.8 $\text{lb}_\text{m}/\text{ft}$ hr. What are its density, viscosity, and kinematic viscosity in SI units?

Solution

$$\rho = \ 53.1 \ \frac{\text{lb}_\text{m}}{\text{ft}^3} \ \times \ 0.4536 \ \frac{\text{kg}}{\text{lb}_\text{m}} \ \times \ \frac{1}{0.3048^3} \ \frac{\text{ft}^3}{\text{m}^3} \ = \ 851 \ \frac{\text{kg}}{\text{m}^3} \ .$$

$$\mu = \ 32.8 \ \frac{\text{lb}_\text{m}}{\text{ft hr}} \ \times \ \frac{1}{2.419} \ \frac{\text{centipoise}}{\text{lb}_\text{m}/\text{ft hr}} \ \times \ 0.01 \ \frac{\text{poise}}{\text{centipoise}} \ = 0.136 \ \text{poise.}$$

Or, converting to SI units, noting that P is the symbol for poise, and evaluating ν:

$$\mu = 0.136 \ \text{P} \times \ 0.1 \ \frac{\text{kg/m s}}{\text{P}} \ = \ 0.0136 \ \frac{\text{kg}}{\text{m s}} \ .$$

$$\nu = \frac{\mu}{\rho} = \frac{0.0136 \ \text{kg/m s}}{851 \ \text{kg/m}^3} \ = \ 1.60 \times 10^{-5} \ \frac{\text{m}^2}{\text{s}} \quad (= 0.160 \ \text{St}). \qquad \square$$

Example 1.2—Mass of Air in a Room   **25**

Table 1.11   *Commonly Used Conversion Factors*

Area	1 mile^2	=	640 acres
	1 acre	=	0.4047 ha
Energy	1 BTU	=	1,055 J
	1 cal	=	4.184 J
	1 J	=	0.7376 ft lb$_f$
	1 erg	=	1 dyne cm
Force	1 lb$_f$	=	4.448 N
	1 N	=	0.2248 lb$_f$
Length	1 ft	=	0.3048 m
	1 m	=	3.281 ft
	1 mile	=	5,280 ft
Mass	1 lb$_m$	=	0.4536 kg
	1 kg	=	2.205 lb$_m$
Power	1 HP	=	550 ft lb$_f$/s
	1 kW	=	737.6 ft lb$_f$/s
Pressure	1 atm	=	14.696 lb$_f$/in^2
	1 atm	=	1.0133 bar
	1 atm	=	1.0133×10^5 Pa
Time	1 day	=	24 hr
	1 hr	=	60 min
	1 min	=	60 s
Viscosity	1 cP	=	2.419 lb$_m$/ft hr
	1 cP	=	0.001 kg/m s
	1 cP	=	0.000672 lb$_m$/ft s
	1 lb$_f$ s/ft^2	=	4.788×10^4 cP
Volume	1 ft^3	=	7.481 U.S. gal
	1 U.S. gal	=	3.785 l
	1 m^3	=	264.2 U.S. gal

Example 1.2—Mass of Air in a Room

Estimate the mass of air in your classroom, which is 80 ft wide, 40 ft deep, and 12 ft high. The gas constant is R = 10.73 psia ft^3/lb-mol °R.

Solution

The volume of the classroom, shown in Fig. E1.2, is:

$$V = 80 \times 40 \times 12 = 3.84 \times 10^4 \text{ ft}^3.$$

Fig. E1.2   Assumed dimensions of classroom.

If the air is approximately 20% oxygen and 80% nitrogen, its mean molecular weight is $M_w = 0.8 \times 28 + 0.2 \times 32 = 28.8$ lb$_m$/lb-mol. From the gas law, assuming an absolute pressure of $p = 14.7$ psia and a temperature of 70 °F = 530 °R, the density is:

$$\rho = \frac{M_w p}{RT} = \frac{28.8 \ (\text{lb}_m/\text{lb mol}) \times 14.7 \ (\text{psia})}{10.73 \ (\text{psia ft}^3/\text{lb mol} \ °R) \times 530 \ (°R)} = 0.0744 \ \text{lb}_m/\text{ft}^3.$$

Hence, the mass of air is:

$$M = \rho V = 0.0744 \ (\text{lb}_m/\text{ft}^3) \times 3.84 \times 10^4 \ (\text{ft}^3) = 2{,}860 \ \text{lb}_m. \qquad \square$$

For the rest of the book, manipulation of units will often be less detailed; the reader should always check if there is any doubt.

1.6 Hydrostatics

Variation of pressure with elevation.   Here, we investigate how the pressure in a stationary fluid varies with elevation z. The result is useful because it can answer questions such as "What is the pressure at the summit of Mt. Annapurna?" or "What forces are exerted on the walls of an oil storage tank?" Consider a hypothetical differential cylindrical element of fluid of cross-sectional area A, height dz, and volume $A\,dz$, which is also surrounded by the same fluid, as shown in Fig. 1.13. Its weight, being the downwards gravitational force on its mass, is $dW = \rho A\,dz\,g$. Two completely equivalent approaches will be presented:

Method 1.   Let p denote the pressure at the base of the cylinder; since p changes at a rate dp/dz with elevation, the pressure is found either from Taylor's expansion or the definition of a derivative to be $p + (dp/dz)dz$ at the top of the cylinder.[7] (Note that we do not anticipate a reduction of pressure with elevation here; hence, the plus sign is used. If, indeed—as proves to be the case—pressure falls with increasing elevation, then the subsequent development will tell us that

---

[7] Further details of this fundamental statement can be found in Appendix A and *must* be fully understood, because similar assertions appear repeatedly throughout the book.

dp/dz is negative.) Hence, the fluid exerts an *upward* force of pA on the base of the cylinder, and a *downward* force of $[p + (dp/dz)dz]A$ on the top of the cylinder.

Next, apply Newton's second law of motion by equating the net upward force to the mass times the acceleration—which is zero, since the cylinder is stationary:

$$pA - \left(p + \frac{dp}{dz}dz\right)A - \underbrace{\rho A\,dz\,g}_{\text{Weight}} = \underbrace{(\rho A\,dz)}_{\text{Mass}} \times 0 = 0. \qquad (1.29)$$

$$\underbrace{\phantom{pA - \left(p + \frac{dp}{dz}dz\right)A}}_{\text{Net pressure force}}$$

Cancellation of pA and division by $A\,dz$ leads to the following *differential equation*, which governs the rate of change of pressure with elevation:

$$\frac{dp}{dz} = -\rho g. \qquad (1.30)$$

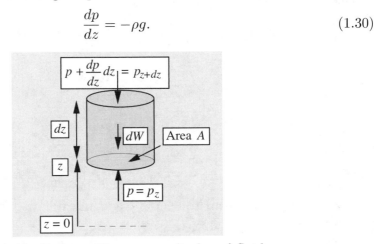

Fig. 1.13   Forces acting on a cylinder of fluid.

Method 2.   Let p_z and p_{z+dz} denote the pressures at the base and top of the cylinder, where the elevations are z and $z+dz$, respectively. Hence, the fluid exerts an *upward* force of $p_z A$ on the base of the cylinder, and a *downward* force of $p_{z+dz}A$ on the top of the cylinder. Application of Newton's second law of motion gives:

$$\underbrace{p_z A - p_{z+dz}A}_{\text{Net pressure force}} - \underbrace{\rho A\,dz\,g}_{\text{Weight}} = \underbrace{(\rho A\,dz)}_{\text{Mass}} \times 0 = 0. \qquad (1.31)$$

Isolation of the two pressure terms on the left-hand side and division by $A\,dz$ gives:

$$\frac{p_{z+dz} - p_z}{dz} = -\rho g. \qquad (1.32)$$

As dz tends to zero, the left-hand side of Eqn. (1.32) becomes the derivative dp/dz, leading to the same result as previously:

$$\frac{dp}{dz} = -\rho g. \qquad (1.30)$$

The same conclusion can also be obtained by considering a cylinder of *finite* height Δz and then letting Δz approach zero.

Note that Eqn. (1.30) predicts a pressure *decrease* in the vertically upward direction at a rate that is proportional to the local density. Such pressure variations can readily be detected by the ear when traveling quickly in an elevator in a tall building, or when taking off in an airplane. The reader must thoroughly understand *both* the above approaches. For most of this book, we shall use Method 1, because it eliminates the steps of taking the limit of $dz \to 0$ and invoking the definition of the derivative.

Pressure in a liquid with a free surface.   In Fig. 1.14, the pressure is p_s at the free surface, and we wish to find the pressure p at a depth H below the free surface—of water in a swimming pool, for example.

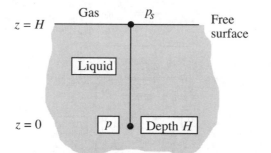

Fig. 1.14   Pressure at a depth H.

Separation of variables in Eqn. (1.30) and integration between the free surface ($z = H$) and a depth H ($z = 0$) gives:

$$\int_{p_s}^{p} dp = -\int_{H}^{0} \rho g \, dz. \tag{1.33}$$

Assuming—quite reasonably—that ρ and g are constants in the liquid, these quantities may be taken outside the integral, yielding:

$$p = p_s + \rho g H, \tag{1.34}$$

which predicts a *linear* increase of pressure with distance downward from the free surface. For large depths, such as those encountered by deep-sea divers, very substantial pressures will result.

Example 1.3—Pressure in an Oil Storage Tank   **29**

Example 1.3—Pressure in an Oil Storage Tank

What is the absolute pressure at the bottom of the cylindrical tank of Fig. E1.3, filled to a depth of H with crude oil, with its free surface exposed to the atmosphere? The specific gravity of the crude oil is 0.846. Give the answers for (a) $H = 15.0$ ft (pressure in lb$_f$/in^2), and (b) $H = 5.0$ m (pressure in Pa and bar). What is the purpose of the surrounding dike?

Fig. E1.3   Crude oil storage tank.

Solution

(a) The pressure is that of the atmosphere, p_a, plus the increase due to a column of depth $H = 15.0$ ft. Thus, setting $p_s = p_a$, Eqn. (1.34) gives:

$$p = p_a + \rho g H$$

$$= 14.7 + \frac{0.846 \times 62.3 \times 32.2 \times 15.0}{144 \times 32.2}$$

$$= 14.7 + 5.49 = 20.2 \text{ psia.}$$

The reader should check the units, noting that the 32.2 in the numerator is g $[=]$ ft/s^2, and that the 32.2 in the denominator is g_c $[=]$ lb$_m$ ft/lb$_f$ s^2.

(b) For SI units, no conversion factors are needed. Noting that the density of water is 1,000 kg/m^3, and that $p_a \doteq 1.01 \times 10^5$ Pa absolute:

$$p = 1.01 \times 10^5 + 0.846 \times 1,000 \times 9.81 \times 5.0 = 1.42 \times 10^5 \text{ Pa} = 1.42 \text{ bar.}$$

In the event of a tank rupture, the dike contains the leaking oil and facilitates prevention of spreading fire and contamination of the environment.

Epilogue

When he arrived at work in an oil refinery one morning, the author saw first-hand the consequences of an inadequately vented oil-storage tank. Rain during the night had caused partial condensation of vapor inside the tank, whose pressure had become sufficiently lowered so that the external atmospheric pressure had crumpled the steel tank just as if it were a flimsy tin can. The refinery manager was not pleased.   □

Example 1.4—Multiple Fluid Hydrostatics

The U-tube shown in Fig. E1.4 contains oil and water columns, between which there is a long trapped air bubble. For the indicated heights of the columns, find the specific gravity of the oil.

Fig. E1.4   Oil/air/water system.

Solution

The pressure p_2 at point 2 may be deduced by starting with the pressure p_1 at point 1 and adding or subtracting, as appropriate, the hydrostatic pressure changes due to the various columns of fluid. Note that the width of the U-tube (2.0 ft) is irrelevant, since there is no change in pressure in the horizontal leg. We obtain:

$$p_2 = p_1 + \rho_o g h_1 + \rho_a g h_2 + \rho_w g h_3 - \rho_w g h_4, \tag{E1.4.1}$$

in which ρ_o, ρ_a, and ρ_w denote the densities of oil, air, and water, respectively. Since the density of the air is very small compared to that of oil or water, the term containing ρ_a can be neglected. Also, $p_1 = p_2$, because both are equal to atmospheric pressure. Equation (E1.4.1) can then be solved for the specific gravity s_o of the oil:

$$s_o = \frac{\rho_o}{\rho_w} = \frac{h_4 - h_3}{h_1} = \frac{3.0 - 1.0}{2.5} = 0.80. \qquad \square$$

Pressure variations in a gas.   For a gas, the density is no longer constant, but is a function of pressure (and of temperature—although temperature variations are usually less significant than those of pressure), and there are two approaches:

1. For *small* changes in elevation, the assumption of constant density can still be made, and equations similar to Eqn. (1.34) are still approximately valid.
2. For moderate or large changes in elevation, the density in Eqn. (1.30) is given by Eqn. (1.7) or (1.8), $\rho = M_w p/RT$ or $\rho = M_w p/ZRT$, depending on whether

Example 1.5—Pressure Variations in a Gas   **31**

the gas is ideal or nonideal. It is understood that *absolute* pressure and temperature must *always* be used whenever the gas law is involved. A separation of variables can still be made, followed by integration, but the result will now be more complicated because the term dp/p occurs, leading—at the simplest (for an isothermal situation)—to a decreasing *exponential* variation of pressure with elevation.

Example 1.5—Pressure Variations in a Gas

For a gas of molecular weight M_w (such as the earth's atmosphere), investigate how the pressure p varies with elevation z if $p = p_0$ at $z = 0$. Assume that the temperature T is constant. What approximation may be made for *small* elevation increases? Explain how you would proceed for the nonisothermal case, in which $T = T(z)$ is a known function of elevation.

Solution

Assuming ideal gas behavior, Eqns. (1.30) and (1.7) give:

$$\frac{dp}{dz} = -\rho g = -\frac{M_w p}{RT}g. \tag{E1.5.1}$$

Separation of variables and integration between appropriate limits yields:

$$\int_{p_0}^{p} \frac{dp}{p} = \ln\frac{p}{p_0} = -\int_{0}^{z}\frac{M_w g}{RT}\,dz = -\frac{M_w g}{RT}\int_{0}^{z}dz = -\frac{M_w g z}{RT}, \tag{E1.5.2}$$

since $M_w g/RT$ is constant. Hence, there is an exponential decrease of pressure with elevation, as shown in Fig. E1.5:

$$p = p_0 \exp\left(-\frac{M_w g}{RT}z\right). \tag{E1.5.3}$$

Since a Taylor's expansion gives $e^{-x} = 1 - x + x^2/2 - \ldots$, the pressure is *approximated* by:

$$p \doteq p_0\left[1 - \frac{M_w g}{RT}z + \left(\frac{M_w g}{RT}\right)^2\frac{z^2}{2}\right]. \tag{E1.5.4}$$

For *small* values of $M_w g z/RT$, the last term is an insignificant second-order effect (compressibility effects are unimportant), and we obtain:

$$p \doteq p_0 - \frac{M_w p_0}{RT}gz = p_0 - \rho_0 gz, \tag{E1.5.5}$$

in which ρ_0 is the density at elevation $z = 0$; this approximation—essentially one of constant density—is shown as the dashed line in Fig. E1.5 and is clearly applicable only for a small change of elevation. Problem 1.19 investigates the upper limit on z for which this linear approximation is realistic. If there are significant elevation changes—as in Problems 1.16 and 1.30—the approximation of Eqn. (E1.5.5) *cannot* be used with any accuracy. Observe with caution that the Taylor's expansion is only a vehicle for demonstrating what happens for small values of $M_w gz/RT$. Actual calculations for larger values of $M_w gz/RT$ should be made using Eqn. (E1.5.3), *not* Eqn. (E1.5.4).

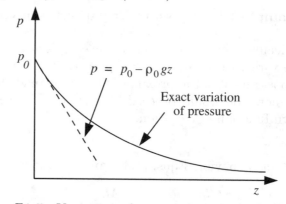

Fig. E1.5 Variation of gas pressure with elevation.

For the case in which the temperature is not constant, but is a known function $T(z)$ of elevation (as might be deduced from observations made by a meteorological balloon), it must be included *inside* the integral:

$$\int_{p_1}^{p_2} \frac{dp}{p} = -\frac{M_w g}{R} \int_0^z \frac{dz}{T(z)}. \tag{E1.5.6}$$

Since $T(z)$ is unlikely to be a simple function of z, a numerical method—such as Simpson's rule in Appendix A—will probably have to be used to approximate the second integral of Eqn. (E1.5.6). □

Total force on a dam or lock gate. Fig. 1.15 shows the side and end elevations of a dam or lock gate of depth D and width W. An expression is needed for the total horizontal force F exerted by the liquid on the dam, so that the latter can be made of appropriate strength. Similar results would apply for liquids in storage tanks. Gauge pressures are used for simplicity, with $p = 0$ at the free surface and in the air outside the dam. Absolute pressures could also be employed, but would merely add a constant atmospheric pressure everywhere, and would eventually be canceled out. If the coordinate z is measured from the bottom of the liquid upward, the corresponding *depth* of a point below the free surface is $D - z$. Hence, from Eqn. (1.34), the differential horizontal force dF on an infinitesimally small rectangular strip of area $dA = W\,dz$ is:

$$dF = pW\,dz = \rho g(D - z)W\,dz. \tag{1.35}$$

(a)   (b)

Fig. 1.15   Horizontal thrust on a dam:
(a) side elevation, (b) end elevation.

Integration from the bottom ($z = 0$) to the top ($z = D$) of the dam gives the total horizontal force:

$$F = \int_0^F dF = \int_0^D \rho g W (D - z)\, dz = \frac{1}{2}\rho g W D^2. \tag{1.36}$$

Horizontal pressure force on an arbitrary plane vertical surface. The preceding analysis was for a regular shape. A more general case is illustrated in Fig. 1.16, which shows a *plane* vertical surface of *arbitrary* shape. Note that it is now slightly easier to work in terms of a *downward* coordinate h.

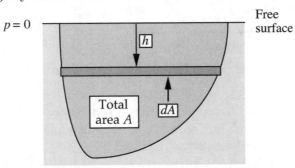

Fig. 1.16   Side view of a pool of liquid
with a submerged vertical surface.

Again taking gauge pressures for simplicity (the gas law is not involved), with $p = 0$ at the free surface, the total horizontal force is:

$$F = \int_A p\, dA = \int_A \rho g h\, dA = \rho g A \frac{\int_A h\, dA}{A}. \tag{1.37}$$

But the depth h_c of the centroid of the surface is defined as:

$$h_c \equiv \frac{\int_A h\, dA}{A}. \tag{1.38}$$

Thus, from Eqns. (1.37) and (1.38), the total force is:

$$F = \rho g h_c A = p_c A, \tag{1.39}$$

in which p_c is the pressure at the centroid.

The advantage of this approach is that the location of the centroid is already known for several geometries. For example, for a rectangle of depth D and width W:

$$h_c = \frac{1}{2}D \quad \text{and} \quad F = \frac{1}{2}\rho g W D^2, \tag{1.40}$$

in agreement with the earlier result of Eqn. (1.36). Similarly, for a vertical circle that is just submerged, the depth of the centroid equals its radius. And, for a vertical triangle with one edge coincident with the surface of the liquid, the depth of the centroid equals one-third of its altitude.

Fig. 1.17   *Thrust on surface of uniform cross-sectional shape.*

Horizontal pressure force on a curved surface.   Fig. 1.17(a) shows the cross section of a submerged surface that is no longer plane. However, the shape *is* uniform normal to the plane of the diagram.

In general, as shown in Fig. 1.17(b), the local pressure force $p\,dA$ on an element of surface area dA does not act horizontally; therefore, its horizontal component must be obtained by projection through an angle of $(\pi/2 - \theta)$, by multiplying by $\cos(\pi/2 - \theta) = \sin\theta$. The total horizontal force F is then:

$$F = \int_A p \sin\theta\, dA = \int_{A^*} p\, dA^*, \tag{1.41}$$

in which $dA^* = dA\sin\theta$ is an element of the projection of A onto the hypothetical vertical plane A*. The integral of Eqn. (1.41) can be obtained readily, as illustrated in the following example.

Example 1.6—Hydrostatic Force on a Curved Surface   **35**

Example 1.6—Hydrostatic Force on a Curved Surface

A submarine, whose hull has a circular cross section of diameter D, is just submerged in water of density ρ, as shown in Fig. E1.6. Derive an equation that gives the total horizontal force F_x on the left half of the hull, for a distance W normal to the plane of the diagram. If $D = 8$ m, the circular cross section continues essentially for the total length $W = 50$ m of the submarine, and the density of sea water is $\rho = 1{,}026$ kg/m^3, determine the total horizontal force on the left-hand half of the hull.

Solution

The force is obtained by evaluating the integral of Eqn. (1.41), which is identical to that for the rectangle in Fig. 1.15:

$$F_x = \int_{A^*} p\,dA = \int_{z=0}^{z=D} \rho g W (D - z)\,dz = \frac{1}{2}\rho g W D^2. \tag{E1.6.1}$$

Insertion of the numerical values gives:

$$F_x = \frac{1}{2} \times 1{,}026 \times 9.81 \times 50 \times 8.0^2 = 1.61 \times 10^7 \text{ N}. \tag{E1.6.2}$$

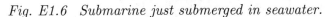

Fig. E1.6   Submarine just submerged in seawater.

Thus, the total force is considerable—about 3.62×10^6 lb$_f$.   □

Archimedes, ca. 287–212 B.C. Archimedes was a Greek mathematician and inventor. He was born in Syracuse, Italy, where he spent much of his life, apart from a period of study in Alexandria. He was much more interested in mathematical research than any of the ingenious inventions that made him famous. One invention was a "burning mirror," which focused the sun's rays to cause intense heat. Another was the rotating Archimedean screw, for raising a continuous stream of water. Presented with a crown supposedly of pure gold, Archimedes tested the possibility that it might be "diluted" by silver by separately immersing the crown and an equal weight of pure gold into his bath, and observed the difference in the overflow. Legend has it that he was so excited by the result that he ran home without his clothes, shouting "$\epsilon \H{v} \rho \eta \kappa \alpha$, $\epsilon \H{v} \rho \eta \kappa \alpha$", "I have found it, I have found it." To dramatize the effect of a lever, he said, "Give me a place to stand, and I will move the earth." He considered his most important intellectual contribution to be the determination of the ratio of the volume of a sphere to the volume of the cylinder that circumscribes it. [Now that calculus has been invented, the reader might like to derive this ratio!] Sadly, Archimedes was killed during the capture of Syracuse by the Romans.

Source: *The Encyclopædia Britannica*, 11th ed., Cambridge University Press (1910–1911).

Buoyancy forces.   If an object is submerged in a fluid, it will experience a net upward or *buoyant* force exerted by the fluid. To find this force, first examine the buoyant force on a submerged circular cylinder of height H and cross-sectional area A, shown in Fig. 1.18.

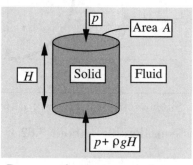

Fig. 1.18 Pressure forces on a submerged cylinder.

The forces on the curved vertical surface act horizontally and may therefore be ignored. Hence, the net *upward* force due to the difference between the opposing pressures on the bottom and top faces is:

$$F = (p + \rho g H - p)A = \rho H A g, \qquad (1.42)$$

Example 1.7—Application of Archimedes' Law   **37**

which is exactly the weight of the displaced liquid, thus verifying *Archimedes' law*, (the buoyant force equals the weight of the fluid displaced) for the cylinder. The same result would clearly be obtained for a cylinder of any uniform cross section.

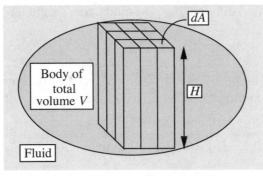

Fig. 1.19   Buoyancy force for an arbitrary shape.

Fig. 1.19 shows a more general situation, with a body of arbitrary shape. However, Archimedes' law still holds since the body can be decomposed into an infinitely large number of vertical rectangular parallelepipeds or "boxes" of infinitesimally small cross-sectional area dA. The effect for one box is then summed or "integrated" over all the boxes, and again gives the net upward buoyant force as the weight of the liquid displaced.

Example 1.7—Application of Archimedes' Law

Consider the situation in Fig. E1.7(a), in which a barrel rests on a raft that floats in a swimming pool. The barrel is then pushed off the raft, and may either float or sink, depending on its contents and hence its mass. The cross-hatching shows the volumes of water that are displaced. For each of the cases shown in Fig. E1.7 (b) and (c), determine whether the water level in the pool will rise, fall, or remain constant, relative to the initial level in (a).

(a) Initial   (b) Final (light barrel)   (c) Final (heavy barrel)

Fig. E1.7   Raft and barrel in swimming pool: (a) initial positions, (b) light barrel rolls off and floats, (c) heavy barrel rolls off and sinks. The cross-hatching shows volumes below the surface of the water.

Solution

Initial state. Let the masses of the raft and barrel be M_r and M_b, respectively. If the volume of displaced water is initially V in (a), Archimedes' law requires that the total weight of the raft and barrel equals the weight of the displaced water, whose density is ρ:

$$(M_r + M_b)g = V\rho g. \tag{E1.7.1}$$

Barrel floats. If the barrel floats, as in (b), with submerged volumes of V_r and V_b for the raft and barrel, respectively, Archimedes' law may be applied to the raft and barrel separately:

$$\text{Raft}: \ M_r g = V_r \rho g, \qquad \text{Barrel}: \ M_b g = V_b \rho g. \tag{E1.7.2}$$

Addition of the two equations (E1.7.2) and comparison with Eqn. (E1.7.1) shows that:

$$V_r + V_b = V. \tag{E1.7.3}$$

Therefore, since the volume of the water is constant, and the total displaced volume does not change, the level of the surface also remains *unchanged*.

Barrel sinks. Archimedes' law may still be applied to the raft, but the weight of the water displaced by the barrel no longer suffices to support the weight of the barrel, so that:

$$\text{Raft}: \ M_r g = V_r \rho g, \qquad \text{Barrel}: \ M_b g > V_b \rho g. \tag{E1.7.4}$$

Addition of the two relations in (E1.7.4) and comparison with Eqn. (E1.7.1) shows that:

$$V_r + V_b < V. \tag{E1.7.5}$$

Therefore, since the volume of the water in the pool is constant, and the total displaced volume is *reduced*, the level of the surface *falls*. This result is perhaps contrary to intuition: since the *whole* volume of the barrel is submerged in (c), it might be thought that the water level will rise above that in (b). However, because the barrel must be heavy in order to sink, the load on the raft and hence V_r are substantially reduced, so that the total displaced volume is also reduced.

This problem illustrates the need for a complete analysis rather than jumping to a possibly erroneous conclusion. $\square$

1.7 Pressure Change Caused by Rotation

Finally, consider the shape of the free surface for the situation shown in Fig. 1.20(a), in which a cylindrical container, partly filled with liquid, is rotated with an angular velocity ω—that is, at $N = \omega/2\pi$ revolutions per unit time. The analysis has applications in fuel tanks of spinning rockets, centrifugal filters, and liquid mirrors.

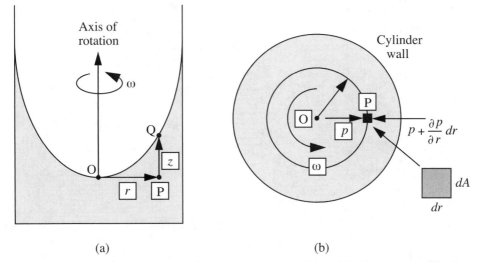

(a)   (b)

Fig. 1.20 Pressure changes for rotating cylinder: (a) elevation, (b) plan.

Point O denotes the origin, where $r = 0$ and $z = 0$. After a sufficiently long time, the rotation of the container will be transmitted by viscous action to the liquid, whose rotation is called a *forced vortex*. In fact, the liquid spins as if it were a *solid body*, rotating with a uniform angular velocity ω, so that the velocity in the direction of rotation at a radial location r is given by $v_\theta = r\omega$. It is therefore appropriate to treat the situation similar to the hydrostatic investigations already made.

Suppose that the liquid element P is essentially a rectangular box with cross-sectional area dA and radial extent dr. (In reality, the element has slightly tapering sides, but a more elaborate treatment taking this into account will yield identical results to those derived here.) The pressure on the inner face is p, whereas that on the outer face is $p + (\partial p/\partial r)dr$. Also, for uniform rotation in a circular path of radius r, the acceleration toward the center O of the circle is $r\omega^2$. Newton's second law of motion is then used for equating the net pressure force toward O to the mass of the element times its acceleration:

$$\underbrace{\left(p + \frac{\partial p}{\partial r}dr - p\right) dA}_{\text{Net pressure force}} = \underbrace{\rho(dA\, dr)}_{\text{Mass}} r\omega^2. \qquad (1.43)$$

Note that the use of a *partial* derivative is essential, since the pressure now varies in both the horizontal (radial) *and* vertical directions. Simplification yields the variation of pressure in the radial direction:

$$\frac{\partial p}{\partial r} = \rho r \omega^2, \tag{1.44}$$

so that pressure *increases* in the radially outward direction.

Observe that the gauge pressure at all points on the interface is zero; in particular, $p_O = p_Q = 0$. Integrating from points O to P (at constant z):

$$\int_{p=0}^{p_P} dp = \rho \omega^2 \int_0^r r \, dr,$$

$$p_P = \frac{1}{2}\rho \omega^2 r^2. \tag{1.45}$$

However, the pressure at P can also be obtained by considering the usual hydrostatic increase in traversing the path QP:

$$p_P = \rho g z. \tag{1.46}$$

Elimination of the intermediate pressure p_P between Eqns. (1.45) and (1.46) relates the elevation of the free surface to the radial location:

$$z = \frac{\omega^2 r^2}{2g}. \tag{1.47}$$

Thus, the free surface is *parabolic* in shape; observe also that the density is not a factor, having been canceled from the equations.

There is another type of vortex—the *free* vortex—that is also important, in cyclone dust collectors and tornadoes, for example, as discussed in Chapters 4 and 7. There, the velocity in the angular direction is given by $v_\theta = c/r$, where c is a constant, so that v_θ is inversely proportional to the radial position.

Example 1.8—Overflow from a Spinning Container

A cylindrical container of height H and radius a is initially half-filled with a liquid. The cylinder is then spun steadily around its vertical axis Z-Z, as shown in Fig. E1.8. At what value of the angular velocity ω will the liquid just start to spill over the top of the container? If $H = 1$ ft and $a = 0.25$ ft, how many rpm (revolutions per minute) would be needed?

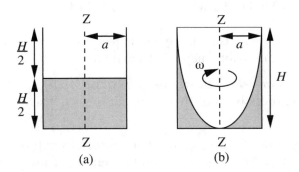

Fig. E1.8 Geometry of a spinning container:
(a) at rest, (b) on the point of overflowing.

Solution

From Eqn. (1.47), the shape of the free surface is a parabola. Therefore, the air inside the rotating cylinder forms a paraboloid of revolution, whose volume is known from calculus to be exactly one-half of the volume of the "circumscribing cylinder," namely, the container.[8] Hence, the liquid at the center reaches the bottom of the cylinder *just* as the liquid at the curved wall reaches the top of the cylinder. In Eqn. (1.47), therefore, set $z = H$ and $r = a$, giving the required angular velocity:

$$\omega = \sqrt{\frac{2gH}{a^2}}.$$

For the stated values:

$$\omega = \sqrt{\frac{2 \times 32.2 \times 1}{0.25^2}} = 32.1 \ \frac{\text{rad}}{\text{s}}, \qquad N = \frac{\omega}{2\pi} = \frac{32.1 \times 60}{2\pi} = 306.5 \ \text{rpm}. \quad \square$$

---

[8] Proof can be accomplished as follows. First, note for the parabolic surface in Fig. E1.8(b), $r = a$ when $z = H$, so, from Eqn. (1.47), $\omega^2/2g = H/a^2$. Thus, Eqn. (1.47) can be rewritten as:

$$z = H \frac{r^2}{a^2}.$$

The volume of the paraboloid of air within the cylinder is therefore:

$$V = \int_{z=0}^{z=H} \pi r^2 \, dz = \int_{z=0}^{z=H} \frac{\pi a^2 z}{H} \, dz = \frac{1}{2}\pi a^2 H,$$

which is exactly one-half of the volume of the cylinder, $\pi a^2 H$. Since the container was initially just half filled, the liquid volume still accounts for the remaining half.

PROBLEMS FOR CHAPTER 1

1. *Units conversion—E.* How many cubic feet are there in an acre-foot? How many gallons? How many cubic meters? How many tonnes of water?

2. *Units conversion—E.* The viscosity μ of an oil is 10 cP and its specific gravity s is 0.8. Reexpress both of these (the latter as density ρ) in both the lb_m, ft, s system and in SI units.

3. *Units conversion—E.* Use conversion factors to express: (a) the gravitational acceleration of 32.174 ft/s^2 in SI units, and (b) a pressure of 14.7 lb_f/in^2 (one atmosphere) in both pascals and bars.

4. *Meteorite density—E.* The Barringer Crater in Arizona was formed 30,000 years ago by a spherical meteorite of diameter 60 m and mass 10^6 t (tonnes), traveling at 15 km/s when it hit the ground.[9] (Clearly, all figures are *estimates*.) What was the mean density of the meteorite? What was the predominant material in the meteorite? Why? If one tonne of the explosive TNT is equivalent to five billion joules, how many tonnes of TNT would have had the same impact as the meteorite?

5. *Reynolds number—E.* What is the mean velocity u_m (ft/s) and the Reynolds number $Re = \rho u_m D / \mu$ for 35 gpm (gallons per minute) of water flowing in a 1.05-in. I.D. pipe if its density is $\rho = 62.3$ lb_m/ft^3 and its viscosity is $\mu = 1.2$ cP? What are the units of the Reynolds number?

6. *Pressure in bubble—E.* Consider a *soap-film bubble* of diameter d. If the external air pressure is p_a, and the surface tension of the soap film is σ, derive an expression for the pressure p_b inside the bubble. *Hint*: Note that there are *two* air/liquid interfaces.

(a)   (b)

Fig. P1.7 Waterflooding of an oil reservoir.

---

[9] Richard A.F. Grieve, "Impact cratering on the earth," *Scientific American*, Vol. 262, No. 4, p. 68 (1990).

7. *Reservoir waterflooding—E.* Fig. P1.7(a) shows how water is pumped down one well, of depth H, into an oil-bearing stratum, so that the displaced oil then flows up through another well. Fig. P1.7(b) shows an enlargement of an idealized pore, of diameter d, at the water/oil interface.

If the water and oil are *just* starting to move, what water inlet pressure p_w is needed if the oil exit pressure is to be p_o? Assume that the oil completely wets the pore (*not* always the case), that the water/oil interfacial tension is σ, and that the densities of the water and oil are ρ_w and ρ_o, respectively.[10]

8. *Barometer reading—M.* In your house (elevation 950 ft above sea level) you have a barometer that registers inches of mercury. On an average day in January, you telephone the weather station (elevation 700 ft) and are told that the exact pressure there is 0.966 bar. What is the correct reading for your barometer, and to how many psia does this correspond? The specific gravity of mercury is 13.57.

Fig. P1.9   Cylinder immersed in water and liquid A.

9. *Two-layer buoyancy—E.* As shown in Fig. P1.9, a layer of an unknown liquid A (immiscible with water) floats on top of a layer of water W in a beaker. A completely submerged cylinder of specific gravity 0.9 adjusts itself so that its axis is vertical and two-thirds of its height projects above the A/W interface and one-third remains below. What is the specific gravity of A? Solve the problem two ways—first using Archimedes' law, and then using a momentum or force balance.

Fig. P1.10   U-tube with immiscible liquids.

10. *Differential manometer—E.* The U-tube shown in Fig. P1.10 has legs of *unequal* internal diameters d_1 and d_2, which are partly filled with immiscible liquids of densities ρ_1 and ρ_2, respectively, and are open to the atmosphere at the top.

---

[10] D.L. Katz et al., *Handbook of Natural Gas Engineering*, McGraw-Hill, New York, 1959, p. 57, indicates a wide range of wettability by water, varying greatly with the particular rock formation.

If an *additional* small volume v_2 of the second liquid is added to the right-hand leg, derive an expression—in terms of ρ_1, ρ_2, v_2, d_1, and d_2—for δ, the amount by which the level at B will fall. If ρ_1 is known, but ρ_2 is unknown, could the apparatus be used for determining the density of the second liquid?

Hints: The lengths h_A, h_B, and h_C have been included just to get started; they must not appear in the final result. After adding the second liquid, consider h_C to have increased by a length Δ—a quantity that must also eventually be eliminated.

Fig. P1.11   Bubble rising in a closed cylinder.

11. *Ascending bubble—E.* As shown in Fig. P1.11, a hollow vertical cylinder with *rigid* walls and of height H is closed at both ends, and is filled with an *incompressible* oil of density ρ. A gauge registers the pressure at the top of the cylinder. When a small bubble of volume v_0 initially adheres to point A at the bottom of the cylinder, the gauge registers a pressure p_0. The gas in the bubble is ideal, and has a molecular weight of M_w. The bubble is liberated by tapping on the cylinder and rises to point B at the top. The temperature T is constant throughout. Derive an expression in terms of any or all of the specified variables for the new pressure-gauge reading p_1 at the top of the cylinder.

12. *Ship passing through locks—M.* A ship of mass M travels uphill through a series of identical rectangular locks, each of equal superficial (bird's-eye view) area A and elevation change h. The steps involved in moving from one lock to the next (1 to 2, for example) are shown as A–B–C in Fig. P1.12. The lock at the top of the hill is supplied by a source of water. The initial depth in lock 1 is H, and the density of the water is ρ.

(a) Derive an expression for the increase in mass of water in lock 1 for the sequence shown in terms of some or all of the variables M, H, h, A, ρ, and g.

(b) If, after reaching the top of the hill, the ship descends through a similar series of locks to its original elevation, again derive an expression for the mass of water gained by a lock from the lock immediately above it.

(c) Does the mass of water to be supplied depend on the mass of the ship if: (i) it travels only uphill, (ii) it travels uphill, then downhill? Explain your answer.

Fig. P1.12 Ship and locks.

13. *Furnace stack—E.* Air ($\rho_a = 0.08$ lb$_m$/ft^3) flows through a furnace where it is burned with fuel to produce a hot gas ($\rho_g = 0.05$ lb$_m$/ft^3) that flows up the stack, as in Fig. P1.13. The pressures in the gas and the immediately surrounding air at the top of the stack at point A are equal.

Fig. P1.13 Furnace stack.

What is the difference Δh (in.) in levels of the water in the manometer connected between the base B of the stack and the outside air at point C? Which side rises? Except for the pressure drop across the furnace (which you need not worry about), treat the problem as one in hydrostatics. That is, ignore any frictional effects and kinetic energy changes in the stack. Also, neglect compressibility effects.

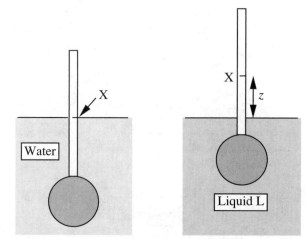

Fig. P1.14   Hydrometer in water and test liquid L.

14. *Hydrometer—E.* When a hydrometer floats in water, its cylindrical stem is submerged so that a certain point X on the stem is level with the free surface of the water, as shown in Fig. P1.14. When the hydrometer is placed in another liquid L of specific gravity s, the stem *rises* so that point X is now a height z above the free surface of L.

Derive an equation giving s in terms of z. If needed, the cross-sectional area of the stem is A, and when in water a total volume V (stem plus bulb) is submerged.

Fig. P1.15   Oil/mercury/water system.

15. *Three-liquid manometer—E.* In the hydrostatic case shown in Fig. P1.15, $a = 6$ ft and $c = 4$ ft. The specific gravities of oil, mercury, and water are $s_o = 0.8$, $s_m = 13.6$, and $s_w = 1.0$. Pressure variations in the air are negligible. What is the difference b in inches between the mercury levels, and which leg of the manometer has the higher mercury level? *Note:* In this latter respect, the diagram may or may not be correct.

16. *Pressure on Mt. Erebus—M*. On page 223 of the biography *Shackleton* (by Roland Huntford, Atheneum, New York, 1986), the Antarctic explorer's colleague, Edward Marshall, is reported as having " . . . fixed the altitude [of Mt. Erebus] by hypsometer. This was simply a small cylinder in which distilled water was boiled and the temperature measured. It was then the most accurate known method of measuring altitude. The summit of Erebus turned out to be 13,500 feet above sea level."[11]

Assuming a uniform (mean) air temperature of $-5\,°F$ (the summer *summit* temperature is $-30\,°F$), and a sea-level pressure of 13.9 psia, at what temperature did the water boil in the hypsometer? At temperatures $T = 160, 170, 180, 190,$ 200, and $210\,°F$, the respective vapor pressures of water are $p_v = 4.741, 5.992,$ 7.510, 9.339, 11.526, and 14.123 psia.

17. *Oil and gas well pressures—M*. A pressure gauge at the top of an oil well 18,000 ft deep registers 2,000 psig. The bottom 4,000-ft portion of the well is filled with oil ($s = 0.70$). The remainder of the well is filled with natural gas ($T = 60\,°F$, compressibility factor $Z = 0.80$, and $s = 0.65$, meaning that the molecular weight is 0.65 times that of air).

Calculate the pressure (psig) at (a) the oil/gas interface, and (b) the bottom of the well.

18. *Thrust on a dam—E*. Concerning the thrust on a rectangular dam, check that Eqn. (1.36) is still obtained if, instead of employing an upward coordinate z, use is made of a *downward* coordinate h (with $h = 0$ at the free surface).

19. *Pressure variations in air—M*. Refer to Example 1.5 concerning the pressure variations in a gas, and assume that you are dealing with air at $40\,°F$. Suppose further that you are using just the linear part of the expansion (up to the term in z) to calculate the absolute pressure at an elevation z above ground level. How large can z be, in miles, with the knowledge that the error amounts to no more than 1% of the exact value?

20. *Grand Coulee dam—E*. The Grand Coulee dam, which first operated in 1941, is 550 ft high and 3,000 ft wide. What is the pressure at the base of the dam, and what is the total horizontal force F lb$_f$ exerted on it by the water upstream?

21. *Force on V-shaped dam—M*. A vertical dam has the shape of a V that is 3 m high and 2 m wide at the top, which is just level with the surface of the water upstream of the dam. Use two different methods to determine the total force (N) exerted by the water on the dam.

22. *Rotating mercury mirror—M*. Physicist Ermanno Borra, of Laval University in Quebec, has made a 40-in. diameter telescopic mirror from a pool of

---

[11] A more recent value is thought to be 12,450 feet.

mercury that rotates at one revolution every six seconds.[12] (Air bearings eliminate vibration, and a thin layer of oil prevents surface ripples.)

By what value Δz would the surface at the center be depressed relative to the perimeter, and what is the focal length (m) of the mirror? The mirror cost Borra $7,500. He estimated that a similar 30-meter mirror could be built for $7.5 million. If the focal length were unchanged, what would be the new value of Δz for the larger mirror? *Hint*: the equation for a parabola of focal length f is $r^2 = 4fz$.

23. *Oil and water in rotating container—E.* A cylindrical container partly filled with immiscible layers of water and oil is placed on a rotating turntable. Develop the necessary equations and prove that the shapes of the oil/air and water/oil interfaces are identical.

24. *Energy to place satellite in orbit—M.* "NASA launched a $195 million astronomy satellite at the weekend to probe the enigmatic workings of neutron stars, black holes, and the hearts of galaxies at the edge of the universe . . . The long-awaited mission began at 8:48 a.m. last Saturday when the satellite's Delta–2 rocket blasted off from the Cape Canaveral Air Station."[13]

This "X-ray Timing Explorer satellite" was reported as having a mass of 6,700 lb$_m$ and being placed 78 minutes after lift-off into a 360-mile-high circular orbit (measured above the earth's surface).

How much energy (J) went directly to the satellite to place it in orbit? What was the corresponding average power (kW)? The force of attraction between a mass m and the mass M_e of the earth is GmM_e/r^2, where r is the distance of the mass from the center of the earth and G is the universal gravitational constant. The value of G is *not* needed in order to solve the problem, as long as you remember that the radius of the earth is 6.37×10^6 m, and that $g = 9.81$ m/s^2 at its surface.

25. *Central-heating loop—M.* Fig. P1.25 shows a piping "loop" that circulates hot water through the system ABCD in order to heat two floors of a house by means of baseboard fins attached to the horizontal runs of pipe (BC and DA). The horizontal and vertical portions of the pipes have lengths L and H, respectively.

The water, which has a mean density of $\bar{\rho}$ and a volume coefficient of expansion α, circulates by the action of natural convection due to a small heater, whose inlet and outlet water temperatures are T_1 and T_2, respectively. The pressure drop due to friction per unit length of piping is cu^2/D, where c is a known constant, u is the mean water velocity, and D is the internal diameter of the pipe. You may assume that the vertical legs AB and CD are insulated, and that equal amounts of heat are dissipated on each floor.

Derive an expression that gives the volumetric circulation rate of water, Q, in terms of c, D, $\bar{\rho}$, α, g, L, H, T_1, and T_2.

---

[12] *Scientific American*, February 1994, pp. 76–81. There is also earlier mention of his work in *Time*, December 15, 1986.

[13] *Manchester Guardian Weekly*, January 7, 1996.

Fins (size exaggerated)

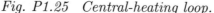

Fig. P1.25   *Central-heating loop.*

26. *Pressure at the center of the earth—M.* Prove that the pressure at the center of the earth is given by $p_c = 3Mg_s/8\pi R^2$, in which g_s is the gravitational acceleration at the surface, M is the mass of the earth, and R is its radius. *Hints:* Consider a small mass m inside the earth, at a radius r from the center. The force of attraction mg_r (where g_r is the *local* gravitational acceleration) between m and the mass M_r enclosed *within* the radius r is GmM_r/r^2, where G is the universal gravitational constant. Repeat for the mass at the surface, and hence show that $g_r/g_s = r/R$. Then invoke hydrostatics.

If the radius of the earth is $R = 6.37 \times 10^6$ m, and its mean density is approximately 5,500 kg/m^3, estimate p_c in Pa and psi.

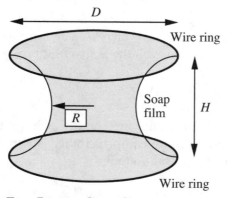

Fig. P1.27   *Soap film on two rings.*

27. *Soap film on wire rings—M.* As shown in Fig. P1.27, a soap film is stretched between two wire rings, each of diameter D and separated by a distance H. Prove that the radius R of the film at its narrowest point is:

$$R = \frac{1}{6}\left(2D + \sqrt{D^2 - 3H^2}\right).$$

You may assume that a section of the soap film is a circular arc, and that $D \geq \sqrt{3}\,H$. What might happen if D is less than $\sqrt{3}\,H$?

Clearly stating your assumptions, derive an expression for the radius, in terms of D and H. Is your expression exact or approximate? Explain.

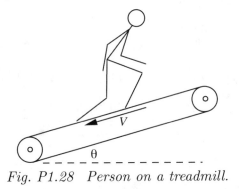

Fig. P1.28 Person on a treadmill.

28. *Treadmill stress test—M.* What power P is needed to resist a force F at a steady velocity V? In a treadmill stress test (Fig. P1.28), you have to keep walking to keep up with a moving belt whose velocity V and angle of inclination θ are steadily increased. Initially, the belt is moving at 1.7 mph and has a grade (defined as $\tan\theta$) of 10%. The test is concluded after 13.3 min, at which stage the belt is moving at 5.0 mph and has a grade of 18%. If your mass is 163 lb$_\mathrm{m}$: (a) how many HP are you exerting at the start of the test, (b) how many HP are you exerting at the end of the test, and (c) how many joules have you expended overall?

29. *Bubble rising in compressible liquid—D.* A liquid of volume V and isothermal compressibility β has its pressure increased by an amount Δp. Explain why the corresponding increase ΔV in volume is given approximately by:

$$\Delta V = -\beta V \, \Delta p.$$

Repeat Problem P1.11, now allowing the oil—whose density and volume are initially ρ_0 and V_0—to have a finite compressibility β. Prove that the ratio of the final bubble volume v_1 to its initial volume v_0 is:

$$\frac{v_1}{v_0} = 1 + \frac{\rho_0 g H}{p_0}.$$

If needed, assume that: (a) the bubble volume is much smaller than the oil volume, and (b) $\beta p_0 V_0 \gg v_1$. If $\rho_0 = 800$ kg/m^3, $\beta = 5.5 \times 10^{-10}$ m^2/N, $H = 1$ m, $p_0 = 10^5$ N/m^2 (initial absolute pressure at the top of the cylinder), $v_0 = 10^{-8}$ m^3, and $V_0 = 0.1$ m^3, evaluate v_1/v_0 and check that assumption (b) above is reasonable.

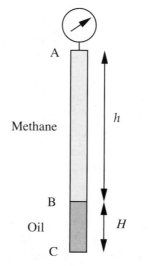

Fig. P1.30 Well containing oil and methane.

30. *Pressures in oil and gas well—M.* Fig. P1.30 shows a well that is 12,000 ft deep. The bottom $H = 2,000$-ft portion is filled with an incompressible oil of specific gravity $s = 0.75$, above which there is an $h = 10,000$-ft layer of methane (CH_4; $C = 12$, $H = 1$) at 100 °F, which behaves as an ideal isothermal gas whose density is *not* constant. The gas and oil are static. The density of water is 62.3 lb_m/ft^3.

(a) If the pressure gauge at the top of the well registers $p_A = 1,000$ psig, compute the *absolute* pressure p_B (psia) at the oil/methane interface. Work in terms of symbols before substituting numbers.

(b) Also compute $(p_C - p_B)$, the additional pressure (psi) in going from the interface B to the bottom of the well C.

Fig. P1.31 Soap film between two disks.

31. *Soap film between disks—E (C).* A circular disk of weight W and radius a is hung from a similar disk by a soap film with surface tension σ, as shown in Fig. P1.31. The gauge pressure inside the film is P.

First, derive an expression for the angle θ in terms of a, P, W, and σ. Then obtain an equation that relates the radius of the neck r to a, P, W, and σ. Assume that: (a) the excess pressure inside a soap film with radii of curvature r_1 and r_2 is $2\sigma(1/r_1 + 1/r_2)$, and (b) the cross section of the film forms a circular arc.

32. *Newspaper statements about the erg—E.* In the *New York Times* for January 18, 1994, the following statement appeared: "An erg is the metric unit scientists use to measure energy. One erg is the amount of energy it takes to move a mass of one gram one centimeter in one second." (This statement related to the earthquake of the previous day, measuring 6.6 on the Richter scale, in the Northridge area of the San Fernando Valley, 20 miles north of downtown Los Angeles.)

Also in the same newspaper, there was a letter of rebuttal on January 30 that stated in part: " . . . This is not correct. The energy required to move a mass through a distance does not depend on how long it takes to accomplish the movement. Thus the definition should not include a unit of time."

A later letter from another reader, on February 10, made appropriate comments about the original article and the first letter. What do you think was said in the second letter?

33. *Centroid of triangle—E.* A triangular plate held vertically in a liquid has one edge (of length B) coincident with the surface of the liquid; the altitude of the plate is H. Derive an expression for the depth of the centroid. What is the horizontal force exerted by the liquid, whose density is ρ, on one side of the plate?

34. *Blake-Kozeny equation—E.* The Blake-Kozeny equation for the pressure drop $(p_1 - p_2)$ in laminar flow of a fluid of viscosity μ through a packed bed of length L, particle diameter D_p and void fraction ε is (Section 4.4):

$$\frac{p_1 - p_2}{L} = 150 \left(\frac{\mu u_0}{D_p^2} \right) \left[\frac{(1 - \varepsilon)^2}{\varepsilon^3} \right].$$

(a) Giving your reasons, suggest appropriate units for ε.
(b) If $p_1 - p_2 = 75$ lb$_f$/in^2, $D_p = 0.1$ in., $L = 6.0$ ft, $\mu = 0.22$ P, and $u_0 = 0.1$ ft/s, compute the value of ε.

35. *Shear stresses for air and water—E.* Consider the situation in Fig. 1.8, with $h = 0.1$ cm and $V = 1.0$ cm/s. The pressure is atmospheric throughout.

(a) If the fluid is air at $20\,^\circ$C, evaluate the shear stress τ_a (dynes/cm^2). Does τ vary across the gap? Explain.
(b) Evaluate τ_w if the fluid is water at $20\,^\circ$C. What is the ratio τ_w/τ_a?
(c) If the temperature is raised to $80\,^\circ$C, does τ_a increase or decrease? What about τ_w?

36. *True/false.* Check *true* or *false*, as appropriate:[14]

(a)   When a fluid is subjected to a steady shear stress, it will reach a state of equilibrium in which no further motion occurs.   T ☐   F ☐

(b)   Pressure and shear stress are two examples of a force per unit area.   T ☐   F ☐

(c)   In fluid mechanics, the basic conservation laws are those of volume, energy, and momentum.   T ☐   F ☐

(d)   Absolute pressures and temperatures must be employed when using the ideal gas law.   T ☐   F ☐

(e)   The density of an ideal gas depends only on its absolute temperature and its molecular weight.   T ☐   F ☐

(f)   Closely, the density of water is 1,000 kg/m^3, and the gravitational acceleration is 9.81 m/s^2.   T ☐   F ☐

(g)   To convert pressure from gauge to absolute, add approximately 1.01 Pa.   T ☐   F ☐

(h)   To convert from psia to psig, add 14.7, approximately.   T ☐   F ☐

(i)   The absolute atmospheric pressure in the classroom is roughly one bar.   T ☐   F ☐

(j)   If ρ is density in g/cm^3 and μ is viscosity in g/cm s, then the kinematic viscosity $\nu = \mu/\rho$ is in stokes.   T ☐   F ☐

(k)   For a given liquid, surface tension and surface energy per unit area have identical numerical values *and* identical units.   T ☐   F ☐

(l)   A force is equivalent to a rate of transfer of momentum.   T ☐   F ☐

(m)   Work is equivalent to a rate of dissipation of power per unit time.   T ☐   F ☐.

(n)   It is possible to have gauge pressures that are as low as −20.0 psig.   T ☐   F ☐

(o)   The density of air in the classroom is roughly 0.08 kg/m^3.   T ☐   F ☐

(p)   Pressure in a static fluid varies in the vertically upward direction z according to $dp/dz = -\rho g_c$.   T ☐   F ☐

---

[14] Solutions to all the true/false assertions are given in Appendix B.

(q)   At any point, the rate of change of pressure with elevation is $dp/dz = -\rho g$, for both incompressible *and* compressible fluids.   T ☐   F ☐

(r)   A vertical pipe full of water, 34 ft high and open at the top, will generate a pressure of about one atmosphere (gauge) at its base.   T ☐   F ☐

(s)   The horizontal force on one side of a vertical circular disc of radius R immersed in a liquid of density ρ, with its center a distance R below the free surface, is $\pi R^3 \rho g$.   T ☐   F ☐

(t)   For a vertical rectangle or dam of width W and depth D, with its top edge submerged in a liquid of density ρ, as in Fig. 1.15, the total horizontal thrust of the liquid can also be expressed as $\int_0^D \rho g h W \, dh$, where h is the coordinate measured *downwards* from the free surface.   T ☐   F ☐

(u)   The horizontal pressure force on a rectangular dam with its top edge in the free surface is F_x. If the dam were made twice as deep, but still with the same width, the total force would be $2F_x$.   T ☐   F ☐

(v)   A solid object completely immersed in oil will experience the same upward buoyant force as when it is immersed in water.   T ☐   F ☐

(w)   Archimedes' law will not be true if the object immersed is hollow (such as an empty box with a tight lid, for example).   T ☐   F ☐

(x)   The rate of pressure change due to centrifugal action is given by $\partial p/\partial r = \rho r^2 \omega$, in which ω is the angular velocity of rotation.   T ☐   F ☐

(y)   To convert radians per second into rpm, divide by 120π.   T ☐   F ☐

(z)   The shape of the free surface of a liquid in a rotating container is a hyperbola.   T ☐   F ☐

(A)   The hydrostatic force exerted on one face of a square plate of side L that is held vertically in a liquid with one edge in the free surface is F. If the plate is lowered vertically by a distance L, the force on one face will be $3F$.   T ☐   F ☐

Chapter 2

MASS, ENERGY, AND MOMENTUM BALANCES

2.1 General Conservation Laws

THE study of fluid mechanics is based, to a large extent, on the conservation laws of three extensive quantities:

1. *Mass*—usually total, but sometimes of one or more individual chemical species.
2. Total *energy*—the sum of internal, kinetic, potential, and pressure energy.
3. *Momentum*, both linear and angular.

For a system viewed as a whole, *conservation* means that there is no net gain nor loss of any of these three quantities, even though there may be some redistribution of them within a system. A general *conservation law* can be phrased relative to the general system shown in Fig. 2.1, in which can be identified:

1. The *system* V.
2. The *surroundings* S.
3. The *boundary* B, also known as the *control surface*, across which the system interacts in some manner with its surroundings.

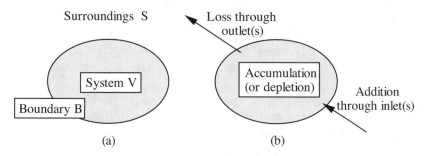

Fig. 2.1 (a) System and its surroundings; (b) transfers to and from a system. For a chemical reaction, creation and destruction terms would also be included inside the system.

The interaction between system and surroundings is typically by one or more of the following mechanisms:

1. A flowing stream, either entering or leaving the system.
2. A "contact" force on the boundary, usually normal or tangential to it, and commonly called a *stress*.
3. A "body" force, due to an external field that acts throughout the system, of which gravity is the prime example.
4. Useful work, such as electrical energy entering a motor or shaft work leaving a turbine.

Let X denote mass, energy, or momentum. Over a finite time period, the general conservation law for X is:

Nonreacting system

$$X_{\text{in}} - X_{\text{out}} = \Delta X_{\text{system}}. \tag{2.1a}$$

For a mass balance on species i in a reacting system

$$X^i_{\text{in}} - X^i_{\text{out}} + X^i_{\text{created}} - X^i_{\text{destroyed}} = \Delta X^i_{\text{system}}. \tag{2.1b}$$

The symbols are defined in Table 2.1. The understanding is that the creation and destruction terms, together with the superscript i, are needed only for mass balances on species i in chemical reactions, which will not be pursued further in this text.

Table 2.1   Meanings of Symbols in Equation (2.1)

Symbol	Meaning
X_{in}	Amount of X brought into the system
X_{out}	Amount of X taken out of the system
X_{created}	Amount of X created within the system
$X_{\text{destroyed}}$	Amount of X destroyed within the system
ΔX_{system}	Increase (accumulation) in the X-content of the system

It is very important to note that Eqn. (2.1a) cannot be applied indiscriminately, and is only observed *in general* for the *three* properties of *mass, energy,* and *momentum*. For example, it is *not* generally true if X is another extensive property such as volume, and is quite meaningless if X is an intensive property such as pressure or temperature.

In the majority of examples in this book, it *is* true that if X denotes mass *and* the density is constant, then Eqn. (2.1a) simplifies to the conservation of *volume,* but this is not the fundamental law. For example, if a gas cylinder is filled up by having nitrogen gas pumped into it, we would very much hope that the volume of the system (consisting of the cylinder and the gas it contains) does *not* increase by the volume of the (compressible) nitrogen pumped into it!

Equation (2.1a) can also be considered on a basis of unit time, in which case all quantities become *rates*; for example, ΔX_{system} becomes the *rate*, dX_{system}/dt, at which the X-content of the system is increasing, x_{in} (note the lower-case "x") would be the *rate* of transfer of X into the system, and so on, as in Eqn. (2.2):

$$x_{\text{in}} - x_{\text{out}} = \frac{dX_{\text{system}}}{dt}. \tag{2.2}$$

2.2 Mass Balances

The general conservation law is typically most useful when *rates* are considered. In that case, if X denotes mass M and x denotes a mass "rate" m (the symbol $\dot{m}$ can also be used) the *transient* mass balance (for a *nonreacting system*) is:

$$m_{\text{in}} - m_{\text{out}} = \frac{dM_{\text{system}}}{dt}, \tag{2.3}$$

in which the symbols have the meanings given in Table 2.2.

Table 2.2   Meanings of Symbols in Equation (2.3)

Symbol	Meaning
m_{in}	Rate of addition of mass into the system
m_{out}	Rate of removal of mass from the system
dM_{system}/dt	Rate of accumulation of mass in the system (will be negative for a depletion of mass)

The majority of the problems in this text will deal with *steady-state* situations, in which the system has the same appearance at all instants of time, as in the following examples:

1. A river, with a flow rate that is constant with time.
2. A tank that is draining through its base, but is also supplied with an identical flow rate of liquid through an inlet pipe, so that the liquid level in the tank remains constant with time.

Steady-state problems are generally easier to solve, because a time derivative, such as dM_{system}/dt, is zero, leading to an *algebraic* equation.

A few problems—such as that in Example 2.1—will deal with *unsteady-state* or *transient* situations, in which the appearance of the system changes with time, as in the following examples:

1. A river, whose level is being raised by a suddenly elevated dam gate downstream.
2. A tank that is draining through its base, but is not being supplied by an inlet stream, so that the liquid level in the tank falls with time.

Transient problems are generally harder to solve, because a time derivative, such as dM_{system}/dt, is retained, leading to a *differential* equation.

Example 2.1—Mass Balance for Tank Evacuation

The tank shown in Fig. E2.1(a) has a volume $V = 1$ m³ and contains air that is maintained at a constant temperature by being in thermal equilibrium with its surroundings.

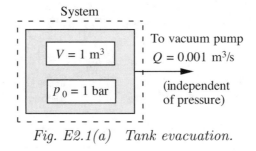

Fig. E2.1(a)   *Tank evacuation.*

If the initial absolute pressure is $p_0 = 1$ bar, how long will it take for the pressure to fall to a final pressure of 0.0001 bar if the air is evacuated at a constant rate of $Q = 0.001$ m³/s, at the pressure prevailing inside the tank at any time?

Solution

First, and fairly obviously, choose the tank as the system, shown by the dashed rectangle. Note that there is no inlet to the system, and just one outlet from it. A mass balance on the air in the system (noting that a rate of loss is negative) gives:

$$- \text{ Rate of loss of mass} = \text{Rate of accumulation of mass}$$

$$-Q\rho = \frac{d}{dt}(V\rho)$$

$$= V\frac{d\rho}{dt} + \rho\frac{dV}{dt} = V\frac{d\rho}{dt}. \qquad (E2.1.1)$$

Note that since the tank volume V is constant, $dV/dt = 0$. For an ideal gas:

$$\rho = \frac{Mp}{RT}, \qquad (E2.1.2)$$

so that:

$$-Q\frac{Mp}{RT} = V\frac{M}{RT}\frac{dp}{dt}. \qquad (E2.1.3)$$

Example 2.1—Mass Balance for Tank Evacuation   **59**

Cancellation of M/RT gives the following *ordinary differential equation*, which governs the variation of pressure p with time t:

$$\frac{dp}{dt} = -\frac{Q}{V} p. \tag{E2.1.4}$$

Separation of variables and integration between $t = 0$ (when the pressure is p_0) and a later time t (when the pressure is p) gives:

$$\int_{p_0}^{p} \frac{dp}{p} = -\frac{Q}{V} \int_{0}^{t} dt \qquad \text{or} \qquad \ln \frac{p}{p_0} = -\frac{Qt}{V}. \tag{E2.1.5}$$

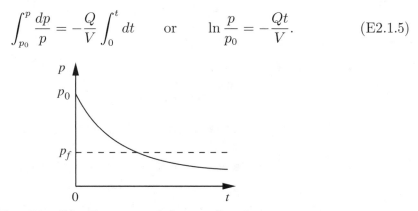

Fig. E2.1(b)   Exponential decay of tank pressure.

The resulting solution shows an exponential decay of the tank pressure with time, also illustrated in Fig. E2.1(b):

$$p = p_0 e^{-Qt/V}. \tag{E2.1.6}$$

Thus, the time t_f taken to evacuate the tank from its initial pressure of 1 bar to a final pressure of $p_f = 0.0001$ bar is:

$$t_f = -\frac{V}{Q} \ln \frac{p_f}{p_0} = -\frac{1}{0.001} \ln \frac{0.0001}{1} = 9{,}210 \text{ s} = 153.5 \text{ min.} \tag{E2.1.7}$$

Problem 2.1 contains a variation of the above, in which air is leaking slowly into the tank from the surrounding atmosphere.   □

Steady-state mass balance for fluid flow.   A particularly useful and simple mass balance—also known as the *continuity equation*—can be derived for the situation shown in Fig. 2.2, where the system resembles a wind sock at an airport. At station 1, fluid flows steadily with density ρ_1 and a uniform velocity u_1 normally across that part of the surface of the system represented by the area A_1. In steady flow, each fluid particle traces a path called a *streamline*. By considering a large number of particles crossing the closed curve C, we have an equally large number of streamlines that then form a surface known as a *stream tube*, across which there is clearly no flow. The fluid then leaves the system with uniform velocity u_2 and density ρ_2 at station 2, where the area normal to the direction of flow is A_2.

Referring to Eqn. (2.3), there is no accumulation of mass because the system is at steady state. Therefore, the only nonzero terms are m_1 (the rate of addition of mass) and m_2 (the rate of removal of mass), which are equal to $\rho_1 A_1 u_1$ and $\rho_2 A_2 u_2$, respectively, so that Eqn. (2.3) becomes:

$$\underbrace{\rho_1 A_1 u_1}_{m_1 \text{ in}} - \underbrace{\rho_2 A_2 u_2}_{m_2 \text{ out}} = 0 \quad \text{(steady state)}, \tag{2.4a}$$

which can be rewritten as:

$$\rho_1 A_1 u_1 = \rho_2 A_2 u_2 = m, \tag{2.4b}$$

where m $(= m_1 = m_2)$ is the mass flow rate entering and leaving the system.

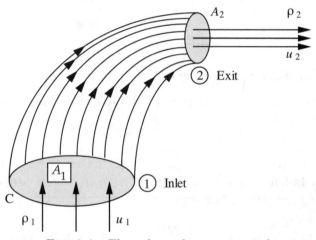

Fig. 2.2   Flow through a stream tube.

For the special but common case of an *incompressible fluid*, $\rho_1 = \rho_2$, so that the steady-state mass balance becomes:

$$A_1 u_1 = A_2 u_2 = \frac{m}{\rho} = Q, \tag{2.5}$$

in which Q is the *volumetric* flow rate.

Equations (2.4a/b) would also apply for nonuniform inlet and exit velocities, if the appropriate *mean* velocities u_{m1} and u_{m2} were substituted for u_1 and u_2. However, we shall postpone the concept of nonuniform velocity distributions to a more appropriate time, particularly to those chapters that deal with *microscopic* fluid mechanics.

2.3 Energy Balances

Equation (2.1) is next applied to the general system shown in Fig. 2.3, it being understood that property X is now *energy*. Observe that there is both flow into and from the system. Also note the quantities defined in Table 2.3.

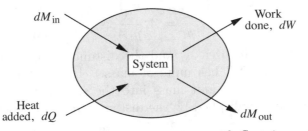

Fig. 2.3 Energy balance on a system with flow in and out.

A differential energy balance results by applying Eqn. (2.1) over a short time period. Observe that there are two transfers into the system (incoming mass and heat) and two transfers out of the system (outgoing mass and work). Since the mass transfers also carry energy with them, there results:

$$dM_{in}\left(e + \frac{p}{\rho} + gz + \frac{u^2}{2}\right)_{in} - dM_{out}\left(e + \frac{p}{\rho} + gz + \frac{u^2}{2}\right)_{out}$$

$$+ dQ - dW = d\left[M\left(e + gz + \frac{u^2}{2}\right)\right]_{system}, (2.6)$$

in which each term has units of energy or work. In the above, the system is assumed for simplicity to be *homogeneous*, so that all parts of it have the same internal, potential, and kinetic energy per unit mass; if such were not the case, integration would be needed throughout the system. Also, multiple inlets and exits could be accommodated by means of additional terms.

Table 2.3 Definitions of Symbols for Energy Balance

Symbol	Definition
dM_{in}	Differential amount of mass entering the system
dM_{out}	Differential amount of mass leaving the system
dQ	Differential amount of heat added *to* the system
dW	Differential amount of useful work done *by* the system
e	Internal energy per unit mass
g	Gravitational acceleration
M	Mass of the system
u	Velocity
ρ	Density

Since the density ρ is the reciprocal of v, the volume per unit mass, $e + p/\rho = e + pv$, which is recognized as the *enthalpy* per unit mass. The *flow energy* term p/ρ in Eqn. (2.6), also known as injection work or flow work, is readily explained by examining Fig. 2.4. Consider unit mass of fluid entering the stream tube under a pressure p_1. The volume of the unit mass is:

$$\frac{1}{\rho_1} = A_1 \frac{1}{\rho_1 A_1}, \tag{2.7}$$

which is the product of the area A_1 and the distance $1/\rho_1 A_1$ through which the mass moves. (Here, the "1" has units of mass.) Hence, the work done on the system by p_1 in pushing the unit mass into the stream tube is the force $p_1 A_1$ exerted by the pressure multiplied by the distance through which it travels:

$$p_1 A_1 \times \frac{1}{\rho_1 A_1} = \frac{p_1}{\rho_1}. \tag{2.8}$$

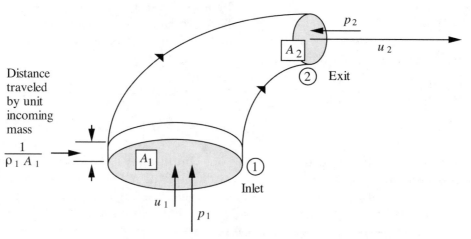

Fig. 2.4   Flow of unit mass to and from stream tube.

Likewise, the work done by the system on the surroundings at the exit is:

$$p_2 A_2 \times \frac{1}{\rho_2 A_2} = \frac{p_2}{\rho_2}. \tag{2.9}$$

Steady-state energy balance.   In the following, all quantities are per unit mass flowing. Referring to the general system shown in Fig. 2.5, the energy entering with the inlet stream plus the heat supplied to the system must equal the energy leaving with the exit stream plus the work done by the system on its surroundings. Therefore, the right-hand side of Eqn. (2.6) is zero under steady-state conditions, and division by $dM_{\text{in}} = dM_{\text{out}}$ gives:

$$e_1 + \frac{u_1^2}{2} + gz_1 + \frac{p_1}{\rho_1} + q = e_2 + \frac{u_2^2}{2} + gz_2 + \frac{p_2}{\rho_2} + w, \tag{2.10}$$

in which each term represents an energy per unit mass flowing.

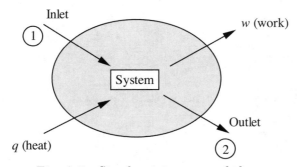

Fig. 2.5   Steady-state energy balance.

For an infinitesimally small system in which *differential* changes are occurring, Eqn. (2.10) may be rewritten as:

$$de + d\left(\frac{u^2}{2}\right) + d(gz) + d(pv) = dq - dw, \qquad (2.11)$$

in which, for example, de is now a *differential* change, and $v = 1/\rho$ is the volume per unit mass.

Now examine the increase in internal energy de, which arises from frictional work $d\mathcal{F}$ dissipated into heat, heat addition dq from the surroundings, less work pdv done by the fluid. That is: $de = d\mathcal{F} + dq - pdv$. Thus, eliminate the change de in the internal energy from Eqn. (2.11), and expand the term $d(pv)$,

$$\underbrace{d\mathcal{F} + dq - pdv}_{de} + d\left(\frac{u^2}{2}\right) + d(gz) + \underbrace{pdv + vdp}_{d(pv)} = dq - dw, \qquad (2.12)$$

which simplifies to the differential form of the *mechanical* energy balance, in which heat terms are absent:

$$d\left(\frac{u^2}{2}\right) + d(gz) + \frac{dp}{\rho} + dw + d\mathcal{F} = 0. \qquad (2.13)$$

For a finite system, for flow from point 1 to point 2, Eqn. (2.13) integrates to:

$$\Delta\left(\frac{u^2}{2}\right) + \Delta(gz) + \int_1^2 \frac{dp}{\rho} + w + \mathcal{F} = 0, \qquad (2.14)$$

in which a finite change is consistently the final minus the initial value, for example:

$$\Delta\left(\frac{u^2}{2}\right) = \frac{u_2^2}{2} - \frac{u_1^2}{2}. \qquad (2.15)$$

An energy balance for an *incompressible* fluid of constant density permits the integral to be evaluated easily, giving:

$$\Delta\left(\frac{u^2}{2}\right) + \Delta(gz) + \frac{\Delta p}{\rho} + w + \mathcal{F} = 0. \tag{2.16}$$

In the majority of cases, g will be virtually constant, in which case there is a further simplification to:

$$\Delta\left(\frac{u^2}{2}\right) + g\Delta z + \frac{\Delta p}{\rho} + w + \mathcal{F} = 0, \tag{2.17}$$

which is a *generalized Bernoulli* equation, *augmented* by two extra terms—the frictional dissipation, $\mathcal{F}$, and the work w done by the system. Note that $\mathcal{F}$ can *never* be negative—it is impossible to convert heat entirely into useful work. The work term w will be positive if the fluid flows through a turbine and performs work on the environment; conversely, it will be negative if the fluid flows through a pump and has work done on it.

Power.   The *rate* of expending energy in order to perform work is known as *power*, with dimensions of ML^2/T^3, typical units being W (J/s) and ft lb$_f$/s. The relations in Table 2.4 are available, depending on the particular context.

Table 2.4   Expressions for Power in Different Systems

System	Expression for P
Flowing stream:	mw (m = mass flow rate, w = work per unit mass)
Force displacement:	Fv (F = force, v = displacement velocity)
Rotating shaft:	$T\omega$ (T = torque, ω = angular velocity of rotation)
Pump:	$Q\Delta p$ (Q = volume flow rate, Δp = pressure increase)

Example 2.2—Pumping *n*-Pentane   **65**

Example 2.2—Pumping *n*-Pentane

Fig. E2.2   Pumping n-pentane.

Fig. E2.2 shows an arrangement for pumping *n*-pentane ($\rho = 39.3$ lb$_\mathrm{m}$/ft^3) at 25 °C from one tank to another, through a vertical distance of 40 ft. All piping is 3-in. I.D. Assume that the overall frictional losses in the pipes are given (by methods to be described in Chapter 3) by:

$$\mathcal{F} = 2.5\ u_\mathrm{m}^2\ \frac{\mathrm{ft}^2}{\mathrm{s}^2} = \frac{2.5 u_\mathrm{m}^2}{g_c}\ \frac{\mathrm{ft\ lb_f}}{\mathrm{lb_m}}. \tag{E2.2.1}$$

For simplicity, however, you may ignore friction in the short length of pipe leading to the pump inlet. Also, the pump and its motor have a combined efficiency of 75%. If the mean velocity u_m is 25 ft/s, determine the following:

(a) The power required to drive the pump.

(b) The pressure at the inlet of the pump, and compare it with 10.3 psia, which is the vapor pressure of *n*-pentane at 25 °C.

(c) The pressure at the pump exit.

Solution

The cross-sectional area of the pipe and the mass flow rate are:

$$A = \frac{\pi}{4}\left(\frac{3}{12}\right)^2 = 0.0491\ \mathrm{ft}^2, \tag{E2.2.2}$$

$$m = \rho u_\mathrm{m} A = 39.3 \times 25 \times 0.0491 = 48.2\ \frac{\mathrm{lb_m}}{\mathrm{s}}. \tag{E2.2.3}$$

Since the supply tank is fairly large, the liquid/vapor interface in it is descending only very slowly, so that $u_1 \doteq 0$. An energy balance between points 1 and 4 (where both pressures are atmospheric, so that $\Delta p = 0$) gives:

$$\Delta\left(\frac{u^2}{2}\right) + g\Delta z + \frac{\Delta p}{\rho} + w + \mathcal{F} = 0, \tag{E2.2.4}$$

$$\frac{25^2 - 0^2}{2} + 32.2 \times 40 + 0 + w + 2.5 \times 25^2 = 0. \tag{E2.2.5}$$

Hence, the work per unit mass flowing is:

$$w = -3{,}163\ \frac{\text{ft}^2}{\text{s}^2} = -\frac{3{,}163\ \text{ft}^2/\text{s}^2}{32.2\ \text{lb}_\text{m}\ \text{ft}/\text{lb}_\text{f}\ \text{s}^2} = -98.3\ \frac{\text{ft lb}_\text{f}}{\text{lb}_\text{m}}, \tag{E2.2.6}$$

in which the minus sign indicates that work is done *on* the liquid. The power required to drive the pump motor is:

$$P = mw = \frac{48.2 \times 98.3}{737.6 \times 0.75} = 8.56\ \text{kW}. \tag{E2.2.7}$$

The pressure at the inlet to the pump is obtained by applying Bernoulli's equation between points 1 and 2:

$$\frac{u_1^2}{2} + \frac{p_1}{\rho} + gz_1 = \frac{u_2^2}{2} + \frac{p_2}{\rho} + gz_2. \tag{E2.2.8}$$

Since the pipe has the same diameter throughout (3 in.), the velocity u_2 entering the pump is the same as that in the vertical section of pipe, 25 ft/s. Solving for the pressure at the pump inlet,

$$p_2 = \rho\left[g(z_1 - z_2) - \frac{u_2^2}{2}\right] = \frac{39.3}{32.2 \times 144}\left(32.2 \times 4.5 - \frac{25^2}{2}\right)$$

$$\doteq -1.42\ \text{psig} = 14.7 - 1.42 = 13.28\ \text{psia}. \tag{E2.2.9}$$

(For better accuracy, friction in the short length of pipe between the tank and the pump inlet should be included, particularly in the context of Chapter 3; however, it will be fairly small and we are justified in ignoring it here.) Note that p_2 is *above* the vapor pressure of n-pentane, which therefore remains as a liquid as it enters the pump. If p_2 were less than 10.3 psia, the n-pentane would tend to vaporize and the pump would not work because of *cavitation*.

The pressure at the pump exit is most readily found by applying an energy balance across the pump. Since there is no change in velocity (the inlet and outlet lines have the same diameter), and no frictional pipeline dissipation:

$$g\Delta z + \frac{\Delta p}{\rho} + w = 0. \tag{E2.2.10}$$

But $p_2 = -1.42$ psig, $w = -3{,}163$ ft^2/s^2, and $\Delta z = 0.5$ ft, so that:

$$p_3 = -1.42 + \frac{39.3}{32.2 \times 144} \times (3{,}163 - 32.2 \times 0.5) = 25.2 \text{ psig.} \qquad \text{(E2.2.11)}$$

The same result could have been obtained by applying the energy balance between points 3 and 4. □

2.4 Bernoulli's Equation

Situations frequently occur in which the following simplifying assumptions can reasonably be made:

1. The flow is steady.
2. There are no work effects; that is, the fluid neither performs work (as in a turbine) nor has work performed on it (as in a pump). Thus, $w = 0$ in Eqn. (2.17).
3. The flow is frictionless, so that $\mathcal{F} = 0$ in Eqn. (2.17). Clearly, this assumption would *not* hold for *long* runs of pipe.
4. The fluid is incompressible; that is, the density is constant. This approximation is excellent for the majority of liquids, and may also be reasonable for some cases of gas flows provided that the pressure variations are moderately small.

Under these circumstances, the general energy balance reduces to:

$$\Delta\left(\frac{u^2}{2}\right) + \Delta(gz) + \frac{\Delta p}{\rho} = 0, \qquad (2.18)$$

which is the famous *Bernoulli's equation*, one of the most important relations in fluid mechanics.

For flow between points 1 and 2 on the *same* streamline, or for any two points in a fluid under static equilibrium (in which case the velocities are zero), Eqn. (2.18) becomes:

$$\underbrace{\frac{u_1^2}{2} + gz_1 + \frac{p_1}{\rho}}_{\text{Total energy at point 1}} = \underbrace{\frac{u_2^2}{2} + gz_2 + \frac{p_2}{\rho}}_{\text{Total energy at point 2}}, \qquad (2.19)$$

which states that although the kinetic, potential, and pressure energies may vary individually, their sum remains constant. Each term in (2.19) must have the same dimensions as the first one, namely, velocity squared or L^2/T^2. Further manipulations in the two principal systems of units yield the following:

1. SI Units

$$\frac{\text{m}^2}{\text{s}^2} = \frac{\text{m}}{\text{kg}} \underbrace{\text{kg} \frac{\text{m}}{\text{s}^2}}_{\text{Force(N)}} = \frac{\text{m N}}{\text{kg}} = \frac{\text{J}}{\text{kg}}, \qquad (2.20)$$

which is readily seen to be energy per unit mass.

Bernoulli, Daniel, born 1700 in Groningen, Holland; died 1782 in Basel, Switzerland. He was the middle of three sons born to Jean Bernoulli, himself chair of mathematics, first at Groningen and later at Basel. The meanness and jealousy of his father discouraged Daniel from continuing his career in medicine, and he became professor of mathematics at St. Petersburg in 1725. In poor health in 1733, he rejoined his family in Basel, where he was appointed professor of anatomy and botany. In 1738 he published his treatise *Hydrodynamica*, which dealt with the interaction between velocities and pressures, and also included the concept of a jet-propelled boat. He received or shared in many prizes from the Academy of Sciences in Paris, including ones related to the measurement of time at sea (important for determining longitude), the inclination of planetary orbits, and tides. He enjoyed a friendly rivalry with the Swiss mathematician Leonhard Euler (1707–1783). Afflicted with asthma in his later life, he devoted much time to the study of probability applied to practical subjects. He recalled with pleasure that when in his youth he introduced himself to a traveling companion by saying "I am Daniel Bernoulli," the reply was "And I am Isaac Newton."

Source: *The Encyclopædia Britannica*, 11th ed., Cambridge University Press (1910–1911).

2. English Units

$$\frac{\text{ft}^2}{\text{s}^2} = \frac{\text{ft}}{\text{lb}_\text{m}} \underbrace{\text{lb}_\text{m} \frac{\text{ft}}{\text{s}^2}}_{\text{poundal}} = \frac{\text{ft poundal}}{\text{lb}_\text{m}}, \tag{2.21}$$

which is again energy per unit mass. However, since the poundal is an archaic unit of force, each term in Bernoulli's equation may be divided by the conversion factor g_c ft $\text{lb}_\text{m}/\text{lb}_\text{f}$ s^2 if the more practical lb_f is required in numerical calculations. For example, the kinetic energy term then becomes:

$$\frac{\text{u}_1^2}{2\text{g}_\text{c}} \; [=] \; \frac{\text{ft lb}_\text{f}}{\text{lb}_\text{m}} \; \text{(energy per unit mass)}. \tag{2.22}$$

As previously indicated, we prefer not to include g_c—or any other conversion factors—directly in these equations.

Head of fluid. A quantity closely related to energy per unit mass may be obtained by dividing (2.19) through by the gravitational acceleration g:

$$\underbrace{\frac{u_1^2}{2g} + z_1 + \frac{p_1}{\rho g}}_{\left(\substack{\text{Velocity + static} \\ \text{+ pressure heads}}\right)} = \underbrace{\frac{u_2^2}{2g} + z_2 + \frac{p_2}{\rho g}}_{\left(\substack{\text{Velocity + static} \\ \text{+ pressure heads}}\right)} = H. \tag{2.23}$$

Each term in (2.23) has dimensions of length, and indeed the terms such as $u_1^2/2g$, z_1, $p_1/\rho g$, and H are called the *velocity* head, *static* head, *pressure* head, and *total* head, respectively.

(a)   (b)

Fig. 2.6 (a) Physical interpretation of velocity, static, and pressure heads for pipe flow, and (b) pressure head increase across a pump.

A physical interpretation of fluid head is readily available by considering the steady flow of a liquid from a tank through the idealized frictionless pipe shown in Fig. 2.6(a). At point 1 (the free surface), the velocity u_1 is virtually zero for a tank of reasonable size, and the pressure p_1 there is also zero (gauge) because the free surface is exposed to the atmosphere. Thus, the velocity and pressure heads are both zero at point 1, so that the total head (H for example) is identical with the static head z_1, namely, the elevation of the free surface relative to some datum level. Hence, Eqn. (2.23) can be rewritten as:

$$z_1 = H = \underbrace{\frac{u_2^2}{2g} + z_2 + \frac{p_2}{\rho g}}. \qquad (2.24)$$

<small>Each of these terms is
interpreted in Fig. 2.6</small>

Looking now at point 2, the static head is simply the elevation z_2 of that point above the datum level; the pressure head is the height above point 2 to which the liquid rises in the manometer—an amount that is just sufficient to balance the pressure p_2; and, by difference, the elevation difference between the top of the liquid in the tube and point 1 must be the velocity head $u_2^2/2g$.

Since the pipe diameter is constant, continuity also requires the velocity and hence the velocity head $u_2^2/2g$ to be constant. Referring to Fig. 2.6(a), since the static head continuously *decreases* along the pipe, and the total head is constant, the pressure head must constantly *increase*. But since the pressure at the exit— or very shortly after it—is atmospheric, the pressure head must again be zero! The reader will doubtless ask: "Is there an anomaly?", and may wish to ponder whether or not the diagram is completely accurate as drawn.

Fig. 2.6(b) shows that the pressure increase Δp across a pump is also equivalent to a head increase $\Delta h = \Delta p/\rho g$, being the increase in liquid levels in piezometric tubes placed at the pump inlet and exit.

Note carefully that the above analysis is for an ideal liquid—one that exhibits no friction. In practice, there would be some loss in total head along the pipe.

2.5 Applications of Bernoulli's Equation

We now apply Eqn. (2.18) to several commonly occurring situations, in which useful relations involving pressures, velocities, and elevations may be obtained. The usual assumptions of steady flow, no external work, no friction, and constant density may reasonably be made in each case.

Fig. 2.7   Tank draining through a rounded orifice.

Tank draining.   First, consider Fig. 2.7, in which a tank is draining through an orifice of cross-sectional area A in its base. If the orifice is *rounded*, the streamlines will be parallel with one another at the exit and the pressure will be uniformly atmospheric there. The elevation of the free surface at point 1 is h above the orifice, where z_2 is taken to be zero. There are no work effects between 1 and 2, and the fluid is incompressible. Also, because the liquid is descending quite slowly, there is essentially no frictional dissipation (unless the liquid is very viscous) and the flow is virtually steady. Hence, Eqn. (2.19) can be applied between 1 and 2, giving:

$$\frac{u_1^2}{2} + gh + \frac{p_1}{\rho} = \frac{u_2^2}{2} + 0 + \frac{p_2}{\rho}. \tag{2.25}$$

But $u_1 \doteq 0$ if the cross-sectional area of the tank is appreciably larger than that of the orifice; also, $p_1 = p_2$, since both are atmospheric pressure. Equation (2.25) then reduces to:

$$u_2 = \sqrt{2gh}, \tag{2.26}$$

so the exit velocity of the liquid is exactly commensurate with a free fall under gravity through a vertical distance h. The corresponding volumetric flow rate is:

$$Q = Au_2 = A\sqrt{2gh}. \tag{2.27}$$

However, if the orifice is *sharp-edged* with area A, as shown in Fig. 2.8, the cross-sectional area of the jet continues to contract after it leaves the orifice because of its inertia to a value a at a location, known as the *vena contracta*,[1] where the streamlines are parallel to one another. In this case, if C_c is the *coefficient of contraction*, the following relations give the area of the *vena contracta* and the total flow rate:

$$a = C_c A, \tag{2.28}$$

$$Q = C_c A\sqrt{2gh}, \tag{2.29}$$

in which the *coefficient of contraction* is found in most instances to have the value:

$$C_c \doteq 0.63. \tag{2.30}$$

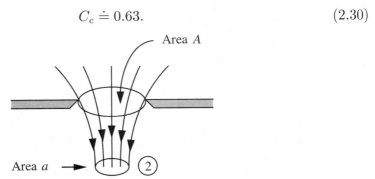

Fig. 2.8 Contraction of the jet through a sharp-edged orifice.

Orifice-plate "meter." The Bernoulli principle—of a decrease in pressure in an accelerated stream—can be employed for the measurement of fluid flow rates in the device shown in Fig. 2.9. There, an *orifice plate* consisting of a circular disc with a central hole of area A_o is bolted between the flanges on two sections of pipe of cross-sectional area A_1.

Bernoulli's equation applies to the fluid as it flows from left to right through the orifice of a reduced area because it is found experimentally that a *contracting* stream is relatively stable, so that frictional dissipation can be ignored, especially over such a short distance. Hence, as the velocity *increases*, the pressure *decreases*. The following theory demonstrates that by measuring the pressure drop $p_1 - p_2$, it is possible to determine the upstream velocity u_1. Let u_2 be the velocity of the jet at the *vena contracta*.

---

[1] Latin for "constricted jet."

Fig. 2.9   Flow through an orifice plate.

Bernoulli's equation applied between points 1 and 2, which have the same elevation ($z_1 = z_2$), gives:

$$\frac{u_1^2}{2} + \frac{p_1}{\rho} = \frac{u_2^2}{2} + \frac{p_2}{\rho}. \tag{2.31}$$

Conservation of mass between points 1 and 2 gives the *continuity equation*:

$$u_1 A_1 = u_2 A_2. \tag{2.32}$$

Elimination of u_2 between Eqns. (2.31) and (2.32) gives:

$$\frac{u_1^2}{2} + \frac{p_1}{\rho} = \frac{u_1^2}{2}\frac{A_1^2}{A_2^2} + \frac{p_2}{\rho}. \tag{2.33}$$

Solution for u_1 yields:

$$u_1 = \sqrt{\frac{2(p_1 - p_2)}{\rho\left(\dfrac{A_1^2}{A_2^2} - 1\right)}}, \tag{2.34}$$

so that the volumetric flow rate Q is:

$$Q = u_1 A_1 = A_1 \sqrt{\frac{2(p_1 - p_2)}{\rho\left(\dfrac{A_1^2}{A_2^2} - 1\right)}} = A_1 \sqrt{\frac{2(p_1 - p_2)}{\rho\left(\dfrac{A_1^2}{C_c^2 A_o^2} - 1\right)}}. \tag{2.35}$$

In Eqn. (2.35), the coefficient of contraction C_c is approximately 0.63 in most cases. However, the following version, which is somewhat less logical than Eqn.

(2.35) and uses a dimensionless *discharge coefficient* C_D, is used in practice instead:

$$Q = C_D A_1 \sqrt{\frac{2(p_1 - p_2)}{\rho \left(\dfrac{A_1^2}{A_o^2} - 1\right)}}.$$

(2.36)

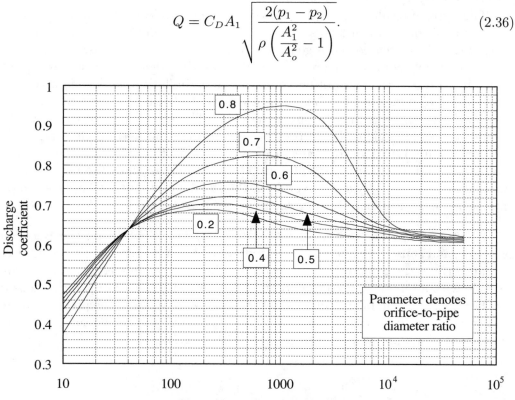

Fig. 2.10   Discharge coefficient for orifice plates.   Based on values from G.G. Brown et al., Unit Operations, *John Wiley & Sons, New York, 1950.*

Fig. 2.10 shows how C_D varies with two additional dimensionless groups, namely, the ratio of the orifice diameter to the pipe diameter, and the *Reynolds number* through the orifice:

$$\frac{D_o}{D_1}, \qquad \mathrm{Re}_o = \frac{u_o D_o \rho}{\mu}.$$

(2.37)

It is easy to show that the Reynolds number at the orifice is given in terms of the upstream Reynolds number Re_1 (which is usually more readily available) by the relation:

$$\mathrm{Re}_o = \frac{D_1}{D_o} \mathrm{Re}_1.$$

(2.38)

Downstream of the *vena contracta*, the jet is found experimentally to become unstable as it expands back again to the full cross-sectional area of the pipe; the subsequent turbulence and loss of useful work means that Bernoulli's equation *cannot* be used in the downstream section.

This type of result—the expression of one dimensionless group in terms of one or more additional dimensionless groups—will occur many times throughout the text, and will be treated more generally in Section 4.5 on *dimensional analysis*. For the moment, however, the reader should note that these dimensionless groups typically represent the ratio of one quantity to a similarly related quantity. The orifice-to-pipe diameter ratio has an obvious geometrical interpretation, stating how large the orifice is in relation to the pipe. The significance of the Reynolds number is not so obvious, but represents the ratio of an inertial effect (given by ρu^2, for example), to a viscous effect (already given in Eqn. (1.14) by the product of viscosity and a velocity gradient—$\mu u/D$, for example); thus, the ratio of these two quantities is $(\rho u^2)/(\mu u/D) = \rho u D/\mu$, the Reynolds number. Chapters 3 and 4 emphasize that turbulence is more likely to occur at higher Reynolds numbers.

Pitot tube.   The device shown in Fig. 2.11 is also based on the Bernoulli principle, and is used for finding the velocity of a moving craft such as a boat or an airplane. Here, the *Pitot tube* is attached to a boat, for example, which is moving steadily with an unknown velocity u_1 through otherwise stagnant water of density ρ. The submerged tip of the tube faces the direction of motion; the pressure at the tip can be found from the height h to which the water rises in the tube or—more practically—by a pressure transducer (see Section 2.7).

Fig. 2.11   Pitot tube.

For simplicity, consider the motion relative to an observer on the boat, in which case the Pitot tube is effectively stationary, with water approaching it with an upstream velocity u_1. Opposite the Pitot tube, the oncoming water decelerates and comes to rest at the *stagnation point* at the tip of the tube.

Application of Bernoulli's equation between points 1 and 2 gives:

$$\frac{p_1}{\rho} + \frac{u_1^2}{2} + 0 = \frac{p_2}{\rho} + \frac{0^2}{2} + g(h+d), \tag{2.39}$$

in which the first zero recognizes the datum level $z_1 = 0$ at point 1, and the second zero indicates that the water is stagnant with $u_2 = 0$ at point 2. But from hydrostatics, the pressure at point 1 is:

$$p_1 = p_2 + \rho g d. \tag{2.40}$$

Subtraction of Eqn. (2.40) from Eqn. (2.39) gives:

$$u_1 = \sqrt{2gh}, \tag{2.41}$$

so that the velocity u_1 of the boat is readily determined from the height of the water in the tube. In practice, a pressure transducer would probably be used for monitoring the excess pressure (corresponding to $\rho g h$) instead of measuring the water level, but we have retained the latter because it is conceptually simpler.

Fig. 2.12   Pitot-static tube.

A very similar device, called the Pitot-*static* tube, is shown in Fig. 2.12, and is employed for measuring the velocity at different radial locations in a pipe. Here, two tubes are involved. The left-hand tube simply measures the pressure and the movable right-hand one is essentially a Pitot tube as before. The velocity u_1 at the particular transverse location where the Pitot tube is placed is given by:

$$u_1 = \sqrt{2gh}. \tag{2.42}$$

Example 2.3—Tank Filling

Fig. E2.3   Tank filling from river: (a) before
pipe is submerged; (b) after pipe is submerged.

Fig. E2.3 shows a concrete tank that is to be filled with water from an adjacent river in order to provide a supply of water for the sprinklers on a golf course. The level of the river is $H = 10$ ft above the base of the tank, and the short connecting pipe, which offers negligible resistance, discharges water at a height $D = 4$ ft above the base of the tank. The inside cross-sectional area of the pipe is $a = 0.1$ ft^2, and that of the tank is $A = 1,000$ ft^2. Derive an algebraic expression for the time t taken to fill the tank, and then evaluate it for the stated conditions.

Solution

The solution is in two parts—before and after the pipe is submerged under the water in the tank, corresponding to (a) and (b) in Fig. E2.3. In both cases, the (gauge) atmospheric pressure is zero.

Part 1. Here, the flow rate is constant, since the water level in the tank is below the pipe outlet and hence offers no resistance to the flow. Bernoulli's equation, applied between the surface of the water in the river and the pipe discharge to the atmosphere, gives:

$$gH = gD + \frac{1}{2}u^2, \tag{E2.3.1}$$

so that:

$$u = \sqrt{2g(H - D)}. \tag{E2.3.2}$$

The volumetric flow rate is constant at ua, and the time for the water to reach the level of the pipe is the gain in volume divided by the flow rate:

$$t_1 = \frac{AD}{a\sqrt{2g(H - D)}}. \tag{E2.3.3}$$

Example 2.3—Tank Filling   **77**

Part 2. For the *submerged* pipe, as in Fig. E2.3(b), the flow rate is now variable, because it is influenced by the level in the tank. The discharge pressure equals the hydrostatic pressure $\rho g(h - D)$ due to the water in the tank above the discharge point. Bernoulli's equation, again applied between the surface of the water in the river and the pipe discharge, yields:

$$gH = gD + \frac{\rho g(h - D)}{\rho} + \frac{1}{2}u^2, \qquad (E2.3.4)$$

or

$$u = \sqrt{2g(H - h)}. \qquad (E2.3.5)$$

A volumetric balance (the density is constant) equates the flow rate into the tank to the rate of increase of volume of water in the tank:

$$ua = A\frac{dh}{dt}. \qquad (E2.3.6)$$

Separation of variables and integration (see Appendix A for a wide variety of standard forms) gives:

$$\frac{a}{A}\int_0^{t_2} dt = \int_D^H \frac{dh}{\sqrt{2g(H - h)}}.$$

The filling time for Part 2 is therefore:

$$t_2 = \frac{A}{a}\left[-\frac{1}{g}\sqrt{2g(H - h)}\right]_D^H = \frac{A}{a}\sqrt{\frac{2(H - D)}{g}}. \qquad (E2.3.7)$$

The total filling time is obtained by adding Eqns. (E2.3.3) and (E2.3.7):

$$
\begin{aligned}
t = t_1 + t_2 &= \frac{A}{a}\left[\frac{D}{\sqrt{2g(H - D)}} + \sqrt{\frac{2(H - D)}{g}}\right] \\
&= \frac{1,000}{0.1}\left[\frac{4}{\sqrt{2 \times 32.2 \times (10 - 4)}} + \sqrt{\frac{2 \times (10 - 4)}{32.2}}\right] \\
&= 10,000 \times (0.203 + 0.610) = 2,030 + 6,100 = 8,130 \text{ s} = 2.26 \text{ hr.} \qquad \square
\end{aligned}
$$

2.6   Momentum Balances

Momentum.   The general conservation law also applies to *momentum* $\mathcal{M}$, which for a mass M moving with a velocity u, as in Fig. 2.13(a), is defined by:

$$\mathcal{M} \equiv Mu \ [=] \ \frac{ML}{T}. \tag{2.43}$$

Strangely, there is no universally accepted symbol for momentum. In this text and elsewhere, the symbols M and m (or $\dot{m}$) frequently denote mass and mass flow rate, respectively. Therefore, we arbitrarily denote momentum and the rate of transfer of momentum due to flow by the symbols $\mathcal{M}$ ("script" M) and $\dot{\mathcal{M}}$, respectively. Momentum is a *vector* quantity, and for the simple case shown has the direction of the velocity u. More generally, there may be velocity components u_x, u_y, and u_z (sometimes also written as v_x, v_y, and v_z, or as u, v, and w) in each of the three coordinate directions, illustrated for Cartesian coordinates in Fig. 2.13(b). In this case, the momentum of the mass M has components Mu_x, Mu_y, and Mu_z in the x, y, and z directions.

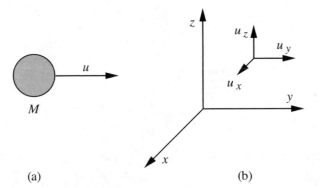

(a)   (b)

Fig. 2.13 (a) Momentum as a product of mass and velocity;
(b) velocity components in the three coordinate directions.

For problems in more than one dimension, conservation of momentum or a *momentum balance* applies in each of the coordinate directions. For example, for the basketball shown in Fig. 2.14, momentum Mu_x in the x direction remains almost constant (drag due to the air would reduce it slightly), whereas the upward momentum Mu_z is constantly diminished—and eventually reversed in sign—by the downward gravitational force.

If a system, such as a river, consists of several parts each moving with different velocities **u** (**boldface** denotes a vector quantity), the total momentum of the system is obtained by integrating over all of its mass:

$$\mathcal{M} = \int_M \mathbf{u} \, dM. \tag{2.44}$$

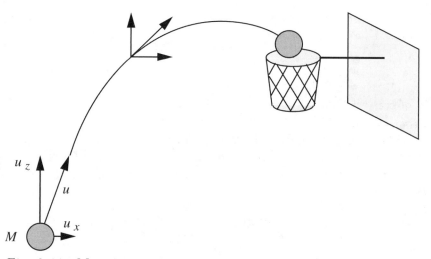

Fig. 2.14   Momentum varies with time in the x and y directions.

Law of momentum conservation.   Following the usual law, Eqn. (2.2), the net rate of transfer of momentum into a system equals the rate of increase of the momentum of the system. The question immediately arises: "How can momentum be transferred?" The answer is that there are two principal modes, by a *force* and by *convection,* as follows in more detail.

1. Momentum Transfer by a Force

A *force* is readily seen to be equivalent to a *rate of transfer of momentum* by examining its dimensions:

$$\text{Force } [=] \ \frac{\text{ML}}{\text{T}^2} = \frac{\text{ML/T}}{\text{T}} = \frac{\text{momentum}}{\text{time}}. \tag{2.45}$$

In fluid mechanics, the most frequently occurring forces are those due to pressure (which acts normal to a surface), shear stress (which acts tangentially to a surface), and gravity (which acts vertically downwards). Pressure and stress are examples of *contact* forces, since they occur over some region of contact with the surroundings of the system. Gravity is also known as a *body* force, since it acts throughout a system.

A momentum balance can be applied to a mass M falling with instantaneous velocity u under gravity in air that offers negligible resistance. Considering momentum as positive downwards, the rate of transfer of momentum to the system (the mass M) is the gravitational force Mg, and is equated to the rate of increase of downwards momentum of the mass, giving:

$$F = Mg = \frac{d}{dt}(\mathcal{M}) = \frac{d}{dt}(Mu) = M\frac{du}{dt}. \tag{2.46}$$

Note that M can be taken outside the derivative *only* if the mass is constant. The acceleration is therefore:

$$\frac{du}{dt} = g, \tag{2.47}$$

and, although this is a familiar result, it nevertheless follows directly from the principle of conservation of momentum.

Another example is provided by the steady flow of a fluid in a pipe of length L and diameter D, shown in Fig. 2.15. The upstream pressure p_1 exceeds the downstream pressure p_2 and thereby provides a driving force for flow from left to right. However, the shear stress τ_w exerted by the wall on the fluid tends to retard the motion.

Fig. 2.15   *Forces acting on fluid flowing in a pipe.*

A steady-state momentum balance to the *right* (note that a *direction* must be specified) is next performed. As the system, choose for simplicity a cylinder that is *moving* with the fluid, since it avoids the necessity of considering flows entering and leaving the system. The result is:

$$\frac{\pi D^2}{4}(p_1 - p_2) - \tau_w \pi D L = \frac{d\mathcal{M}}{dt} = 0. \tag{2.48}$$

Here, the first term is the rate of addition of momentum to the system resulting from the net pressure difference $p_1 - p_2$, which acts on the *circular* area $\pi D^2 / 4$. The second term is the rate of subtraction of momentum from the system by the wall shear stress, which acts to the *left* on the *cylindrical* area $\pi D L$. Since the flow is steady, there is no change of momentum of the system with time, and $d\mathcal{M}/dt = 0$. Simplification gives:

$$\tau_w = \frac{(p_1 - p_2)D}{4L}, \tag{2.49}$$

which is an important equation from which the wall shear stress can be obtained from the pressure drop (which is easy to determine experimentally) independently of any constitutive equation relating the stress to a velocity gradient.

2. Momentum Transfer by Convection or Flow

The *convective transfer* of momentum by flow is more subtle, but can be appreciated with reference to Fig. 2.16, in which water from a hose of cross-sectional area A impinges with velocity u on the far side of a trolley of mass M with (for simplicity) frictionless wheels. The dotted box delineates a stationary system within which the momentum Mv is increasing to the right, because the trolley clearly tends to accelerate in that direction. The reason is that momentum is being transferred across the surface BC into the system by the *convective action* of the jet. The rate of transfer is the mass flow rate $m = \rho u A$ times the velocity u, namely:

$$\dot{\mathcal{M}} = mu = \rho A u^2 \;\; [=] \;\; \frac{\text{M}}{\text{T}}\frac{\text{L}}{\text{T}} = \frac{\text{ML/T}}{\text{T}}, \tag{2.50}$$

which is again momentum per unit time.

$$m = \rho u A$$

Nozzle of area A

Jet
u
M
v

Fig. 2.16   Convection of momentum.

The acceleration of the trolley will now be found by applying momentum balances in two different ways, depending on whether the control volume is stationary or moving. In each case, the water leaves the nozzle of cross-sectional area A with velocity u and the trolley has a velocity v. Both velocities are relative to the nozzle. The reader should make a determined effort to understand both approaches, since momentum balances are conceptually more difficult than mass and energy balances. It is also essential to perform the momentum balance in an *inertial frame of reference*—one that is either stationary or moving with a uniform velocity, but *not accelerating*.

1. Control surface moving with trolley. As shown in Fig. 2.17, the control surface delineating the system is moving to the right at the same velocity v as the trolley. The observer perceives water entering the system across BC not with velocity u but with a *relative* velocity $(u - v)$, so that the rate m of convection of *mass* into the control volume is:

$$m = \rho A(u - v). \tag{2.51}$$

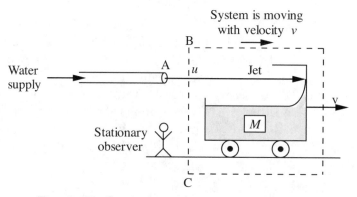

Fig. 2.17 Stationary observer—moving system.

A momentum balance (positive direction to the right) gives:

Rate of addition of momentum = Rate of increase of momentum,   (2.52)

$$\rho A(u - v)u = mu = \frac{d}{dt}(Mv) = M\frac{dv}{dt} + v\frac{dM}{dt}.$$ (2.53)

Note that to obtain the momentum flux, m is multiplied by the absolute velocity u [not the relative velocity $(u - v)$, which has already been accounted for in the mass flux m]. Also, the mass of the system is not constant, but is increasing at a rate given by:

$$\frac{dM}{dt} = m = \rho A(u - v).$$ (2.54)

It follows from the last three equations that the acceleration $a = dv/dt$ of the trolley to the right is:

$$a = \frac{\rho A}{M}(u - v)^2.$$ (2.55)

2. Control volume fixed. In Fig. 2.18, the control surface is now fixed in space, so that the trolley is moving within it. Also—and not quite so obviously, that part of the jet of length L inside the control surface is lengthening at a rate $dL/dt = v$ and increasing its momentum ρALu, and this must of course be taken into account. A mass balance first gives:

Rate of addition of mass = Rate of increase of mass,

$$m = \rho Au = \frac{dM}{dt} + \frac{d}{dt}(\rho AL) = \frac{dM}{dt} + \rho Av.$$ (2.56)

Likewise, a momentum balance gives:

Rate of addition = Rate of increase.

$$mu = \rho Au^2 = \frac{d}{dt}(Mv + \rho ALu) = Ma + v\frac{dM}{dt} + \rho Avu.$$ (2.57)

Example 2.4—Impinging Jet of Water   **83**

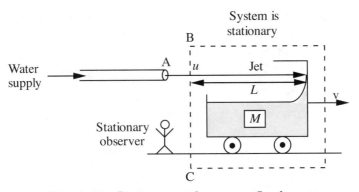

Fig. 2.18 Stationary observer—fixed system.

By eliminating dM/dt between Eqns. (2.56) and (2.57) and rearranging, the acceleration becomes identical with that in (2.55), thus verifying the equivalence of the two approaches:

$$a = \frac{\rho A}{M}(u - v)^2. \tag{2.58}$$

Example 2.4—Impinging Jet of Water

Fig. E2.4 shows a plan of a jet of water impinging against a shield that is held stationary by a force F opposing the jet, which divides into several radially outward streams, each leaving at right angles to the jet. If the total water flow rate is $Q = 1$ ft^3/s and its velocity is $u = 100$ ft/s, find F (lb$_f$).

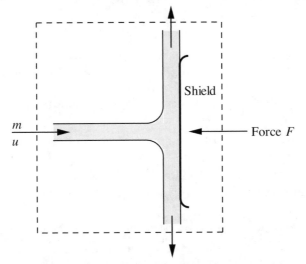

Fig. E2.4   Jet impinging against shield.

Solution

   Perform a momentum balance to the *right* on the system bounded by the dotted control surface. If the mass flow rate is m, the rate of transfer of momentum into the system by convection is mu. The exiting streams have no momentum to the right. Also, the opposing force amounts to a rate of addition of momentum F to the left. Hence, at steady state:

$$mu - F = 0, \tag{E2.4.1}$$

so that:

$$F = mu = \rho Qu = \frac{62.4 \times 1 \times 100}{32.2} = 193.8 \text{ lb}_f. \qquad \square$$

Example 2.5—Velocity of Wave on Water

   As shown in Fig. E2.5(a), a small disturbance in the form of a wave of slightly increased depth $D + dD$ travels with velocity u along the surface of a layer of otherwise stagnant water of depth D. Note that no part of the water itself moves with velocity u—just the dividing line between the stagnant and disturbed regions; in fact, the water velocity just upstream of the front is a small quantity du, and downstream it is zero. If the viscosity is negligible, find the wave velocity in terms of the depth.

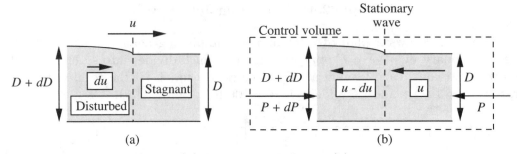

Fig. E2.5   *(a) Moving wave front; (b) stationary wave front as seen by observer traveling with wave.*

   Estimate the warning times available to evacuate communities that are threatened by the following "avalanches" of water:

(a) A tidal wave (closely related to a *tsunami*[2]), generated by an earthquake 1,000 miles away, traveling across an ocean of average depth 2,000 ft.

(b) An onrush, caused by a failed dam 50 miles away, traveling down a river of depth 12 ft that is also flowing towards the community at 4 mph.

---

[2] According to *The New York Times* of December 27, 2004, a *tsunami* is a *series* of waves generated by underwater seismic disturbances. In the disaster of the previous day, the Indian tectonic plate slipped under (and raised) the Burma plate off the coast of Sumatra. The article reported that the ocean floor could rise dozens of feet over a distance of hundreds of miles, displacing unbelievably enormous quantities of water. The waves reached Indonesia, Thailand, Sri Lanka, and elsewhere, causing immense destruction and the eventual loss of more than 225,000 lives.

Example 2.5—Velocity of Wave on Water   **85**

Solution

As shown in Fig. E2.5(a), the problem is a transient one, since the picture changes with time as the wave front moves to the right. The solution is facilitated by superimposing a velocity u to the left, as in Fig. E2.5(b)—that is, by taking the viewpoint of an observer traveling *with* the wave, who now "sees" water coming from the right with velocity u and leaving to the left with velocity $u - du$.

All forces, flow rates, and momentum fluxes will be based on unit width normal to the plane of the figure. Because the disturbance is *small*, second-order differentials such as $(dD)^2$ and $du\,dD$ can be neglected. Referring to Fig. E2.5(b), the total downstream and upstream pressure forces are obtained by integration:

$$P = \int_0^D \rho g h\,dh = \frac{1}{2}\rho g D^2, \quad P + dP = \int_0^{D+dD} \rho g h\,dh = \frac{1}{2}\rho g (D+dD)^2. \quad \text{(E2.5.1)}$$

A *mass* balance on the indicated control volume relates the downstream and upstream velocities and their depths (remember that calculations are based on unit width, so $D \times 1 = D$ is really an area):

$$\rho u D = \rho(u - du)(D + dD), \qquad \text{or} \qquad dD = D\,\frac{du}{u}. \quad \text{(E2.5.2)}$$

Since the viscosity is negligible, there is no shear stress exerted by the floor on the water above. Therefore, a steady-state *momentum* balance to the left on the control volume yields:

$$\underbrace{\frac{1}{2}\rho g D^2}_{P} + \underbrace{(\rho u D)u}_{\substack{\text{Convected} \\ \text{in}}} - \underbrace{\frac{1}{2}\rho g(D+dD)^2}_{P+dP} - \underbrace{(\rho u D)(u - du)}_{\substack{\text{Convected} \\ \text{out}}} = 0, \quad \text{(E2.5.3)}$$

which simplifies to:

$$(\rho u D)\,du = \rho g D\,dD, \qquad \text{or} \qquad u\,du = g\,dD. \quad \text{(E2.5.4)}$$

Substitution for dD from the mass balance yields the velocity u of the wave:

$$u\,du = g D\,\frac{du}{u}, \qquad \text{or} \qquad u = \sqrt{gD}. \quad \text{(E2.5.5)}$$

Note that the wave velocity increases in proportion to the square root of the depth of the water. Another viewpoint is that the *Froude* number, $\text{Fr} = u^2/gD$, being the ratio of inertial (ρu^2) to hydrostatic ($\rho g D$) effects, is unity.

The calculations for the water "avalanches" now follow:

(a) *Ocean wave*:

$$u = \sqrt{32.2 \times 2{,}000} = 254 \text{ ft/s}, \tag{E2.5.6}$$

so that the warning time is:

$$t = \frac{1{,}000 \times 5{,}280}{254 \times 3{,}600} = 5.78 \text{ hr.} \tag{E2.5.7}$$

(a) *River wave*:

$$u = \sqrt{32.2 \times 12} = 19.7 \text{ ft/s} \; = \frac{19.7 \times 3{,}600}{5{,}280} = 13.4 \text{ mph.} \tag{E2.5.8}$$

Since the river is itself flowing at 4 mph, the total wave velocity is $13.4 + 4 = 17.4$ mph. Thus, the warning time is:

$$t = \frac{50}{17.4} = 2.87 \text{ hr.} \tag{E2.5.9}$$

The velocity of a *sinusoidally* varying wave traveling on deep water is discussed in Section 7.10. □

Further momentum balances for an orifice plate and sudden expansion. This section concludes with two examples in which pressure changes in *turbulent* zones, where Bernoulli's equation *cannot* be applied, are determined from momentum balances.

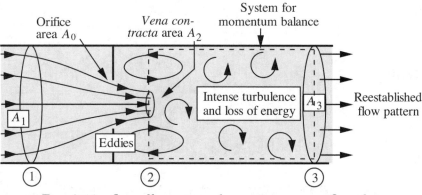

Fig. 2.19 *Overall pressure drop across an orifice plate.*

Fig. 2.19 shows the flow through an orifice plate. Mass balances yield the following (note that $A_1 = A_3$):

$$m = \rho u_1 A_1 = \rho u_2 A_2 = \rho u_3 A_3, \tag{2.59}$$

or:

$$u_3 = u_1 \quad \text{and} \quad u_2 = \frac{u_1 A_1}{A_2}. \tag{2.60}$$

In the *upstream* section, where the streamlines are converging and the flow is relatively frictionless, Bernoulli's equation has already been used to relate changes in pressure to changes in velocity:

$$\frac{u_1^2}{2} + \frac{p_1}{\rho} = \frac{u_2^2}{2} + \frac{p_2}{\rho}. \tag{2.61}$$

Elimination of u_2 between (2.60) and (2.61) gives the pressure drop in the upstream section:

$$\frac{p_1 - p_2}{\rho} = \frac{u_2^2}{2} - \frac{u_1^2}{2} = \frac{u_1^2}{2}\left(\frac{A_1^2}{A_2^2} - 1\right). \tag{2.62}$$

In the *downstream* section, it is characteristic of a jet of expanding cross-sectional area that it generates excessive turbulence, so that the frictional dissipation term $\mathcal{F}$ is no longer negligible and Bernoulli's equation *cannot* be applied. However, the length of pipe between 2 and 3 is sufficiently small so that the wall shear stress imposed by it can be ignored. A momentum balance (positive to the right) on the dotted control volume between locations 2 and 3 gives:

$$mu_2 - mu_3 + (p_2 - p_3)A_1 = 0. \tag{2.63}$$

Here, the first two terms correspond to convection of momentum into and from the control volume; the second two terms reflect the pressure forces on the two ends, both of area A_1. Although there will be some fluctuations of momentum within the system because of the turbulence, these variations will essentially cancel out over a sufficiently long time period; this *quasi* steady-state viewpoint says that the flow is *steady in the mean*. The eddies surrounding the jet at the *vena contracta* keep recirculating and the net momentum transfer from them across the control surface is zero.

The pressure drop in the downstream section is obtained by eliminating u_3 between Eqns. (2.60) and (2.63):

$$\frac{p_2 - p_3}{\rho} = \frac{m}{\rho A_1}(u_3 - u_2) = u_1(u_1 - u_2) = u_1^2\left(1 - \frac{A_1}{A_2}\right). \tag{2.64}$$

The *overall* pressure drop is found by adding Eqns. (2.62) and (2.64):

$$\frac{p_1 - p_3}{\rho} = \frac{u_1^2}{2}\left(\frac{A_1^2}{A_2^2} - 2\frac{A_1}{A_2} + 1\right) = \frac{u_1^2}{2}\left(\frac{A_1}{A_2} - 1\right)^2. \tag{2.65}$$

Observe that the last term in Eqn. (2.65), being the product of two squares, must always be *positive*. Hence, p_3 is always less than p_1, and there is overall a *loss* of useful work. Indeed, the corresponding frictional dissipation term $\mathcal{F}$ can be found by employing the overall *energy* balance, Eqn. (2.17), between points 1 and 3:

$$\Delta\left(\frac{u^2}{2}\right) + g\Delta z + \frac{\Delta p}{\rho} + w + \mathcal{F} = 0. \tag{2.66}$$

Here, the first, second, and fourth terms are zero because overall there is no change in velocity, no elevation change, and zero work performed. The frictional dissipation per unit mass flowing is therefore:

$$\mathcal{F} = -\frac{\Delta p}{\rho} = \frac{p_1 - p_3}{\rho} = \frac{u_1^2}{2}\left(\frac{A_1}{A_2} - 1\right)^2, \tag{2.67}$$

which value can then be used for an *inclined* or even vertical pipe, the effect of elevation being accounted for by the $g\Delta z$ term.

The following questions also deserve reflection:

1. Is p_3 less than, equal to, or greater than p_2? Why?
2. What is involved in performing a momentum balance between points 1 and 2?

A CFD (*computational fluid dynamics*) simulation of highly turbulent flow through an orifice plate is presented in Example 9.4 in Chapter 9. To show the power of CFD, the lower half of a small part of that simulation is shown in Fig. 2.20. Note the remarkable similarity to the earlier qualitative sketch of Fig. 2.19.

Fig. 2.20   *Streamlines computed by COMSOL for flow from left to right through an orifice plate.*

A *sudden expansion* in a pipeline shown in Fig. 2.21 occurs when two pipes of different diameters are joined together. The corresponding $\mathcal{F}$ term is obtained by a momentum balance similar to that in the downstream section of the orifice plate, and is left as an exercise (see Problem P2.22).

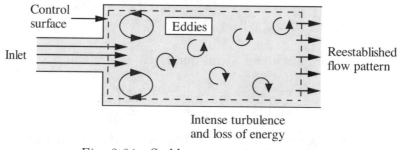

Fig. 2.21   *Sudden expansion in a pipe.*

Example 2.6—Flow Measurement by a Rotameter   **89**

Example 2.6—Flow Measurement by a Rotameter

Fig. E2.6(a) shows a flow-measuring instrument called a *rotameter*, consisting typically of a solid "float" (made of an inert material such as stainless steel, glass, or tantalum) inside a gradually *tapered* glass tube. As the fluid flows upward, the float reaches a position of equilibrium, from which the flow rate can be read from the adjacent scale, which is often etched on the glass tube.

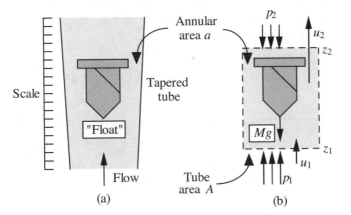

Fig. E2.6 (a) Section of rotameter, and (b) control volume.

The float is often stabilized by helical grooves incised into it, which induce *rotation*—hence the name. Other shapes of floats—including spheres in the smaller instruments—may be employed. The problem is to analyze the flow and the equilibrium position of the float. Assume that viscous effects are unimportant, which is true for the majority of industrially important fluids as they pass around the float.

Solution

Some flow-measuring devices, such as the orifice plate and Venturi meter, incorporate a *fixed* reduction of area, which produces a pressure drop that depends on the flow rate. In contrast, the rotameter depends on the change of an annular area a between the float and the tube, which is a function of the vertical location of the float, to yield an essentially fixed pressure drop at all flow rates. In other words, the annular area functions as an orifice of variable area.

Referring to Fig. E2.6(b), mass, energy, and upward momentum balances yield:

Continuity:
$$m = \rho u_1 A = \rho u_2 a, \tag{E2.6.1}$$

Bernoulli:
$$\tfrac{1}{2}u_1^2 + \frac{p_1}{\rho} + gz_1 = \tfrac{1}{2}u_2^2 + \frac{p_2}{\rho} + gz_2, \tag{E2.6.2}$$

$$\text{Momentum:}\quad \underbrace{(p_1 - p_2)A}_{\text{Pressure}} + \underbrace{mu_1 - mu_2}_{\text{Convection}} - \underbrace{\left[\overbrace{(z_2 - z_1)A - \frac{M}{\rho_f}}^{\text{Volume of fluid}} \right] \rho g - Mg}_{\text{Gravity}} = 0, \quad \text{(E2.6.3)}$$

in which ρ and ρ_f are the densities of the fluid and the float, respectively. In the momentum balance, the three types of terms are indicated by underbraces, and the bracketed expression is the volume of fluid in the control volume. Starting with Eqn. (E2.6.3), terms involving the pressures and elevations can be eliminated by using Eqn. (E2.6.2) and terms involving m and u_2 can be eliminated by using Eqn. (E2.6.1), leading after some algebra to:

$$\frac{1}{2}\rho u_1^2 A \left(\frac{A}{a} - 1 \right)^2 = Mg \left(1 - \frac{\rho}{\rho_f} \right). \quad \text{(E2.6.4)}$$

The flow rate is then:

$$Q = Au_1 = A \sqrt{\frac{2Mg \left(1 - \frac{\rho}{\rho_f} \right)}{\rho A \left(\frac{A}{a} - 1 \right)^2}} \doteq a\sqrt{\frac{2Mg}{\rho A}}, \quad \text{(E2.6.5)}$$

in which the approximate form holds for $a \ll A$ and $\rho \ll \rho_f$. Since $a = a(z)$ and $A = A(z)$ depend on the location z of the float, Eqn. (E2.6.5) demonstrates that the flow rate can be determined from the equilibrium position of the float. In practice, since a is not known precisely (there may be a *vena contracta* effect), Eqn. (E2.6.5) can be used for a preliminary design, the rotameter being subsequently *calibrated* before accurate use. □

Angular momentum. So far, we have been concerned with the conservation of momentum in a specified *linear* direction. However, it is also possible to consider *angular momentum*, which is introduced by examining Fig. 2.22(a), in which a differential mass dM is rotating with angular velocity ω rad/s in an arc having a radius of curvature r. The angular momentum dA of the mass is defined as:

$$dA = ru\, dM = \omega r^2 dM. \quad (2.68)$$

For a finite *distributed* mass, the total angular momentum is obtained by integration over the entire mass:

$$A = \int_M \omega r^2 dM = \omega I, \quad (2.69)$$

in which the *moment of inertia I* is defined as:

$$I = \int_M r^2\, dM. \quad (2.70)$$

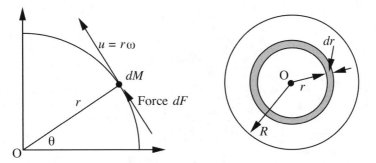

Fig. 2.22   (a) Mass moving in arc of radius r;
(b) integration to find the total angular momentum.

For example, for the flywheel of width W, radius R, and mass M, whose cross section is shown in Fig. 2.21(b):

$$I = \int_0^R r^2 \underbrace{2\pi r dr W \rho}_{dM} = \underbrace{\rho W \pi R^2}_{M} \frac{R^2}{2} = M \frac{R^2}{2}. \tag{2.71}$$

Also, *torque*, being the product of a force times a radius, is given in differential and integrated form by:

$$dT = r\,dF, \qquad T = \int_F r\,dF, \tag{2.72}$$

and is analogous to force in linear momentum problems and will result in a transfer of angular momentum. In the same way that the application of a force F to a constant mass M moving with velocity v leads to the law $F = M dv/dt$, one form of an angular momentum balance is:

$$T = I \frac{d\omega}{dt}. \tag{2.73}$$

Here, T is the torque applied to the axle of the flywheel whose moment of inertia is I, resulting in an angular acceleration $d\omega/dt$.

As a final example, consider the impeller of a centrifugal pump, whose cross section is shown in Fig. 2.23. (For practical reasons, the vanes are usually curved, not straight, as discussed further in Section 4.2.)

The impeller rotates with angular velocity ω, and its rotation causes the fluid to be thrown radially outwards between the vanes by centrifugal action (again, see Section 4.2). The fluid enters at a radial position r_1 and leaves at a radial position r_2; the corresponding tangential velocities u_1 and u_2 denote the inlet and exit liquid velocities relative to a stationary observer. The arrows A, B, C, and D illustrate the liquid velocity relative to the rotating impeller.

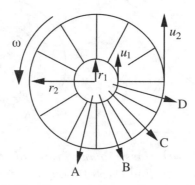

Fig. 2.23   Cross section of pump impeller.

An angular momentum balance on the impeller gives:

$$\frac{dA}{dt} = (ru)_{\text{in}}m_{\text{in}} - (ru)_{\text{out}}m_{\text{out}} + T = (ru)_1 m_1 - (ru)_2 m_2 + T. \qquad (2.74)$$

For steady operation, the derivative is zero, and the torque required to rotate the impeller is:

$$T = m[(ru)_2 - (ru)_1] = m\omega(r_2^2 - r_1^2). \qquad (2.75)$$

Here, m is the mass flow rate through the pump, subscripts 1 and 2 denote inlet and exit conditions, and $u = r\omega$ denotes the corresponding tangential velocities. The corresponding *power* needed to drive the pump is the product of the angular velocity and the torque (see also Table 2.4):

$$P = \omega T. \qquad (2.76)$$

2.7   Pressure, Velocity, and Flow Rate Measurement

Previous sections have alluded to various ways of measuring the most important quantities associated with static and flowing fluids. This section reviews and amplifies such methods. Excellent and comprehensive surveys are available in a book and on a CD.[3]

Pressure.   A *piezometer* is the generic name given to a pressure-measuring device. Three basic types are recognized here; in each case, learning from the principle of the Pitot tube, it is important that the opening to the device be tangential to any fluid motion, otherwise an erroneous reading will result.

---

[3] See R.J. Goldstein, ed., *Fluid Mechanics Measurements*, 2nd ed., Hemisphere Publishing Corporation, New York, 1996; and *Material Balances & Visual Equipment Encyclopedia of Chemical Engineering Equipment*, CD produced by the Multimedia Education Laboratory, Department of Chemical Engineering, University of Michigan, Susan Montgomery (Director), 2002.

Fig. 2.24 (a) piezometric tube, and (b) manometer.

1. Transparent tubes can be connected to the point(s) where the pressure is needed, two examples being given in Fig. 2.24:

   (a) In the *piezometric tube*, the height to which the liquid rises is observed, and the required gauge pressure at point 1 in the pipe or vessel is $p_1 = \rho g h$. Fairly obviously, this arrangement can only be used if the fluid is a liquid and not a gas.

   (b) For measuring a pressure *difference*, such as occurs across an orifice plate, a differential "U" manometer can be employed, using a liquid that is immiscible with the one whose pressure is desired (and which may now be either a liquid *or* a gas). The arrangement shown is for $\rho_B > \rho_A$, and the required pressure difference is $p_1 - p_2 = (\rho_B - \rho_A)g\Delta h$. For $\rho_B < \rho_A$, the manometer tube would be an inverted "U" and would be placed above the pipe. The manometer can also be employed in situation (a) if the right-hand leg is vented to the atmosphere, in which case the fluid whose pressure is needed can then be a gas.

2. The *Bourdon gauge*, of which the essentials are illustrated in Fig. 2.25 (the linkage to the needle is actually more intricate), employs a curved, flattened metal tube, open at one end and closed at the other. Pressure applied to the open end tends to cause the tube to straighten out, and this motion is transmitted to a needle that rotates and points to the appropriate reading on a circular scale.

3. The pressure *transducer* employs a small thin diaphragm, usually contained in a hollow metallic cylinder that is "plugged" into the location where the pressure is needed. Variations of pressure cause the diaphragm to bend, and the extent of such distortion can be determined by different methods. For example, piezoresistive and piezoelectric material properties of either the diaphragm or another sensing element bonded to it will cause variations in the strain to be translated into variations in the electrical resistance or electric polarization,

either of which can be converted into an electrical signal for subsequent processing. Advances in semiconductor fabrication have allowed a wide variety of pressure transducers to become readily available, and these are clearly the preferred instruments if the pressure is to be recorded or transmitted to a process-control computer.

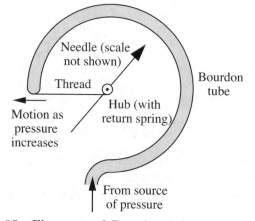

Fig. 2.25   Elements of Bourdon tube pressure gauge.

Velocity.   Fluid velocities can be determined by several methods, some of the more important being:

1. *Tracking* of small particles or bubbles moving with the fluid. By photographing over a known short time period, the velocity can be deduced from the distance traveled by a particle. For flow inside a transparent pipe, the distance of the particle from the wall can be determined either by focusing closely on a known location or, better, by using stereoscopic photography with two cameras and comparing the apparent particle locations on the resulting two negatives. Laser-Doppler velocimetry (LDV) determines the velocity by measuring the Doppler shift of a laser beam caused by the moving microscopic particle. Infrared lasers are used for assessing upper-atmosphere turbulence.

2. *Pitot tube* measurements, as already discussed.

3. *Hot-wire anemometry*, in which a very small and thin electrically heated wire suspended between two supports is introduced into the moving stream. Measurements are made of the resistance of the wire, which depends on its temperature, which is governed by the local heat-transfer coefficient, which in turn depends on the local velocity.

Flow rate.   The following is a representative selection of the wide variety of methods and instruments for measuring flow rates of fluids.

1. *Discharge* of the flowing stream into a receptacle so that the volume or mass of fluid flowing during a known period of time can be measured directly.

2. Devices based on the *Bernoulli* principle, in which the pressure decreases because of an increased velocity through a restriction. Examples are the orifice plate and rotameter already discussed, and the Venturi (see Problem 2.12).

3. Flow of a liquid over a *weir* or *notch*, in which the depth of the liquid depends on the flow rate.

4. *Turbine meter*, consisting of a small in-line turbine placed inside a section of pipe; the rotational speed, which can be transmitted electrically to a recorder, depends on the flow rate.

5. *Thermal flow meter*, in which a small heater is located between two temperature detectors—one (A) upstream and the other (B) downstream of the heater. For a given power input, the temperature difference $(T_B - T_A)$ is measured and will vary inversely with the flow rate, which can then be determined.

6. *Target meter*, typically consisting of a disk mounted on a flexible arm and placed normal to the flow in a pipe. The displacement of the disk, and hence the flow rate, is determined from the output of a strain gauge attached to the arm.

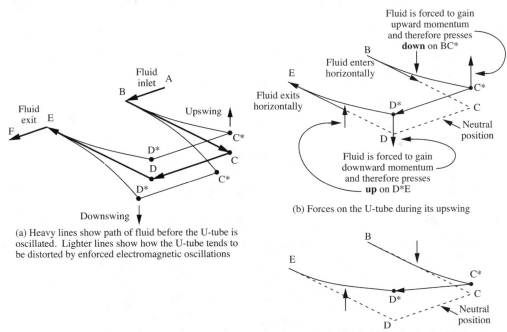

(a) Heavy lines show path of fluid before the U-tube is oscillated. Lighter lines show how the U-tube tends to be distorted by enforced electromagnetic oscillations

(b) Forces on the U-tube during its upswing

(c) The U-tube is therefore twisted during its upswing (and twists in the opposite direction during its downswing)

Fig. 2.26   Principles of a Coriolis mass-flow meter.

7. *Coriolis flow meter.* Here—see Fig. 2.26(a)—the fluid is forced to flow through a slightly flexible U-tube BCDE (with suitably rounded corners) that is vibrated up and down by a magnetic driver. Fig. 2.26(b) examines what happens

when the U-tube is in the "upswing" position. Suppose that the orientation is such that the fluid enters *horizontally*. Because the fluid is forced to conform to the path BC* it acquires some vertically upward momentum and therefore exerts an opposite or *downward* force on the leg BC*. And because it must lose this vertical momentum before it leaves the U-tube at E, the fluid exerts an *upward* force on the leg D*E. The net result (Fig. 2.26(c)) is that—seen from the CD end—the U-tube twists *clockwise* while in the upswing mode. And when it is in the downswing mode, exactly the opposite rotation occurs. Electromagnetic sensors near the two long arms of the U-tube detect the magnitude of the twisting motion, which depends on the *mass* flow rate of the fluid. The instrument is therefore useful for measuring mass flow rates even though the fluid may be non-Newtonian and possibly contain some gas bubbles. It is called a Coriolis meter because it can also be analyzed by considering the forces on a moving fluid in a rotating or oscillating system of coordinates.

8. *Vortex-shedding meter.* An obstacle is placed in the flowing stream, and vortices are shed on alternate sides of the wake. The frequency of shedding, and hence the flow rate, is determined by various methods, one of which measures the oscillations of a flexible strip placed in the wake.

PROBLEMS FOR CHAPTER 2

1. *Evacuation of leaking tank—M*. This problem concerns the tank shown in Fig. P2.1 and is the same as Example 2.1, except that there is now a small leak into the tank from the outside air, whose pressure is 1 bar. The mass flow rate of the leak equals $c(\rho_0 - \rho)$, where $c = 10^{-5}$ m^3/s is a constant, ρ_0 is the density of the ambient air, and ρ is the density of the air inside the tank. The initial density of air inside the tank is also ρ_0.

What is the lowest attainable pressure p^* inside the tank? How long will it take the pressure to fall halfway from its initial value to p^*?

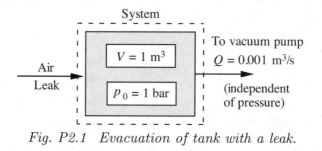

Fig. P2.1   Evacuation of tank with a leak.

2. *Slowed traffic—E*. A two-lane highway carries cars traveling at an average speed of 60 mph. In a construction zone, where the cars have merged into one lane, the average speed is 20 mph and the average distance between front bumpers of successive cars is 25 ft. What is the average distance between front bumpers in *each* lane of the two-lane section? Why? How many cars per hour are passing through the construction zone?

3. *"Density" of cars—E.* While driving down the expressway at 65 mph, you count an average of 18 cars going in the other direction for each mile you travel. At any instant, how many cars are there per mile traveling in the opposite direction to you?

4. *Ethylene pipeline—E.* Twenty lb_m/s of ethylene gas (assumed to behave ideally) flow steadily at $60\,°F$ in a pipeline whose internal diameter is 8 in. The pressure at upstream location 1 is $p_1 = 60$ psia, and has fallen to $p_2 = 25$ psia at downstream location 2. Why does the pressure fall, and what are the velocities of the ethylene at the two locations?

5. *Transient behavior of a stirred tank—E.* The well-stirred tank of volume $V = 2$ m^3 shown in Fig. P2.5 is initially filled with brine, in which the initial concentration of sodium chloride at $t = 0$ is $c_0 = 1$ kg/m^3. Subsequently, a flow rate of $Q = 0.01$ m^3/s of pure water is fed steadily to the tank, and the same flow rate of brine leaves the tank through a drain. Derive an expression for the subsequent concentration of sodium chloride c in terms of c_0, t, Q, and V. Make a sketch of c versus t and label the main features. How long (minutes and seconds) will it take for the concentration of sodium chloride to fall to a final value of $c_f = 0.0001$ kg/m^3?

Inlet

Stirred tank

Outlet

Fig. P2.5   Stirred tank with continuous flow.

6. *Stirred tank with crystal dissolution—M.* A well-stirred tank of volume $V = 2$ m^3 is filled with brine, in which the initial concentration of sodium chloride at $t = 0$ is $c_0 = 1$ kg/m^3. Subsequently, a flow rate of $Q = 0.01$ m^3/s of pure water is fed steadily to the tank, and the *same* flow rate of brine leaves the tank through a drain. Additionally, there is an ample supply of sodium chloride crystals in the bottom of the tank, which dissolve at a uniform rate of $m = 0.02$ kg/s.

Why is it reasonable to suppose that the volume of brine in the tank remains constant? Derive an expression for the subsequent concentration c of sodium chloride in terms of c_0, m, t, Q, and V. Make a sketch of c versus t and label the main features. Assuming an inexhaustible supply of crystals, what will the concentration c of sodium chloride in the tank be at $t = 0$, 10, 100, and ∞ s? Carefully define the system on which you perform a transient mass balance.

7. *Soaker garden hose—D.* A "soaker" garden hose with porous canvas walls is shown in Fig. P2.7. Water is supplied at a pressure p_0 at $x = 0$, and the far end at $x = L$ is blocked off with a cap. At any intermediate location x, water leaks out through the wall at a volumetric rate $q = \beta(p - p_a)$ per unit length, where p is the local pressure inside the hose and the constant β and the external air pressure p_a are both known.

Fig. P2.7 *A garden hose with porous walls.*

You may assume that the volumetric flow rate of water inside the hose is proportional to the negative of the pressure gradient:

$$Q = -\alpha\frac{dp}{dx}, \tag{P2.7.1}$$

where the constant α is also known. (Strictly speaking, Eqn. (P2.7.1) holds only for *laminar* flow, which may or may not be the case—see Chapter 3 for further details.) By means of a mass balance on a differential length dx of the hose, prove that the variations of pressure obey the differential equation:

$$\frac{d^2P}{dx^2} = \gamma^2 P, \tag{P2.7.2}$$

where $P = (p - p_a)$ and $\gamma^2 = \beta/\alpha$. Show that:

$$P = A\sinh\gamma x + B\cosh\gamma x \tag{P2.7.3}$$

satisfies Eqn. (P2.7.2), and determine the constants A and B from the boundary conditions. Hence, prove that the variations of pressure are given by:

$$\frac{p - p_a}{p_0 - p_a} = \frac{\cosh\gamma(L - x)}{\cosh\gamma L}. \tag{P2.7.4}$$

Then show that the total rate of loss of water from the hose is given by:

$$Q_{\text{loss}} = \alpha\gamma(p_0 - p_a)\tanh\gamma L. \tag{P2.7.5}$$

Finally, sketch $(p - p_a)/(p_0 - p_a)$ versus x for low, intermediate, and high values of γ. Point out the main features of your sketch.

8. *Performance of a siphon—E.* As shown in Fig. P2.8, a pipe of cross-sectional area $A = 0.01$ m^2 and total length 5.5 m is used for siphoning water from a tank. The discharge from the siphon is 1.0 m below the level of the water in the tank. At its highest point, the pipe rises 1.5 m above the level in the tank. What is the water velocity v (m/s) in the pipe? What is the lowest pressure in bar (gauge), and where does it occur? Neglect pipe friction. Are your answers reasonable?

1.5 m   Siphon

1.0 m   Tank

Discharge

Fig. P2.8 Siphon for draining tank.

If the siphon reaches virtually all the way to the bottom of the tank (but is not blocked off), is the time taken to drain the tank equal to $t = V/vA$, where V is the initial volume of water in the tank, and v is still the velocity as computed above when the tank is full? Explain your answer.

9. *Pitot tube—E.* The speed of a boat is measured by a Pitot tube. When traveling in seawater ($\rho = 64$ lb$_m$/ft^3), the tube measures a pressure of 2.5 lb$_f$/in^2 due to the motion. What is the speed (mph) of the boat? What would be the speed in fresh water ($\rho = 62.4$ lb$_m$/ft^3), also for a pressure of 2.5 lb$_f$/in^2?

10. *Leaking carbon dioxide—M.* A long vertical tube is open at the top and contains a small orifice in its base, which is otherwise closed. The lower 10-m section of the tube is filled with carbon dioxide (MW = 44). Otherwise, there is air (MW = 28.8) above the carbon dioxide in the tube and also outside the tube. Calculate the velocity (m/s) of carbon dioxide that issues from the orifice. In which direction does it flow?

11. *Two pressure gauges—E.* Fig. P2.11 shows two pressure gauges that are mounted on a vertical water pipe 40 ft apart, yet they read exactly the same pressure, 100 psig.

(a) Is the water flowing? Why?

(b) If so, in which direction is it flowing? Why? *Hint*: You may wish to use an overall energy balance in the pipe between the two gauges to check your answer.

Fig. P2.11   Two pressure gauges.

12. *Venturi "meter"—M.* A volumetric flow rate Q of liquid of density ρ flows through a pipe of cross-sectional area A, and then passes through the Venturi "meter" shown in Fig. P2.12, whose "throat" cross-sectional area is a. A manometer containing mercury (ρ_m) is connected between an upstream point (station 1) and the throat (station 2), and registers a difference in mercury levels of Δh. Derive an expression giving Q in terms of A, a, g, Δh, ρ, ρ_m, and C_D.

Fig. P2.12   Venturi "meter."

If the diameters of the pipe and the throat of the Venturi are 6 in. and 3 in. respectively, what flow rate (gpm) of iso-pentane ($\rho = 38.75$ lb$_m$/ft^3) would register a Δh of 20 in. on a mercury ($s = 13.57$) manometer? What is the corresponding pressure drop in psi? Assume a discharge coefficient of $C_D = 0.98$. *Note:* Those parts of the manometer not occupied by the mercury are filled with iso-pentane, since there is free communication with the pipe via the pressure tappings.

13. *Orifice plate—M.* A horizontal 2-in. I.D. pipe carries kerosene at 100 °F, with density 50.5 lb$_m$/ft^3 and viscosity 3.18 lb$_m$/ft hr. In order to measure the flow rate, the line is to be fitted with a sharp-edged orifice plate, with pressure tappings that are connected to a mercury manometer that reads up to a 15-in. difference in the mercury levels. If the largest flow rate of kerosene is expected to be 560 lb$_m$/min, specify the diameter of the orifice plate that would then just register the full 15-in. difference between the mercury levels.

14. *Tank draining—M.* A cylindrical tank of diameter 1 m has a well-rounded orifice of diameter 2 cm in its base. How long will it take an initial depth of acetone equal to 2 m to drain completely from the tank? How long would it take if the orifice were sharp-edged? If needed, the density of acetone is 49.4 lb_m/ft^3.

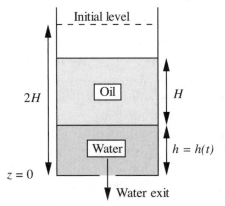

Fig. P2.15   Draining tank with oil and water.

15. *Draining immiscible liquids from a tank—M*. Fig. P2.15 shows a tank of cross-sectional area A that initially ($t = 0$) contains two layers, each of depth H: oil (density ρ_o), and water (ρ_w). A sharp-edged orifice of cross-sectional area a and coefficient of contraction 0.62 in the base of the tank is then opened. Derive an expression for the time t taken for the water to drain from the tank, in terms of H, g, A, a, ρ_o, and ρ_w. Neglect friction and assume that $A \gg a$. You should need *one* of the integrals given in Appendix A.

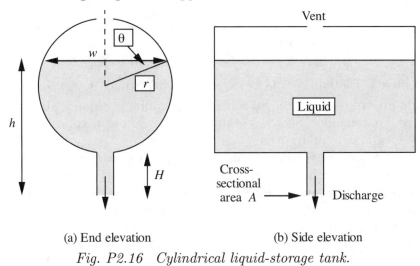

(a) End elevation   (b) Side elevation

Fig. P2.16   Cylindrical liquid-storage tank.

16. *Draining a horizontal cylindrical tank—D*. A cylindrical tank of radius r and length L, vented at the top to the atmosphere, is shown in side and end elevations in Fig. P2.16. Initially ($t = 0$), it is full of a low-viscosity liquid, which is then allowed to drain through an exit pipe of length H and cross-sectional area A.

Prove that the time taken to drain just the tank (excluding the exit pipe) is:

$$t = \frac{2Lr^2}{A\sqrt{2g}} \int_0^\pi \frac{\sin^2 \theta \, d\theta}{\sqrt{H + r(1 + \cos\theta)}}.$$

If $L = 5$ m, $r = 1$ m, $H = 1$ m, what value of A will enable the tank to drain in one hour? If necessary, use Simpson's rule to approximate the integral.

17. *Lifting silicon wafers—M*. Your supervisor has suggested the device shown in Fig. P2.17 for picking up delicate circular silicon wafers without touching them. A mass flow rate m of air of constant density ρ at a pressure p_0 is blown down a central tube of radius r_1, which is connected to a flange of external radius r_2. The device is positioned at a distance H above the silicon wafer, also of radius r_2, that is supposedly to be picked up. The air flows radially outwards with radial velocity $u_r(r)$ in the gap and is eventually discharged to the atmosphere at a pressure p_2, which is somewhat lower than p_0.

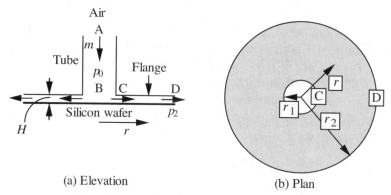

(a) Elevation   (b) Plan

Fig. P2.17   Picking up a silicon wafer.

If the flow is frictionless and gravitational effects may be neglected:

(a) Obtain an expression for u_r in terms of the radial position r and any or all of the given parameters. Sketch a graph that shows how u_r varies with r between radial locations r_1 and r_2.

(b) What equation relates the velocity and pressure in the air stream to each other? Sketch another graph that shows how the pressure $p(r)$ in the gap between the flange and wafer varies with radial position between r_1 and r_2. *Hint*: it may be best to work backwards, starting with a pressure p_2 at radius r_2. Clearly indicate the value p_0 on your graph.

(c) Comment on the likely merits of the device for picking up the wafer.

18. *Rocket performance—M*. Fig. P2.18 shows a rocket whose mass is M lb$_m$ and whose vertical velocity at any instant is v ft/s. A propellant is being ejected downwards through the exhaust nozzle with a mass flow rate m lb$_m$/s and a velocity u ft/s relative to the rocket. The gravitational acceleration is g ft/s^2 and the rocket experiences a drag force $F = kv^2$ lb$_m$ft/s^2, where k is a known constant.

Fig. P2.18   *Rocket traveling upward.*

(a) What is the absolute downward velocity of the propellant leaving the exhaust nozzle?
(b) What is the flux of momentum downward through the nozzle?
(c) Is the mass of the rocket constant?
(d) Perform a momentum balance on the moving rocket as the system. Make it quite clear whether your viewpoint is that of an observer on the ground or one traveling with the rocket. Hence, derive a formula for the rocket acceleration dv/dt, in terms of M, m, v, u, and k.

19. *Jet airplane—E.* A small jet airplane of mass 10,000 kg is in steady level flight at a velocity of $u = 250$ m/s through otherwise stationary air ($\rho_a = 1.0$ kg/m^3). The outside air enters the engines, where it is burned with a negligible amount of fuel to produce a gas of density $\rho_g = 0.40$ kg/m^3, which is discharged at atmospheric pressure with velocity $v = 600$ m/s relative to the airplane through exhaust ports with a total cross-sectional area of 1 m^2. Calculate:

(a) The mass flow rate (kg/s) of gas.
(b) The frictional drag force (N) due to air resistance on the external surfaces of the airplane.
(c) The power (W) of the engine.

20. *Jet-propelled boat—M.* As shown in Fig. P2.20, a boat of mass $M = 1,000$ lb$_m$ is propelled on a lake by a pump that takes in water and ejects it, at a constant velocity of $v = 30$ ft/s relative to the boat, through a pipe of cross-sectional area $A = 0.2$ ft^2. The resisting force F of the water is proportional to the square of the boat velocity u, which has a maximum value of 20 ft/s.

What is the acceleration of the boat when its velocity is $u = 10$ ft/s? For a stationary observer on the lakeshore, use two approaches for a momentum balance and check that they give the same result. (Take a control volume large enough so that the water ahead of the boat is undisturbed.)

(a) A control surface moving with the boat.

(b) A control surface fixed in space, within which the boat is moving.

   The power for propelling the boat is known to be:

$$P = \frac{m}{2}(v^2 - u^2),$$

where m is the mass flow rate of water through the pump. Explain this result as simply as possible from the viewpoint of an observer traveling with the boat. Give a reason why the power P *decreases* as the velocity u of the boat *increases!*

Fig. P2.20   *Jet-propelled boat.*

   21. *Branch pipe—M.* The system shown in Fig. P2.21 carries crude oil of specific gravity 0.8. The total volumetric flow rate at point 1 is 10 ft³/s. Branches 1, 2, and 3 are 5, 4, and 3 inches in diameter, respectively. All three branches are in a horizontal plane, and friction is negligible. The pressure gauges at points 1 and 3 read $p_1 = 25$ psig and $p_3 = 20$ psig, respectively. What is the pressure at point 2? (*Hint:* Note that Bernoulli's equation can only be applied to one streamline at a time!)

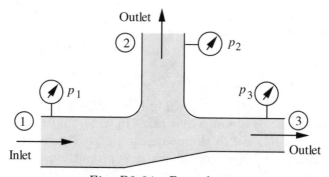

Fig. P2.21   *Branch pipe.*

   22. *Sudden expansion in a pipe—M.* Fig. P2.22 shows a sudden expansion in a pipe. Why can't Bernoulli's equation be applied between points 1 and 2, even though the flow pattern is fully established at both locations?

Prove that there is an overall *increase* of pressure equal to:

$$p_2 - p_1 = \rho u_2^2 \left(\frac{A_2}{A_1} - 1 \right).$$

Where does the energy come from to provide this pressure increase? Also prove that the frictional dissipation term is:

$$\mathcal{F} = \frac{1}{2} u_2^2 \left(\frac{A_2}{A_1} - 1 \right)^2.$$

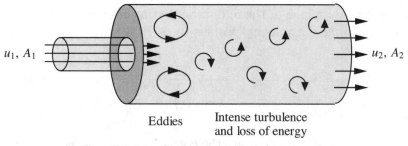

u_1, A_1 $\qquad$ u_2, A_2

Eddies $\qquad$ Intense turbulence and loss of energy

Fig. P2.22   Sudden expansion in a pipe.

23. *Force on a return elbow—M*. Fig. P2.23 shows an idealized view of a return elbow or "U-bend," which is connected to two pipes by flexible hoses that transmit no forces. Water flows at a velocity 10 m/s through the pipe, which has an internal diameter of 0.1 m. The gauge pressures at points 1 and 2 are 3.0 and 2.5 bar, respectively. What horizontal force F (N, to the left) is needed to keep the return elbow in position?

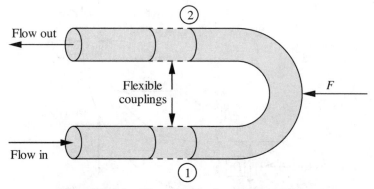

Flow out

Flexible couplings

F

Flow in

Fig. P2.23   Flow around a return elbow.

24. *Hydraulic jump—M*. Fig. P2.24 shows a *hydraulic jump*, which sometimes occurs in open channel or river flow. Under the proper conditions, a rapidly flowing stream of liquid suddenly changes to a slowly flowing or *tranquil* stream with an attendant *rise* in the elevation of the liquid surface.

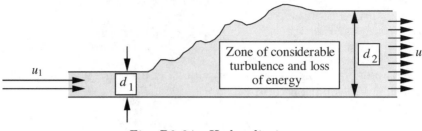

Fig. P2.24   Hydraulic jump.

By using mass and momentum balances on control volumes that extend unit distance normal to the plane of the figure, neglecting friction at the bottom of the channel, and assuming the velocity profiles to be flat, prove that the downstream depth d_2 is given by:

$$d_2 = -\frac{1}{2}d_1 + \sqrt{\left(\frac{d_1}{2}\right)^2 + \frac{2u_1^2 d_1}{g}}.$$

If $u_1 = 5$ m/s and $d_1 = 0.2$ m, what are the corresponding downstream values? (An overall energy balance—*not* required here—will show that the viscous or frictional loss $\mathcal{F}$ per unit mass flowing is positive only for the direction of the jump as shown; hence, a deep stream cannot spontaneously change to a shallow stream.)

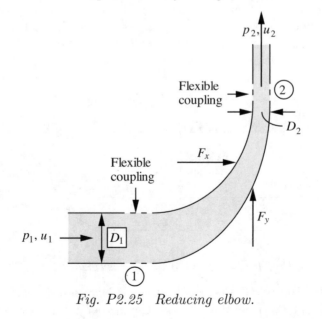

Fig. P2.25   Reducing elbow.

25.   *Reducing elbow—M.* Fig. P2.25 shows a reducing elbow located in a horizontal plane (gravitational effects are unimportant), through which a liquid of constant density is flowing. The flexible connections, which do not exert any forces on the elbow, serve only to delineate the system that is to be considered; they would not be used in practice, because the retaining forces F_x and F_y would be provided by the walls of the pipe.

Neglecting frictional losses, derive expressions for the following, in terms of any or all of the known inlet gauge pressure p_1, inlet velocity u_1, inlet and exit diameters D_1 and D_2 (and hence the corresponding cross-sectional areas A_1 and A_2), and liquid density ρ:

(a) The exit velocity u_2 and pressure p_2.
(b) The retaining forces F_x and F_y needed to hold the elbow in position.

Calculate the two forces F_x and F_y if $D_1 = 0.20$ m, $D_2 = 0.15$ m, $p_1 = 1.5$ bar (gauge), $u_1 = 5.0$ m/s, and $\rho = 1,000$ kg/m^3 (the liquid is water).

26. *Jet-ejector pump—M*. Water from a supply having a total head of 100 ft is used in the jet of an ejector, shown in Fig. P2.26, in order to lift 10 ft^3/s of water from another source (at a lower level) that has a total head of -5 ft.

Fig. P2.26   Ejector "pump."

The nozzle area of the jet pipe is 0.05 ft^2 and the annular area of the suction line at section 1 is 0.5 ft^2. Determine the volumetric flow rate from the 100-ft head supply and the total delivery head for the ejector at section 2.

27. *Speed of a sound wave—M*. As shown in Fig. P2.27(a), a sound wave is a very small disturbance that propagates through a medium; its *wave front*, which moves with velocity c, separates the disturbed and undisturbed portions of the medium. (No part of the fluid itself actually moves with a velocity c.) In the undisturbed medium, the velocity is $u = 0$, the pressure is p, and the density is ρ. Just behind the wave front, in the disturbed medium, these values have been changed by infinitesimally small amounts, so that the corresponding values are du, $p + dp$, and $\rho + d\rho$. The system is best analyzed from the viewpoint of an observer traveling with the wave front, as shown in Fig. P2.27(b), which is obtained by superimposing a velocity c to the left, so the wave front is now stationary, and the flow is steady, from right to left.

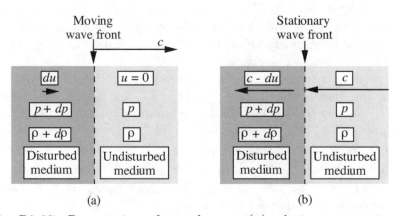

(a)   (b)

Fig. P2.27   Propagation of sound wave; (a) relative to a stationary observer; (b) relative to an observer traveling with the wave front.

By appropriate steady-state mass and momentum balances on a control volume, which you should clearly indicate, prove that the velocity of sound is:

$$c = \sqrt{\frac{dp}{d\rho}}.$$

(Although not needed here, it may further be shown that the above derivative should be evaluated at constant entropy.)

28. *Car retarding force—M. Experimentally,* how would you *most easily* determine the total retarding force (wind, road, etc.) on a car at various speeds?

Fig. P2.29   Plan of garden sprinkler.

29. *Garden sprinkler, arms rotating—E.* Fig. P2.29 shows an idealized plan of a garden sprinkler. The central bearing is well lubricated, so the arms are quite free to rotate about the central pivot. Each of the two nozzles has a cross-sectional area of 5 sq mm, and each arm is 20 cm long. Why is there rotation in the indicated direction? If the water supply rate to the sprinkler is 0.0001 m³/s, determine:

(a) The velocity u (m/s) of the water jets relative to the nozzles.
(b) The angular velocity of rotation, ω, of the arms, in both rad/s and rps.

30. *Garden sprinkler, arms fixed—E.* Consider again the sprinkler of Problem 2.29. For the same water flow rate, what applied torque is needed to prevent the arms from rotating?

31. *Oil/water separation—M.* Design a piece of equipment, as simple as possible, that will separate, by settling, a stream of an oil/water mixture into two individual streams of oil and water, as sketched in Fig. P2.31 (where the relative levels are not necessarily correct).

Fig. P2.31   *Unit for separating oil and water.*

Adhere to the following specifications:

(a) The exit streams are at atmospheric pressure.
(b) Anticipate that the inlet stream will fluctuate with time in both its oil and water content, with upper flow rate limits of 10 and 40 gpm, respectively.
(c) To insure proper separation, the residence time in the equipment of both the oil and water should be, on average, at least 20 minutes. For each phase, the residence time is defined as $t = V/Q$, the volume V occupied by the phase divided by its volumetric flow rate Q.
(d) A small relative water flow rate should not cause oil to leave through the water outlet, nor should a small relative oil flow rate cause water to leave through the oil outlet.

Your answer should specify the following details, at least:

(a) The volume and shape of the equipment.
(b) The locations of the inlet and exit pipes.
(c) The internal configuration.

32. *Slowing down of the earth—D.* If a and b are the semimajor and semiminor axes, respectively, consider the ellipse whose equation is:

$$\frac{x^2}{a^2} + \frac{y^2}{b^2} = 1. \tag{P2.32.1}$$

Consider the *ellipsoid*, shown in Fig. P2.32, formed by rotating the ellipse about its minor axis. Write down expressions for the mass dM and moment of inertia dI (about the semiminor axis) for the disk of radius x and thickness dy shown in the figure. Hence, prove that the mass and moment of inertia of the ellipsoid are:

$$M = \frac{4}{3}\pi\rho a^2 b, \qquad I = \frac{8}{15}\pi\rho a^4 b. \tag{P2.32.2}$$

Fig. P2.32   Rotating ellipsoid.

At midnight on December 31, 1995, an extra second was added to all cesium-133 atomic clocks to align them more closely with time as measured by the earth's rotational speed, which had been declining. Twenty such "leap" seconds were added in the twenty-four years from 1972 to 1995. One theory for the slowing down is that the earth is gradually bulging at the equator and flattening at the poles.[4] Assuming that the earth was a perfect sphere in 1972, and was in 1995 an ellipsoid, by what fraction of its original radius had it changed at the equator and at the poles? If the earth's radius is 6.37×10^6 m, what are the corresponding distances?

33. *Head for a real liquid—M*. Fig. 2.6 gave an interpretation of fluid heads for an ideal, *frictionless* liquid. For a *real* liquid, in which pipe friction is significant, redraw the diagram and explain all of its features. Incorporate four equally spaced manometer locations, starting at the entrance to the pipe and finishing at its exit.

34. *Acceleration of a jet airplane—M*. Fig. P2.34 shows an airplane of mass M that is flying horizontally with velocity u through otherwise stationary air. The airplane is propelled by a jet engine that ejects hot gas of density ρ_g at atmospheric pressure with velocity v *relative to the engine*, through exhaust ports of total cross-sectional area A. The drag force F is cu^2, where c is a known constant.

Control volume moving with airplane

Fig. P2.34   Airplane in horizontal flight.

Answer the following, using any or all of the symbols u, v, ρ_g, A, M, and c, as already defined:

(a) Write down an expression for m, the mass flow rate of hot gas leaving the engines.

---

[4] Curt Suplee, "Setting Earth's clock," *Washington Post,* December 31, 1990. See also Dava Sobel, "The 61-second minute," *New York Times*, December 31, 1995.

(b) As "seen" by a stationary observer on the ground, what is the velocity and direction of the exhaust gases?

(c) Perform a momentum balance on a control surface moving with the airplane, and derive an expression for its acceleration a. Assume that the mass of fuel consumed is negligibly small in comparison with that of the air passing through the engine.

(d) Compute a (m/s^2) if $u = 100$ m/s, $v = 300$ m/s, $\rho_g = 0.4$ kg/m^3, $A = 2$ m^2, $M = 10^4$ kg, and $c = 0.60$ kg/m.

35. *Performance of a "V-notch"—M*. Fig. P2.35(a) shows a "V-notch," which is a triangular opening of half-angle θ cut in the end wall of a tank or channel carrying a liquid. We wish to deduce the volumetric flow rate Q of liquid spilling over the notch from the height H of the free surface above the bottom of the notch.

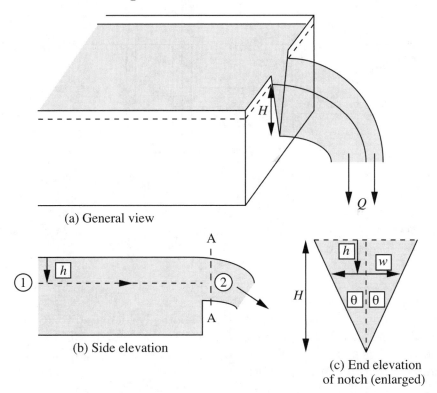

(a) General view

(b) Side elevation

(c) End elevation of notch (enlarged)

Fig. P2.35   *Discharge over a V-notch.*

The side elevation, Fig. P2.35(b), shows a representative streamline, flowing horizontally at a constant depth h below the free surface, between an upstream point 1 (where the pressure is hydrostatic and the velocity negligibly small) to a downstream point 2 at the notch exit, where the pressure is atmospheric (all the way across the section A–A). What is the reason for supposing that the velocity of the liquid as it spills over the notch is: $u_2 = \sqrt{2gh}$?

By integrating over the total depth of the notch, determine Q in terms of H, g, and θ. *Hint*: First determine the width w of the notch at a depth h in terms of H, h, and θ. (In practice, there is a significant further contraction of the stream after it leaves the notch, and the theoretical value obtained for Q has to be multiplied by a coefficient of discharge, whose value is typically $C_D \doteq 0.62$.)

36. *Tank-draining with a siphon—M.* Fig. P2.36 shows a tank of horizontal cross-sectional area A that is supplied with a steady volumetric flow rate Q of liquid, and that is simultaneously being drained through a siphon of cross-sectional area a, whose discharge is at the same level as the base of the tank. Frictional dissipation in the siphon may be neglected.

Fig. P2.36   *Siphon for draining tank.*

(a) Assuming that pipe friction is negligible, prove that the steady-state level in the tank is given by:

$$h^* = \frac{Q^2}{2ga^2}.$$

(b) If the incoming flow rate of liquid is now suddenly increased to $2Q$, prove that the time taken for the depth of liquid to rise to $2h^*$ (shown by the uppermost dashed line) is given by the following expression, in which you are expected to compute the value of the coefficient α:

$$t = \frac{\alpha A Q}{ga^2}.$$

Hints: You should need one of the integrals given in Appendix A. Start by performing a transient mass balance on the tank when the depth of liquid is h, between h^* and $2h^*$. Comment on the fact that the *higher* the flow rate, the *longer* the time it will take!

37. *Fluid "jetser"—M.* The October 1993 *Journal of Engineering Education* gives an account of a competition in the Civil Engineering department at Oregon State University. There, "Students design and build a fluid jetser powered solely by water from a reservoir connected to a nozzle and mounted on a wheeled platform." A possible configuration for such a "fluid jetser" is shown in Fig. P2.37.

Fig. P2.37   A fluid jetser.

Neglecting fluid friction, use Bernoulli's equation to derive an expression for v (the exit velocity of the water, relative to the nozzle) in terms of the current value of the height differential h. If the cross-sectional area of the nozzle is A, and the density of water is ρ, give an expression for the mass flow rate m that is leaving the nozzle.

Then perform a momentum balance on the moving jetser, as seen by a stationary observer. Hence, derive an expression for the acceleration du/dt in terms of any or all of h, A, F (the drag force), u, M (the current mass of the vehicle), ρ, and any physical constant(s) that you think are necessary.

38. *Force to hold orifice pipe in river—D (C).* A thin-walled pipe, of diameter 4 in., is held stationary in a river flowing at 6 ft/s parallel to its axis. The pipe contains a sharp-edged orifice, of diameter 3 in., with a contraction coefficient of 0.60 based on the area. Neglecting entry losses and shear forces on the pipe, and assuming that the orifice is distant from either end of the pipe, calculate the flow rate through the pipe and the force required to hold it in the stream.

39. *Froth on sieve tray—D (C).* Fig. P2.39 shows the flow pattern on a sieve tray operating at a high liquid flow rate and with a gas that is negligibly absorbed in the liquid. The liquid flows from left to right across the tray; AB is a region of clear liquid where the velocity is u_L and the density ρ_L; the perforations begin at B, and BC is a froth zone, the mean density of the froth being ρ_F. Show by a momentum balance that the fluid depths in the two regions are related by the equation:

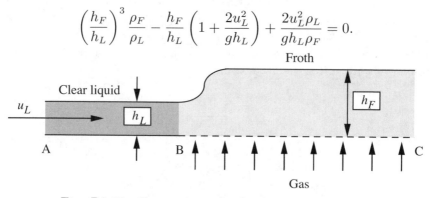

$$\left(\frac{h_F}{h_L}\right)^3 \frac{\rho_F}{\rho_L} - \frac{h_F}{h_L}\left(1 + \frac{2u_L^2}{gh_L}\right) + \frac{2u_L^2\rho_L}{gh_L\rho_F} = 0.$$

Fig. P2.39 *Formation of a froth on a sieve tray.*

Calculate h_F when $h_L = 0.2$ ft, $u_L = 0.5$ ft/s, $\rho_F/\rho_L = 0.5$. Calculate the ratio p_L/p_F of the pressure increases in the liquid and froth layers.

40. *Multiple orifice plates—M (C).* What is meant by the *pressure recovery* behind an orifice plate? Show that the overall pressure drop across an orifice plate, allowing for partial pressure recovery downstream, is kQm, where Q is the volumetric flow rate, m is the mass flow rate, and k is a function of the geometry of the system. Assume that k is constant, but comment on this assumption.

Fig. P2.40 *Four orifice plates in a Wheatstone-bridge arrangement.*

Four identical orifice plates are connected together in an arrangement analogous to a Wheatstone bridge, as shown in Fig. P2.40. The pump delivers a constant mass flow rate m_2 of an incompressible liquid from D to B. Show that, provided m_2 is large, the pressure drop between A and C is directly proportional to the liquid mass flow rate m_1 in the pipe XY.

What is the critical value of m_2, below which this expression for the pressure drop ceases to hold?

41. *Pressure recovery in sudden expansion—M (C).* Show that at a sudden enlargement in a pipe, the pressure rise is a maximum when the diameter of the larger pipe is $\sqrt{2}$ times the fixed diameter of the smaller pipe, and that the rate of energy dissipation is then $\rho a u^3/8$, in which u is the velocity in the smaller pipe of cross-sectional area a.

42. *Manifold in reactor base—D (C).* Fig. P2.42 shows a manifold MMM for injecting liquid into the base of a reactor. The manifold is formed from a pipe of cross-sectional area A. Liquid enters with velocity v through a pipe P of cross-sectional area a, and leaves uniformly around the circumference of the manifold via a large number of small holes. The rate of recirculation is indicated by the velocity V at section BB just upstream of the injection point. By assuming no loss of energy between sections B and A, but assuming that momentum is conserved from section A to section B, show that, when wall friction is negligible:

$$V = v(A - a)\sqrt{\frac{2}{Aa}}.$$

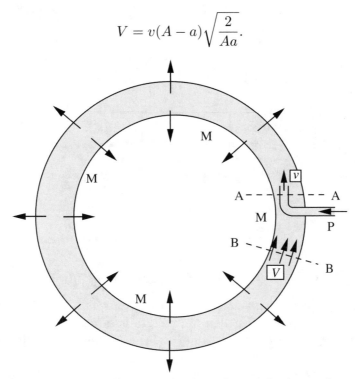

Fig. P2.42   Manifold for injecting liquid into the base of a reactor.

43. *Leaking ventilation duct—M (C).* A ventilation duct of square cross section operating just above atmospheric pressure has a damaged joint, the longitudinal section being shown in Fig. P2.43. The cross-sectional area open to the atmosphere is 10% of the total cross-sectional area (measured normal to the flow direction).

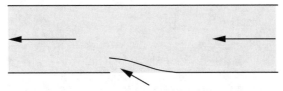

Fig. P2.43   Ventilation duct with damaged joint.

If the velocity upstream of the damaged section is 3 m/s, obtain an expression for the inward leakage rate as a function of the pressure drop along the duct. Neglect pressure drop due to friction. If friction were taken into account, what additional information would be required to solve the problem?

44. *Packed-column flooding—M (C).* Fig. P2.44 shows an apparatus for investigating the mechanism of flooding in packed columns. A is a vertical cylinder covered by a liquid film that runs down steadily and continuously under gravity. The nozzle B, shown in section, is symmetrical about the centerline of A, and at a typical section X–X the area between the wall of the nozzle and the liquid surface is $A = a + bx^2$.

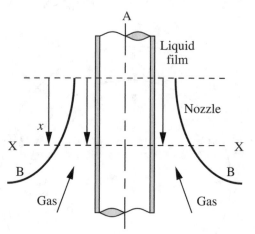

Fig. P2.44   Apparatus for investigating "flooding."

Show that if the liquid surface remains cylindrical, and the gas velocity is uniform across any section X–X, the pressure gradient dp/dx in the gas stream will be a maximum at $x = \sqrt{a/5b}$, and will arrest the downward flow of liquid at a gas flow rate Q given by:

$$Q^2 = \frac{108 g \rho_L a^3}{125 \rho_G b} \sqrt{\frac{5b}{a}},$$

in which ρ_G and ρ_L are the gas and liquid densities, respectively.

45. *Interpretation of pressure head—M.* Analyze the situation in Fig. 2.6(a), and comment critically as to whether the liquid level in the manometer is reasonable as drawn.

46. *Velocities and pressures—M.* Water flows from point A in the reservoir through a circular channel of varying diameter to its final discharge at point B. Sketch clearly the general trends in the velocity and gauge pressure along the streamline from A to B.

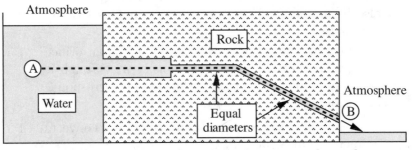

Fig. P2.46   Water flow through varying channel.

47. *True/false.* Check *true* or *false*, as appropriate:

(a) Heat terms are present in the mechanical energy balance, and are absent in Bernoulli's equation.   T ☐   F ☐

(b) If pure water flows into a tank of brine of constant volume, the concentration of salt in the exit stream decreases exponentially with time.   T ☐   F ☐

(c) Streamlines can cross one another if the fluid has a sufficiently high velocity.   T ☐   F ☐

(d) As gasoline flows steadily upward in a pipe of uniform diameter, its velocity decreases because of the negative influence of gravity.   T ☐   F ☐

(e) In the energy balance or Bernoulli's equation, the term $\Delta p/\rho$ has the same units as the square of a velocity.   T ☑   F ☑

(f) If Bernoulli's equation is divided through by g_c, then each term becomes a fluid head, and has the units of length, such as feet or meters.   T ☐   F ☐

(g) Consider an initial ($t = 0$) liquid level of $h = h_0$ in Fig. 2.7 (tank draining), with a corresponding *initial* exit flow rate of $Q_0 = A\sqrt{2gh_0}$. The time taken to drain the tank will be the initial volume of liquid in the tank divided by Q_0.   T ☐   F ☐

(h) A Pitot tube works on the principle of converting kinetic energy into potential energy.   T ☐   F ☐

(i)   As fluid flows through an orifice plate to the location   T ☐   F ☐
        of the *vena contracta,* its pressure will rise, because
        it is going faster there.

(j)   If a value for the frictional dissipation term $\mathcal{F}$ is   T ☐   F ☐
        known for an orifice plate in a horizontal pipe, it can
        also be applied to situations that are not necessarily
        horizontal.

(k)   If a mercury manometer connected across an orifice   T ☐   F ☐
        plate with kerosene flowing of density ρ_k through it
        registers a difference of Δh in levels, the correspond-
        ing pressure drop is $p_1 - p_2 = \rho_{\text{Hg}} g \Delta h$.

(l)   A stream of mass flow rate m and velocity u convects   T ☐   F ☐
        momentum at a rate mu^2.

(m)   At any instant, the rate of increase of momentum   T ☐   F ☐
        of a liquid-fueled rocket is $M dv/dt$, where M is the
        current mass of the rocket and v is its velocity.

(n)   The thrust of a jet engine is $m(v-u)$, where m is the   T ☐   F ☐
        mass flow rate through the engine, v is the velocity
        of the exhaust relative to the engine, and u is the
        velocity of the craft on which the engine is mounted.

(o)   There is always a pressure *decrease* across a sudden   T ☐   F ☐
        expansion in a pipeline.

(p)   A hydraulic jump is irreversible, and can occur only   T ☐   F ☐
        when a relatively deep stream of liquid suddenly be-
        comes a relatively shallow stream.

(q)   The momentum of a baseball decreases between the   T ☐   F ☐
        mound and plate because of gravitational forces.

(r)   A boat, traveling forwards at a velocity u on a lake,   T ☐   F ☐
        has a pump on board that takes water from the lake
        and ejects it to the rear at a velocity v relative to the
        boat. The velocity of the water as seen by an observer
        on the shore is $v - u$ in the forwards direction.

(s)   After the fluid downstream of an orifice plate has   T ☐   F ☐
        reestablished its flow pattern, it will return to the
        same pressure that it had upstream of the orifice
        plate.

(t)   Bernoulli's equation (with $w = \mathcal{F} = 0$) holds across a   T ☐   F ☐
        sudden expansion in a pipe.

(u)   A momentum balance can be used to determine the   T ☐   F ☐
       frictional dissipation term, $\mathcal{F}$, for a sudden expansion
       in a pipeline.

(v)   A pirouetting (spinning) ice skater, with her arms   T ☐   F ☐
       outstretched, rotates faster when she brings her hands
       together above her head mainly because of the re-
       duced air drag.

Chapter 3

FLUID FRICTION IN PIPES

3.1 Introduction

$\mathbf{I}$N chemical engineering process operations, fluids are typically conveyed through pipelines, in which viscous action—with or without accompanying turbulence—leads to "friction" and a dissipation of useful work into heat. Such friction is normally overcome either by means of the pressure generated by a pump or by the fluid falling under gravity from a higher to a lower elevation. In both instances, it is usually necessary to know what flow rate and velocity can be expected for a given driving force. This topic will now be discussed.

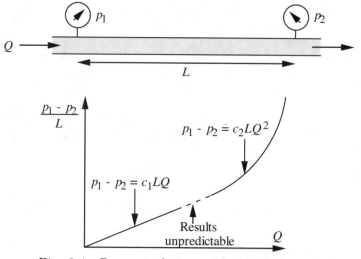

Fig. 3.1 *Pressure drop in a horizontal pipe.*

Fig. 3.1 shows a pipe fitted with pressure gauges that record the pressures p_1 and p_2 at the beginning and end of a test section of length L. A *horizontal* pipe is intentionally chosen because the observations are not then complicated by the effect of gravity. In addition, for good results, it is desirable that a substantial length of straight pipe should precede the first pressure gauge, in order that the flow pattern is *fully developed* (no longer varying with distance along the pipe) at that location.

Reynolds, Osborne, born 1842 in Belfast, Ireland; died 1912 in Somerset, England. Born into an Anglican clerical family, Reynolds entered an apprenticeship in 1861 with Edward Hayes, a mechanical engineer, before obtaining a degree in mathematics at Cambridge in 1867. After brief employment as a civil engineer, he competed successfully the next year for a newly created professorship at Owens College, Manchester, holding this position for the next 37 years. He worked on a wide range of topics in engineering and physics. He demonstrated the significance of the dimensionless group that now bears his name in his paper, "An experimental investigation of the circumstances which determine whether the motion of water in parallel channels shall be direct or sinuous and of the law of resistance in parallel channels," *Philosophical Transactions of the Royal Society*, Vol. 174, pp. 935–982 (1883). His analogy between heat and momentum transport (see Chapter 9) was published in his paper "On the extent and action of the heating of steam boilers," *Proceedings of the Manchester Literary and Philosophical Society*, Vol. 14, p. 8 (1874–1875). In 1885, he attributed the name *dilatancy* to the ability of closely packed granules to *increase* the volume of their interstices (the void fraction) when disturbed. He was elected a Fellow of the Royal Society in 1877.

Source: *Dictionary of Scientific Biography*, Charles Scribner's Sons, New York, 1975.

For a given flow rate, repetition of the experiment for different lengths demonstrates that the pressure drop $(p_1 - p_2)$ is directly proportional to L. Hence, it is appropriate to plot the pressure drop per unit length $(p_1 - p_2)/L$ (the negative of the *pressure gradient* dp/dz, where z denotes distance along the pipe) against the volumetric flow rate Q. There are three distinct flow regimes in the resulting graph:

1. For flow rates that are low (in a sense to be defined shortly), the pressure gradient is directly proportional to the flow rate.

2. For intermediate flow rates, the results are irreproducible, and alternate seemingly randomly between extensions of regimes 1 and 3.

3. For high flow rates, the pressure gradient is closely proportional to the *square* of the flow rate.

These regimes are known as the *laminar, transition,* and *turbulent* zones, respectively. The situation is further illuminated by the famous 1883 experiment of Sir Osborne Reynolds (see box above), similar to that illustrated in Fig. 3.2, where a liquid flows in a transparent tube. A fine steady filament of a dye is introduced by a hypodermic needle into the center of the flowing liquid stream, care being taken to ensure that there is no instability due to an imbalance of velocities. (For gases, the flow can be visualized by injecting a filament of smoke, such as kerosene vapor.) Again, three distinct flow regimes are found, which correspond exactly to those already encountered above:

1. For low flow rates, Fig. 3.2(a), the injected dye jet maintains its integrity as a long filament that travels along with the liquid. (The jet actually broadens gradually, due to diffusion.)
2. For intermediate flow rates, the results are irreproducible, and seem to alternate between extensions of regimes 1 and 3.
3. For high flow rates, Fig. 3.2(b), the jet of dye mixes very rapidly with the surrounding liquid and becomes highly diluted, so that it soon becomes invisible. The reason is that the liquid flow in the pipe is *unstable*, consisting of random turbulent motions superimposed on the bulk flow to the right.

Fig. 3.2   The Reynolds experiment.

Table 3.1   Dependence of Pipe Flow Regime on the Reynolds Number

Approximate Value of Reynolds Number	Flow Regime	Pressure Gradient Is Proportional to
$< 2{,}000$	Laminar	Q
$2{,}000$—$4{,}000$	Transition	Variable
$> 4{,}000$	Turbulent	$Q^{1.8}$—Q^2

Further experiments show that the three regimes do not depend solely on the flow rate, but on a dimensionless *combination* of the mean fluid velocity u_m, its density ρ and viscosity μ, and the diameter D of the pipe. The combination or *dimensionless group* is defined by:

$$\mathrm{Re} = \frac{\rho u_\mathrm{m} D}{\mu},\qquad (3.1)$$

and is called the *Reynolds number*, and indicates the relative importance of *iner-tial* effects (as measured by ρu_{m}^2—see Eqn. (3.4), for example) to *viscous* effects $(\mu u_{\mathrm{m}}/D)$. Table 3.1 shows which regime can be expected for a given Reynolds number. The exponent on Q is 1.8 or 2, depending on whether the pipe is hydraulically *smooth* or *rough*, respectively, in a sense to be defined later. In Sections 3.2 and 3.3 we shall study flow in the laminar and turbulent regimes more closely.

3.2 Laminar Flow

In order to avoid the additional complication of gravity (which *will* be included later), consider flow in the *horizontal* cylindrical pipe of radius a shown in Fig. 3.3.

Fig. 3.3 *Forces acting on a cylindrical fluid element.*

Consider further a moving cylinder of fluid of radius r and length L. In this case, there is zero *convective* transport of momentum across the two circular ends of the cylinder, and the analysis is simplified.[1] Because of the retarding action of the pipe wall, there will be a shear stress τ exerted to the left on the curved surface of the cylinder by the fluid between it and the pipe wall.[2] The net pressure force acting on the circular area πr^2 of the two ends is exactly counterbalanced by the shear stress acting on the curved surface, of area $2\pi r L$.

Thus, a steady-state momentum balance to the right gives:

$$p\pi r^2 - \left(p + \frac{dp}{dz}L\right)\pi r^2 - \tau 2\pi r L = 0. \qquad (3.2)$$

Simplifying,

$$\tau = \frac{r}{2}\left(-\frac{dp}{dz}\right). \qquad (3.3)$$

Since the pressure gradient is readily determined experimentally (Section 3.1), Eqn. (3.3) may be used for finding the shear stress at any radial location. Observe

---

[1] If a cylinder *fixed in space* is considered instead, there will be a convective transfer of z momentum in through the left-hand end, and out through the right-hand end. However, because the velocity profile is fully developed, these rates of transport are equal in magnitude and—since one is positive (an addition to the system) and the other negative (a loss from the system), they cancel each other and the same result is obtained.

[2] A more sophisticated definition of the shear stress, involving a sign convention and two subscripts, is intentionally being postponed, to Section 5.6.

that since dp/dz is negative, τ is indeed positive in the direction shown in Fig. 3.3. Equation (3.3) predicts a linear variation of τ with r, shown in the left part of Fig. 3.4. The shear stress is zero at the centerline, rising to a maximum value of τ_w at the pipe wall.

Fig. 3.4   *Shear stress and velocity distributions for laminar flow.*

The analysis so far holds equally well for laminar *or* turbulent flow. However, specializing now to the case of *laminar* flow of a *Newtonian* fluid, the shear stress is directly proportional to the velocity gradient:

$$\tau = -\mu \frac{du}{dr},\tag{3.4}$$

in which μ is the viscosity of the fluid. A justification for Eqn. (3.4) will be given in Section 3.3, on the basis of momentum transport on a molecular scale. Referring to Fig. 3.3, note that τ is positive as drawn and that du/dr is negative (since the velocity decreases to zero as the wall is approached). Elimination of τ between Eqns. (3.3) and (3.4) gives:

$$-\mu \frac{du}{dr} = \frac{r}{2}\left(-\frac{dp}{dz}\right),\tag{3.5}$$

which integrates to:

$$\int_0^u du = -\frac{1}{2\mu}\left(-\frac{dp}{dz}\right)\int_a^r r\,dr,\tag{3.6}$$

in which the boundary condition of $u = 0$ at $r = a$ reflects a condition of *no slip* at the pipe wall.

The final expression for the velocity u at any radial location r is:

$$u = \frac{1}{4\mu}\left(-\frac{dp}{dz}\right)(a^2 - r^2).\tag{3.7}$$

Observe that the pressure gradient can be expressed in terms of the total pressure drop $-\Delta p$ over a finite length L, as follows:

$$\frac{dp}{dz} = \frac{p_2 - p_1}{L} = -\frac{p_1 - p_2}{L} = \frac{\Delta p}{L}.\tag{3.8}$$

Hence, the velocity profile can also be rewritten as:

$$u = \frac{-\Delta p}{4\mu L}(a^2 - r^2).\qquad(3.9)$$

As shown in the right part of Fig. 3.4, note that the velocity profile is *parabolic* in shape, and that the velocity is directly proportional to the pressure gradient and inversely proportional to the viscosity.

Fig. 3.5   Flow through a differential annulus.

The total volumetric flow rate Q is obtained by integration over the cross section of the pipe. Consider the differential annulus of internal radius r and width dr shown in Fig. 3.5. Its area may be obtained in either of two ways: (a) as the difference in areas between two circles of radii $r + dr$ and r, neglecting a term in $(dr)^2$; or (b) by "unwinding" the annulus and regarding it as a rectangular strip of length $2\pi r$ and width dr. The area dA and the corresponding flow rate dQ through it are:

$$dA = \pi(r + dr)^2 - \pi r^2 = \pi r^2 + 2\pi r\, dr + \pi(dr)^2 - \pi r^2 = 2\pi r\, dr,\quad(3.10a)$$

$$dA = 2\pi r\, dr,\qquad(3.10b)$$

$$dQ = u\, dA = 2\pi r u\, dr.\qquad(3.11)$$

The total flow rate is therefore:

$$Q = \int_0^Q dQ = 2\pi \int_0^a \underbrace{\frac{1}{4\mu}\left(-\frac{dp}{dz}\right)}_{\text{Constant}}(a^2 - r^2)r\, dr$$

$$= \frac{\pi a^4}{8\mu}\left(-\frac{dp}{dz}\right) = \frac{\pi a^4}{8\mu}\frac{p_1 - p_2}{L},\qquad(3.12)$$

a relation for pipe flow known as the *Hagen-Poiseuille law*.

The above analysis can also be extended to the case of an *inclined* pipe, again by performing a momentum balance, but now including a gravitational force. The pressure gradient would be supplemented by the term $\pm \rho g \sin \theta$, and the result for the velocity profile, for example, would be:

$$u = \frac{1}{4\mu}\left(-\frac{dp}{dz} \pm \rho g \sin \theta\right)(a^2 - r^2),\qquad(3.13)$$

where θ is the (positive) angle of inclination to the horizontal. The plus sign in Eqn. (3.13) holds for downhill flow, and the minus sign for uphill flow. The velocity profile will still be parabolic, regardless of the orientation of the pipe.

Frictional dissipation term $\mathcal{F}$. An alternative and generally more useful approach is first to establish the *frictional dissipation* term for a horizontal pipe, shown in Fig. 3.6.

Fig. 3.6   *Flow in a horizontal pipe.*

The Hagen-Poiseuille law can be rephrased as:

$$Q = \frac{\pi a^4 (-\Delta p)}{8 \mu L}, \tag{3.14}$$

where $\Delta p = p_2 - p_1$ (as always, "Δ" denotes the final minus the initial value). But an overall energy balance, Eqn. (2.17), gives:

$$\alpha \Delta \left(\frac{u^2}{2} \right) + g \Delta z + \frac{\Delta p}{\rho} + w + \mathcal{F} = 0, \tag{3.15}$$

in which the factor α is unimportant at present but will be explained shortly. The first, second, and fourth terms are zero because there is no change in velocity, no change in elevation, and no work performed. Solution for $\mathcal{F}$ and elimination of Δp using Eqn. (3.14) gives:

$$\mathcal{F} = -\frac{\Delta p}{\rho} = \frac{8 \mu L Q}{\pi a^4 \rho} = \frac{8 \mu u_{\mathrm{m}} L}{\rho a^2}. \tag{3.16}$$

This expression for $\mathcal{F}$ is "universal" in the sense that it does not depend on the inclination of the pipe; it may therefore be used in the energy balance for flow in *horizontal, inclined,* or *vertical* pipes. The reason is that the frictional dissipation depends on the magnitude and shape of the velocity profile, which is parabolic in all instances, and does *not* depend on the pipe orientation.

The following relations are also available:

Maximum velocity

The fluid velocity is greatest at the centerline, and substitution of $r = 0$ into Eqn. (3.7) gives:

$$u_{\mathrm{max}} = \frac{a^2}{4\mu} \left(-\frac{dp}{dz} \right). \tag{3.17}$$

Mean velocity

The mean velocity of the fluid is obtained by dividing the total flow rate from Eqn. (3.12) by the cross-sectional area, and equals half the maximum velocity:

$$u_{\mathrm{m}} = \frac{Q}{\pi a^2} = \frac{a^2}{8\mu}\left(-\frac{dp}{dz}\right) = \frac{1}{2}u_{\max}. \tag{3.18}$$

Kinetic energy per unit mass

For purposes of overall energy balances, we need to find the kinetic energy associated with a unit mass of fluid as it traverses a particular axial location z. It will be seen that this is *not* simply half the square of the mean velocity, $u_{\mathrm{m}}^2/2$.

For any general property ψ per unit mass ($\psi = u^2/2$ in the case of kinetic energy), define an average value $\overline{\psi}$ per unit mass flowing as:

$$\overline{\psi} = \frac{\displaystyle\int_0^a \underbrace{2\pi r\,dr}_{dA}\,(\rho u \psi)}{\underbrace{\text{Total amount of }\psi\text{ flowing}}}\Bigg/\frac{}{\displaystyle\int_0^a \underbrace{2\pi r\,dr}_{dA}\,(\rho u)}. \tag{3.19}$$

$$\underbrace{}_{\text{Total mass flow rate}}$$

Substitution of $\psi = u^2/2$ into Eqn. (3.19) and recognizing that the denominator is simply ρQ gives the kinetic energy per unit mass flowing:

$$\overline{KE} = \frac{1}{\rho Q}\int_0^a 2\pi r\,dr\left(\rho u\frac{u^2}{2}\right) = \left[\frac{a^2}{8\mu}\left(-\frac{dp}{dz}\right)\right]^2 = u_{\mathrm{m}}^2 \quad \left(\text{not }\frac{u_{\mathrm{m}}^2}{2}\right). \tag{3.20}$$

Thus, the kinetic energy per unit mass in laminar pipe flow is:

$$\overline{KE} = \alpha\,\frac{u_{\mathrm{m}}^2}{2}, \tag{3.21}$$

in which $\alpha = 2$.

For *turbulent* velocity profiles, the integration of (3.19) is still performed with $\psi = u^2/2$, but now with a velocity profile that is appropriate for turbulent flow (see Section 3.5). Compared to their laminar counterparts, turbulent velocity profiles are much flatter near the centerline and steeper near the wall, and the result is $\alpha \doteq 1.07$. Further, since the kinetic energy term is frequently small when compared with other terms in the energy balance, α can usually safely be taken as unity for turbulent flow, and will therefore often be omitted.

Example 3.1—Polymer Flow in a Pipeline

A polymer flows steadily in the horizontal pipe of Fig. E3.1(a) under the following conditions: $\rho = 900$ kg/m^3, $\mu = 0.01$ Pa s (kg/m s), $D = 0.02$ m, and $u_m = 0.5$ m/s. Evaluate the following, clearly indicating the units:

(a) The Reynolds number.
(b) The frictional dissipation per meter per kg flowing.
(c) The pressure drop per meter.
(d) The elevation *decrease* for every meter of length if the polymer were to flow steadily at the same rate without any pumping needed. In this case, gravity provides the necessary energy for flow (i.e., $\Delta p = 0$), as in Fig. E3.1(b).

Fig. E3.1   *Polymer flow in pipeline: (a) horizontal, (b) inclined.*

Solution

(a) A check on the (dimensionless) Reynolds number gives:

$$\text{Re} = \frac{\rho u_m D}{\mu} = \frac{900 \times 0.5 \times 0.02}{0.01} = 900. \qquad \text{(E3.1.1)}$$

(b) Since the Reynolds number is below the critical value of 2,000, the flow is *laminar*, and the frictional dissipation term per unit length ($L = 1$ m) can be computed from Eqn. (3.16):

$$\mathcal{F} = \frac{8\mu u_m L}{\rho a^2} = \frac{8 \times 0.01 \times 0.5 \times 1}{900 \times (0.01)^2} = 0.444 \ \frac{\text{m}^2}{\text{s}^2} = 0.444 \ \frac{\text{J}}{\text{kg}}. \qquad \text{(E3.1.2)}$$

(c) An energy balance between any two points in the pipeline gives:

$$\alpha \Delta \left(\frac{u^2}{2} \right) + g\Delta z + \frac{\Delta p}{\rho} + w + \mathcal{F} = 0. \qquad \text{(E3.1.3)}$$

However, the first, second, and fourth terms are zero (constant velocity, no elevation change for a horizontal pipe, and no work done), so the pressure drop per meter is:

$$-\Delta p = \rho \mathcal{F} = 900 \times 0.444 \ \frac{\text{kg m/s}^2}{\text{m}^2} = 400 \text{ Pa} = 0.0040 \text{ bar}. \qquad \text{(E3.1.4)}$$

(d) To find the elevation decrease for no pumping, again apply the overall energy balance with zero kinetic energy change and no work done. Additionally, there is no change in the pressure, since the loss in potential energy is completely balanced by pipe friction, so that:

$$\Delta z = -\frac{\mathcal{F}}{g} = -\frac{0.444}{9.81} = -0.0453 \text{ m.} \qquad \text{(E3.1.5)}$$

That is, the elevation must drop by 0.0453 m for every meter of pipe length. □

3.3 Models for Shear Stress

Newton's law of viscosity relating the shear stress to the velocity gradient has a ready interpretation based on momentum transport resulting from molecular diffusion. As an introduction, consider first the situation in Fig. 3.7, which shows a plan of two trolleys on frictionless tracks.

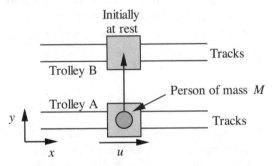

Fig. 3.7 Lateral transport of momentum between two trolleys on frictionless tracks.

A person of mass M jumps from trolley A, which is moving to the right with velocity u, onto trolley B, which is hitherto stationary. The person clearly transports an amount of momentum $\mathcal{M} = Mu$ from A to B, with the result that B *accelerates*; note that the momentum is in the x direction, but that it is transported in the *transverse* or y direction. If the transfer were *continuous*, with several people jumping successively—assuming that the trolley is large enough to accommodate them—the net effect would be a *steady* force in the x direction exerted by A on B, analogous to a shear force.

Momentum transport in laminar flow.   A similar phenomenon occurs for laminar flow, except that the lateral transport is now due to random molecular movement, known as *Brownian* motion.[3]

---

[3] G. Stix, *Scientific American*, Vol. 291, No. 3, September 2004, p. 46, says: "[In 1905] Albert Einstein published a paper in *Annalen*, 'On the motion of small particles suspended in liquids at rest required by

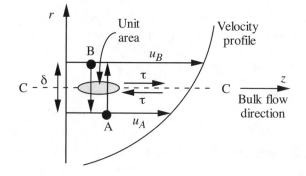

Fig. 3.8   Molecular diffusion model for shear stress in laminar flow.

Fig. 3.8 shows part of the velocity profile for flow in the axial or z direction of a pipe. Consider a plane C–C at any radial location. Because of Brownian motion, molecules such as those at A and B will constantly be crossing C–C from below and above, at a mass rate m per unit area per unit time. A molecule such as B, whose z component of velocity is u_B, will bring down with it a slightly larger amount of z momentum than that taken upward by a molecule such as A, whose z component of velocity is only u_A. For a velocity profile as shown, there is a net *positive* rate of transfer of z momentum downward across C–C, and this is manifest as a shear stress given by:

$$\tau = m(u_B - u_A). \tag{3.22}$$

On the average, the molecules will travel a small distance δ, known as their *mean free path*, before they collide with other molecules and surrender their momentum. The velocity discrepancy $u_B - u_A$ is therefore the product of δ and the local velocity gradient, yielding Newton's law of viscosity:

$$\tau \doteq m\delta \left(\frac{du}{dr}\right) = \mu \frac{du}{dr}, \tag{3.23}$$

where $\mu = m\delta$ is the *viscosity* of the fluid. Since both m and δ are independent of the flow rate in the z direction, the viscosity is constant, independent of the bulk motion. Note that we now have a *physical basis* for the similar result of (3.4), in which the minus sign merely reflects an alternative viewpoint for the direction of the shear stress.

---

the molecular-kinetic theory of heat,' that supplied a prediction of the number and mass of molecules in a given volume of liquid—and how these molecules would flit around. The erratic movements were known as Brownian motion, after the observation by botanist Robert Brown in 1827 of the irregular zigs and zags of particles inside pollen grains in water. Einstein suggested that the movements of the water molecules would be so great that they would jostle suspended particles, a dance that could be witnessed under a microscope."

Newton, Sir Isaac, born 1642 in Lincolnshire, England; died 1727 in Kensington, buried in Westminster Abbey, London. He was an unsurpassed scientific genius. After a brief interruption of his education to help with the family farm, he entered Trinity College Cambridge as a student in 1661, obtained his B.A. degree in mathematics in 1665, and was elected a fellow of his college in 1667. Newton's research in optics embraced reflection, refraction, and polarization of light, and he invented a reflecting telescope. About 1666, he deduced from Kepler's laws of planetary motion that the gravitational attraction between the sun and a planet must vary inversely as the square of the distance between them. Newton was appointed Lucasian professor of mathematics in 1669, and elected a fellow of the Royal Society in 1671. His *magnum opus* was *Philosophiæ Naturalis Principia Mathematica*; published in three volumes in 1687, "*Principia*" included the laws of mechanics, celestial motions, hydrodynamics, wave motion, and tides. At this stage, despite all of his accomplishments, Newton had neither been rewarded monetarily nor with a position of national prominence. However, matters improved when he was appointed Master of the Royal Mint in London in 1699, a position he held with distinction until his death.

Source: *The Encyclopædia Britannica*, 11th ed., Cambridge University Press (1910–1911).

For *gases*, the mean free path can be predicted from kinetic theory, leading to the following theoretical expression for the viscosity, which is found to agree well in most cases with experiment, provided the pressure is less than approximately 10 atm:

$$\mu = \frac{1}{\pi r^2}\sqrt{\frac{mkT}{3}}. \tag{3.24}$$

Here, r is the effective radius of a gas molecule and m is its mass; k is Boltzmann's constant (1.380×10^{-16} ergs/molecule K) and T is the absolute temperature. Note the predictions that: (a) μ is independent of pressure, and (b) μ *rises* with increasing temperature.

For *liquids,* there is no simple corresponding expression for their viscosities. In general, however—and in contrast to gases—the viscosity of liquids *falls* with increasing temperature. Note also the *empirical* correlations of Eqns. (1.17) and (1.18) for the viscosities of liquids and gases, which are based on experimental observations.

Momentum transport in turbulent flow.   An analogous situation holds for turbulent flow. However, the random molecular motion is now substantially *augmented* by a turbulent eddy motion, which is on a *much* larger scale. Turbulent flow can be described by the concept of the *eddy kinematic viscosity*, ν_T, giving:

$$\tau = \rho(\nu + \nu_T)\frac{du}{dr} \doteq \rho\nu_T\frac{du}{dr}, \tag{3.25}$$

it being noted—from experiment—that $\nu_T \gg \nu$. The eddy kinematic viscosity ν_T is *not* constant, but is found experimentally to be approximately directly proportional to $u_\mathrm{m}D$, where D is the pipe diameter; this result is also plausible on qualitative grounds—an increase in u_m causes more turbulence and a larger D permits eddies to travel further. The momentum transport is also proportional to the fluid density. Since du/dr is also roughly proportional to u_m (a doubling of u_m also approximately doubles the velocity gradient at any position), and inversely proportional to D, it follows that:

$$\tau \doteq c\rho u_\mathrm{m}D\left(\frac{u_\mathrm{m}}{D}\right) \doteq c\rho u_\mathrm{m}^2, \tag{3.26}$$

in which c is a constant, as yet unknown, and which depends on the particular radial location we are examining. A more detailed discussion of the eddy viscosity is given in Chapter 9.

Introduction to dimensional analysis.   By focusing on the shear stress τ_w at the wall, Eqn. (3.26) can be rephrased by asserting that in turbulent flow, the dimensionless wall shear stress or *friction factor*:

$$f = \frac{\tau_\mathrm{w}}{\rho u_\mathrm{m}^2}, \tag{3.27}$$

is a constant. The friction factor is another *dimensionless group*, being the ratio of the wall shear stress to the inertial force per unit area that would result from the impingement of a stream of density ρ and velocity u_m normally against a wall, as in Example 2.4. However, because the assumptions made above are not entirely correct, the friction factor is not quite constant in practice, and will be seen to depend on two more dimensionless groups—the Reynolds number and the relative roughness of the pipe wall.

Other versions in more common use are the *Fanning* friction factor:

$$f_\mathrm{F} = \frac{\tau_\mathrm{w}}{\frac{1}{2}\rho u_\mathrm{m}^2}, \tag{3.28}$$

and the *Moody* friction factor:

$$f_\mathrm{M} = \frac{\tau_\mathrm{w}}{\frac{1}{8}\rho u_\mathrm{m}^2}. \tag{3.29}$$

In this text, we shall mainly use f_F; most of the time we shall drop the appellation "Fanning" and simply call it *the* friction factor.

3.4 Piping and Pumping Problems

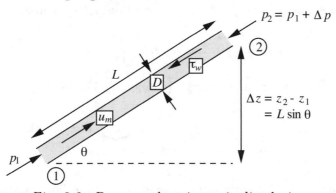

Fig. 3.9 *Pressure drop in an inclined pipe.*

Upward flow in an inclined pipe is depicted in Fig. 3.9. A steady-state momentum balance in the direction of flow on the fluid in the pipe gives:

$$\underbrace{(p_1 - p_2)\frac{\pi D^2}{4}}_{\left(\substack{\text{Net pres-}\\\text{sure force}}\right)} - \underbrace{\tau_{\text{w}}\pi DL}_{\left(\substack{\text{Wall shear}\\\text{stress force}}\right)} - \underbrace{\frac{\pi D^2}{4}\rho L g \sin\theta}_{\left(\substack{\text{Gravitational}\\\text{force}}\right)} = 0. \qquad (3.30)$$

Here, the three terms are the net pressure force that drives the flow, the retarding shear force exerted by the pipe wall, and the retarding gravitational weight of the fluid. Equation (3.30) can be rewritten as:

$$-\Delta p = p_1 - p_2 = 4\tau_{\text{w}}\frac{L}{D} + \rho g \Delta z. \qquad (3.31)$$

But, from the definition of the Fanning friction factor, Eqn. (3.28):

$$\tau_{\text{w}} = \frac{1}{2}f_{\text{F}}\rho u_{\text{m}}^2. \qquad (3.32)$$

Elimination of the wall shear stress between Eqns. (3.31) and (3.32) gives:

$$-\Delta p = p_1 - p_2 = \underbrace{2f_{\text{F}}\rho u_{\text{m}}^2\frac{L}{D}}_{\text{Friction}} + \underbrace{\rho g \Delta z}_{\text{Gravity}} = \frac{32f_{\text{F}}\rho Q^2 L}{\pi^2 D^5} + \rho g \Delta z. \qquad (3.33)$$

Here, the second form is in terms of the volumetric flow rate Q instead of the mean velocity. Thus, the pressure drop is clearly seen to depend on two factors—pipe friction and gravity. Although the preceding analysis was based on *upward* flow, the reader should double-check that Eqn. (3.33) also holds for *downward* flow.

An alternative and somewhat more generally useful form of Eqn. (3.33) is obtained by investigating the *frictional dissipation* per unit mass. Rearrangement of Eqn. (3.33) gives:

$$g\Delta z + \frac{\Delta p}{\rho} + 2f_F u_m^2 \frac{L}{D} = 0. \qquad (3.34)$$

But the overall (incompressible) energy balance, (2.17), is:

$$\underbrace{\Delta\left(\frac{u^2}{2}\right)}_{\text{zero}} + g\Delta z + \frac{\Delta p}{\rho} + \underbrace{w}_{\text{zero}} + \mathcal{F} = 0. \qquad (3.35)$$

Note that there is zero change in kinetic energy for pipe flow because the velocity is constant, and also that there is no work in the absence of a pump or turbine. A comparison of Eqns. (3.34) and (3.35) immediately gives the frictional dissipation per unit mass:

$$\mathcal{F} = 2f_F u_m^2 \frac{L}{D} = \frac{32f_F Q^2 L}{\pi^2 D^5}, \qquad (3.36)$$

a result that is valid for laminar *or* turbulent flow.

However, for the special case of *laminar* flow, we know from Eqn. (3.16) that:

$$\mathcal{F} = 2f_F u_m^2 \frac{L}{D} = \frac{8\mu u_m L}{\rho a^2}. \qquad (3.37)$$

Since $a = D/2$, the friction factor for *laminar* flow is therefore:

$$f_F = \frac{16\mu}{\rho u_m D} = \frac{16}{\text{Re}}. \qquad (3.38)$$

Experimentally, the friction factor f_F depends on the Reynolds number Re and—if the flow is turbulent—on the pipe roughness ratio ε/D. The salient features are shown in the *friction-factor plot* of Fig. 3.10.

Note the following:

1. For *laminar flow* (Re $\leq$ 2,000 approximately), Eqn. (3.38) is obeyed.
2. In the *transition* region (2,000 $<$ Re $\leq$ 4,000 approximately), there is no definite correlation, and f_F cannot be given a value. Because of this uncertainty, pipe designs in this region should be avoided.
3. In *turbulent* flow (Re $>$ 4,000), there is a family of curves, with f_F increasing as the relative roughness ε/D increases. Except for smooth pipes (see below), f_F becomes independent of Re at high Reynolds numbers.

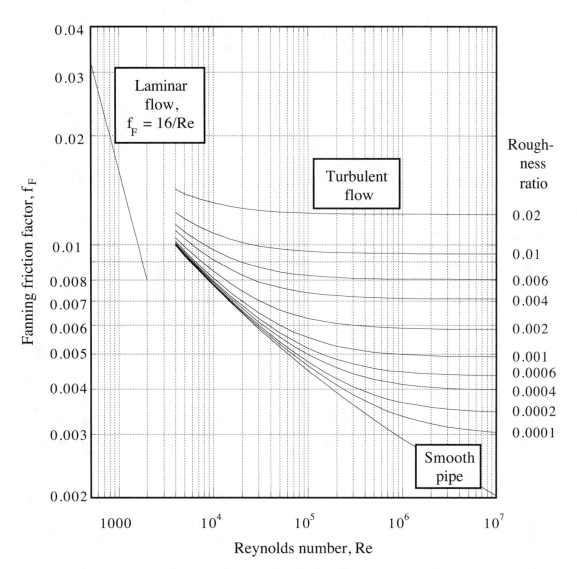

Fig. 3.10 Fanning friction factor for flow in pipes. The turbulent region is based on the Colebrook and White equation.

Pipe roughness. The question immediately arises as to how a *roughness* length scale ε can be assigned to the surface of a particular wall material. Initially, Nikuradse conducted pressure-drop measurements on pipes that he had artificially roughened by coating their inside walls with glue and sprinkling on a single layer of sand grains of known diameter.[4] The situation is idealized in Fig. 3.11, where ε is the diameter of the sand grains.

---

[4] J. Nikuradse, "Strōmungsgesetze in rauhen Rohren," *VDI-Forschungsarbeiten,* No. 362 (1933).

Fig. 3.11   Artificially roughened wall.

Nikuradse then deduced the friction factor for a large number of cases and was able to build up a friction-factor plot for *artificially* roughened pipes that was very similar to those shown in Fig. 3.10 for "real" surfaces. By comparing the two plots, it is then simple to assign an effective value of ε/D and, hence, ε for a "real" surface. Representative values are given in Table 3.2 for a variety of surfaces.

Table 3.2   Effective Surface Roughnesses[5]

Surface	ε (ft)	ε (mm)
Concrete	0.001–0.01	0.3–3.0
Cast iron	0.00085	0.25
Galvanized iron	0.0005	0.15
Commercial steel	0.00015	0.046
Drawn tubing	0.000005	0.0015

Friction-factor formulas.   The following formula, known as the *Colebrook and White* equation, gives a good representation of the experimentally determined friction factor in the turbulent region, and is the basis for that part of Fig. 3.10:

$$\frac{1}{\sqrt{f_\mathrm{F}}} = -4.0 \log_{10}\left(\frac{\varepsilon}{D} + \frac{4.67}{\mathrm{Re}\sqrt{f_\mathrm{F}}}\right) + 2.28$$

$$= -1.737 \ln\left(0.269\frac{\varepsilon}{D} + \frac{1.257}{\mathrm{Re}\sqrt{f_\mathrm{F}}}\right), \tag{3.39}$$

in which the logarithm to base e is intended in the second version. Also, for the special case of a *hydraulically smooth* surface, the following relation, known as the *Blasius* equation, correlates experimental observations for turbulent flow at Reynolds numbers below 100,000:

$$f_\mathrm{F} = 0.0790\,\mathrm{Re}^{-1/4}. \tag{3.40}$$

Hydraulically smooth means that the surface irregularities do not protrude beyond the laminar boundary layer immediately adjacent to the wall (see the last part of

---

[5] See for example, G.G. Brown et al., *Unit Operations*, Wiley & Sons, New York, 1950, p. 140.

Section 3.5). For this reason, the roughness is unimportant—not the case if the irregularities are large enough to extend into the turbulent core, in which case they would enhance the degree of turbulence and therefore influence the friction factor.

Unfortunately, Eqn. (3.39) is not explicit in the friction factor; that is, for a given Reynolds number, f_F cannot be computed directly because it appears on both sides of the equation. However, an evaluation of the right-hand side that incorporates a first *estimate* for f_F will give $1/\sqrt{f_F}$, from which a second estimate can be obtained. This second estimate can be "recycled" back into the right-hand side and the process repeated. This iterative process—known as *successive substitution*—converges very quickly to the desired value for the friction factor.

In an article that discusses easy-to-use formulas for the friction factor, Olujić notes that Shacham has pointed out that, starting with a first estimate of $f_F = 0.0075$, the procedure converges with an average accuracy of less than one percent within just *one* iteration, for a very wide range of Reynolds numbers and roughness ratios.[6] Therefore, it is reasonable to incorporate this starting estimate directly into Eqn. (3.39), which then gives the following *explicit* form for the friction factor:

$$f_F = \left\{ -1.737 \ln \left[0.269 \frac{\varepsilon}{D} - \frac{2.185}{\text{Re}} \ln \left(0.269 \frac{\varepsilon}{D} + \frac{14.5}{\text{Re}} \right) \right] \right\}^{-2}. \qquad (3.41)$$

In order to obtain the friction factor for rough pipes, we recommend:

1. For hand calculations, use Fig. 3.10.
2. For computer programs and spreadsheet calculations, use Eqn. (3.41) for the turbulent region ($\text{Re} > 4{,}000$) and $f_F = 16/\text{Re}$ for the laminar region ($\text{Re} \leq 2{,}000$). Avoid designs in the uncertain transition region ($2{,}000 < \text{Re} \leq 4{,}000$).

Commercial steel pipe is manufactured in *standard sizes*, a selection of which is shown in Table 3.3. Note in Table 3.3 that the nominal size is roughly the same as the inside diameter, and that the wall thickness depends on the *schedule number* n, defined as:

$$n = 1{,}000 \, \frac{p_{max}}{S_a}, \qquad (3.42)$$

in which p_{max} is the maximum allowable pressure in the pipe and S_a is the allowable tensile stress in the pipe wall.

---

[6] Ž. Olujić, "Compute friction factors fast for flow in pipes," *Chemical Engineering* (December 14, 1981), pp. 91–94.

Table 3.3   *Representative Pipe Sizes*[7]

Nominal Size (in.)	Outside Diameter (in.)	Schedule Number	Wall Thickness (in.)	Inside Diameter (in.)
1/2	0.840	40	0.109	0.622
		80	0.147	0.546
3/4	1.050	40	0.113	0.824
		80	0.154	0.742
1	1.315	40	0.133	1.049
		80	0.179	0.957
2	2.375	40	0.154	2.067
		80	0.218	1.939
3	3.500	40	0.216	3.068
		80	0.300	2.900
		160	0.437	2.626
4	4.500	40	0.237	4.026
		80	0.337	3.826
		160	0.531	3.438
6	6.625	40	0.280	6.065
		80	0.432	5.761
		160	0.718	5.189
8	8.625	40	0.322	7.981
		80	0.500	7.625
		160	0.906	6.813
10	10.75	40	0.365	10.020
		80	0.593	9.564
		160	1.125	8.500
12	12.75	40	0.406	11.938
		80	0.687	11.376
		160	1.312	10.126
16	16.00	40	0.500	15.000
		80	0.843	14.314
		160	1.562	12.876
24	24.00	40	0.687	22.626
		80	1.218	21.564
		160	2.312	19.376

---

[7] See, for example, J.H. Perry, ed., *Chemical Engineers' Handbook*, 3rd ed., McGraw-Hill, New York, 1950, pp. 415, 416.

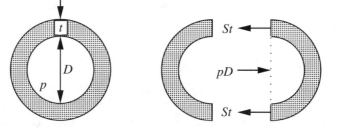

Fig. 3.12 Hoop stress in a pipe wall.

An interpretation of Eqn. (3.42) is readily apparent by studying Fig. 3.12, in which the hoop stress S in a pipe of diameter D subject to a pressure p is found by imagining the pipe to be split into two halves. The pressure force per unit length tending to blow away the right (or left) half is pD. Also, if t is the wall thickness, there is a total restoring force $2St$ within the pipe wall. Equating the two forces when p and S have reached their respective limits p_{max} and S_a:

$$p_{max}D = 2S_a t, \tag{3.43}$$

or,

$$\frac{t}{D} = \frac{p_{max}}{2S_a}. \tag{3.44}$$

Thus, the schedule number of (3.42) is 2,000 times the ratio of the necessary wall thickness to the pipe diameter. (The above treatment is most accurate for *thin-walled pipes*.)

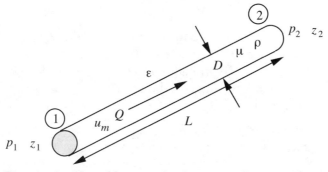

Fig. 3.13 Variables involved in pipe-flow problems.

Solution of "simple" piping problems.   Consider the transport of a fluid in a pipe from point 1 to point 2, as shown in Fig. 3.13, in which the following are assumed to be *known*—as will usually be the case in practice:

1. The elevation increase, $\Delta z = z_2 - z_1$, which may be positive, zero, or negative.
2. The length L of the pipe.
3. The material of construction of the pipe, and hence the pipe wall roughness ε.
4. The density ρ and viscosity μ of the fluid.

The following additional variables are of prime importance:

1. The volumetric flow rate, Q.
2. The internal pipe diameter, D.
3. The pressure drop, $-\Delta p = p_1 - p_2$.

The specification of any *two* of Q, D, and $-\Delta p$ (there are three cases) then enables the *remaining one* of them to be determined. The following algorithms are equally applicable to solutions by hand calculator or by a computer application such as a spreadsheet.

The calculations are not equally straightforward in each of the three cases just identified, and we distinguish between the following possibilities and approaches, the second and third of which require *iterative* types of solution:

1. Known flow rate and diameter. The combination of specified Q and D represents the easiest situation, in which the following direct (noniterative) steps are needed to determine the required pressure drop—which would then typically enable the size of an accompanying pump to be found:

(a) Compute the mean velocity $u_m = 4Q/\pi D^2$, the Reynolds number Re $= \rho u_m D/\mu$, and the roughness ratio ε/D.
(b) Based on the values of Re and ε/D, determine the friction factor f_F from either the friction factor chart or from the equations that represent it.
(c) Compute the pressure drop from Eqn. (3.33):

$$-\Delta p = p_1 - p_2 = 2f_F \rho u_m^2 \frac{L}{D} + \rho g \Delta z. \tag{3.33}$$

2. Known diameter and pressure drop. Given D and $-\Delta p$, the problem is now to find the flow rate Q. There is an immediate difficulty because—in the absence of the flow rate—the Reynolds number cannot be determined directly, so the friction factor is also unknown, even though the roughness ratio *is* known. The following steps are typically needed:

(a) Compute the roughness ratio ε/D.
(b) *Assume* or make a reasonable guess as to the first estimate for the Reynolds number Re. Since the majority of pipe-flow problems are in turbulent flow, a value such as Re = 10,000 or 100,000 should be considered. If the flow is of a viscous polymer, then the flow is probably laminar, in which case a value such as Re = 1,000 could be appropriate.
(c) Based on the values of ε/D and Re, compute the friction factor f_F from either the friction factor chart or from the equations that represent it.
(d) Compute the mean velocity u_m from:

$$-\Delta p = p_1 - p_2 = 2f_F \rho u_m^2 \frac{L}{D} + \rho g \Delta z, \tag{3.33}$$

which can be rearranged to give u_{m} explicitly:

$$u_{\mathrm{m}} = \sqrt{\frac{D}{2 f_{\mathrm{F}} \rho L} [(p_1 - p_2) - \rho g \Delta z]}. \qquad (3.45)$$

(e) Compute the Reynolds number from $\mathrm{Re} = \rho u_{\mathrm{m}} D / \mu$. If this is acceptably close to the value used in Step (c), then the problem is essentially solved, in which case proceed with Step (f). If not, return to Step (c).

(f) Compute the flow rate from $Q = u_{\mathrm{m}} \pi D^2 / 4$.

3. Known flow rate and pressure drop. Given Q and $-\Delta p$, the problem is to find the diameter D. The immediate difficulty is that—in the absence of the diameter—neither the Reynolds number *nor* the roughness ratio is known, so the friction factor is also unknown. The following steps are typically needed (other approaches are possible, but the formula given in Step (e) for the diameter is likely to lead to the quickest convergence):

(a) *Estimate* or guess the pipe diameter D. Hence, compute the corresponding mean velocity, $u_{\mathrm{m}} = 4Q / \pi D^2$.

(b) Compute the Reynolds number, $\mathrm{Re} = \rho u_{\mathrm{m}} D / \mu$.

(c) Compute the roughness ratio ε / D.

(d) Based on the available values of ε / D and Re, compute the friction factor f_{F} from either the friction factor chart or from the equations that represent it.

(e) Compute the diameter from:

$$D = \left[\frac{32 f_{\mathrm{F}} \rho Q^2 L}{\pi^2 (p_1 - p_2 - \rho g \Delta z)} \right]^{1/5}. \qquad (3.46)$$

(Note that this equation results by eliminating the mean velocity between Eqn. (3.33) and $Q = u_{\mathrm{m}} \pi D^2 / 4$.)

(f) Compute the mean velocity, $u_{\mathrm{m}} = 4Q / \pi D^2$.

(g) Compute the Reynolds number, $\mathrm{Re} = \rho u_{\mathrm{m}} D / \mu$. If this is acceptably close to the value used in Step (d), then the problem is essentially solved. If not, return to Step (c).

The above three situations will be fully explored in Examples 3.2, 3.3, and 3.4, which address Cases 1, 2, and 3, respectively. Additionally, Example 3.5 will solve a few loose ends not covered in Examples 3.2–3.4. We have intentionally opted to involve the same physical situation throughout, so the reader can concentrate on the different solution procedures.

Example 3.2—Unloading Oil from a Tanker
Specified Flow Rate and Diameter

General. The following statements apply equally to Examples 3.2, 3.3, 3.4, and 3.5. Fig. E3.2 shows a pump that transfers a steady stream of 35°API crude oil from an oil tanker to a refinery storage tank, both free surfaces being open to the atmosphere. The effective length—including fittings—of the commercial steel pipe is 6,000 ft. The discharge at point 4 is 200 ft above the pump exit, which is level with the free surface of oil in the tanker. However, because of an intervening hill, point 3 is at a higher altitude than point 4. Losses between points 1 and 2 may be ignored.

The crude oil has the following properties: $\rho = 53$ lb$_\mathrm{m}$/ft^3; $\mu = 13.2$ cP; vapor pressure $p_v = 4.0$ psia.

Specific to Example 3.2. Implement the algorithm for a Case 1–type problem. If the pipeline is Schedule 40 with a nominal diameter of 6 in., and the required flow rate is 506 gpm, what pressure p_2 is needed at the pump exit? Solve the problem first by hand calculations, and then by a spreadsheet.

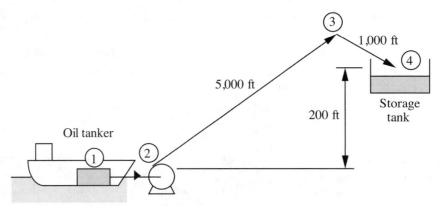

Fig. E3.2   Unloading tanker with intervening hill (vertical scale exaggerated).

Solution

(a) The main body of calculations is performed consistently in the ft, lb$_\mathrm{m}$, s, system. The flow rate of 506 gpm is first converted into cubic feet per second (the reader should by now be readily able to identify all of the conversion factors used):

$$Q = \frac{506}{7.48 \times 60} = 1.127 \; \frac{\mathrm{ft}^3}{\mathrm{s}}. \tag{E3.2.1}$$

From Table 3.3, for nominal 6-in. Schedule 40 pipe, the actual internal diameter is $D = 6.065$ in. $= 0.5054$ ft. The mean velocity is obtained by dividing the flow rate by the cross-sectional area:

$$u_\mathrm{m} = \frac{Q}{\pi D^2/4} = \frac{4Q}{\pi D^2} = \frac{4 \times 1.127}{\pi \times 0.5054^2} = 5.62 \; \frac{\mathrm{ft}}{\mathrm{s}}, \tag{E3.2.2}$$

Example 3.2—Unloading Oil, Specified Flow Rate and Diameter   **143**

from which the Reynolds number can be obtained:

$$\text{Re} = \frac{\rho u_m D}{\mu} = \frac{53 \times 5.62 \times 0.5054}{13.2 \times 0.000672} = 16,790. \tag{E3.2.3}$$

Table E3.2   Spreadsheet Solution of a Case 1–Type Problem

Flow in a Pipeline: Case 1, Specified Diameter and Flow Rate

Constants

g =	32.2	ft/s^2	pi =		3.1416
(lbm/ft s)/cp =	0.000672		gal/ft^3 =		7.48
g_c =	32.2	lbm ft/lbf s^2	s/min =		60
in/ft =	12		in^2/ft^2 =		144

Input data

D =	6.065	in	Q =	506	gpm
rho =	53	lbm/ft^3	mu =	13.2	cp
deltaz =	200	ft	L =	6,000	ft
eps =	0.00015	ft			

Intermediate values

D =	0.5054	ft	Q =	1.127	ft^3/s
mu =	0.00887	lbm/ft s	u_m =	5.620	ft/s
eps/D =	0.000297		Re =	16,970	

Pressure-drop calculations

fF =	0.00690		
2(fF)(rho)(u_m^2)L/D =	59.1	psi	
(rho)g(deltaz) =	73.6	psi	

Final solution

(-) deltap =	132.7	psi
p_2 =	132.7	psig

From Table 3.2, the roughness of commercial steel is $\varepsilon = 0.00015$ ft, so the pipe roughness ratio is:

$$\frac{\varepsilon}{D} = \frac{0.00015}{0.5054} = 0.000297 \tag{E3.2.4}$$

(b) Based on the above roughness ratio and Reynolds number (which clearly shows that the flow is turbulent), the Fanning friction factor is found from Fig. 3.10 to be approximately $f_F = 0.0070$, or somewhat more accurately from the Shacham equation as:

$$f_F = \left\{ -1.737 \ln \left[0.269 \times 0.000297 - \frac{2.185}{16,790} \ln \left(0.269 \times 0.000297 + \frac{14.5}{16,790} \right) \right] \right\}^{-2}$$

$$= 0.00692. \tag{E3.2.5}$$

(c) Now apply the overall energy balance between points 2 and 4. There is no change in kinetic energy and no work term, so:

$$p_2 - p_4 = \underbrace{2f_F \rho u_m^2 \frac{L}{D}}_{\text{Friction}} + \underbrace{\rho g(z_4 - z_2)}_{\text{Hydrostatic}}$$

$$= \frac{1}{32.2 \times 144} \left(\underbrace{2 \times 0.00692 \times 53 \times 5.620^2 \times \frac{6,000}{0.5054}}_{275,043} + \underbrace{53 \times 32.2 \times 200}_{341,268} \right)$$

$$= 132.9 \text{ psi.} \tag{E3.2.6}$$

The same sequence of calculations can also be performed by an Excel or other type of spreadsheet, as shown in Table E3.2. Note the organization of the spreadsheet, in which all values are clearly identified by algebraic symbols and their corresponding units. Observe also that *all* necessary constants and conversion factors are clearly displayed, rather than being "buried" in the various formulas in which they are used. The trifling difference between the computed friction factor in Eqn. (E3.2.5) and that in the spreadsheet, which used the logarithm to base 10 version, is because of truncation error.

If a single problem of this type is to be solved, the author finds that the hand-calculation method is quicker than setting up a new spreadsheet for its implementation. However, there should be no question of the superiority of the spreadsheet approach if the same type of calculation is to be repeated with different data. □

Example 3.3—Unloading Oil from a Tanker
Specified Diameter and Pressure Drop

Still consider the situation described at the beginning of Example 3.2, but now implement the algorithm for problems of type Case 2. If the pipeline is now specified to be of Schedule 40 with a nominal diameter of 6 in., and the available pressure at the pump exit is $p_2 = 132.7$ psig, what flow rate Q (gpm) can be expected?

Solution

(a) As in Example E3.2, the pipe diameter is $D = 0.5054$ ft and the roughness ratio is $\varepsilon/D = 0.000297$.

(b) Assume a first estimate of the Reynolds number, Re $= 100,000$, for example (a value that is intentionally quite high, in order to demonstrate the rapid convergence of the method).

(c) Based on the above roughness ratio and assumed Reynolds number (which corresponds to turbulent flow), the Fanning friction factor is found from Fig. 3.10

Example 3.3—Unloading Oil, Specified Diameter and Pressure Drop   **145**

to be approximately $f_F = 0.005$, or somewhat more accurately from the Shacham equation as:

$$f_F = \left\{-1.737\ln\left[0.269 \times 0.000297 - \frac{2.185}{10^5}\ln\left(0.269 \times 0.000297 + \frac{14.5}{10^5}\right)\right]\right\}^{-2}$$

$$= 0.00488. \tag{E3.3.1}$$

(d) The corresponding first estimate of the mean velocity is:

$$u_m = \sqrt{\frac{D}{2f_F\rho L}\left[(p_1 - p_2) - \rho g\Delta z\right]} \tag{E3.3.2}$$

$$= \sqrt{\frac{0.5054}{2 \times 0.00488 \times 53 \times 6{,}000}\left[132.7 \times 32.2 \times 144 - 53 \times 32.2 \times 200\right]}$$

$$= 6.681 \text{ ft/s.}$$

(e) Reevaluate the Reynolds number:

$$\text{Re} = \frac{53 \times 6.681 \times 0.5054}{13.2 \times 0.000672} = 20{,}175. \tag{E3.3.3}$$

Since this is different from the value of 100,000 assumed in (c), repeat steps (c), (d), and (e) until there is no further significant change. The successive computed values are summarized in Table E3.3.1.

Table E3.3.1   *Values as the Solution Converges*

Re	f_F	u_m(ft/s)
100,000	0.00488	6.681
20,175	0.00663	5.729
17,301	0.00687	5.630
17,002	0.00690	5.619
16,968	0.00690	5.617

(f) Finally, the required flow rate is computed:

$$Q = \frac{\pi D^2}{4}u_m = 7.48 \times 60 \times \frac{\pi \times 0.5054^2}{4} \times 5.617 = 506 \text{ gpm.} \tag{E3.3.4}$$

The spreadsheet solution is given in Table E3.3.2. In the "Iterated Values" section, the cells containing the values of the Reynolds number, mean velocity, and friction factor constitute a "circular reference," because the formula for Re involves u_m, the formula for f_F involves Re, and the formula for u_m involves f_F. However, the "iterate" feature of the Excel spreadsheet very quickly brings these values into consistency with one another, the final converged values being shown. The exact sequence of values computed by Excel is unknown.

Table E3.3.2   Spreadsheet Solution of a Case 2–Type Problem

Flow in a Pipeline: Case 2, Specified Diameter and Pressure Drop

Constants

g =	32.2	ft/s^2	pi =	3.1416	
(lbm/ft s)/cp =	0.000672		gal/ft^3 =	7.48	
g_c =	32.2	lbm ft/lbf s^2	s/min =	60.0	
in/ft =	12.0		in^2/ft^2 =	144	

Input data

D =	6.065	in	(-) deltap =	132.7	psi
rho =	53	lbm/ft^3	mu =	13.2	cp
deltaz =	200	ft	L =	6,000	ft
eps =	0.00015	ft			

Intermediate values

D =	0.5054	ft	(-) deltap =	615,303	lbm/ft s^2
mu =	0.00887	lbm/ft s	eps/D =	0.000297	

Iterated values

Re =	16,963		fF =	0.00690
u_m =	5.617	ft/s		

Final solution

Q =	1.127	ft^3/s	Q =	506 gpm

Table E3.4   Spreadsheet Solution of a Case 3–Type Problem

Flow in a Pipeline: Case 3, Specified Flow Rate and Pressure Drop

Constants

g =	32.2	ft/s^2	pi =	3.1416	
(lbm/ft s)/cp =	0.000672		gal/ft^3 =	7.48	
g_c =	32.2	lbm ft/lbf s^2	s/min =	60.0	
in/ft =	12		in^2/ft^2 =	144	

Input data

Q =	506	gpm	(-) deltap =	132.7	psi
rho =	53	lbm/ft^3	mu =	13.2	cp
deltaz =	200	ft	L =	6,000	ft
eps =	0.00015	ft			

Intermediate values

Q =	1.127	ft^3/s	(-) deltap =	615,303	lbm/ft s^2
mu =	0.00887	lbm/ft s			

Iterated values

D =	0.5055	ft	u_m =	5.618	ft/s
Re =	16,968		eps/D =	0.000297	
fF =	0.00690				

Final solution

D =	6.066	in

Example 3.4—Unloading Oil, Specified Flow Rate and Pressure Drop   **147**

Example 3.4—Unloading Oil from a Tanker
Specified Flow Rate and Pressure Drop

Consider again the situation described at the beginning of Example 3.2, but now implement the algorithm for problems of type Case 3. If the flow rate is specified as $Q = 506$ gpm, and the available pressure at the pump exit is $p_2 = 132.7$ psig, what pipe diameter D (in.) is needed?

A hand calculation will not be presented here, for two reasons:

1. The reader should have understood the general idea from Cases 1 and 2, as exemplified by the detailed hand calculations in Examples 3.2 and 3.3.
2. More quantities are involved in the iterative calculations for Case 3, and there is much to be said for using spreadsheet calculations exclusively.

Therefore, the final spreadsheet solution is given in Table E3.4. Observe that there are now *five* mutually dependent quantities: D, u_{m}, Re, ε/D, and f_{F}. The Excel "iterate" feature is again used in order to converge rapidly on the final indicated values. □

Example 3.5—Unloading Oil from a Tanker
Miscellaneous Additional Calculations

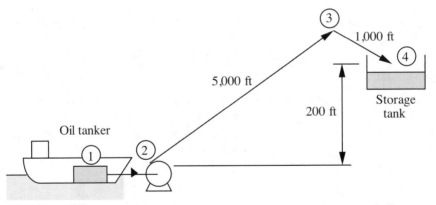

Fig. E3.5 (E3.2)   Unloading tanker with intervening hill.

As shown in Fig. E3.5, still consider the situation studied in Examples 3.2, 3.3, and 3.4, for which $D = 0.5054$ ft, $p_2 = 132.7$ psig, and $Q = 506$ gpm. Answer the following additional questions:

(a) If the combination of pump and motor is 80% efficient, how much electrical power (kW) is needed to drive the pump?
(b) If, in order to avoid vapor lock, the pressure in the pipeline must always be above the vapor pressure of the crude oil, what is the maximum permissible elevation of point 3 relative to point 4?

(c) If the flow in the pipeline were at the upper limit of being laminar, what pump exit pressure would then be needed? (Answer this part without using the friction factor plot.)

Solution

(a) To obtain the necessary pumping power, first apply the energy balance across the pump, between points 1 and 2 (with virtually no change in kinetic energy[8] or elevation, and no explicit representation of friction):

$$\frac{p_2 - p_1}{\rho} + w = 0. \tag{E3.5.1}$$

The work performed *on* the crude oil, per unit mass, is:

$$-w = 132.7 \ \frac{\text{lb}_f}{\text{in}^2} \times 144 \ \frac{\text{in}^2}{\text{ft}^2} \times \frac{1}{53} \ \frac{\text{ft}^3}{\text{lb}_m} = 360 \ \frac{\text{ft lb}_f}{\text{lb}_m}. \tag{E3.5.2}$$

The mass flow rate of the oil is:

$$m = \frac{506 \times 53}{7.48 \times 60} = 59.75 \ \frac{\text{lb}_m}{\text{s}}. \tag{E3.5.3}$$

Bearing in mind the efficiency of 80%, the required electrical power to be delivered to the pump is:

$$P = \frac{360 \times 59.75}{737 \times 0.80} = 36.48 \ \text{kW}. \tag{E3.5.4}$$

(b) Note that vapor lock is most likely to occur at the highest elevation—namely, at point 3. Therefore, to find the maximum elevation of point 3 *without* causing vapor lock, apply the energy equation between points 3 and 4, again with zero kinetic-energy change and no work term:

$$g(z_4 - z_3) + \frac{p_4 - p_3}{\rho} + 2f_F u_m^2 \frac{L}{D} = 0, \tag{E3.5.5}$$

$$32.2(z_4 - z_3) + \underbrace{\frac{(14.7 - 4) \times 32.2 \times 144}{53}}_{936.1} + \underbrace{2 \times 0.00692 \times 5.617^2 \times \frac{1,000}{0.5054}}_{864.0} = 0.$$

$$\tag{E3.5.6}$$

Solving for the elevation difference:

$$z_3 - z_4 = 55.9 \ \text{ft.} \tag{E3.5.7}$$

---

[8] The kinetic energy term $\Delta(u^2)/2 = 0.490$ ft lb$_f$/lb$_m$ is small compared to $\Delta p/\rho = 363.3$ ft lb$_f$/lb$_m$.

Example 3.4—Unloading Oil, Specified Flow Rate and Pressure Drop   **149**

That is, the highest point in the pipeline is limited to 55.9 ft elevation above the final discharge at point 4. If it were any higher, the pressure would fall to the vapor pressure of the oil (4.0 psia) and the oil would *start* to vaporize; the extent of vaporization would be limited by the amount of heat available to supply the necessary latent heat of vaporization.

(c) If the flow were at the upper limit of the laminar range, the Reynolds number would be Re $= \rho u_\mathrm{m} D/\mu = 2,000$, corresponding to a mean velocity of:

$$u_\mathrm{m} = \frac{2{,}000 \times 13.2 \times 0.000672}{53 \times 0.5054} = 0.662 \ \frac{\mathrm{ft}}{\mathrm{s}}. \qquad (E3.5.8)$$

The corresponding frictional dissipation per unit mass is:

$$\mathcal{F} = \frac{8\mu L Q}{\pi a^4 \rho} = \frac{8\mu L u_\mathrm{m}}{a^2 \rho} = \frac{8 \times 13.2 \times 0.000672 \times 6{,}000 \times 0.662}{\left(\dfrac{0.5054}{2}\right)^2 \times 53} = 83.32 \ \frac{\mathrm{ft}^2}{\mathrm{s}^2}.$$

$$(E3.5.9)$$

Application of the energy balance between points 2 and 4, with $p_4 = 0$ and $\Delta z = 200$ ft, gives:

$$-\frac{p_2 \times 144 \times 32.2}{53} + 200 \times 32.2 + 83.32 = 0, \qquad (E3.5.10)$$

so that the required pump exit pressure is:

$$p_2 = 74.6 \ \mathrm{psig}. \qquad (E3.5.11) \ \square$$

Alternative treatment as simultaneous nonlinear equations.   A different but equivalent approach to simple piping problems of the nature just discussed in Examples 3.2, 3.3, and 3.4 is to recognize that the situation—whether Case 1, 2, or 3 is involved—is governed by the following system of *simultaneous nonlinear equations*:

Pressure drop:

$$-\Delta p = p_1 - p_2 = 2 f_\mathrm{F} \rho u_\mathrm{m}^2 \frac{L}{D} + \rho g \Delta z. \qquad (3.47)$$

Flow rate:

$$Q = \frac{\pi D^2}{4} u_\mathrm{m}. \qquad (3.48)$$

Reynolds number:

$$\mathrm{Re} = \frac{\rho u_\mathrm{m} D}{\mu}. \qquad (3.49)$$

Equations representing the friction-factor plot (avoid $2{,}000 < \mathrm{Re} \leq 4{,}000$):

$$\mathrm{Re} \leq 2{,}000: \quad f_\mathrm{F} = \frac{16}{\mathrm{Re}},$$

$$\mathrm{Re} > 4{,}000: \quad f_\mathrm{F} = \left\{ -1.737 \ln \left[0.269\frac{\varepsilon}{D} - \frac{2.185}{\mathrm{Re}} \ln \left(0.269\frac{\varepsilon}{D} + \frac{14.5}{\mathrm{Re}} \right) \right] \right\}^{-2}.$$

$$(3.50)$$

Depending on the particular case for which the solution is required, there will be different sets of known quantities and unknown quantities. However, with the ready availability of spreadsheets such as Excel (with its "Solver" feature) and equation solvers such as Polymath, the above equations can be solved fairly easily.

3.5   Flow in Noncircular Ducts

The cross section of a pipe is most frequently circular, but other shapes may be encountered. For example, the rectangular cross section of many domestic hot-air heating ducts should be apparent to most people living in the United States. The situation for a *horizontal* duct is illustrated in Fig. 3.14; the cross-sectional shape is quite arbitrary—it doesn't have to be rectangular as shown—as long as it is uniform at all locations. There, A is the cross-sectional area and P is the *wetted perimeter*—defined as the length of wall that is actually in contact with the fluid. For the flow of a gas, P will always be the length of the complete periphery of the duct; for liquids, however, it will be somewhat less than the periphery if the liquid has a free surface and incompletely fills the total cross section.

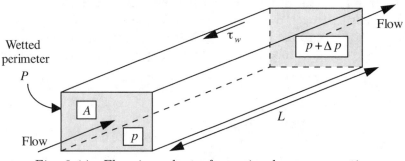

Fig. 3.14 *Flow in a duct of noncircular cross section.*

If τ_w is the wall shear stress and there is a pressure drop $-\Delta p$ over a length L, a momentum balance in the direction of flow yields:

$$[p - (p + \Delta p)]A - \tau_\mathrm{w}PL = 0. \tag{3.51}$$

The pressure drop is therefore:

$$-\Delta p = \tau_{\mathrm{w}} \frac{PL}{A} = \underbrace{\frac{\tau_{\mathrm{w}}}{\frac{1}{2}\rho u_{\mathrm{m}}^2}}_{f_{\mathrm{F}}} \frac{1}{2} \rho u_{\mathrm{m}}^2 \frac{PL}{A} \tag{3.52}$$

$$= 2 f_{\mathrm{F}} \rho u_{\mathrm{m}}^2 \frac{L}{4A/P} = 2 f_{\mathrm{F}} \rho u_{\mathrm{m}}^2 \frac{L}{D_{\mathrm{e}}}. \tag{3.53}$$

Thus, the equation for the pressure drop is identical with that of Eqn. (3.33) for a *circular* pipe provided that D is replaced by the *hydraulic mean diameter* D_e, defined by:

$$D_{\mathrm{e}} = \frac{4A}{P}. \tag{3.54}$$

The reader may wish to check that $D_e = D$ for a circular duct. Following similar lines as those used previously, the frictional dissipation per unit mass can be deduced as:

$$\mathcal{F} = 2 f_{\mathrm{F}} u_{\mathrm{m}}^2 \frac{L}{D_{\mathrm{e}}}, \tag{3.55}$$

and this expression can then be employed for *inclined* ducts of noncircular cross section.

Steady flow in open channels. A similar treatment follows for a liquid flowing steadily down a channel inclined at an angle θ to the horizontal, such as a river or irrigation ditch, shown in Fig. 3.15. Again, as long as the cross section is uniform along the channel, it can be quite arbitrary in shape, not necessarily rectangular. The driving force is now gravity, there being no variation of pressure because the free surface is uniformly exposed to the atmosphere.

If the wetted perimeter is again P and the cross-sectional area occupied by the liquid is A, a steady-state momentum balance in the direction of flow gives:

$$\rho A L g \sin\theta - \tau_{\mathrm{w}} P L = 0. \tag{3.56}$$

Noting that:

$$L\sin\theta = z_1 - z_2 = -\Delta z, \tag{3.57}$$

division of Eqn. (3.56) by $-\rho A$ gives:

$$g\Delta z + \frac{\tau_{\mathrm{w}} P L}{\rho A} = 0, \tag{3.58}$$

in which the second term can be rearranged as:

$$2 \frac{\tau_{\mathrm{w}}}{\frac{1}{2}\rho u_{\mathrm{m}}^2} u_{\mathrm{m}}^2 \frac{L}{4A/P} = 2 f_{\mathrm{F}} u_{\mathrm{m}}^2 \frac{L}{4A/P}. \tag{3.59}$$

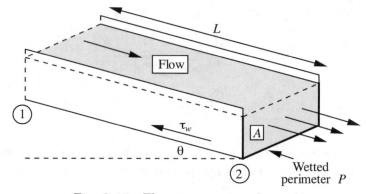

Fig. 3.15   Flow in an open channel.

Comparison with the overall energy balance:

$$g\Delta z + \mathcal{F} = 0, \tag{3.60}$$

gives the frictional dissipation per unit mass as:

$$\mathcal{F} = 2f_{\mathrm{F}} u_{\mathrm{m}}^2 \frac{L}{D_{\mathrm{e}}}, \quad \text{where} \quad D_{\mathrm{e}} = \frac{4A}{P}, \tag{3.61}$$

which has exactly the same form as Eqns. (3.54) and (3.55).

Example 3.6—Flow in an Irrigation Ditch

The irrigation ditch shown in Fig. E3.6 has a cross section that is 6 ft wide ×
6 ft deep. It conveys water from location 1 to location 2, between which there is
a certain drop in elevation.

Fig. E3.6   Cross section of irrigation ditch.

With a flow rate of $Q = 72$ ft³/s of water, the ditch is filled to a depth of
4 ft. If the same ditch, transporting water between the same two locations, were
completely filled to a depth of 6 ft, by what percentage would the flow rate increase?
Start by applying the overall energy balance between points 1 and 2, and assume
that the friction factor remains constant.

Example 3.6—Flow in an Irrigation Ditch   **153**

Solution

Apply the energy equation between two points separated by a distance L:

$$\underbrace{\Delta \frac{u^2}{2}}_{\text{Zero}} + g\Delta z + \underbrace{\frac{\Delta p}{\rho} + w}_{\text{Zero}} + \underbrace{2f_F u_{\mathrm{m}}^2 \frac{L}{D_e}}_{\mathcal{F}} = 0,$$

or,

$$g\Delta z + 2f_F u_{\mathrm{m}}^2 \frac{L}{D_e} = 0. \tag{E3.6.1}$$

Whether the ditch is filled to 4 ft or 6 ft, all quantities in Eqn. (E3.6.1) remain the same except u_{m} and D_e, so that:

$$u_{\mathrm{m}}^2 = cD_e, \tag{E3.6.2}$$

in which c is a factor that incorporates everything that remains constant.

Case 1. When the ditch is filled to a depth of only 4 ft, the hydraulic mean diameter is:

$$D_e = \frac{4A}{P} = \frac{4 \times 24}{14} = 6.86 \text{ ft.} \tag{E3.6.3}$$

But the mean velocity is:

$$u_{\mathrm{m}} = \frac{Q}{A} = \frac{72}{24} = 3.0 \ \frac{\text{ft}}{\text{s}}, \tag{E3.6.4}$$

so that the value of the constant in Eqn. (E3.6.2) is:

$$c = \frac{3^2}{6.86} = 1.31 \ \frac{\text{ft}}{\text{s}^2}. \tag{E3.6.5}$$

Case 2. When the ditch is filled to a depth of 6 ft, the hydraulic mean diameter is:

$$D_e = \frac{4A}{P} = \frac{4 \times 36}{18} = 8.00 \text{ ft.} \tag{E3.6.6}$$

From Eqns. (E3.6.2) and (E3.6.5), the mean velocity is:

$$u_{\mathrm{m}} = \sqrt{1.31 \times 8} = 3.24 \ \frac{\text{ft}}{\text{s}}, \tag{E3.6.7}$$

so that the flow rate is now:

$$Q = u_{\mathrm{m}}A = 3.24 \times 36 = 116.7 \ \frac{\text{ft}^3}{\text{s}}. \tag{E3.6.8}$$

The percentage increase in flow rate is therefore:

$$\frac{116.7 - 72}{72} \times 100 = 62\%. \tag{E3.6.9}$$

Thus, the increase in flow rate is somewhat *more* than the 50% increase in the depth of water. The reason is that the increased area for flow is accompanied by a somewhat lower increase in the length of the wetted perimeter.   □

Pressure drop across pipe fittings.   A variety of auxiliary hardware such as valves and elbows is associated with most piping installations. These *fittings* invariably cause the flow to deviate from its normal straight course and hence induce additional turbulence and frictional dissipation. Indeed, the resulting additional pressure drop is sometimes comparable to that in the pipeline itself.

The basic procedure is to recognize that the fitting causes an additional pressure drop that would be produced by a certain length of pipe into which the fitting is introduced. Therefore, we substitute for the fitting an extra contribution to the length of the pipe, based on the *equivalent length* $(L/D)_e$ of the fitting. For example, referring to Table 3.4, three standard 90° elbows in a 6–in.-diameter line cause a pressure drop that is equivalent to an extra 45 ft of pipe.

The gate valve uses a retractable circular plate that normally has one of two extreme positions: (a) complete obstruction of the flow, or (b) essentially no obstruction. The gate valve cannot be used for fine control of the flow rate, for which the globe or needle valve, with an adjustable plug or needle partly obstructing a smaller orifice, is more effective.

Table 3.4   Equivalent Lengths of Pipe Fittings[9,10]

Type of Fitting	$(L/D)_e$
Angle valve (open)	160
Close return bend	75
Gate valve (open)	6.5
Globe valve (open)	330
Square 90° elbow	70
Standard 90° elbow	30
Standard "T" (through side outlet)	70
45° elbow	15
Sudden contraction, 4:1	15
Sudden contraction, 2:1	11
Sudden contraction, 4:3	6.5
Sudden expansion, 1:4	30
Sudden expansion, 1:2	20
Sudden expansion, 3:4	6.5

Laminar and turbulent velocity profiles.   The parabolic velocity profile already encountered in *laminar* flow in a pipe is again illustrated on the left of Fig.

---

[9]   See for example, G.G. Brown et al., *Unit Operations*, Wiley & Sons, New York, 1950, p. 140.

[10]   For sudden expansions and contractions, the *diameter ratio* is given in Table 3.4; also, the equivalent length in these cases is based on the smaller diameter.

3.16. On the right, we see for the first time the general shape of the velocity profile for *turbulent* flow. Chapter 9 shows that although in turbulent flow the velocities exhibit random fluctuations, it is still possible to work in terms of a *time-averaged* axial velocity. For simplicity at this stage, we shall still use u to denote such a quantity, although in Chapter 9 it will be replaced with a symbol such as $\bar{v}_z$.

Fig. 3.16 *Laminar and turbulent velocity profiles.*

In addition to the generally higher velocities, note that the overall turbulent velocity profile consists of two profoundly different zones:

1. A very thin region, known as the *laminar sublayer*, in which turbulent effects are essentially absent, the shear stress is virtually constant, and there is an extremely steep velocity gradient. The following equation is derived in Section 9.7 for the thickness δ of the laminar sublayer relative to the pipe diameter D as a function of the Reynolds number in the pipe:

$$\frac{\delta}{D} \doteq 62 \, \text{Re}^{-7/8}. \tag{3.62}$$

Observe that the laminar sublayer becomes thinner as the Reynolds number increases, because the greater intensity of turbulence extends closer to the wall.

2. A *turbulent* core, which extends over nearly the whole cross section of the pipe. Here, the velocity profile is relatively flat because rapid turbulent radial momentum transfer tends to "iron out" any differences in velocity. A representative equation for the ratio of the velocity u at a distance y from the wall to the centerline velocity u_c is:

$$\frac{u}{u_c} = \left(\frac{y}{a}\right)^{1/n}, \tag{3.63}$$

in which n in the exponent varies somewhat with the Reynolds number, as in Table 3.5. For $n = 1/7$, Eqn. (3.63) is plotted in Fig. 9.11.

Table 3.5 Exponent n for Equation (3.63)[11]

Re	n
4.0×10^3	6
2.3×10^4	6.6
1.1×10^5	7
1.1×10^6	8.8
2.0×10^6	10
3.2×10^6	10

3.6 Compressible Gas Flow in Pipelines

A complete treatment of compressible gas flow is beyond the scope of this text, and two important examples—one here and one in the next section—must suffice.

Most of the discussion so far has involved *incompressible* fluids—primarily liquids, although gases may be included in approximate treatments if the changes in pressure, and hence in the density, are mild. However, there are situations involving gases in which the density variations caused by changes in pressure (and possibly temperature) *must* be taken into account if completely erroneous results are to be avoided. An example of this type of flow is closely approximated by the interstate underground pipeline transport of natural gas, in which the surrounding ground maintains essentially isothermal conditions.

Therefore, consider the steady flow of an ideal gas of molecular weight M_w in a long-distance horizontal pipeline of length L and diameter D, as shown in Fig. 3.17. The inlet and exit pressures and densities are p_1, ρ_1 and p_2, ρ_2, respectively. The pipeline is assumed to be sufficiently long in relation to its diameter that it comes into thermal equilibrium with its surroundings; thus, the flow is *isothermal*, at an absolute temperature T.

Fig. 3.17 Isothermal flow of gas in a pipeline.

If the mean velocity is u and the pressure is p, a differential energy balance over a length dx gives:

$$d\left(\frac{u^2}{2}\right) + \frac{dp}{\rho} + d\mathcal{F} = 0. \tag{3.64}$$

---

[11] H. Schlichting, *Boundary-Layer Theory*, McGraw-Hill, New York, 1955, p. 403.

Note that the change in kinetic energy cannot necessarily be ignored, since fluctuations in density will cause the gas either to accelerate or decelerate. Expansion of the differential, substitution of an alternative expression for $d\mathcal{F}$, and division by u^2 produces:

$$\frac{du}{u} + \frac{dp}{\rho u^2} + 2f_{\mathrm{F}}\frac{dx}{D} = 0. \tag{3.65}$$

Because of continuity, the *mass velocity* $G = \rho u$ is constant:

$$dG = 0 = \rho\, du + u\, d\rho, \tag{3.66}$$

or, since the density is proportional to the absolute pressure:

$$\frac{du}{u} = -\frac{d\rho}{\rho} = -\frac{dp}{p}. \tag{3.67}$$

Also note that:

$$\frac{1}{\rho u^2} = \frac{\rho}{G^2} = \frac{\rho_1}{p_1}\frac{p}{G^2}. \tag{3.68}$$

From Eqns. (3.65), (3.67), and (3.68), there results:

$$-\int_{p_1}^{p_2} \frac{dp}{p} + \frac{\rho_1}{p_1 G^2}\int_{p_1}^{p_2} p\,dp + \frac{2f_{\mathrm{F}}}{D}\int_0^L dx = 0. \tag{3.69}$$

Note that since G is constant, and the viscosity of a gas is virtually independent of pressure, the Reynolds number, $\mathrm{Re} = \rho u D/\mu = GD/\mu$, is essentially constant. Hence, the friction factor f_{F}, which depends only on the Reynolds number and the roughness ratio, is justifiably taken outside the integral in Eqn. (3.69).

Performing the integration:

$$\frac{4f_{\mathrm{F}}L}{D} = (p_1^2 - p_2^2)\frac{\rho_1}{p_1 G^2} + \ln\left(\frac{p_2}{p_1}\right)^2. \tag{3.70}$$

The mass velocity in the pipeline is therefore:

$$G^2 = \frac{\rho_1}{p_1}\frac{p_1^2 - p_2^2}{\dfrac{4f_{\mathrm{F}}L}{D} - \ln\left(\dfrac{p_2}{p_1}\right)^2} = \frac{M_{\mathrm{w}}}{RT}\frac{p_1^2 - p_2^2}{\dfrac{4f_{\mathrm{F}}L}{D} - \ln\left(\dfrac{p_2}{p_1}\right)^2}. \tag{3.71}$$

The last term in the denominator, being derived from the kinetic-energy term of Eqn. (3.64), is typically relatively small; if indeed it can be ignored (such an assumption should be checked with numerical values), the *Weymouth* equation results:

$$G^2 = \frac{M_{\mathrm{w}}}{RT}\frac{D}{4f_{\mathrm{F}}L}(p_1^2 - p_2^2). \tag{3.72}$$

However, if the ratio of the absolute pressures p_2/p_1 is significantly less than unity, the last term in the denominator of Eqn. (3.71) cannot be ignored. Consider the situation in which the exit pressure p_2 is progressively reduced below the inlet pressure, as shown in Fig. 3.18. As expected, equation (3.71) predicts an initial increase in the mass velocity G as p_2 is reduced below p_1. However, a *maximum* value of G is eventually reached when p_2 has fallen to a critical value p_2^*; a further reduction in the exit pressure then *apparently* leads to a *reduction* in G.

The critical exit pressure is obtained by noting that at the maximum,

$$\frac{dG^2}{dp_2} = 0, \tag{3.73}$$

which, when applied to Eqn. (3.71), gives, after some algebra:

$$\left(\frac{p_1}{p_2^*}\right)^2 - \ln\left(\frac{p_1}{p_2^*}\right)^2 = 1 + \frac{4f_F L}{D}, \tag{3.74}$$

which can be solved for p_2^*. The corresponding maximum mass velocity $G_{\max}$ can be shown to obey the equation[12]:

$$G_{\max}^2 = \frac{\rho_1 p_1}{1 + \dfrac{4f_F L}{D} + \ln\dfrac{p_1 \rho_1}{G_{\max}^2}}, \tag{3.75}$$

which can be solved for $G_{\max}$ by successive substitution or Newton's method.

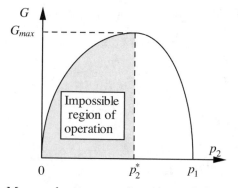

Fig. 3.18   Mass velocity as a function of the exit pressure.

---

[12]   The derivation of Eqn. (3.75) is somewhat tricky. First eliminate $\ln(p_1/p_2^*)^2$ between Eqns. (3.71) and (3.74), giving:

$$G_{\max}^2 = \frac{\rho_1}{p_1}(p_2^*)^2. \tag{A}$$

Then substitute for $(p_2^*)^2$ from Eqn. (A) in both the numerator and denominator of Eqn. (3.71). Rearrangement then yields Eqn. (3.75).

The curve in Fig. 3.18 in the region $0 < p_2 < p_2^*$ is actually an illusion because it would involve a decrease of entropy, and a further reduction of p_2 below p_2^* does *not* decrease the flow rate. Instead, the exit pressure remains at p_2^* and there is a sudden irreversible expansion or *shock wave* at the pipe exit from p_2^* down to the pressure p_2 ($< p_2^*$) just outside the exit. The shock occurs over an extremely narrow region of a few molecules in thickness, in which there are abrupt changes in pressure, temperature, density, and velocity.

It may also be shown that:

$$G_{\max} = \sqrt{p_2^* \rho_2^*} = \rho_2^* u_2^*, \qquad (3.76)$$

where an asterisk denotes conditions under the maximum mass flow rate. Also, the corresponding exit velocity:

$$u_2^* = \sqrt{\frac{p_2^*}{\rho_2^*}}, \qquad (3.77)$$

can be interpreted as the velocity of a hypothetical *isothermal* sound wave at the exit conditions, since we have the following relations for the velocity of sound and an ideal gas:

$$c = \sqrt{\frac{dp}{d\rho}}, \qquad \frac{p}{\rho} = \frac{RT}{M_{\mathrm{w}}}. \qquad (3.78)$$

Thus, the velocity of a hypothetical isothermal sound wave is given by:

$$\text{Isothermally}: \quad c = \sqrt{\frac{RT}{M_{\mathrm{w}}}}. \qquad (3.79)$$

In practice, however, sound waves travel nearly *isentropically*, and the sonic velocity is then:

$$\text{Isentropically}: \quad c = \sqrt{\frac{\gamma RT}{M_{\mathrm{w}}}}, \qquad (3.80)$$

in which $\gamma = c_p/c_v$ is the ratio of the specific heat at constant pressure to that at constant volume.

3.7 Compressible Flow in Nozzles

Another case involving compressible flow occurs with the discharge of a gas from a high-pressure reservoir through a nozzle consisting of a converging section that narrows to a "throat," possibly followed by a diverging section or "diffuser." Representative applications occur in the flow of combustion gases from rockets, steam-jet ejectors (used for creating partial vacuums by entrainment), the emergency escape of gas through a rupture disk in a high-pressure reactor, and the generation of supersonic flows.

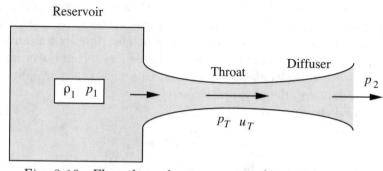

Fig. 3.19   Flow through a converging/diverging nozzle.

As shown in Fig. 3.19, the gas in the reservoir has an absolute pressure p_1 and a density ρ_1; the final discharge is typically to the atmosphere, at an absolute pressure p_2. The transfer of gas is rapid and there is little chance of heat transfer to the wall of the nozzle, so the flow is adiabatic. Furthermore, since only short lengths are involved, friction may be neglected, so the expansion is *isentropic*, being governed by the equation:

$$\frac{p}{\rho^\gamma} = c = \frac{p_1}{\rho_1^\gamma}, \quad \text{or} \quad \rho = \rho_1 \left(\frac{p}{p_1}\right)^{1/\gamma}, \tag{3.81}$$

in which c is a constant and $\gamma = c_p/c_v$, the ratio of specific heats. An ideal gas and constant specific heat are assumed.[13]

---

[13] Proof of Eqn. (3.81) is not immediately obvious, so here is one version. A differential change in the internal energy u (per mole) is:

$$du = Tds - Pdv. \tag{1}$$

But for an ideal gas the internal energy depends only on the temperature, so that:

$$c_v = \left(\frac{\partial u}{\partial T}\right)_v = \frac{du}{dT}. \tag{2}$$

From Eqns. (1) and (2) we obtain:

$$c_v dT = Tds - Pdv. \tag{3}$$

Also, for an ideal gas:

$$Pv = RT. \tag{4}$$

But for frictionless adiabatic flow, there is no change in the entropy. Thus, from Eqns. (3) and (4), assuming constant c_v:

$$c_v dT = -RT \frac{dv}{v} \quad \text{or} \quad c_v \int_{T_1}^{T_2} \frac{dT}{T} = -R \int_{v_1}^{v_2} \frac{dv}{v}, \tag{5}$$

so that:

$$c_v \ln \frac{T_2}{T_1} = -R \ln \frac{v_2}{v_1} \quad \text{or} \quad \frac{v_2}{v_1} = \left(\frac{T_2}{T_1}\right)^{-c_v/R}. \tag{6}$$

For horizontal flow between the reservoir and some downstream position where the velocity is u and the pressure is p, Bernoulli's equation gives:

$$\frac{u^2}{2} - \frac{u_1^2}{2} + \int_{p_1}^{p} \frac{dp}{\rho} = 0. \tag{3.82}$$

in which the integral is:

$$\int_{p_1}^{p} \frac{dp}{\rho} = \frac{p_1^{1/\gamma}}{\rho_1} \int_{p_1}^{p} \frac{dp}{p^{1/\gamma}} = \frac{\gamma}{\gamma - 1} \frac{p_1}{\rho_1} \left[\left(\frac{p}{p_1}\right)^{(\gamma-1)/\gamma} - 1 \right]. \tag{3.83}$$

Because the velocity u_1 in the reservoir is essentially zero, the last two equations yield a relation between the velocity and pressure at any point in the flowing gas:

$$u^2 = \frac{2\gamma}{\gamma - 1} \frac{p_1}{\rho_1} \left[1 - \left(\frac{p}{p_1}\right)^{(\gamma-1)/\gamma} \right]. \tag{3.84}$$

Since $m = \rho u A$ at any location, where m is the mass flow rate and A is the cross-sectional area of the nozzle, Eqn. (3.84) may be rewritten as:

$$\left(\frac{m}{A}\right)^2 = \frac{2\gamma}{\gamma - 1} p_1 \rho_1 \left[1 - \left(\frac{p}{p_1}\right)^{(\gamma-1)/\gamma} \right] \left(\frac{p}{p_1}\right)^{\frac{2}{\gamma}}. \tag{3.85}$$

The mass velocity m/A is clearly a maximum at the throat, where it has the value m/A_{T}. However, the pressure at the throat is still a variable, and a maximum of m/A_{T} occurs with respect to p when:

$$\frac{d(m/A_{\mathrm{T}})}{d(p/p_1)} = 0. \tag{3.86}$$

Using Eqn. (4), the temperature ratio may be reexpressed, giving:

$$\frac{v_2}{v_1} = \left(\frac{P_2 v_2}{P_1 v_1}\right)^{-c_v/R}, \tag{7}$$

or:

$$\frac{P_2}{P_1} = \left(\frac{v_2}{v_1}\right)^{\left(1+\frac{c_v}{R}\right)\left(-\frac{R}{c_v}\right)}. \tag{8}$$

But since $c_p - c_v = R$ for an ideal gas, the exponent in Eqn. (8) simplifies to:

$$\left(1 + \frac{c_v}{R}\right)\left(-\frac{R}{c_v}\right) = -\left(\frac{R}{c_v} + 1\right) = -\frac{c_p}{c_v} = -\gamma, \tag{9}$$

in which $\gamma = c_p/c_v$. Thus, the final relation is:

$$P_1 v_1^{\gamma} = P_2 v_2^{\gamma}. \tag{10}$$

After some algebra, these last two equations give the critical pressure ratio at the throat, corresponding to the maximum possible mass flow rate:

$$\frac{p_{cT}}{p_1} = \left(\frac{2}{\gamma + 1}\right)^{\gamma/(\gamma-1)}. \tag{3.87}$$

After several lines of algebra involving Eqns. (3.81), (3.84), and (3.87), the corresponding velocity u_{cT} at the throat is found to be:

$$u_{cT}^2 = \left(\frac{2\gamma}{\gamma + 1}\right)\frac{p_1}{\rho_1} = \frac{\gamma p_{cT}}{\rho_{cT}} = \frac{\gamma R T_{cT}}{M_w}, \tag{3.88}$$

in which the subscript c denotes critical conditions. That is, the gas velocity at the throat equals the local sonic velocity, as in Eqn. (3.80). Under these conditions, known as *choking at the throat*, the critical mass flow rate m_c is:

$$m_c = A_T \sqrt{\gamma p_1 \rho_1 \left(\frac{2}{\gamma + 1}\right)^{(\gamma+1)/(\gamma-1)}}. \tag{3.89}$$

Now consider what happens for various exit pressures p_2, decreasing progressively from the reservoir pressure p_1:

(A) If the exit pressure is slightly below the reservoir pressure, there will be a small flow rate, which can be computed from Eqn. (3.85) by substituting $A = A_2$ (the exit area) and $p = p_2$. Equation (3.85) then gives the variation of pressure in the nozzle. The flow is always subsonic.

(B) The same as A, except the mass flow rate is higher.

(C) If the exit pressure is reduced sufficiently, the velocity at the throat increases to the critical value given by Eqn. (3.88). In the diverging section, the pressure increases and the flow is subsonic.

(D) For an exit pressure lying between C and E, no continuous solution is possible. The flow, which is critical, is supersonic for a certain distance beyond the throat, but there is then a *very* sudden increase of pressure, known as a *shock*, and the flow is thereafter subsonic. The shock is an irreversible phenomenon, resulting in abrupt changes in velocity, pressure, and temperature over an extremely short distance of a few molecules in thickness.

(E) For the same critical mass flow rate as in C, Eqn. (3.85) has a second root, corresponding to an exit pressure at E in Fig. 3.20. In this case, however, there is a continuous *decrease* of pressure in the diffuser, where the flow is now *supersonic*.

(F) For an exit pressure lower than E, a further irreversible expansion occurs just outside the nozzle.

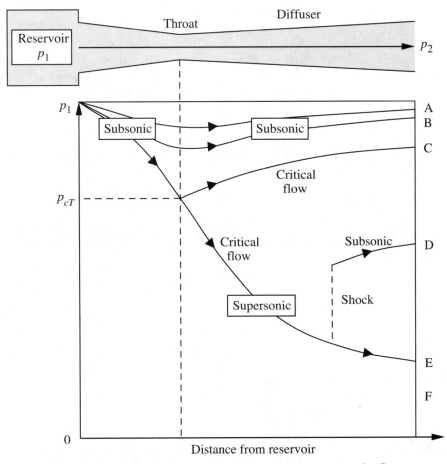

Fig. 3.20   Effect of varying exit pressure on nozzle flow.

If the diffuser section is absent, then the flow is essentially that through an orifice in a high-pressure reservoir. The flow will be subsonic if the exit pressure exceeds p_{cT}. If the exit pressure equals p_{cT}, then critical flow occurs with the sonic velocity through the orifice. And if it falls below this value, critical flow will still occur, but with a further irreversible expansion just outside the orifice.

3.8   Complex Piping Systems

The chemical engineer should be prepared to cope with pumping and piping systems that are far more complicated than the examples discussed thus far in this chapter. The only such type of a more complex system to be considered here is the simplest, which involves steady, incompressible flow. A modest complication of the systems already studied can be seen by glancing at the scheme in Fig. E3.7, in which water is pumped from a low-lying reservoir into a pipe that subsequently divides into two *branches* in order to feed two elevated tanks.

Although the details of solution of these problems depends on the particular situation, the general approach is first to subdivide the system into a number of discrete elements or units such as tanks, pipes, and pumps. Each such element typically lies between two junction points or *nodes*, at each of which it is connected to one or more other adjoining elements. A system of simultaneous equations based on the following principles is then developed:

(a) Continuity of mass (or volume, for an incompressible fluid): at any node, the sum of the incoming flow rates must equal the sum of the outgoing flow rates. For example, if pipe A leads into a node, and pipes B and C leave from it:

$$Q_A = Q_B + Q_C. \tag{3.90}$$

(b) An energy balance for every segment of pipe that connects two nodes. It is customary to replace the mean velocity with the volumetric flow rate:

$$Q = \frac{\pi D^2}{4} u_m, \tag{3.91}$$

so that a representative energy balance is:

$$g\Delta z + \frac{\Delta p}{\rho} + \frac{32 f_F Q^2 L}{\pi^2 D^5} = 0. \tag{3.92}$$

(c) An equation for each pump, such as:

$$\Delta p = a - bQ^2 \qquad \text{or} \qquad \frac{\Delta p}{\rho} + w = 0, \tag{3.93}$$

in which a and b are coefficients that depend on the particular pump. The second version in Eqn. (3.93) would only be used if any two of the following variables were specified: (a) the pump inlet pressure, (b) the pump discharge pressure, and (c) the work performed per unit mass flowing.

The system of simultaneous equations will be nonlinear, because of the Q^2 terms appearing in the pipe and pump equations, and can be solved for the unknown pressures and flow rates by methods that are largely governed by the complexity of the system:

(a) For systems with a modest number of nodes—say no more than 20—standard software such as Excel or Polymath can be employed. Each such solution, similar to that explained in Example 3.7 below, will be specific to the particular problem at hand, and cannot be generalized to other situations.

(b) For larger systems, especially when many piping and pumping problems are to be investigated, a general-purpose computer program can be written, which will accomplish the following steps:

Example 3.7—Solution of a Piping/Pumping Problem   **165**

 (i) Read information about the various elements, their characteristic parameters (such as the length of a pipe), and how they are connected to one another.
 (ii) Read information about the required system performance. For example, the pressure at the free surface of water in a tower would normally be set to zero gauge, and a delivery rate of 1,000 gpm might be needed from a certain fire hydrant.
 (iii) Apply the continuity principle to each node at which the pressure is not specified, leading to a system of simultaneous nonlinear equations in the unknown nodal pressures.
 (iv) Call on a standard general-purpose nonlinear simultaneous equation solving routine, typically involving the Newton-Raphson method, to calculate the unknown nodal pressures.
 (v) When all the pressures have been calculated, solve for all the unknown flow rates.
 (vi) Present the results in a useful format.

These steps are illustrated for a simple system in Example 3.7.

Example 3.7—Solution of a Piping/Pumping Problem

Consider the installation shown in Fig. E3.7, in which friction for the short run of pipe between the supply tank and pump can be ignored, and in which nodes 1 and 2 are essentially at the same elevation. The equation for the pressure increase (psi) across the centrifugal pump is (see Eqn. (4.4)):

$$\Delta p = p_2 - p_1 = a - bQ_A^2, \qquad \text{(E3.7.1)}$$

in which a $= 72$ psi and b $= 0.0042$ psi/(gpm)^2. The friction factors in the three pipes are $f_{FA} = 0.00523$, $f_{FB} = 0.00584$, and $f_{FC} = 0.00556$, and the density of the water being pumped is $\rho = 62.4$ lb$_m$/ft^3.

Other parameters are given in Tables E3.7.1 and E3.7.2. Assume that the pipe lengths have already included the equivalent lengths of all fittings and valves.

Table E3.7.1   Nodal Elevations

Node	Elevation, ft
1	0
2	0
3	25
4	80
5	60

Table E3.7.2   Pipeline Parameters

Pipe	D, in.	L, ft
A	3.068	80
B	2.067	300
C	2.067	500

Fig. E3.7   Pumping and piping installation.

Solution

By dividing Eqn. (3.92) by g, inserting the value for π, and substituting appropriate conversion factors, the energy balance for a pipe element becomes:

$$\Delta z + 4.636 \times 10^3 \frac{\Delta p}{\rho g} + 4.01 \frac{f_F Q^2 L}{g D^5} = 0, \qquad \text{(E3.7.2)}$$

in which the variables have the following units: z, ft; p, psi; ρ, $\mathrm{lb_m/ft^3}$; Q, gpm; L, ft; g, $\mathrm{ft/s^2}$; and D, in. The energy balance is now applied three times:

Between points 2 and 3

$$f_1 = z_3 - z_2 + 4.636 \times 10^3 \frac{p_3 - p_2}{\rho g} + 4.01 \frac{f_{FA} Q_A^2 L_A}{g D_A^5} = 0. \qquad \text{(E3.7.3)}$$

Between points 3 and 4

$$f_2 = z_4 - z_3 + 4.636 \times 10^3 \frac{p_4 - p_3}{\rho g} + 4.01 \frac{f_{FB} Q_B^2 L_B}{g D_B^5} = 0. \qquad \text{(E3.7.4)}$$

Between points 3 and 5

$$f_3 = z_5 - z_3 + 4.636 \times 10^3 \frac{p_5 - p_3}{\rho g} + 4.01 \frac{f_{FC} Q_C^2 L_C}{g D_C^5} = 0. \qquad \text{(E3.7.5)}$$

Considerations of continuity at the branch point 3 and the performance of the pump give:

Example 3.7—Solution of a Piping/Pumping Problem   **167**

Continuity at point 3

$$f_4 = Q_A - Q_B - Q_C = 0, \tag{E3.7.6}$$

Between points 1 and 2

$$f_5 = p_2 - (p_1 + a - bQ_A^2) = 0. \tag{E3.7.7}$$

Table E3.7.3   Results from the Spreadsheet

Example 3.7: Solution of Pumping/Piping System

A. Input data

Elevations

			Pipe parameters		
z_1 =	0	ft	D_A =	3.068	in
z_2 =	0	ft	D_B =	2.067	in
z_3 =	25	ft	D_C =	2.067	in
z_4 =	80	ft	L_A =	80.0	ft
z_5 =	60	ft	L_B =	300.0	ft
			L_C =	500.0	ft

Pump coefficients

			Constants		
a =	72	psi	rho =	62.4	lbm/ft^3
b =	0.0042	psi/(gpm)^2	g =	32.2	ft/s^2
			c_1 =	4.636E+03	in^2 lbm/ft lbf s^2

Friction factors

			c_2 =	4.01	in^5 ft/gpm^2 s^2
f_FA =	0.00523	f_FB =	0.00584	f_FC =	0.00556

B. Converged solution

Pressures

			Flow rates		
p_1 =	0.0	psi	Q_A =	89.42	gpm
p_2 =	38.4	psi	Q_B =	35.06	gpm
p_3 =	26.9	psi	Q_C =	54.36	gpm
p_4 =	0.0	psi			
p_5 =	0.0	psi			

Functional values

f_1 =	0.000
f_2 =	0.000
f_3 =	0.000
f_4 =	0.000
f_5 =	0.000
Sum =	0.001 (of absolute functional values)

These last relationships represent five simultaneous nonlinear equations. The values of the five unknowns p_2, p_3, Q_A, Q_B, and Q_C are found by using the Excel

Solver, for which the results are shown in Table E3.7.3. The strategy is governed by noting that although *each* of the $f_i, i = 1, \ldots, 5$ should equal zero at convergence, the Solver can only minimize a *single* cell. Therefore, the cell identified as "Sum" is the sum of the *absolute* values, $\sum_{i=1}^{5} |f_i|$, and if this is brought close to zero, then all the individual f_i values must also be essentially zero.

The same final values for the two unknown pressures and the three flow rates are obtained for different starting estimates of these five variables. Observe that each of the f_i values is essentially zero, and that continuity is observed: $Q_A = Q_B + Q_C$. Solutions for the same basic configuration of tanks, pump, and pipes, but with, for example, different elevations and pump characteristics, could easily be obtained by changing the appropriate input cells and repeating the solution. However, if a different system were to be investigated, the *structure* of the spreadsheet would have to be changed.

For simplicity in this example, constant yet realistic values were specified for the friction factors in the pipes A, B, and C. A simple extension would be to incorporate extra cells to allow for variability with the Reynolds number and roughness ratio, as in Eqn. (3.50). □

PROBLEMS FOR CHAPTER 3

Unless otherwise stated, all piping is Schedule 40 commercial steel; for water: $\rho = 62.3$ lb$_m$/ft^3 = 1,000 kg/m^3, $\mu = 1.0$ cP.

1. *Momentum flux in laminar flow—M.* For laminar flow in a pipe, derive an expression for the total momentum flux per unit mass flowing in terms of the mean velocity u_m.

2. *Switching oil colors—M (C).* For the laminar velocity profile $u = \alpha(a^2 - r^2)$, prove that the *fraction* f of the total volumetric flow rate that occurs between the wall and a radial location $r = R$ is given by:

$$f = \left(1 - \frac{R^2}{a^2}\right)^2 .$$

Oil colored with fluorescein is in laminar flow with mean velocity u_m in a long pipe of length L, when the stream at the inlet is suddenly switched to colorless oil. Draw a diagram showing a representative location of the interface between the colorless and colored oil, *after* the colorless oil has started appearing at the pipe exit. How long a time t will it take, as a multiple of L/u_m, for the flow at the exit to consist of 99% colorless oil?

3. *Laminar tank draining—M.* The tank and pipe shown in Fig. P3.3 are initially filled with a liquid of viscosity μ and density ρ. Assuming laminar flow, taking pipe friction to be the only resistance, and ignoring exit kinetic-energy effects, prove that the time taken to drain just the tank is:

$$t = \frac{8\mu L R^2}{\rho g r^4} \ln\left(1 + \frac{H}{L}\right).$$

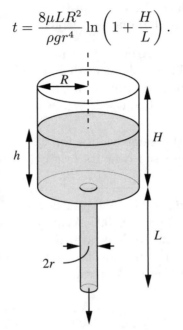

Fig. P3.3   Tank draining in laminar flow.

4. *Friction factor plot—E.* On your friction factor plot, check the following concerning the Fanning friction factor:

(a) That $f_F = 16/\text{Re}$ for the laminar-flow regime.
(b) That the label $\mathcal{F}/[(4\Delta x/D)(V^2/2)]$, which appears in some other published friction factor plots, is the same as $f_F = \tau_w/(\frac{1}{2}\rho u_m^2)$.
(c) The accuracy of the Blasius equation in the region $5{,}000 < \text{Re} < 100{,}000$.

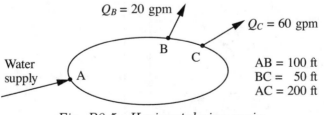

Fig. P3.5   Horizontal ring main.

5. *Flow in a ring main—M.* Fig. P3.5 shows a *horizontal* ring main, which consists of a continuous loop of pipe, such as might be used for supplying water to various points on one floor of a building. Water enters the main at A at 50 psig, and is discharged at B and C at rates of 20 gpm and 60 gpm, respectively. Tests on the particular pipe forming the main show that the frictional pressure drop (psi) is given by $-\Delta p = 0.0002\, Q^2 L$, where L is the length of pipe in feet and Q is the flow rate in gpm. Estimate the flow rate in the section AC and the pressure at B. Was the flow during the tests laminar or turbulent?

6. *Pipeline corrosion—E.* A horizontal pipeline is designed for a given ε, L, Q, ρ, μ, and Δp, where all symbols have their usual meanings, and the resulting diameter is calculated to be a certain value D. It is then found that scaling or corrosion is likely to occur, and that ε may rise *tenfold*, giving f_F about twice as large as originally thought.

In order to maintain the same values for Q and Δp, by what ratio should the design diameter be increased over its original value to allow for scaling and corrosion?

7. *Two Reynolds numbers—E.* A liquid flows turbulently through a smooth horizontal glass tube with Re = 10,000. A value of $f_F = 0.0080$ is indicated by the friction-factor diagram. If the flow rate through the same tubing is increased tenfold, with Re = 100,000, the friction factor falls to 0.0045. At the higher flow rate, would you expect the frictional pressure drop per unit length to increase or decrease? By what factor?

8. *Erroneous friction factor—E.* Water is flowing turbulently at a mean velocity of $u_m = 10$ ft/s in a 1.0-in. I.D. horizontal pipe, and the Fanning friction factor is $f_F = 0.0060$. What error in the pressure drop would ensue if (erroneously) the assumption were made that the flow was laminar, abandoning the previous value of the friction factor?

Fig. P3.9 "Pumping" kerosene.

9. *Pumping kerosene—M*. Fig. P3.9 shows how nitrogen gas under a pressure $p_n = 15.0$ psig can be used for "pumping" kerosene at 75 °F ($\rho = 51.0$ lb$_m$/ft^3, $\mu = 4.38$ lb$_m$/ft hr) through an elevation increase of 20 ft. If there is an effective length (including fittings) of 150 ft of nominal 2-in. pipe between the two tanks, what is the flow rate of kerosene in gpm? Neglect exit kinetic-energy effects.

10. *Lodge water supply—M*. A mountain water reservoir in a national park is to provide water at a flow rate of $Q = 200$ gpm and a minimum pressure of 40 psig to the lodge in the valley 200 ft below the reservoir. If the effective length of pipe is 2,000 ft, what is the minimum standard pipe size that is needed? Neglect exit kinetic-energy effects.

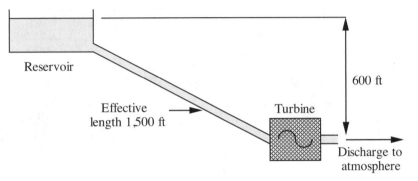

Fig. P3.11   Hydroelectric installation.

11. *Hydroelectric installation—M*. Fig. P3.11 shows a small hydroelectric installation. Water from the large mountain reservoir flows steadily through a steel pipe of 12 in. nominal diameter and effective length 1,500 ft into a turbine. The water is ultimately discharged through a short length of 12-in. pipe to the atmosphere, where the elevation is 600 ft below the surface of the reservoir.

If the water flow rate is 12.5 ft^3/s, and the turbine/generator produces 425 kW of useful electrical power, estimate the following:

(a) The efficiency of the turbine/generator combination.
(b) The pressure (psig) at the inlet to the turbine.
(c) The power (kW) dissipated by pipe friction.
(d) The velocity gradient (s^{-1}) at the pipe wall.

Also, what would probably happen if during an emergency a large gate valve at the turbine inlet were suddenly closed during a period of a *few* seconds?

Data: Internal diameter of pipe $= 11.939$ in.
Effective roughness $= 0.00015$ ft.
Viscosity of water $= 1.475$ cP.

12. *Drilling mud circulation—M*. Prove that the equivalent or hydraulic mean diameter for flow in the annular space between two concentric cylinders of diameters d_1 and d_2 is given by $D_e = d_1 - d_2$.

Fig. P3.12   Drilling mud circulation system.

Fig. P3.12 illustrates the mud-circulation system on an oil-well drilling rig. Drilling mud from a mixing tank T flows to the inlet of the pump P, which discharges through BD to the inside of the drill pipe DE. During drilling, the mud flow is to be steady at $Q = 100$ gpm. The mud is a Newtonian liquid with $\mu = 5.0$ cP at the average flowing temperature of 70 °F, and its density, due to weighing agents and other additives, is $\rho = 67$ lb$_m$/ft^3.

The drill pipe DE, of depth 10,000 ft, is surrounded by the casing C. At the bottom, the mud jets out through the drill bit and recirculates back through the annular space to F, where it is piped back to the tank T. The surface piping has a total equivalent length (including all valves, elbows, etc.) of 1,000 ft. The mild steel piping has a roughness $\varepsilon = 0.00015$ ft. Other properties of the piping are given in Table P3.12.

Table P3.12   Data for pipes (all Schedule 80)

Pipe	Nominal Size	O.D. (in.)	I.D. (in.)	Cross-Sectional Area (sq in.)
Surface piping	2 in.	2.375	1.939	2.953
Drill pipe DE	2 in.	2.375	1.939	2.953
Drill casing C	6 in.	6.625	5.761	26.07

Calculate:

(a) The flow rate Q in ft^3/s throughout the system.

(b) The mean velocities (ft/s) in the surface piping, the drill pipe, and the annular space between the casing and the tubing.

Then, assuming for the moment that all friction factors are the same, show that the frictional dissipation $\mathcal{F}$ for the annular space is likely to contribute only on the order of 1% to $\mathcal{F}$ for the surface and drill pipe, and may therefore be reasonably neglected.

Finally, if the pump is running at 79% overall efficiency, compute the required pumping horsepower, within 2%.

13. *Pumping and piping—M.* Fig. P3.13 shows a centrifugal pump that is used for pumping water from one tank to another through a 1,000 ft (including fittings) nominal 4-in. I.D. pipeline.

Prove that the pressure drop (psi) in the pipeline between points 2 and 3 is given closely in terms of the flow rate Q (ft³/s) by:

$$p_2 - p_3 = 10.83 + 10,265 f_F Q^2,$$

in which f_F is the Fanning friction factor.

Fig. P3.13   Pumping installation.

The performance curve for the pump has been determined, and relates the pressure increase Δp (psi) across the pump to the flow rate Q (ft³/s) through it:

$$\Delta p = 19.2 - 133.4 Q^{4.5}.$$

If the viscosity of the water is 1 cP, determine:

    (a) The flow rate Q (ft³/s).
    (b) The pressure increase across the pump, Δp (psi).
    (c) The Reynolds number Re in the pipeline.
    (d) The Fanning friction factor, f_F.

14. *Ring main for fire hydrants—D*. Prove for flow of water in an inclined pipe that:

$$32.2\Delta z + 74.3\Delta p + 4.00 f_F Q^2 \frac{L}{D^5} = 0.$$

Here, the symbols have their usual meanings, but the following units have been used: Δz (ft), Δp (psi), Q (gpm), L (ft), and D (in.).

Fig. P3.14   Ring main for feeding fire hydrants.

Fig. P3.14 shows a pump that takes its suction from a pond and discharges water into a ring main that services the fire hydrants for a chemical plant. All pipe is nominal 6-in. I.D. At the pump exit, the pipe immediately divides into two branches. Points 1 and 2 are essentially at the same elevation, and losses before the pump may be neglected.

If the pump exit pressure—which is also the pressure p_2 at the dividing point 2—is 85 psig, determine the total flow rate coming from *both* branches through a fire hydrant at point 3, whose elevation is 100 ft above the pump exit, if the delivery pressure is $p_3 = 20$ psig. The effective distances (including all fittings) between points 2 and 3 are 1,000 ft and 2,000 ft for the shorter and longer paths, respectively. Take only a *single* estimate of the Fanning friction factor—don't spend time refining it by iteration. Also, if the pump and its motor have a combined efficiency of 75%, what power (HP) is needed to drive the pump?

15. *Optimum pipe diameter—D*. A pump delivers a power P kW to transfer Q m³/s of crude oil of density ρ kg/m³ through a long-distance horizontal pipeline of length L m, with a friction factor f_F. The installed cost of the pipeline is $\$c_1 D^m L$ (where $m = 1.4$) and that of the pumping station is $\$(c_2 + c_3 P)$; both these costs are amortized over n years. Electricity costs $\$c_4$ per kWh and the pump has an efficiency η. The values of c_1, c_2, c_3, and c_4 are known. The pump inlet and pipeline exit pressure are the same. If there are N hours in a year, prove that the optimum pipe diameter giving the lowest total annual cost is:

$$D_{\text{opt}} = \left(\frac{5\alpha}{\beta m}\right)^{1/(m+5)}, \quad \text{where} \quad \alpha = \left(\frac{32 f_F \rho Q^3 L}{\pi^2}\right)\left(\frac{N c_4}{1{,}000\eta} + \frac{c_3}{n}\right), \quad \beta = \frac{c_1 L}{n}.$$

If $c_1 = 2{,}280$, $c_2 = 95{,}000$, $c_3 = 175$, $c_4 = 0.11$, $\rho = 850$, $L = 50{,}000$, $\eta = 0.75$, and $f_F = 0.0065$, all in units consistent with the above, evaluate D_{opt} for all six combinations of $Q = 0.05$, 0.2, and 0.5 m^3/s, with $n = 10$ and 20 years.

16. *Replacement of ventilation duct—M.* An existing horizontal ventilation duct of length L has a square cross section of side d. It is to be replaced with a new duct of rectangular cross section, $d \times 2d$. Due to complications of installing the larger duct, its length will be $2L$. If the overall pressure drop is unchanged, what percentage improvement in volumetric flow rate may be expected with the new duct? Neglect all losses except wall friction, and assume for simplicity that the dimensionless wall shear stress $f_F = \tau_w/\frac{1}{2}\rho u_m^2$ has the same value in both cases.

17. *Flow in a concrete aqueduct—M.* An open concrete aqueduct of surface roughness $\varepsilon = 0.01$ ft has a rectangular cross section. The aqueduct is 10 ft wide, and falls 10.5 ft in elevation for each mile of length. It is to carry 150,000 gpm of water at 60 °F. If $f_F = 0.0049$, what is the minimum depth needed if the aqueduct is not to overflow?

18. *Natural gas pipeline—M.* Natural gas (methane, assumed ideal) flows steadily at 55 °F in a nominal 12-in. diameter horizontal pipeline that is 20 miles long, with $f_F = 0.0035$. If the inlet pressure is 100 psia, what exit pressure would correspond to the maximum flow rate through the pipeline? If the actual exit pressure is 10 psia, what is the mass flow rate of the gas (lb$_m$/hr)?

19. *Pumping ethylene—D.* Ethylene gas is to be pumped along a 6-in. I.D. pipe for a distance of 5 miles at a mass flow rate of 2.0 lb$_m$/s. The delivery pressure at the end of the pipe is to be 2.0 atm absolute, and the flow may be considered isothermal, at 60 °F. If $f_F = 0.0030$, calculate the required inlet pressure. Assume ideal gas behavior, and justify any further assumptions.

20. *Fluctuations in a surge tank—D.* The installation shown in Fig. P3.20 delivers water from a reservoir of constant elevation H to a turbine. The surge tank of diameter D is intended to prevent excessive pressure rises in the pipe whenever the valve is closed quickly during an emergency.

Assuming constant density, neglecting (because of the relatively large diameter of the surge tank) the effects of acceleration and friction of water in the surge tank, and allowing for possible negative values of u, prove that a momentum balance on the water in the pipe leads to:

$$g(H - h) - 2f_F \frac{L}{d} u|u| = L\frac{du}{dt}. \qquad \text{(P3.20.1)}$$

Here, h = height of water in the surge tank (h_0 under steady conditions), f_F = Fanning friction factor, u = mean velocity in the pipe, g = gravitational acceleration, and t = time.

Fig. P3.20   *Surge tank for turbine installation.*

Also, prove that:

$$\frac{\pi d^2}{4} u = \frac{\pi D^2}{4}\frac{dh}{dt} + Q_v. \tag{P3.20.2}$$

The flow rate Q_v through the valve in the period $0 \le t \le t_c$, during which it is being closed, is approximated by:

$$Q_v = k\left(1 - \frac{t}{t_c}\right)\sqrt{h - h^*}, \tag{P3.20.3}$$

where the constant k depends on the particular valve and the downstream head h^* depends on the particular emergency at the turbine.

What is the physical interpretation of h^*? During and after an emergency shutdown of the valve, compute the variations with time of the velocity in the pipe and the level in the surge tank. Euler's method for solving ordinary differential equations is suggested, either embodied in a computer program or in a spreadsheet. Plot u and h against time for $0 \le t \le t_{max}$, and give a physical explanation of the results.

Test Data

$g = 32.2$ ft/s^2, $H = 100$ ft, $h_0 = 88$ ft, $f_F = 0.0060$, $L = 2{,}000$ ft, $d = 2$ ft, $t_c = 6$ s, $k = 21.4$ ft$^{2.5}$/s, $t_{max} = 500$ s, and $D = 4$ ft. Take two extreme values for h^*: (a) its original steady value, $h_0 - Q_{v0}^2/k^2$, where Q_{v0} is the original steady flow rate, and (b) zero.

21. *Laminar sublayer in turbulent flow—E*. Fig. P3.21 shows a highly idealized view of the velocity profile for turbulent pipe flow of a fluid. Assume that a central turbulent core of uniform velocity u_m occupies virtually all of the cross section, and that there is a very thin laminar sublayer of thickness δ between it and the wall.

Fig. P3.21   *Turbulent velocity profile with laminar sublayer.*

If the viscosity of the fluid is μ, write down a formula for the wall shear stress τ_w in terms of μ, u_{m}, and δ . If the Fanning friction factor is given by the Blasius equation, derive a formula for the dimensionless ratio δ/D (where D is the pipe diameter) in terms of the Reynolds number, Re. Evaluate this ratio for Re $= 10^4$, 10^5, and 10^6, and comment on the results.

22. *Reservoir and ring main—M.* (a) For fluid flow in a pipeline, starting from Eqn. (3.36), prove that the frictional dissipation per unit mass is given by:

$$\mathcal{F} = cLQ^2, \quad \text{where} \quad c = \frac{32f_F}{\pi^2 D^5}.$$

Here, Q is the volumetric flow rate, and all other symbols have their usual meanings.

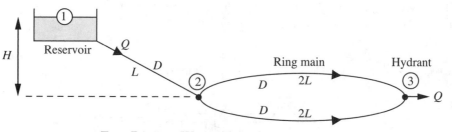

Fig. P3.22   *Water distribution system.*

Fig. P3.22 shows a horizontal ring main that is supplied with water from a reservoir of elevation H, via a pipeline of diameter D and length L. Between nodes 2 and 3, the ring main has two equal legs, each of diameter D and length $2L$. Kinetic-energy changes may be ignored.

(b) Derive an expression for the flow rate Q from a hydrant at node 3, where the required gauge delivery pressure is p_3. Your answer should give Q in terms of L, H, c, g, p_3, and ρ. *Hint*: use an energy balance in two stages, first between nodes 1 and 2, and *then* between nodes 2 and 3, along just *one* of the two identical legs.

(c) If the pipeline is Schedule 40 commercial steel with a nominal diameter of 12 in., and if the Reynolds number is very high, what is a good estimate for f_F? For $H = 250$ ft, $p_3 = 20$ psig, and $L = 2,500$ ft, determine Q in ft^3/s. Use just one value for f_F—do not iterate.

23. *Another ring main—M (C).* Cooling water is supplied to a factory by a horizontal ring main ABCDEA of 1-ft-diameter pipe. Water at 100 psig is fed into the main at point A. Table P3.23 shows (a) where the water is subsequently withdrawn, and (b) the distances along each leg of the main.

Table P3.23   Withdrawal Points and Leg Lengths

Withdrawal Point	Flow Rate (ft^3/s)	Leg	Length (Yards)
B	10	AB	200
C	5	BC	400
D	8	CD	600
E	8	DE	400
		EA	400

At which point does the flow reverse in the main? Calculate the least pressure in the system. Assume $f_F = 0.0040$ throughout.

24. *Leaking flange—D (C).* Prove that for unidirectional viscous flow between parallel plates separated by a distance h, the pressure gradient is $-12\mu u_m/h^2$, in which μ is the fluid viscosity and u_m is the mean velocity.

Fig. P3.24   Liquid leakage in a flange.

Viscous liquid is contained, at gauge pressure P, within a pipe of diameter $D = 2R$ having flanges of outer diameter $2mR$, as indicated in Fig. P3.24. A leak develops, so that the fluid flows radially outward through the gap h between the flanges. Neglecting inertial effects, show that the flow rate Q due to leakage and the axial thrust F on one flange are:

$$Q = \frac{\pi h^3 P}{6\mu \ln m}, \qquad F = P\pi R^2 \left(\frac{m^2 - 1}{2 \ln m} - 1 \right).$$

25. *Comparison of friction factor formulas—M.* The Shacham explicit formula for the Fanning friction factor, f_F, was presented in Eqn. (3.41) as an accurate alternative to the implicit Colebrook and White formula of Eqn. (3.39). Use a spreadsheet to compare the values given by both formulas for all 36 combinations of $\varepsilon/D = 0.05$, 0.01, 0.001, 10^{-4}, 10^{-5}, and 10^{-6}, and Re = 4,000, 10^4, 10^5, 10^6, 10^7, and 10^8. What are (a) the average, and (b) the maximum percentage deviations of the values given by the Shacham formula as compared to those from the Colebrook and White formula?

26. *General purpose spreadsheet for simple piping problems—D.* Write a general purpose *single* spreadsheet that will accommodate the solution for as wide a variety as possible of simple piping problems. Make sure that it can handle Case 1–, 2–, and 3–type problems, for which the spreadsheet should be printed. Give some accompanying text, describing your method.

27. *Rise of liquid in a capillary tube—M.* A long vertical tube of very narrow internal diameter d is dipped just below the surface of a liquid of density ρ, viscosity μ, and surface tension σ. Assuming that the liquid wets the tube with zero contact angle, derive an expression for the time t taken for the liquid to rise to a height H in the tube. Assume laminar flow and neglect kinetic-energy effects.

Hint: First consider the situation where the liquid has risen to an intermediate height h; at this moment, derive an expression for the mean upward velocity u_m in terms of ρ, g, σ, μ, h, and d.

Fig. P3.28   *Two pumps feeding to an upper tank.*

28. *Complex piping system—D.* Consider the piping system shown in Fig. P3.28. The pressures p_1 and p_5 are both essentially atmospheric (0 psig); there is an increase in elevation between points 4 and 5, but pipes C and D are horizontal.

The head-discharge curves for the centrifugal pumps can be represented by:

$$\Delta p = a - bQ^2,$$

in which Δp is the pressure increase in psig across the pump, Q is the flow rate in gpm, and a and b are constants depending on the particular pump.

Use a spreadsheet that will accept values for a_A, b_A, a_B, b_B, $z_5 - z_4$, D_C, L_C, D_D, L_D, D_E, L_E, and the Fanning friction factor (assumed constant throughout). Then solve for the unknowns Q_C, Q_D, Q_E, p_2, p_3, and p_4. Take $z_5 - z_4 = 70$ ft, $f_F = 0.00698$, with other parameters given in Tables P3.28.1 and P3.28.2.

Table P3.28.1   Pump Parameters

Pump	a, psi	b, psi/(gpm)^2
A	156.6	0.00752
B	117.1	0.00427

Table P3.28.2   Pipe Parameters

Pipe	D, in.	L, ft
C	1.278	125
D	2.067	125
E	2.469	145

Assume that the above pipe lengths have already included the equivalent lengths of all fittings and valves.

29. *Liquid oscillations in U-tube—M.* Your supervisor has proposed that the density of a liquid may be determined by placing it in a glass U-tube and observing the period of oscillations when one side is momentarily subjected to an excess pressure that is then released. She suggests that the longer periods will correspond to the denser liquids. Conduct a thorough analysis to determine the validity of the proposed method. For simplicity, neglect friction.

30. *Energy from a warm mountain?—D.* Consider a tunnel or shaft entering the base of a mountain and leaving at its summit. If the desert sun maintains the mountain at a warm temperature, but the outside air cools significantly at night, evaluate the prospects of generating power at night by installing a turbine in the tunnel.

31. *Laminar flow in a vertical pipe—E.* Repeat the analysis in Section 3.2, but now for upward flow in a vertical pipe. Prove that the velocity is:

$$u = \frac{1}{4\mu}\left(-\frac{dp}{dz} - \rho g\right)(a^2 - r^2).$$

32. *Aspects of laminar pipe flow—M.* A polymer of density $\rho = 0.80$ g/cm^3 and viscosity $\mu = 230$ cP flows at a rate $Q = 1{,}560$ cm^3/s in a horizontal pipe of diameter 10 cm. Evaluate the following, all in CGS units: (a) the mean velocity, u_m; (b) the Reynolds number Re, hence verifying that the flow is laminar; (c) the maximum velocity, u_{max}; (d) the pressure drop per unit length, $-dp/dz$; (e) the wall shear stress, τ_w; (f) the Fanning friction factor, f_F; and (g) the frictional dissipation $\mathcal{F}$ for 100 cm of pipe.

33. *Viscous flow in a plunger—M.* A tube of diameter $D = 2.0$ cm and length $L = 100$ cm is initially filled with a liquid of density $\rho = 1.0$ g/cm^3 and viscosity $\mu = 100$ P. It is then drained by the application of a constant force $F = 10^5$ dynes to a plunger, as shown in Fig. P3.33.

Fig. P3.33   *Draining a tube with a plunger.*

Assuming laminar flow, compute the time to expel one-half of the liquid. Then check the laminar flow assumption.

34. *True/false.* Check *true* or *false*, as appropriate:

(a)   The Reynolds number is a measure of the ratio of inertial forces to viscous forces.   T ☐   F ☐

(b)   The distribution of shear stress for laminar flow in a pipe varies parabolically with the radius.   T ☐   F ☐

(c)   For laminar flow in a pipe, the shear stress τ varies linearly with distance from the centerline, whereas for turbulent flow it varies as the square of the distance from the centerline.   T ☐   F ☐

(d)   The Hagen-Poiseuille law predicts how the shear stress varies with radial location in laminar pipe flow.   T ☐   F ☐

(e)   For laminar flow in a horizontal pipeline under a constant pressure gradient, a doubling of the diameter results in a doubling of the flow rate.   T ☐   F ☐

(f)   For laminar flow in a pipe with mean velocity u_m, the kinetic energy per unit mass is one-half the square of the mean velocity, namely, $u_\mathrm{m}^2/2$.   T ☐   F ☐

(g)   The kinetic energy per unit mass flowing is approximately $u_\mathrm{m}^2/2$ for turbulent flow in a pipe.   T ☐   F ☐

(h)   Referring to Fig. 3.7, when the person jumps to trolley B, trolley A will decelerate because it is losing momentum.   T ☐   F ☐

(i)   Newton's law relating shear stress and viscosity can be related to the transfer of momentum on a molecular scale.   T ☐   F ☐

(j)   A friction factor is a dimensionless wall shear stress.   T ☐   F ☐

(k)   The viscosity of an ideal gas increases as the pressure increases, because the molecules are closer together and offer more resistance.   T ☐   F ☐

(l)   In turbulent flow, the eddy viscosity ε is usually of comparable magnitude to the molecular viscosity μ.   T ☐   F ☐

(m)   A simple eddy transport model for the turbulent shear stress predicts a constant friction factor.   T ☐   F ☐

(n)   When a frictional dissipation term $\mathcal{F}$ has been obtained for horizontal flow, it may then be used for flow in an inclined pipe for the same flow rate.   T ☐   F ☐

(o)   The Shacham equation is *explicit* in the friction factor.   T ☐   F ☐

(p)   The frictional dissipation term for pipe flow is given by $2f_F \rho u_m^2 L/D$ (energy/unit mass).   T ☐   F ☐

(q)   For a rough pipe, the Fanning friction factor keeps on decreasing as the Reynolds number increases.   T ☐   F ☐

(r)   If in a piping problem the diameter of the pipe, its roughness, the flow rate, and the properties of the fluid are given, a few iterations will generally be needed in order to converge on the proper value of the friction factor and hence the pressure drop.   T ☐   F ☐

(s)   If the friction factor and pressure gradient in a horizontal pipeline remain constant, a doubling of the diameter will cause a 16-fold increase in the flow rate.   T ☐   F ☐

(t)   For a given volumetric flow rate Q, the pressure drop for turbulent flow in a pipe is approximately proportional to $1/D^5$, where D is the pipe diameter.   T ☐   F ☐

(u)   For incompressible pipe flow from points 1 to 2, $\mathcal{F}$ is never negative in the relation $E_2 - E_1 + \mathcal{F} = 0$, where E is the sum of the kinetic, potential, and pressure energy. (Assume $w = 0$.)   T ☐   F ☐

(v)   For pipe flow, the friction factor varies *gradually* as   T ☐ F ☐
        the Reynolds number increases from laminar flow to
        turbulent flow.

(w)   The Colebrook and White equation is *explicit* in the   T ☐ F ☐
        friction factor.

(x)   A hydraulically smooth pipe is one in which the wall   T ☐ F ☐
        surface irregularities do not protrude beyond the lam-
        inar boundary layer next to the wall.

(y)   The hydraulic mean diameter for an open rectangular   T ☐ F ☐
        ditch of depth D and width $3D$ is $1.5D$. (Assume that
        the ditch is full of water.)

(z)   The hydraulic mean diameter for a ventilation duct   T ☐ F ☐
        of depth D and breadth $3D$ is $2D$.

(A)   For a slightly inclined pipe of internal diameter D that   T ☐ F ☐
        is running half full of liquid, the equivalent diameter
        is also D.

(B)   The effective length of a close return bend in a 6-in.   T ☐ F ☐
        nominal diameter pipe is about 37.5 ft.

(C)   The effective length of an open globe valve in a 12-in.   T ☐ F ☐
        nominal diameter pipe is about 100 ft.

(D)   For turbulent flow, the thickness of the laminar sub-   T ☐ F ☐
        layer increases as the Reynolds number increases.

(E)   At a Reynolds number of 100,000, the thickness of the   T ☐ F ☐
        laminar sublayer for pipe flow is roughly one-tenth the
        diameter of the pipe.

(F)   In turbulent flow, the laminar sublayer is an ex-   T ☐ F ☐
        tremely thin region next to the wall, across which
        there is a significant change in the velocity.

(G)   For steady flow of a compressible gas in a pipeline,   T ☐ F ☐
        the mass flow rate is the same at any location.

(H)   For steady isothermal flow of a compressible gas in a   T ☐ F ☐
        pipeline, the Weymouth equation is valid if the pipe
        friction is neglected.

(I)   For steady isothermal flow of a compressible gas in   T ☐ F ☐
        a pipeline, the mass flow rate is proportional to the
        pressure drop.

(J)   For isothermal flow of a compressible gas in a horizontal pipeline, some pressure energy is consumed in overcoming friction *and* in changing the kinetic energy of the gas.   T ☐   F ☐

Chapter 4

FLOW IN CHEMICAL ENGINEERING EQUIPMENT

4.1 Introduction

THIS chapter concludes the presentation of *macroscopic* topics, by discussing important applications of fluid mechanics to several chemical engineering processing operations. Since the variety of such operations is fairly large, it will be impossible to cover everything; therefore, the focus will be on a representative set of topics in which the application of fluid mechanics plays a fundamental role in chemical processing. In fact, the general theme is the basic theory that underlies a selection of the so-called "unit operations." Certain other applications—including those involved in polymer processing, two-phase flow, and bubbles in fluidized beds—depend more on *microscopic* fluid mechanics for their interpretation, and will be postponed until Chapters 6–11.

In most cases the theory is necessarily simplified, sometimes leading to approximate predictions. However, the reader should thereby gain a knowledge of some of the important issues, which will then enable him or her to make a critical examination of articles in equipment handbooks, process design software, etc., which will generally be needed if serious designs of chemical plants are to be made.

The design and use of process equipment lies at the heart of chemical engineering. Students are encouraged to take every opportunity to see the wide variety of such equipment firsthand, by visiting chemical engineering laboratories, chemical plants, oil refineries, sugar mills, paper mills, glass bottle plants, polymer processing operations, pharmaceutical production facilities, breweries, and waste-treatment plants, etc. Until such visits can be made, an excellent substitute is available on a compact disk produced by Dr. Susan Montgomery and her coworkers.[1] The CD can be used on either a PC or a Macintosh computer and consists of many photographs with accompanying descriptions of equipment, arranged under the following headings: materials transport, heat transfer, separations, process vessels, mixing, chemical reactors, process parameters, and process control. The visual encyclopedia is very easy to navigate, and four representative windows are reproduced, with permission, in Figs. 4.1–4.4.

---

[1] *Material Balances & Visual Equipment Encyclopedia of Chemical Engineering Equipment*, CD produced by the Multimedia Education Laboratory, Department of Chemical Engineering, University of Michigan, Susan Montgomery (Director), 2002.

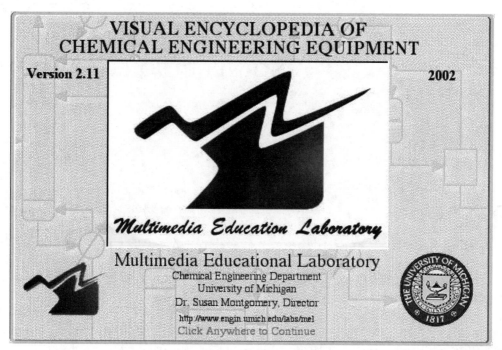

Fig. 4.1   Title window of the visual equipment encyclopedia.

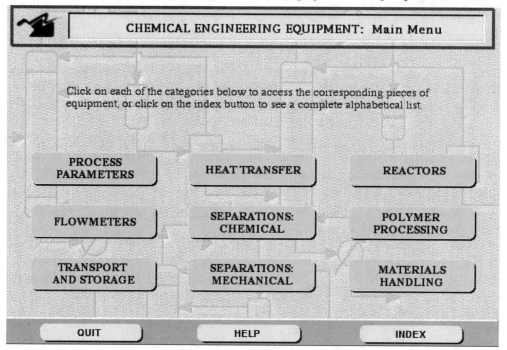

Fig. 4.2   Main menu of the visual equipment encyclopedia.

CYCLONES / HYDROCYCLONES: Equipment Design

Overflow

Feed inlet

Inlet chamber

Vortex finder

Cone

Underflow

Spigot

The picture shows the parts of a cyclone. For more information on a given part, click on its name.

For a given particle size, a portion of the particles will exit out the bottom with the underflow, and the rest will exit with the overflow. The heavier the particle, the greater the chance that it exits out the bottom.

The cutpoint size, a measure of a cyclone's performance, is defined as the particle size for which half the particles exit at the top and half at the bottom. A coarser (larger) cutpoint means that only larger particles can be separated. A finer (smaller) cutpoint means that smaller particles can be removed.

(Courtesy of Lakos Filtration Systems, Fresno, CA)

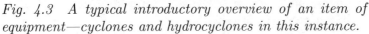

BACK TO PREVIOUS MENU   LAST   PAGE 1 OF 2

Fig. 4.3   A typical introductory overview of an item of equipment—cyclones and hydrocyclones in this instance.

CYCLONES / HYDROCYCLONES: Usage Examples

Hydrocyclones are used in desliming, degritting, concentration, recovery of solids, clarification, open circuit classification and closed-circuit grinding. The hydrocyclone below right is used to remove polypropylene powder from a solvent recovery unit.

(Courtesy of Clean Gas Systems, Inc., Ronkonkoma, NY)

Cyclones can recover powder coatings, dusts, chemicals and catalysts. The quad cyclone above is used to collect coal dust. The cyclones on the near right are used in the powder coating process.

(Courtesy ITW Gema, Indianapolis, IN)

(Courtesy of Richard Mozley Limited, Cornwall, UK)

BACK TO PREVIOUS MENU   LAST   PAGE 1 OF 1   NEXT

Fig. 4.4   A typical window showing further equipment details.

4.2 Pumps and Compressors

A fluid may be transferred from one location to another in either of two basic ways:

1. If it is a liquid and there is a drop in elevation, allowing it to fall under gravity.
2. Passing it through a machine such as a pump or compressor that imparts energy to it, typically increasing its pressure (sometimes its velocity), which then enables it to overcome the resistance of the pipe through which it subsequently flows.

Devices that increase the pressure of a flowing fluid usually fall into one of the following two main categories, the first of which is subdivided into two subcategories:

1. Positive displacement pumps, whose nature is either *reciprocating* or *rotary*.
2. Centrifugal pumps, fans, and blowers.

Reciprocating positive displacement pumps.   Fig. 4.5 shows how a piston moving to and fro in a cylinder is used for pumping a fluid. The pump is double-acting—the four valves allow fluid to be pumped continuously, whether the piston is moving to the right or the left. The particular instant shown is when the piston is moving to the right. Valve D is closed and fluid is being pumped from the right-hand side of the cylinder through the open valve C to the outlet; simultaneously, valve B is open and fluid is being sucked from the inlet into the left-hand side of the cylinder. In the return stroke, only valves A and D would be open, pumping fluid from the left-hand side to the outlet while filling up the right-hand side from the inlet.

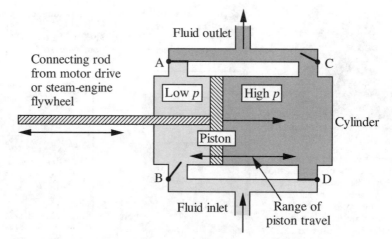

Fig. 4.5 Idealized reciprocating positive displacement pump.

Reciprocating pumps may be used for both liquids and gases, and are excellent for generating high pressures. In the case of gases—for which the pump is called a *compressor*—there is a significant temperature rise, and intercoolers will be needed if several pumps are used in series in order to produce very high pressures. A variation that avoids friction between the piston and cylinder in order to make a tight seal is to have a pulsating flexible diaphragm. To avoid damage to reciprocating pumps, a provision must be made for automatic opening of a relief valve or recycle line if a valve on the outlet side is inadvertently closed.

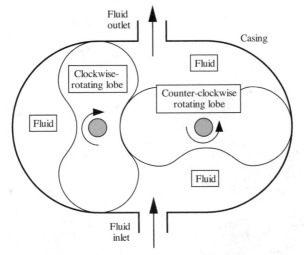

Fig. 4.6   Idealized rotary positive displacement pump.

Rotary positive displacement pumps.   Fig. 4.6 shows how two counter-rotating double lobes inside a casing can be used for boosting the pressure of a fluid between inlet and outlet. A variety of other configurations is possible, such as two triple lobes or two intermeshed gears. The rotary pump is good for handling viscous liquids, but because of the close tolerances needed, it cannot be manufactured large enough to compete with centrifugal pumps for coping with very high flow rates.

Centrifugal pumps.   As shown in Fig. 4.7, the centrifugal pump typically resembles a hair drier without the heating element. The impeller usually consists of two flat disks, separated by a distance d by a number of curved vanes, that rotate inside the stationary housing. Fluid enters the impeller through a hole (location "1") or "eye" at its center, and is flung outward by centrifugal force into the periphery of the housing ("2") and from there to the volute chamber and pump exit ("3"). Centrifugal pumps are particularly suitable for handling large flow rates, and also for liquids containing suspended solids. The following is only an *approximate* analysis, the key to which lies in understanding the various velocities at the impeller exit, as follows:

1. The impeller, which has an outer radius r_2 and rotates with an angular velocity ω, has a linear velocity $u_2 = \omega r_2$ at its periphery.

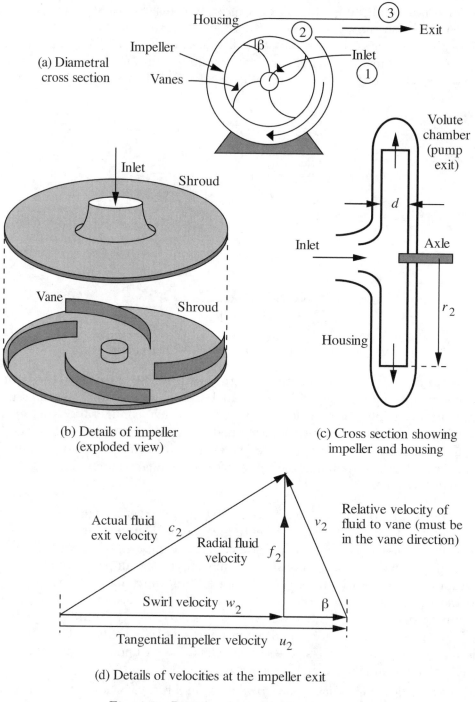

(a) Diametral cross section

(b) Details of impeller (exploded view)

(c) Cross section showing impeller and housing

(d) Details of velocities at the impeller exit

Fig. 4.7   Details of a centrifugal pump.

2. Because the fluid is guided by the vanes, which make an angle β with the periphery of the impeller, the *relative* velocity of the fluid to the impeller, v_2, must also be in this direction.

3. The actual fluid exit velocity, as would be seen by a stationary observer, is c_2, the resultant of u_2 and v_2.

4. This fluid velocity, c_2, has a radially outward component f_2, related to the flow rate through the pump by $Q = 2\pi r_2 df_2$.

5. Further, the actual velocity c_2 has a component w_2 tangential to the impeller, and this is known as the "swirl" velocity.

There is a similar set of velocities at the impeller inlet, but they are significantly smaller and may be disregarded in an introductory analysis.

From Eqn. (2.75), the torque needed to drive the impeller and hence the power transmitted to the fluid are:

$$T = mr_2w_2, \qquad P = \omega T = m(\omega r_2)w_2 = mu_2w_2, \tag{4.1}$$

where m is the mass flow rate through the pump; note the use of the swirl velocity. The additional head Δh imparted to the fluid is the energy it gains per unit mass, divided by the acceleration of gravity:

$$\Delta h = \frac{P}{mg} = \frac{u_2w_2}{g} = \frac{u_2(u_2 - f_2\cot\beta)}{g}. \tag{4.2}$$

Within the impeller, this increased head is reflected largely by an increase in the fluid velocity from its entrance value to c_2. However, in the volute chamber, there is subsequently a *decrease* in the velocity, so that the kinetic energy just gained is converted to pressure energy. Thus, the overall pressure increase is:

$$\Delta p = p_3 - p_1 = \rho g\Delta h = \rho u_2w_2 \doteq \rho u_2^2, \tag{4.3}$$

in which the last approximation—assuming $w_2 \doteq u_2$—will be seen from Fig 4.3(d) to be realistic at low flow rates, for which f_2 is small. Note that Eqn. (4.2) predicts that Δh should decline linearly with increasing flow rate, which is proportional to f_2. This declining head/discharge characteristic is a direct result of the "swept-back" vanes ($\beta < 90°$), and is a desirable feature in preserving stability in some pumping and piping schemes; swept-forward vanes are generally undesirable.

In practice, because of increased turbulence and other losses, Δh is found not to decline linearly with increasing flow rate Q, but in the manner shown in Fig. 4.8. In many cases, the curve is satisfactorily represented by the following relation, where a, b, and n are constants, with n often approximately equal to two:

$$\Delta h = a - bQ^n, \tag{4.4}$$

Fig. 4.8   Head/discharge curve for centrifugal pump.

In addition, the above simplified analysis suggests two *dimensionless groups* that can be used for all pumps of a given design that are geometrically similar—that is, apart from size, they look alike. If N denotes the rotational speed of the impeller:

$$\omega = 2\pi N, \quad u_2 = \omega r_2 = \omega \frac{D}{2} = \pi N D. \tag{4.5}$$

The pressure increase at low flow rates is then approximately:

$$\Delta p \doteq \rho u_2^2 = \rho \left(\frac{\omega D}{2} \right)^2 = \rho \pi^2 D^2 N^2, \tag{4.6}$$

so that the dimensionless group $\Delta p/(\rho D^2 N^2)$ should be roughly constant at low flow rates.

The volumetric flow rate is obtained by multiplying the area $\pi D c_1 D$ between the disks of the impeller (the gap width $c_1 D$ increases linearly with D for a given pump design) by the *radially* outward velocity $c_2 u_2$ (this simple theory proposes that the flow rate is roughly proportional to the tangential velocity of the impeller), where c_1 and c_2 are constants, so that:

$$Q = (\pi D)(c_1 D)(c_2 u_2). \tag{4.7}$$

Substitution of u_2 from (4.5) gives:

$$Q = c_1 c_2 \pi^2 D^3 N. \tag{4.8}$$

Thus, the dimensionless group $Q/(N D^3)$ should be roughly constant at low flow rates. In practice, the assumptions made above fail progressively as the flow rate increases. Nevertheless, the two dimensionless groups derived above—for the pressure increase and flow rate—are usually adequate to characterize all pumps of a given design, no matter what the flow rate. Thus, all such pumps can be characterized by the *single* curve shown in Fig. 4.9; it is the values of the two *groups* that count, not the individual values of Δp, ρ, D, Q, and N.

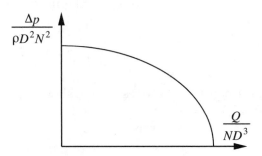

Fig. 4.9   General characteristic curve for centrifugal pump.

Example 4.1—Pumps in Series and Parallel

For a certain type of centrifugal pump, the head increase Δh (ft) is closely related to the flow rate Q (gpm) by the equation:

$$\Delta h = a - bQ^2, \tag{E4.1.1}$$

where a and b are constants that have been determined by tests on the pump. Two identical such pumps are now connected together, as shown in Fig. E4.1(a), either in series or in parallel.

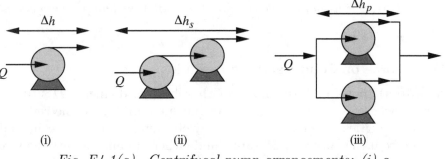

Fig. E4.1(a)   Centrifugal pump arrangements: (i) a single pump, (ii) two in series, (iii) two in parallel.

Derive expressions for the head increases Δh_s and Δh_p for these two arrangements in terms of the total flow rate Q through them. Also, display the results graphically for the values $a = 25$ ft and $b = 0.0025$ ft/(gpm)^2.

Solution

When the pumps are in *series*, the head increases are additive, and the total increase is double that for the single pump:

$$\Delta h_s = 2\left(a - bQ^2\right). \tag{E4.1.2}$$

For the *parallel* configuration, the flow through each pump is only $Q/2$. The overall head increase is the same as that for either pump singly:

$$\Delta h_p = a - b \left(\frac{Q}{2} \right)^2 . \tag{E4.1.3}$$

For the given values of the constants, the results are shown in Fig. E4.1(b). Observe that the series and parallel configurations are useful for allowing operation with increased head and flow rate, respectively.

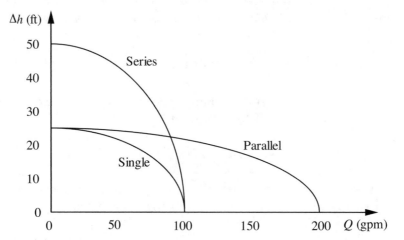

Fig. E4.1(b)   Performance curves for three different pump arrangements.

4.3   Drag Force on Solid Particles in Fluids

Fig. 4.10(a) shows a stationary smooth sphere of diameter D situated in a fluid stream, whose velocity far away from the sphere is u_∞ to the right. Except at very low velocities, when the flow is entirely laminar, the wake immediately downstream from the sphere is unstable, and either laminar or turbulent vortices will constantly be shed from various locations around the sphere. Because of the turbulence, the pressure on the downstream side of the sphere will never fully recover to that on the upstream side, and there will be a net *form drag* to the right on the sphere. (For purely stable laminar flow, the pressure recovery is complete, and the form drag is zero.) In addition, because of the velocity gradients that exist near the sphere, there will also be a net *viscous drag* to the right. The sum of these two effects is known as the (total) drag force, F_D. A similar drag occurs for spheres and other objects *moving* through an otherwise stationary fluid—it is the *relative* velocity that counts. And if *both* the fluid and the solid object are moving with respective velocities V_f and V_s, the drag force is in the direction of the relative velocity $V_f - V_s$.

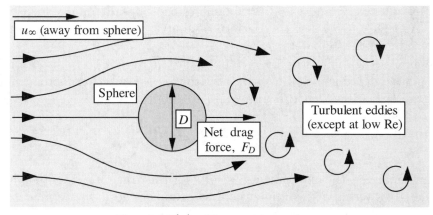

Fig. 4.10(a)   Flow past a sphere.

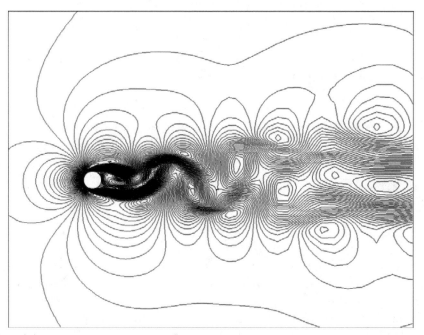

Fig. 4.10(b)   Instantaneous velocity contour plots for flow past a cylinder, produced by the FlowLab computational fluid dynamics program at Re = 100. After being shed from the cylinder, the vortices move alternately clockwise (top row) and counterclockwise (bottom row). Courtesy Chi-Yang Cheng, Fluent, Inc.

In Part II of this book we shall include several examples in which flow patterns can be simulated on the computer—a field known as *computational fluid dynamics* or CFD. One of these examples appears at the end of Chapter 13, but since it involves flow past a cylinder (somewhat similar to a sphere) it is appropriate to reproduce part of it here, in Fig. 4.10(b). For the case here (Re = 100) the flow is laminar but unsteady, and the diagram shows contours of equal velocities at one instant. As time progresses, the pattern oscillates as the vortices are shed from alternate sides of the cylinder.

The analysis is facilitated by recalling that there is a correlation for flow in smooth pipes between two dimensionless groups—the friction factor (or dimensionless wall shear stress) and the Reynolds number:

$$f_F = \frac{\tau_w}{\frac{1}{2}\rho u_m^2}, \qquad \text{Re} = \frac{\rho u_m D}{\mu}. \tag{4.9}$$

In the same manner, the experimental results for the drag on a smooth sphere may be correlated in terms of two dimensionless groups—the *drag coefficient* C_D and the Reynolds number:

$$C_D = \frac{F_D/A_p}{\frac{1}{2}\rho u_\infty^2}, \qquad \text{Re} = \frac{\rho u_\infty D}{\mu}, \tag{4.10}$$

in which $A_p = \pi D^2/4$ is the *projected area* of the sphere in the direction of motion, and ρ and μ are the properties of the fluid.

Table 4.1 Drag Force on a Sphere

Range of Re	Type of Flow in the Wake	Correlation for C_D
Re < 1	Laminar	$C_D = 24/\text{Re}$
$1 < \text{Re} < 10^3$	Transition	$C_D \doteq 18\text{Re}^{-0.6}$
$10^3 < \text{Re} < 2 \times 10^5$	Turbulent	$C_D \doteq 0.44$

The resulting correlation for the "standard drag curve" (SDC) is shown in Fig. 4.11 for the curve marked $\psi = 1$ (the other curves will be explained later), in a graph that has certain resemblances to the friction factor diagram of Fig. 3.11. There are at least three distinct regions, as shown in Table 4.1.

Also, D.G. Karamanev has recalled[2] that the entire curve up to the "critical" Reynolds number of about 200,000 can be represented quite accurately by the *single* equation of Turton and Levenspiel[3]:

$$C_D = \frac{24}{\text{Re}}\left(1 + 0.173\,\text{Re}^{0.657}\right) + \frac{0.413}{1 + 16{,}300\,\text{Re}^{-1.09}}. \tag{4.11}$$

---

[2] Private communication, March 18, 2002.

[3] R. Turton and O. Levenspiel, "A short note on the drag correlation for spheres," *Powder Technology*, Vol. 47, p. 83 (1986).

Karamanev has also developed an ingenious alternative approach for determining the velocity at which a particle settles in a fluid—presented near the end of this section, together with an important observation on the *rise* velocities of spheres whose density is significantly less than that of the surrounding liquid.

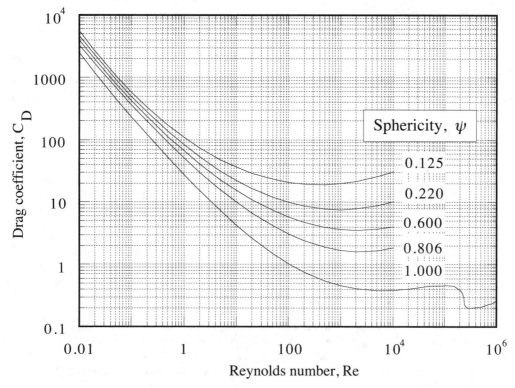

Fig. 4.11   Drag coefficients for objects with different values of the sphericity ψ; the curve for $\psi = 1$ corresponds to a sphere.[4]

In connection with Fig. 4.11, note:

1. The transition from laminar to turbulent flow is much more gradual than that for pipe flow. Because of the confined nature of pipe flow, it is possible for virtually the entire flow field to become turbulent; however, for a sphere in an essentially infinite "sea" of fluid, it would require an impossibly large amount of energy to render the fluid turbulent everywhere, so the transition to turbulence proceeds only by degrees.

2. The upper limit for purely laminar flow is about Re = 1, in contrast to Re = 2,000 for pipe flow. A prime reason for this is the highly unstable nature of flow in a sudden expansion, which is essentially occurring in the wake of the sphere.

---

[4] Based on values published in G.G. Brown et al., *Unit Operations*, Wiley & Sons, New York, 1950, p. 76.

3. There is a fairly sudden downward "blip" in the drag coefficient at about Re = 200,000, because the boundary layer on the sphere suddenly changes from laminar to turbulent. Dimples on a golf ball encourage this type of transition to occur at even lower Reynolds numbers. (Also see Section 8.7 for a more complete explanation.)

For the laminar flow region, the law $C_D = 24/\text{Re}$ easily rearranges to:

$$F_D = 3\pi\mu u_\infty D, \tag{4.12}$$

which is known as *Stokes' law,*[5] which can also be proved theoretically (but not easily!), starting from the microscopic equations of motion (the Navier-Stokes equations).

Stokes, Sir George Gabriel, born 1819 in County Sligo, Ireland; died 1903 in Cambridge, England. He entered Pembroke College, Cambridge, graduated with the highest honors in mathematics in 1841, and was elected a fellow of his college the same year. Stokes was appointed Lucasian professor of mathematics in 1849, and was president of the Royal Society from 1885 to 1890. He excelled in both theoretical and experimental mathematical physics. His early papers, published from 1842 to 1850, dealt mainly with hydrodynamics, the motion of fluids with friction, waves, and the drag on ships. Later work centered on a wide variety of topics, including the propagation of sound waves, diffraction and polarization of light, fluorescence of certain materials subject to ultraviolet light, optical properties of glass and improvements in telescopes, Röntgen rays, and differential equations relating to the stresses and strains in railway bridges.

Source: *The Encyclopædia Britannica*, 11th ed., Cambridge University Press (1910–1911).

Nonspherical particles. For particles that are not spheres, two quantities must first be defined:

1. The *sphericity* ψ of the particle:

$$\psi = \frac{\text{Surface area of a sphere having the same volume as the particle}}{\text{Surface area of the particle}}. \tag{4.13}$$

It is easy to show that $\psi = 1$ corresponds to a sphere. Further, sphericities of all other particles must be less than one, because for a given volume a sphere has the minimum possible surface area.

2. The *equivalent particle diameter*, D_p, defined as the diameter of a sphere having the same volume as the particle.

---

[5] G.G. Stokes, "On the effect of the internal friction of fluids on the motion of pendulums," *Cambridge Philosophical Transactions*, Part II, Vol. IX, p. 8 (1850).

The corresponding drag coefficient, again defined by Eqn. (4.10), can then be obtained from Fig. 4.11, in which D_p is involved in the Reynolds number, and ψ is the parameter on a family of curves. The value of A_p is nominally the projected area normal to the direction of flow, but this definition lacks precision in the case of a moving particle, which could rotate and lack a constant orientation.

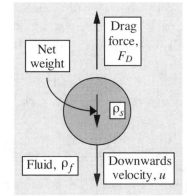

Fig. 4.12 Settling of a sphere under gravity.

Settling under gravity. Consider the spherical particle of diameter D and density ρ_s shown in Fig. 4.12, which is settling under gravity in a fluid of density ρ_f and viscosity μ_f. A downward momentum balance equates the downward weight of the sphere minus the upward buoyant force, minus the upward drag force, to the downward rate of increase of momentum of the sphere:

$$\underbrace{\frac{\pi D^3}{6}(\rho_s - \rho_f)g}_{\text{Net Weight}} - \underbrace{F_D}_{\text{Drag}} = \underbrace{\frac{d}{dt}\left(\frac{\pi D^3 \rho_s}{6}u\right)}_{\substack{\text{Rate of increase} \\ \text{of momentum}}}$$

$$= \frac{\pi D^3 \rho_s}{6}\frac{du}{dt} \quad \text{(for constant mass).} \qquad (4.14)$$

The simplification of constant mass holds in many situations—but not, for example, for a liquid sphere that is evaporating. Integration of (4.14) enables the velocity u to be obtained as a function of time.

Terminal velocity. An important case of Eqn. (4.14) occurs when the sphere is traveling at its steady *terminal velocity* u_t. In this case, the net weight of the sphere is exactly counterbalanced by the drag force and there is no acceleration, so that $du/dt = 0$ and:

$$F_D = \frac{\pi D^3}{6}(\rho_s - \rho_f)g. \qquad (4.15)$$

The corresponding drag coefficient is:

$$C_{\mathrm{D}} = \frac{F_D/\left(\frac{\pi D^2}{4}\right)}{\frac{1}{2}\rho_f u_t^2} = \frac{4}{3}\frac{gD}{u_t^2}\frac{\rho_s - \rho_f}{\rho_f}. \tag{4.16}$$

A typical problem will specify the value of u_t and ask what is the corresponding value of D, or *vice versa*. Equation (4.16) is not particularly useful, since both the left- and right-hand sides contain unknowns. Alternative forms, whose validity the reader should check, are:

$$C_{\mathrm{D}}Re^2 = \frac{4}{3}\frac{g\rho_f D^3}{\mu^2}(\rho_s - \rho_f). \tag{4.17}$$

$$\frac{C_{\mathrm{D}}}{Re} = \frac{4}{3}\frac{g\mu}{\rho_f^2 u_t^3}(\rho_s - \rho_f). \tag{4.18}$$

Clearly, the right-hand side of Eqn. (4.17) is independent of u_t, and this equation will be useful if u_t is sought. In such an event, the product $C_D Re^2$ is known and the drag coefficient and Reynolds number (and hence u_t) can then be computed with reference to Fig. 4.11. Likewise, the right-hand side of Eqn. (4.18) is independent of D, and this equation will be useful if D is sought. Appropriate values for C_D can be obtained from Table 4.1, Fig. 4.11, or Eqn. (4.11).

Karamanev method for rapid calculation of falling and rising terminal velocities. Additionally, D.G. Karamanev has developed a method based on the *Archimedes* number[6]:

$$\mathrm{Ar} = \frac{D^3 g|(\rho_s - \rho_f)|\rho_f}{\mu^2} = \frac{3}{4}\mathrm{Re}^2 C_D, \tag{4.19}$$

in which the absolute value $|(\rho_s - \rho_f)|$ allows for the sphere to be either heavier (downwards terminal velocity) or lighter (upwards terminal velocity) than the surrounding fluid. Karamanev shows that the entire curve up to the "critical" Reynolds number of about 200,000 can be represented quite accurately by the *single* equation:

$$C_D = \frac{432}{\mathrm{Ar}}\left(1 + 0.0470\,\mathrm{Ar}^{2/3}\right) + \frac{0.517}{1 + 154\,\mathrm{Ar}^{-1/3}}. \tag{4.20}$$

Thus, if the particle diameter and density and fluid properties are known, the Archimedes number and hence the drag coefficient can be calculated explicitly and quickly, in a form ideally suited for spreadsheet application.

---

[6] D.G. Karamanev, "Equations for calculation of the terminal velocity and drag coefficient of solid spheres and gas bubbles," *Chemical Engineering Communications*, Vol. 147, pp. 75–84 (1996).

Downward terminal velocity. After Ar and C_D have been computed, the terminal velocity for *downward settling*—that is, when $\rho_s > \rho_f$, is given by:

$$u_t = \sqrt{\frac{4gD|\rho_s - \rho_f|}{3\rho_f C_D}}. \tag{4.21}$$

Upward terminal velocity. For the case in which the solid density is significantly less than that of the surrounding liquid, that is, $\rho_s \ll \rho_f$, Karamanev summarizes the results of extensive experiments that show:

1. For Re < 135 (corresponding to Ar < 13,000), the SDC (standard drag curve) still applies, and the upward terminal velocity can still be computed from Eqns. (4.20) and (4.21).

2. For Re > 135 (corresponding to Ar > 13,000), the SDC *no longer* applies. Instead, the drag coefficient shows a sudden upward "jump" to a constant value of $C_D = 0.95$, and this value is then used in Eqn. (4.21) to obtain the upward terminal velocity. At this higher value for C_D, the spheres followed a perfect spiral trajectory; the angle between the spiral tangent and the horizontal plane was always very close to 60°. A similar result holds for *bubbles* rising in liquids.

Applications. Four representative applications of the above theory of *particle mechanics* are sketched in Fig. 4.13, and are explained as follows:

(a) *Separation between particles* of different size and density may be achieved by introducing the particles into a stream of liquid that flows down a slightly inclined channel. Depending on the relative rates of settling, different types of particles may be collected in compartments A, B, etc. Clearly, the interaction between particles complicates the issue, but the simple theory presented above should be adequate to make a preliminary design.

(b) *Electrostatic precipitators* cause fine dust particles or liquid droplets (positively charged, for example) in a fast-moving stream of stack gas to be attracted to a negatively charged electrode. Whether or not the particles actually reach the electrode, from which they can be collected, depends on the drag exerted on them by the surrounding gas.

(c) *Spray driers* are used for making dried milk, detergent powders, fertilizers, some instant coffees, and many other granular materials. In each case, a solution of the solid is introduced as a spray into the top of a column. Hot air is blown up through the column in order to evaporate the water from the droplets, so that the pure solid can be recovered at the bottom of the column. In this case, the diameter of the particles is constantly changing, and considerations of mass and heat transfer are also needed for a full analysis.

Fig. 4.13 Applications of drag theory: (a) particle separation, (b) electrostatic precipitator, (c) spray drier, and (d) falling-sphere viscometer.

(d) *Falling-sphere viscometers* can be used for determining the viscosity of a polymeric liquid. By timing the fall of a sphere, chosen to be sufficiently small so that the Stokes' law regime is observed, the viscosity can be deduced from Eqn. (4.11).

Example 4.2—Manufacture of Lead Shot

Lead shot of diameter d and density ρ is manufactured by spraying molten lead from the top of a "shot tower," in which the hot lead spheres are cooled by the surrounding air as they fall through a height H, solidifying by the time they reach the cushioning pool of water at the base of the tower. To assist your colleague, who is an expert in heat transfer, derive an expression for the time of fall t of the shot, as a function of its diameter. High accuracy is not needed—make any plausible simplifying assumptions.

Example 4.2—Manufacture of Lead Shot   **203**

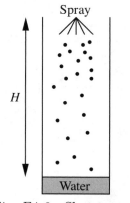

Fig. E4.2   Shot tower.

Solution

Start with Eqn. (4.12) divided through by $\pi D^3 \rho_s/6$, and neglect ρ_f in comparison with ρ_s:

$$g - \frac{6F_D}{\pi D^3 \rho_s} = \frac{du}{dt}. \tag{E4.2.1}$$

But, from Eqn. (4.10), since $A_p = \pi D^2/4$:

$$F_D = \frac{1}{8}\pi D^2 \rho_f u^2 C_D,$$

so that Eqn. (E4.2.1) becomes:

$$g - \frac{3\rho_f C_D u^2}{4\rho_s D} = g - cu^2 = \frac{du}{dt}, \tag{E4.2.2}$$

in which $c = 3\rho_f C_D/4\rho_s D$.

Assume that after a short initial period the drag coefficient is uniform, so that c is approximately constant. Separation of variables and integration between the spray nozzle and the bottom of the tower gives:

$$\int_0^t dt = \int_0^u \frac{du}{g - cu^2}, \tag{E4.2.3}$$

or, using Appendix A to determine the integral:

$$t = \frac{1}{2\sqrt{gc}} \ln \frac{\sqrt{g} + u\sqrt{c}}{\sqrt{g} - u\sqrt{c}}. \tag{E4.2.4}$$

Solution of (E4.2.4) for the velocity gives:

$$u = \frac{dx}{dt} = b\,\frac{e^{at} - 1}{e^{at} + 1}, \tag{E4.2.5}$$

in which:

$$a = 2\sqrt{gc}, \qquad b = \sqrt{\frac{g}{c}}. \tag{E4.2.6}$$

Integration of Eqn. (E4.2.5) yields:

$$\int_0^x dx = b\int_0^t \frac{e^{at}}{e^{at} + 1}\,dt - b\int_0^t \frac{dt}{e^{at} + 1},$$

$$\begin{aligned}
x &= \frac{b}{a}\ln\left(e^{at} + 1\right)\Big|_0^t - \frac{b}{a}\left[at - \ln\left(1 + e^{at}\right)\right]_0^t \\
&= \frac{1}{2c}\left[2\ln\left(\frac{e^{at} + 1}{2}\right) - at\right].
\end{aligned} \tag{E4.2.7}$$

After substituting $x = H$, Eqn. (E4.2.7) gives the time t taken for the spheres to fall through a vertical distance H. By using standard expansions for e^x and $\ln(1+x)$, several lines of algebra show that in the limit as c and/or t become small, Eqn. (E4.2.7) gives:

$$x \doteq \frac{1}{2}gt^2 - \frac{1}{12}cg^2t^4, \tag{E4.2.8}$$

in which the first term corresponds to a free fall in the absence of any drag, and the second term accounts for the drag.[7]   □

4.4 Flow Through Packed Beds

Flow through *packed beds* occurs in several areas of chemical engineering. Examples are the flow of gas through a tubular reactor containing catalyst particles, and the flow of water through cylinders packed with ion-exchange resin in order to produce deionized water. The flow of oil through porous rock formations is a closely related phenomenon; in this case, the individual particles are essentially fused together. In all cases, it is usually necessary for a certain flow rate to be able to predict the corresponding pressure drop, which may be substantial, especially if the particles are small.

The analysis is performed for the case of a *horizontal* packed bed, shown in Fig. 4.14, in order to avoid the complicating effect of gravity. Table 4.2 lists the relevant notation.

---

[7] I thank my colleague Robert Ziff for using Mathematica to check Eqn. (E4.2.8).

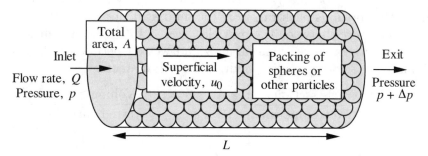

Fig. 4.14 Flow through a packed bed.

Table 4.2 Notation for Flow Through Packed Beds

Symbol	Meaning
A	Cross-sectional area of bed
a_v	Surface area of a particle divided by its volume
D_p	Effective particle diameter, $6/a_v$
L	Bed length
Q	Volumetric flow rate
u_0	Superficial fluid velocity, Q/A
ε	Fraction void (not occupied by particles)
ρ, μ	Fluid density and viscosity

The reader should check that D_p, as defined in Table 4.2, reproduces the actual diameter for the special case of a spherical particle.

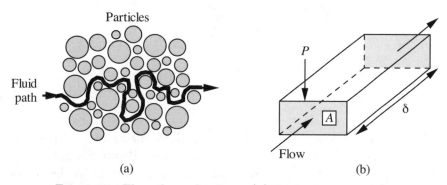

Fig. 4.15 Flow through pores: (a) the tortuous path between particles; (b) an idealized pore (the cross section can be any shape—not necessarily rectangular).

The situation may be analyzed to a certain extent by referring to Fig. 4.15(a), which shows the tortuous path taken by the fluid as it negotiates its way through the interstices or *pores* between the particles. Fig. 4.15(b) shows unit length of an idealized pore, with cross-sectional area A and wetted perimeter P. For a given total volume V, the corresponding hydraulic mean diameter is:

$$D_{\mathrm{e}} = 4 \; \frac{(\text{Cross sectional area A}) \times \delta}{(\text{Wetted perimeter P}) \times \delta} = 4 \; \frac{\text{Volume of voids}}{\text{Wetted surface area}}$$

$$= 4 \; \frac{\varepsilon V}{V(1-\varepsilon)a_{\mathrm{v}}} = \frac{4\varepsilon}{a_{\mathrm{v}}(1-\varepsilon)}. \tag{4.22}$$

For a horizontal pore, the pressure drop is therefore:

$$-\Delta p = 2f_{\mathrm{F}}\rho u_{\mathrm{m}}^2 \frac{L}{D_{\mathrm{e}}} = 2f_{\mathrm{F}}\rho\left(\frac{u_0}{\varepsilon}\right)^2 L \, \frac{a_{\mathrm{v}}(1-\varepsilon)}{4\varepsilon} \tag{4.23}$$

$$= 3f_{\mathrm{F}}\rho u_0^2 \frac{(1-\varepsilon)}{\varepsilon^3} \frac{L}{D_p}. \tag{4.24}$$

Rearrangement of (4.24) yields:

$$\frac{-\Delta p}{\rho u_0^2} \frac{D_p}{L} \frac{\varepsilon^3}{1-\varepsilon} = 3f_{\mathrm{F}} = 1.75 \; (\text{experimentally}). \tag{4.25}$$

Thus, theory indicates for turbulent flow, in which f_{F} is essentially constant, that the somewhat unusual dimensionless group on the left-hand side of Eqn. (4.25) should be constant. This prediction is completely substantiated by experiment, and the value of the constant is 1.75.

More generally, however, allowance should be made for a laminar contribution, which will prevail at low Reynolds numbers. The resulting *Ergun equation*, which is one of the most successful correlations in chemical engineering, is:

$$\frac{-\Delta p}{\rho u_0^2} \frac{D_p}{L} \frac{\varepsilon^3}{1-\varepsilon} = \underbrace{\frac{150}{Re}}_{\text{Laminar}} + \underbrace{1.75}_{\text{Turbulent}}, \tag{4.26}$$

in which the Reynolds number is:

$$Re = \frac{\rho u_0 D_p}{(1-\varepsilon)\mu}. \tag{4.27}$$

The Ergun equation is shown in Fig. 4.16; the limiting cases for low and high Reynolds numbers are called the Blake-Kozeny and Burke-Plummer equations, respectively. Observe that these two forms (one proportional to the reciprocal of the Reynolds number, and the other a constant) are analogous to our previous experience for the friction factor in pipes, first in laminar and then in highly turbulent flow.

The y-axis is labeled: Dimensionless pressure drop, $\dfrac{-\Delta p}{\rho u_0^2}\dfrac{D_p}{L}\dfrac{\varepsilon^3}{1-\varepsilon}$

The x-axis is labeled: Reynolds number Re, $\dfrac{\rho u_0 D_p}{(1-\varepsilon)\mu}$

Labels within the figure: Blake-Kozeny equation; Ergun equation; Burke-Plummer equation

Fig. 4.16   The Ergun equation.

Frictional dissipation term for packed beds.   So far, we have been concerned only with *horizontal* beds, for which the overall energy balance is:

$$\frac{\Delta p}{\rho} + \mathcal{F} = 0, \qquad (4.28)$$

Thus, from Eqn. (4.22), the frictional dissipation term per unit mass flowing is:

$$\mathcal{F} = -\frac{\Delta p}{\rho} = \frac{150 u_0 \mu L (1-\varepsilon)^2}{\rho D_p^2 \varepsilon^3} + 1.75\,\frac{u_0^2 L (1-\varepsilon)}{D_p \varepsilon^3}. \qquad (4.29)$$

Although $\mathcal{F}$ from Eqn. (4.29) has been derived for the horizontal bed (in order to isolate the purely frictional effect), this relation can then be substituted into the appropriate energy balance for an inclined or vertical packed bed.

d'Arcy's law.   The above theory can be applied to a *consolidated* or *porous* medium, in which the particles are fused together, such as would occur in a sandstone rock formation through which oil is flowing. Since the flow rate is likely to be small, and again considering *horizontal* flow (that is, ignoring changes in pressure caused by hydrostatic effects), the turbulent term in Eqn. (4.22) can be neglected as a good approximation, giving:

$$\frac{-\Delta p}{\rho u_0^2} \frac{D_p}{L} \frac{\varepsilon^3}{1 - \varepsilon} = \frac{150(1 - \varepsilon)\mu}{\rho u_0 D_p}. \tag{4.30}$$

Rearranging, the superficial velocity is given by *d'Arcy's law:*

$$u_0 = \frac{-\Delta p}{L\mu} \underbrace{\frac{D_p^2 \varepsilon^3}{150(1 - \varepsilon)^2}}_{\kappa} = -\frac{\kappa}{\mu} \frac{\Delta p}{L}. \tag{4.31}$$

Note that since the concept of an individual particle diameter no longer exists, the fraction κ shown in Eqn. (4.31) with an underbrace is collectively considered to be another physical property of the porous medium, known as its *permeability*. If u_0 is measured in cm/s, Δp in atm, μ in cP, and L in cm, the unit of permeability is known as the *darcy*, which is equivalent to:

$$1 \text{ darcy} = 1 \, \frac{\text{cm/s cP}}{\text{atm/cm}} \doteq 0.986 \times 10^{-8} \text{ cm}^2 = 1.06 \times 10^{-11} \text{ ft}^2. \tag{4.32}$$

The *differential* form of d'Arcy's law in one dimension is:

$$u_0 = -\frac{\kappa}{\mu} \frac{dp}{dx}, \tag{4.33}$$

which is a classical type of relation, in which a *flux* (here a volumetric flow rate per unit area) is proportional to a *conductivity* (κ/μ) times a negative *gradient* of a potential driving force (dp/dx). A more general, vector form of d'Arcy's law, $\mathbf{v} = -(\kappa/\mu)\nabla p$, is given in Eqn. (7.76).

Example 4.3—Pressure Drop in a Packed-Bed Reactor

A liquid reactant is pumped through the catalytic reactor shown in Fig. E4.3, which consists of a horizontal cylinder packed with catalyst spheres of diameter d_1 = 2.0 mm. Tests summarized in Table E4.3 show the pressure drops $-\Delta p$ across the reactor at two different volumetric flow rates Q.

If the maximum pressure drop is limited by the pump to 50 psi, what is the upper limit on the flow rate? After the existing catalyst is spent, a similar batch is unfortunately unavailable, and the reactor has to be packed with a second batch whose diameter is now d_2 = 1.0 mm. What is the new maximum allowable flow rate if the pump is still limited to 50 psi?

Example 4.3—Pressure Drop in a Packed-Bed Reactor   **209**

Table E4.3 Packed-Bed Pressure Drop

Q, ft^3/hr	$-\Delta p$, psi
12.0	9.6
24.0	24.1

Fig. E4.3 Packed-bed reactor.

Solution

From Eqn. (4.29), for constant μ, L, ε, and ρ, noting that u_0 is proportional to Q and that all conversion factors can be absorbed into the constants a and b:

$$-\Delta p = \frac{aQ}{D_p^2} + \frac{bQ^2}{D_p}. \tag{E4.3.1}$$

Inserting values from Table E4.3:

$$9.6 = \frac{12a}{2^2} + \frac{144b}{2} = 3a + 72b, \tag{E4.3.2}$$

$$24.1 = \frac{24a}{2^2} + \frac{576b}{2} = 6a + 288b. \tag{E4.3.3}$$

Solution of these two simultaneous equations gives $a = 2.38$ and $b = 0.0340$, so the maximum flow rate Q_{max} is given by the quadratic equation:

$$50 = \frac{2.38\,Q_{max}}{2^2} + \frac{0.0340\,Q_{max}^2}{2}, \tag{E4.3.4}$$

from which $Q_{max} = 39.5$ ft^3/hr.

For the new catalyst, D_p is now only 1.0, so the new maximum flow rate obeys the equation:

$$50 = \frac{2.38 Q_{max}}{1^2} + \frac{0.0340 Q_{max}^2}{1}, \tag{E4.3.5}$$

yielding $Q_{max} = 16.9$ ft^3/hr. Note that the flow rate declines appreciably for the finer size of packing. □

4.5   Filtration

Introduction and plate-and-frame filters.   A filter is a device for removing solid particles from a fluid stream (often from a liquid). Examples are:

1. In the paper industry, to separate paper-pulp from a water/pulp suspension.
2. In sugar refining, either to clarify sugar solutions or to remove wanted saccharates from a slurry.
3. In the recovery of magnesium from seawater, to separate out the insoluble magnesium hydroxide.
4. In metallurgical extraction, to remove the unwanted mineral residues from which silver and gold have been extracted by a cyanide solution.
5. In automobiles, to clean oil and air.
6. In municipal domestic water plants, to purify water.

The basic elements of a filter are shown in Fig. 4.17.

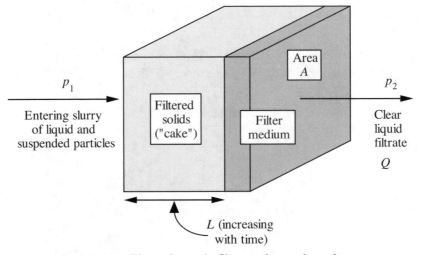

Fig. 4.17   Flow through filter cake and medium.

A *slurry*, containing liquid and suspended particles at an inlet pressure p_1, flows through the filter *medium*, such as a cloth, gauze, or layer of very fine particles. The clear liquid or *filtrate* passes at a volumetric rate Q through the medium into a region where the pressure is p_2, whereas the suspended particles form a porous semisolid *cake* of ever-increasing thickness L.

A *plate-and-frame filter* consists of several such devices operating in parallel. The cloth is supported on a porous metal plate, and successive plates are separated by a frame, which also incorporates various channels to supply the slurry and remove the filtrate. When the cake has built up to occupy the entire space between successive plates, the filter must be dismantled in order to remove the cake, wash the filter, and restart the operation. Detailed views are given in Fig. 4.18.

In many cases, the resistance of the filter medium can be neglected. If A is the area of the filter, and if V denotes the total volume of filtrate passed since starting with $L = 0$ at $t = 0$, d'Arcy's law gives:

$$u_0 = \frac{Q}{A} = \frac{1}{A}\frac{dV}{dt} = \frac{\kappa(p_1 - p_2)}{\mu L}. \qquad (4.34)$$

Fig. 4.18 *The elements of a plate-and-frame filter.*

But the thickness of the cake increases linearly with the volume of filtrate, so that:

$$L = \frac{\alpha V}{A},$$

(4.35)

in which α is the volume of cake deposited by unit volume of filtrate. Hence,

$$\frac{1}{A}\frac{dV}{dt} = (p_1 - p_2)\frac{\kappa A}{\alpha \mu V}.$$

(4.36)

Depending largely on the characteristics of the pump supplying the slurry under pressure, two principal modes of operation are now recognized.

1. *Constant-pressure* operation occurs approximately if a centrifugal pump, not operating near its maximum flow rate, is employed. With the pressure drop $(p_1 - p_2)$ thereby held constant, integration of Eqn. (4.36) up to a time t yields:

$$\frac{\alpha \mu}{\kappa A^2}\int_0^V V\,dV = (p_1 - p_2)\int_0^t dt.$$

(4.37)

That is, the volume of filtrate varies with the square root of elapsed time according to:

$$V = \sqrt{\frac{2\kappa A^2(p_1 - p_2)t}{\alpha \mu}}.$$

(4.38)

2. *Constant flow-rate* operation occurs when a positive displacement pump is used, in which case the inlet pressure simply adjusts to whatever is needed to maintain the flow rate Q at a steady value. Since $V = Qt$, differentiation yields:

$$\frac{dV}{dt} = Q.$$

(4.39)

Substitution for dV/dt from Eqn. (4.36) then gives the relation between the pressure drop and the flow rate:

$$p_1 - p_2 = \frac{\alpha \mu Q^2 t}{\kappa A^2}.$$

(4.40)

Rotary vacuum filters. A disadvantage of the plate-and-frame filter is its *intermittent* operation—it must be dismantled and cleaned when the cake has built up to occupy the entire space between the plates. Generally, chemical engineers prefer *continuous* processing operations, which in the case of filtration can be achieved by the rotary vacuum filter shown in Fig. 4.19.

The slurry to be filtered is supplied continuously to a large bath, in which a partly submerged perforated drum is rotating slowly at an angular velocity ω.

The drum is divided internally into several separate longitudinal segments, and by a complex set of valves (not shown here) each segment can be maintained either below or above atmospheric pressure. Thus, a partial vacuum applied to the submerged segments causes filtration to occur, the cake building up on the surface of the drum, and the filtrate passing inside the drum, where it is removed at one end by piping (also not shown). The partial vacuum also causes the wash water to pass through the cake, and it too is collected by additional piping at one end of the drum. The washed cake is finally detached by a scraper or "doctor knife," assisted by a small positive pressure inside the segment just approaching the scraper.

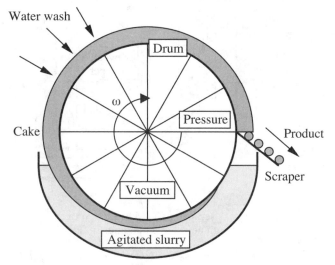

Fig. 4.19   Cross section of a rotary vacuum filter.

The analysis of the rotary vacuum filter is similar to that of the plate-and-frame filter, in which the time of operation is the period for one complete rotation $(2\pi/\omega)$ multiplied by the fraction of segments under vacuum that are in contact with the slurry. The operation is essentially constant pressure, because of the steady vacuum inside the drum relative to the atmosphere.

Centrifugal filters.   One type of *centrifugal filter* is shown in Fig. 4.20. It consists of a cylindrical basket with a perforated vertical surface (as in a washing machine), covered with filter cloth, that is rotated at a high speed. Slurry sprayed on the inside is flung outwards by centrifugal action and soon starts to deposit a lining of cake on the inside of the wall. The filtrate discharges through the perforations and is collected in an outer casing. After a suitable amount of cake has been deposited, the slurry feed is stopped and the basket is slowed down, during which period the cake is washed and scraped off the wall. The cake is then deposited into a receptacle through openable doors in the base.

The analysis of the centrifugal filter follows standard lines. In the *slurry*, the pressure increases because of centrifugal action, from atmospheric pressure at $r = r_1$ to a maximum at $r = a$. In the *cake*, centrifugal action again tends to increase the pressure, but friction dissipates this effect, so that the discharge at $r = r_2$ has reverted to atmospheric pressure. The liquid will "back up" to the appropriate radius r_1 that suffices to provide the necessary driving force to overcome friction in the cake.

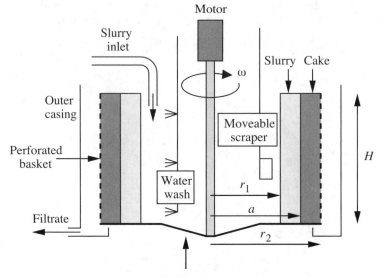

Fig. 4.20 Idealized cross section of a centrifugal filter.

The basic equations governing pressure in the slurry and in the cake are:

$$\text{Slurry:} \quad \frac{dp}{dr} = \rho_S \omega^2 r; \qquad \text{Cake:} \quad \frac{dp}{dr} = \rho_F \omega^2 r - \frac{\mu Q}{2\pi r \kappa H}. \qquad (4.41)$$

Here, κ is the cake permeability, and the superficial velocity for d'Arcy's law has been recognized as $u_0 = Q/(2\pi r H)$, where Q is the filtrate flow rate. Unless otherwise stated, the resistance of the filter medium is usually neglected, and the slurry and filtrate densities ρ_S and ρ_F have essentially the same value, ρ. The slurry equation can be integrated *forward*, from r_1, where $p = 0$, to give the pressure in the slurry. The cake equation can be integrated *backward*, from r_2, where $p = 0$, to give the pressure in the cake. The two expressions for the pressure must match, of course, at the slurry/cake interface, $r = a$.

Considerations of the rate of cake deposition show that the inner radius a of the cake gradually *decreases* as solids are deposited, according to:

$$\alpha Q = -2\pi a H \frac{da}{dt}, \qquad (4.42)$$

in which α is the volume of cake per unit volume of filtrate.

4.6 Fluidization

Fig. 4.21(a) illustrates upwards flow of a fluid through a bed of initial height h_0 that is packed with particles of diameter D_p. Fig. 4.21(b) shows the relation between the actual bed height h and the superficial velocity u.

For low u, h is almost unchanged from its initial value. However, as u is increased, the pressure drop $p_1 - p_2$ also increases, and will eventually build up to a value that suffices to counterbalance the downward weight of the particles. At this point, when u has reached the *incipient fluidizing velocity* u_0, the particles are essentially weightless and will start circulating virtually as though they were a liquid. Further increases in u will cause the bed to expand (still in a fluidized state), whereas the pressure drop now increases only slightly.

Fluidized beds are excellent for providing good contact and mixing between fluid and solid, as is required in some catalytic reactors. (See also Section 10.6 for further details of fluidized beds—particularly relating to *particulate* and *aggregative* modes of operation.) The value of the incipient fluidizing velocity may be obtained by the following treatment. An energy balance applied between the bed inlet and exit gives:

$$gh_0 + \frac{p_2 - p_1}{\rho_f} + \mathcal{F} = 0. \tag{4.43}$$

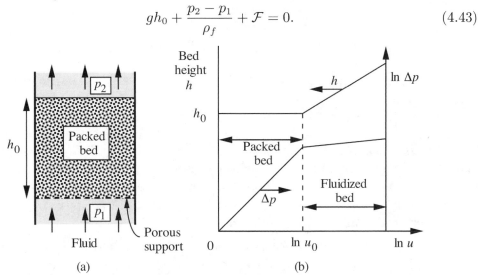

Fluid   Porous support   0

(a)   (b)

Fig. 4.21   Fluidization: (a) upward flow through a packed bed, and (b) variation of bed height with superficial velocity.

The pressure drop can now be extracted and equated to the downward weight of particles and fluid per unit area:

$$p_1 - p_2 = \rho_f \mathcal{F} + gh_0 \rho_f = \underbrace{gh_0[(1 - \varepsilon_0)\rho_s + \varepsilon_0 \rho_f]}_{\substack{\text{Total downward weight} \\ \text{of particles and fluid}}}, \tag{4.44}$$

where ε_0 is the void fraction when the bed is on the verge of fluidization. Isolation of the frictional dissipation term gives:

$$\rho_f \mathcal{F} = gh_0(1 - \varepsilon_0)(\rho_s - \rho_f). \tag{4.45}$$

But $\mathcal{F}$ is given by the right-hand side of Eqn. (4.29):

$$\mathcal{F} = \frac{150 u_0 \mu h_0 (1 - \varepsilon_0)^2}{\rho_f D_p^2 \varepsilon_0^3} + 1.75 \frac{u_0^2 h_0 (1 - \varepsilon_0)}{D_p \varepsilon_0^3}. \tag{4.46}$$

Thus, from Eqns. (4.45) and (4.46), after canceling $(1 - \varepsilon_0)h_0$, the incipient fluidizing velocity u_0 is given in terms of all other known quantities by:

$$\frac{150(1 - \varepsilon_0)\mu u_0}{D_p^2 \varepsilon_0^3} + 1.75 \frac{\rho_f u_0^2}{D_p \varepsilon_0^3} = g(\rho_s - \rho_f). \tag{4.47}$$

A much more complete discussion of fluidized beds is given in the latter part of Chapter 10.

4.7 Dynamics of a Bubble-Cap Distillation Column

The basic principles learned so far can be employed to insure the satisfactory operation of a distillation column, a cross section of which is shown in Fig. 4.22(a). Such a column is used to separate or *fractionate* a mixture, based on the different boiling points or volatilities of the components in a feed stream, which could typically consist of ethanol and water, or a mixture of "light" (low boiling point) and "heavy" (high boiling point) hydrocarbons. The most volatile components become concentrated toward the top of the column, and the least volatile toward the bottom.

The essential parts of the column are:

1. A *reboiler*, typically steam-heated, that boils the liquid from the bottom of the column, part of which is withdrawn as the bottoms product, rich in the heavy components. The rest of the liquid is vaporized and returned to the column.

2. A series of *trays* on which the liquid mixture is boiling. Vapor somewhat enriched in the lighter components rises to the tray above, and liquid somewhat enriched in the heavier components falls to the tray below.

3. A *condenser*, typically cooled by water, liquefies the vapor from the top tray. Part of the liquid is returned to the top tray as *reflux*, and the rest is withdrawn as the overhead product, rich in the light components.

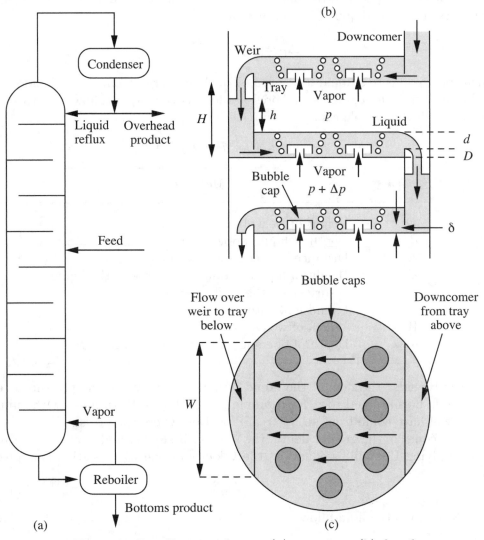

Fig. 4.22 Distillation column: (a) overview; (b) detail of liquid on three successive trays; (c) plan of one tray.

As seen from Fig. 4.22(b), boiling liquid is continuously spilling over a weir at one side of every tray, from there flowing via a *downcomer* and through a constriction to the tray below. Vapor, boiling from the liquid on each tray, is simultaneously flowing upward through *bubble caps* (which act as liquid seals, and only a few of which are shown in the diagram) into the boiling liquid on the tray immediately above. We wish to insure that the arrangement is stable.

Adopt the notation shown in Table 4.3. The pressure of the vapor leaving a tray must be high enough to overcome the hydrostatic pressure of the liquid on the tray above, and hence to enable the vapor to flow through the bubble caps. The excess pressure required is approximately:

$$\Delta p = \rho_L g (d + D), \tag{4.48}$$

which corresponds to a liquid head of $(d+D)$. A further refinement, not made here, would be to include a small extra pressure loss as the vapor follows the tortuous path through the bubble cap.

Table 4.3   Notation for Distillation Column

Symbol	Meaning
D	Height of weir above tray
d	Depth of liquid above weir
H	Distance between successive trays
h	Height of liquid in downcomer above the liquid surface on the next tray below
L	Liquid flow rate
W	Width of weir
δ	Depth of opening at bottom of downcomer

Next consider the flow of the liquid over the weir, where the pressure is everywhere uniform (equal to p, for example). Liquid at a depth y below the upper surface will have come from an upstream location where the pressure is $p + \rho_L g y$ and, on account of the greater depth, the velocity is smaller and may be neglected. Thus, applying Bernoulli's equation, the velocity at the weir location is approximately:

$$u = \sqrt{2gy}. \tag{4.49}$$

Integration gives the liquid flow rate:

$$L = C_D \int_0^d W \sqrt{2gy}\, dy = \frac{2}{3} C_D \sqrt{2g}\, W d^{3/2}. \tag{4.50}$$

Here, a coefficient of discharge C_D, typically about 0.62, has been introduced to allow for deviations from the theory, mainly because of a further contraction of the liquid stream as it spills over the weir.

The available driving head h has to overcome two resistances:

1. The head $(d + D)$ needed to cause the gas to flow.
2. The loss of kinetic energy as the liquid jet at the bottom of the downcomer is dissipated as it enters the tray.

Thus, equating these two effects:

$$h = d + D + \frac{1}{2g} \left(\frac{L}{W\delta} \right)^2 .$$ (4.51)

For sufficiently high liquid flow rates, the level in any downcomer can only back up to the level of the tray above before it starts interfering with the flow from the weir above. In such event, the tray spacing is the sum of the three individual heights shown in Fig. 4.22(b):

$$H = h + d + D.$$ (4.52)

Under these conditions, also using Eqn. (4.47):

$$H = 2(d + D) + \frac{1}{2g} \left(\frac{L}{W\delta} \right)^2 .$$ (4.53)

Note from Eqn. (4.50) that:

$$d = \left(\frac{9L^2}{8gC_D^2 W^2} \right)^{1/3} ,$$ (4.54)

and eliminate the unknown d from Eqns. (4.53) and (4.54), finally giving:

$$D + \left(\frac{9L_{\max}^2}{8gC_D^2 W^2} \right)^{1/3} + \frac{1}{4g} \left(\frac{L_{\max}}{W\delta} \right)^2 = \frac{1}{2} H.$$ (4.55)

Hence, the maximum liquid flow rate $L_{\max}$ under which the column can operate successfully is given by Eqn. (4.55). Any attempt to increase the liquid flow rate beyond $L_{\max}$ will cause liquid to occupy the entire column, a phenomenon known as "flooding." Under these circumstances, there is no space left for the vapor, and normal operation as an effective distillation column ceases.

4.8 Cyclone Separators

Solid particles—even dust—may be separated from a fluid stream—usually a gas—by means of a *cyclone separator*, the elements of which are shown in Fig. 4.23. The notation for the analysis appears in Table 4.4.

A volumetric flow rate Q of particle-containing gas enters tangentially through an inlet port, of cross-sectional area A, into the top of a virtually empty cylinder, of radius r_2. The swirling motion tends to cause the particles to be flung out to the wall of the cyclone, from which they subsequently fall by gravity into the lower conical portion for collection. The particle-free gas discharges upward through a large pipe of radius r_1.

The effectiveness of the cyclone may be *approximated* by a simple analysis, by considering both the centrifugal force and fluid drag acting on the particles. The velocity of the gas in the cylindrical portion has three components:

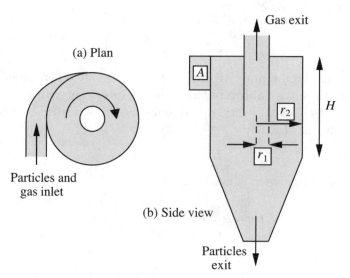

(a) Plan

Particles and
gas inlet

Gas exit

A

r_2

H

r_1

(b) Side view

Particles
exit

Fig. 4.23 Cyclone separator.

Table 4.4   Notation for Cyclone Separator

Symbol	Meaning
A	Cross-sectional area of inlet port
D	Particle diameter (critical value D^*)
H	Height of cylindrical part of cyclone
m	Particle mass
Q	Volumetric flow rate of gas
r	Radial coordinate
r_1	Radius of gas exit pipe
r_2	Radius of cylindrical part of cyclone
v_r	Radially inward velocity component
v_θ	Tangential velocity component
μ	Viscosity of gas
ρ_{p}	Particle density

1. An inward radial velocity, v_r, as the gas travels from the inlet to the exit. For a cylinder height H, continuity gives as a first approximation:

$$v_r \doteq \frac{Q}{2\pi r H}. \tag{4.56}$$

2. A tangential velocity, v_θ, which—again to a first approximation—is inversely proportional to the radius, as in a free vortex (see Section 7.2). Since the inlet

value of v_θ is Q/A at a radius r_2, its value at any smaller radius r is:

$$v_\theta = \frac{Qr_2}{Ar}. \tag{4.57}$$

Equation (4.57) also follows from the principle of conservation of angular momentum (see Section 2.6), in which ωr^2 is constant, where $\omega = v_\theta/r$ is the angular velocity.

3. A vertical component, v_z, first descending from the inlet (even into the conical portion) and eventually changing direction and rising into the gas exit. In the present simplified analysis, v_z will be ignored, but it could be accommodated by treating the situation as a potential flow problem according to the methods given in Chapter 7, in which case a relatively complex computer-assisted numerical solution would be needed to obtain a proper description of the entire flow pattern.

Consider the forces acting on a representative particle of density $\rho_{\rm p}$, which is assumed to be so small that Stokes' law applies. (For larger particles, an appropriate drag coefficient would have to be incorporated into the analysis.) A particle will remain at a radial location r when the radially *outward* centrifugal force is counterbalanced by an *inward* drag, giving:

$$\frac{mv_\theta^2}{r} = 3\pi\mu v_r D. \tag{4.58}$$

Substitute for the two velocity components from Eqns. (4.56) and (4.57), and note the particle mass is $m = \rho_{\rm p}\pi D^3/6$. The inward drag and outward centrifugal force will then be in balance at a radius r for a particle of diameter:

$$D = \frac{r}{r_2}\sqrt{\frac{9\mu A^2}{\pi p_P Q H}}. \tag{4.59}$$

Taking the most pessimistic view, the diameter D^* of the largest particle that we are sure will not be attracted toward the gas exit is obtained by setting $r = r_2$:

$$D^* = \sqrt{\frac{9\mu A^2}{\pi p_P Q H}}. \tag{4.60}$$

A particle whose diameter equals or exceeds D^* will be trapped at the outer wall and will fall by gravity so that it is separated from the gas. However, a particle with $D < D^*$ will tend to be dragged toward the exit tube and may not be separated from the gas. Observe that small values of A and large values of Q will serve to reduce D^* and hence enable smaller particles to be collected.

If we set $r = r_1$ in Eqn. (4.59), we obtain a diameter D_{min} below which all particles will tend to reach the exit tube:

$$D_{min} = \frac{r_1}{r_2} \sqrt{\frac{9\mu A^2}{\pi p_P Q H}}. \tag{4.61}$$

We emphasize that motion in the vertical direction has been neglected and that the above results are only approximate.

4.9 Sedimentation

Section 4.2 dealt with the relative motion of a *single* particle in a fluid. In particular, a method was discussed for obtaining the *terminal* velocity u_t of a single particle settling under gravity in a fluid. Some chemical engineering operations involve *many* such particles that are sufficiently close together so that the previous theory no longer applies.

Table 4.5   Values of the Richardson-Zaki Exponent

Re	n
Re < 0.2	4.65
0.2 < Re < 1	$4.35 \text{Re}^{-0.03}$
1 < Re < 500	$4.45 \text{Re}^{-0.1}$
Re > 500	2.39

Fortunately, the Richardson/Zaki[8] correlation is available to give the settling velocity u of a group of particles as a function of the void fraction ε (the fraction of the total volume that is occupied by the fluid):

$$u = u_t \varepsilon^n. \tag{4.62}$$

Here, the value of exponent n is primarily a function of the Reynolds number $\text{Re} = \rho_f u D / \mu$, as shown in Table 4.5. (Richardson and Zaki also found that n depends to a much smaller extent on the ratio of the particle diameter to that of the containing vessel—a fact that can reasonably be ignored here.) Note that Eqn. (4.62) correctly reduces to $u = u_t$ (the terminal velocity of a single particle) when the void fraction is $\varepsilon = 1$.

---

[8]   J.F. Richardson and W.N. Zaki, "Sedimentation and fluidisation," *Transactions of the Institution of Chemical Engineers*, Vol. 32, p. 35 (1954).

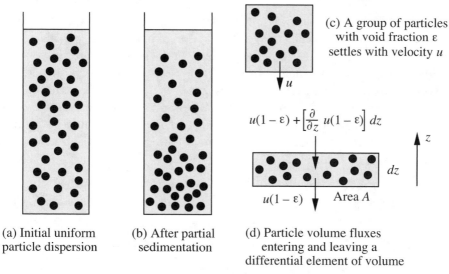

(a) Initial uniform particle dispersion

(b) After partial sedimentation

(c) A group of particles with void fraction ε settles with velocity u

(d) Particle volume fluxes entering and leaving a differential element of volume

Fig. 4.24 Particle sedimentation.

Now examine the sedimentation in a liquid of a large cluster of particles in a container, as shown in Fig. 4.24. Initially, the particles are uniformly distributed throughout the liquid, as in (a). At some later time, as in (b), they will tend to congregate toward the bottom of the container. That is, the void fraction will vary from a relatively high value at the top of the container to a relatively low value at the bottom. We wish to derive the differential equation that governs the variation of the void fraction ε with both height z and time t.

Consider a differential element of cross-sectional area A and height dz, as in Fig. 4.24(d). Since the fraction of volume occupied by the particles is $(1 - \varepsilon)$, the rate at which *particle* volume leaves the element through its lower surface is $u(1-\varepsilon)$ per unit area. The rate at which particle volume enters the element through its upper surface is differentially greater, as shown in the diagram. Next, equate the net rate of particle volume entering the element to the rate of increase of particle volume inside the element, giving:

$$A\frac{\partial}{\partial z}[u(1 - \varepsilon)]dz = \frac{\partial}{\partial t}\left[(1 - \varepsilon)A\,dz\right]. \tag{4.63}$$

Substitution of the particle velocity from Eqn. (4.62) yields the following differential equation:

$$u_t\varepsilon^{n-1}[n - (n + 1)\varepsilon]\frac{\partial \varepsilon}{\partial z} + \frac{\partial \varepsilon}{\partial t} = 0. \tag{4.64}$$

Unfortunately, there is no ready analytical solution of Eqn. (4.64), but—starting at $t = 0$ with a known initial distribution of ε—numerical methods could be employed to determine how the void fraction ε varies with elevation z and time t.

4.10 Dimensional Analysis

Dimensional analysis is important because it enables us to express relations between variables—whether analytical solutions or experimental correlations—very concisely. Instead of attempting to establish a relation that may involve *several* variables *independently*, a much simpler relation is typically established between a relatively *small* number of *groups* of variables. With proper planning, the technique also often reduces the number of experiments that are needed in certain investigations. There are two main approaches in determining the appropriate dimensionless groups, depending whether or not an analytical solution or similar model is already available.

1. Four examples of the first approach, which rearranges an existing solution into dimensionless groups, are available from material already studied:

(a) Consider Eqn. (4.38), for the constant-pressure operation of a batch filter:

$$V = \sqrt{\frac{2\kappa A^2 (p_1 - p_2)t}{\alpha \mu}}. \tag{4.38}$$

Rearrangement in terms of a dimensionless group Π, gives:

$$\Pi = \frac{\alpha \mu V^2}{2\kappa A^2 (p_1 - p_2)t} = 1, \tag{4.65}$$

which states that no matter what the individual values of V, μ, etc., the performance of *all* filters can be expressed by the equation:

$$\Pi = 1. \tag{4.66}$$

(b) The shape of the free surface of a rotating liquid is:

$$\frac{gz}{\omega^2 r^2} = \frac{1}{2}. \tag{1.47}$$

Observe that both the group of variables on the left-hand side, and the fraction on the right-hand side, are dimensionless.

(c) For the evacuation of the tank in Example 2.1, the dimensionless pressure ratio p/p_0 is a function of the dimensionless time vt/V.

$$\frac{p}{p_0} = e^{-vt/V}. \tag{E2.1.6}$$

(d) Finally, consider Eqn. (3.40), the semi-empirical Blasius relation that expresses the dimensionless wall shear stress for turbulent flow in a smooth pipe in terms of the (dimensionless) Reynolds number:

$$f_F = 0.0790 \, \mathrm{Re}^{-1/4}, \tag{3.40}$$

which can be written out fully as:

$$\frac{\tau_w}{\frac{1}{2}\rho u_m^2} = 0.0790 \left(\frac{\mu}{\rho u_m D}\right)^{1/4}.$$ (4.67)

Again note that the correlation in terms of the dimensionless groups is much more concise than that with the individual variables.

2. The second approach, in which appropriate dimensionless groups have to be found, is useful when there is no existing model or solution. First, four *fundamental dimensions* are identified, as shown in Table 4.6; temperature is given for completeness, but is generally unnecessary for most fluid mechanics work.

Table 4.6   Fundamental Dimensions

Fundamental Dimension	*Symbol*
Length	L
Mass	M
Time	T
Temperature	deg

Now express the dimensions of other variables, known as *derived quantities*, in terms of the fundamental dimensions, as shown in Table 4.7. There, "deg" denotes Kelvins, degrees Fahrenheit, etc., as appropriate.

Wall shear stress for pipe flow.   The general approach is illustrated for a specific case. Assume, based on experience, that the wall shear stress τ_w in pipe flow is likely to depend only on the distance x from the inlet, D, u_m, ρ, μ, and the wall roughness ε. That is,

$$\tau_w = \psi(x, \ D, \ u_m, \ \rho, \ \mu, \ \varepsilon),$$ (4.68)

where ψ denotes a functional dependency—as yet unknown. Hence, there exists some relation between all the seven variables, written as:

$$\psi(\tau_w, \ x, \ D, \ u_m, \ \rho, \ \mu, \ \varepsilon) = 0.$$ (4.69)

Observe that the functions ψ are different in (4.68) and (4.69); since both are unknown, it is pointless to use a different symbol each time, so ψ is really a "generic" functional dependency. The dimensions of the various quantities are given in Table 4.8.

Table 4.7   Dimensions of Derived Quantities

Quantity	Representative Symbol	Dimensions
Angular acceleration	$\dot{\omega}$	T^{-2}
Angular velocity	ω	T^{-1}
Area	A	L^2
Density	ρ	M/L^3
Energy	E	ML^2/T^2
Force	F	ML/T^2
Heat-transfer coefficient	h	M/T^3 deg
Kinematic viscosity	ν	L^2/T
Linear acceleration	a	L/T^2
Linear velocity	u	L/T
Mass diffusivity	$\mathcal{D}$	L^2/T
Mass flow rate	m	M/T
Mass transfer coefficient	h_d	L/T
Momentum	$\mathcal{M}$	ML/T
Power	P	ML^2/T^3
Pressure	p	M/LT^2
Rotational speed	ω	T^{-1}
Sonic velocity	c	L/T
Specific heat	c_p	L^2/T^2 deg
Stress	τ	M/LT^2
Surface tension	σ	M/T^2
Thermal conductivity	k	ML/T^3 deg
Thermal diffusivity	α	L^2/T
Viscosity	μ	M/LT
Volume	V	L^3

Table 4.8   Dimensions of Quantities for Pipe Flow

Quantity:	τ_w	x	D	u_m	ρ	μ	ε
Dimensions:	M/LT^2	L	L	L/T	M/L^3	M/LT	L

Next, choose as many *primary quantities* as there are fundamental dimensions (three in this case), as long as they contain all the relevant fundamental dimensions explicitly among them. For example, we can choose D, ρ, and u_m (L, M/L^3, and L/T) as primary quantities because the dimensions L, M, and T are available in D, ρD^3, and D/u_m. However, x, D, and μ (L, L, M/LT) would be an improper choice because there is no combination of these variables that generates M and T individually (the ratio M/T persists). Having chosen D, ρ, and u_m, form dimensionless ratios, given the general symbol Π, for the remaining quantities τ_w, μ, x, and ε, as follows:

For τ_w, the first group is formulated as:

$$\Pi_1 = \frac{\tau_w}{D^a \rho^b u_m^c}. \tag{4.70}$$

Next, choose the exponents a, b, and c so that Π_1 is dimensionless:

$$\text{Dimensions of } \tau_w = \text{Dimensions of } D^a \rho^b u_m^c$$

$$\text{M L}^{-1}\text{T}^{-2} = (\text{L})^a \left(\frac{\text{M}}{\text{L}^3}\right)^b \left(\frac{\text{L}}{\text{T}}\right)^c. \tag{4.71}$$

Considering each of the fundamental dimensions M, L, and T in turn, three simultaneous equations result:

$$\begin{aligned}
\text{M}: & \quad 1 = b, \\
\text{L}: & \quad -1 = a - 3b + c, \\
\text{T}: & \quad -2 = -c,
\end{aligned} \tag{4.72}$$

for which the solution is $a = 0$, $b = 1$, and $c = 2$. The first dimensionless group is therefore:

$$\Pi_1 = \frac{\tau_w}{\rho u_m^2}. \tag{4.73}$$

A similar procedure for μ, x, and ε gives the following dimensionless groups:

$$\Pi_2 = \frac{\mu}{D\rho u_m}, \qquad \Pi_3 = \frac{x}{D}, \qquad \Pi_4 = \frac{\varepsilon}{D}. \tag{4.74}$$

It is then a fundamental postulate, known as the *Buckingham Pi Theorem*, that there exists some relation among the dimensionless groups thus formed:

$$\psi(\Pi_1, \Pi_2, \Pi_3, \Pi_4) = 0, \tag{4.75}$$

Note that the original problem of developing a correlation among *seven* variables has now been simplified enormously to that of finding a correlation among just *four* dimensionless groups.

Substituting for the Π groups in (4.75):

$$\psi\left(\frac{\tau_w}{\rho u_m^2}, \frac{\mu}{D\rho u_m}, \frac{x}{D}, \frac{\varepsilon}{D}\right) = 0. \qquad (4.76)$$

Since the wall shear stress is the quantity of greatest interest, an obvious rearrangement of (4.76) is:

$$\frac{\tau_w}{\rho u_m^2} = \psi\left(\frac{\mu}{D\rho u_m}, \frac{x}{D}, \frac{\varepsilon}{D}\right) = \psi\left(\frac{\rho u_m D}{\mu}, \frac{x}{D}, \frac{\varepsilon}{D}\right). \qquad (4.77)$$

That is, given experimental values of τ_w, x, D, u_m, ρ, μ, and ε, dimensional analysis indicates that all pipe flow data can be correlated by a series of plots, each for a known value of x/D, of $\tau_w/\rho u_m^2$ versus $\rho u_m D/\mu$, with ε/D as a parameter. This conclusion is completely verified and quantified by experiment. In practice, except near the pipe entrance in laminar flow, x/D is found to be of secondary importance only and is often ignored, in which case the friction factor plot of Fig. 3.10 is obtained.

Table 4.9   Important Dimensionless Numbers,
Being the Ratio of Two Physical Effects, $\mathcal{N}/\mathcal{D}$

Dimensionless Number	Symbol	Typical Formula	$\mathcal{N}$	$\mathcal{D}$
Capillary number	Ca	$\mu u/\sigma$	Viscous forces	Surface forces
Drag coefficient	C_D	$F_D/\frac{1}{2}\rho u_\infty^2 A_p$	Total drag force	Inertial force
Friction factor	f_F	$\tau_w/\frac{1}{2}\rho u_m^2$	Shear force	Inertial force
Froude number	Fr	u^2/gD	Inertial force	Gravitational force
Grashof number	Gr	$\beta g\Delta T L^3 \rho^2/\mu^2$	Buoyancy force	Viscous force
Mach number	M	u/c	Local velocity	Sonic velocity
Nusselt number	Nu	hD/k	Total heat transfer	Conductive transfer
Peclet number	Pe	Du/α	Convective transfer	Conductive transfer
Prandtl number	Pr	$\mu c_p/k$	Kinematic viscosity	Thermal diffusivity
Rayleigh number	Ra	$Gr \times Pr$	—	—
Relative roughness	ε/D	ε/D	Wall roughness	Pipe diameter
Reynolds number	Re	$\rho u D/\mu$	Inertial force	Viscous force
Schmidt number	Sc	$\mu/\rho\mathcal{D}$	Kinematic viscosity	Mass diffusivity
Sherwood number	Sh	$h_d D/\mathcal{D}$	Total mass transfer	Diffusional transfer
Stokes number	St	$\rho g D^2/\mu u$	Gravitational force	Viscous force
Weber number	We	$u^2 \rho D/\sigma$	Inertial forces	Surface forces

Table 4.9 gives the dimensionless groups (including those involved in heat transfer) of greatest interest to the chemical engineer. In almost all cases, each group corresponds to the ratio of two competing effects, $\mathcal{N}$ (numerator) divided by $\mathcal{D}$ (denominator). For example, the Reynolds number is the ratio of an inertial effect, ρu_m^2, to a viscous effect, $\mu u_m/D$.

The following terms are often used in dimensional analysis:

Example 4.4—Thickness of the Laminar Sublayer   **229**

1. *Dynamical similarity*, which indicates equality of the appropriate dimensionless groups in two cases that are being compared. Translated in terms of a Reynolds number, for example, this means that the balance between inertial and viscous forces is the same in the two situations.
2. *Geometrical similarity*, which means that except for size, the geometrical appearance is the same. For example, if the performance of a full-size oceangoing oil tanker were to be predicted by performing tests on a scale model, then— fairly obviously—the model should be that of an oil tanker and not a rowing boat!

Example 4.4—Thickness of the Laminar Sublayer

Fig. E4.4 shows an idealized version of the velocity profile for turbulent flow in a pipe of diameter D. The central turbulent core, which occupies almost all of the cross section, has a uniform (time-averaged) velocity u. Additionally, there is a thin laminar sublayer of thickness δ, adjacent to the wall, in which the velocity builds up linearly from zero at the wall to u at the junction with the turbulent core.

From experiment, the dimensionless shear stress (friction factor) is found to be constant at high Reynolds numbers, independent of the flow rate:

$$\frac{\tau_w}{\frac{1}{2}\rho u^2} = c, \tag{E4.4.1}$$

Determine, in dimensionless terms, how the thickness δ of the laminar sublayer varies with the velocity u.

Fig. E4.4   *Idealized turbulent velocity profile.*

Solution

The wall shear stress is the product of the viscosity and the velocity gradient in the laminar sublayer:

$$\tau_w = \mu \frac{u}{\delta}. \tag{E4.4.2}$$

Elimination of τ_w between Eqns. (E4.4.1) and (E4.4.2) gives:

$$\frac{1}{2}c\rho u^2 = \mu \frac{u}{\delta}. \tag{E4.4.3}$$

By solving for δ and dividing by D, Eqn. (E4.4.3) yields the desired result, as a relation between two dimensionless groups:

$$\frac{\delta}{D} = \frac{2\mu}{c\rho u D} = \frac{2}{c\mathrm{Re}}, \tag{E4.4.4}$$

in which Re is the Reynolds number, $\rho u D / \mu$. Thus, the thickness of the laminar sublayer is inversely proportional to the Reynolds number. That is, as Re increases, the flow in the central portion becomes more turbulent, confining the laminar sublayer to a thinner region next to the wall. Eqn. (E4.4.4) enables δ/D to be estimated: for example, taking $c = f_\mathrm{F} = 0.00325$ and $\mathrm{Re} = 50,000$ as representative values, we find that $\delta/D = 0.0123$.

Finally, note that the present result compares favorably with Eqn. (3.62), which holds if a more sophisticated model (the one-seventh power law) is taken for the velocity profile in the turbulent core:

$$\frac{\delta}{D} \doteq 62 \; Re^{-7/8}. \tag{3.62} \quad \square$$

PROBLEMS FOR CHAPTER 4

Unless otherwise stated, all piping is Schedule 40 commercial steel, and the properties of water are: $\rho = 62.3 \; lb_\mathrm{m}/ft^3$, $\mu = 1.0 \; cP$.

1. *Pumping air and oil—E.* Ideally, what pressure increases (psi) could be expected across centrifugal pumps of 6-in. and 12-in. impeller diameters when pumping air ($\rho = 0.075 \; lb_\mathrm{m}/ft^3$) and oil ($\rho = 50 \; lb_\mathrm{m}/ft^3$)? Four answers are expected. The impellers run at 1,200 rpm.

2. *Pump and pipeline—M.* The head/discharge curve of a centrifugal pump is shown in Fig. P4.2. The exit of the pump is connected to 1,000 ft of nominal 2-in. diameter horizontal pipe ($D = 2.067$ in.). What flow rate (ft^3/s) of water can be expected? Assume atmospheric pressure at the pump inlet and pipe exit, and take $f_\mathrm{F} = 0.00475$.

3. *Pump scale model—E.* A centrifugal pump operating at 1,800 rpm is to be designed to handle a liquid hydrocarbon of specific gravity 0.95. To check its performance, a half-scale model is to be tested, operating at 1,200 rpm, pumping water. The scale model is found to deliver 200 gpm with a head increase of 22.6 ft. Assuming dynamical similarity (equality of the appropriate dimensionless groups), what will be the corresponding flow rate (gpm) and head increase (ft) for the full-size pump?

Fig. P4.2 Pump head/discharge characteristic curve.

4. *Dimensional analysis of pumps—M*. Across a centrifugal pump, the increase in energy per unit mass of liquid is $g\Delta h$, where g is the gravitational acceleration and Δh is the increase in head. This quantity $g\Delta h$ (treat the combination as a single entity) may be a function of the impeller diameter D, the rotational speed N, and liquid density ρ (but not the viscosity), and the flow rate Q. Perform a dimensional analysis from first principles, in which $g\Delta h$ and Q are embodied in two separate dimensionless groups.

A centrifugal pump operating at 1,450 rpm is to be designed to handle a liquid hydrocarbon of s.g. (specific gravity) 0.95. To predict its performance, a half-scale model is to be tested, operating at 725 rpm, pumping a light oil of s.g. 0.90. The scale model is found to deliver 200 gpm with a head increase of 20 ft. Assuming dynamical similarity, what will be the corresponding flow rate and head increase for the full-size pump?

5. *Pumps in series and parallel—M*. Two different series/parallel arrangements of three identical centrifugal pumps are shown in Fig. P4.5. The head increase Δh across a single such pump varies with the flow rate Q through it according to:

$$\Delta h = a - bQ^2.$$

Derive expressions for the head increases $\Delta h_{(a)}$ and $\Delta h_{(b)}$, in terms of a and b and the total flow rate Q, for these two configurations. Sketch your results on a graph, also including the performance curve for the single pump. *Note:* In case (b), it might be possible for the two pumps in series in the one branch to overpower the third pump and cause a *reversal* of flow through it. Such a possibility is prevented by the check valve, which permits only *forward* flow.

(a)   (b)

Fig. P4.5   Series/parallel pump arrangements.

6. *Solar-car performance—E.* A solar car has a frontal area of 12.5 ft^2 and a drag coefficient of 0.106.[9] If the electric motor is delivering 1.2 kW of useful power to the driving wheels, estimate the corresponding maximum speed near Alice Springs in the Australian "outback."

7. *Terminal velocity of hailstones—M.* Occasionally, 1.0-in. diameter hailstones fall in Ann Arbor. What is their terminal velocity in ft/s? Take $C_D = 0.40$ as a first approximation and assume anything else that is reasonable. *Data:* densities (lb$_m$/ft^3): ice, 57.2; air, 0.0765; viscosity of air: 0.0435 lb$_m$/ft hr.

8. *Hot-air balloon emergency—E.* Uninflated, the total mass of a hot-air balloon, including all accessories and the balloonist, is 500 lb$_m$. Inflated, it behaves virtually as a perfect sphere of diameter 80 ft, and rises in air at 50 °F, whose density is 0.0774 lb$_m$/ft^3. Unfortunately, the burner fails, the air in the balloon cools, the balloon loses virtually all of its buoyancy, and soon reaches a steady downward terminal velocity, fortunately still retaining its spherical shape.

The balloonist is yourself. Having taken a fluid mechanics course, you are accustomed to making quick calculations, and are prepared to endure a 10 ft/s crash, but will use a parachute otherwise. You know that for very high Reynolds numbers the drag coefficient is constant at about 0.44. Based on the above, would you use the parachute? Why?

9. *Viscosity determination—M.* This problem relates to finding the viscosity of a liquid by observing the time taken for a small ball bearing to fall steadily through a known vertical distance in the liquid.

---

[9] This problem was inspired by the success of the *Sunrunner* solar-powered car designed and built in 12 months by University of Michigan engineering students. In General Motors *Sunrayce USA* in July 1990 against 31 other university teams, *Sunrunner* placed first in an 11-day race from Epcot Center, Florida, to the General Motors Technical Center in Warren, Michigan. In October 1990, *Sunrunner* was again first among university teams (and third overall among 36 contestants, including professional teams) in the World Solar Challenge race from Darwin to Adelaide. *Sunrunner's* motor was rated at 3 HP, so it could actually deliver more than 1.2 kW if it drew additional electricity from the storage battery.

A stainless steel sphere of diameter $d = 1$ mm and density $\rho_s = 7,870$ kg/m³ falls steadily under gravity through a polymeric fluid of density $\rho_f = 1,052$ kg/m³ and viscosity $\mu = 0.1$ kg/m s (Pa s). What is the downward terminal velocity (cm/s) of the sphere?

10. *Ascending hot-air balloon—M.* A spherical hot-air balloon of diameter 40 ft and deflated mass 500 lb$_\text{m}$ is released from rest in still air at 50 °F. The gas inside the balloon is effectively air at 200 °F. Assuming a constant drag coefficient of $C_D = 0.60$, estimate:

(a) The steady upward terminal velocity of the balloon.

(b) The time in seconds it takes to attain 99% of this velocity.

The density of air (lb$_\text{m}$/ft³) is 0.0774 at 50 °F and 0.0598 at 200 °F. The table of integrals in Appendix A should be helpful.

11. *Baseball travel—M.* The diameter of a baseball is 2.9 in, and its mass is 0.32 lb$_\text{m}$. If the drag coefficient is $C_D = 0.50$, obtain an expression in terms of u^2 for the drag force of the air ($\rho = 0.075$ lb$_\text{m}$/ft³) on the baseball. If it leaves the pitcher on the mound at 100 mph, estimate its velocity by the time it crosses the plate, 60 ft away. *Hint:* Eventually use the relation $u = dx/dt$ to change the standard momentum balance into a differential equation in which the variables are u and x (*not* u and t), before integrating.

12. *Packed bed flow—M.* Outline briefly the justification for supposing that the energy-equation frictional dissipation term for flow with superficial velocity u_0 through a packed bed of length L is of the form:

$$\mathcal{F} = (au_0 + bu_0^2)L, \tag{P4.12.1}$$

in which a and b are constants that depend on the nature of the packing and the properties of the fluid flowing through the bed.

As shown in Fig. P4.12, a bed of ion-exchange resin particles of depth $L = 2$ cm is supported by a metal screen that offers negligible resistance to flow at the bottom of a cylindrical container. Liquid (which is essentially water with $\mu = 1$ cP and $\rho = 1$ g/cm³) flows steadily *down* through the bed. The pressures at both the free surface of the water and at the exit from the bed are both atmospheric.

The following results are obtained for the liquid height H as a function of superficial velocity u_0:

H (cm):	2.5	75.4
u_0 (cm/s):	0.1	1.0

First, obtain the values of the constants a and b for the packed bed. (*Hint:* Perform overall energy balances between the liquid entrance and the packing exit, ignoring any exit kinetic energy effects.) Second, what is the d'Arcy's law permeability (cm²) for the packed bed at *very low* flow rates?

Fig. P4.12   Flow through a packed bed.

Third, a prototype apparatus is to be constructed in which the same type of ion-exchange particles are contained between two metal screens in the form of a hollow cylinder of outer radius 5 cm and inner radius 0.5 cm. What pressure difference (bar) is needed to effect a steady flow rate of 10 cm^3/s of water per cm length of the hollow cylinder? (If needed, assume the flow is from the outside to the inside.) *Hint:* Start from equation (P4.12.1) and obtain a differential equation that gives dp/dr.

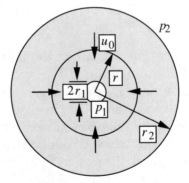

Fig. P4.13   Horizontal cross section of a well.

13. *Performance of a water well—M*. Fig. P4.13 shows the horizontal cross section of a well of radius r_1 in a bed of fine sand that produces water at a volumetric flow rate Q per unit depth and at a pressure p_1. The water flows radially inwards from the outlying region, with symmetry about the axis of the well. A pressure transducer enables the pressure p_2 to be monitored at a radial distance r_2.

Prove that the superficial velocity u_0 radially inwards of the water varies with radial position r according to:

$$u_0 = \frac{Q}{2\pi r}.$$

The following data were obtained for $r_1 = 3$ in. and $r_2 = 300$ in.:

Q (gpm/ft):	100	200
$p_2 - p_1$ (psi):	20	50

Calculate $p_2 - p_1$ for $Q = 300$ gpm/ft. Also, for $Q = 100$ gpm/ft and $p_1 = 0$ psi, what would the pressure be at $r = 6$ in?

Start from the Ergun equation, which has the following appropriate form:

$$\frac{dp}{dr}\frac{d_p}{\rho u_0^2}\frac{\varepsilon^3}{1-\varepsilon} = 1.75 + \frac{150(1-\varepsilon)\mu}{\rho u_o d_p}.$$

14. *Pressure drop in an ion-exchange bed—E.* A horizontal water purification unit consists of a hollow cylinder that is packed with ion-exchange resin particles. Tests with water flowing through the unit gave the following results:

Flow rate Q (gallons/hr):	10.0	20.0
Pressure drop, $-\Delta p$ (psi):	4.0	10.0

If the available pump limits the pressure drop over the unit to a maximum of 54 psi, what is the maximum flow rate of water that can be pumped through it?

15. *Plate-and-frame filter—M.* The following data were obtained for a plate-and-frame filter of total area $A = 500$ cm² operating under a constant pressure drop of $\Delta p = 0.1$ atm:

Time t (min):	50	75	100
Volume of filtrate V (liter):	9.7	12.3	14.0

The filtrate is essentially water. The volume of the cake is one-tenth the volume of the filtrate passed. The resistance of the filter medium may be neglected.

(a) Make an appropriate plot and estimate the permeability κ (darcies) of the cake.
(b) At a certain time t after start-up, the filter is shut down. There follows a cleaning time t_c, in which the accumulated cake is removed, the cloth is cleaned, and the filter is reassembled. This cyclical pattern of productive operation, followed by cleaning, etc., is continued indefinitely. If $t_c = 30$ min, what value of t will maximize the average volume of filtrate produced per unit time?
(c) What is the value (liter/min) of this maximum average volumetric flow rate of filtrate?

16. *Dimensional analysis for ship model—M.* The total force F resisting the motion of a ship or its scale model depends on the density ρ of the liquid, its viscosity μ, the gravitational acceleration g, the length L of the ship, and its velocity u. In what phenomenon that accompanies ship motion is g involved? If tests are to be performed to determine F for several models (each with a different L) of a ship of given design, using several different liquids, show that $F/\rho u^2 L^2$ is one of the appropriate groups for correlating the results. What are the other groups?

In practice, what liquid is likely to be used for the tests? Is it feasible to maintain the values of *all* the dimensionless groups constant between the models and the full-size ship? What would you recommend if the effect of viscosity were of secondary importance? Explain your answers.

17. *Dimensional analysis for pump power—E.* The power P needed to drive a particular type of centrifugal pump depends to a first approximation only on the rotational speed N, the volumetric flow rate Q, the impeller diameter D, and the density ρ of the fluid being pumped, but not its viscosity. What dimensionless groups could be used for correlating P in terms of Q?

18. *Dimensional analysis for disk torque—E.* The torque T required to rotate a disc submerged in a large volume of liquid depends on the liquid density ρ, its viscosity μ, the angular velocity ω, and the disc diameter D. Use dimensional analysis to find dimensionless groups that will serve to correlate experimental data for the torque as a function of viscosity. Choose D, ρ, and ω as primary variables. Which, if any, of the groups is equivalent to a Reynolds number?

19. *Power of automobile—E.* On the I-68 highway just west of Cumberland, MD, there are two inclines, one uphill and the other downhill, both with a 5.5% grade.[10] Your car, which has a mass of 1,800 lb_m, achieves a steady speed of 63 mph when freewheeling downhill, and (under power) can ascend the incline at a steady speed of 57 mph. What is the effective HP of your car?

20. *Drainage ditch capacity—M.* A drainage ditch alongside a highway with a 3% grade has a rectangular cross section of depth 4 ft and width 8 ft, and—to prevent soil erosion—is fully packed with rock fragments of effective diameter 5 in., sphericity 0.8, and porosity 0.42. During a rainstorm, what is the maximum capacity of the ditch (gpm) if the water just reaches the top of the ditch?

21. *Dimensional analysis of centrifugal pumps—M.* For centrifugal pumps of a given design (that is, those that are *geometrically similar*), there exists a functional relationship of the form:

$$\psi(Q, P, \rho, N, D) = 0,$$

---

[10] I-68 opened on August 2, 1991.

where P is the power required to drive the pump (with dimensions ML^2/T^3), Q is the volumetric flow rate (L^3/T), ρ is the density of the fluid being pumped (M/L^3), N is the rotational speed of the impeller (T^{-1}), and D is the impeller diameter (L).

By choosing ρ, N, and D as the primary quantities, we wish to establish two groups, one for Q, and the other for P, that can be used for representing data on all pumps of the given design. Verify that the group for Q is $\Pi_1 = Q/ND^3$, and determine the group Π_2 involving P.

A one-third scale model pump ($D_1 = 0.5$ ft) is to be tested when pumping $Q_1 = 100$ gpm of water ($\rho_1 = 62.4$ lb$_m$/ft^3) in order to predict the performance of a proposed full-size pump ($D_2 = 1.5$ ft.) that is intended to operate at $N_2 = 750$ rpm with a flow rate of $Q_2 = 1,000$ gpm when pumping an oil of density $\rho_2 = 50$ lb$_m$/ft^3.

If dynamical similarity is to be preserved (equality of dimensionless groups):

(a) At what rotational speed N_1 rpm should the scale model be driven?
(b) If under these conditions the scale model needs $P_1 = 1.20$ HP to drive it, what power P_2 will be needed for the full-size pump?

22. *Burning fuel droplet—D.* A droplet of liquid fuel has an initial diameter of D_0. As it burns in air, it loses mass at a rate proportional to its current surface area. If the droplet takes a time t_b to burn completely, prove that its diameter D varies with time according to:

$$D = D_0 \left(1 - \frac{t}{t_b}\right).$$

If the droplet falls in laminar flow under gravity, prove that the distance x it has descended is governed by the differential equation:

$$\frac{d^2x}{dt^2} + \frac{1}{D^2}\left(\frac{18\mu}{\rho} - \frac{3DD_0}{t_b}\right)\frac{dx}{dt} - g = 0,$$

where ρ is the droplet density (much greater than that of air) and μ is the viscosity of the air. (Since D depends on t, this differential equation is fairly complicated, and its solution would most readily be obtained by a numerical method.)

If the droplet is always essentially at its terminal velocity, prove that the distance L it will fall before complete combustion is given by:

$$L = \frac{D_0^2 \rho g t_b}{54\mu}.$$

23. *Particle ejection from fluidized bed—M (C).* Fig. P4.23 shows a particle of mass M that is ejected vertically upward from the surface of a fluidized bed with an initial velocity v_s. The velocity of the fluidizing gas above the bed is v_g, and the resulting drag force on the particle is $D = c(v_g - v)$, where c is a constant and v is the current velocity of the particle.

Fig. P4.23 Departure of particle from the top of a fluidized bed.

Prove that the maximum height h to which the particle can be entrained above the bed is given by:

$$h = \frac{v_g - v_b}{g} \left[v_s + v_b \ln \left(1 - \frac{v_s}{v_b} \right) \right],$$

in which v_b is the steady upward velocity of the particle when the drag and gravitational forces are balanced. *Hint*: You can take either of two approaches to develop the necessary differential equation in v and z:

(a) Perform a conventional momentum balance and then involve the identity $v = dz/dt$.

(b) Realize that a differential decrease in the kinetic energy of the particle equals the work done by the downward forces on the particle as it travels through a differential distance dz.

24. *Plate-and-frame and rotary vacuum filters—D (C).* Filtrate production from a compressible sludge is maintained by using both a plate-and-frame filter and a continuously operating rotary vacuum filter. The plate-and-frame filter produces 50 gpm of filtrate averaged over a day, using a slurry pump with a capacity of 100 gpm fitted with a relief valve to insure that the pressure does not exceed 60 psig. The time needed to clean the filter is 25 min.

The rotary vacuum filter has a diameter of 5 ft and a width of 5 ft, and the internal segments are arranged so that 20% of the total filtering surface is effective at any time. The filter drum rotates at half a revolution each minute, and the filtrate is produced at 25 gpm when the pressure inside the drum equals 8.2 psia, and at 30 gpm when the drum pressure is reduced to 3.2 psia. The slurry trough is at atmospheric pressure.

What is the minimum surface area required for the plate-and-frame filter? Assume that the rate of filtration is given by:

$$\frac{dV}{dt} = \frac{c(\Delta p)^n}{V},$$

where V is the amount of filtrate produced per unit area of filter in time t, Δp is the pressure difference across the filter, and c and n are constants.

25. *Pressure in a centrifugal filter—D.* Consider the filter shown in Fig. 4.18, with the notation and equations given there. By starting from Eqns. (4.41), prove that the pressure in the cake and slurry are given in dimensionless form by:

$$\text{Slurry: } P = R^2 - R_1^2; \qquad \text{Cake: } P = R^2 - 1 + \frac{(1 - R_1^2) \ln R}{\ln R_a}.$$

Here, $P = 2p/(\rho \omega^2 r_2^2)$, $R = r/r_2$, and subscripts 1 and a correspond to radii of r_1 and a, respectively.

For a filter operating at $R_a = 0.8$, plot P versus R on a single graph for $R_1 = 0.7$, 0.75, 0.78, and 0.8. Comment on your findings.

26. *Transient effects in a centrifugal filter—D.* Consider the centrifugal filter shown in Fig. 4.18, with the notation and equations given there.

If the flow rate of filtrate is steady and there is initially no cake, prove that after a time t the inner radius of the slurry is given by:

$$r_1 = \sqrt{r_2^2 - \frac{\mu Q}{\pi \kappa H \rho \omega^2} \ln \frac{r_2}{\sqrt{r_2^2 - (Q \varepsilon t / \pi H)}}}.$$

Hint: Start by using the relations for dp/dr in Eqn. (4.41) to obtain two expressions for the pressure p_a at the slurry/cake interface; then eliminate p_a and obtain an expression for r_1.

A centrifugal filter operates with the following values: $r_2 = 0.5$ m, $\rho = 1,000$ kg/m^3, $\omega = 40\pi$ rad/s, $\kappa = 3.2 \times 10^{-13}$ m^2, $H = 0.5$ m, $Q = 0.005$ m^3/s, $\varepsilon = 0.1$, and $\mu = 0.001$ kg/m s. After five minutes, what are the values of r_1 and a?

Fig. P4.27 Apparatus for studying flow in a porous medium.

27. *Porous medium flow—D (C).* Fig. P4.27 shows an apparatus for studying flow in a porous medium. Fluid of high viscosity μ flows between two parallel plates PP under the influence of a uniform pressure gradient dp/dx. Midway between the plates is a slab S of porous material of void fraction ε and permeability κ. The diagram shows the velocity profile in the fluid, the velocity within the slab being the interstitial velocity:

$$u_i = -\frac{\kappa}{\mu}\frac{dp}{dx}.$$

Explain why the velocity profile is of the form indicated, with particular attention to the boundary conditions at the slab surfaces. Show that the total flow rate Q per unit width is:

$$Q = \frac{1}{\mu}\left(-\frac{dp}{dx}\right)\left[\frac{a^3}{6} + \kappa(a + \varepsilon h)\right].$$

28. *Plate-and-frame filtration—M (C)*. Derive the general filtration equation for a constant-rate period followed by a constant-pressure period:

$$\frac{t - t_r}{V - V_r} = \frac{r}{A\Delta p}\left[\frac{V_c}{2A}(V - V_r) + \frac{V_c V_r}{A} + \frac{c}{r}\right],$$

where t is the total filtration time, t_r is the time of filtration at constant rate, V is the total filtrate volume, V_r is the filtrate volume collected in the constant-rate period, A is the total cross-sectional area of the filtration path, r is the specific resistance of the filter cake, c is the resistance coefficient of the filter cloth, V_c is the volume of filter cake formed per unit volume of filtrate collected, and Δp is the pressure drop across the filter.

A plate-and-frame filter is to be designed to filter 300 m^3 of slurry in each cycle of operation. A test on a small filter of area 0.1 m^2 at a pressure drop of 1 bar gave the following results:

Time (s):	250	500	750	1,000
Volume of filtrate collected (m^3):	0.0906	0.1285	0.1570	0.1815

Assuming negligible filter-cloth resistance, estimate the filtration area required in the full-scale filter when the cycle of operation consists of half an hour at a constant rate of 2×10^{-3} m^3/s m^2, followed by one hour at the pressure attained at the end of the constant-rate period. Also evaluate this pressure.

29. *Centrifugal pump efficiency—M (C)*. A single-stage centrifugal pump has swept-back vanes, which at the outlet make an angle β with the tangent to the outer diameter, whose value is D. The axial width at the outer periphery is b, and the rotational speed is N revolutions per unit time. There is no recovery of kinetic energy in the volute chamber, and the actual whirl velocity is the ideal value. Assuming the flow to be radial at the entry, show that the output head is:

$$H = \frac{u^2 - f^2\text{cosec}^2\beta}{2g},$$

in which $u = \pi DN$ and f is the radial velocity of flow at the exit. Derive an expression for the power input, and show that the maximum efficiency is $1/(1 + \sin\beta)$.

30. *Distillation column flooding—E.* Following the notation of Section 4.7, a bubble-cap distillation column has values $H = 1$ ft, $C_D = 0.62$, $D = 0.1$ ft, $W = 2$ ft, and $\delta = 0.1$ ft. If flooding is just to be avoided, what is the maximum liquid flow rate L (ft^3/s) that can be accommodated? What are then the liquid velocity through the opening at the bottom of a downcomer and the height of the liquid above the weir?

31. *Cyclone separator performance—E.* Following the notation of Section 4.8, a cyclone separator has values $A = 0.2$ ft^2, $H = 2$ ft, $Q = 5$ ft^3/s, $\mu = 0.034$ lb$_m$/ft hr, and $\rho_p = 120$ lb$_m$/ft^3. What is the diameter of the smallest spherical particle that can be separated by the cyclone? Express your answer in both feet and microns (μm).

32. *How fast did that dinosaur go?—M.* Chemical engineers are often expected to be versatile, as in this problem, suggested by an article in the April 1991 issue of *Scientific American*, "How Dinosaurs Ran," by R. McNeill Alexander. Assume that the speed u of a walking, running, or galloping animal depends on its leg length ℓ, its stride length s, and g (this because such locomotion involves some up-and-down motion). Perform standard dimensional analysis to show that the ratio s/ℓ should be a function of the Froude number $u^2/g\ell$.

Based on observations of different quadrupeds from cats to rhinoceroses, and two species of bipeds (humans and kangaroos), the article showed that with some experimental scatter, $\ln(s/\ell) \doteq 0.618 + 0.308 \ln(\mathrm{Fr}) + 0.035\,[\ln(\mathrm{Fr})]^2$.[11] For most animals, the leg length is approximately four times the foot length or diameter.

Fossilized dinosaur tracks in a Texan gully showed footprints with an approximate diameter of 0.95 m and a stride length of 4.5 m. What was the speed of the dinosaur on that day?

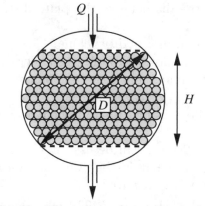

Fig. P4.33   Flow through a spherical reactor.

---

[11] The article also showed a transition from a trotting to a galloping gait at Fr = 2.55, equally true whether the animal was a ferret or a rhinoceros!

33. *Pressure drop in spherical reactor—M.* Fig. P4.33 shows a spherical reactor of internal diameter D that is packed to a height H (symmetrically disposed about the "equator") with spherical catalyst particles of diameter d and void fraction ε. A volumetric flow rate Q of a liquid of density ρ and viscosity μ flows through the packing.

How would you determine the resulting pressure drop? Give sufficient detail so that somebody else could perform all necessary calculations based on your plan.

34. *Pumping into a filter—E.* Fig. P4.34 shows a centrifugal pump with pressure increase $p_2 - p_1 = a - bQ^2$ atm, where a and b are known constants, pumping a slurry into a plate-and-frame filter. The filter has a cross-sectional area A cm^2, cake permeability κ darcies [(cm/s) cP/(atm/cm)], filtrate viscosity μ cP, and cake-to-filtrate volumetric ratio ε. A total volume V cm^3 of filtrate has been passed. The pump inlet and filter exit pressures are equal.

Fig. P4.34 Pump feeding into a filter.

If $a = 0.2$, $b = 10^{-5}$, $A = 100$, $\kappa = 100$, $\mu = 1$, $\varepsilon = 0.1$, and $V = 10^4$, all in units consistent with the above definitions, calculate the current volumetric flow rate Q cm^3/s. Ignore the slight difference between slurry and filtrate volumes.

35. *Equipment encyclopedia—M.* Examine the *Material Balances & Visual Equipment Encyclopedia of Chemical Engineering Equipment CD*, discussed in Section 4.1, to answer the following ten questions:

(a) In the animation of the plate distillation column, how many trays are there?
(b) Above what value of the diameter is the plate distillation column the most cost efficient?
(c) What is the fluid medium in a hydrocyclone?
(d) In a cyclone, does the overflow contain the smaller or the larger particles?
(e) In a filter, what is the maximum ratio of volume of wash liquid to filtrate volume?
(f) In a rotary vacuum filter, approximately what fraction of the drum is submersed in the slurry?
(g) How many shapes of collectors can be used in electrostatic precipitators?
(h) In a packed-bed reactor, what is the two-word phrase for the variation of product concentration in the reactor?
(i) In the centrifugal pump section, how many head/flow rate curves are shown on a single graph?
(j) In a fluidized-bed reactor, what is the typical range in size of the catalyst particles?

36. *Approximation for falling sphere—M.* Referring to Example 4.2, prove for small times that Eqn. (E4.2.7) can be approximated by Eqn. (E4.2.8).

37. *True/false.* Check *true* or *false*, as appropriate:

(a) In a centrifugal pump, the major portion of the pressure increase occurs as the fluid passes through the vanes of the impeller. T □ F □

(b) For most designs of centrifugal pumps, the head increases as the flow rate increases because of the greater kinetic energy. T □ F □

(c) Data for centrifugal pumps of a given design may be correlated by plotting $\Delta p/(\rho D^2 N^2)$ against $Q/(ND^3)$. T □ F □

(d) If two centrifugal pumps are placed in parallel, the overall pressure increase for a given total flow rate Q will be twice what it is for a single pump with the same flow rate Q. T □ F □

(e) Rotational speed N is related to the angular velocity of rotation ω by the equation $N = 2\pi\omega$. T □ F □

(f) The drag coefficient is a dimensionless force per unit area. T □ F □

(g) Flow around a sphere is laminar for a Reynolds number of 1,000. T □ F □

(h) Stokes' law for the drag force on a sphere is $F = 3\pi\mu u_\infty D$. T □ F □

(i) For flow around a sphere, there is a transition region from laminar to turbulent flow in which the value of the drag coefficient is quite uncertain. T □ F □

(j) Consider a sphere settling at its terminal velocity in a fluid. If the following dimensionless group is known to have the value of 10,000:

$$\frac{4}{3}\frac{g\rho_f D^3}{\mu^2}(\rho_s - \rho_f),$$

then the corresponding value of the drag coefficient is approximately 2.5. T □ F □

(k) When a descending sphere has reached its terminal velocity, Stokes' law is always satisfied. T □ F □

(l)   Analysis of the performance of a spray-drying column   T ☐   F ☐
         involves a knowledge of the drag on particles.

(m)   The sphericity of a cube is greater than that of a   T ☐   F ☐
         sphere of the same volume because the cube has more
         surface area.

(n)   Turbulent flow through a packed bed can be modeled   T ☐   F ☐
         as flow through a noncircular duct.

(o)   The Ergun equation applies for both laminar and tur-   T ☐   F ☐
         bulent flow.

(p)   The Burke-Plummer and Blake-Kozeny equations are   T ☐   F ☐
         limiting forms of the Ergun equation, for low and high
         values of the Reynolds number, respectively.

(q)   A tubular reactor is packed with catalyst particles,   T ☐   F ☐
         each of which is a cylinder of diameter 1 cm and length
         2 cm. The effective particle diameter is 0.5 cm.

(r)   A high permeability means that a porous medium   T ☐   F ☐
         offers a high resistance to flow through it.

(s)   For flow through a porous material, the pressure drop   T ☐   F ☐
         is usually proportional to the square of the flow rate.

(t)   After a frictional dissipation term $\mathcal{F}$ has been estab-   T ☐   F ☐
         lished for flow in a packed bed, it may be used in an
         energy balance for flow in either horizontal, vertical,
         or inclined directions, provided the flow rate is not
         changed.

(u)   To convert from darcies to cm^2, multiply by 1.06   T ☐   F ☐
         $\times 10^{-11}$ approximately.

(v)   In constant-pressure operation of a filter, the amount   T ☐   F ☐
         of filtrate passed is directly proportional to the
         elapsed time.

(w)   If a plate-and-frame filter is operated at a constant   T ☐   F ☐
         flow rate, the pressure drop across it is proportional
         to the square root of the elapsed time.

(x)   Fluidized particles behave essentially as though they   T ☐   F ☐
         were a liquid.

(y)   A person who can swim in water could probably also   T ☐   F ☐
         swim in a fluidized bed of sand particles, assuming
         the bed were large enough.

(z) Dynamical similarity means that a scale model *looks* T ☐ F ☐
exactly like a full-size object (pump, ship, etc.), except only for differences in size.

(A) Geometrical similarity means that a scale model (of a T ☐ F ☐
pump, ship, etc.) has exactly the same values of the
dimensionless groups (Reynolds number, for example)
as the corresponding full-size object.

(B) A friction factor is essentially a ratio of shear forces T ☐ F ☐
to viscous forces.

(C) The Stokes number is essentially a ratio of gravita- T ☐ F ☐
tional forces to inertial forces.

(D) The Grashof, Nusselt, Peclet, and Prandtl numbers T ☐ F ☐
are related to heat transfer.

(E) The Froude number is a measure of the ratio of iner- T ☐ F ☐
tial forces to gravitational forces.

(F) The Froude number is important in determining the T ☐ F ☐
drag on ships because it represents the ratio of viscous
forces to gravitational forces.

(G) A plate-and-frame filter has the disadvantage that it T ☐ F ☐
is subject to intermittent operation.

(H) For constant-pressure operation of a plate-and-frame T ☐ F ☐
filter, the volume of filtrate is proportional to the
elapsed time.

(I) Progressing radially outward in a centrifugal filter, T ☐ F ☐
the pressure typically increases in the slurry and decreases in the cake.

(J) There is a definite upper limit on the liquid through- T ☐ F ☐
put that a bubble-cap distillation column can handle.

(K) Between the gas inlet and exit in a cyclone, a forced T ☐ F ☐
vortex exists, in which the tangential velocity v_θ is
proportional to the radial distance from the centerline.

(L) A cyclone can separate particles from a gas more eas- T ☐ F ☐
ily at low flow rates.

PART II

MICROSCOPIC

FLUID MECHANICS

Chapter 5

DIFFERENTIAL EQUATIONS
OF FLUID MECHANICS

5.1 Introduction to Vector Analysis

THE various quantities used in fluid mechanics may be subdivided into the categories of *scalars*, *vectors*, and *tensors*. At any point in space and time, a *scalar* needs only a single number to represent it, examples being temperature, volume, and density. A *vector*, however, needs for its description both a *magnitude* and a *direction*, examples being force, velocity, and momentum; for example, the gravitational force on one kilogram is g newtons, *vertically downward*. A *tensor* is more complicated, and its discussion will be postponed until the shear-stress tensor is introduced in Section 5.7. Appendix C gives details of a few important vector and tensor operations not covered in the present chapter.

Vectors.   A vector, such as **v** shown in Fig. 5.1, is determined by both its magnitude or length, and by its direction. Also shown are the *unit vectors* $\mathbf{e}_x$, $\mathbf{e}_y$, and $\mathbf{e}_z$, which are of length unity and point in each of the respective coordinate directions.

Vectors are usually designated in bold-faced type. In handwriting, however, it is more practical to use lightface forms such as $\vec{v}$, $\overline{v}$, or $\underline{v}$ instead.

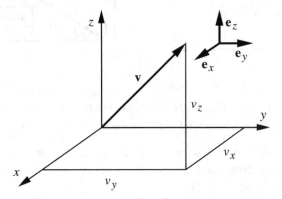

Fig. 5.1   A vector **v***, and the three unit vectors* $\mathbf{e}_x$*,* $\mathbf{e}_y$*, and* $\mathbf{e}_z$ *in the rectangular Cartesian coordinate system.*

It is also essential in most fluid mechanics problems to consider a coordinate system, of which the simplest is the RCCS or *rectangular Cartesian coordinate system*, employing *three* distance coordinates, x, y, and z. In it, a vector such as $\mathbf{v}$ can be expressed as a linear combination of the unit vectors $\mathbf{e}_x$, $\mathbf{e}_y$, and $\mathbf{e}_z$:

$$\mathbf{v} = v_x \mathbf{e}_x + v_y \mathbf{e}_y + v_z \mathbf{e}_z, \tag{5.1}$$

where v_x, v_y, and v_z are the *components* of the vector in the x, y, and z directions, respectively, as shown in Fig. 5.1.

5.2 Vector Operations

Vector addition and subtraction.   If a vector $\mathbf{u}$ is added to another vector $\mathbf{v}$, the result is a vector $\mathbf{w}$ whose components equal the sums of the corresponding components of $\mathbf{u}$ and $\mathbf{v}$:

$$w_x = u_x + v_x, \qquad w_y = u_y + v_y, \qquad w_z = u_z + v_z. \tag{5.2}$$

A similar result holds for subtraction of one vector from another.

Vector products.   Multiplication of a vector $\mathbf{v}$ by a scalar s gives a vector $s\mathbf{v}$ whose components are the same as those of $\mathbf{v}$, except that each is multiplied by s. Two types of *vector multiplication* are particularly important.

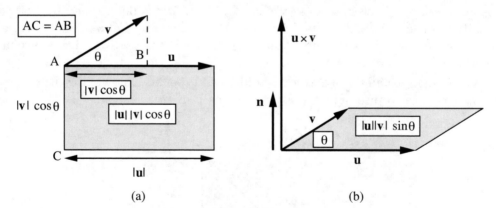

(a)   (b)

Fig. 5.2   Geometrical representations of vector multiplication: (a) dot product; (b) cross product.

1. The *dot product* $\mathbf{u} \cdot \mathbf{v}$ of two vectors $\mathbf{u}$ and $\mathbf{v}$ is a *scalar*, defined as the product of their individual magnitudes $|\mathbf{u}|$ and $|\mathbf{v}|$ and the cosine of the angle θ between the vectors (Fig. 5.2(a)):

$$\mathbf{u} \cdot \mathbf{v} = |\mathbf{u}||\mathbf{v}| \cos \theta, \tag{5.3}$$

which is the area of the rectangle with sides $|\mathbf{u}|$ and $|\mathbf{v}|\cos\theta$. The dot product of two unit vectors $\mathbf{e}_i$ and $\mathbf{e}_j$ may also be expressed concisely as:

$$\mathbf{e}_i \cdot \mathbf{e}_j = \delta_{ij}, \tag{5.4}$$

in which the *Kronecker delta* symbol $\delta_{ij} = 0$ for $i \neq j$ and $\delta_{ii} = 1$ for $i = j$. For example,

$$\mathbf{e}_x \cdot \mathbf{e}_x = 1, \qquad \mathbf{e}_x \cdot \mathbf{e}_y = 0, \qquad \mathbf{e}_x \cdot \mathbf{e}_z = 0. \tag{5.5}$$

With this information, dot products of vectors may be reexpressed in forms involving the individual components, such as:

$$\mathbf{u} \cdot \mathbf{v} = u_x v_x + u_y v_y + u_z v_z. \tag{5.6}$$

2. The *cross product* $\mathbf{u} \times \mathbf{v}$ of two vectors $\mathbf{u}$ and $\mathbf{v}$ is a *vector*, whose *magnitude* is the area of the parallelogram with adjacent sides $\mathbf{u}$ and $\mathbf{v}$, namely, the product of the individual magnitudes and the sine of the angle θ between them. The *direction* of $\mathbf{u} \times \mathbf{v}$ is along the unit vector $\mathbf{n}$, normal to the plane of $\mathbf{u}$ and $\mathbf{v}$, such that $\mathbf{u}$, $\mathbf{v}$, and $\mathbf{n}$ form a right-handed set, as shown in Fig. 5.2(b):

$$\mathbf{u} \times \mathbf{v} = |\mathbf{u}||\mathbf{v}|\mathbf{n}\sin\theta. \tag{5.7}$$

Representative nonzero cross products of the unit vectors are:

$$\mathbf{e}_x \times \mathbf{e}_y = \mathbf{e}_z, \quad \mathbf{e}_y \times \mathbf{e}_z = \mathbf{e}_x, \quad \mathbf{e}_z \times \mathbf{e}_x = \mathbf{e}_y, \quad \mathbf{e}_x \times \mathbf{e}_z = -\mathbf{e}_y, \quad \mathbf{e}_x \times \mathbf{e}_x = \mathbf{0}, \tag{5.8}$$

where $\mathbf{0}$ is the null vector. The cross product may conveniently be reformulated in terms of the individual vector components and also as a determinant:

$$\mathbf{u} \times \mathbf{v} = (u_y v_z - u_z v_y)\mathbf{e}_x + (u_z v_x - u_x v_z)\mathbf{e}_y + (u_x v_y - u_y v_x)\mathbf{e}_z = \begin{vmatrix} \mathbf{e}_x & \mathbf{e}_y & \mathbf{e}_z \\ u_x & u_y & u_z \\ v_x & v_y & v_z \end{vmatrix}. \tag{5.9}$$

The more complicated double-dot and dyadic products are discussed in Sections 11.3 and 11.4.

Vector differentiation. The subsequent development is greatly facilitated by the introduction of the vector differential operator ∇ (*nabla* or *del*), defined in the RCCS as:

$$\nabla = \mathbf{e}_x \frac{\partial}{\partial x} + \mathbf{e}_y \frac{\partial}{\partial y} + \mathbf{e}_z \frac{\partial}{\partial z}, \tag{5.10}$$

in which the unit vectors are weighted according to the corresponding derivatives.

Gradient.   Fig. 5.3 shows that in the RCCS, if a scalar $s = s(x, y, z)$ is a function of position, then a constant value of s, such as $s = s_1$, defines a surface. The *gradient* of a scalar s at a point P, designated grad s (to be shown shortly to be equal to ∇s), is a *directional derivative*. It is defined as a *vector* in the *direction* in which s increases most rapidly with distance, and whose *magnitude* equals that rate of increase.

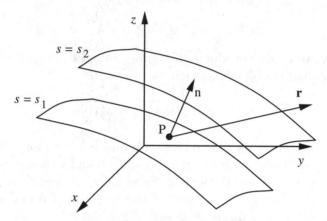

*Fig. 5.3   The direction in which s increases most rapidly with distance is **n**, a unit vector normal to the surface of constant s at point P, and pointing in the direction of increasing s.*

An expression for the gradient is obtained by first considering a unit vector **r** in any direction:

$$\mathbf{r} = \frac{dx}{dr}\mathbf{e}_x + \frac{dy}{dr}\mathbf{e}_y + \frac{dz}{dr}\mathbf{e}_z, \tag{5.11}$$

(whose magnitude can readily be shown to be unity from geometrical considerations), together with the vector ∇s:

$$\nabla s = \frac{\partial s}{\partial x}\mathbf{e}_x + \frac{\partial s}{\partial y}\mathbf{e}_y + \frac{\partial s}{\partial z}\mathbf{e}_z. \tag{5.12}$$

The component of ∇s in the direction of **r** is the dot product:

$$\nabla s \cdot \mathbf{r} = \frac{\partial s}{\partial x}\frac{dx}{dr} + \frac{\partial s}{\partial y}\frac{dy}{dr} + \frac{\partial s}{\partial z}\frac{dz}{dr}. \tag{5.13}$$

Also note that a differential change in s, and hence its rate of change in the **r** direction, are given by:

$$ds = \frac{\partial s}{\partial x}dx + \frac{\partial s}{\partial y}dy + \frac{\partial s}{\partial z}dz, \qquad \frac{ds}{dr} = \frac{\partial s}{\partial x}\frac{dx}{dr} + \frac{\partial s}{\partial y}\frac{dy}{dr} + \frac{\partial s}{\partial z}\frac{dz}{dr}. \tag{5.14}$$

Hence, from Eqns. (5.13) and (5.14):

$$\nabla s \cdot \mathbf{r} = \frac{ds}{dr}. \tag{5.15}$$

When $\mathbf{r}$ coincides with $\mathbf{n}$, the rate of change of s with distance, ds/dr, is greatest, equaling $ds/dn = \nabla s \cdot \mathbf{n}$. But this greatest value can only occur if ∇s is collinear with $\mathbf{n}$ (when $\cos\theta$ in the dot product equals unity and $\nabla s \cdot \mathbf{n} = |\nabla s|$). Therefore, ∇s as defined in Eqn. (5.12) equals grad s, the gradient of s.

Example 5.1—The Gradient of a Scalar[1]

A scalar c varies with position according to:

$$c = x^2 + 4y^2. \tag{E5.1.1}$$

That is, surfaces of constant values of c are ellipses, two of which (for $c = 4$ and $c = 16$) are shown in Fig. E5.1. Evaluate the gradient of c at four representative points, A, B, C, and D on the ellipse for which $c = 16$. The coordinates of these points are $(-4, 0)$, $(4, 0)$, $(2, \sqrt{3})$, and $(0, 2)$.

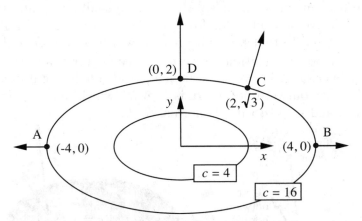

Fig. E5.1   *Gradients of c at four representative points.*

Solution

The derivatives of c with respect to the two coordinate directions, and the corresponding gradient are:

$$\frac{\partial c}{\partial x} = 2x, \qquad \frac{\partial c}{\partial y} = 8y, \qquad \nabla c = 2x\,\mathbf{e}_x + 8y\,\mathbf{e}_y. \tag{E5.1.2}$$

---

[1] For Examples 5.1–5.4, two-dimensional situations have been chosen because: (a) they are easily visualized, and (b) this text is largely confined to no more than two-dimensional problems.

Table E5.1   Gradients ∇c and Magnitudes $|\nabla c|$ at Four Points

Point	x	y	∇c	$|\nabla c|$
A	-4	0	$-8\,\mathbf{e}_x$	8
B	4	0	$8\,\mathbf{e}_x$	8
C	2	$\sqrt{3}$	$4\,\mathbf{e}_x + 8\sqrt{3}\,\mathbf{e}_y$	$4\sqrt{13}$
D	0	2	$16\,\mathbf{e}_y$	16

The individual gradients and their magnitudes at the four points are given in Table E5.1. Observe that each gradient is pointing normally outwards from the ellipse, in the direction of increasing c, and that the magnitude of ∇c is greatest at point D on the y-axis, where the ellipses are spaced most closely. $\square$

Flux.   The *flux* $\mathbf{v}$ of an extensive quantity, X for example, is a vector that denotes the direction and the rate at which X is being transported (by flow, diffusion, conduction, etc.), typically per unit area. Examples are fluxes of mass, momentum, energy, and volume. The last of these is given a special and well-known name, *velocity*, for which typical units are $(\mathrm{m}^3/\mathrm{s})/\mathrm{m}^2$ (volume transported per unit time per unit area) or m/s. Although we shall eventually focus on $\mathbf{v}$ as the *velocity*, the development holds for any other vector flux. If the *total* flux is intended—as opposed to that per unit area—it will be so designated in this text.

Fig. 5.4(a) shows the relation between the vector flux $\mathbf{v}$ and the element of a surface that it is crossing. Note that the magnitude of the total flux of X across the surface dS, with outward normal $\mathbf{n}$, is $\mathbf{v} \cdot \mathbf{n}\, dS$, which reduces to zero if $\mathbf{v}$ is in the plane of dS, and to $|\mathbf{v}|\, dS$ if $\mathbf{v}$ is normal to dS.

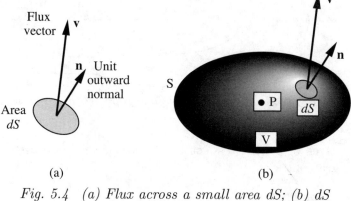

(a)   (b)

*Fig. 5.4   (a) Flux across a small area dS; (b) dS
as part of the boundary surface S of the volume V.*

Divergence.   The *divergence* of a vector $\mathbf{v}$ at a point P, designated div $\mathbf{v}$ (later to be shown equal to $\nabla \cdot \mathbf{v}$), is now introduced. As shown in Fig. 5.4(b), consider a small volume V enveloping P, and designate the corresponding boundary surface as S. The divergence of $\mathbf{v}$ is then defined as:

$$\operatorname{div} \mathbf{v} = \lim_{V \to 0} \frac{1}{V} \iint_S \mathbf{v} \cdot \mathbf{n} \, dS. \tag{5.16}$$

If dS is an element of the surface, $\mathbf{v} \cdot \mathbf{n} \, dS$ can be recognized as the magnitude of the *outward* flux of X across dS. The divergence is therefore the net rate of outflow of X per unit volume, as the volume becomes vanishingly small at P.

An expression for $\operatorname{div} \mathbf{v}$ can be derived in the RCCS, for example, by referring to the differential rectangular parallelepiped element, of volume $dx \, dy \, dz$, shown in Fig. 5.5.

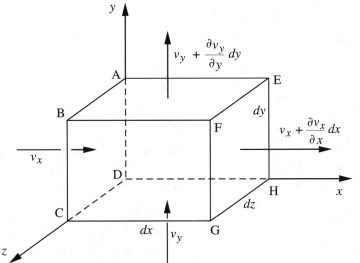

Fig. 5.5   Fluxes across the four faces normal to the x and y axes; those across the other two faces are omitted for clarity.

Consider the rate of transport *into* the element across the face ABCD. Since this face is normal to the x axis, the rate is v_x (the x component of the vector flux) times $dy \, dz$ (the area of the face):

$$v_x \, dy \, dz. \tag{5.17}$$

Likewise, the rate of transport *out of* the element across face EFGH is:

$$\left(v_x + \frac{\partial v_x}{\partial x} \, dx \right) dy \, dz, \tag{5.18}$$

because the rate v_x is now augmented by the rate $\partial v_x / \partial x$ at which it changes in the x direction, times the small increment dx. The net rate of *outflow* across these two faces is obtained by subtracting Eqn. (5.17) from Eqn. (5.18):

$$\left(v_x + \frac{\partial v_x}{\partial x} \, dx \right) dy \, dz - v_x \, dy \, dz = \left(\frac{\partial v_x}{\partial x} \, dx \right) dy \, dz. \tag{5.19}$$

Similar arguments hold for the fluxes across the other four faces, so that the total *outflow* rate for all *six* faces is:

$$\left(\frac{\partial v_x}{\partial x}\,dx\right)dy\,dz + \left(\frac{\partial v_y}{\partial y}\,dy\right)dz\,dx + \left(\frac{\partial v_z}{\partial z}\,dz\right)dx\,dy. \tag{5.20}$$

Dividing by the element volume $dx\,dy\,dz$, and using the definition of Eqn. (5.16):

$$\operatorname{div}\mathbf{v} = \frac{\partial v_x}{\partial x} + \frac{\partial v_y}{\partial y} + \frac{\partial v_z}{\partial z}. \tag{5.21}$$

The alternative notation $\nabla \cdot \mathbf{v}$ (which will be used hereafter) is now easy to comprehend:

$$\begin{aligned}
\nabla \cdot \mathbf{v} &= \left(\mathbf{e}_x\frac{\partial}{\partial x} + \mathbf{e}_y\frac{\partial}{\partial y} + \mathbf{e}_z\frac{\partial}{\partial z}\right) \cdot (v_x\mathbf{e}_x + v_y\mathbf{e}_y + v_z\mathbf{e}_z) \\
&= \frac{\partial v_x}{\partial x} + \frac{\partial v_y}{\partial y} + \frac{\partial v_z}{\partial z} = \operatorname{div}\mathbf{v}.
\end{aligned} \tag{5.22}$$

An example of a vector flux is $\mathbf{q}$, the rate (and direction) at which heat is conducted in a medium because of gradients of the temperature T. In this case, experiments show that, at the simplest, Fourier's law is obeyed:

$$\mathbf{q} = -k\nabla T, \tag{5.23}$$

in which k is the *thermal conductivity* of the medium. The minus sign indicates that the heat is flowing in the direction of *decreasing* temperature.

We turn now to the most commonly occurring case in fluid mechanics, in which $\mathbf{v}$ is indeed the velocity—that is, the flux of volume. For an incompressible fluid, the density is constant; hence, there cannot be any net outflow (or inflow) of volume from an infinitesimally small fluid element. If $\nabla \cdot \mathbf{v} = 0$, which is true for an incompressible fluid, there is no depletion or accumulation of volume, and the flow is called *solenoidal*. However, positive (or negative) values of the divergence imply such a depletion (or accumulation). For example, a positive value of $\nabla \cdot \mathbf{v}$ could occur for a compressible gas whose density is decreasing, caused by an outflow of volume from an infinitesimally small element, with a corresponding diminution of pressure in that element.

The following identity holds for any scalar c and any vector $\mathbf{v}$, and will occasionally be useful:

$$\nabla \cdot c\mathbf{v} = c\nabla \cdot \mathbf{v} + \mathbf{v} \cdot \nabla c. \tag{5.24}$$

For the particular case in which $c = \rho$ (the fluid density), a physical interpretation of Eqn. (5.24) is given in Example 5.6.

Example 5.3—An Alternative to the Differential Element   **257**

Example 5.2—The Divergence of a Vector

Evaluate the divergence of the vector $\mathbf{v} = v_x\,\mathbf{e}_x + v_y\,\mathbf{e}_y$, which represents the velocity of a traveling wave (see Section 7.10), and whose components are:

$$v_x = -ce^{ky}\sin(kx - \omega t), \qquad v_y = ce^{ky}\cos(kx - \omega t). \qquad \text{(E5.2.1)}$$

Solution

The divergence is obtained by summing the appropriate derivatives:

$$\nabla \cdot \mathbf{v} = \frac{\partial v_x}{\partial x} + \frac{\partial v_y}{\partial y} = \frac{\partial}{\partial x}\left[-ce^{ky}\sin(kx - \omega t)\right] + \frac{\partial}{\partial y}\left[ce^{ky}\cos(kx - \omega t)\right]$$
$$= -cke^{ky}\cos(kx - \omega t) + cke^{ky}\cos(kx - \omega t) = 0. \qquad \text{(E5.2.2)}$$

In common with most problems studied in this book, the flow is incompressible and $\nabla \cdot \mathbf{v} = 0$. Note that although time t was one of the three independent variables $(x, y, \text{ and } t)$, it did not enter the determination of the divergence. $\square$

Example 5.3—An Alternative to the Differential Element

When performing operations similar to the above, some people prefer first to consider a *finite* volume, such as one of dimensions $\Delta x \times \Delta y \times \Delta z$, and then let this shrink to one of differential size. Investigate this alternative.

Solution

The net rate of *outflow* across the faces normal to the x axis is now given by the difference of the velocities at locations x and $x + \Delta x$, multiplied by the area $\Delta y \Delta z$ across which the transports occur:

$$(v_x|_{x+\Delta x} - v_x|_x)\,\Delta y \Delta z. \qquad \text{(E5.3.1)}$$

By following similar arguments for the outflow across the other four faces, the total outflow rate for all six faces is:

$$(v_x|_{x+\Delta x} - v_x|_x)\,\Delta y \Delta z + (v_y|_{y+\Delta y} - v_y|_y)\,\Delta z \Delta x + (v_z|_{z+\Delta z} - v_z|_z)\,\Delta x \Delta y. \quad \text{(E5.3.2)}$$

Division by the volume $\Delta x \Delta y \Delta z$ then gives the rate of outflow per unit volume:

$$\frac{v_x|_{x+\Delta x} - v_x|_x}{\Delta x} + \frac{v_y|_{y+\Delta y} - v_y|_y}{\Delta y} + \frac{v_z|_{z+\Delta z} - v_z|_z}{\Delta z}. \qquad \text{(E5.3.3)}$$

Each dimension is now allowed to shrink to a *differential* value. By definition of the derivative:

$$\lim_{\Delta x \to 0} \frac{v_x|_{x+\Delta x} - v_x|_x}{\Delta x} = \frac{\partial v_x}{\partial x}. \qquad \text{(E5.3.4)}$$

Likewise, the second and third terms in Eqn. (E5.3.3) yield $\partial v_y/\partial y$ and $\partial v_z/\partial z$, so that the total rate of outflow per unit volume, which by definition is the divergence of the velocity, becomes identical to that obtained previously:

$$\nabla \cdot \mathbf{v} = \frac{\partial v_x}{\partial x} + \frac{\partial v_y}{\partial y} + \frac{\partial v_z}{\partial z}. \qquad (E5.3.5)$$

The preference of whether to start with a differential element, or with a finite element that is allowed to shrink to a differential one and then invoke the definition of a derivative, is a personal one. We have opted for the former (and have therefore generally followed this approach throughout), but recognize that the alternative may be preferred by others. The alternative approach may be adopted when performing both mass balances and momentum balances, the latter of which is discussed in Section 5.6. □

Angular velocity. Fig. 5.6 shows a fluid (or solid) particle P that rotates in a circular path with angular velocity $\boldsymbol{\omega}$ about the axis OQ. Note that the angular velocity (radians per unit time) is a vector whose direction is the same as the axis of rotation. Let $\mathbf{r}$ be the vector OP that locates P relative to an origin O *anywhere* on the axis of rotation.

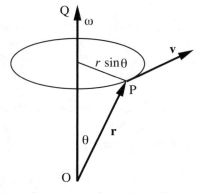

Fig. 5.6 Rotation of P about the axis OQ.

If ω and r denote the magnitudes of $\boldsymbol{\omega}$ and $\mathbf{r}$, respectively, then the *magnitude* of the *velocity* of P is ω times $r \sin \theta$ (the distance of P from the axis). The *direction* of the velocity is such that $\boldsymbol{\omega}$, $\mathbf{r}$, and $\mathbf{v}$ form a right-handed set. Based on this information, the reader may wish to check that the linear velocity of P is the vector $\boldsymbol{\omega} \times \mathbf{r}$, which can also be expressed as a determinant:

$$\mathbf{v} = \boldsymbol{\omega} \times \mathbf{r} = \begin{vmatrix} \mathbf{e}_x & \mathbf{e}_y & \mathbf{e}_z \\ \omega_x & \omega_y & \omega_z \\ r_x & r_y & r_z \end{vmatrix}. \qquad (5.25)$$

This result is *independent* of the location of the origin O, as long as it lies on the axis of rotation.

Curl.   Refer first to the area A in Fig. 5.7(a), which has a positive normal direction **n** and boundary curve C. The *circulation* $\mathcal{C}$ in the direction **n** of a vector **v** at a point P is defined as:

$$\mathcal{C} = \lim_{A \to 0} \frac{1}{A} \int_C \mathbf{v} \cdot \mathbf{ds}, \tag{5.26}$$

whose value will, in general, depend on the orientation of A and hence on the direction of its normal **n**. Observe that the path taken is *clockwise* as seen by an observer facing in the direction of positive **n**, and that this is also the direction of a vector element d**s** of the boundary C. The circulation is important because if it is zero for all orientations—approximately the case for low-viscosity flows—then the flow pattern may be amenable to the simpler treatment discussed in Chapter 7.

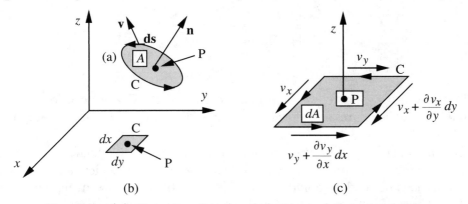

Fig. 5.7   (a) Notation for the definition of the circulation of a vector **v**; *(b) a small element* $dx\,dy$ *for investigating the z component of curl* **v**; *(c) the same, magnified, showing velocity components for integration around the boundary C.*

The *curl* of a vector **v** at a point P, designated curl **v** (later to be shown equal to $\nabla \times \mathbf{v}$), is a *vector*, whose three components are equal to the circulation for each of the x, y, and z directions. When curl **v** is resolved in any arbitrary direction **n**, the resulting component (curl **v**) $\cdot$ **n** gives the circulation in that direction.

First, investigate the z component of curl **v**, defined as:

$$\mathrm{curl}_z \mathbf{v} = \lim_{A \to 0} \frac{1}{A} \int_C \mathbf{v} \cdot \mathbf{ds}. \tag{5.27}$$

Here, A is an area in the x/y plane that includes point P, as in Fig. 5.7 (b) and (c), and C is its boundary; d**s** denotes a (vector) element of C, considered clockwise when looking at the plane in the z direction (from below). The definition actually holds for any shape of area that shrinks to zero, but which—for our purposes—is most conveniently taken as a rectangle of area $dxdy$, as in Fig. 5.7.

By performing the integration around C, and noting that the appropriate four magnitudes of $d\mathbf{s}$ are dy, $-dx$, $-dy$, and dx, Eqn. (5.27) gives:

$$
\begin{aligned}
\text{curl}_z\mathbf{v} &= \frac{1}{dx\,dy}\left[\left(v_y + \frac{\partial v_y}{\partial x}\,dx\right)dy - \left(v_x + \frac{\partial v_x}{\partial y}\,dy\right)dx - v_y\,dy + v_x\,dx\right] \\
&= \frac{\partial v_y}{\partial x} - \frac{\partial v_x}{\partial y}.
\end{aligned}
\tag{5.28}
$$

After obtaining the x and y components similarly, curl $\mathbf{v}$ is then:

$$
\begin{aligned}
\text{curl}\,\mathbf{v} &= \left(\frac{\partial v_z}{\partial y} - \frac{\partial v_y}{\partial z}\right)\mathbf{e}_x + \left(\frac{\partial v_x}{\partial z} - \frac{\partial v_z}{\partial x}\right)\mathbf{e}_y + \left(\frac{\partial v_y}{\partial x} - \frac{\partial v_x}{\partial y}\right)\mathbf{e}_z \\
&= \begin{vmatrix} \mathbf{e}_x & \mathbf{e}_y & \mathbf{e}_z \\ \dfrac{\partial}{\partial x} & \dfrac{\partial}{\partial y} & \dfrac{\partial}{\partial z} \\ v_x & v_y & v_z \end{vmatrix} = \nabla \times \mathbf{v}.
\end{aligned}
\tag{5.29}
$$

Having proved its validity, we shall henceforth use the $\nabla \times \mathbf{v}$ notation for curl $\mathbf{v}$. Also note that the expressions for $\nabla \times \mathbf{v}$ in the CCS and SCS are more complicated, being given in Table 5.4.

For the special—and most commonly occurring case—in which $\mathbf{v}$ is the velocity vector, we can now deduce the physical meaning of $\nabla \times \mathbf{v}$, which is also known as the *vorticity*. The reader should discover (see Problem 5.2) by using Eqns. (5.25) and (5.29) that, for a constant angular velocity, as in rigid-body rotation:

$$
\begin{aligned}
\nabla \times \mathbf{v} &= \left(\mathbf{e}_x\frac{\partial}{\partial x} + \mathbf{e}_y\frac{\partial}{\partial y} + \mathbf{e}_z\frac{\partial}{\partial z}\right) \times (\boldsymbol{\omega} \times \mathbf{r}) \\
&= 2\left(\omega_x\mathbf{e}_x + \omega_y\mathbf{e}_y + \omega_z\mathbf{e}_z\right) = 2\boldsymbol{\omega}.
\end{aligned}
\tag{5.30}
$$

That is, $\nabla \times \mathbf{v}$ is *twice* the angular velocity of the fluid at a point.

Irrotationality.   Consider the special situation in which the fluid velocity $\mathbf{v}$ (the volumetric flux per unit area) is given by the gradient of a scalar ϕ:

$$
\mathbf{v} = \nabla\phi,
\tag{5.31}
$$

in much the same way that the conductive heat flux $\mathbf{q}$ per unit area is given by Fourier's law, $\mathbf{q} = -k\nabla T$, where k is the thermal conductivity and T is the temperature. It is fairly easy to show (see Problem 5.3) that, provided $\mathbf{v} = \nabla\phi$, the curl of the velocity is everywhere zero:

$$
\nabla \times \mathbf{v} = 0,
\tag{5.32}
$$

so from Eqn. (5.30) there is no angular velocity. In such event, the flow is termed *irrotational* or *potential*, and the scalar function ϕ is called the *velocity potential*. Such flows arise approximately at relatively high Reynolds numbers, when the effects of viscosity are small.

In the limiting case of zero viscosity, there would be no shear stresses, so an element of flowing fluid initially possessing zero angular velocity could never start rotating. An *approximate* physical example is given by a cup of coffee, containing cream, so that its surface can be visualized. If the cup is rotated—through half a revolution, for example—there is almost no corresponding rotation of the coffee, because it has a relatively small viscosity. However, if the cup is *continuously* rotated, the effect of viscosity will eventually cause all the liquid to rotate with the same angular velocity as the cup.

In irrotational flow, a fluid element can still *deform*, but cannot rotate. In two dimensions, for example, a small square element could deform into a rhombus, but the diagonals of the latter, even though one would be elongated and the other shortened, would still maintain the same *directions* as those of the square element. Such a situation is shown in Fig. 5.8.

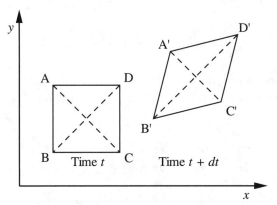

Fig. 5.8   A fluid element that moves and deforms, but does not rotate.

Further, for the usual case of an incompressible fluid:

$$\nabla \cdot \mathbf{v} = 0, \tag{5.33}$$

Substitution of $\mathbf{v}$ from Eqn. (5.31) immediately leads to:

$$\nabla^2 \phi = \nabla \cdot \nabla \phi = \left(\mathbf{e}_x \frac{\partial}{\partial x} + \mathbf{e}_y \frac{\partial}{\partial y} + \mathbf{e}_z \frac{\partial}{\partial z} \right) \cdot \left(\mathbf{e}_x \frac{\partial \phi}{\partial x} + \mathbf{e}_y \frac{\partial \phi}{\partial y} + \mathbf{e}_z \frac{\partial \phi}{\partial z} \right)$$

$$= \frac{\partial^2 \phi}{\partial x^2} + \frac{\partial^2 \phi}{\partial y^2} + \frac{\partial^2 \phi}{\partial z^2} = 0, \tag{5.34}$$

which is the famous equation of *Laplace*, governing the velocity potential in the RCCS; ∇^2 is called the *Laplacian* operator. As seen in Chapter 7, Laplace's equation is very important in the study of irrotational flow.

Example 5.4—The Curl of a Vector

Evaluate the curl of the velocity vector $\mathbf{v} = v_x \, \mathbf{e}_x + v_y \, \mathbf{e}_y$ for the following two cases ($v_z = 0$ in both instances):

$$\text{(a)} \ \ v_x = -\omega y, \quad v_y = \omega x; \qquad \text{(b)} \ \ v_x = -\frac{cy}{x^2 + y^2}, \quad v_y = \frac{cx}{x^2 + y^2}. \qquad \text{(E5.4.1)}$$

Solution

In both cases, $v_z = 0$ and the remaining components v_x and v_y do not depend on the coordinate z; therefore, the x and y components of $\nabla \times \mathbf{v}$ (which would multiply the unit vectors $\mathbf{e}_x$ and $\mathbf{e}_y$) are—from Eqn. (5.29)—both zero. Thus, we are concerned only with the z components, which multiply the unit vector $\mathbf{e}_z$.

Case (a)

$$\text{curl}_z \mathbf{v} = \frac{\partial v_y}{\partial x} - \frac{\partial v_x}{\partial y} = \frac{\partial(\omega x)}{\partial x} - \frac{\partial(-\omega y)}{\partial y} = \omega - (-\omega) = 2\omega. \qquad \text{(E5.4.2)}$$

Thus, the flow represented by the velocity has a vorticity $2\omega \mathbf{e}_z$ and an angular velocity $\omega \mathbf{e}_z$. It corresponds to a *forced vortex*, discussed further in Section 7.2.

Case (b)

$$\text{curl}_z \mathbf{v} = \frac{\partial v_y}{\partial x} - \frac{\partial v_x}{\partial y} = \frac{\partial}{\partial x}\left(\frac{cx}{x^2 + y^2}\right) - \frac{\partial}{\partial y}\left(-\frac{cy}{x^2 + y^2}\right)$$

$$= \frac{c(y^2 - x^2)}{(x^2 + y^2)^2} - \frac{c(y^2 - x^2)}{(x^2 + y^2)^2} = 0. \qquad \text{(E5.4.3)}$$

Thus, the flow represented by the velocity has zero vorticity and zero angular velocity. It corresponds to a *free vortex*, to be discussed further in Section 7.2.

As can be discovered from Problem 5.12, the above two cases may be examined somewhat more easily in cylindrical coordinates, which is one of the topics of the next section. $\qquad \square$

Example 5.5—The Laplacian of a Scalar

Evaluate the Laplacian of the following scalar:

$$\phi = ce^{ky} \cos(kx - \omega t). \qquad \text{(E5.5.1)}$$

Solution

The necessary derivatives for forming the Laplacian, $\nabla^2\phi$, are:

$$\frac{\partial\phi}{\partial x} = -cke^{ky}\sin(kx - \omega t), \qquad \frac{\partial^2\phi}{\partial x^2} = -ck^2e^{ky}\cos(kx - \omega t),$$

$$\frac{\partial\phi}{\partial y} = cke^{ky}\cos(kx - \omega t), \qquad \frac{\partial^2\phi}{\partial y^2} = ck^2e^{ky}\cos(kx - \omega t). \qquad (E5.5.2)$$

The Laplacian of ϕ is therefore:

$$\frac{\partial^2\phi}{\partial x^2} + \frac{\partial^2\phi}{\partial y^2} = -ck^2e^{ky}\cos(kx - \omega t) + ck^2e^{ky}\cos(kx - \omega t) = 0. \qquad (E5.5.3)$$

The scalar ϕ is closely related to the potential function for an irrotational wave traveling on the surface of deep water, to be discussed further in Section 7.11. Note that although time t is one of the three independent variables, it is not involved in the formation of the Laplacian. $\qquad\square$

5.3 Other Coordinate Systems

The coordinate systems most frequently used in fluid mechanics problems, which should already be somewhat familiar, are:

1. The *rectangular Cartesian coordinate system* (RCCS), already discussed.
2. The *cylindrical coordinate system* (CCS).
3. The *spherical coordinate system* (SCS).

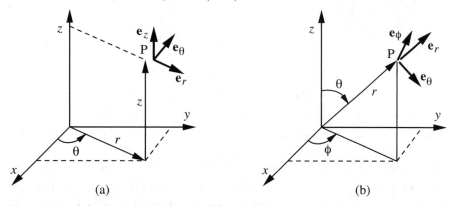

Fig. 5.9 (a) Cylindrical and (b) spherical coordinate systems (note that in some texts, θ and ϕ are interchanged from those shown above).

The cylindrical and spherical systems are shown in Fig. 5.9. Note that the RCCS is superimposed in both cases, since we sometimes need to convert from one system to another. As far as possible, the system chosen should be most suited to the geometry of the problem at hand. As an obvious illustration, the CCS is employed when studying flow in a pipe, because the pipe wall is described by a constant value of the radial coordinate ($r = a$, for example).

Cylindrical coordinate system. To locate the point P, the CCS employs *two* distance coordinates, r and z, and *one* angle, θ (restricted in range from 0 to 2π). In the CCS, a vector such as **v** can be expressed in terms of the unit vectors $\mathbf{e}_r$, $\mathbf{e}_\theta$, and $\mathbf{e}_z$, by:

$$\mathbf{v} = v_r\mathbf{e}_r + v_\theta\mathbf{e}_\theta + v_z\mathbf{e}_z, \tag{5.35}$$

where v_r, v_θ, and v_z are the components of the vector in the r, θ, and z coordinate directions, respectively. The relations between the RCCS and CCS are:

$$x = r\cos\theta, \qquad y = r\sin\theta, \qquad z = z,$$
$$r = \sqrt{x^2 + y^2}, \qquad \theta = \tan^{-1}\frac{y}{x}, \qquad z = z. \tag{5.36}$$

$$\mathbf{e}_r = \cos\theta\,\mathbf{e}_x + \sin\theta\,\mathbf{e}_y, \quad \mathbf{e}_\theta = -\sin\theta\,\mathbf{e}_x + \cos\theta\,\mathbf{e}_y, \quad \mathbf{e}_z = \mathbf{e}_z. \tag{5.37}$$

Spherical coordinate system. To locate the point P, the SCS employs *one* distance coordinate, r, and *two* angles, θ (ranging from 0 to π) and ϕ (0 to 2π). In the SCS, a vector such as **v** can be expressed in terms of the unit vectors $\mathbf{e}_r$, $\mathbf{e}_\theta$, and $\mathbf{e}_\phi$:

$$\mathbf{v} = v_r\mathbf{e}_r + v_\theta\mathbf{e}_\theta + v_\phi\mathbf{e}_\phi, \tag{5.38}$$

where v_r, v_θ, and v_ϕ are the components of the vector in the r, θ, and ϕ coordinate directions, respectively. The relations between the RCCS and SCS are:

$$x = r\sin\theta\cos\phi, \qquad y = r\sin\theta\sin\phi, \qquad z = r\cos\theta,$$
$$r = \sqrt{x^2 + y^2 + z^2}, \qquad \theta = \tan^{-1}\frac{\sqrt{x^2 + y^2}}{z}, \qquad \phi = \tan^{-1}\frac{y}{x}. \tag{5.39}$$

$$\mathbf{e}_r = \sin\theta\cos\phi\,\mathbf{e}_x + \sin\theta\sin\phi\,\mathbf{e}_y + \cos\theta\,\mathbf{e}_z,$$
$$\mathbf{e}_\theta = \cos\theta\cos\phi\,\mathbf{e}_x + \cos\theta\sin\phi\,\mathbf{e}_y - \sin\theta\,\mathbf{e}_z, \tag{5.40}$$
$$\mathbf{e}_\phi = -\sin\phi\,\mathbf{e}_x + \cos\phi\,\mathbf{e}_y.$$

The form of the del operator in the CCS and SCS depends on the context in which it is being used, and "del" is therefore not a particularly useful concept in cylindrical and spherical coordinates. However, note that ∇s, $\nabla \cdot \mathbf{v}$, $\nabla \times \mathbf{v}$, and $\nabla^2 s$ still remain as *convenient notations* for the gradient, divergence, curl, and Laplacian in all coordinate systems.

Expressions for the gradient, divergence, curl, and Laplacian in the CCS and SCS can be derived from the corresponding expressions in the RCCS. The *derivations*, which involve coordinate transformations, are laborious and beyond the main theme of this section. One of the difficulties that arises is with the unit vectors; whereas those for the RCCS are invariant, the directions of some of these unit vectors in the CCS and SCS are functions of the coordinates themselves, necessitating considerable care in differentiation. The reader will need some of the *results*, however, which are given (with those for the RCCS) in Tables 5.1–5.5.

Table 5.1   Expressions for the Del Operator

Rectangular: $\quad \nabla = \mathbf{e}_x \dfrac{\partial}{\partial x} + \mathbf{e}_y \dfrac{\partial}{\partial y} + \mathbf{e}_z \dfrac{\partial}{\partial z}.$

Cylindrical: $\quad$ Form depends on usage and will not be pursued.

Spherical: $\quad$ Form depends on usage and will not be pursued.

Table 5.2   Expressions for the Gradient of a Scalar

Rectangular: $\quad \nabla s = \dfrac{\partial s}{\partial x} \mathbf{e}_x + \dfrac{\partial s}{\partial y} \mathbf{e}_y + \dfrac{\partial s}{\partial z} \mathbf{e}_z.$

Cylindrical: $\quad \nabla s = \dfrac{\partial s}{\partial r} \mathbf{e}_r + \dfrac{1}{r}\dfrac{\partial s}{\partial \theta} \mathbf{e}_\theta + \dfrac{\partial s}{\partial z} \mathbf{e}_z.$

Spherical: $\quad \nabla s = \dfrac{\partial s}{\partial r} \mathbf{e}_r + \dfrac{1}{r}\dfrac{\partial s}{\partial \theta} \mathbf{e}_\theta + \dfrac{1}{r \sin \theta}\dfrac{\partial s}{\partial \phi} \mathbf{e}_\phi.$

Table 5.3   Expressions for the Divergence of a Vector

Rectangular: $\quad \nabla \cdot \mathbf{v} = \dfrac{\partial v_x}{\partial x} + \dfrac{\partial v_y}{\partial y} + \dfrac{\partial v_z}{\partial z}.$

Cylindrical: $\quad \nabla \cdot \mathbf{v} = \dfrac{1}{r}\dfrac{\partial (r v_r)}{\partial r} + \dfrac{1}{r}\dfrac{\partial v_\theta}{\partial \theta} + \dfrac{\partial v_z}{\partial z}.$

Spherical: $\quad \nabla \cdot \mathbf{v} = \dfrac{1}{r^2}\dfrac{\partial (r^2 v_r)}{\partial r} + \dfrac{1}{r \sin \theta}\dfrac{\partial (v_\theta \sin \theta)}{\partial \theta} + \dfrac{1}{r \sin \theta}\dfrac{\partial v_\phi}{\partial \phi}.$

Table 5.4   Expressions for the Curl of a Vector

Rectangular: $\nabla \times \mathbf{v} = \left(\dfrac{\partial v_z}{\partial y} - \dfrac{\partial v_y}{\partial z} \right) \mathbf{e}_x + \left(\dfrac{\partial v_x}{\partial z} - \dfrac{\partial v_z}{\partial x} \right) \mathbf{e}_y + \left(\dfrac{\partial v_y}{\partial x} - \dfrac{\partial v_x}{\partial y} \right) \mathbf{e}_z.$

Cylindrical: $\nabla \times \mathbf{v} = \left(\dfrac{1}{r}\dfrac{\partial v_z}{\partial \theta} - \dfrac{\partial v_\theta}{\partial z} \right) \mathbf{e}_r + \left(\dfrac{\partial v_r}{\partial z} - \dfrac{\partial v_z}{\partial r} \right) \mathbf{e}_\theta$

$$+ \frac{1}{r} \left(\frac{\partial (r v_\theta)}{\partial r} - \frac{\partial v_r}{\partial \theta} \right) \mathbf{e}_z.$$

Spherical: $\nabla \times \mathbf{v} = \dfrac{1}{r \sin \theta} \left(\dfrac{\partial (v_\phi \sin \theta)}{\partial \theta} - \dfrac{\partial v_\theta}{\partial \phi} \right) \mathbf{e}_r$

$$+ \left(\frac{1}{r \sin \theta} \frac{\partial v_r}{\partial \phi} - \frac{1}{r} \frac{\partial (r v_\phi)}{\partial r} \right) \mathbf{e}_\theta + \frac{1}{r} \left(\frac{\partial (r v_\theta)}{\partial r} - \frac{\partial v_r}{\partial \theta} \right) \mathbf{e}_\phi.$$

Table 5.5   Expressions for the Laplacian of a Scalar

Rectangular: $\nabla^2 s = \dfrac{\partial^2 s}{\partial x^2} + \dfrac{\partial^2 s}{\partial y^2} + \dfrac{\partial^2 s}{\partial z^2}.$

Cylindrical: $\nabla^2 s = \dfrac{1}{r}\dfrac{\partial}{\partial r}\left(r \dfrac{\partial s}{\partial r} \right) + \dfrac{1}{r^2}\dfrac{\partial^2 s}{\partial \theta^2} + \dfrac{\partial^2 s}{\partial z^2}.$

Spherical: $\nabla^2 s = \dfrac{1}{r^2}\dfrac{\partial}{\partial r}\left(r^2 \dfrac{\partial s}{\partial r} \right) + \dfrac{1}{r^2 \sin \theta}\dfrac{\partial}{\partial \theta}\left(\sin \theta \dfrac{\partial s}{\partial \theta} \right) + \dfrac{1}{r^2 \sin^2 \theta}\dfrac{\partial^2 s}{\partial \phi^2}.$

5.4   The Convective Derivative

The development and understanding of the differential equations of fluid mechanics are facilitated by the introduction of a new type of derivative that follows the *Lagrangian* viewpoint, and is symbolized for a property or variable X by the notation:

$$\frac{\mathcal{D}X}{\mathcal{D}t}. \tag{5.41}$$

This derivative is defined as the rate of change of X with respect to time, *following the path taken by the fluid*, and is known as the *convective* or *substantial* derivative.

To illustrate, focus on something that will be of immediate use in Section 5.5— the convective derivative of the density, $\mathcal{D}\rho/\mathcal{D}t$. In general, density will depend on both time and position; that is, $\rho = \rho(t, x, y, z)$. Its total differential is therefore:

$$d\rho = \frac{\partial \rho}{\partial t}\, dt + \frac{\partial \rho}{\partial x}\, dx + \frac{\partial \rho}{\partial y}\, dy + \frac{\partial \rho}{\partial z}\, dz. \tag{5.42}$$

Over the time increment dt, take the space increments dx, dy, and dz to correspond to the distances traveled by a fluid particle, as shown in Fig. 5.10. That is:

$$dx = v_x\,dt, \qquad dy = v_y\,dt, \qquad dz = v_z\,dt. \tag{5.43}$$

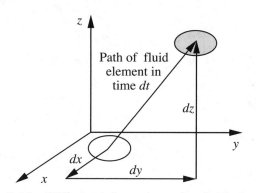

Fig. 5.10 Differential motion of a fluid element.

By dividing Eqn. (5.42) by dt and introducing dx, dy, and dz from Eqn. (5.43), the following relation is obtained for the time variation of density for an observer moving with the fluid:

$$\frac{\mathcal{D}\rho}{\mathcal{D}t} = \left(\frac{d\rho}{dt}\right)_{\substack{\text{Moving} \\ \text{with fluid}}} = \frac{\partial\rho}{\partial t} + \underbrace{v_x\frac{\partial\rho}{\partial x} + v_y\frac{\partial\rho}{\partial y} + v_z\frac{\partial\rho}{\partial z}}_{\substack{\text{Rate of change of density with} \\ \text{time due to fluid motion}}} = \frac{\partial\rho}{\partial t} + \underbrace{\mathbf{v}\cdot\nabla\rho}_{\text{See text}}. \tag{5.44}$$

The physical significance of the last term in Eqn. (5.44), $\mathbf{v}\cdot\nabla\rho$, will be explained shortly, in Example 5.6.

From Eqn. (5.44), the convective derivative consists of two parts:

1. The rate of change $\partial\rho/\partial t$ of the density with time at a *fixed point*, corresponding to the *Eulerian* viewpoint.
2. An additional contribution, emphasized by the underbrace in Eqn. (5.44), due to the fluid motion.

The convective differential operator is therefore defined as:

$$\frac{\mathcal{D}}{\mathcal{D}t} = \frac{\partial}{\partial t} + v_x\frac{\partial}{\partial x} + v_y\frac{\partial}{\partial y} + v_z\frac{\partial}{\partial z} = \frac{\partial}{\partial t} + \mathbf{v}\cdot\nabla. \tag{5.45}$$

5.5 Differential Mass Balance

A differential mass balance, also known as the *continuity equation*, is readily derived by accounting for the change of mass inside the *fixed*, constant-volume element dV, shown in Fig. 5.11.

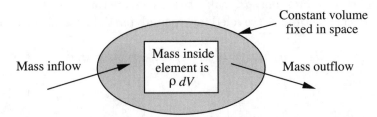

Fig. 5.11   Control volume for differential mass balance.

Assuming that dV is sufficiently small so that at any instant the density ρ is uniform within it, its mass is ρdV. Due to fluid flow, there will be transfers of mass into and from the element, the local mass flux being $\rho\mathbf{v}$. Recalling the definition of the divergence just after Eqn. (5.16), the *net* rate of *outflow* of mass per unit volume is $\nabla \cdot \rho\mathbf{v}$. Equating the net rate of *inflow* of mass (the negative of the outflow) to the rate of accumulation in the element:

$$-(\nabla \cdot \rho\mathbf{v})\, dV = \frac{\partial(\rho dV)}{\partial t}. \tag{5.46}$$

Cancellation of the constant volume dV and rearrangement gives:

$$\frac{\partial \rho}{\partial t} + \nabla \cdot \rho\mathbf{v} = 0. \tag{5.47}$$

In the RCCS, this differential mass balance appears as:

$$\frac{\partial \rho}{\partial t} + \frac{\partial(\rho v_x)}{\partial x} + \frac{\partial(\rho v_y)}{\partial y} + \frac{\partial(\rho v_z)}{\partial z} = 0. \tag{5.48}$$

Complete forms for the three principal coordinate systems are given in Table 5.6. Each of these states that the density and velocity components cannot change arbitrarily, but are always constrained in their variations so that mass is conserved.

Table 5.6   Differential Mass Balances

Rectangular Cartesian Coordinates:

$$\frac{\partial \rho}{\partial t} + \frac{\partial(\rho v_x)}{\partial x} + \frac{\partial(\rho v_y)}{\partial y} + \frac{\partial(\rho v_z)}{\partial z} = 0. \tag{5.48}$$

Cylindrical Coordinates:

$$\frac{\partial \rho}{\partial t} + \frac{1}{r}\frac{\partial(\rho r v_r)}{\partial r} + \frac{1}{r}\frac{\partial(\rho v_\theta)}{\partial \theta} + \frac{\partial(\rho v_z)}{\partial z} = 0. \tag{5.49}$$

Spherical Coordinates:

$$\frac{\partial \rho}{\partial t} + \frac{1}{r^2}\frac{\partial(\rho r^2 v_r)}{\partial r} + \frac{1}{r \sin\theta}\frac{\partial(\rho v_\theta \sin\theta)}{\partial \theta} + \frac{1}{r \sin\theta}\frac{\partial(\rho v_\phi)}{\partial \phi} = 0. \tag{5.50}$$

Example 5.6—Physical Interpretation of the Net Rate of Mass Outflow   **269**

Example 5.6—Physical Interpretation of the Net Rate of Mass Outflow

Following Eqn. (5.24), the divergence term in Eqn. (5.47) can be expanded to yield:

$$\underbrace{\nabla \cdot \rho\mathbf{v}}_{1} = \underbrace{\rho\nabla \cdot \mathbf{v}}_{2} + \underbrace{\mathbf{v} \cdot \nabla\rho}_{3}. \tag{E5.6.1}$$

Give the physical significance of each of the three numbered terms in Eqn. (E5.6.1).

Solution

Each term corresponds to a rate per unit volume, with the following physical significance.

1. As explained above, the first term is the *overall* or net rate of loss of mass.
2. Note first, from the definition of the divergence, that $\nabla \cdot \mathbf{v}$ is the rate of outflow of volume (per unit volume). This situation could occur if the fluid—probably a gas—were expanding due to a decrease in pressure, in which case there would be an outflow of volume across the boundaries of a fixed unit volume. Multiplication by the density then gives the second term as the rate of loss of mass due to *expansion*.
3. To illustrate, first imagine the case of a rectangular fluid element, as in Fig. 5.5, but with the only nonzero velocity component being v_x *uniformly* in the x direction. Suppose further that the density increases in the x direction, so that $\partial\rho/\partial x$ is positive and hence $\nabla\rho \neq 0$. The flow into the left-hand face will bring less mass into the element than the flow out of the right-hand face takes from the element, so there will be a net loss of mass due to the *flow*. The accommodation of similar losses due to flow in the other two coordinate directions then leads to the third term in Eqn. (E5.6.1). □

Of course, the continuity equation simplifies enormously if the density is constant, which is approximately true for the majority of liquids. In this event, the term $\partial\rho/\partial t$ is zero, and ρ can be canceled from all the remaining terms, giving:

$$\nabla \cdot \mathbf{v} = 0. \tag{5.51}$$

The corresponding simplified forms in the three coordinate systems can be obtained either from Table 5.6, or by equating to zero the forms of the divergence given in Table 5.3. Of these, the most familiar form is that in the RCCS, namely:

$$\frac{\partial v_x}{\partial x} + \frac{\partial v_y}{\partial y} + \frac{\partial v_z}{\partial z} = 0. \tag{5.52}$$

Example 5.7—Alternative Derivation of the Continuity Equation

There are other, equivalent ways of deriving the continuity equation. To illustrate, derive the continuity equation in vector form by following the motion of a fluid element of *constant mass,* whose volume may be changing, as shown in Fig. E5.7, along the path A–B–C.

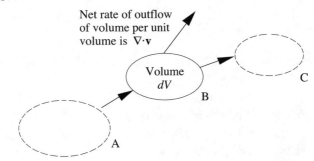

Net rate of outflow
of volume per unit
volume is $\nabla \cdot \mathbf{v}$

Volume
dV

A

B

C

Fig. E5.7 Motion of a fluid element along the path A–B–C.

Solution

Focus attention when the element is at location B, where its volume is dV. By definition of the divergence of the velocity vector as the net rate of outflow of volume per unit volume, the fractional rate at which the element is *increasing* in volume is $\nabla \cdot \mathbf{v}$. (A positive value of the divergence means that the constant mass is now occupying a larger volume, because this consists of the original volume plus the outflow or expansion.) But since the mass of the element is constant, its density must be *decreasing* at the same fractional rate. Since we are following the path taken by the fluid, the latter is given by $(\mathcal{D}\rho/\mathcal{D}t)/\rho$. The relationship between the two quantities is therefore:

$$\frac{1}{\rho}\frac{\mathcal{D}\rho}{\mathcal{D}t} = -\nabla \cdot \mathbf{v}, \qquad (E5.7.1)$$

or,

$$\frac{\mathcal{D}\rho}{\mathcal{D}t} + \rho\nabla \cdot \mathbf{v} = 0. \qquad (E5.7.2)$$

The reader can show (see Problem 5.6) that Eqns. (5.47) and (E5.7.2) are equivalent. As a further example, Problem 5.7 illustrates yet another way of deriving the continuity equation. □

5.6 Differential Momentum Balances

Section 5.5 derived the differential mass balance, or continuity equation, using both vector notation and the three usual coordinate systems. Now proceed to the differential *momentum* balances, of which there will be *three*, one for each coordinate direction. The following derivation is fairly involved, and requires concentration. Nevertheless, these momentum balances are essential in order to comprehend the basic equations of fluid mechanics. Similar equations can also be derived for energy and mass transfer. Thus, the concepts presented here will also be important in other areas of chemical engineering.

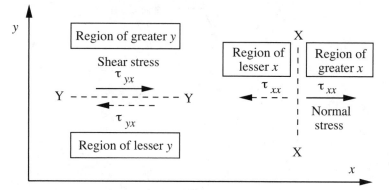

Fig. 5.12   *Tangential and normal stresses τ_{yx} and τ_{xx} acting on surfaces of constant y and x, respectively.*

Sign convention for stresses. We start by establishing the notation and sign convention for tangential and normal stresses due to viscous action, for which representative components appear as τ_{yx} and τ_{xx} in Fig. 5.12. Concentrate first on the stresses represented by the *full* arrows. Note that each is denoted by the symbol "τ," with two subscripts.

The *first* subscript denotes the *surface* on which the stress acts. For example, the stress τ_{yx} acts on a surface of constant y (shown as Y–Y in the figure), whereas τ_{xx} acts on a surface of constant x (X–X).

The *second* subscript denotes the *direction* of the stress, considered to be positive when exerted by the fluid in the region of greater "first subscript" on the fluid of lesser "first subscript." For example, τ_{yx} acts in the x direction (corresponding to its second subscript) for the fluid of greater first subscript, y, (above the line Y–Y) acting on the region of lesser y (below the line Y–Y). Likewise, τ_{xx} also acts in the x direction (corresponding to its second subscript) for the fluid of greater first subscript, x (to the right of the line X–X), acting on the region of lesser x (to the left of the line X–X).

Observe that stresses are always perfectly balanced across surfaces such as those represented by X–X and Y–Y. Thus, the stress τ_{yx} *below* the surface Y–Y (shown by the dashed arrow) is exactly the same as that above Y–Y, but in the *opposite* direction. The same can be said for τ_{xx}. If the stresses were unequal, there would be a finite force acting on the surface, which has zero mass, resulting in an infinite acceleration. Such is not the case!

Fig. 5.13 Tangential and normal stresses acting on a rectangular parallelepiped element. The stresses acting on the faces of constant z are omitted for clarity.

Differential momentum balance.   Consider now the tangential and normal stresses acting on the differential rectangular parallelepiped of fluid, of dimensions $dx \times dy \times dz$, shown in Fig. 5.13. There are six normal stresses (one for each face) and twelve tangential or shear stresses (two for each face), for a total of eighteen stresses. However, to avoid encumbering the diagram with an excessive number of symbols, the six stresses on the faces normal to the z axis are omitted for clarity. Observe that the directions of all the stresses follow the sign convention just described.

The usual symbol τ is used to denote the *tangential* or viscous shear stresses. However, the *normal* stresses are denoted by a different symbol, σ, because it will be seen later that a normal stress such as σ_{xx} can be decomposed into two parts, $\sigma_{xx} = -p + \tau_{xx}$, namely:

1. The fluid pressure which, being *compressive*, acts in the opposite direction to that of the convention, and is given a negative sign, as in $-p$.

2. An additional contribution, due to viscous action, such as τ_{xx}.

As usual, stresses change by a differential amount from one face to the opposite face. For example, τ_{yx}, acting in the negative x direction on the bottom face, is augmented to $\tau_{yx} + (\partial \tau_{yx}/\partial y)dy$, acting in the positive x direction, on the top face.

Also denote the body forces per unit mass acting in the three coordinate directions as g_x, g_y, and g_z. The symbol "g" is used because gravity is the obvious body force. The understanding is that it would be replaced or augmented by additional body forces—such as those due to electromagnetic action—whenever appropriate. In the usual case, with the z axis pointing vertically upward, $g_x = g_y = 0$, and $g_z = -g$, where g is the gravitational acceleration.

Now perform three successive momentum balances, one for each coordinate direction, on an element of fixed mass ($\rho\, dx\, dy\, dz$) that is *moving with the fluid*. This viewpoint simplifies the derivation, in that we do not have to consider convective transports of momentum across each of the six faces. (Problem 5.10 checks that a momentum balance on an element *fixed in space*—in which convective transports have to be considered—leads to exactly the same result.)

A momentum balance in the x direction gives:

$$\underbrace{\rho\, dx\, dy\, dz \frac{\mathcal{D}v_x}{\mathcal{D}t}}_{\substack{\text{Rate of increase} \\ \text{of x momentum}}} = \underbrace{\left(\frac{\partial \sigma_{xx}}{\partial x}\, dx \right) dy\, dz + \left(\frac{\partial \tau_{yx}}{\partial y}\, dy \right) dz\, dx + \left(\frac{\partial \tau_{zx}}{\partial z}\, dz \right) dx\, dy}_{\substack{\text{Net viscous and pressure} \\ \text{force in the x direction}}}$$

$$+ \underbrace{g_x \rho\, dx\, dy\, dz}_{\text{Body force}}. \tag{5.53}$$

Observe that the *convective* derivative must be employed on the left-hand side, because we have chosen to perform a momentum balance on an element moving with the fluid.

Division of Eqn. (5.53) by the common volume $dx\, dy\, dz$, and repetition for the other two coordinate directions gives:

$$\begin{aligned}
\rho \frac{\mathcal{D}v_x}{\mathcal{D}t} &= \frac{\partial \sigma_{xx}}{\partial x} + \frac{\partial \tau_{yx}}{\partial y} + \frac{\partial \tau_{zx}}{\partial z} + \rho g_x, \\
\rho \frac{\mathcal{D}v_y}{\mathcal{D}t} &= \frac{\partial \tau_{xy}}{\partial x} + \frac{\partial \sigma_{yy}}{\partial y} + \frac{\partial \tau_{zy}}{\partial z} + \rho g_y, \\
\rho \frac{\mathcal{D}v_z}{\mathcal{D}t} &= \frac{\partial \tau_{xz}}{\partial x} + \frac{\partial \tau_{yz}}{\partial y} + \frac{\partial \sigma_{zz}}{\partial z} + \rho g_z.
\end{aligned} \tag{5.54}$$

Bear in mind that a "shorthand" notation appears on the left-hand side of Eqn. (5.54); the convective derivatives can be written out more fully according to the definition given in Eqn. (5.45). Although Eqn. (5.54), with *stress* components on the right-hand sides, may be useful in certain circumstances, particularly when special formulas for the stresses are available for non-Newtonian fluids (see Chapter 11), it is usually preferable to cast them in terms of *velocities* for the special but very important case of *Newtonian* fluids, as will be done in the next section.

5.7 Newtonian Stress Components in Cartesian Coordinates

In general, there are nine stress components acting at any point in a fluid, illustrated in Fig. 5.14(a).

The assembly of all nine components in the form of a matrix is called a *tensor*. Thus, both the viscous-stress and the total-stress tensors are expressed as:

$$\boldsymbol{\tau} = \begin{pmatrix} \tau_{xx} & \tau_{xy} & \tau_{xz} \\ \tau_{yx} & \tau_{yy} & \tau_{yz} \\ \tau_{zx} & \tau_{zy} & \tau_{zz} \end{pmatrix}, \qquad \boldsymbol{\sigma} = \begin{pmatrix} \sigma_{xx} & \tau_{xy} & \tau_{xz} \\ \tau_{yx} & \sigma_{yy} & \tau_{yz} \\ \tau_{zx} & \tau_{zy} & \sigma_{zz} \end{pmatrix}. \qquad (5.55)$$

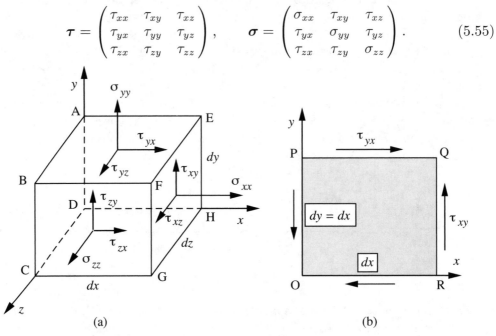

Fig. 5.14 (a) The nine normal and shear stress components;
(b) shear stresses acting on an element in the x/y plane.

(Actually, there is somewhat more to the definition of a tensor. There is a rule by which its elements are transformed when proceeding to a different frame of coordinates, but it is unnecessary to pursue this aspect here.) It will now be shown that, of the *six* shear stress components, only *three* are independent. Consider a rectangular parallelepiped element of fluid extending for any distance L in the z direction, and whose square cross section $dx \times dx$ in the x/y plane is shown in Fig. 5.14(b). There are four relevant shear stresses, indicated by the arrows, which, if not in balance, may cause the element to rotate. Equating the net clockwise moment about the origin O to the product of the moment of inertia I of the element times its rate of increase of angular velocity $d\omega/dt$:

$$\underbrace{\underbrace{\underbrace{L\,dx}_{\text{Area}}\,(\tau_{yx} - \tau_{xy})\,dx}_{\text{Net force}}}_{\text{Clockwise moment}} = I\,\frac{d\omega}{dt}. \qquad (5.56)$$

On the left-hand side, the second dx is the radius arm for multiplying the net force in order to obtain the torque. For the element, I can be shown to be proportional to $(dx)^4$. Cancellation of $(dx)^2$ leaves a factor of $(dx)^2$ on the right-hand side. Since $d\omega/dt$ remains finite as dx approaches zero, this can only occur if:

$$\tau_{yx} = \tau_{xy}. \qquad (5.57)$$

Similarly, $\tau_{zy} = \tau_{yz}$ and $\tau_{xz} = \tau_{zx}$. That is, there are only *three* independent shear-stress components.

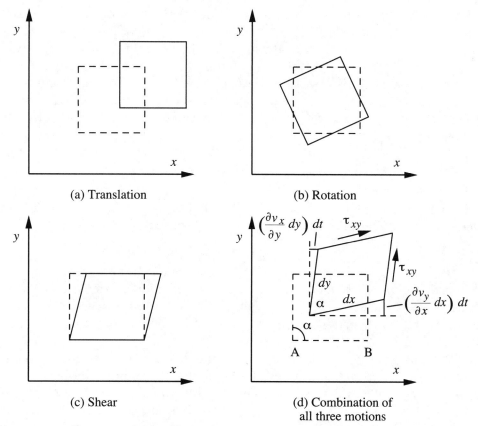

(a) Translation   (b) Rotation

(c) Shear   (d) Combination of all three motions

Fig. 5.15   Basic modes of fluid motion.   The dashed outline shows the initial location of an element.   The full outline shows its location after translation, rotation, shear, or a combination of all three of these motions.

Basic relation for a Newtonian fluid.   Fig. 5.15(a–c) shows three basic ways in which a fluid may move. In each case, the dashed outline indicates the initial shape of a fluid element in the x/y plane. Shortly afterward, it will have assumed a new configuration, denoted by the full outline. The following can occur:

(a) *Translation*—the element moves to a new location without changing its shape.

(b) *Rotation*—the element turns, without moving its center of gravity.

(c) *Shear*—the element deforms into a parallelogram.

Fig. 5.15(d) illustrates a combination of all three motions, of which we want to determine just the *shear* component, which can be discovered by finding the rate of change of the angle α, which starts as a right angle, but is slightly diminished after a small time interval dt has elapsed.

Consider first the bottom side AB of the element. The rate at which any point on it moves in the y direction depends on the local value of v_y, and should this vary with x, then the side will be tilted one way or the other and will no longer be parallel to the x axis. Hence, during the time dt, point B will have advanced by an amount $(\partial v_y/\partial x)\, dx\, dt$ relative to point A, and the side AB will therefore be tilted counterclockwise by an angle $(\partial v_y/\partial x)\, dt$. By a similar argument, the left-hand side of the element will be tilted clockwise by an angle $(\partial v_x/\partial y)\, dt$. Both of these actions (as drawn) tend to *diminish* the angle α; summation of the effects and division by dt gives:

$$\frac{d\alpha}{dt} = -\left(\frac{\partial v_x}{\partial y} + \frac{\partial v_y}{\partial x} \right). \tag{5.58}$$

For a *Newtonian* fluid, it is a fundamental assumption, borne out by experiment, that the shear stress causing the deformation is proportional to the rate at which α is decreasing (also known as the *strain rate*):

$$\tau_{xy} \ (= \tau_{yx}) = -\mu\, \frac{d\alpha}{dt} = \mu \left(\frac{\partial v_x}{\partial y} + \frac{\partial v_y}{\partial x} \right). \tag{5.59}$$

The constant of proportionality μ is the *viscosity*. Similar relations can be deduced for the other shear stresses. The complete set of equations, which collectively express *Newton's law of viscosity*, is given for rectangular coordinates in the first half of Table 5.7.

Although it follows along similar lines, the derivation of the relations for the normal stress is *considerably* more involved and will not be given here. Instead, the end result is given in the second half of Table 5.7. Thus, as hinted previously, the normal stresses (which, by the sign convention, are considered positive when causing tension) consist of two parts: the fluid pressure p (a compression—hence the negative sign) and components caused by viscous action. The latter are given by:

$$\tau_{xx} = 2\mu \frac{\partial v_x}{\partial x} - \frac{2}{3}\mu \nabla \cdot \mathbf{v},$$

$$\tau_{yy} = 2\mu \frac{\partial v_y}{\partial y} - \frac{2}{3}\mu \nabla \cdot \mathbf{v}, \tag{5.62}$$

$$\tau_{zz} = 2\mu \frac{\partial v_z}{\partial z} - \frac{2}{3}\mu \nabla \cdot \mathbf{v}.$$

There is a ready qualitative physical explanation of the first terms on the right-hand sides of Eqn. (5.62). For example, if the fluid is being stretched in the x direction, $\partial v_x / \partial x$ will be positive, resulting in a positive value of τ_{xx}—that is, a *tension*—in order to do so.

Table 5.7 Stress Components in Rectangular Coordinates[2]

$$\tau_{xy} = \tau_{yx} = \mu \left(\frac{\partial v_x}{\partial y} + \frac{\partial v_y}{\partial x} \right),$$

$$\tau_{yz} = \tau_{zy} = \mu \left(\frac{\partial v_y}{\partial z} + \frac{\partial v_z}{\partial y} \right), \qquad (5.60)$$

$$\tau_{zx} = \tau_{xz} = \mu \left(\frac{\partial v_z}{\partial x} + \frac{\partial v_x}{\partial z} \right).$$

$$\sigma_{xx} = -p + 2\mu \frac{\partial v_x}{\partial x} - \frac{2}{3} \mu \nabla \cdot \mathbf{v},$$

$$\sigma_{yy} = -p + 2\mu \frac{\partial v_y}{\partial y} - \frac{2}{3} \mu \nabla \cdot \mathbf{v}, \qquad (5.61)$$

$$\sigma_{zz} = -p + 2\mu \frac{\partial v_z}{\partial z} - \frac{2}{3} \mu \nabla \cdot \mathbf{v}.$$

By similar arguments, the rate of *rotation* of the fluid element in Fig. 5.15 could also be expressed in terms of derivatives of the velocity components, and this will be pursued in Section 7.2, when angular velocity and vorticity are studied in more detail.

For completeness, the stress components in cylindrical and spherical coordinates are given in Tables 5.8 and 5.9.

The shear- and normal-stress relations, (5.60) and (5.61), may now be substituted into the three momentum balances of Eqn. (5.54), as detailed in Example 5.8 below. Clearly, the resulting relations will be quite complicated, and at an introductory level it is often appropriate to make the simplification of *constant viscosity*. The final result is:

Momentum balance for constant-viscosity Newtonian fluid

$$\rho \left(\frac{\partial v_x}{\partial t} + v_x \frac{\partial v_x}{\partial x} + v_y \frac{\partial v_x}{\partial y} + v_z \frac{\partial v_x}{\partial z} \right) = -\frac{\partial p}{\partial x} + \mu \left(\frac{\partial^2 v_x}{\partial x^2} + \frac{\partial^2 v_x}{\partial y^2} + \frac{\partial^2 v_x}{\partial z^2} \right)$$

---

[2] Equation numbers also appear in Tables 5.7–5.9. Since tables often have to be moved around during page composition, this has led to an apparent out-of-sequence order for some equations on this and the following two pages.

$$+ \rho g_x + \frac{1}{3} \mu \frac{\partial}{\partial x} \nabla \cdot \mathbf{v},$$

$$\rho \left(\frac{\partial v_y}{\partial t} + v_x \frac{\partial v_y}{\partial x} + v_y \frac{\partial v_y}{\partial y} + v_z \frac{\partial v_y}{\partial z} \right) = -\frac{\partial p}{\partial y} + \mu \left(\frac{\partial^2 v_y}{\partial x^2} + \frac{\partial^2 v_y}{\partial y^2} + \frac{\partial^2 v_y}{\partial z^2} \right)$$

$$+ \rho g_y + \frac{1}{3} \mu \frac{\partial}{\partial y} \nabla \cdot \mathbf{v},$$

$$\rho \left(\frac{\partial v_z}{\partial t} + v_x \frac{\partial v_z}{\partial x} + v_y \frac{\partial v_z}{\partial y} + v_z \frac{\partial v_z}{\partial z} \right) = -\frac{\partial p}{\partial z} + \mu \left(\frac{\partial^2 v_z}{\partial x^2} + \frac{\partial^2 v_z}{\partial y^2} + \frac{\partial^2 v_z}{\partial z^2} \right)$$

$$+ \rho g_z + \frac{1}{3} \mu \frac{\partial}{\partial z} \nabla \cdot \mathbf{v}. \qquad (5.67)$$

The three individual relationships in Eqn. (5.67) can be viewed as the x, y, and z components of a single *vector* equation:

$$\rho \frac{D\mathbf{v}}{Dt} = -\nabla p + \mu \nabla^2 \mathbf{v} + \rho \mathbf{g} + \frac{1}{3} \mu \nabla (\nabla \cdot \mathbf{v}). \qquad (5.68)$$

As shown in Appendix C, the term $\nabla^2 \mathbf{v}$ has components in the x, y, and z directions, each of which amounts to the Laplacian of the individual velocity components v_x, v_y, and v_z; also, $\mathbf{g}$ is a vector whose components are g_x, g_y, and g_z; and $\nabla (\nabla \cdot \mathbf{v})$, having components $\partial (\nabla \cdot \mathbf{v}) / \partial x$, $\partial (\nabla \cdot \mathbf{v}) / \partial y$, and $\partial (\nabla \cdot \mathbf{v}) / \partial z$, is the gradient of the divergence of the velocity vector.

Table 5.8   Stress Components in Cylindrical Coordinates

$$\tau_{r\theta} = \tau_{\theta r} = \mu \left(r \frac{\partial}{\partial r} \left(\frac{v_\theta}{r} \right) + \frac{1}{r} \frac{\partial v_r}{\partial \theta} \right),$$

$$\tau_{\theta z} = \tau_{z\theta} = \mu \left(\frac{\partial v_\theta}{\partial z} + \frac{1}{r} \frac{\partial v_z}{\partial \theta} \right), \qquad (5.63)$$

$$\tau_{zr} = \tau_{rz} = \mu \left(\frac{\partial v_z}{\partial r} + \frac{\partial v_r}{\partial z} \right).$$

$$\sigma_{rr} = -p + 2\mu \frac{\partial v_r}{\partial r} - \frac{2}{3} \mu \nabla \cdot \mathbf{v},$$

$$\sigma_{\theta\theta} = -p + 2\mu \left(\frac{1}{r} \frac{\partial v_\theta}{\partial \theta} + \frac{v_r}{r} \right) - \frac{2}{3} \mu \nabla \cdot \mathbf{v}, \qquad (5.64)$$

$$\sigma_{zz} = -p + 2\mu \frac{\partial v_z}{\partial z} - \frac{2}{3} \mu \nabla \cdot \mathbf{v}.$$

Table 5.9 Stress Components in Spherical Coordinates

$$\tau_{r\theta} = \tau_{\theta r} = \mu \left(r \frac{\partial}{\partial r} \left(\frac{v_\theta}{r} \right) + \frac{1}{r} \frac{\partial v_r}{\partial \theta} \right),$$

$$\tau_{\theta\phi} = \tau_{\phi\theta} = \mu \left(\frac{\sin\theta}{r} \frac{\partial}{\partial \theta} \left(\frac{v_\phi}{\sin\theta} \right) + \frac{1}{r\sin\theta} \frac{\partial v_\theta}{\partial \phi} \right), \tag{5.65}$$

$$\tau_{\phi r} = \tau_{r\phi} = \mu \left(\frac{1}{r\sin\theta} \frac{\partial v_r}{\partial \phi} + r \frac{\partial}{\partial r} \left(\frac{v_\phi}{r} \right) \right).$$

$$\sigma_{rr} = -p + 2\mu \frac{\partial v_r}{\partial r} - \frac{2}{3} \mu \nabla \cdot \mathbf{v},$$

$$\sigma_{\theta\theta} = -p + 2\mu \left(\frac{1}{r} \frac{\partial v_\theta}{\partial \theta} + \frac{v_r}{r} \right) - \frac{2}{3}\mu \nabla \cdot \mathbf{v}, \tag{5.66}$$

$$\sigma_{\phi\phi} = -p + 2\mu \left(\frac{1}{r\sin\theta} \frac{\partial v_\phi}{\partial \phi} + \frac{v_r}{r} + \frac{v_\theta \cot\theta}{r} \right) - \frac{2}{3}\mu \nabla \cdot \mathbf{v}.$$

Three important tensors. For an incompressible Newtonian fluid, the relation between the shear stresses and the rates of strain may be expressed more concisely in terms of tensors. The *viscous-stress* and *total-stress* tensors have already been defined:

$$\boldsymbol{\tau} = \begin{pmatrix} \tau_{xx} & \tau_{xy} & \tau_{xz} \\ \tau_{yx} & \tau_{yy} & \tau_{yz} \\ \tau_{zx} & \tau_{zy} & \tau_{zz} \end{pmatrix}, \qquad \boldsymbol{\sigma} = \begin{pmatrix} \sigma_{xx} & \tau_{xy} & \tau_{xz} \\ \tau_{yx} & \sigma_{yy} & \tau_{yz} \\ \tau_{zx} & \tau_{zy} & \sigma_{zz} \end{pmatrix}. \tag{5.69}$$

A *rate-of-deformation*, *rate-of-strain*, or simply *strain-rate* tensor can also be defined:

$$\dot{\boldsymbol{\gamma}} = \begin{pmatrix} 2\dfrac{\partial v_x}{\partial x} & \left(\dfrac{\partial v_x}{\partial y} + \dfrac{\partial v_y}{\partial x} \right) & \left(\dfrac{\partial v_x}{\partial z} + \dfrac{\partial v_z}{\partial x} \right) \\ \left(\dfrac{\partial v_y}{\partial x} + \dfrac{\partial v_x}{\partial y} \right) & 2\dfrac{\partial v_y}{\partial y} & \left(\dfrac{\partial v_y}{\partial z} + \dfrac{\partial v_z}{\partial y} \right) \\ \left(\dfrac{\partial v_z}{\partial x} + \dfrac{\partial v_x}{\partial z} \right) & \left(\dfrac{\partial v_z}{\partial y} + \dfrac{\partial v_y}{\partial z} \right) & 2\dfrac{\partial v_z}{\partial z} \end{pmatrix} = \nabla\mathbf{v} + (\nabla\mathbf{v})^T, \tag{5.70}$$

in which the *dyadic product* $\nabla\mathbf{v}$ and its transpose $(\nabla\mathbf{v})^T$ are defined in Appendix C. Since $\nabla \cdot \mathbf{v} = 0$ for an incompressible fluid, Eqns. (5.60)–(5.62) can be reexpressed as:

$$\boldsymbol{\tau} = \mu\dot{\boldsymbol{\gamma}}, \qquad \boldsymbol{\sigma} = -p\mathbf{I} + \mu\dot{\boldsymbol{\gamma}}, \tag{5.71}$$

where $\mathbf{I}$ is the *unit tensor*, having diagonal elements equal to unity and off-diagonal elements equal to zero.

The viscous-stress and strain-rate tensors will find important applications when non-Newtonian fluids are considered in Chapter 11.

Example 5.8—Constant-Viscosity Momentum
Balances in Terms of Velocity Gradients

Verify the first of the three momentum balances of Eqn. (5.67) by starting from the x momentum balance of Eqn. (5.54) and substituting for the stresses σ_{xx}, τ_{yx}, and τ_{zx} from Eqns. (5.60) and (5.61).

Solution

The indicated substitution leads to:

$$\rho \frac{Dv_x}{Dt} = \frac{\partial \sigma_{xx}}{\partial x} + \frac{\partial \tau_{yx}}{\partial y} + \frac{\partial \tau_{zx}}{\partial z} + \rho g_x$$

$$= \frac{\partial}{\partial x}\left(-p + 2\mu \frac{\partial v_x}{\partial x} - \frac{2}{3}\mu \nabla \cdot \mathbf{v}\right) + \frac{\partial}{\partial y}\left[\mu\left(\frac{\partial v_x}{\partial y} + \frac{\partial v_y}{\partial x}\right)\right]$$

$$+ \frac{\partial}{\partial z}\left[\mu\left(\frac{\partial v_x}{\partial z} + \frac{\partial v_z}{\partial x}\right)\right] + \rho g_x. \tag{E5.8.1}$$

For the case of constant viscosity, collection of terms gives:

$$\rho \frac{Dv_x}{Dt} = -\frac{\partial p}{\partial x} + \mu\left(\frac{\partial^2 v_x}{\partial x^2} + \frac{\partial^2 v_x}{\partial y^2} + \frac{\partial^2 v_x}{\partial z^2}\right) + \rho g_x$$

$$+ \mu \underbrace{\left[\frac{\partial}{\partial x}\left(\frac{\partial v_x}{\partial x}\right) + \frac{\partial}{\partial y}\left(\frac{\partial v_y}{\partial x}\right) + \frac{\partial}{\partial z}\left(\frac{\partial v_z}{\partial x}\right)\right]}_{\text{See text}} - \frac{2}{3}\mu \frac{\partial}{\partial x}\nabla \cdot \mathbf{v}. \tag{E5.8.2}$$

But, since the order of partial differentiation can be reversed, the expression indicated by the underbrace is the x derivative of $\nabla \cdot \mathbf{v}$ and can be combined with the subsequent term. Expansion of the convective derivative on the left-hand side then gives the desired result for the x momentum balance:

$$\rho \underbrace{\left(\frac{\partial v_x}{\partial t} + v_x \frac{\partial v_x}{\partial x} + v_y \frac{\partial v_x}{\partial y} + v_z \frac{\partial v_x}{\partial z}\right)}_{\text{Inertial or acceleration terms}} = \underbrace{-\frac{\partial p}{\partial x}}_{\substack{\text{Pressure} \\ \text{term}}} + \underbrace{\mu\left(\frac{\partial^2 v_x}{\partial x^2} + \frac{\partial^2 v_x}{\partial y^2} + \frac{\partial^2 v_x}{\partial z^2}\right)}_{\text{Primary viscous terms}}$$

$$+ \underbrace{\rho g_x}_{\substack{\text{Body} \\ \text{force} \\ \text{term}}} + \underbrace{\frac{1}{3}\mu \frac{\partial}{\partial x}\nabla \cdot \mathbf{v}}_{\substack{\text{Secondary} \\ \text{viscous term}}}. \tag{E5.8.3}$$

Here, the nature of the various terms is also indicated. The derivations of the y and z equations follow on similar lines.   □

Frequently—as occurs for liquids in most circumstances—the fluid is essentially incompressible, in which case the term $\nabla \cdot \mathbf{v}$ can be neglected. In such event, the momentum balances are called the *Navier-Stokes equations*.

This chapter concludes with a display of the momentum balances (in terms of stresses) and the Navier-Stokes equations in the three main coordinate systems (Tables 5.10–5.12), together with a short example for the case of variable viscosity.

Table 5.10   Momentum Equations in Rectangular Cartesian Coordinates

In terms of stresses (under all circumstances):   (5.72)

$$\rho \left(\frac{\partial v_x}{\partial t} + v_x \frac{\partial v_x}{\partial x} + v_y \frac{\partial v_x}{\partial y} + v_z \frac{\partial v_x}{\partial z} \right) = -\frac{\partial p}{\partial x} + \frac{\partial \tau_{xx}}{\partial x} + \frac{\partial \tau_{yx}}{\partial y} + \frac{\partial \tau_{zx}}{\partial z} + \rho g_x,$$

$$\rho \left(\frac{\partial v_y}{\partial t} + v_x \frac{\partial v_y}{\partial x} + v_y \frac{\partial v_y}{\partial y} + v_z \frac{\partial v_y}{\partial z} \right) = -\frac{\partial p}{\partial y} + \frac{\partial \tau_{xy}}{\partial x} + \frac{\partial \tau_{yy}}{\partial y} + \frac{\partial \tau_{zy}}{\partial z} + \rho g_y,$$

$$\rho \left(\frac{\partial v_z}{\partial t} + v_x \frac{\partial v_z}{\partial x} + v_y \frac{\partial v_z}{\partial y} + v_z \frac{\partial v_z}{\partial z} \right) = -\frac{\partial p}{\partial z} + \frac{\partial \tau_{xz}}{\partial x} + \frac{\partial \tau_{yz}}{\partial y} + \frac{\partial \tau_{zz}}{\partial z} + \rho g_z.$$

In terms of velocities, for a Newtonian fluid with constant ρ and μ: (5.73)

$$\rho \left(\frac{\partial v_x}{\partial t} + v_x \frac{\partial v_x}{\partial x} + v_y \frac{\partial v_x}{\partial y} + v_z \frac{\partial v_x}{\partial z} \right) = -\frac{\partial p}{\partial x} + \mu \left(\frac{\partial^2 v_x}{\partial x^2} + \frac{\partial^2 v_x}{\partial y^2} + \frac{\partial^2 v_x}{\partial z^2} \right) + \rho g_x,$$

$$\rho \left(\frac{\partial v_y}{\partial t} + v_x \frac{\partial v_y}{\partial x} + v_y \frac{\partial v_y}{\partial y} + v_z \frac{\partial v_y}{\partial z} \right) = -\frac{\partial p}{\partial y} + \mu \left(\frac{\partial^2 v_y}{\partial x^2} + \frac{\partial^2 v_y}{\partial y^2} + \frac{\partial^2 v_y}{\partial z^2} \right) + \rho g_y,$$

$$\rho \left(\frac{\partial v_z}{\partial t} + v_x \frac{\partial v_z}{\partial x} + v_y \frac{\partial v_z}{\partial y} + v_z \frac{\partial v_z}{\partial z} \right) = -\frac{\partial p}{\partial z} + \mu \left(\frac{\partial^2 v_z}{\partial x^2} + \frac{\partial^2 v_z}{\partial y^2} + \frac{\partial^2 v_z}{\partial z^2} \right) + \rho g_z.$$

Navier, Claude-Louis-Marie-Henri, born 1785 in Dijon, France, died 1836 in Paris. Navier became a protégé and friend of Fourier, who was one of his professors at the Ecole Polytechnique in Paris. Much of his early efforts from 1807 to 1813 went into editing and publishing the works of two others: (a) the manuscripts of his great-uncle, Emiland Gauthey, a notable civil engineer; and (b) a revised edition of Bélidor's *Science des ingénieurs*. The greater part of Navier's own engineering work centered on the mechanics of structural materials—wood, stone, and iron—thereby furnishing important analytical tools to civil engineers. He designed and almost completed a suspension bridge spanning the river Seine, but flooding caused by the accidental breakage of a sewer resulted in the listing of the bridge. Although repairs were probably quite feasible, political forces led to the demolition of the bridge. Some of Navier's theoretical work studied the motion of solids and liquids, leading to the famous equations identified with his name and that of Stokes. After the French Revolution, Navier became an important consultant to the state, recommending policies for road construction and traffic, and for laying out a national railway system.

Source: *Dictionary of Scientific Biography*, Charles Scribner's Sons, New York, 1975.

$$\textit{Table 5.11} \quad \textit{Momentum Equations in Cylindrical Coordinates}$$

In terms of stresses (under all circumstances):   (5.74)

$$\rho\left(\frac{\partial v_r}{\partial t} + v_r\frac{\partial v_r}{\partial r} + \frac{v_\theta}{r}\frac{\partial v_r}{\partial \theta} - \frac{v_\theta^2}{r} + v_z\frac{\partial v_r}{\partial z}\right)$$

$$= -\frac{\partial p}{\partial r} + \frac{1}{r}\frac{\partial(r\tau_{rr})}{\partial r} + \frac{1}{r}\frac{\partial \tau_{r\theta}}{\partial \theta} - \frac{\tau_{\theta\theta}}{r} + \frac{\partial \tau_{rz}}{\partial z} + \rho g_r,$$

$$\rho\left(\frac{\partial v_\theta}{\partial t} + v_r\frac{\partial v_\theta}{\partial r} + \frac{v_\theta}{r}\frac{\partial v_\theta}{\partial \theta} + \frac{v_r v_\theta}{r} + v_z\frac{\partial v_\theta}{\partial z}\right)$$

$$= -\frac{1}{r}\frac{\partial p}{\partial \theta} + \frac{1}{r^2}\frac{\partial(r^2\tau_{r\theta})}{\partial r} + \frac{1}{r}\frac{\partial \tau_{\theta\theta}}{\partial \theta} + \frac{\partial \tau_{\theta z}}{\partial z} + \rho g_\theta,$$

$$\rho\left(\frac{\partial v_z}{\partial t} + v_r\frac{\partial v_z}{\partial r} + \frac{v_\theta}{r}\frac{\partial v_z}{\partial \theta} + v_z\frac{\partial v_z}{\partial z}\right)$$

$$= -\frac{\partial p}{\partial z} + \frac{1}{r}\frac{\partial(r\tau_{rz})}{\partial r} + \frac{1}{r}\frac{\partial \tau_{\theta z}}{\partial \theta} + \frac{\partial \tau_{zz}}{\partial z} + \rho g_z.$$

In terms of velocities, for a Newtonian fluid with constant ρ and μ: (5.75)

$$\rho\left(\frac{\partial v_r}{\partial t} + v_r\frac{\partial v_r}{\partial r} + \frac{v_\theta}{r}\frac{\partial v_r}{\partial \theta} - \frac{v_\theta^2}{r} + v_z\frac{\partial v_r}{\partial z}\right)$$

$$= -\frac{\partial p}{\partial r} + \mu\left[\frac{\partial}{\partial r}\left(\frac{1}{r}\frac{\partial(rv_r)}{\partial r}\right) + \frac{1}{r^2}\frac{\partial^2 v_r}{\partial \theta^2} - \frac{2}{r^2}\frac{\partial v_\theta}{\partial \theta} + \frac{\partial^2 v_r}{\partial z^2}\right] + \rho g_r,$$

$$\rho\left(\frac{\partial v_\theta}{\partial t} + v_r\frac{\partial v_\theta}{\partial r} + \frac{v_\theta}{r}\frac{\partial v_\theta}{\partial \theta} + \frac{v_r v_\theta}{r} + v_z\frac{\partial v_\theta}{\partial z}\right)$$

$$= -\frac{1}{r}\frac{\partial p}{\partial \theta} + \mu\left[\frac{\partial}{\partial r}\left(\frac{1}{r}\frac{\partial(rv_\theta)}{\partial r}\right) + \frac{1}{r^2}\frac{\partial^2 v_\theta}{\partial \theta^2} + \frac{2}{r^2}\frac{\partial v_r}{\partial \theta} + \frac{\partial^2 v_\theta}{\partial z^2}\right] + \rho g_\theta,$$

$$\rho\left(\frac{\partial v_z}{\partial t} + v_r\frac{\partial v_z}{\partial r} + \frac{v_\theta}{r}\frac{\partial v_z}{\partial \theta} + v_z\frac{\partial v_z}{\partial z}\right)$$

$$= -\frac{\partial p}{\partial z} + \mu\left[\frac{1}{r}\frac{\partial}{\partial r}\left(r\frac{\partial v_z}{\partial r}\right) + \frac{1}{r^2}\frac{\partial^2 v_z}{\partial \theta^2} + \frac{\partial^2 v_z}{\partial z^2}\right] + \rho g_z.$$

Table 5.12 Momentum Equations in Spherical Coordinates

In terms of stresses (under all circumstances): (5.76)

$$
\rho\left(\frac{\partial v_r}{\partial t} + v_r \frac{\partial v_r}{\partial r} + \frac{v_\theta}{r}\frac{\partial v_r}{\partial \theta} + \frac{v_\phi}{r\sin\theta}\frac{\partial v_r}{\partial \phi} - \frac{v_\theta^2 + v_\phi^2}{r}\right) = -\frac{\partial p}{\partial r}
$$
$$
+ \frac{1}{r^2}\frac{\partial}{\partial r}(r^2\tau_{rr}) + \frac{1}{r\sin\theta}\frac{\partial(\tau_{r\theta}\sin\theta)}{\partial \theta} + \frac{1}{r\sin\theta}\frac{\partial\tau_{r\phi}}{\partial \phi} - \frac{\tau_{\theta\theta} + \tau_{\phi\phi}}{r} + \rho g_r,
$$

$$
\rho\left(\frac{\partial v_\theta}{\partial t} + v_r \frac{\partial v_\theta}{\partial r} + \frac{v_\theta}{r}\frac{\partial v_\theta}{\partial \theta} + \frac{v_\phi}{r\sin\theta}\frac{\partial v_\theta}{\partial \phi} + \frac{v_r v_\theta}{r} - \frac{v_\phi^2\cot\theta}{r}\right) = -\frac{1}{r}\frac{\partial p}{\partial \theta}
$$
$$
+ \frac{1}{r^2}\frac{\partial(r^2\tau_{r\theta})}{\partial r} + \frac{1}{r\sin\theta}\frac{\partial(\tau_{\theta\theta}\sin\theta)}{\partial \theta} + \frac{1}{r\sin\theta}\frac{\partial\tau_{\theta\phi}}{\partial \phi} + \frac{\tau_{r\theta}}{r} - \frac{\cot\theta}{r}\tau_{\phi\phi} + \rho g_\theta,
$$

$$
\rho\left(\frac{\partial v_\phi}{\partial t} + v_r \frac{\partial v_\phi}{\partial r} + \frac{v_\theta}{r}\frac{\partial v_\phi}{\partial \theta} + \frac{v_\phi}{r\sin\theta}\frac{\partial v_\phi}{\partial \phi} + \frac{v_\phi v_r}{r} + \frac{v_\theta v_\phi\cot\theta}{r}\right) = -\frac{1}{r\sin\theta}\frac{\partial p}{\partial \phi}
$$
$$
+ \frac{1}{r^2}\frac{\partial(r^2\tau_{r\phi})}{\partial r} + \frac{1}{r}\frac{\partial\tau_{\theta\phi}}{\partial \theta} + \frac{1}{r\sin\theta}\frac{\partial\tau_{\phi\phi}}{\partial \phi} + \frac{\tau_{r\phi}}{r} + \frac{2\cot\theta}{r}\tau_{\theta\phi} + \rho g_\phi.
$$

In terms of velocities, for a Newtonian fluid with constant ρ and μ: (5.77)

$$
\rho\left(\frac{\partial v_r}{\partial t} + v_r \frac{\partial v_r}{\partial r} + \frac{v_\theta}{r}\frac{\partial v_r}{\partial \theta} + \frac{v_\phi}{r\sin\theta}\frac{\partial v_r}{\partial \phi} - \frac{v_\theta^2 + v_\phi^2}{r}\right)
$$
$$
= -\frac{\partial p}{\partial r} + \mu\left[\frac{1}{r^2}\frac{\partial}{\partial r}\left(r^2\frac{\partial v_r}{\partial r}\right) + \frac{1}{r^2\sin\theta}\frac{\partial}{\partial \theta}\left(\sin\theta\frac{\partial v_r}{\partial \theta}\right)\right.
$$
$$
\left. + \frac{1}{r^2\sin^2\theta}\frac{\partial^2 v_r}{\partial \phi^2} - \frac{2v_r}{r^2} - \frac{2}{r^2}\frac{\partial v_\theta}{\partial \theta} - \frac{2v_\theta\cot\theta}{r^2} - \frac{2}{r^2\sin\theta}\frac{\partial v_\phi}{\partial \phi}\right] + \rho g_r,
$$

$$
\rho\left(\frac{\partial v_\theta}{\partial t} + v_r \frac{\partial v_\theta}{\partial r} + \frac{v_\theta}{r}\frac{\partial v_\theta}{\partial \theta} + \frac{v_\phi}{r\sin\theta}\frac{\partial v_\theta}{\partial \phi} + \frac{v_r v_\theta}{r} - \frac{v_\phi^2\cot\theta}{r}\right)
$$
$$
= -\frac{1}{r}\frac{\partial p}{\partial \theta} + \mu\left[\frac{1}{r^2}\frac{\partial}{\partial r}\left(r^2\frac{\partial v_\theta}{\partial r}\right) + \frac{1}{r^2\sin\theta}\frac{\partial}{\partial \theta}\left(\sin\theta\frac{\partial v_\theta}{\partial \theta}\right)\right.
$$
$$
\left. + \frac{1}{r^2\sin^2\theta}\frac{\partial^2 v_\theta}{\partial \phi^2} + \frac{2}{r^2}\frac{\partial v_r}{\partial \theta} - \frac{v_\theta}{r^2\sin^2\theta} - \frac{2\cos\theta}{r^2\sin^2\theta}\frac{\partial v_\phi}{\partial \phi}\right] + \rho g_\theta,
$$

$$
\rho\left(\frac{\partial v_\phi}{\partial t} + v_r \frac{\partial v_\phi}{\partial r} + \frac{v_\theta}{r}\frac{\partial v_\phi}{\partial \theta} + \frac{v_\phi}{r\sin\theta}\frac{\partial v_\phi}{\partial \phi} + \frac{v_\phi v_r}{r} + \frac{v_\theta v_\phi\cot\theta}{r}\right)
$$
$$
= -\frac{1}{r\sin\theta}\frac{\partial p}{\partial \phi} + \mu\left[\frac{1}{r^2}\frac{\partial}{\partial r}\left(r^2\frac{\partial v_\phi}{\partial r}\right) + \frac{1}{r^2\sin\theta}\frac{\partial}{\partial \theta}\left(\sin\theta\frac{\partial v_\phi}{\partial \theta}\right)\right.
$$
$$
\left. + \frac{1}{r^2\sin^2\theta}\frac{\partial^2 v_\phi}{\partial \phi^2} - \frac{v_\phi}{r^2\sin^2\theta} + \frac{2}{r^2\sin\theta}\frac{\partial v_r}{\partial \phi} + \frac{2\cos\theta}{r^2\sin^2\theta}\frac{\partial v_\theta}{\partial \phi}\right] + \rho g_\phi.
$$

Example 5.9—Vector Form of Variable-Viscosity Momentum Balance

Occasionally, even though the density is essentially constant, there may be significant viscosity variations—especially for non-Newtonian and turbulent flows—in which case the simplifications of the previous example cannot be made. We start again from the first equation in Example 5.8:

$$
\rho \frac{Dv_x}{Dt} = \frac{\partial \sigma_{xx}}{\partial x} + \frac{\partial \tau_{yx}}{\partial y} + \frac{\partial \tau_{zx}}{\partial z} + \rho g_x
$$

$$
= \frac{\partial}{\partial x}\left(-p + 2\mu\frac{\partial v_x}{\partial x} - \frac{2}{3}\mu \nabla \cdot \mathbf{v}\right) + \frac{\partial}{\partial y}\left[\mu\left(\frac{\partial v_x}{\partial y} + \frac{\partial v_y}{\partial x}\right)\right]
$$

$$
+ \frac{\partial}{\partial z}\left[\mu\left(\frac{\partial v_x}{\partial z} + \frac{\partial v_z}{\partial x}\right)\right] + \rho g_x. \tag{E5.9.1}
$$

For an incompressible fluid, which frequently occurs, $\nabla \cdot \mathbf{v} = 0$ and we obtain, for the x-momentum balance:

$$
\rho\left(\frac{\partial v_x}{\partial t} + v_x\frac{\partial v_x}{\partial x} + v_y\frac{\partial v_x}{\partial y} + v_z\frac{\partial v_x}{\partial z}\right) = -\frac{\partial p}{\partial x} + \frac{\partial}{\partial x}\left[2\mu\frac{\partial v_x}{\partial x}\right]
$$

$$
+ \frac{\partial}{\partial y}\left[\mu\left(\frac{\partial v_x}{\partial y} + \frac{\partial v_y}{\partial x}\right)\right] + \frac{\partial}{\partial z}\left[\mu\left(\frac{\partial v_z}{\partial x} + \frac{\partial v_x}{\partial z}\right)\right] + \rho g_x. \tag{E5.9.2}
$$

In vector form this last equation, together with similar y and z momentum balances, can be summarized as:

$$
\rho\frac{\partial \mathbf{v}}{\partial t} + \rho\mathbf{v} \cdot \nabla\mathbf{v} = -\nabla p + \nabla \cdot \mu(\nabla\mathbf{v} + (\nabla\mathbf{v})^T) + \mathbf{F}. \tag{E5.9.3}
$$

Here, $\mathbf{F}$ includes the three individual gravitational terms. Appendix C defines the dyadic product $\nabla\mathbf{v}$ and its transpose, together with $\nabla \cdot \boldsymbol{\tau}$, from which it follows that $\nabla \cdot \mu(\nabla\mathbf{v} + (\nabla\mathbf{v})^T)$ expands into three sets of viscous terms, as in Eqn. (E5.9.2) and its counterparts in the y and z directions.

The compact form of Eqn. (E5.9.3) will be used in Section 9.5, when the k/ε method for computing turbulent flows is discussed. □.

PROBLEMS FOR CHAPTER 5

1. *Linear velocity in terms of angular velocity—E.* By focusing on the meaning of $\boldsymbol{\omega} \times \mathbf{r}$, both regarding its magnitude and direction, show that it equals $\mathbf{v}$ in Fig. 5.6. Also prove that $\mathbf{v} = \boldsymbol{\omega} \times \mathbf{r}$ can be expressed as the determinant in Eqn. (5.25).

2. *Angular velocity and the curl—E.* Verify Eqn. (5.30), namely, that the curl of the velocity equals twice the angular velocity, which is assumed to be constant, as in rigid-body rotation. *Hint*: Note that the components of any position vector $\mathbf{r}$ in the RCCS are $r_x = x$, $r_y = y$, and $r_z = z$.

3. *Curl of the gradient of a scalar—E.* If a vector $\mathbf{v}$ is given by the gradient of a scalar, as in $\mathbf{v} = \nabla\phi$, prove that its curl is zero, as in Eqn. (5.32). What is the significance of this result?

4. *Cylindrical coordinates—E.* In the CCS, what types of surfaces are described by constant values of r, θ, and z?

5. *Spherical coordinates—E.* In the SCS, what types of surfaces are described by constant values of r, θ, and ϕ?

6. *Alternative form of the continuity equation—E.* By using the vector identity $\nabla \cdot \rho\mathbf{v} = \mathbf{v} \cdot \nabla\rho + \rho\nabla \cdot \mathbf{v}$, prove that the continuity equation:

$$\frac{\partial\rho}{\partial t} + \nabla \cdot \rho\mathbf{v} = 0, \tag{5.47}$$

can be reexpressed in the equivalent form,

$$\frac{\mathcal{D}\rho}{\mathcal{D}t} + \rho\nabla \cdot \mathbf{v} = 0. \tag{E5.7.2}$$

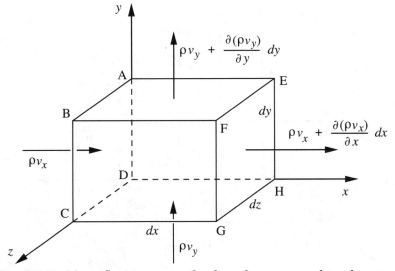

Fig. P5.7 Mass fluxes across the four faces normal to the x and y axes; those across the other two faces are omitted for clarity.

7. *Alternative derivation of continuity equation—M.* By performing a transient mass balance on the *stationary* control volume ABCDEFGH shown in Fig. P5.7, check that you again obtain the continuity equation in RCCS, Eqn. (5.48).

8. *Pressure in terms of normal stresses—E.* Prove from Eqns. (5.61) that the pressure is the negative of the mean of the three normal stresses.

9. *Transverse velocity in a boundary layer (I)—M.* A highly simplified viewpoint of incompressible fluid flow in a boundary layer on a flat plate (see Section 8.2) gives the x velocity component v_x and the boundary-layer thickness δ in terms of the following dimensionless groups:

$$\frac{v_x}{v_{x\infty}} = \frac{y}{\delta}, \qquad \text{where} \qquad \frac{\delta}{x} = 3.46\sqrt{\frac{\nu}{v_{x\infty}x}}.$$

Here, $v_{x\infty}$ is the x velocity well away from the plate, x and y are coordinates along and normal to the plate, respectively, and ν is the kinematic viscosity of the fluid.

If $v_y = 0$ at $y = 0$ (the plate), derive two equations that relate the following two dimensionless velocities (involving v_x and v_y) to the dimensionless distance ζ:

$$\frac{v_x}{v_{x\infty}}, \qquad v_y\sqrt{\frac{x}{\nu v_{x\infty}}}, \qquad \zeta = y\sqrt{\frac{v_{x\infty}}{\nu x}}.$$

Investigate the extent to which the dimensionless x and y velocities agree with those in Fig. 8.5 (which is based on the accurate Blasius solution), for $\zeta = 0, 1, 2,$ and 3.

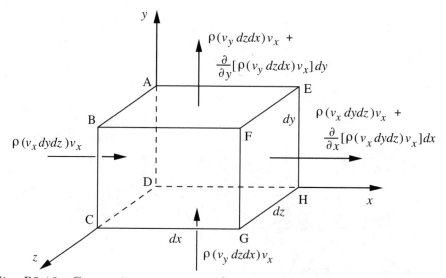

Fig. P5.10   Convective momentum fluxes across the four faces normal to the x and y axes; those across the other two faces are omitted for clarity.

10. *Momentum balance on a fixed control volume—M.* Repeat the differential x momentum balance in Section 5.6, but now assume that the parallelepiped-shaped control volume shown in Fig. 5.13 is *fixed* in space. Observe that there are now *six* convective momentum fluxes to be considered, four of which are shown in Fig. P5.10. All the terms on the right-hand side of Eqn. (5.53) will be unchanged, but you will no longer be using the convective derivative on the left-hand side. You will also need to involve the continuity equation, (5.48).

11. *Transverse velocity in a boundary layer (II)—M.* A simplified viewpoint of incompressible fluid flow in a boundary layer on a flat plate (see Section 8.2) gives the x velocity component v_x and the boundary-layer thickness δ in terms of the following dimensionless groups:

$$\frac{v_x}{v_{x\infty}} = \sin\frac{\pi y}{2\delta}, \qquad \text{where} \qquad \frac{\delta}{x} = 4.79\sqrt{\frac{\nu}{v_{x\infty}x}}.$$

Here, $v_{x\infty}$ is the x velocity well away from the plate, x and y are coordinates along and normal to the plate, respectively, and ν is the kinematic viscosity of the fluid.

If $v_y = 0$ at $y = 0$ (the plate), derive two equations that relate the following two dimensionless velocities (involving v_x and v_y) to the dimensionless distance ζ:

$$\frac{v_x}{v_{x\infty}}, \qquad v_y\sqrt{\frac{x}{\nu v_{x\infty}}}, \qquad \zeta = y\sqrt{\frac{v_{x\infty}}{\nu x}}.$$

Explain the extent to which the dimensionless x and y velocities agree with those in Fig. 8.5 (which is based on the accurate Blasius solution), for $\zeta = 0, 1, 2,$ and 3.

12. *Vortices and angular velocity—E.* Two types of vortices—forced and free—will be described in Chapter 7. In cylindrical coordinates, the velocity components for a *forced* vortex [Fig. 7.1(a)] are:

$$v_r = 0, \quad v_\theta = r\omega, \quad v_z = 0,$$

whereas those for a *free* vortex [Fig. 7.1(b)] are:

$$v_r = 0, \quad v_\theta = \frac{c}{r}, \quad v_z = 0.$$

Determine the z component of the vorticity, and hence the angular velocity, in each case.

13. *Flow past a cylinder—M.* In a certain incompressible flow, the vector velocity $\mathbf{v}$ is the gradient of a scalar ϕ. That is, $\mathbf{v} = \nabla\phi$, where, in cylindrical coordinates, the flow is such that:

$$\phi = U\left(r + \frac{a^2}{r}\right)\cos\theta.$$

(The flow is that of a uniform stream past a cylinder, as shown in Fig. 7.8.)

(a) Prove that the two nonzero velocity components are:

$$v_r = U\left(1 - \frac{a^2}{r^2}\right)\cos\theta, \qquad v_\theta = -U\left(1 + \frac{a^2}{r^2}\right)\sin\theta.$$

(b) Verify that these velocities satisfy the continuity equation.

(c) Evaluate the z component of the vorticity, $\nabla \times \mathbf{v}$, and hence of the angular velocity $\boldsymbol{\omega}$, as functions of position.

14. *Flow past a sphere—E.* For incompressible flow past a sphere (Fig. 7.17), the velocity components in spherical coordinates are here presumed to be:

$$v_r = U\cos\theta\left(1 - \frac{a^3}{r^3}\right), \qquad v_\theta = U\sin\theta\left(1 + \frac{a^3}{2r^3}\right), \qquad v_\phi = 0.$$

Unfortunately, the above expression for v_θ contains a sign error. Use the continuity equation to locate the error and correct it. Then prove that the angular velocity of the flow is zero everywhere.

15. *Vorticity in pipe flow—E.* For laminar viscous flow in a pipe of radius a, the z velocity component at any radius r $(\leq a)$ is:

$$v_z = \frac{1}{4\mu}\left(-\frac{\partial p}{\partial z}\right)(a^2 - r^2),$$

where μ is the viscosity and $\partial p/\partial z$ is the pressure gradient. The other two velocity components, v_r and v_θ, are both zero. Evaluate the three components of the vorticity $\nabla \times \mathbf{v}$ as functions of position.

16. *Propagation of a sound wave—D.* As a one-dimensional sound wave propagates through an otherwise stationary inviscid fluid that is compressible, the density ρ and velocity v_x deviate by *small* amounts ρ' and v_x' from their initial values of ρ_0 and 0, respectively:

$$\rho = \rho_0 + \rho',$$

$$v_x = 0 + v_x' = v_x'.$$

From mass and momentum balances, simplified to one dimension, prove that the deviations ρ' obey:

$$\frac{\partial^2 \rho'}{\partial t^2} = c^2\frac{\partial^2 \rho'}{\partial x^2},$$

where $c = \sqrt{dp/d\rho}$ is a property of the fluid (see Problem 2.27), being about 1,000 and 4,000 ft/s for air and water, respectively. Ignore any small second-order quantities.

Verify by substitution into the above differential equation that a possible solution for the density fluctuations is:

$$\rho' = A\sin\left[\omega\left(\frac{x}{c} - t\right)\right],$$

in which A is a constant amplitude. Also answer the following:

(a) Prove that $\lambda = 2\pi c/\omega$ is the *wavelength* by verifying that $\rho'(x) = \rho'(x + \lambda)$.

(b) What is the velocity of the wave? *Hint*: At what later value of time, $t + T$, does $\rho'(t) = \rho'(t + T)$? Note that the wave has traveled a distance λ in this additional time T.

(c) What is the corresponding function for $v_x'(x, t)$?

17. *Continuity equation for a sound wave—M.* A one-dimensional sound wave of wavelength λ travels with a velocity c. The velocity and density can be shown to vary as:

$$v_x = \frac{Ac}{\rho_0} \sin\left[\omega\left(\frac{x}{c} - t\right)\right],$$

$$\rho = \rho_0 + A\sin\left[\omega\left(\frac{x}{c} - t\right)\right].$$

For waves of small amplitude, $(A \ll \rho_0)$, derive an expression for the divergence of ρv_x, and verify that the continuity equation, (5.48), is satisfied.

18. *Sound waves in a conical organ pipe—M.* Analyze the pressure fluctuations in the conical organ pipe shown in Fig. P5.18, at the apex of which $(r = 0)$ a vibrating reed provides an oscillating pressure, with the other end open to the atmosphere $(p = 0)$. Because of the conical shape, spherical coordinates are appropriate, with $v_\theta = v_\phi = 0$. The density, velocity, and pressure are of the form $\rho = \rho_0 + \rho'$, $v_r = 0 + v_r'$, and $p = p_0 + p'$, in which ρ', v_r', and p' denote *small* fluctuations from the steady values ρ_0, 0, and p_0. Viscosity and gravity are insignificant. The velocity of sound is given by $c^2 = dp/d\rho$.

Fig. P5.18   *Conical organ pipe with oscillating pressure at $r = 0$.*

(a) Simplify the continuity and r momentum equations in spherical coordinates.

(b) By substituting for ρ, v_r, and p in terms of fluctuating components, and ignoring all terms in which primed quantities appear twice, prove that the pressure fluctuations obey:

$$\frac{\partial^2 p'}{\partial t^2} = \frac{c^2}{r^2}\frac{\partial}{\partial r}\left(r^2\frac{\partial p'}{\partial r}\right). \tag{P5.18.1}$$

(c) You may assume *without proof* that a solution to Eqn. (P5.18.1) is:

$$p' = \frac{1}{r}(A\cos kr + B\sin kr)\sin \omega t, \tag{P5.18.2}$$

in which $\omega = kc$ and k is yet to be determined.

(i) Why is A zero?

(ii) In terms of the length L, what values of k are permissible?

19. *Vector operations—M*. If $\mathbf{u} = 3y^2\mathbf{e}_x - x\mathbf{e}_y + 5\mathbf{e}_z$, $\mathbf{v} = x\mathbf{e}_x + (y - z)\mathbf{e}_y$, and $a = 5x^3 - y + yz$, evaluate:

(a)	$\mathbf{u} \cdot \mathbf{v}$	(b)	$\mathbf{u} \times \mathbf{v}$	(c)	$\nabla \times \mathbf{v}$
(d)	∇a	(e)	$\nabla^2 a$	(f)	$\nabla \cdot \mathbf{u}$

Based on your answers above, evaluate the following true/false assertions:

(a) The dot product of two vectors is a scalar.

(b) The cross product of two vectors is a scalar.

(c) The curl of a vector is a vector.

(d) The gradient of a scalar is a scalar.

(e) The Laplacian of a scalar is a vector.

(f) The divergence of a vector is a scalar.

20. *True/false.* Check *true* or *false*, as appropriate:

(a) The dot product of a unit vector with itself is zero.   T ☐   F ☐

(b) The magnitude of the cross product of two vectors is the area of a parallelogram with adjacent sides equal to the two vectors.   T ☐   F ☐

(c) If you wanted to walk to the summit of a hill in the fewest number of (equal) steps, you should always follow the direction of the gradient of the elevation at all stages of your path.   T ☐   F ☐

(d) Velocity is another way of expressing a mass flux.   T ☐   F ☐

(e) The divergence of the velocity vector is the net rate of outflow of volume per unit volume.   T ☐   F ☐

(f) If the density is constant, the microscopic mass balance reduces to $\nabla \times \mathbf{v} = 0$.   T ☐   F ☐

(g) Angular velocity is equal to twice the vorticity.   T ☐   F ☐

(h) The Laplacian operator is equivalent to the divergence of the gradient operator.   T ☐   F ☐

(i) In spherical coordinates, the three unit vectors are $\mathbf{e}_z$, $\mathbf{e}_\theta$, and $\mathbf{e}_\phi$.   T ☐   F ☐

(j) The convective derivative gives the rate of change of a property, following the path taken by a fluid.   T ☐   F ☐

(k)   Newton's law of viscosity relates to the rate of change   T ☐   F ☐
of an angle of a deforming fluid element under the
influence of applied shear stresses.

(l)   In the Navier-Stokes equations, there are just three   T ☐   F ☐
principal types of terms: inertial, viscous, and gravitational.

(m)   In cylindrical coordinates, the location of a point P is   T ☐   F ☐
specified by one distance and two angles.

(n)   The sign convention for a normal stress is that it is   T ☐   F ☐
considered positive if it is a compressive stress, such
as pressure.

(o)   The basic modes of fluid motion are translation, ro-   T ☐   F ☐
tation, and shear.

(p)   The unit vector $\mathbf{e}_r$ in cylindrical coordinates always   T ☐   F ☐
points in the same direction, for all values of r, θ, and
z.

(q)   The term $\rho \nabla \cdot \mathbf{v}$ represents a rate of increase of mass,   T ☐   F ☐
such as could occur if the pressure of a small element
of gas increases.

(r)   For steady flow, the continuity equation always re-   T ☐   F ☐
duces to $\nabla \cdot \mathbf{v} = 0$.

(s)   The curl of the velocity vector equals the angular ve-   T ☐   F ☐
locity.

(t)   Of the nine components of the total-stress tensor, σ_{xx},   T ☐   F ☐
τ_{xy}, τ_{yx}, etc., only six are independent.

Chapter 6

SOLUTION OF VISCOUS-FLOW PROBLEMS

6.1 Introduction

THE previous chapter contained derivations of the relationships for the conservation of mass and momentum—the *equations of motion*—in rectangular, cylindrical, and spherical coordinates. All the experimental evidence indicates that these are indeed the most fundamental equations of fluid mechanics, and that in principle they govern any situation involving the flow of a Newtonian fluid. Unfortunately, because of their all-embracing quality, their solution in analytical terms is difficult or impossible except for relatively simple situations. However, it is important to be aware of these "Navier-Stokes equations," for the following reasons:

1. They lead to the analytical and exact solution of some simple, yet important problems, as will be demonstrated by examples in this chapter.
2. They form the basis for further work in other areas of chemical engineering.
3. If a few realistic *simplifying assumptions* are made, they can often lead to approximate solutions that are eminently acceptable for many engineering purposes. Representative examples occur in the study of boundary layers, waves, lubrication, coating of substrates with films, and inviscid (irrotational) flow.
4. With the aid of more sophisticated techniques, such as those involving power series and asymptotic expansions, and *particularly* computer-based numerical methods (as implemented by CFD or computational fluid dynamics software such as FlowLab and COMSOL), they can lead to the solution of moderately or highly advanced problems, such as those involving injection-molding of polymers and even the incredibly difficult problem of weather prediction.

The following sections present exact solutions of the equations of motion for several relatively simple problems in rectangular, cylindrical, and spherical coordinates, augmented by a couple of CFD examples. Throughout, unless otherwise stated, the flow is assumed to be *steady, laminar,* and *Newtonian,* with *constant* density and viscosity. Although these assumptions are necessary in order to obtain solutions, they are nevertheless realistic in several instances.

The reader is cautioned that probably a majority of industrial processes involve *turbulent* flow, which is more complicated, and for this reason sometimes receives scant attention in university courses. The author has seen instances in which erroneous assumptions of laminar flow have led to answers that are wildly inaccurate.

All of the examples in this chapter are characterized by low Reynolds numbers. That is, the viscous forces are much more important than the inertial forces, and are usually counterbalanced by pressure or gravitational effects. Typical applications occur in microfluidics (involving tiny channels—see Chapter 12) and in the flow of high-viscosity polymers. Situations in which viscous effects are relatively *unimportant* will be discussed in Chapter 7.

Solution procedure.   The general *procedure* for solving each problem involves the following steps:

1. Make reasonable simplifying assumptions. Almost all of the cases treated here will involve *steady incompressible* flow of a *Newtonian* fluid in a *single* coordinate direction. Further, *gravity* may or may not be important, and a certain amount of *symmetry* may be apparent.

2. Write down the *equations of motion*—both mass (continuity) and momentum balances—and simplify them according to the assumptions made previously, striking out terms that are zero. Typically, only a very few terms—perhaps only one in some cases—will remain in each differential equation. The simplified continuity equation usually yields information that can subsequently be used to simplify the momentum equations.

3. *Integrate* the simplified equations in order to obtain expressions for the dependent variables such as velocities and pressure. These expressions will usually contain some, as yet, arbitrary constants—typically *two* for the velocities (since they appear in second-order derivatives in the momentum equations) and *one* for the pressure (since it appears only in a first-order derivative).

4. Invoke the *boundary conditions* in order to evaluate the constants appearing in the previous step. For *pressure,* such a condition usually amounts to a specified pressure at a certain location—at the inlet of a pipe, or at a free surface exposed to the atmosphere, for example. For the *velocities*, these conditions fall into either of the following classifications:

   (a) Continuity of the velocity, amounting to a *no-slip* condition. Thus, the velocity of the fluid in contact with a solid surface typically equals the velocity of that surface—zero if the surface is stationary.[1] And, for the few cases in which one fluid (A, for example) is in contact with another immiscible fluid (B), the velocity in fluid A equals the velocity in fluid B at the common interface.

---

[1] In a few exceptional situations there may be lack of adhesion between the fluid and surface, in which case *slip* can occur. Also see Example 12.4, in which electroosmosis gives the illusion of slip.

(b) Continuity of the shear stress, usually between two fluids A and B, leading to the product of viscosity and a velocity gradient having the same value at the common interface, whether in fluid A or B. If fluid A is a liquid, and fluid B is a relatively stagnant gas, which—because of its low viscosity—is incapable of sustaining any significant shear stress, then the common shear stress is effectively zero.

5. At this stage, the problem is essentially solved for the pressure and velocities. Finally, if desired, shear-stress distributions can be derived by differentiating the velocities in order to obtain the velocity *gradients*; numerical predictions of process variables can also be made.

Types of flow.   Two broad classes of viscous flow will be illustrated in this chapter:

1. *Poiseuille flow,* in which an applied pressure difference causes fluid motion between stationary surfaces.
2. *Couette flow,* in which a moving surface drags adjacent fluid along with it and thereby imparts a motion to the rest of the fluid.

Occasionally, it is possible to have both types of motion occurring simultaneously, as in the screw extruder analyzed in Example 6.4.

6.2   Solution of the Equations of Motion in Rectangular Coordinates

The remainder of this chapter consists almost entirely of a series of worked examples, illustrating the above steps for solving viscous-flow problems.

Example 6.1—Flow Between Parallel Plates

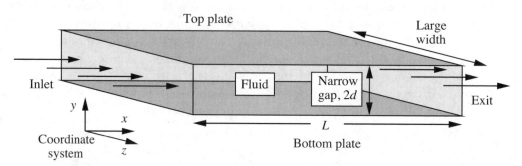

Fig. E6.1.1   Geometry for flow through a rectangular duct. The spacing between the plates is exaggerated in relation to their length.

Fig. E6.1.1 shows a fluid of viscosity μ that flows in the x direction between two rectangular plates, whose width is very large in the z direction when compared to their separation in the y direction. Such a situation could occur in a die when

Example 6.1—Flow Between Parallel Plates   **295**

a polymer is being extruded at the exit into a sheet, which is subsequently cooled and solidified. Determine the relationship between the flow rate and the pressure drop between the inlet and exit, together with several other quantities of interest.

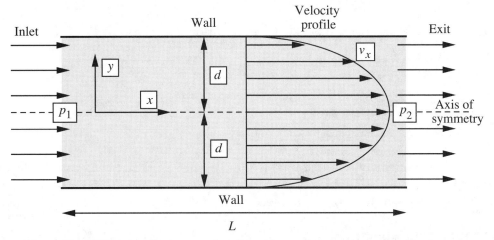

Fig. E6.1.2   Geometry for flow through a rectangular duct.

Simplifying assumptions. The situation is analyzed by referring to a cross section of the duct, shown in Fig. E6.1.2, taken at any fixed value of z. Let the depth be $2d$ ($\pm d$ above and below the centerline or axis of symmetry $y = 0$), and the length L. Note that the motion is of the *Poiseuille* type, since it is caused by the applied pressure difference $(p_1 - p_2)$. Make the following fairly realistic *assumptions* about the flow:

1. As already stated, it is steady and Newtonian, with constant density and viscosity. (These assumptions will often be taken for granted, and not restated, in later problems.)

2. Entrance effects can be neglected, so that the flow is fully developed, in which case there is only one nonzero velocity component—that in the direction of flow, v_x. Thus, $v_y = v_z = 0$.

3. Since, in comparison with their spacing, $2d$, the plates extend for a very long distance in the z direction, all locations in this direction appear essentially identical to one another. In particular, there is no variation of the velocity in the z direction, so that $\partial v_x / \partial z = 0$.

4. Gravity acts vertically downward; hence, $g_y = -g$ and $g_x = g_z = 0$.

5. The velocity is zero in contact with the plates, so that $v_x = 0$ at $y = \pm d$.

Continuity.   Start by examining the general continuity equation, (5.48):

$$\frac{\partial \rho}{\partial t} + \frac{\partial(\rho v_x)}{\partial x} + \frac{\partial(\rho v_y)}{\partial y} + \frac{\partial(\rho v_z)}{\partial z} = 0, \tag{5.48}$$

which, in view of the constant-density assumption, simplifies to Eqn. (5.52):

$$\frac{\partial v_x}{\partial x} + \frac{\partial v_y}{\partial y} + \frac{\partial v_z}{\partial z} = 0. \tag{5.52}$$

But since $v_y = v_z = 0$:

$$\frac{\partial v_x}{\partial x} = 0, \tag{E6.1.1}$$

so v_x is independent of the distance from the inlet, and the velocity profile will appear the same for *all* values of x. Since $\partial v_x/\partial z = 0$ (assumption 3), it follows that $v_x = v_x(y)$ is a function of y only.

Momentum balances.   With the stated assumptions of a Newtonian fluid with constant density and viscosity, Eqn. (5.73) gives the x, y, and z momentum balances:

$$\rho\left(\frac{\partial v_x}{\partial t} + v_x\frac{\partial v_x}{\partial x} + v_y\frac{\partial v_x}{\partial y} + v_z\frac{\partial v_x}{\partial z}\right) = -\frac{\partial p}{\partial x} + \mu\left(\frac{\partial^2 v_x}{\partial x^2} + \frac{\partial^2 v_x}{\partial y^2} + \frac{\partial^2 v_x}{\partial z^2}\right) + \rho g_x,$$

$$\rho\left(\frac{\partial v_y}{\partial t} + v_x\frac{\partial v_y}{\partial x} + v_y\frac{\partial v_y}{\partial y} + v_z\frac{\partial v_y}{\partial z}\right) = -\frac{\partial p}{\partial y} + \mu\left(\frac{\partial^2 v_y}{\partial x^2} + \frac{\partial^2 v_y}{\partial y^2} + \frac{\partial^2 v_y}{\partial z^2}\right) + \rho g_y,$$

$$\rho\left(\frac{\partial v_z}{\partial t} + v_x\frac{\partial v_z}{\partial x} + v_y\frac{\partial v_z}{\partial y} + v_z\frac{\partial v_z}{\partial z}\right) = -\frac{\partial p}{\partial z} + \mu\left(\frac{\partial^2 v_z}{\partial x^2} + \frac{\partial^2 v_z}{\partial y^2} + \frac{\partial^2 v_z}{\partial z^2}\right) + \rho g_z.$$

With $v_y = v_z = 0$ (from assumption 2), $\partial v_x/\partial x = 0$ [from the simplified continuity equation, (E6.1.1)], $g_y = -g$, $g_x = g_z = 0$ (assumption 4), and steady flow (assumption 1), these momentum balances simplify enormously, to:

$$\mu\frac{\partial^2 v_x}{\partial y^2} = \frac{\partial p}{\partial x}, \tag{E6.1.2}$$

$$\frac{\partial p}{\partial y} = -\rho g, \tag{E6.1.3}$$

$$\frac{\partial p}{\partial z} = 0. \tag{E6.1.4}$$

Pressure distribution.   The last of the simplified momentum balances, Eqn. (E6.1.4), indicates no variation of the pressure across the width of the system (in the z direction), which is hardly a surprising result. When integrated, the

Example 6.1—Flow Between Parallel Plates   **297**

second simplified momentum balance, Eqn. (E6.1.3), predicts that the pressure varies according to:

$$p = -\rho g \int dy + f(x) = -\rho g y + f(x).$$ (E6.1.5)

Observe carefully that since a *partial* differential equation is being integrated, we obtain not a constant of integration, but a *function* of integration, $f(x)$.

Assume—to be verified later—that $\partial p/\partial x$ is constant, so that the centerline pressure (at $y = 0$) is given by a linear function of the form:

$$p_{y=0} = a + bx.$$ (E6.1.6)

The constants a and b may be determined from the inlet and exit centerline pressures:

$$x = 0: \qquad p = p_1 = a,$$ (E6.1.7)

$$x = L: \qquad p = p_2 = a + bL,$$ (E6.1.8)

leading to:

$$a = p_1, \qquad b = -\frac{p_1 - p_2}{L}.$$ (E6.1.9)

Thus, the centerline pressure falls *linearly* from p_1 at the inlet to p_2 at the exit:

$$f(x) = p_1 - \frac{x}{L}(p_1 - p_2),$$ (E6.1.10)

so that the complete pressure distribution is

$$p = p_1 - \frac{x}{L}(p_1 - p_2) - \rho g y.$$ (E6.1.11)

That is, the pressure declines linearly, both from the bottom plate to the top plate, and *also* from the inlet to the exit. In the majority of applications, $2d \ll L$, and the relatively small pressure variation in the y direction is usually ignored. Thus, p_1 and p_2, although strictly the *centerline* values, are typically referred to as *the* inlet and *the* exit pressures, respectively.

Velocity profile.   Since, from Eqn. (E6.1.1), v_x does not depend on x, $\partial^2 v_x/\partial y^2$ appearing in Eqn. (E6.1.2) becomes a *total* derivative, so this equation can be rewritten as:

$$\mu \frac{d^2 v_x}{dy^2} = \frac{\partial p}{\partial x},$$ (E6.1.12)

which is a second-order ordinary differential equation, in which the pressure gradient will be shown to be *uniform* between the inlet and exit, being given by:

$$-\frac{\partial p}{\partial x} = \frac{p_1 - p_2}{L}.$$ (E6.1.13)

A minus sign is used on the left-hand side, since $\partial p/\partial x$ is negative, thus rendering both sides of Eqn. (E6.1.13) as positive quantities.

Equation (E6.1.12) can be integrated twice, in turn, to yield an expression for the velocity. After multiplication through by dy, a first integration gives:

$$\int \frac{d^2 v_x}{dy^2}\, dy = \int \frac{d}{dy}\left(\frac{dv_x}{dy}\right) dy = \int \frac{1}{\mu}\left(\frac{\partial p}{\partial x}\right) dy,$$

$$\frac{dv_x}{dy} = \frac{1}{\mu}\left(\frac{\partial p}{\partial x}\right) y + c_1. \qquad (E6.1.14)$$

A second integration, of Eqn. (E6.1.14), yields:

$$\int \frac{dv_x}{dy}\, dy = \int \left[\frac{1}{\mu}\left(\frac{\partial p}{\partial x}\right) y + c_1\right] dy,$$

$$v_x = \frac{1}{2\mu}\left(\frac{\partial p}{\partial x}\right) y^2 + c_1 y + c_2. \qquad (E6.1.15)$$

The two constants of integration, c_1 and c_2, are determined by invoking the boundary conditions:

$$y = 0: \qquad \frac{dv_x}{dy} = 0, \qquad (E6.1.16)$$

$$y = d: \qquad v_x = 0, \qquad (E6.1.17)$$

leading to:

$$c_1 = 0, \qquad c_2 = -\frac{1}{2\mu}\left(\frac{\partial p}{\partial x}\right) d^2. \qquad (E6.1.18)$$

Eqns. (E6.1.15) and (E6.1.18) then furnish the velocity profile:

$$v_x = \frac{1}{2\mu}\left(-\frac{\partial p}{\partial x}\right)(d^2 - y^2), \qquad (E6.1.19)$$

in which $-\partial p/\partial x$ and $(d^2 - y^2)$ are both *positive* quantities. The velocity profile is *parabolic* in shape, and is shown in Fig. E6.1.2.

Alternative integration procedure.   Observe that we have used *indefinite* integrals in the above solution, and have employed the boundary conditions to determine the constants of integration. An alternative approach would again be to integrate Eqn. (E6.1.12) twice, but now to involve *definite* integrals by inserting the boundary conditions as limits of integration.

Example 6.1—Flow Between Parallel Plates   **299**

Thus, by separating variables, integrating once, and noting from symmetry about the centerline that $dv_x/dy = 0$ at $y = 0$, we obtain:

$$\mu \int_0^{dv_x/dy} d\left(\frac{dv_x}{dy}\right) = \frac{\partial p}{\partial x} \int_0^y dy, \qquad \text{(E6.1.20)}$$

or:

$$\frac{dv_x}{dy} = \frac{1}{\mu}\left(\frac{\partial p}{\partial x}\right) y. \qquad \text{(E6.1.21)}$$

A second integration, noting that $v_x = 0$ at $y = d$ (zero velocity in contact with the upper plate—the no-slip condition) yields:

$$\int_0^{v_x} dv_x = \frac{1}{\mu}\left(\frac{\partial p}{\partial x}\right) \int_d^y y\, dy. \qquad \text{(E6.1.22)}$$

That is:

$$v_x = \frac{1}{2\mu}\left(-\frac{\partial p}{\partial x}\right)(d^2 - y^2), \qquad \text{(E6.1.23)}$$

in which two minus signs have been introduced into the right-hand side in order to make quantities in both pairs of parentheses positive. This result is identical to the earlier Eqn. (E6.1.19). The student is urged to become familiar with both procedures, before deciding on the one that is individually best suited.

Also, the reader who is troubled by the assumption of symmetry of v_x about the centerline (and by never using the fact that $v_x = 0$ at $y = -d$), should be reassured by an alternative approach, starting from Eqn. (E6.1.15):

$$v_x = \frac{1}{2\mu}\left(\frac{\partial p}{\partial x}\right) y^2 + c_1 y + c_2. \qquad \text{(E6.1.24)}$$

Application of the *two* boundary conditions, $v_x = 0$ at $y = \pm d$, gives

$$c_1 = 0, \qquad c_2 = -\frac{1}{2\mu}\left(\frac{\partial p}{\partial x}\right) d^2, \qquad \text{(E6.1.25)}$$

leading again to the velocity profile of Eqn. (E6.1.19) *without* the assumption of symmetry.

Volumetric flow rate. Integration of the velocity profile yields an expression for the volumetric flow rate Q per unit width of the system. Observe first that the differential flow rate through an element of depth dy is $dQ = v_x dy$, so that:

$$Q = \int_0^Q dQ = \int_{-d}^d v_x\, dy = \int_{-d}^d \frac{1}{2\mu}\left(-\frac{\partial p}{\partial x}\right)(d^2 - y^2)\, dy = \frac{2d^3}{3\mu}\left(-\frac{\partial p}{\partial x}\right). \qquad \text{(E6.1.26)}$$

Since from an overall macroscopic balance Q is constant, it follows that $\partial p / \partial x$ is also constant, independent of distance x; the assumptions made in Eqns. (E6.1.6) and (E6.1.13) are therefore verified. The *mean velocity* is the total flow rate per unit depth:

$$v_{xm} = \frac{Q}{2d} = \frac{d^2}{3\mu} \left(-\frac{\partial p}{\partial x} \right),$$ (E6.1.27)

and is therefore two-thirds of the maximum velocity, $v_{x\mathrm{max}}$, which occurs at the centerline, $y = 0$.

Fig. E6.1.3   *Pressure and shear-stress distributions.*

Shear-stress distribution.   Finally, the shear-stress distribution is obtained by employing Eqn. (5.60):

$$\tau_{yx} = \mu \left(\frac{\partial v_x}{\partial y} + \frac{\partial v_y}{\partial x} \right).$$ (5.60)

By substituting for v_x from Eqn. (E6.1.15) and recognizing that $v_y = 0$, the shear stress is:

$$\tau_{yx} = -y \left(-\frac{\partial p}{\partial x} \right).$$ (E6.1.28)

Referring back to the sign convention expressed in Fig. 5.12, the first minus sign in Eqn. (E6.1.28) indicates for positive y that the fluid in the region of greater y is acting on the region of lesser y in the *negative* x direction, thus trying to retard the fluid between it and the centerline, and acting against the pressure gradient. Representative distributions of pressure and shear stress, from Eqns. (E6.1.11) and (E6.1.28), are sketched in Fig. E6.1.3. More precisely, the arrows at the left and right show the external pressure forces acting on the fluid contained between $x = 0$ and $x = L$.   □

Example 6.2—Shell Balance for Flow Between Parallel Plates   **301**

6.3 Alternative Solution Using a Shell Balance

Because the flow between parallel plates was the first problem to be examined, the analysis in Example 6.1 was purposely very thorough, extracting the last "ounce" of information. In many other applications, the velocity profile and the flow rate may be the only quantities of prime importance. On the average, therefore, subsequent examples in this chapter will be shorter, concentrating on certain features and ignoring others.

The problem of Example 6.1 was solved by starting with the completely general equations of motion and then simplifying them. An alternative approach involves a direct momentum balance on a differential element of fluid—a "shell"—as illustrated in Example 6.2.

Example 6.2—Shell Balance for Flow Between Parallel Plates

Employ the shell-balance approach to solve the same problem that was studied in Example 6.1.

Assumptions.   The necessary "shell" is in reality a differential element of fluid, as shown in Fig. E6.2. The element, which has dimensions of dx and dy in the plane of the diagram, extends for a depth of dz (any other length may be taken) normal to the plane of the diagram.

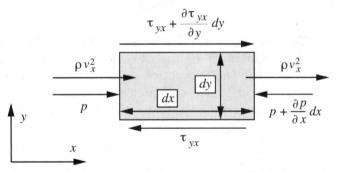

Fig. E6.2   Momentum balance on a fluid element.

If, for the present, the element is taken to be a system that is fixed in space, there are three different types of rate of x momentum transfer to it:

1. A convective transfer of $\rho v_x^2\, dy\, dz$ *in* through the left-hand face, and an identical amount *out* through the right-hand face. Note here that we have implicitly assumed the consequences of the continuity equation, expressed in Eqn. (E6.1.1), that v_x is constant along the duct.
2. Pressure forces on the left- and right-hand faces. The latter will be smaller, because $\partial p/\partial x$ is negative in reality.

3. Shear stresses on the lower and upper faces. Observe that the directions of the arrows conform strictly to the sign convention established in Section 5.6. A momentum balance on the element, which is not accelerating, gives:

$$\underbrace{\underbrace{\rho v_x^2 \, dy \, dz}_{\text{In}} - \underbrace{\rho v_x^2 \, dy \, dz}_{\text{Out}}}_{\text{Net convective transfer}} + \underbrace{\underbrace{p \, dy \, dz}_{\text{Left}} - \underbrace{\left(p + \frac{\partial p}{\partial x} \, dx\right) dy \, dz}_{\text{Right}}}_{\text{Net pressure force}}$$

$$+ \underbrace{\underbrace{\left(\tau_{yx} + \frac{\partial \tau_{yx}}{\partial y} \, dy\right) dx \, dz}_{\text{Upper}} - \underbrace{\tau_{yx} \, dx \, dz}_{\text{Lower}}}_{\text{Net shear force}} = 0. \quad \text{(E6.2.1)}$$

The usual cancellations can be made, resulting in:

$$\frac{d\tau_{yx}}{dy} = \frac{\partial p}{\partial x}, \quad \text{(E6.2.2)}$$

in which the total derivative recognizes that the shear stress depends only on y and not on x. Substitution for τ_{yx} from Eqn. (5.60) with $v_y = 0$ gives:

$$\mu \frac{d^2 v_x}{dy^2} = \frac{\partial p}{\partial x}, \quad \text{(E6.2.3)}$$

which is identical with Eqn. (E6.1.12) that was derived from the simplified Navier-Stokes equations. The remainder of the development then proceeds as in the previous example. Note that the convective terms can be sidestepped entirely if the momentum balance is performed on an element that is chosen to be *moving* with the fluid, in which case there is no flow either into or out of it. ☐

The choice of approach—simplifying the full equations of motion, or performing a shell balance—is very much a personal one, and we have generally opted for the former. The application of the Navier-Stokes equations, which are admittedly rather complicated, has the advantages of not "reinventing the (momentum balance) wheel" for each problem, and also of assuring us that no terms are omitted. Conversely, a shell balance has the merits of relative simplicity, although it may be quite difficult to perform convincingly for an element with *curved* sides, as would occur for the problem in spherical coordinates discussed in Example 6.8.

This section concludes with another example problem, which illustrates the application of two further boundary conditions for a liquid, one involving it in contact with a *moving* surface, and the other at a gas/liquid interface where there is a condition of *zero shear*.

Example 6.3—Film Flow on a Moving Substrate   **303**

Example 6.3—Film Flow on a Moving Substrate

Fig. E6.3.1 shows a coating experiment involving a flat photographic film that is being pulled up from a processing bath by rollers with a steady velocity U at an angle θ to the horizontal. As the film leaves the bath, it entrains some liquid, and in this particular experiment it has reached the stage where: (a) the velocity of the liquid in contact with the film is $v_x = U$ at $y = 0$, (b) the thickness of the liquid is constant at a value δ, and (c) there is no net flow of liquid (as much is being pulled up by the film as is falling back by gravity). (Clearly, if the film were to retain a permanent coating, a net upwards flow of liquid would be needed.)

Fig. E6.3.1   Liquid coating on a photographic film.

Perform the following tasks:

1. Write down the differential mass balance and simplify it.
2. Write down the differential momentum balances in the x and y directions. What are the values of g_x and g_y in terms of g and θ? Simplify the momentum balances as much as possible.
3. From the simplified y momentum balance, derive an expression for the pressure p as a function of y, ρ, δ, g, and θ, and hence demonstrate that $\partial p / \partial x = 0$. Assume that the pressure in the surrounding air is zero everywhere.
4. From the simplified x momentum balance, assuming that the air exerts a negligible shear stress τ_{yx} on the surface of the liquid at $y = \delta$, derive an expression for the liquid velocity v_x as a function of U, y, δ, and α, where $\alpha = \rho g \sin \theta / \mu$.
5. Also derive an expression for the net liquid flow rate Q (per unit width, normal to the plane of Fig. E6.3.1) in terms of U, δ, and α. Noting that $Q = 0$, obtain an expression for the film thickness δ in terms of U and α.
6. Sketch the velocity profile v_x, labeling all important features.

Assumptions and continuity.   The following assumptions are reasonable:

1. The flow is steady and Newtonian, with constant density ρ and viscosity μ.
2. The z direction, normal to the plane of the diagram, may be disregarded entirely. Thus, not only is v_z zero, but all derivatives with respect to z, such as $\partial v_x / \partial z$, are also zero.

3. There is only one nonzero velocity component, namely, that in the direction of motion of the photographic film, v_x. Thus, $v_y = v_z = 0$.
4. Gravity acts vertically downwards.

Because of the constant-density assumption, the continuity equation, (5.48), simplifies, as before, to:

$$\frac{\partial v_x}{\partial x} + \frac{\partial v_y}{\partial y} + \frac{\partial v_z}{\partial z} = 0. \qquad (E6.3.1)$$

But since $v_y = v_z = 0$, it follows that

$$\frac{\partial v_x}{\partial x} = 0. \qquad (E6.3.2)$$

so v_x is independent of distance x along the film. Further, v_x does not depend on z (assumption 2); thus, the velocity profile $v_x = v_x(y)$ depends only on y and will appear the same for *all* values of x.

Momentum balances.   With the stated assumptions of a Newtonian fluid with constant density and viscosity, Eqn. (5.73) gives the x and y momentum balances:

$$\rho\left(\frac{\partial v_x}{\partial t} + v_x\frac{\partial v_x}{\partial x} + v_y\frac{\partial v_x}{\partial y} + v_z\frac{\partial v_x}{\partial z}\right) = -\frac{\partial p}{\partial x} + \mu\left(\frac{\partial^2 v_x}{\partial x^2} + \frac{\partial^2 v_x}{\partial y^2} + \frac{\partial^2 v_x}{\partial z^2}\right) + \rho g_x,$$

$$\rho\left(\frac{\partial v_y}{\partial t} + v_x\frac{\partial v_y}{\partial x} + v_y\frac{\partial v_y}{\partial y} + v_z\frac{\partial v_y}{\partial z}\right) = -\frac{\partial p}{\partial y} + \mu\left(\frac{\partial^2 v_y}{\partial x^2} + \frac{\partial^2 v_y}{\partial y^2} + \frac{\partial^2 v_y}{\partial z^2}\right) + \rho g_y,$$

Noting that $g_x = -g\sin\theta$ and $g_y = -g\cos\theta$, these momentum balances simplify to:

$$\frac{\partial p}{\partial x} + \rho g\sin\theta = \mu\frac{\partial^2 v_x}{\partial y^2}, \qquad (E6.3.3)$$

$$\frac{\partial p}{\partial y} = -\rho g\cos\theta. \qquad (E6.3.4)$$

Integration of Eqn. (E6.3.4), between the free surface at $y = \delta$ (where the gauge pressure is zero) and an arbitrary location y (where the pressure is p) gives:

$$\int_0^p dp = -\rho g\cos\theta\int_\delta^y dy + f(x), \qquad (E6.3.5)$$

so that:

$$p = \rho g\cos\theta(\delta - y) + f(x). \qquad (E6.3.6)$$

Note that since a *partial* differential equation is being integrated, a *function* of integration, $f(x)$, is again introduced. Another way of looking at it is to observe

Example 6.3—Film Flow on a Moving Substrate   **305**

that if Eqn. (E6.3.6) is differentiated with respect to y, we would recover the original equation, (E6.3.4), because $\partial f(x)/\partial y = 0$.

However, since $p = 0$ at $y = \delta$ (the air/liquid interface) for all values of x, the function $f(x)$ must be zero. Hence, the pressure distribution:

$$p = (\delta - y)\rho g \cos \theta, \tag{E6.3.7}$$

shows that p is not a function of x.

In view of this last result, we may now substitute $\partial p/\partial x = 0$ into the x-momentum balance, Eqn. (E6.3.3), which becomes:

$$\frac{d^2 v_x}{dy^2} = \frac{\rho g}{\mu} \sin \theta = \alpha, \tag{E6.3.8}$$

in which the constant α has been introduced to denote $\rho g \sin \theta/\mu$. Observe that the second derivative of the velocity now appears as a *total* derivative, since v_x depends on y only.

A first integration of Eqn. (E6.3.8) with respect to y gives:

$$\frac{dv_x}{dy} = \alpha y + c_1. \tag{E6.3.9}$$

The boundary condition of zero shear stress at the free surface is now invoked:

$$\tau_{yx} = \mu \left(\frac{\partial v_y}{\partial x} + \frac{\partial v_x}{\partial y} \right) = \mu \frac{dv_x}{dy} = 0. \tag{E6.3.10}$$

Thus, from Eqns. (E6.3.9) and (E6.3.10) at $y = \delta$, the first constant of integration can be determined:

$$\frac{dv_x}{dy} = \alpha\delta + c_1 = 0, \qquad \text{or} \qquad c_1 = -\alpha\delta. \tag{E6.3.11}$$

A second integration, of Eqn. (E6.3.9) with respect to y, gives:

$$v_x = \alpha \left(\frac{y^2}{2} - y\delta \right) + c_2. \tag{E6.3.12}$$

The second constant of integration, c_2, can be determined by using the boundary condition that the liquid velocity at $y = 0$ equals that of the moving photographic film. That is, $v_x = U$ at $y = 0$, yielding $c_2 = U$; thus, the final velocity profile is:

$$v_x = U - \alpha y \left(\delta - \frac{y}{2} \right). \tag{E6.3.13}$$

Observe that the velocity profile, which is parabolic, consists of two parts:

1. A constant and positive part, arising from the film velocity, U.
2. A variable and negative part, caused by gravity, which reduces v_x at increasing distances y from the film and eventually causes it to become negative.

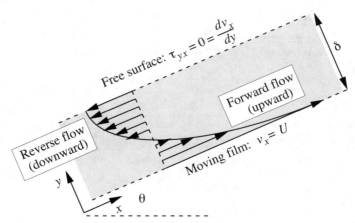

*Fig. E6.3.2   Velocity profile in thin liquid layer on moving
photographic film for the case of zero net liquid flow rate.*

Exactly how much of the liquid is flowing upwards, and how much downwards, depends on the values of the variables U, δ, and α. However, we are asked to investigate the situation in which there is no net flow of liquid—that is, as much is being pulled up by the film as is falling back by gravity. In this case:

$$Q = \int_0^\delta v_x \, dy = \int_0^\delta \left[U - \alpha y \left(\delta - \frac{y}{2} \right) \right] dy = U\delta - \frac{1}{3}\alpha\delta^3 = 0, \qquad \text{(E6.3.14)}$$

giving the thickness of the liquid film as:

$$\delta = \sqrt{\frac{3U}{\alpha}}. \qquad \text{(E6.3.15)}$$

The velocity profile for this case of $Q = 0$ is shown in Fig. E6.3.2.   □

Example 6.4—Transient Viscous Diffusion of Momentum (COMSOL)   **307**

Example 6.4—Transient Viscous Diffusion of Momentum (COMSOL)

This problem is one of the few instances in this book in which we investigate *transient* or unsteady-state fluid flow. It illustrates how momentum is transferred under the diffusive influence of viscosity.

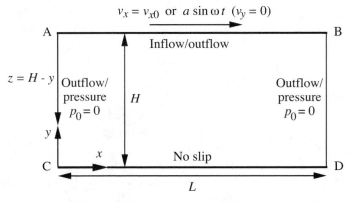

Fig. E6.4.1   *Region of liquid being studied, with the four COMSOL boundary conditions.*

As shown in Fig. E6.4.1, consider a liquid of density $\rho = 1$ g/cm^3 and viscosity μ (whose values will be given later), contained between two parallel plates AB and CD of length $L = 2$ cm and separation $H = 1$ cm. The lower plate CD is stationary, and the upper plate AB can undergo two types of movement:

1. At time $t = 0$, it is suddenly set in motion with a constant x velocity $v_x = v_{x0}$. In our problem, $v_{x0} = 1$ cm/s and $\mu = 0.5$ g/cm s $= 0.5$ P.

2. At time $t = 0$, it is oscillated to the right and left in such a way that its x velocity varies with time according to $v_{x0} = a \sin \omega t$. In our problem the amplitude $a = 1$ cm/s, the angular velocity $\omega = 2\pi$, and the viscosity may be either $\mu = 0.1$ or 0.5 P.

Very similar types of motion occur in concentric-cylinder viscometers, as in Figs. 1.1 and 11.13(b), although with smaller moving-surface separations.

Use COMSOL to investigate how v_x varies between the midpoints of the lower and upper plates at different times, and interpret the results. If needed, further details of the implementation of COMSOL are given in Chapter 14.

Note that COMSOL will employ the full Navier-Stokes and continuity equations in its solution. However, because the only nonzero velocity component is v_x, the following familiar diffusion-type equation is effectively being solved:

$$\rho \frac{\partial v_x}{\partial t} = \mu \frac{\partial^2 v_x}{\partial y^2}. \tag{E6.4.1}$$

Solution—Case 1 (Constant Velocity)

1. In COMSOL, choose the sequence **New, Chemical Engineering Module, Momentum Balance**, and **Incompressible Navier-Stokes**.

2. Draw the rectangle shown in Fig. E6.4.1, with opposite vertices (0, 0) and (2, 1).

3. Under **Options/Constants**, set vx0 = 1, rho = 1, and mu = 0.5.

4. Under **Physics/Subdomain Settings**, indicate that the density is rho and the viscosity is mu.

5. Under **Physics/Boundary Settings**, establish the boundary conditions shown in Fig. E6.4.1. On the upper plate (an "Inflow/Outflow" boundary), set the x and y velocities to be u0 = vx0 and v0 = 0.

6. Since the problem is a transient one, under **Solve/Solver Parameters** choose **Solver/Time dependent** and for **General/Time Stepping/Times**, insert 0:0.01:0.1 0.2:0.1:0.5 1:0.5:2—meaning that solutions will be available for subsequent plotting at $t = 0, 0.01, 0.02, \ldots 0.1, 0.2, 0.3, 0.4, 0.5, 1.0, 1.5, 2.0$ s.

7. Initialize the mesh without refinement and solve the problem. Observe, as the solution progresses, that COMSOL starts cautiously with a very small time step, which it realizes can rapidly be increased without compromising the accuracy of the solution.

8. Finally, plot v_x against distance y between the midpoints of the lower and upper plates, for selected values of time. Under **Postprocessing/Cross Section Plot Parameters**, make the following selections:

(a) Under the **General** tab, for **Solutions to use**, select **stored output times** of 0.01, 0.02, 0.05, 0.1, 0.2, 0.5, and 2.0 s. On a Macintosh computer, this is achieved by holding down the Command key (Control key for a PC) and clicking successively on each time.

(b) Under the **Line/Extrusion** tab, choose **Line plot, x-velocity** as the predefined quantity, **y** as the **x-axis data**. For **Cross-section line data**, indicate x0 = 1, x1 = 1, y0 = 0, and y1 = 1, followed by **OK**.

Comments on the results. Fig. E6.4.2 shows how the velocity v_x varies with distance y from the lower plate at various times. The result is a classical case of diffusion—in this instance, of x momentum (generated by the motion of the upper plate, which moves at 1 cm/s) into the liquid below it. For the smallest time plotted ($t = 0.01$ s), most of the liquid, from $y = 0$ to 0.7 cm, is essentially undisturbed, followed by a rising v_x between $y = 0.7$ and 1 cm. For early times (or for an infinitely deep liquid), before the influence of the lower plate is felt, there is an analytical solution:

$$\frac{v_x}{v_{x0}} = \operatorname{erfc} \frac{z}{2\sqrt{\nu t}}, \qquad (E6.4.2)$$

in which $z = H - y$ is the downward distance from the upper plate, $\nu = \mu/\rho$ is the kinematic viscosity, and erfc denotes the complementary error function.

Example 6.4—Transient Viscous Diffusion of Momentum (COMSOL)   **309**

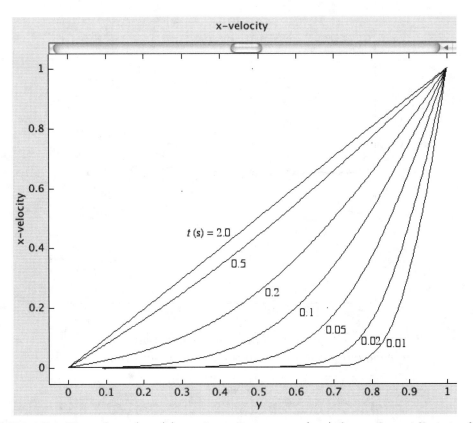

Fig. E6.4.2   Plot of v_x (cm/s) against distance y (cm) from the midpoint of the lower plate to the midpoint of the upper plate, for the indicated values of time (s).

As time increases, more and more of the liquid is set into motion by the diffusion of x momentum from the upper plate. Eventually—theoretically at $t = \infty$ but effectively at a time of about 2 s—a steady state is reached, in which the shear stress exerted by the upper plate is counterbalanced by the shear stress in the opposite direction from the stationary lower plate. Since the velocity profile is linear, dv_x/dy is constant; thus, these shear stresses are equal (and opposite) because they are both given by $\tau_{yx} = \mu dv_x/dy$.

Solution—Case 2 (Oscillating Velocity)

Since the problem is very similar to that of Case 1, only the differences between the two solution procedures will be noted:

1. Under **Constants**, replace vx0 with the amplitude a = 1, and add omega = 2*pi = 6.283185. To start, set the viscosity to be mu = 0.1.

2. Under **Boundary Settings**, the velocity of the upper boundary u0 will now be a*sin(omega*t).

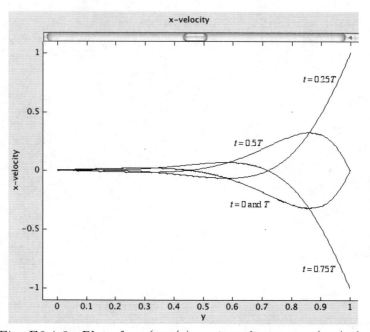

Fig. E6.4.3   Plot of v_x (cm/s) against distance y (cm), for one complete cycle of the upper plate. Viscosity $\mu = 0.1$ P.

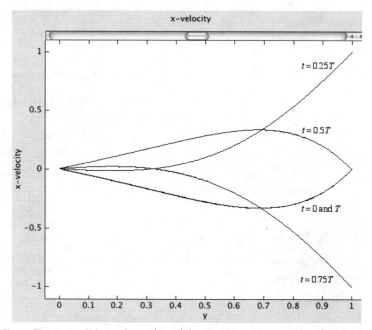

Fig. E6.4.4   Plot of v_x (cm/s) against distance y (cm), for one complete cycle of the upper plate. Viscosity $\mu = 0.5$ P.

Example 6.4—Transient Viscous Diffusion of Momentum (COMSOL)   **311**

3. Under **Times**, insert `0:2:18 19:0.25:20`, so solutions will be available at 2-s intervals up to 18 s and at 0.25-s intervals between 19 and 20 s.

4. For plotting, under **Solutions to use**, select **stored output times** of `19, 19.25, 19.5, 19.75` and 20 s. Such large values of time will insure that the cyclical velocities are repeating themselves almost exactly—indeed, they are only marginally different from the solutions between 2 and 3 s.

5. Proceeding as before, obtain a second plot for a viscosity of $\mu = 0.5$ P.

Comments on the results. The results for the two values of the viscosity are shown in Figs. E6.4.3 and E6.4.4. Since $\omega = 2\pi$, the period of the oscillations is $T = 1$ s. Thus, velocity profiles separated by 0.25 s will occur over one complete cycle of $t = 0, 0.25T, 0.5T, 0.75T$, and T. Note two main effects:

(a) *Penetration depth.* For the lower viscosity, $\mu = 0.1$ P, the surface oscillations do not penetrate as far into the liquid as those for the higher viscosity of $\mu = 0.5$ P. In other words, the lower viscosity does not permit as rapid a diffusion of momentum as the higher viscosity.

(b) *Phase change.* Note that for $t = T$, when the upper plate has a zero velocity, much of the liquid has a *negative* velocity, largely due to the diffusional effects of the negative surface velocity at an earlier time such as $t = 0.75T$. Similar observations can be made for other values of time.

Both effects have applications in the testing of non-Newtonian fluids, for which the phase difference between stress and strain is illustrated in terms of *displacements* in Fig. 11.11. Since the velocity v_x is the derivative of the distance moved, the displacement D of the upper plate is:

$$D = -\frac{a}{\omega} \cos \omega t. \qquad (E6.4.3)$$

For an infinitely deep liquid, there is again an exact solution:

$$v_x = ae^{-\beta z} \sin(\omega t - \beta z), \qquad (E6.4.4)$$

in which z is the depth and:

$$\beta^2 = \frac{\omega}{2\nu}. \qquad (E6.4.5)$$

The attenuation and phase change with depth z are very apparent from Eqn. (E6.4.4). $\qquad \square$

6.4 Poiseuille and Couette Flows in Polymer Processing

The study of *polymer processing* falls into the realm of the chemical engineer. First, the polymer, such as nylon, polystyrene, or polyethylene, is produced by a chemical reaction—either as a liquid or solid. (In the latter event, it would subsequently have to be melted in order to be processed further.) Second, the polymer must be formed by suitable equipment into the desired final shape, such as a film, fiber, bottle, or other molded object. The procedures listed in Table 6.1 are typical of those occurring in polymer processing.

Table 6.1   Typical Polymer-Processing Operations

Operation	*Description*
Extrusion	Molten polymer is forced, either by a single or twin screw, or other means such as a piston (ram) or gear pump, through a narrow opening (called a *die*) to form a continuous pipe, tube, profile, sheet, film, or filament. The "extrudate" from the die is then cooled—usually by air or water—and the product collected (wound up, sawn to length, etc.). Typical examples are PVC pipe (for drainage and sewage), PE film (for plastic bags), and nylon filaments (for toothbrush bristles).
Drawing or "spinning"	Molten polymer flows through narrow openings, either in the form of film or filaments, and is pulled by a roller that stretches (draws) and orients the film or filaments in the "direction of draw." Examples are oriented polypropylene (OPP) films used for making woven plastic bags (e.g., for packaging fertilizer), and plastic fibers such as nylon and polyester for making fabrics and garments (blouses, socks, etc.).
Injection molding	Molten polymer (normally a thermoplastic) is forced under high pressure into a cold mold where it cools and solidifies into the shape of the mold cavity(ies). The process is used for making a multitude of products, such as buckets and bowls (HDPE), domestic telephones (ABS), bottle crates (PP), pipe fittings (rigid PVC), and CD, DVD, and videotape boxes (in PP or GPPS).

Example 6.5—The Single-Screw Extruder   **313**

Calendering	Molten polymer is forced through the gaps of successive highly polished rolls where it is shaped into a uniformly thick film or sheet. The nature of the surfaces of the rolls strongly influences the final appearance of the film or sheet, which may be smooth, matte, or embossed (e.g., simulated leather). The main product of this process is calendered flexible PVC sheet used for the backing of files (stationery) and for automobile seat covers and internal linings (leather is used for the more expensive cars!). The rubber industry also uses calendering extensively.
Blow molding	An extruded tube (parison) of molten polymer is clamped between two mold halves and inflated by injecting compressed air into the center of the parison. The parison expands, like blowing up a balloon, until it fills the mold cavity where it cools and hardens into a solid product—usually a bottle made of HDPE. Rigid PVC is also commonly blow molded (e.g., into clear shampoo and shower-gel bottles).
Coating or "spreading"	A low viscosity polymer is applied as a thin film by a steel blade ("doctor" blade) onto a substrate such as paper, fabric, or even another polymer sheet or film. PVC "leathercloth" is made by this process— heavily plasticized PVC paste is coated onto a cotton or synthetic fiber fabric, followed by heating in an oven to "fuse" (solidify) the PVC (used a lot for car upholstery).

Since polymers are generally highly viscous, their flows can be obtained by solving the equations of viscous motion. In this chapter, we cover the rudiments of extrusion, die flow, and drawing or spinning. The analysis of calendering and coating is considerably more complicated, but can be rendered tractable if reasonable simplifications, known collectively as the *lubrication approximation,* are made, as discussed in Chapter 8. The treatments of injection molding and blow molding are beyond the scope of this book.

Example 6.5—The Single-Screw Extruder

Because molten polymers are usually very viscous, they often need very high pressures to push them through dies. One such "pump" for achieving this is the *screw extruder,* shown in Fig. E6.5.1. The polymer enters the feed hopper as pellets, falls into the screw channels in the feed section, and is pushed forward by the screw, which rotates at an angular velocity ω, clockwise as seen by an observer looking along the axis from the inlet to the exit (notice that the screw is left-handed). The heated barrel, together with shear effects, melts the pellets, which then become fluid prior to entering the metering section.

Fig. E6.5.1 Screw extruder.

There are three zones in the extruder:

1. In the *feed* zone the granules are transported into the barrel, where they melt just before entering the compression zone. The large constant depth of the channel in the feed zone means that there will be negligible pressure exerted on the downstream melt in the metering zone.

2. In the *compression* or "transition" zone the channel depth decreases from the feed zone to the metering zone. Therefore, the melt will gradually increase in pressure as it travels along the compression zone, reaching a maximum pressure at the beginning of the metering zone.

3. In the *metering* zone, length L_0, the screw channel has a constant shallow depth h (with $h \ll r$) and the melt in the channel should now be homogeneous or uniform. Thus, the metering zone of the screw acts like a constant-delivery pump since the screw is rotating at a constant speed. The melt pressure uniformly increases as it passes along the metering zone. Therefore, calculations of *extruder output* are based on the *metering zone* of the screw. Extruder output calculations are relatively simple due to the uniformity of conditions existing in the metering zone of the screw.

The preliminary analysis given here neglects any heat-transfer effects in the metering section, and also assumes that the polymer has a constant Newtonian viscosity μ.

The investigation is facilitated by taking the viewpoint of a hypothetical observer *located on the screw,* in which case the screw surface and the flights appear to be stationary, with the barrel moving with velocity $V = r\omega$ at a *helix angle θ* to

Example 6.5—The Single-Screw Extruder **315**

the flight axis, as shown in Fig. E6.5.2. The alternative viewpoint of an observer *located on the inside surface of the barrel* is not very fruitful, because not only are the flights seen as *moving* boundaries, but the observations would be periodically *blocked* as the flights passed over the observer! The width of the screw channel measured perpendicularly to the vertical sides of the flight flanks is designated W.

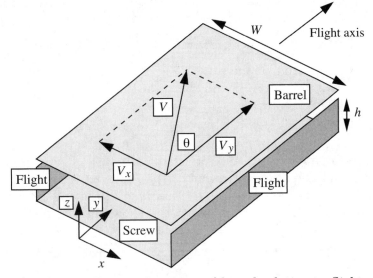

Fig. E6.5.2   Diagonal motion of barrel relative to flights.

Solution

Motion in two principal directions is considered:

1. Flow *parallel* to the flight axis, caused by a barrel velocity of $V_y = V \cos \theta = r\omega \cos \theta$ relative to the (now effectively stationary) flights and screw.
2. Flow *normal* to the flight axis, caused by a barrel velocity of $V_x = -V \sin \theta = -r\omega \sin \theta$ relative to the (stationary) flights and screw.

In each case, the flow is considered one-dimensional, with "end-effects" caused by the presence of the flights being unimportant. A glance at Fig. E6.5.3(b) will give the general idea. Although the flow in the x-direction must reverse itself as it nears the flights, it is reasonable to assume for $h \ll W$ that there is a substantial central region in which the flow is essentially in the positive or negative x direction.

1. Motion parallel to the flight axis.   The reader may wish to investigate the additional simplifying assumptions that give the y momentum balance as:

$$\frac{\partial p}{\partial y} = \mu \frac{d^2 v_y}{dz^2}. \tag{E6.5.1}$$

Integration twice yields the velocity profile as:

$$v_y = \frac{1}{2\mu}\left(\frac{\partial p}{\partial y}\right)z^2 + c_1 z + c_2 = \underbrace{\frac{1}{2\mu}\left(-\frac{\partial p}{\partial y}\right)(hz - z^2)}_{\text{Poiseuille flow}} + \underbrace{\frac{z}{h}r\omega\cos\theta}_{\text{Couette flow}}. \qquad (E6.5.2)$$

Here, the integration constants c_1 and c_2 have been determined in the usual way by applying the boundary conditions:

$$z = 0: \quad v_y = 0; \qquad z = h: \quad v_y = V_y = V\cos\theta = r\omega\cos\theta. \qquad (E6.5.3)$$

Note that the negative of the pressure gradient is given in terms of the inlet pressure p_1, the exit pressure p_2, and the total length $L \; (= L_0/\sin\theta)$ measured along the screw flight axis by:

$$-\frac{\partial p}{\partial y} = -\frac{p_2 - p_1}{L} = \frac{p_1 - p_2}{L}, \qquad (E6.5.4)$$

and is a negative quantity since the screw action builds up pressure and $p_2 > p_1$. Thus, Eqn. (E6.5.2) predicts a Poiseuille-type *backflow* (caused by the adverse pressure gradient) and a Couette-type *forward flow* (caused by the relative motion of the barrel to the screw). The combination is shown in Fig. E6.5.3(a).

(a) Cross section along flight axis, showing velocity profile.

(b) Cross section normal to flight axis, showing streamlines.

Fig. E6.5.3 *Fluid motion (a) along, and (b) normal to the flight axis, as seen by an observer on the screw.*

The total flow rate Q_y of polymer melt in the direction of the flight axis is obtained by integrating the velocity between the screw and barrel, and recognizing that the width between flights is W:

$$Q_y = W\int_0^h v_y\,dz = \underbrace{\frac{Wh^3}{12\mu}\left(\frac{p_1 - p_2}{L}\right)}_{\text{Poiseuille}} + \underbrace{\frac{1}{2}Whr\omega\cos\theta}_{\text{Couette}}. \qquad (E6.5.5)$$

Example 6.5—The Single-Screw Extruder   **317**

The actual value of Q_y will depend on the resistance of the die located at the extruder exit. In a hypothetical case, in which the die offers no resistance, there would be no pressure increase in the extruder ($p_2 = p_1$), leaving only the Couette term in Eqn. (E6.5.5). For the practical situation in which the die offers significant resistance, the Poiseuille term would serve to *diminish* the flow rate given by the Couette term.

2. *Motion normal to the flight axis.* By a development very similar to that for flow parallel to the flight axis, we obtain:

$$\frac{\partial p}{\partial x} = \mu \frac{d^2 v_x}{dz^2}, \tag{E6.5.6}$$

$$v_x = \underbrace{\frac{1}{2\mu}\left(-\frac{\partial p}{\partial x}\right)(hz - z^2)}_{\text{Poiseuille flow}} - \underbrace{\frac{z}{h}r\omega\sin\theta}_{\text{Couette flow}}, \tag{E6.5.7}$$

$$Q_x = \int_0^h v_x\,dz = \underbrace{\frac{h^3}{12\mu}\left(-\frac{\partial p}{\partial x}\right)}_{\text{Poiseuille}} - \underbrace{\frac{1}{2}hr\omega\sin\theta}_{\text{Couette}} = 0. \tag{E6.5.8}$$

Here, Q_x is the flow rate in the x direction, per unit depth along the flight axis, and must equal zero, because the flights at either end of the path act as barriers. The negative of the pressure gradient is therefore:

$$-\frac{\partial p}{\partial x} = \frac{6\mu r\omega\sin\theta}{h^2}, \tag{E6.5.9}$$

so that the velocity profile is given by:

$$v_x = \frac{z}{h}r\omega\left(2 - 3\frac{z}{h}\right)\sin\theta. \tag{E6.5.10}$$

Note from Eqn. (E6.5.10) that v_x is zero when either $z = 0$ (on the screw surface) or $z/h = 2/3$. The reader may wish to sketch the general appearance of $v_z(z)$. □

Representative values. A COMSOL example follows shortly, and it is appropriate to consider typical measurements and polymer properties for screw extruders[2]:

---

[2] I heartily thank Mr. John W. Ellis, of the Petroleum and Petrochemical College at Chulalongkorn University in Bangkok, who is an expert in polymer processing, for supplying these values and most of the information in Table 6.1. Previously he wrote *Polymer Products: Design, Materials and Processing*, with co-author David Morton-Jones, published by Chapman and Hall (1986). His latest book, *Introduction to Plastic Foams*, co-authored by Dr. Duanghathai Pentrakoon (of the Faculty of Science, Chulalongkorn University, Bangkok), is published by the Chulalongkorn University Press (2005).

1. The size of the screw varies from a laboratory diameter of 20 mm up to large production machines having diameters of 150 mm (very common) or much larger—up to 600 mm.

2. The helix angle is almost always $\theta = \tan^{-1}(1/\pi) = 17.66°$, corresponding to a "square thread" screw profile in which the pitch P equals the outside diameter $D = 2r$ of the screw.

3. For a 100-mm-diameter screw, a typical channel depth (metering section) would be about $h = 5$ or 6 mm.

4. The screw rpm ranges from a slow speed of about 30 rpm to a high speed of 150 rpm. The large extruders run at a slower rpm. A typical screw surface speed (the barrel velocity relative to the screw) for extruding polyethylene is 0.5 m/s. For a 100-mm screw, a typical screw speed would be 65 rpm.

5. The *metering section* length L_0 depends on the type of polymer being used—amorphous, semicrystalline, etc. This length is usually expressed as a multiple of the screw diameter, such as $L_0 = 6.5D$. Typical extruder screws have an *overall* length of $25D$.

6. Polymers generally "shear thin" with increasing shear rate, so the viscosity will depend on the prevailing shear rate (related to rpm) of the melt in the screw channel at the melt temperature being used. But for extrusion processes we can assume a melt viscosity of anything between 200 and 1,000 N s/m² (i.e., Pa s) for extrusion grades of polyolefins (LDPE, HDPE, and PP).

Example 6.6—Flow Patterns in a Screw Extruder (COMSOL)

Consider the screw extruder just discussed in Example 6.5. Solve the equations of motion to determine as precisely as possible the flow pattern given approximately in Fig. E6.5.3(b), normal to the flight axis. Note that the pressure cannot be allowed to "float," but must be specified to some arbitrary value at any one point in the cross section shown in Fig. E6.6.1, such as point 4.

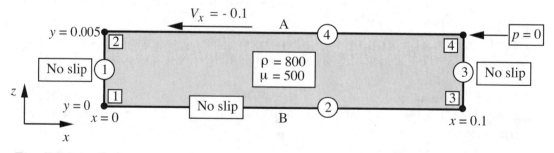

Fig. E6.6.1 Cross section of screw extruder normal to the flight axis. Numerals 1, 2, 3, and 4 in circles denote the four boundary segments, and those in boxes denote the four points where these segments join one another. Eventually, a velocity cross plot will be made between midpoints A and B of the top and bottom surfaces.

Example 6.6—Flow Patterns in a Screw Extruder (COMSOL)   **319**

Based on the comments at the end of Example 6.5, use the following values, which relate to the extrusion of a polyolefin:

1. Polymer properties: $\rho = 800$ kg/m^3, $\mu = 500$ Pa s (N s/m^2).
2. Rotational speed $N = 65$ rpm, so that $\omega = 2\pi \times 65/60 = 6.8$ rad/s.
3. Channel depth $h = 5$ mm $= 0.005$ m.
4. For a square pitch, the pitch P and screw diameter D are identical, taken to be 100 mm $= 0.1$ m. The distance between flights is $W = P\cos\theta = 0.1 \times \cos 17.66° = 0.0953$ m. For simplicity in our example, we shall round up slightly and take $W = 0.1$ m.
5. The relative velocity is $V = \omega r = \omega D/2 = 6.8 \times 0.1/2 = 0.340$ m/s, with x component $V_x = -0.340 \times \sin\theta = -0.340 \times \sin 17.66° = -0.103 \doteq -0.1$ m/s.

Solution

1. In COMSOL, choose the sequence **New, Chemical Engineering Module, Momentum Balance**, and **Incompressible Navier-Stokes**.

2. Under **Options/Constants** set rho = 800, mu = 500, and Vx = -0.1.

2. Draw the rectangle shown in Fig. E6.6.1, with opposite vertices (0, 0) and (0.1, 0.005).

3. Under **Physics/Subdomain Settings**, indicate that the density is rho and the viscosity is mu.

4. Under **Physics/Boundary Settings**, set boundaries 1, 2, 3 to be no-slip and boundary 4 to have u0 = Vx and v0 = 0.

5. Under **Physics/Point Settings**, check **Point constraint** for point 4, at which p0 = 0. At points 1, 2, and 3 there is no such constraint.

6. Initialize and refine the mesh *twice*, giving 1,024 elements and 4,951 degrees of freedom. Then solve the problem.

7. Under **Postprocessing**, make appropriate choices in the **Plot Parameters** pulldown menu to generate the following:

(a) An arrow plot of the x velocities.

(b) A picture that shows 20 streamlines.

(c) A contour plot of the pressure distribution. Obtain values for the pressure at the midpoints of the two ends by first double-clicking on "Snap" at the bottom of the screen, which enables the mouse to point freely and not be constrained to the nearest grid lines. Then click on each midpoint in turn and observe the value of the pressure in the lower left-hand corner of the window.

(d) A cross-section plot of the pressure between the midpoints A (0.05, 0.005) of the upper moving surface and B (0.05, 0) of the lower stationary surface. Here, you will need to choose the **Line/Extrusion** option in the **Cross-Section Plot Parameters** window.

Discussion of Results

The COMSOL results are shown in Figs. E6.6.2–E6.6.6. The arrow plot of Fig. E6.6.2 clearly shows that the velocity profile is virtually unchanged over the whole width of the cross section. Near the top surface the flow is induced by the moving surface to be in the negative x direction. Toward the fixed bottom surface there is a flow reversal, so that at any value of x the net flow rate $Q = \int_0^h v_x \, dy$ is zero.

Fig. E6.6.2   Arrow plot, in which the arrow length is proportional to the local velocity. To improve legibility, only the left portion has been reproduced and enlarged.

Fig. E6.6.3   Streamlines, in which the aspect ratio is the same as in the problem statement. The two arrows have been added to emphasize the counterclockwise direction of flow.

Fig. E6.6.4   Streamlines, in which the height, from $y = 0$ to 0.005 has been "stretched" to show more detail. The two arrows have been added to emphasize the counterclockwise direction of flow.

Example 6.6—Flow Patterns in a Screw Extruder (COMSOL)   **321**

Fig. E6.6.3, which contains 20 streamlines, now indicates that the essential flow reversal occurs very close to each end of the cross section. To make the flow pattern more clear, the streamlines are repeated in Fig. E6.6.4, in which a larger scale is employed in the y direction—that is, the x/y geometry is no longer to scale.

Fig. E6.6.5 Isobars, with the pressure varying from p = 0 at the right-hand end to about p = 1.1 × 10⁶ N/m² at the left-hand end. Note that except for the two upper corners, where there is a conflict in velocities between the moving and no-slip boundaries, the pressure depends only on the horizontal coordinate x.

Fig. E6.6.6 A cross plot that shows how the horizontal velocity v_x varies from the midpoint A of the top surface to the midpoint B of the bottom surface.

The pressure distribution is shown in Fig. E6.6.5 as a series of isobars. Note that the isobars are spaced very evenly, and that the pressure increase from right to left is from $p = 0$ to $p = 1.1 \times 10^6$ N/m². Recall that the simplified theory gave the pressure gradient as:

$$-\frac{\partial p}{\partial x} = \frac{6\mu r\omega \sin\theta}{h^2}. \tag{E6.5.9}$$

Thus, the theoretical increase in pressure from right to left should be:

$$0.1\left(\frac{6 \times 500 \times 0.05 \times 6.8 \times \sin 16.77°}{0.005^2}\right) = 1.18 \times 10^6 \text{ N/m}^2, \tag{E6.6.1}$$

in good agreement with the more accurate two-dimensional solution of the Navier-Stokes equations shown in Fig. E6.6.5.

Finally, Fig. E6.6.6 shows how v_x varies from the midpoint A of the top surface to the midpoint B of the bottom surface. As explained in Eqn. (E6.5.7), there are two contributions to this velocity profile:

1. A Couette flow from right to left, driven by the leftward-moving top surface.
2. A Poiseuille flow from left to right, driven by the excess pressure at the left.

$\square$

6.5   Solution of the Equations of Motion in Cylindrical Coordinates

Several chemical engineering operations exhibit symmetry about an axis z and involve one or more surfaces that can be described by having a constant radius for a given value of z. Examples are flow in pipes, extrusion of fibers, and viscometers that involve flow between concentric cylinders, one of which is rotating. Such cases lend themselves naturally to solution in *cylindrical* coordinates, and two examples will now be given.

Example 6.7—Flow Through an Annular Die

Following the discussion of polymer processing in the previous section, now consider flow through a die that could be located at the exit of the screw extruder of Example 6.4. Consider a die that forms a *tube* of polymer (other shapes being sheets and filaments). In the die of length D shown in Fig. E6.7, a pressure difference $p_2 - p_3$ causes a liquid of viscosity μ to flow steadily from left to right in the annular area between two fixed concentric cylinders. Note that p_2 is chosen for the inlet pressure in order to correspond to the extruder exit pressure from Example 6.5. The inner cylinder is solid, whereas the outer one is hollow; their radii are r_1 and r_2, respectively. The problem, which could occur in the extrusion of plastic tubes, is to find the velocity profile in the annular space and the total volumetric flow rate Q. Note that *cylindrical* coordinates are now involved.

Assumptions and continuity equation.   The following assumptions are realistic:

1. There is only one nonzero velocity component, namely that in the direction of flow, v_z. Thus, $v_r = v_\theta = 0$.
2. Gravity acts vertically downward, so that $g_z = 0$.
3. The axial velocity is independent of the angular location; that is, $\partial v_z / \partial \theta = 0$.

To analyze the situation, again start from the continuity equation, (5.49):

$$\frac{\partial \rho}{\partial t} + \frac{1}{r}\frac{\partial(\rho r v_r)}{\partial r} + \frac{1}{r}\frac{\partial(\rho v_\theta)}{\partial \theta} + \frac{\partial(\rho v_z)}{\partial z} = 0, \qquad (5.49)$$

Example 6.7—Flow Through an Annular Die   **323**

which, for constant density and $v_r = v_\theta = 0$, reduces to:

$$\frac{\partial v_z}{\partial z} = 0, \tag{E6.7.1}$$

verifying that v_z is independent of distance from the inlet, and that the velocity profile $v_z = v_z(r)$ appears the same for all values of z.

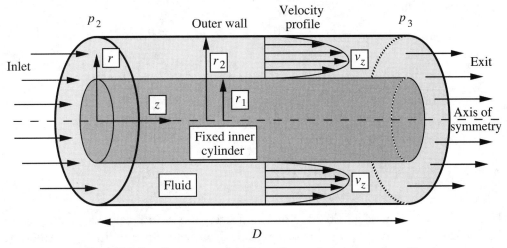

Fig. E6.7   Geometry for flow through an annular die.

Momentum balances.   There are again three momentum balances, one for each of the r, θ, and z directions. If explored, the first two of these would ultimately lead to the pressure variation with r and θ at any cross section, which is of little interest in this problem. Therefore, we extract from Eqn. (5.75) only the z momentum balance:

$$\rho\left(\frac{\partial v_z}{\partial t} + v_r\frac{\partial v_z}{\partial r} + \frac{v_\theta}{r}\frac{\partial v_z}{\partial \theta} + v_z\frac{\partial v_z}{\partial z}\right)$$
$$= -\frac{\partial p}{\partial z} + \mu\left[\frac{1}{r}\frac{\partial}{\partial r}\left(r\frac{\partial v_z}{\partial r}\right) + \frac{1}{r^2}\frac{\partial^2 v_z}{\partial \theta^2} + \frac{\partial^2 v_z}{\partial z^2}\right] + \rho g_z. \tag{E6.7.2}$$

With $v_r = v_\theta = 0$ (from assumption 1), $\partial v_z/\partial z = 0$ [from Eqn. (E6.7.1)], $\partial v_z/\partial \theta = 0$ (assumption 3), and $g_z = 0$ (assumption 2), this momentum balance simplifies to:

$$\mu\left[\frac{1}{r}\frac{d}{dr}\left(r\frac{dv_z}{dr}\right)\right] = \frac{\partial p}{\partial z}, \tag{E6.7.3}$$

in which total derivatives are used because v_z depends only on r.

Shortly, we shall prove that the pressure gradient is uniform between the die inlet and exit, being given by:

$$-\frac{\partial p}{\partial z} = \frac{p_2 - p_3}{D}, \tag{E6.7.4}$$

in which both sides of the equation are positive quantities. Two successive integrations of Eqn. (E6.7.3) may then be performed, yielding:

$$v_z = -\frac{1}{4\mu}\left(-\frac{\partial p}{\partial z}\right)r^2 + c_1\ln r + c_2. \tag{E6.7.5}$$

The two constants may be evaluated by applying the boundary conditions of zero velocity at the inner and outer walls,

$$r = r_1: v_z = 0, \qquad r = r_2: v_z = 0, \tag{E6.7.6}$$

giving:

$$c_1 = \frac{1}{4\mu}\left(-\frac{\partial p}{\partial z}\right)\frac{r_2^2 - r_1^2}{\ln(r_2/r_1)}, \qquad c_2 = \frac{1}{4\mu}\left(-\frac{\partial p}{\partial z}\right)r_2^2 - c_1\ln r_2. \tag{E6.7.7}$$

Substitution of these values for the constants of integration into Eqn. (E6.7.5) yields the final expression for the velocity profile:

$$v_z = \frac{1}{4\mu}\left(-\frac{\partial p}{\partial z}\right)\left[\frac{\ln(r/r_2)}{\ln(r_2/r_1)}(r_2^2 - r_1^2) + (r_2^2 - r^2)\right], \tag{E6.7.8}$$

which is sketched in Fig. E6.7. Note that the maximum velocity occurs somewhat *before* the halfway point in progressing from the inner cylinder to the outer cylinder.

Volumetric flow rate. The final quantity of interest is the volumetric flow rate Q. Observing first that the flow rate through an annulus of internal radius r and external radius $r + dr$ is $dQ = v_z 2\pi r\, dr$, integration yields:

$$Q = \int_0^Q dQ = \int_{r_1}^{r_2} v_z 2\pi r\, dr. \tag{E6.7.9}$$

Since $r\ln r$ is involved in the expression for v_z, the following indefinite integral is needed:

$$\int r\ln r\, dr = \frac{r^2}{2}\ln r - \frac{r^2}{4}, \tag{E6.7.10}$$

giving the final result:

$$Q = \frac{\pi(r_2^2 - r_1^2)}{8\mu}\left(-\frac{\partial p}{\partial z}\right)\left[r_2^2 + r_1^2 - \frac{r_2^2 - r_1^2}{\ln(r_2/r_1)}\right]. \tag{E6.7.11}$$

Since Q, μ, r_1, and r_2 are constant throughout the die, $\partial p/\partial z$ is also constant, thus verifying the hypothesis previously made. Observe that in the limiting case of $r_1 \to 0$, Eqn. (E6.7.11) simplifies to the Hagen-Poiseuille law, already stated in Eqn. (3.12).

This problem may also be solved by performing a momentum balance on a shell that consists of an *annulus* of internal radius r, external radius $r + dr$, and length dz. □

Example 6.8—Spinning a Polymeric Fiber   **325**

Example 6.8—Spinning a Polymeric Fiber

A Newtonian polymeric liquid of viscosity μ is being "spun" (drawn into a fiber or filament of small diameter before solidifying by pulling it through a chemical setting bath) in the apparatus shown in Fig. E6.8.

The liquid volumetric flow rate is Q, and the filament diameters at $z = 0$ and $z = L$ are D_0 and D_L, respectively. To a first approximation, neglect the effects of gravity, inertia, and surface tension (this last effect is examined in detail in Problem 6.26). Derive an expression for the tensile force F needed to pull the filament downward. Assume that the axial velocity profile is "flat" at any vertical location, so that v_z depends only on z, which is here most conveniently taken as positive in the *downward* direction. Also derive an expression for the downward velocity v_z as a function of z. The inset of Fig. E6.8 shows further details of the notation concerning the filament.

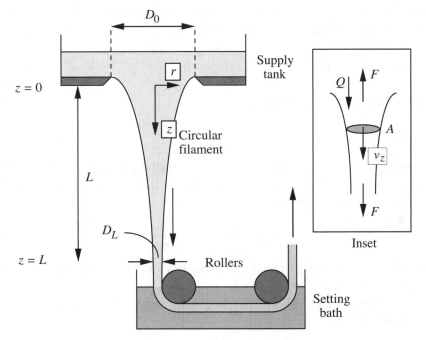

Fig. E6.8   *"Spinning" a polymer filament, whose diameter in relation to its length is exaggerated in the diagram.*

Solution

It is first necessary to determine the radial velocity and hence the pressure inside the filament. From continuity:

$$\frac{1}{r}\frac{\partial(rv_r)}{\partial r} + \frac{\partial v_z}{\partial z} = 0. \tag{E6.8.1}$$

Since v_z depends only on z, its axial derivative is a function of z only, or $dv_z/dz = f(z)$, so that Eqn. (E6.8.1) may be rearranged and integrated at constant z to give:

$$\frac{\partial(rv_r)}{\partial r} = -rf(z), \qquad rv_r = -\frac{r^2 f(z)}{2} + g(z). \qquad \text{(E6.8.2)}$$

But to avoid an infinite value of v_r at the centerline, g(z) must be zero, giving:

$$v_r = -\frac{rf(z)}{2}, \qquad \frac{\partial v_r}{\partial r} = -\frac{f(z)}{2}. \qquad \text{(E6.8.3)}$$

To proceed with reasonable expediency, it is necessary to make some simplification. After accounting for a primary effect (the difference between the pressure in the filament and the surrounding atmosphere), we assume that a secondary effect (variation of pressure across the filament) is negligible; that is, the pressure does not depend on the radial location (see Problem 6.27 for a more accurate investigation). Note that the external (gauge) pressure is zero everywhere, and apply the first part of Eqn. (5.64) at the free surface:

$$\sigma_{rr} = -p + 2\mu\frac{\partial v_r}{\partial r} = -p - \mu f(z) = 0, \quad \text{or} \quad p = -\mu\frac{dv_z}{dz}. \qquad \text{(E6.8.4)}$$

The axial stress is therefore:

$$\sigma_{zz} = -p + 2\mu\frac{dv_z}{dz} = 3\mu\frac{dv_z}{dz}. \qquad \text{(E6.8.5)}$$

It is interesting to note that the same result can be obtained with an alternative assumption.[3] The axial tension in the fiber equals the product of the cross-sectional area and the local axial stress:

$$F = A\sigma_{zz} = 3\mu A\frac{dv_z}{dz}. \qquad \text{(E6.8.6)}$$

Since the effect of gravity is stated to be insignificant, F is a constant, regardless of the vertical location.

At any location, the volumetric flow rate equals the product of the cross-sectional area and the axial velocity:

$$Q = Av_z. \qquad \text{(E6.8.7)}$$

A differential equation for the velocity is next obtained by dividing one of the last two equations by the other, and rearranging:

$$\frac{1}{v_z}\frac{dv_z}{dz} = \frac{F}{3\mu Q}. \qquad \text{(E6.8.8)}$$

---

[3] See S. Middleman's *Fundamentals of Polymer Processing*, McGraw-Hill, New York, 1977, p. 235. There, the author assumes $\sigma_{rr} = \sigma_{\theta\theta} = 0$, followed by the identity: $p = -(\sigma_{zz} + \sigma_{rr} + \sigma_{\theta\theta})/3$.

Integration, noting that the inlet velocity at $z = 0$ is $v_{z0} = Q/(\pi D_0^2/4)$, gives:

$$\int_{v_{z0}}^{v_z} \frac{dv_z}{v_z} = \frac{F}{3\mu Q} \int_0^z dz, \qquad \text{where} \quad v_{z0} = \frac{4Q}{\pi D_0^2}, \tag{E6.8.9}$$

so that the axial velocity increases exponentially with distance:

$$v_z = v_{z0} e^{Fz/3\mu Q}. \tag{E6.8.10}$$

The tension is obtained by applying Eqns. (E6.8.7) and (E6.8.10) just before the filament is taken up by the rollers:

$$v_{zL} = \frac{4Q}{\pi D_L^2} = v_{z0} e^{FL/3\mu Q}. \tag{E6.8.11}$$

Rearrangement yields:

$$F = \frac{3\mu Q}{L} \ln \frac{v_{zL}}{v_{z0}}, \tag{E6.8.12}$$

which predicts a force that *increases* with higher viscosities, flow rates, and draw-down ratios (v_{zL}/v_{z0}), and that *decreases* with longer filaments.

Elimination of F from Eqns. (E6.8.10) and (E6.8.12) gives an expression for the velocity that depends only on the variables specified originally:

$$v_z = v_{z0} \left(\frac{v_{zL}}{v_{z0}} \right)^{z/L} = v_{z0} \left(\frac{D_0}{D_L} \right)^{2z/L}, \tag{E6.8.13}$$

a result that is independent of the viscosity. $\qquad\qquad\square$

6.6   Solution of the Equations of Motion in Spherical Coordinates

Most of the introductory viscous-flow problems will lend themselves to solution in either rectangular or cylindrical coordinates. Occasionally, as in Example 6.7, a problem will arise in which *spherical* coordinates should be used. It is a fairly advanced problem! Try first to appreciate the broad steps involved, and then peruse the fine detail at a second reading.

Example 6.9—Analysis of a Cone-and-Plate Rheometer

The problem concerns the analysis of a cone-and-plate rheometer, an instrument developed and perfected in the 1950s and 1960s by Prof. Karl Weissenberg, for measuring the viscosity of liquids, and also known as the "Weissenberg rheogoniometer."[4] A cross section of the essential features is shown in Fig. E6.9, in which the liquid sample is held by surface tension in the narrow opening between a rotating lower circular plate, of radius R, and an upper cone, making an angle of β with the vertical axis. The plate is rotated steadily in the ϕ direction with an angular velocity ω, causing the liquid in the gap to move in concentric circles with a velocity v_ϕ. (In practice, the tip of the cone is slightly truncated, to avoid friction with the plate.) Observe that the flow is of the *Couette* type.

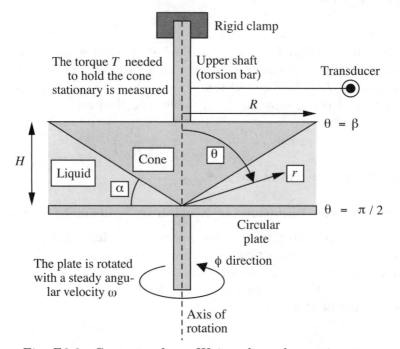

Fig. E6.9   Geometry for a Weissenberg rheogoniometer.
(The angle between the cone and plate is exaggerated.)

The top of the upper shaft—which acts like a torsion bar—is clamped rigidly. However, viscous friction will twist the cone and the lower portions of the upper shaft very slightly; the amount of motion can be detected by a light arm at the extremity of which is a transducer, consisting of a small piece of steel, attached to

---

[4] Professor Weissenberg (1893–1976) once related to the author that he (Prof. Weissenberg) was attending an instrument trade show in London. There, the rheogoniometer was being demonstrated by a young salesman who was unaware of Prof. Weissenberg's identity. Upon inquiring how the instrument worked, the salesman replied: "I'm sorry, sir, but it's quite complicated, and I don't think you will be able to understand it."

Example 6.9—Analysis of a Cone-and-Plate Rheometer   **329**

the arm, and surrounded by a coil of wire; by monitoring the inductance of the coil, the small angle of twist can be obtained; a knowledge of the elastic properties of the shaft then enables the restraining torque T to be obtained. From the analysis given below, it is then possible to deduce the viscosity of the sample. The instrument is so sensitive that if no liquid is present, it is capable of determining the viscosity of the air in the gap!

The problem is best solved using *spherical* coordinates, because the surfaces of the cone and plate are then described by *constant* values of the angle θ, namely β and $\pi/2$, respectively.

Assumptions and the continuity equation.   The following realistic assumptions are made:

1. There is only one nonzero velocity component, namely that in the ϕ direction, v_ϕ. Thus, $v_r = v_\theta = 0$.
2. Gravity acts vertically downward, so that $g_\phi = 0$.
3. We do not need to know how the pressure varies in the liquid. Therefore, the r and θ momentum balances, which would supply this information, are not required.

The analysis starts once more from the continuity equation, (5.50):

$$\frac{\partial \rho}{\partial t} + \frac{1}{r^2}\frac{\partial(\rho r^2 v_r)}{\partial r} + \frac{1}{r\sin\theta}\frac{\partial(\rho v_\theta \sin\theta)}{\partial \theta} + \frac{1}{r\sin\theta}\frac{\partial(\rho v_\phi)}{\partial \phi} = 0, \tag{5.50}$$

which, for constant density and $v_r = v_\theta = 0$ reduces to:

$$\frac{\partial v_\phi}{\partial \phi} = 0, \tag{E6.9.1}$$

verifying that v_ϕ is independent of the angular location ϕ, so we are correct in examining just one representative cross section, as shown in Fig. E6.9.

Momentum balances.   There are again three momentum balances, one for each of the r, θ, and ϕ directions. From the third assumption above, the first two such balances are of no significant interest, leaving, from Eqn. (5.77), just that in the ϕ direction:

$$\rho\left(\frac{\partial v_\phi}{\partial t} + v_r\frac{\partial v_\phi}{\partial r} + \frac{v_\theta}{r}\frac{\partial v_\phi}{\partial \theta} + \frac{v_\phi}{r\sin\theta}\frac{\partial v_\phi}{\partial \phi} + \frac{v_\phi v_r}{r} + \frac{v_\theta v_\phi \cot\theta}{r}\right)$$

$$= -\frac{1}{r\sin\theta}\frac{\partial p}{\partial \phi} + \mu\left[\frac{1}{r^2}\frac{\partial}{\partial r}\left(r^2\frac{\partial v_\phi}{\partial r}\right) + \frac{1}{r^2\sin\theta}\frac{\partial}{\partial \theta}\left(\sin\theta\frac{\partial v_\phi}{\partial \theta}\right)\right.$$

$$\left. + \frac{1}{r^2\sin^2\theta}\frac{\partial^2 v_\phi}{\partial \phi^2} - \frac{v_\phi}{r^2\sin^2\theta} + \frac{2}{r^2\sin\theta}\frac{\partial v_r}{\partial \phi} + \frac{2\cos\theta}{r^2\sin^2\theta}\frac{\partial v_\theta}{\partial \phi}\right] + \rho g_\phi. \tag{E6.9.2}$$

With $v_r = v_\theta = 0$ (assumption 1), $\partial v_\phi / \partial \phi = 0$ [from Eqn. (E6.9.1)], and $g_\phi = 0$ (assumption 2), the momentum balance simplifies to:

$$\frac{\partial}{\partial r} \left(r^2 \frac{\partial v_\phi}{\partial r} \right) + \frac{1}{\sin \theta} \frac{\partial}{\partial \theta} \left(\sin \theta \frac{\partial v_\phi}{\partial \theta} \right) - \frac{v_\phi}{\sin^2 \theta} = 0. \qquad \text{(E6.9.3)}$$

Determination of the velocity profile.   First, seek an expression for the velocity in the ϕ direction, which is expected to be proportional to both the distance r from the origin and the angular velocity ω of the lower plate. However, its variation with the coordinate θ is something that has to be discovered. Therefore, postulate a solution of the form:

$$v_\phi = r\omega f(\theta), \qquad \text{(E6.9.4)}$$

in which the function $f(\theta)$ is to be determined. Substitution of v_ϕ from Eqn. (E6.9.4) into Eqn. (E6.9.3), and using f' and f'' to denote the first and second *total* derivatives of f with respect to θ, gives:

$$f'' + f' \cot \theta + f(1 - \cot^2 \theta)$$

$$\equiv \frac{1}{\sin^2 \theta} \frac{d}{d\theta} (f' \sin^2 \theta - f \sin \theta \cos \theta)$$

$$= \frac{1}{\sin^2 \theta} \frac{d}{d\theta} \left[\sin^3 \theta \frac{d}{d\theta} \left(\frac{f}{\sin \theta} \right) \right] = 0. \qquad \text{(E6.9.5)}$$

The reader is encouraged, as always, to check the missing algebraic and trigonometric steps, although they are rather tricky here![5]

By multiplying Eqn. (E6.9.5) through by $\sin^2 \theta$, it follows after integration that the quantity in brackets is a constant, here represented as $-2c_1$, the reason for the "-2" being that it will cancel with a similar factor later on:

$$\sin^3 \theta \frac{d}{d\theta} \left(\frac{f}{\sin \theta} \right) = -2c_1. \qquad \text{(E6.9.6)}$$

Separation of variables and indefinite integration (without specified limits) yields:

$$\int d \left(\frac{f}{\sin \theta} \right) = \frac{f}{\sin \theta} = -2c_1 \int \frac{d\theta}{\sin^3 \theta}. \qquad \text{(E6.9.7)}$$

---

[5] Although our approach is significantly different from that given on page 98 *et seq.* of R.B. Bird, W.E. Stewart, and E.N. Lightfoot, *Transport Phenomena*, Wiley & Sons, New York, 1960, we are indebted to these authors for the helpful hint they gave in establishing the equivalency expressed in our Eqn. (E6.9.5).

Example 6.9—Analysis of a Cone-and-Plate Rheometer   **331**

To proceed further, we need the following two standard indefinite integrals and one trigonometric identity[6]:

$$\int \frac{d\theta}{\sin^3 \theta} = -\frac{1}{2} \frac{\cot \theta}{\sin \theta} + \frac{1}{2} \int \frac{d\theta}{\sin \theta}, \tag{E6.9.8}$$

$$\int \frac{d\theta}{\sin \theta} = \ln\left(\tan \frac{\theta}{2} \right) = \frac{1}{2} \ln\left(\tan^2 \frac{\theta}{2} \right), \tag{E6.9.9}$$

$$\tan^2 \frac{\theta}{2} = \frac{1 - \cos \theta}{1 + \cos \theta}. \tag{E6.9.10}$$

Armed with these, Eqn. (E6.9.7) leads to the following expression for f:

$$f = c_1 \left[\cot \theta + \frac{1}{2} \left(\ln \frac{1 + \cos \theta}{1 - \cos \theta} \right) \sin \theta \right] + c_2, \tag{E6.9.11}$$

in which c_2 is a second constant of integration.

Implementation of the boundary conditions.   The constants c_1 and c_2 are found by imposing the two boundary conditions:

1. At the lower plate, where $\theta = \pi/2$ and the expression in parentheses in Eqn. (E6.9.11) is zero, so that $f = c_2$, the velocity is simply the radius times the angular velocity:

$$v_\phi \equiv r\omega f = r\omega c_2 = r\omega, \quad \text{or} \quad c_2 = 1. \tag{E6.9.12}$$

2. At the surface of the cone, where $\theta = \beta$, the velocity $v_\phi = r\omega f$ is zero. Hence $f = 0$, and Eqn. (E6.9.11) leads to:

$$c_1 = -c_2 g(\beta) = -g(\beta), \quad \text{where} \quad \frac{1}{g(\beta)} = \cot \beta + \frac{1}{2} \left(\ln \frac{1 + \cos \beta}{1 - \cos \beta} \right) \sin \beta. \tag{E6.9.13}$$

Substitution of these expressions for c_1 and c_2 into Eqn. (E6.9.11), and noting that $v_\phi = r\omega f$, gives the final (!) expression for the velocity:

$$v_\phi = r\omega \left[1 - \frac{\cot \theta + \frac{1}{2} \left(\ln \frac{1 + \cos \theta}{1 - \cos \theta} \right) \sin \theta}{\cot \beta + \frac{1}{2} \left(\ln \frac{1 + \cos \beta}{1 - \cos \beta} \right) \sin \beta} \right]. \tag{E6.9.14}$$

As a partial check on the result, note that Eqn. (E6.9.14) reduces to $v_\phi = r\omega$ when $\theta = \pi/2$ and to $v_\phi = 0$ when $\theta = \beta$.

---

[6] From pages 87 (integrals) and 72 (trigonometric identity) of J.H. Perry, ed., *Chemical Engineers' Handbook*, 3rd ed., McGraw-Hill, New York, 1950.

Shear stress and torque.   Recall that the primary goal of this investigation is to determine the torque T needed to hold the cone stationary. The relevant shear stress exerted by the liquid on the surface of the cone is $\tau_{\theta\phi}$—that exerted on the under surface of the cone (of constant first subscript, $\theta = \beta$) in the positive ϕ direction (refer again to Fig. 5.12 for the sign convention and notation for stresses). Since this direction is the same as that of the rotation of the lower plate, we expect that $\tau_{\theta\phi}$ will prove to be positive, thus indicating that the liquid is trying to turn the cone in the same direction in which the lower plate is rotated.

From the second of Eqn. (5.65), the relation for this shear stress is:

$$\tau_{\theta\phi} = \mu \left[\frac{\sin\theta}{r} \frac{\partial}{\partial\theta} \left(\frac{v_\phi}{\sin\theta} \right) + \frac{1}{r\sin\theta} \frac{\partial v_\theta}{\partial\phi} \right]. \tag{E6.9.15}$$

Since $v_\theta = 0$, and recalling Eqns. (E6.9.4), (E6.9.6), and (E6.9.13), the shear stress on the cone becomes:

$$(\tau_{\theta\phi})_{\theta=\beta} = \mu \left[\frac{\sin\theta}{r} \frac{\partial}{\partial\theta} \left(\frac{r\omega f}{\sin\theta} \right) \right]_{\theta=\beta} = -\left(\frac{2c_1\omega\mu}{\sin^2\theta} \right)_{\theta=\beta} = \frac{2\omega\mu g(\beta)}{\sin^2\beta}. \tag{E6.9.16}$$

One importance of this result is that it is *independent* of r, giving a constant stress and strain throughout the liquid, a significant simplification when deciphering the experimental results for non-Newtonian fluids (see Chapter 11). In effect, the increased velocity differences between the plate and cone at the greater values of r are offset in exact proportion by the larger distances separating them.

The torque exerted by the liquid on the cone (in the positive ϕ direction) is obtained as follows. The surface area of the cone between radii r and $r + dr$ is $2\pi r \sin\beta\, dr$, and is located at a lever arm of $r\sin\beta$ from the axis of symmetry. Multiplication by the shear stress and integration gives:

$$T = \int_0^{R/\sin\beta} \underbrace{(2\pi r \sin\beta\, dr)}_{\text{Area}} \underbrace{r\sin\beta}_{\substack{\text{Lever} \\ \text{arm}}} \underbrace{(\tau_{\theta\phi})_{\theta=\beta}}_{\text{Stress}}, \tag{E6.9.17}$$

Substitution of $(\tau_{\theta\phi})_{\theta=\beta}$ from Eqn. (E6.9.16) and integration gives the torque as:

$$T = \frac{4\pi\omega\mu g(\beta)R^3}{3\sin^3\beta}. \tag{E6.9.18}$$

Problem 6.15 reaches essentially the same conclusion much more quickly for the case of a small angle between the cone and the plate. The torque for holding the cone stationary has the same value, but is, of course, in the *negative* ϕ direction.

Since R and $g(\beta)$ can be determined from the radius and the angle β of the cone in conjunction with Eqn. (E6.9.13), the viscosity μ of the liquid can finally be determined.   □

PROBLEMS FOR CHAPTER 6

*Unless otherwise stated, all flows are steady state, for
a Newtonian fluid with constant density and viscosity.*

1. *Stretching of a liquid film—M.* In broad terms, explain the meanings of the
following two equations, paying attention to any sign convention:

$$\sigma_{xx} = -p + 2\mu\,\frac{\partial v_x}{\partial x}, \qquad \sigma_{yy} = -p + 2\mu\,\frac{\partial v_y}{\partial y}.$$

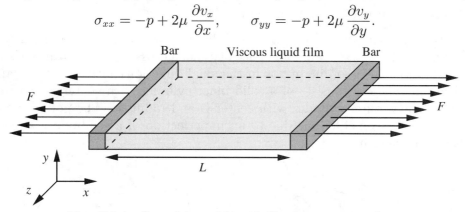

Fig. P6.1   Stretching of liquid film between two bars.

Fig. P6.1 shows a film of a viscous liquid held between two bars spaced a distance
L apart. If the film thickness is uniform, and the total volume of liquid is V, show
that the force necessary to separate the bars with a relative velocity dL/dt is:

$$F = \frac{4\mu V}{L^2}\frac{dL}{dt}.$$

Fig. P6.2   Coating of wire drawn through a die.

2. *Wire coating—M.* Fig. P6.2 shows a rodlike wire of radius r_1 that is being
pulled steadily with velocity V through a horizontal die of length L and internal
radius r_2. The wire and the die are coaxial, and the space between them is filled
with a liquid of viscosity μ. The pressure at both ends of the die is atmospheric.
The wire is coated with the liquid as it leaves the die, and the thickness of the
coating eventually settles down to a uniform value, δ.

Neglecting end effects, use the equations of motion in cylindrical coordinates
to derive expressions for:

(a) The velocity profile within the annular space. Assume that there is only one nonzero velocity component, v_z, and that this depends only on radial position.

(b) The total volumetric flow rate Q through the annulus.

(c) The limiting value for Q if r_1 approaches zero.

(d) The final thickness, δ, of the coating on the wire. (Here, only an equation is needed from which δ could be found.)

(e) The force F needed to pull the wire.

3. *Off-center annular flow—D (C).* A liquid flows under a pressure gradient $\partial p/\partial z$ through the *narrow* annular space of a die, a cross section of which is shown in Fig. P6.3(a). The coordinate z is in the axial direction, normal to the plane of the diagram. The die consists of a solid inner cylinder with center P and radius b inside a hollow outer cylinder with center O and radius a. The points O and P were *intended* to coincide, but due to an imperfection of assembly are separated by a small distance δ.

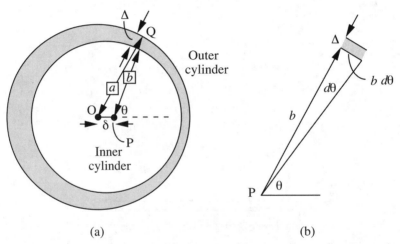

(a)   (b)

Fig. P6.3   Off-center cylinder inside a die (gap width exaggerated): (a) complete cross section; (b) effect of incrementing θ.

By a simple geometrical argument based on the triangle OPQ, show that the gap width Δ between the two cylinders is given approximately by:

$$\Delta \doteq a - b - \delta \cos \theta,$$

where the angle θ is defined in the diagram.

Now consider the radius arm b swung through an angle $d\theta$, so that it traces an arc of length $b\,d\theta$. The flow rate dQ through the shaded element in Fig. P6.3(b) is approximately that between parallel plates of width $b\,d\theta$ and separation Δ. Hence, prove that the flow rate through the die is given approximately by:

$$Q = \pi bc(2\alpha^3 + 3\alpha\delta^2),$$

in which:

$$c = \frac{1}{12\mu}\left(-\frac{\partial p}{\partial z}\right), \qquad \text{and} \qquad \alpha = a - b.$$

Assume from Eqn. (E6.1.26) that the flow rate per unit width between two flat plates separated by a distance h is:

$$\frac{h^3}{12\mu}\left(-\frac{\partial p}{\partial z}\right).$$

What is the ratio of the flow rate if the two cylinders are touching at one point to the flow rate if they are concentric?

4. *Compression molding—M.* Fig. P6.4 shows the (a) beginning, (b) intermediate, and (c) final stages in the compression molding of a material that behaves as a liquid of high viscosity μ, from an initial cylinder of height H_0 and radius R_0 to a final disk of height H_1 and radius R_1.

Fig. P6.4   Compression molding between two disks.

In the molding operation, the upper disk A is squeezed with a uniform velocity V toward the stationary lower disk B.

(a) Derive expressions for H and R as functions of time t, H_0, and R_0.

(b) Consider the radially outward volumetric flow rate Q per unit perimeter, crossing a cylinder of radius r, as in Fig. P6.4(b). Obtain a relation for Q in terms of V and r.

(c) Assume by analogy from Eqn. (E6.1.26) that per unit width of a channel of depth H, the volumetric flow rate is:

$$Q = \frac{H^3}{12\mu}\left(-\frac{\partial p}{\partial r}\right).$$

(d) Ignoring small variations of pressure in the z direction, derive an expression for the radial variation of pressure.

(e) Prove that the total compressive force F that must be exerted downward on the upper disk is:

$$F = \frac{3\pi\mu V R^4}{2H^3}.$$

(f) Draw a sketch that shows how F varies with time.

5. *Film draining—M.* Fig. P6.5 shows an idealized view of a liquid film of viscosity μ that is draining under gravity down the side of a flat vertical wall. Such a situation would be *approximated* by the film left on the wall of a tank that was suddenly drained through a large hole in its base.

Fig. P6.5   Liquid draining from a vertical wall.

What are the justifications for assuming that the velocity profile at any distance x below the top of the wall is given by:

$$v_x = \frac{\rho g}{2\mu} y(2h - y),$$

where $h = h(x)$ is the local film thickness? Derive an expression for the corresponding downward mass flow rate m per unit wall width (normal to the plane of the diagram).

Perform a *transient* mass balance on a differential element of the film and prove that h varies with time and position according to:

$$\frac{\partial h}{\partial t} = -\frac{1}{\rho}\frac{\partial m}{\partial x}.$$

Now substitute your expression for m, to obtain a partial differential equation for h. Try a solution of the form:

$$h = c\,t^p x^q,$$

and determine the unknowns c, p, and q. Discuss the limitations of your solution.

Note that a similar situation occurs when a substrate is suddenly lifted from a bath of coating fluid.

6. *Sheet "spinning"—M.* A Newtonian polymeric liquid of viscosity μ is being "spun" (drawn into a sheet of small thickness before solidifying by pulling it through a chemical setting bath) in the apparatus shown in Fig. P6.6.

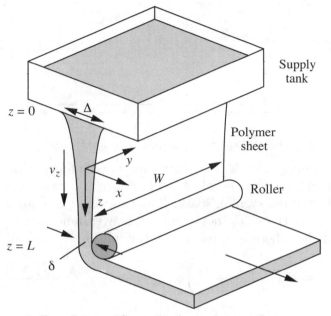

Fig. P6.6   "Spinning" a polymer sheet.

The liquid volumetric flow rate is Q, and the sheet thicknesses at $z = 0$ and $z = L$ are Δ and δ, respectively. The effects of gravity, inertia, and surface tension are negligible. Derive an expression for the tensile force needed to pull the filament downward. *Hint*: Start by assuming that the vertically downward velocity v_z depends only on z and that the lateral velocity v_y is zero. Also derive an expression for the downward velocity v_z as a function of z.

7. *Details of pipe flow—M.* A fluid of density ρ and viscosity μ flows from left to right through the horizontal pipe of radius a and length L shown in Fig. P6.7. The pressures at the centers of the inlet and exit are p_1 and p_2, respectively. You may assume that the only nonzero velocity component is v_z, and that this is not a function of the angular coordinate, θ.

Stating any further necessary assumptions, derive expressions for the following, in terms of any or all of a, L, p_1, p_2, ρ, μ, and the coordinates r, z, and θ:

(a) The velocity profile, $v_z = v_z(r)$.
(b) The total volumetric flow rate Q through the pipe.
(c) The pressure p at any point (r, θ, z).
(d) The shear stress, τ_{rz}.

Fig. P6.7 *Flow of a liquid in a horizontal pipe.*

8. *Natural convection—M.* Fig. P6.8 shows two infinite parallel vertical walls that are separated by a distance $2d$. A fluid of viscosity μ and volume coefficient of expansion β fills the intervening space. The two walls are maintained at uniform temperatures T_1 (cold) and T_2 (hot), and you may assume (to be proved in a heat-transfer course) that there is a linear variation of temperature in the x direction. That is:

$$T = \overline{T} + \frac{x}{d}\left(\frac{T_2 - T_1}{2}\right), \qquad \text{where} \qquad \overline{T} = \frac{T_1 + T_2}{2}.$$

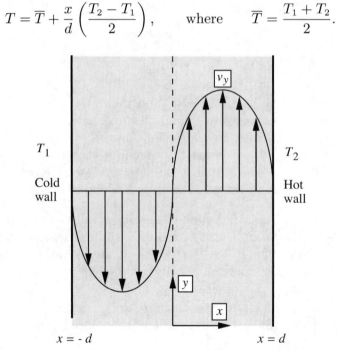

Fig. P6.8 *Natural convection between vertical walls.*

The density is *not* constant, but varies according to:

$$\rho = \overline{\rho}\left[1 - \beta\left(T - \overline{T}\right)\right],$$

where $\bar{\rho}$ is the density at the mean temperature $\overline{T}$, which occurs at $x = 0$.

If the resulting natural-convection flow is steady, use the equations of motion to derive an expression for the velocity profile $v_y = v_y(x)$ between the plates. Your expression for v_y should be in terms of any or all of x, d, T_1, T_2, $\bar{\rho}$, μ, β, and g.

Hints: In the y momentum balance, you should find yourself facing the following combination:

$$-\frac{\partial p}{\partial y} + \rho g_y,$$

in which $g_y = -g$. These two terms are *almost* in balance, but not quite, leading to a small—but important—buoyancy effect that "drives" the natural convection. The variation of pressure in the y direction may be taken as the normal hydrostatic variation:

$$\frac{\partial p}{\partial y} = -g\bar{\rho}.$$

We then have:

$$-\frac{\partial p}{\partial y} + \rho g_y = g\bar{\rho} + \bar{\rho}\left[1 - \beta\left(T - \overline{T}\right)\right](-g) = \bar{\rho}\beta g\left(T - \overline{T}\right),$$

and this will be found to be a vital contribution to the y momentum balance.

9. *Square duct velocity profile—M.* A certain flow in rectangular Cartesian coordinates has only one nonzero velocity component, v_z, and this does not vary with z. If there is no body force, write down the Navier-Stokes equation for the z momentum balance.

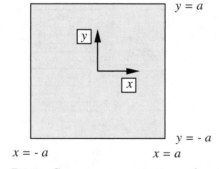

Fig. P6.9   Square cross section of a duct.

One-dimensional, fully developed steady flow occurs under a pressure gradient $\partial p/\partial z$ in the z direction, parallel to the axis of a square duct of side $2a$, whose cross section is shown in Fig. P6.9. The following equation has been proposed for the velocity profile:

$$v_z = \frac{1}{2\mu}\left(-\frac{\partial p}{\partial z}\right)a^2\left[1 - \left(\frac{x}{a}\right)^2\right]\left[1 - \left(\frac{y}{a}\right)^2\right].$$

Without attempting to integrate the momentum balance, investigate the possible merits of this proposed solution for v_z. Explain whether or not it is correct.

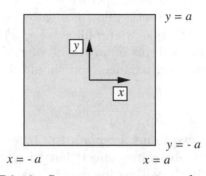

Fig. P6.10   Square cross section of a duct.

10. *Poisson's equation for a square duct—E.* A polymeric fluid of uniform viscosity μ is to be extruded after pumping it through a long duct whose cross section is a square of side $2a$, shown in Fig. P6.10. The flow is parallel everywhere to the axis of the duct, which is in the z direction, normal to the plane of the diagram.

If $\partial p/\partial z$, μ, and a are specified, show that the problem of obtaining the axial velocity distribution $v_z = v_z(x, y)$, amounts to solving Poisson's equation—of the form $\nabla^2 \phi = f(x, y)$, where f is specified and ϕ is the unknown. Also note that Poisson's equation can be solved numerically by the COMSOL computational fluid dynamics program introduced in Chapter 14 (see also Example 7.4).

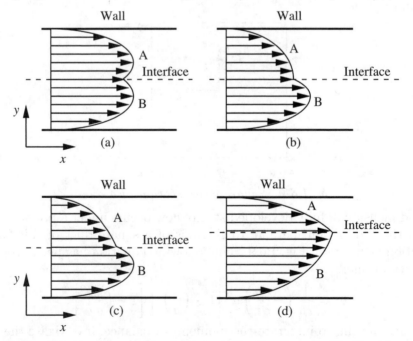

Fig. P6.11   Proposed velocity profiles for immiscible liquids.

11. *Permissible velocity profiles—E.* Consider the shear stress τ_{yx}; why must it be *continuous*—in the y direction, for example—and *not* undergo a sudden step-change in its value? Two *immiscible* Newtonian liquids A and B are in steady laminar flow between two parallel plates. Which—if any—of the velocity profiles shown in Fig. P6.11 are impossible? Profile A meets the interface normally in (b), but at an angle in (c). Any apparent location of the interface at the centerline is coincidental and should be ignored. Explain your answers carefully.

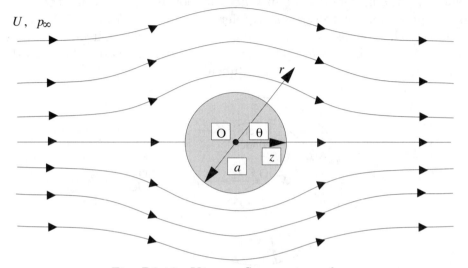

Fig. P6.12   Viscous flow past a sphere.

12. *"Creeping" flow past a sphere—D.* Figure P6.12 shows the steady, "creeping" (very slow) flow of a fluid of viscosity μ past a sphere of radius a. Far away from the sphere, the pressure is p_∞ and the undisturbed fluid velocity is U in the positive z direction. The following velocity components and pressure have been proposed in spherical coordinates:

$$p = p_\infty - \frac{3\mu U a}{2r^2}\cos\theta,$$

$$v_r = U\left(1 - \frac{3a}{2r} + \frac{a^3}{2r^3}\right)\cos\theta,$$

$$v_\theta = -U\left(1 - \frac{3a}{4r} - \frac{a^3}{4r^3}\right)\sin\theta,$$

$$v_\phi = 0.$$

Assuming the velocities are sufficiently small so that terms such as $v_r(\partial v_r/\partial r)$ can be neglected, and that gravity is unimportant, prove that these equations do indeed satisfy the following conditions, and therefore are the solution to the problem:

(a) The continuity equation.

(b) The r and θ momentum balances.

(c) A pressure of p_∞ and a z velocity of U far away from the sphere.

(d) Zero velocity components on the surface of the sphere.

Also derive an expression for the net force exerted in the z direction by the fluid on the sphere, and compare it with that given by Stokes' law in Eqn. (4.11).

Note that the problem is one in *spherical* coordinates, in which the z axis has no formal place, except to serve as a reference direction from which the angle θ is measured. There is also symmetry about this axis, such that any derivatives in the ϕ direction are zero. *Note*: The actual *derivation* of these velocities, *starting* from the equations of motion, is fairly difficult!

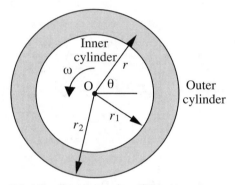

Fig. P6.13   Section of a Couette viscometer.

13. *Torque in a Couette viscometer—M.* Fig. P6.13 shows the horizontal cross section of a concentric cylinder or "Couette" viscometer, which is an apparatus for determining the viscosity μ of the fluid that is placed between the two vertical cylinders. The inner and outer cylinders have radii of r_1 and r_2, respectively. If the inner cylinder is rotated with a steady angular velocity ω, and the outer cylinder is stationary, derive an expression for v_θ (the θ velocity component) as a function of radial location r.

If, further, the torque required to rotate the *inner* cylinder is found to be T per unit length of the cylinder, derive an expression whereby the unknown viscosity μ can be determined, in terms of T, ω, r_1, and r_2. *Hint*: You will need to consider one of the shear stresses given in Table 5.8.

14. *Wetted-wall column—M.* Fig. P6.14 shows a "wetted-wall" column, in which a thin film of a reacting liquid of viscosity μ flows steadily down a plane wall, possibly for a gas-absorption study. The volumetric flow rate of liquid is specified as Q per unit width of the wall (normal to the plane of the diagram).

Assume that there is only one nonzero velocity component, v_x, and that this does not vary in the x direction, and that the gas exerts negligible shear stress on the liquid film. Starting from the equations of motion, derive an expression for the "profile" of the velocity v_x (as a function of ρ, μ, g, y, and δ), and also for the film thickness, δ (as a function of ρ, μ, g, and Q).

Flow rate
Q

Wall

Liquid
film

y

x

δ

Fig. P6.14   Wetted-wall column.

15. *Simplified view of a Weissenberg rheogoniometer—M.* Consider the Weissenberg rheogoniometer with a very shallow cone; thus, referring to Fig. E6.7, $\beta = \pi/2 - \alpha$, where α is a small angle

(a) Without going through the complicated analysis presented in Example 6.7, outline your reasons for supposing that the shear stress at any location on the cone is:

$$(\tau_{\theta\phi})_{\theta=\beta} = \frac{\omega\mu}{\alpha}.$$

(b) Hence, prove that the torque required to hold the cone stationary (or to rotate the lower plate) is:

$$T = \frac{2}{3}\frac{\pi\omega\mu R^4}{H}.$$

(c) By substituting $\beta = \pi/2 - \alpha$ into Eqn. (E6.7.13) and expanding the various functions in power series (only a very few terms are needed), prove that $g(\beta) = 1/(2\alpha)$, and that Eqn. (E6.7.18) again leads to the expression just obtained for the torque in part (b) above.

16. *Parallel-disk rheometer—M.* Fig. P6.16 shows the diametral cross section of a viscometer, which consists of two opposed circular horizontal disks, each of radius R, spaced by a vertical distance H; the intervening gap is filled by a liquid of constant viscosity μ and constant density. The upper disk is stationary, and the lower disk is rotated at a steady angular velocity ω in the θ direction.

There is only *one* nonzero velocity component, v_θ, so the liquid everywhere moves in circles. Simplify the general continuity equation in cylindrical coordinates, and hence deduce those coordinates ($r?$, $\theta?$, $z?$) on which v_θ may depend.

Now consider the θ-momentum equation, and simplify it by eliminating all zero terms. Explain briefly: (a) why you would expect $\partial p/\partial \theta$ to be zero, and (b) why you *cannot* neglect the term $\partial^2 v_\theta/\partial z^2$.

Also explain briefly the logic of supposing that the velocity in the θ direction is of the form $v_\theta = r\omega f(z)$, where the function $f(z)$ is yet to be determined. Now substitute this into the simplified θ-momentum balance and determine $f(z)$, using the boundary conditions that v_θ is zero on the upper disk and $r\omega$ on the lower disk.

Why would you designate the shear stress exerted by the liquid on the lower disk as $\tau_{z\theta}$? Evaluate this stress as a function of radius.

17. *Screw extruder optimum angle—M.* Note that the flow rate through the die of Example 6.7, given in Eqn. (E6.7.11), can be expressed as:

$$Q = \frac{c(p_2 - p_3)}{\mu D},$$

in which c is a factor that accounts for the geometry.

Suppose that this die is now connected to the exit of the extruder studied in Example 6.5, and that $p_1 = p_3 = 0$, both pressures being atmospheric. Derive an expression for the optimum flight angle θ_{opt} that will *maximize* the flow rate Q_y

through the extruder and die. Give your answer in terms of any or all of c, D, h, L_0, r, W, μ, and ω.

For what value of c would the pressure at the exit of the extruder approach its largest possible value p_{2max}? Derive an expression for p_{2max}.

18. *Annular flow in a die—E.* Referring to Example 6.7, concerning annular flow in a die, answer the following questions, giving your explanation in both cases:

(a) What form does the velocity profile, $v_z = v_z(r)$, assume as the radius r_1 of the inner cylinder becomes vanishingly small?

(b) Does the maximum velocity occur halfway between the inner and outer cylinders, or at some other location?

19. *Rotating rod in a fluid—M.* Fig. P6.19(a) shows a horizontal cross section of a long vertical cylinder of radius a that is rotated steadily counterclockwise with an angular velocity ω in a very large volume of liquid of viscosity μ. The liquid extends effectively to infinity, where it may be considered at rest. The axis of the cylinder coincides with the z axis.

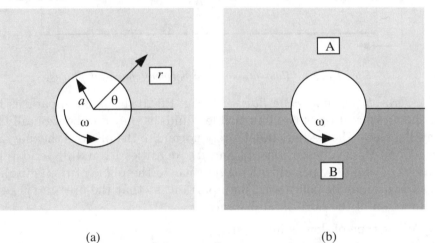

(a)   (b)

Fig. P6.19   *Rotating cylinder in (a) a single*
liquid, and (b) two immiscible liquids.

(a) What type of flow is involved? What coordinate system is appropriate?

(b) Write down the differential equation of mass and that one of the three general momentum balances that is most applicable to the determination of the velocity v_θ.

(c) Clearly stating your assumptions, simplify the situation so that you obtain an ordinary differential equation with v_θ as the dependent variable and r as the independent variable.

(d) Integrate this differential equation, and introduce any boundary condition(s), and prove that $v_\theta = \omega a^2/r$.

(e) Derive an expression for the shear stress $\tau_{r\theta}$ at the surface of the cylinder. Carefully explain the plus or minus sign in this expression.

(f) Derive an expression that gives the torque T needed to rotate the cylinder, per unit length of the cylinder.

(g) Derive an expression for the vorticity component $(\nabla \times \mathbf{v})_z$. Comment on your result.

(h) Fig. P6.19(b) shows the initial condition of a mixing experiment in which the cylinder is in the middle of two immiscible liquids, A and B, of *identical* densities and viscosities. After the cylinder has made one complete rotation, draw a diagram that shows a representative location of the interface between A and B.

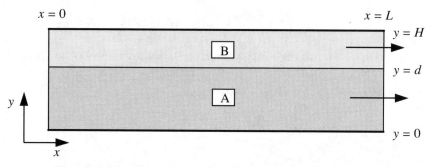

Fig. P6.20   *Two-phase flow between parallel plates.*

20. *Two-phase immiscible flow—M.* Fig. P6.20 shows an apparatus for measuring the pressure drop of two immiscible liquids as they flow horizontally between two parallel plates that extend indefinitely normal to the plane of the diagram. The liquids, A and B, have viscosities μ_A and μ_B, densities ρ_A and ρ_B, and volumetric flow rates Q_A and Q_B (per unit depth normal to the plane of the figure), respectively. Gravity may be considered unimportant, so that the pressure is essentially only a function of the horizontal distance, x.

(a) What type of flow is involved?

(b) Considering layer A, start from the differential equations of mass and momentum, and, clearly stating your assumptions, simplify the situation so that you obtain a differential equation that relates the horizontal velocity v_{xA} to the vertical distance y.

(c) Integrate this differential equation so that you obtain v_{xA} in terms of y and any or all of d, dp/dx, μ_A, ρ_A, and (assuming the pressure gradient is uniform) two arbitrary constants of integration, say, c_{1A} and c_{2A}. Assume that a similar relationship holds for v_{xB}.

(d) Clearly state the four boundary and interfacial conditions, and hence derive expressions for the four constants, thus giving the velocity profiles in the two layers.

(e) Sketch the velocity profiles and the shear-stress distribution for τ_{yx} between the upper and lower plates.

(f) Until now, we have assumed that the interface level $y = d$ is known. In reality, however, it will depend on the relative flow rates Q_A and Q_B. Show clearly how this dependency could be obtained, but do not actually carry the analysis through to completion.

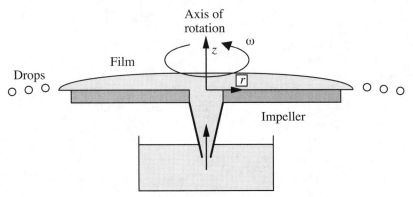

Fig. P6.21   Rotating-impeller humidifier.

21. *Room humidifier—M.* Fig. P6.21 shows a room humidifier, in which a circular impeller rotates about its axis with angular velocity ω. A conical extension dips into a water-bath, sucking up the liquid (of density ρ and viscosity μ), which then spreads out over the impeller as a thin laminar film that rotates everywhere with an angular velocity ω, eventually breaking into drops after leaving the periphery.

(a) Assuming incompressible steady flow, with symmetry about the vertical (z) axis, and a relatively small value of v_z, what can you say from the continuity equation about the term:

$$\frac{\partial(rv_r)}{\partial r}\;?$$

(b) If the pressure in the film is everywhere atmospheric, the only significant inertial term is v_θ^2/r, and information from (a) can be used to neglect one particular term, to what two terms does the r momentum balance simplify?

(c) Hence prove that the velocity v_r in the radial direction is a half-parabola in the z direction.

(d) Derive an expression for the total volumetric flow rate Q, and hence prove that the film thickness at any radial location is given by:

$$\delta = \left(\frac{3Q\nu}{2\pi r^2\omega^2}\right)^{1/3},$$

where ν is the kinematic viscosity.

Gap width is exaggerated here

Fig. P6.22   Transport of inner cylinder.

22. *Transport of inner cylinder—M (C).* As shown in Fig. P6.22, a long solid cylinder of radius r_c and length L is being transported by a viscous liquid of the same density down a pipe of radius a, which is much smaller than L. The annular gap, of extent $a - r_c$, is much smaller than a. Assume: (a) the cylinder remains concentric within the pipe, (b) the flow in the annular gap is laminar, (c) the shear stress is essentially constant across the gap, and (d) entry and exit effects can be neglected. Prove that the velocity of the cylinder is given fairly accurately by:

$$v_c = \frac{2Q}{\pi(a^2 + r_c^2)},$$

where Q is the volumetric flow rate of the liquid upstream and downstream of the cylinder. *Hint:* Concentrate first on understanding the physical situation. Don't rush headlong into a lengthy analysis with the Navier-Stokes equations!

Flow

v_z

Fig. P6.23   Flow between inclined planes.

23. *Flow between inclined planes—M (C).* A viscous liquid flows between two infinite planes inclined at an angle 2α to each other. Prove that the liquid velocity, which is everywhere parallel to the line of intersection of the planes, is given by:

$$v_z = \frac{r^2}{4\mu}\left(\frac{\cos 2\theta}{\cos 2\alpha} - 1\right)\left(-\frac{\partial p}{\partial z}\right),$$

where z, r, θ are cylindrical coordinates. The z axis is the line of intersection of the planes and the r axis at $\theta = 0$ bisects the angle between the planes. Assume laminar flow, with:

$$\frac{\partial p}{\partial z} = \mu \nabla^2 v_z = \mu \left[\frac{1}{r} \frac{\partial}{\partial r} \left(r \frac{\partial v_z}{\partial r} \right) + \frac{1}{r^2} \frac{\partial^2 v_z}{\partial \theta^2} \right],$$

and start by proposing a solution of the form $v_z = r^n g(\theta)(-\partial p/\partial z)/\mu$, where the exponent n and function $g(\theta)$ are to be determined.

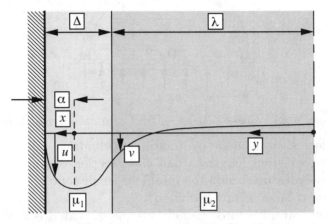

Fig. P6.24   Flow of two immiscible liquids in a pipe.
The thickness Δ of the film next to the wall is exaggerated.

24. *Immiscible flow inside a tube—D (C).* A film of liquid of viscosity μ_1 flows down the inside wall of a circular tube of radius $(\lambda + \Delta)$. The central core is occupied by a second immiscible liquid of viscosity μ_2, in which there is no net vertical flow. End effects may be neglected, and steady-state circulation in the core liquid has been reached. If the flow in both liquids is laminar, so that the velocity profiles are parabolic as shown in Fig. P6.24, prove that:

$$\alpha = \frac{2\mu_2 \Delta^2 + \lambda \mu_1 \Delta}{4\mu_2 \Delta + \lambda \mu_1},$$

where Δ is the thickness of the liquid film, and α is the distance from the wall to the point of maximum velocity in the film. Fig. P6.24 suggests notation for solving the problem. To save time, assume parabolic velocity profiles without proof: $u = a + bx + cx^2$ and $v = d + ey + fy^2$.

Discuss what happens when $\lambda/\Delta \to \infty$; and also when $\mu_2/\mu_1 \to 0$; and when $\mu_1/\mu_2 \to 0$.

25. *Blowing a polyethylene bubble—D.* For an incompressible fluid in cylindrical coordinates, write down:

(a) The continuity equation, and simplify it.
(b) Expressions for the viscous normal stresses σ_{rr} and $\sigma_{\theta\theta}$, in terms of pressure, viscosity, and strain rates.

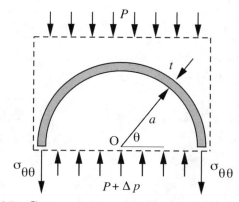

Fig. P6.25   Cross section of half of a cylindrical bubble.

A polyethylene sheet is made by inflating a cylindrical bubble of molten polymer effectively at constant length, a half cross section of which is shown in Fig. P6.25. The excess pressure inside the bubble is small compared with the external pressure P, so that $\Delta p \ll P$ and $\sigma_{rr} \doteq -P$.

By means of a suitable force balance on the indicated control volume, prove that the circumferential stress is given by $\sigma_{\theta\theta} = -P + a\Delta p/t$. Assume pseudo-steady state—that is, the circumferential stress just balances the excess pressure, neglecting any acceleration effects.

Hence, prove that the expansion velocity v_r of the bubble (at $r = a$) is given by:

$$v_r = \frac{a^2 \Delta p}{4\mu t},$$

and evaluate it for a bubble of radius 1.0 m and film thickness 1 mm when subjected to an internal gauge pressure of $\Delta p = 40$ N/m². The viscosity of polyethylene at the appropriate temperature is 10^5 N s/m².

26. *Surface-tension effect in spinning—M.* Example 6.8 ignored surface-tension effects, which would increase the pressure in a filament of radius R approximately by an amount σ/R, where σ is the surface tension. Compare this quantity with the reduction of pressure, $\mu dv_z/dz$, caused by viscous effects, for a polymer with $\mu = 10^4$ P, $\sigma = 0.030$ kg/s², $L = 1$ m, R_L (exit radius) $= 0.0002$ m, $v_{z0} = 0.02$ m/s, and $v_{zL} = 2$ m/s. Consider conditions both at the beginning and end of the filament. Comment briefly on your findings.

27. *Radial pressure variations in spinning—M.* Example 6.8 assumed that the variation of pressure across the filament was negligible. Investigate the validity of this assumption by starting with the suitably simplified momentum balance:

$$\frac{\partial p}{\partial r} = \mu \left[\frac{\partial}{\partial r} \left(\frac{1}{r} \frac{\partial (r v_r)}{\partial r} \right) + \frac{\partial^2 v_r}{\partial z^2} \right].$$

If v_z is the local axial velocity, prove that the corresponding increase of pressure from just inside the free surface (p_R) to the centerline (p_0) is:

$$p_0 - p_R = \frac{\mu v_z (\ln \beta)^3 R^2}{4L^3},$$

where $\beta = v_{zL}/v_{z0}$. Obtain an expression for the ratio $\xi = (p_R - p_0)/(\mu dv_z/dz)$, in which the denominator is the pressure decrease due to viscosity when crossing the interface from the air into the filament, and which *was* accounted for in Example 6.6.

Estimate ξ at the beginning of the filament for the situation in which $\mu = 10^4$ P, $L = 1$ m, R_L (exit radius) $= 0.0002$ m, $v_{z0} = 0.02$ m/s, and $v_{zL} = 2$ m/s. Comment briefly on your findings.

28. *Condenser with varying viscosity—M.* In a condenser, a viscous liquid flows steadily under gravity as a uniform laminar film down a vertical cooled flat plate. Due to conduction, the liquid temperature T varies linearly across the film, from T_0 at the cooled plate to T_1 at the hotter liquid/vapor interface, according to $T = T_0 + y(T_1 - T_0)/h$. The viscosity of the liquid is given approximately by $\mu = \mu^*(1 - \alpha T)$, where μ^* and α are constants.

Prove that the viscosity at any location can be reexpressed as:

$$\mu = \mu_0 + cy, \qquad \text{in which} \qquad c = \frac{\mu_1 - \mu_0}{h},$$

where μ_0 and μ_1 are the viscosities at temperatures T_0 and T_1, respectively.

What is the expression for the shear stress τ_{yx} for a liquid for steady flow in the x direction? Derive an expression for the velocity v_x as a function of y. Make sure that you *do not* base your answer on any equations that assume *constant* viscosity.

Sketch the velocity profile for both a small and a large value of α.

29. *Rotating sphere—M.* A sphere of radius R, immersed in a liquid of viscosity μ of infinite extent, rotates about its z axis (see Fig. 5.9(b)) with a steady angular velocity ω in the ϕ direction. You may assume that the only nonzero liquid velocity component is that in the ϕ direction, and is of the form $v_\phi = c \sin \theta / r^2$.

(a) Verify that the proposed form for v_ϕ satisfies the appropriate momentum balance, and determine c in terms of R and ω.

(b) Prove that the shear stress at any radial location in the liquid is $\tau_{r\phi} = -3\mu c \sin \theta / r^3$, and carefully explain the minus sign.

(c) Obtain an expression for the torque T needed to rotate the sphere, as a function of R, ω, and μ. You should need the indefinite integral:

$$\int \sin^3 \theta \, d\theta = -\cos \theta + \frac{1}{3} \cos^3 \theta.$$

30. *Moving plate—M.* Similar to the first part of Example 6.4, consider a viscous fluid of infinite extent and kinematic viscosity ν with a long flat plate in its surface. At $t = 0$, the plate suddenly starts moving with a steady velocity v_{x0} in the x direction. Prove, by checking $\partial v_x/\partial t = \nu \partial^2 v_x/\partial z^2$ and the boundary conditions, that the fluid velocity at a depth z is subsequently given by:

$$\frac{v_x}{v_{x0}} = \text{erfc}\frac{z}{2\sqrt{\nu t}}. \tag{E6.4.2}$$

You may assume that:

$$\text{erfc } y = \frac{2}{\sqrt{\pi}} \int_y^\infty e^{-s^2}\, ds, \quad \frac{d}{dy}\text{erfc } y = -\frac{2}{\sqrt{\pi}}e^{-y^2}, \quad \text{erfc}(0) = 1. \tag{P6.30.1}$$

31. *Oscillating plate—M.* Similar to the second part of Example 6.4, consider a viscous fluid of infinite extent and kinematic viscosity ν with a long flat plate in its surface. The plate oscillates backwards and forwards in the x direction with velocity $v_{x0} = a\sin\omega t$. After any initial transients have died out, prove, by checking $\partial v_x/\partial t = \nu \partial^2 v_x/\partial z^2$ and the boundary conditions, that the fluid velocity at a depth z is subsequently given by:

$$v_x = ae^{-\beta z}\sin(\omega t - \beta z), \tag{E6.4.4}$$

in which

$$\beta^2 = \frac{\omega}{2\nu}. \tag{E6.4.5}$$

32. *True/false.* Check *true* or *false*, as appropriate:

(a)   For horizontal flow of a liquid in a rectangular duct between parallel plates, the pressure varies linearly both in the direction of flow and in the direction normal to the plates.   T ☐ F ☐

(b)   For horizontal flow of a liquid in a rectangular duct between parallel plates, the boundary conditions can be taken as zero velocity at one of the plates and *either* zero velocity at the other plate *or* zero velocity gradient at the centerline.   T ☐ F ☐

(c)   For horizontal flow of a liquid in a rectangular duct between parallel plates, the shear stress varies from zero at the plates to a maximum at the centerline.   T ☐ F ☐

(d)   For horizontal flow of a liquid in a rectangular duct between parallel plates, a measurement of the pressure gradient enables the shear-stress distribution to be found.   T ☐ F ☐

(e)   In fluid mechanics, when integrating a partial differ-   T ☐   F ☐
       ential equation, you get one or more *constants* of in-
       tegration, whose values can be determined from the
       boundary condition(s).

(f)   For flows occurring between $r = 0$ and $r = a$ in cylin-   T ☐   F ☐
       drical coordinates, the term $\ln r$ may appear in the
       final expression for one of the velocity components.

(g)   For flows in ducts and pipes, the volumetric flow rate   T ☐   F ☐
       can be obtained by differentiating the velocity profile.

(h)   Natural convection is a situation whose analysis de-   T ☐   F ☐
       pends on *not* taking the density as constant every-
       where.

(i)   A key feature of the Weissenberg rheogoniometer is   T ☐   F ☐
       the fact that a conical upper surface results in a uni-
       form velocity gradient between the cone and the plate,
       for all values of radial distance.

(j)   If, in three dimensions, the pressure obeys the equa-   T ☐   F ☐
       tion $\partial p/\partial y = -\rho g$, and both $\partial p/\partial x$ and $\partial p/\partial z$ are
       nonzero, then integration of this equation gives the
       pressure as $p = -\rho g y + c$, where c is a constant.

(k)   If two immiscible liquids A and B are flowing in the x T ☐   F ☐
       direction between two parallel plates, both the veloc-
       ity v_x and the shear stress τ_{yx} are continuous at the
       interface between A and B, where the coordinate y is
       normal to the plates.

(l)   In compression molding of a disk between two plates,   T ☐   F ☐
       the force required to squeeze the plates together de-
       creases as time increases.

(m)   For flow in a wetted-wall column, the pressure in-   T ☐   F ☐
       creases from atmospheric pressure at the gas/liquid
       interface to a maximum at the wall.

(n)   For one-dimensional flow in a pipe—either laminar   T ☐   F ☐
       or turbulent—the shear stress τ_{rz} varies linearly from
       zero at the wall to a maximum at the centerline.

(o)   In Example 6.1, for flow between two parallel plates,   T ☐   F ☐
       the shear stress τ_{yx} is negative in the upper half
       (where $y > 0$), meaning that physically it acts in
       the opposite direction to that indicated by the con-
       vention.

Chapter 7

LAPLACE'S EQUATION, IRROTATIONAL AND POROUS-MEDIA FLOWS

7.1 Introduction

THE previous chapter dealt with fluid motions in which the viscosity always played a key role. However, away from a solid boundary, the effect of viscosity is frequently small and may be neglected. The example was given at the end of Section 5.2 of the rotation of a cup containing coffee, in which there was little perceptible rotation of the liquid itself. For a cup of molasses, the situation would obviously be different.

There are several practical situations, in which the fluid can be treated as essentially *inviscid*, occurring for example in the flow of air relative to an airplane, flow of water in lakes and harbors, surface waves on water, air motion in tornadoes, etc., and these will be discussed in this chapter. It is understood that we are not inquiring about the motion in either of the following two cases:

1. Very close to a solid boundary, where the action of viscosity *is* important, and which will be discussed in Chapter 8.

2. In the wake of a solid obstacle, where laminar instabilities or turbulence can occur at high Reynolds numbers—for illustrations see the applications of computational fluid dynamics using COMSOL and FlowLab in Examples 9.4 (flow through an orifice plate) and 13.4 (flow in the wake of a cylinder).

Section 7.3 will demonstrate that these inviscid flows are governed by *Laplace's* equation. Somewhat paradoxically, Section 7.9 will show that Laplace's equation also applies to a phenomenon that is apparently at the other end of the spectrum—to the flow of *viscous* fluids in porous media, which is of considerable practical significance in the production of oil, the underground storage of natural gas, and the flow of groundwater. And, as will be seen in Chapter 10, Laplace's equation plays an important role in the rise velocity of large bubbles in liquids and fluidized beds, and also governs the pressure distribution in the particulate phase of fluidized beds. Finally, Example 12.3 shows that potential-flow theory can be applied to electroosmotic flow around a particle in a microchannel.

Motion of an inviscid fluid. We start by rederiving Bernoulli's equation for *inviscid* fluids, but this time in any number of space dimensions. In such cases, with $\mu = 0$, Eqn. (5.68) yields for the usual case of an incompressible fluid the so-called *Euler equation*:

$$\frac{\mathcal{D}\mathbf{v}}{\mathcal{D}t} = -\frac{1}{\rho}\nabla p + \mathbf{F}, \tag{7.1}$$

in which the body force vector $\mathbf{F}$ is typically the gradient of a scalar Φ (known as the *body-force potential*):

$$\mathbf{F} = -\nabla\Phi = -\nabla gz, \tag{7.2}$$

where gravity is assumed to be the body force and z is oriented vertically upwards. Since ρ is constant, Eqns. (7.1) and (7.2) yield:

$$\frac{\mathcal{D}\mathbf{v}}{\mathcal{D}t} = -\nabla\left(\frac{p}{\rho} + gz\right). \tag{7.3}$$

The following vector identity is quoted without proof:

$$\frac{\mathcal{D}\mathbf{v}}{\mathcal{D}t} = \frac{\partial\mathbf{v}}{\partial t} + \nabla\left(\frac{1}{2}\mathbf{v}^2\right) - \mathbf{v}\times\boldsymbol{\zeta}, \tag{7.4}$$

in which $\mathbf{v}^2 = \mathbf{v}\cdot\mathbf{v}$, and Eqn. (5.30) has already shown the vorticity $\boldsymbol{\zeta}$ to equal:

$$\boldsymbol{\zeta} = \left(\frac{\partial v_z}{\partial y} - \frac{\partial v_y}{\partial z}\right)\mathbf{e}_x + \left(\frac{\partial v_x}{\partial z} - \frac{\partial v_z}{\partial x}\right)\mathbf{e}_y + \left(\frac{\partial v_y}{\partial x} - \frac{\partial v_x}{\partial y}\right)\mathbf{e}_z = \nabla\times\mathbf{v} = 2\boldsymbol{\omega}, \tag{5.29}$$

where $\boldsymbol{\omega}$ is the angular velocity of the fluid at a point.

Hence, from Eqns. (7.3) and (7.4),

$$\frac{\partial\mathbf{v}}{\partial t} - \mathbf{v}\times\boldsymbol{\zeta} = -\nabla\left(\frac{p}{\rho} + \frac{1}{2}\mathbf{v}^2 + gz\right), \tag{7.5}$$

which, for *steady flow*, reduces to

$$\mathbf{v}\times\boldsymbol{\zeta} = \nabla\left(\frac{p}{\rho} + \frac{1}{2}\mathbf{v}^2 + gz\right). \tag{7.6}$$

That is, the vector $\mathbf{v}\times\boldsymbol{\zeta}$ is normal to the surfaces

$$\frac{p}{\rho} + \frac{1}{2}\mathbf{v}^2 + gz = c, \tag{7.7}$$

where c is a constant for a particular surface, which must therefore contain $\mathbf{v}$ (that is, *streamlines*) and $\boldsymbol{\zeta}$ (*vortex lines*). The tangent to a streamline has the direction of the velocity; that to a vortex line has the direction of vorticity. For *irrotational flows*, which have zero vorticity ($\boldsymbol{\zeta} = \mathbf{0}$), the constant c is the same throughout the fluid, giving *Bernoulli's equation*, of considerable importance for the motion of ideal *frictionless* fluids.

Another special case of Eqn. (7.5) occurs for *transient* irrotational flows:

$$\frac{\partial\mathbf{v}}{\partial t} = -\nabla\left(\frac{p}{\rho} + \frac{1}{2}\mathbf{v}^2 + gz\right), \tag{7.8}$$

which finds application in the study of surface waves in Section 7.11.

7.2 Rotational and Irrotational Flows

Fig. 7.1 shows examples of both irrotational and rotational flows, in each of which the streamlines, viewed here from above the free surface of the liquid in a container, are concentric circles. To visualize the flows, a cork is floating in the liquid, and is here designated as a square with (for easy visualization) diagonals and a protruding arrow. Although both flows consist of a *vortex*, in which fluid particles are following circular paths, the two cases are quite different:

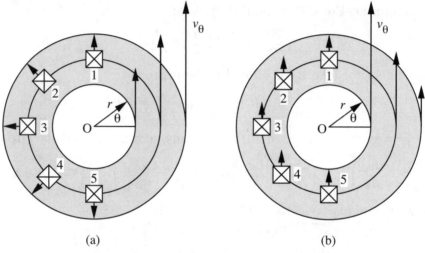

(a)   (b)

Fig. 7.1   (a) Forced vortex (rotational)
and (b) free vortex (irrotational) flows.

(a) The first case corresponds to a *forced* vortex in which (mainly due to a rotating container and the action of viscosity) the liquid is turning with a constant angular velocity ω, just as if it were a solid body. Thus, the sole velocity component is given by $v_\theta = r\omega$, which *increases* with radius. As the path 1–2–3–4–5 is followed, the cork clearly rotates counterclockwise, thus indicating a counterclockwise angular velocity throughout. The flow is therefore *rotational*.

(b) The second case corresponds to a *free vortex*, which can occur in essentially inviscid liquids, the sole velocity component now being *inversely* proportional to the radius; that is, $v_\theta = c/r$. As the path 1–2–3–4–5 is followed, the cork maintains its orientation, indicating that the vorticity is zero. The flow is therefore *irrotational*.

The cyclone separator, discussed in Section 4.8, presents an example of a free vortex in the central core of the equipment; however, in a small region near the entrance of the separator, the flow approximates a forced vortex, because of the driving effect of the inlet gas. The draining of a bathtub or sink offers another example of a free vortex, since the velocity in the angular direction speeds up as the drain hole is approached. The question of the *direction* of rotation into the drain has been nicely answered by Cope.[1]

Vorticity and angular velocity. We have already seen that the angular velocity at a point is one-half of the vorticity, giving (in two dimensions):

$$\omega = \frac{1}{2}\left(\frac{\partial v_y}{\partial x} - \frac{\partial v_x}{\partial y}\right), \tag{7.9}$$

which will now be interpreted and verified for two-dimensional flow.

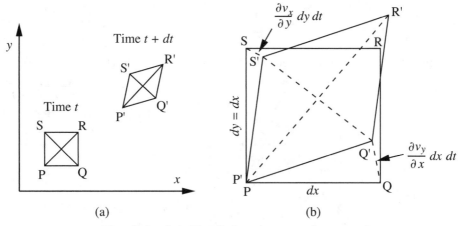

(a)   (b)

Fig. 7.2 (a) Fluid element translates and rotates; (b) determination of angular velocity.

Consider the situation in Fig. 7.2(a), in which, over a small time step dt, an initially square fluid element PQRS moves to a new location P′Q′R′S′. Observe that the element translates (moves), deforms, and possibly rotates, as discussed earlier following Fig. 5.15. Fig. 7.2(b) shows both elements enlarged and superimposed to have a common lower left-hand corner, with P and P′ coincident. The angular velocity of the element is determined by finding out how much the diagonal PR has rotated when it assumes its new position P′R′. Observe that PQ has rotated counterclockwise to its new position P′Q′. The y velocity v_y of Q exceeds that of P by an amount $(\partial v_y/\partial x)\,dx$; thus, in a time dt, the distance QQ′ will be $(\partial v_y/\partial x)\,dx\,dt$.

---

[1] Yes, it *is* true that—if the water is *perfectly* still before the drain plug is pulled—the direction of rotation of the bathtub vortex *does* depend on whether it is in the northern or southern hemisphere. However, any quite small initial rotation of the water would dominate this natural effect caused by the earth's rotation. In a letter to the editor of the *American Scientist*, Vol. 71 (Nov/Dec 1983), Winston Cope stated: "I spent most of 1974 with the U.S. Navy at the South Pole, where one of my projects was to demonstrate the Coriolis effect. At the poles the Coriolis effect is greatest, and we used a smaller tank than described by Dr. Sibulkin: half of a 50-gallon drum. Because the temperature of the room was below freezing for water, a solution of ethylene glycol was used. It took 3–5 days for the rotation effects due to filling to subside, but a small, consistently clockwise vortex resulted as the fluid drained. The motion was easily detected by means of talc particles on the surface. The rotation of the vortex was the same whether the fluid drained through a nozzle toward the floor, or from a nozzle through a siphon from above."

The angle Q'PQ is therefore $(\partial v_y/\partial x)\,dt$, so that the (counterclockwise) angular velocity of PQ is $\partial v_y/\partial x$. By a similar argument, the clockwise angular velocity of PS is $\partial v_x/\partial y$.

The counterclockwise angular velocity of the diagonal PR is the mean of the angular velocities of PQ and PS:

$$\omega = \frac{1}{2}\left(\frac{\partial v_y}{\partial x} - \frac{\partial v_x}{\partial y}\right). \qquad (7.10)$$

A comparison with the z component of vorticity given in Eqn. (5.29):

$$\frac{\partial v_y}{\partial x} - \frac{\partial v_x}{\partial y}, \qquad (7.11)$$

reveals again that the angular velocity is one-half of the vorticity.

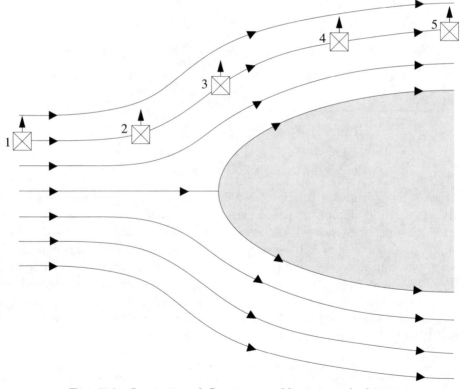

Fig. 7.3   Irrotational flow past a blunt-nosed object.

An additional example of irrotational flow, viewed from above a free surface, is given in Fig. 7.3, for liquid motion past a blunt-nosed solid. Again, the floating cork maintains a constant orientation. Observe in our idealization that the fluid in contact with the solid object violates the "no-slip" boundary condition, and is

Example 7.1—Forced and Free Vortices   **359**

allowed to have a *finite* velocity, because there are no viscous stresses to retard it. In practice, of course, the velocity at the boundary would be zero, but would quickly build up to the value in the "mainstream" across a very thin boundary layer. Thus, *apart* from this boundary layer, the irrotational flow solution can still give a fairly accurate picture of the overall flow.

Example 7.1—Forced and Free Vortices

The flow pattern in a liquid stirred in a large closed cylindrical tank can be approximated by a *forced* vortex of radius a and angular velocity ω (corresponding to the location of the stirrer), surrounded by a *free* vortex that extends indefinitely radially outward, as shown in Fig. E7.1.1.

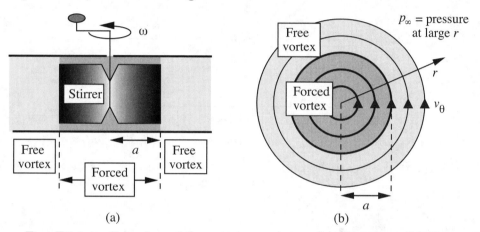

Fig. E7.1.1   Forced- and free-vortex regions: (a) elevation; (b) plan.

If the pressure is p_∞ for large values of r, derive and plot expressions for the tangential velocity v_θ and pressure p as functions of radial location r. If the top of the tank is removed, so that the liquid is allowed to have a free surface, comment on its shape. Also, discuss how the free vortex is generated in this example.

Solution

First, consider the tangential *velocity* v_θ in the two regions:

$$\text{Forced vortex } (r \le a): \quad v_\theta = r\omega; \qquad \text{free vortex } (r \ge a): \quad v_\theta = \frac{c}{r}. \quad \text{(E7.1.1)}$$

At the junction between the two regions, where $r = a$, the velocities must be identical, so that $(v_\theta)_{r=a} = a\omega = c/a$; thus, the constant is $c = \omega a^2$ and the velocities in the two regions are:

$$\text{Forced vortex}: v_\theta = r\omega; \qquad \text{free vortex}: v_\theta = \frac{\omega a^2}{r}, \quad \text{(E7.1.2)}$$

representative profiles being shown in Fig. E7.1.2.

Second, since there is zero vorticity in the *free* vortex, Bernoulli's equation applies, it being noted that the velocity declines to zero for large r:

$$\text{Free vortex}: \quad \frac{p}{\rho} + \frac{v_\theta^2}{2} = \frac{p_\infty}{\rho}, \quad \text{or} \quad p = p_\infty - \frac{\rho\omega^2 a^4}{2r^2}, \tag{E7.1.3}$$

so that the pressure at the junction between the two vortices is:

$$p_a = p_\infty - \frac{1}{2}\rho\omega^2 a^2. \tag{E7.1.4}$$

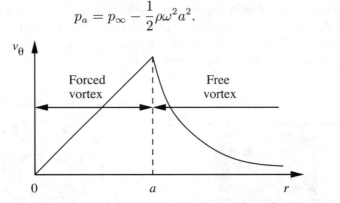

Fig. E7.1.2   *Forced- and free-vortex velocity distributions.*

For the *forced* vortex, the radial variation of pressure is already given in Eqn. (1.44):

$$\frac{dp}{dr} = \rho r \omega^2, \tag{E7.1.5}$$

a result that can also be obtained from the full r momentum balance in cylindrical coordinates, Eqn. (5.75), by making appropriate simplifications and substituting $v_\theta = r\omega$. Integration of Eqn. (E7.1.5), noting that $p = p_a$ at $r = a$, gives:

$$\text{Forced vortex}: \quad p = p_a - \frac{1}{2}\rho\omega^2(a^2 - r^2) = p_\infty - \rho\omega^2\left(a^2 - \frac{1}{2}r^2\right). \tag{E7.1.6}$$

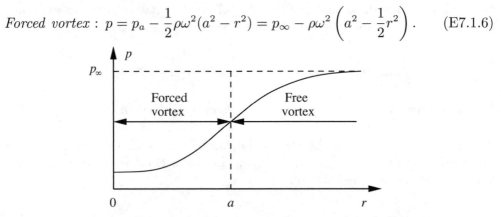

Fig. E7.1.3   *Forced- and free-vortex pressure distributions.*

A representative complete pressure distribution is given in Fig. E7.1.3. Note that the overall pressure changes in the two vortices are identical ($\rho\omega^2 a^2/2$). If the liquid has a free surface, the pressure toward the bottom of its container must be balanced hydrostatically by a column of liquid that extends to the free surface. Thus, the free surface must have a shape that resembles the curve in Fig. E7.1.3, and this is substantiated by experiment.

Regarding the generation of the free vortex, consider the liquid in the region $r > a$ to be initially at rest when the stirrer is turned on. With appropriate simplifications, the θ momentum balance becomes:

$$\rho\frac{\partial v_\theta}{\partial t} = \mu\frac{\partial}{\partial r}\left(\frac{1}{r}\frac{\partial(rv_\theta)}{\partial r}\right), \tag{E7.1.7}$$

and will govern how v_θ builds up to its final value as shown in Fig. E7.1.2, at which stage the right-hand side of Eqn. (E7.1.7) is zero. Somewhat paradoxically in this example, viscosity is important in determining the final velocity profile, even though the resulting free vortex region has zero vorticity. □

7.3 Steady Two-Dimensional Irrotational Flow

Rectangular Cartesian coordinates.   Most of the development in this section will be in terms of x/y coordinates, after which the important relationships will be restated for two-dimensional cylindrical coordinates.

For inviscid fluids, Bernoulli's equation, (7.7), is already available for relating changes in pressure, velocity, and elevation. Here, we use the *continuity equation*,

$$\frac{\partial v_x}{\partial x} + \frac{\partial v_y}{\partial y} = 0, \tag{7.12}$$

and the *irrotationality condition* to show that the flow can be represented by a *velocity potential* ϕ that obeys Laplace's equation.

Note first that the angular velocity of a fluid element is caused by *shear stresses*, which act tangentially to the surface of the element, and may have a net resulting moment about its center of gravity, thus causing it to rotate. (Pressure forces act normal to the surface of an element, and have no resulting moment, so cannot induce rotation of the element.) In the absence of viscosity, there can be no shear stresses. Hence, an element of fluid starting with zero angular velocity cannot acquire any. Therefore, from Eqn. (7.10), zero angular velocity implies that:

$$\frac{\partial v_y}{\partial x} = \frac{\partial v_x}{\partial y}. \tag{7.13}$$

Velocity potential.   Now consider a new function, the *velocity potential* ϕ, which is such that the velocity components v_x and v_y are obtained by differentiation of ϕ in the x and y directions:

$$v_x = \frac{\partial \phi}{\partial x}, \qquad v_y = \frac{\partial \phi}{\partial y}. \tag{7.14}$$

That is, the vector velocity is the *gradient* of the velocity potential[2]:

$$\mathbf{v} = v_x \mathbf{e}_x + v_y \mathbf{e}_y = \nabla \phi. \tag{7.15}$$

The reader can check that the concept of the velocity potential *automatically* satisfies the irrotationality condition, Eqn. (7.13). Substitution of the velocity components from Eqn. (7.14) into the continuity equation, (7.12), gives *Laplace's equation* for the *velocity potential*:

$$\frac{\partial^2 \phi}{\partial x^2} + \frac{\partial^2 \phi}{\partial y^2} = 0. \tag{7.16}$$

Laplace, Pierre Simon, Marquis de, born 1749 at Beaumont-en-Auge in Normandy, died 1827 at Arcueil, France. Laplace was the son of a farmer in Normandy; his subsequent career led him to become known as "the Newton of France." He excelled in mathematics and astronomy, and some of his brilliant early work was instrumental in demonstrating the stability of the solar system. His *magnum opus* was *Mécanique céleste*, published in five volumes between 1799 and 1825. He became a member of the Academy of Sciences in 1785. His concept of a potential function was fundamental to subsequent theories of electricity, magnetism, and heat. Extending previous work of Gauss and Legendre, he established the statistical basis for the method of least squares. He made significant contributions to an incredibly wide range of other scientific topics, including the stability of Saturn's rings, the velocity of sound, specific heats, capillary action, numerical interpolation, and Laplace transforms. Modesty was not Laplace's strong suit, and he exhibited a predilection for positions of political power and the titles that accompanied them. His last words before his death were: "Ce que nous connaissons est peu de chose, ce que nous ignorons est immense." [What we know is little enough; what we don't know is enormous.]

Source: *The Encyclopædia Britannica*, 11th ed., Cambridge University Press (1910–1911).

Stream function.   As an alternative, we can define a *stream function*, ψ, such that the velocity components v_x and v_y are obtained by differentiation of ψ in the y and $-x$ directions, respectively:

$$v_x = \frac{\partial \psi}{\partial y}, \qquad v_y = -\frac{\partial \psi}{\partial x}. \tag{7.17}$$

---

[2] Some writers define the potential with a negative sign: $\mathbf{v} = -\nabla \phi$. The reader should therefore be alert as to which convention is being used.

Compared with Eqn. (7.14), note the reversed order of x and y, and also the minus sign in the expression for v_y.

The reader can also check that the concept of the stream function *automatically* satisfies the continuity equation, (7.12). Substitution of the velocity components from Eqn. (7.17) into the irrotationality condition, (7.13), gives *Laplace's equation* for the *stream function* in two-dimensional irrotational flows:

$$\frac{\partial^2 \psi}{\partial x^2} + \frac{\partial^2 \psi}{\partial y^2} = 0. \tag{7.18}$$

Table 7.1 summarizes the above facts about the velocity potential and the stream function. It will also be shown in the next section that a contour of constant ψ represents the path followed by the fluid.

Table 7.1 Basic Properties of the Velocity Potential and Stream Function

Quantity	Automatically Satisfies	To Obtain Laplace's Equation, Combine It With
ϕ	Irrotationality	Continuity
ψ	Continuity	Irrotationality

Fig. 7.4   Cylindrical (r/θ) and rectangular (x/y) coordinates.

Two-dimensional cylindrical coordinates.   First, examine the more commonly occurring case for which $v_z = 0$. For r/θ coordinates—shown in relation to x/y coordinates in Fig. 7.4—the corresponding relationships between the velocity components and the potential and stream function can be proved:

$$v_r = \frac{\partial \phi}{\partial r}, \qquad v_\theta = \frac{1}{r}\frac{\partial \phi}{\partial \theta}, \tag{7.19a}$$

$$v_r = \frac{1}{r}\frac{\partial \psi}{\partial \theta}, \qquad v_\theta = -\frac{\partial \psi}{\partial r}. \tag{7.19b}$$

From Eqn. (5.49) and Table 5.4, the continuity equation and irrotationality condition are:

$$\text{Continuity :} \quad \frac{\partial(rv_r)}{\partial r} + \frac{\partial v_\theta}{\partial \theta} = 0. \qquad \text{Irrotationality :} \quad \frac{\partial v_r}{\partial \theta} = \frac{\partial(rv_\theta)}{\partial r}. \qquad (7.20)$$

Appropriate substitutions again lead to Laplace's equation in the potential function and stream function:

$$\frac{1}{r}\frac{\partial}{\partial r}\left(r\frac{\partial \phi}{\partial r}\right) + \frac{1}{r^2}\frac{\partial^2 \phi}{\partial \theta^2} = 0, \qquad \frac{1}{r}\frac{\partial}{\partial r}\left(r\frac{\partial \psi}{\partial r}\right) + \frac{1}{r^2}\frac{\partial^2 \psi}{\partial \theta^2} = 0. \qquad (7.21)$$

Second, note that two-dimensional flows may also exist in cylindrical coordinates with $v_\theta = 0$, in which v_r and v_z are the velocity components of interest. In this case, the appropriate relations for the velocities, potential function, and stream function are:

$$v_r = \frac{\partial \phi}{\partial r} = \frac{1}{r}\frac{\partial \psi}{\partial z}, \qquad v_z = \frac{\partial \phi}{\partial z} = -\frac{1}{r}\frac{\partial \psi}{\partial r}. \qquad (7.22a)$$

$$\frac{\partial^2 \phi}{\partial r^2} + \frac{1}{r}\frac{\partial \phi}{\partial r} + \frac{\partial^2 \phi}{\partial z^2} = 0, \qquad \frac{\partial^2 \psi}{\partial r^2} - \frac{1}{r}\frac{\partial \psi}{\partial r} + \frac{\partial^2 \psi}{\partial z^2} = 0. \qquad (7.22b)$$

Note that the second-order equation for the stream function is no longer Laplace's equation.

Flow in porous media.   The treatment in Section 7.9 will show that the pressure for the flow of a viscous fluid, such as oil, in a porous medium, such as a permeable rock formation, also obeys Laplace's equation (in two-dimensional rectangular coordinates, for example):

$$\nabla^2 p = \frac{\partial^2 p}{\partial x^2} + \frac{\partial^2 p}{\partial y^2} = 0. \qquad (7.23)$$

Thus, much of the discussion in this chapter applies not only to flow of essentially *inviscid* fluids, but *also* to the flow of *viscous* fluids in porous media. Obvious important applications are to the recovery of oil from underground rock formations, to the motion of groundwater in soil, and the underground storage of natural gas. A representative example appears in Section 7.9.

7.4 Physical Interpretation of the Stream Function

For steady two-dimensional flow in x/y coordinates, an alternative and equivalent definition to that of Eqn. (7.17) for the stream function is given by reference to Fig. 7.5(a), in which a family of *streamlines* (lines of constant ψ) is shown. Consider the flow rate Q, per unit depth normal to the plane of the diagram, *between* streamlines that have values of the stream function equal to ψ_1 and ψ_2.

The equivalent viewpoint to that of Eqn. (7.17) is to define the flow rate Q as the *difference* between the two values of the stream function, namely, $Q = \psi_2 - \psi_1$. The sign convention is such that Q is taken to be positive if flowing from left to right across a path such as PS *from ψ_1 to ψ_2*. That is, if the flow is physically in the indicated direction, then $\psi_2 > \psi_1$; if it is in the opposite direction, then $\psi_2 < \psi_1$.

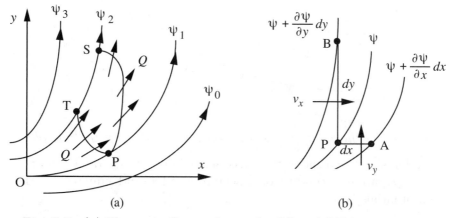

Fig. 7.5 *(a) Flow rate Q crossing paths PS and PT joining streamlines with values ψ_1 and ψ_2; (b) velocity components v_x and v_y crossing segments dy and dx between neighboring streamlines.*

Since the value of ψ is constant along a streamline, the flow rate Q will still be the same across any other path such as PT. It immediately follows that the flow across ST is zero, and that no flow can occur *across* a streamline; hence, the flow is always in the direction of the streamline passing through any point. Representative units for Q and ψ are m^2/s.

To verify that this viewpoint is completely consistent with that of Eqn. (7.17), consider Fig. 7.5(b), which shows three points, P, A, and B, on three streamlines that are only differentially separated. Across PB and PA, paying attention to the direction of flow as defined above, the flow rates dQ_x and dQ_y per unit depth are:

$$dQ_x = v_x\,dy = \frac{\partial \psi}{\partial y}\,dy, \qquad dQ_y = -v_y\,dx = \frac{\partial \psi}{\partial x}\,dx. \qquad (7.24)$$

Thus, from Eqn. (7.24), the velocity components are identical with those given previously:

$$v_x = \frac{\partial \psi}{\partial y}, \qquad v_y = -\frac{\partial \psi}{\partial x}. \qquad (7.17)$$

Since we have already established that the direction of the flow is normal to the equipotentials, the streamlines and equipotentials must everywhere be *orthogonal* to one another, and a representative situation is shown in Fig. 7.6.

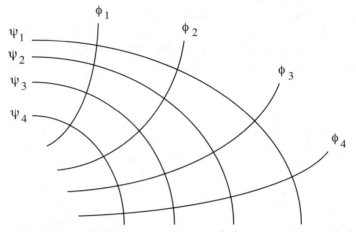

Fig. 7.6   Orthogonality of streamlines (ψ) and equipotentials (ϕ).

The reader should consider the *direction* of flow if the values of the stream function $\psi_1 \dots \psi_4$ are in ascending or descending order, and similarly for the values of the equipotentials $\phi_1 \dots \phi_4$.

7.5   Examples of Planar Irrotational Flow

Here, we examine a few two-dimensional irrotational flows, in which it is necessary to become accustomed to working in either rectangular (x/y) or cylindrical (r/θ) coordinates, depending on the problem at hand.

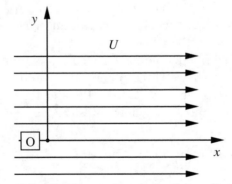

Fig. 7.7   Uniform stream in the x direction.

Uniform stream.   One of the simplest flows is that in which the two velocity components are $v_x = U$ (a specified value) and $v_y = 0$, corresponding to a uniform stream in the x direction, as shown in Fig. 7.7. The reader can check that the velocity potential and stream function are given by:

$$\phi = Ux, \tag{7.25}$$

$$\psi = Uy. \tag{7.26}$$

Observe also that if the function for *either* ϕ or ψ is known, then the *other one* can usually be readily deduced. For example, suppose that ϕ is given from Eqn. (7.25). By using the definitions of Eqn. (7.14), the x and y velocity components can be obtained and then equated to the corresponding derivatives of the stream function from Eqn. (7.17):

$$v_x = \frac{\partial \phi}{\partial x} = U = \frac{\partial \psi}{\partial y}, \qquad v_y = \frac{\partial \phi}{\partial y} = 0 = -\frac{\partial \psi}{\partial x}. \tag{7.27}$$

Integration of the two relations in Eqn. (7.27) gives two expressions for the stream function:

$$\psi = Uy + f(x), \qquad \psi = g(y), \tag{7.28}$$

in which $f(x)$ and $g(y)$ are arbitrary functions of integration. (Remember, we are integrating *partial* differential equations, not *ordinary* differential equations, and therefore obtain *functions* of integration, not just *constants* of integration.) However, the only way the two expressions for ψ in Eqn. (7.28) can be mutually consistent is if $f(x) = c$ and $g(y) = Uy + c$, in which c is a constant, conveniently— but not necessarily—taken as zero. Therefore, we conclude that $\psi = Uy$, in agreement with the stated function in Eqn. (7.26).

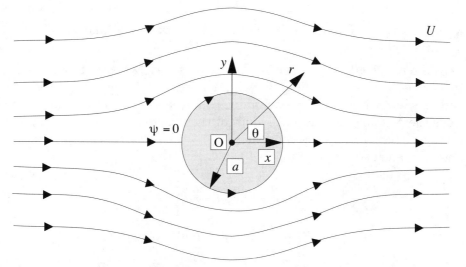

Fig. 7.8 Streamlines for flow past a cylinder.

Flow past a cylinder. Fig. 7.8 illustrates the steady flow of an inviscid fluid past a cylinder of radius a; far away from the cylinder, the velocity is U in the x direction and zero in the y direction. In this particular situation, it is convenient to introduce polar coordinates, r and θ, in addition to x and y, such that:

$$x = r \cos \theta, \qquad y = r \sin \theta. \tag{7.29}$$

The velocity potential and stream function are:

$$\phi = U \left(r + \frac{a^2}{r} \right) \cos \theta, \tag{7.30}$$

$$\psi = U \left(r - \frac{a^2}{r} \right) \sin \theta. \tag{7.31}$$

These functions are verified if they satisfy the following three conditions:

1. *Laplace's equations.* It will be left as an exercise to prove compatibility with Eqn. (7.21):

$$\frac{1}{r} \frac{\partial}{\partial r} \left(r \frac{\partial \phi}{\partial r} \right) + \frac{1}{r^2} \frac{\partial^2 \phi}{\partial \theta^2} = 0, \qquad \frac{1}{r} \frac{\partial}{\partial r} \left(r \frac{\partial \psi}{\partial r} \right) + \frac{1}{r^2} \frac{\partial^2 \psi}{\partial \theta^2} = 0. \tag{7.32}$$

2. *Uniform x velocity far away from the cylinder.* As the radial coordinate becomes large ($r \to \infty$), the potential and stream function become:

$$\phi = Ur \cos \theta = Ux, \tag{7.33}$$

$$\psi = Ur \sin \theta = Uy, \tag{7.34}$$

which are readily shown to be consistent with $v_x = U$ and $v_y = 0$ far away from the cylinder.

3. *Zero radial velocity at the surface of the cylinder.* Because in our idealization there is no viscosity, no boundary layer forms on the surface of the cylinder, and this surface is itself a streamline—more correctly, a *divided* streamline, since there is flow around both sides of the cylinder. It therefore follows that there can be no radial component of the velocity at $r = a$. For the velocity potential and stream function, we have:

$$\left(\frac{\partial \phi}{\partial r} \right)_{r=a} = \left[U \left(1 - \frac{a^2}{r^2} \right) \cos \theta \right]_{r=a} = 0, \tag{7.35}$$

$$\psi = 0 \text{ (surface of cylinder is a streamline).} \tag{7.36}$$

Example 7.2—Stagnation Flow   **369**

Example 7.2—Stagnation Flow

Investigate the potential flow whose stream function in x/y coordinates is:

$$\psi = cxy. \tag{E7.2.1}$$

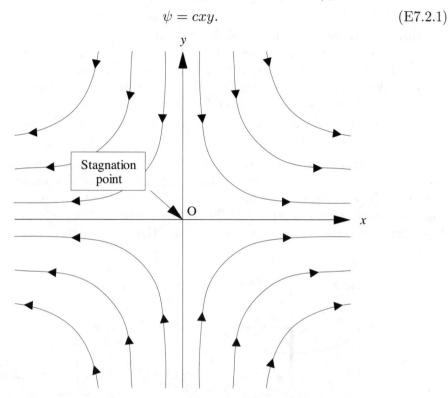

Fig. E7.2.1   *Potential stagnation flow.*

Solution

From Eqn. (7.17), the two velocity components are:

$$v_x = \frac{\partial \psi}{\partial y} = cx, \qquad v_y = -\frac{\partial \psi}{\partial x} = -cy, \tag{E7.2.2}$$

leading to the following observations (c is assumed to be positive):

1. Along the x-axis ($y = 0$), $v_y = 0$, and the flow is parallel to the x-axis. For positive values of x, the flow is to the right, and for negative values of x, it is to the left.
2. Along the y-axis ($x = 0$), $v_x = 0$, and the flow is parallel to the y-axis. For positive values of y, the flow is downward, and for negative values of y, it is upwards.
3. The equation for the stream function, $\psi = cxy$, is that of a family of hyperbolas.
4. There is a stagnation point at the origin, O.

With above in mind, it is now possible to deduce the flow pattern, which is shown in Fig. E7.2.1. It could occur when two streams—one coming from the $+y$ direction and the other coming from the $-y$ direction—meet each other. Observe that the streams are deflected so that they leave in the $+x$ and $-x$ directions.

However, there are other possibilities. We could simply choose to ignore the lower half of the region, for negative values of y. In this case, the flow would essentially be that of a downward wind against a horizontal plane, such as the ground. We could also focus exclusively on the upper right-hand quadrant, for example, in which case the flow would be that in a corner.

The derivation of the corresponding velocity potential is left as an exercise. $\square$

Line source. Fig. 7.9(a) shows a "line" source that extends indefinitely along an axis of symmetry, from which it emits a flow uniformly in all radial directions. The strength m of the source is defined such that the total volumetric flow rate emitted by unit length of the source is $2\pi m$. (The inclusion of the factor 2π is merely a convenience, since it simplifies the subsequent equations.)

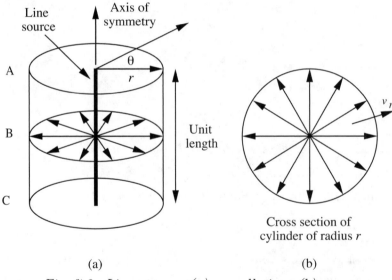

(a)   (b)

Fig. 7.9   Line source: (a) overall view; (b) cross section, showing the velocity v_r at a radius r.

A view along the axis of a circular section of radius r at any location—such as at B—is shown in Fig. 7.9(b). If the outwards radial velocity is v_r, the total volumetric flow rate per unit depth is the velocity v_r times the surface area $2\pi r$ of the imaginary cylindrical shell through which the flow is passing:

$$2\pi m = 2\pi r v_r, \tag{7.37}$$

Example 7.3—Combination of a Uniform Stream and a Line Sink (C)   **371**

so that:

$$v_r = \frac{m}{r} = \frac{\partial \phi}{\partial r} = \frac{1}{r}\frac{\partial \psi}{\partial \theta}, \qquad v_\theta = \frac{1}{r}\frac{\partial \phi}{\partial \theta} = -\frac{\partial \psi}{\partial r} = 0. \qquad (7.38)$$

Since there is no swirl, note that $v_\theta = 0$. Thus, the potential and stream functions corresponding to the line source are, apart from a possible constant:

$$\phi = m \ln r, \qquad \psi = m\theta. \qquad (7.39)$$

A line source is a useful "building block" in constructing two-dimensional flow patterns, as will be seen in Example 7.3. A line source of negative strength is called a *line sink*, and resembles a vertical well into which a liquid such as oil or water is flowing from the surrounding earth or porous rock.

Example 7.3—Combination of a Uniform Stream and a Line Sink (C)

An inviscid fluid flows with velocity U parallel to a wall, as shown in Fig. E7.3.1. The narrow slot S is long in the direction normal to the diagram, and a volumetric flow rate Q per unit slot length is withdrawn. (The particle shown at $x = X$ is not needed in this example, but will be important in Problem 7.13.)

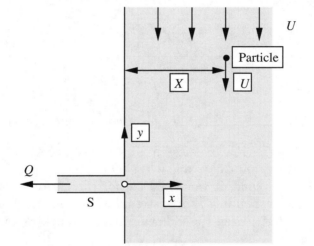

Fig. E7.3.1   Partial diversion of a stream through a slot.

Derive an expression for the stream function, and sketch several representative streamlines. At a large value of y, what is the value of x for the dividing streamline? Discuss a practical application of this type of flow.

Solution

The overall stream function is the sum of the stream functions for the following two flows:

1. A uniform stream in the direction of the negative y axis. By analogy with Eqn. (7.26):

$$\psi = Ux, \quad \text{since this gives}: v_x = \frac{\partial \psi}{\partial y} = 0, \quad v_y = -\frac{\partial \psi}{\partial x} = -U. \qquad \text{(E7.3.1)}$$

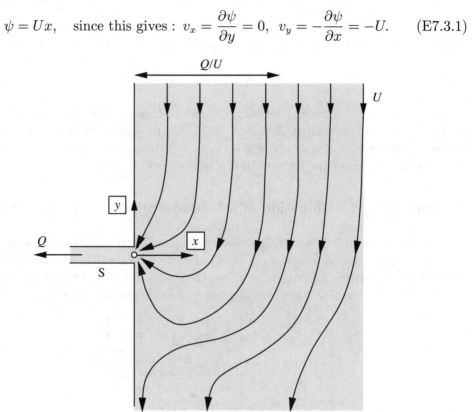

Fig. E7.3.2 *Streamlines for diversion through slot.*

2. Because the slot is narrow, it is essentially a line sink with its axis normal to the plane of the diagram. Since a flow rate Q per unit depth is withdrawn from just *half* the total plane, it would be $2Q$ for the entire plane of Fig. 7.9(b). Therefore, the corresponding *source* strength for use in conjunction with Eqn. (7.39) is $m = -Q/\pi$, giving the following stream function for the slot:

$$\psi = -\frac{Q}{\pi} \theta = -\frac{Q}{\pi} \tan^{-1} \frac{y}{x}. \qquad \text{(E7.3.2)}$$

The stream function for the combination of uniform stream and line sink is:

$$\psi = Ux - \frac{Q}{\pi} \tan^{-1} \frac{y}{x}. \qquad \text{(E7.3.3)}$$

Representative streamlines are shown in Fig. E7.3.2. Note that the width of the incoming stream that is diverted through the slot is Q/U—this value, when multiplied by the velocity U, gives the required slot flow rate Q.

Example 7.4—Flow Patterns in a Lake (COMSOL)   **373**

The flow is important because it corresponds approximately to the diversion of part of a large river into a small side channel, or of a gas into a sampling port. And—if the direction of Q were reversed—the flow would be that of a small tributary *into* a river.   □

Example 7.4—Flow Patterns in a Lake (COMSOL)

Fig. E7.4.1 shows a bird's-eye view of a lake whose shape is approximated by merging two ellipses and two rectangles. These rectangles protrude to the "west" at AB and to the "south" at CD, and can serve as an inlet and an exit to and from the lake, respectively.

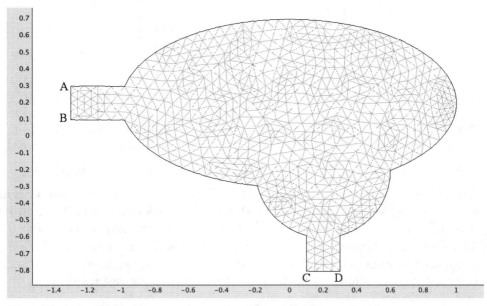

Fig. E7.4.1   A bird's-eye view of a lake, which results from merging two ellipses and two rectangles. The finite-element mesh has been refined once.

Note that the inviscid flow (possibly with rotation) is governed by Poisson's equation:

$$-\nabla^2\psi = -\left(\frac{\partial^2\psi}{\partial x^2} + \frac{\partial^2\psi}{\partial y^2}\right) = f, \tag{E7.4.1}$$

in which ψ is the stream function. In COMSOL, Eqn. (E7.4.1) is represented as $-\nabla \cdot (c\nabla u) = f$, with $c = 1$.

For two-dimensional flow, it can readily be shown that $-\nabla^2\psi = \zeta$, where $\zeta = (\partial v_y/\partial x - \partial v_x/\partial y)$ is the *vorticity*, so that f acts as a source term for the vorticity. If $f = 0$, the vorticity will be zero and there will be no rotation. But a positive value of f would mean that the vorticity is positive, corresponding to a counterclockwise rotation. And a negative value of f would indicate a clockwise rotation. Such rotations could result from the shear stresses induced on the surface of the lake by a *nonuniform* wind, whose velocity v_y varies with x as shown in Fig. E7.4.2. Note that in the eastern region ($x > 0$) the intensity increases toward the east, and in the western region ($x < 0$) the intensity increases toward the west.

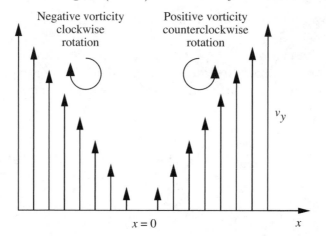

Fig. E7.4.2   *Velocities v_y corresponding to a nonuniform wind, shearing the surface of the water in the lake and generating vorticity as indicated.*

Solve for the pattern of streamlines for each of the following three cases:

1. The stream function equals $\psi = 0$ along the near shore between points B and C and equals $\psi = 100$ along the far shore between points D and A. From B to A and from C to D, ψ varies linearly from zero to 100. Everywhere, $f = 0$.

2. The rivers are "turned off," so the lake is now isolated and its boundary becomes the single streamline $\psi = 0$. However, the vorticity source term is now $f = 2{,}500x$. Thus, we would expect a counterclockwise rotation for $x > 0$ and a clockwise rotation for $x < 0$.

3. Finally, consider the case of both river flow *and* the nonuniform wind. The boundary conditions will be the same as for Case 1, but the expression for f will be the same as for Case 2.

Solution

1. In COMSOL, choose the sequence **New, PDE Modes, Classical PDEs,** and **Poisson's Equation.**

2. After the sequence **Options, Axes/Grid Settings, Grid,** uncheck **Auto,** and set both the x and y spacings to be 0.1.

3. Draw the following two rectangles and ellipses:

(a) R1, with width 0.4, height 0.2, and corner at $(-1.3, 0.1)$.
(b) R2, with width 0.2, height 0.3, and corner at $(0.1, -0.8)$.

Example 7.4—Flow Patterns in a Lake (COMSOL)   **375**

(c) **E1**, with A-semiaxes 1.0, B-semiaxes 0.5, and center at $(0, 0.2)$.

(d) **E2**, with A-semiaxes 0.4, B-semiaxes 0.4, and center at $(0.2, -0.2)$.

4. Form the composite object **CO1** from **R1+R2+E1+E2**, deleting interior boundaries by *unchecking* the **Keep interior boundaries** default case.

5. Don't forget that clicking on the **Zoom Extents** button will give an image that fills the screen as much as possible.

6. Draw the mesh and refine it just once. In **Mesh Statistics** under the **Mesh** pulldown menu, observe, on a Macintosh computer, 2,939 degrees of freedom and 1,420 elements (on a PC, 2,955 and 1,428, respectively).

Table E7.4.1 Subdomain and Boundary Settings for the Three Cases

Case	Subdomain Coefficients	Dirichlet Boundary-Settings Coefficients ($hu = r$ in COMSOL)
1	$c = 1, f = 0$	Near shore (B to C): $h = 1, r = 0$ Far shore (D to A): $h = 1, r = 100$ AB: $h = 1, r = 500(y - 0.1)$ CD: $h = 1, r = 500(x - 0.1)$
2	$c = 1, f = 2{,}500x$	All boundary segments: $h = 1, r = 0$
3	$c = 1, f = 2{,}500x$	Same as for Case 1

7. For each of the three options in turn, under the **Physics/Subdomain** and **Physics/Boundary Settings** pulldown menus, implement the values shown in Table E7.4.1, and solve the problem. Note that an arithmetic expression such as $500(y - 0.1)$ will be entered as `500*(y - 0.1)` in COMSOL. Also in this approach, all boundary segments have Dirichlet conditions imposed on them (specified value of the stream function ψ, known as u in COMSOL). For an alternative approach along AB, see the discussion of results for Case 1 below.

8. Under **Postprocessing**, make appropriate choices in the **Plot Parameters** pulldown menu to generate a plot showing 20 streamlines in each case. The **Contour** option will be needed.

Discussion of Results

Case 1. The streamlines are shown in Fig. E7.4.3. The COMSOL results do not show the direction of flow, but it can easily be ascertained as follows. Observe over most of the lake that $v_x = \partial\psi/\partial y$ is positive and that $v_y = -\partial\psi/\partial x$ is negative. Clearly, the flow enters at AB and leaves at CD, and arrows to that effect have been added.

The direction of flow could easily be reversed, by assigning $\psi = 100$ to the near shore BC and $\psi = 0$ to the far shore DA. Note that a Dirichlet boundary condition was used for the inlet, with the stream function varying linearly from zero at B to 100 at A. However, the same results would have been obtained if the Neumann condition, amounting to $\partial u(= \psi)/\partial n = 0$, had been used instead, because it would imply $v_y = 0$ and no crossflow at the inlet. (And similarly for the exit between C and D.)

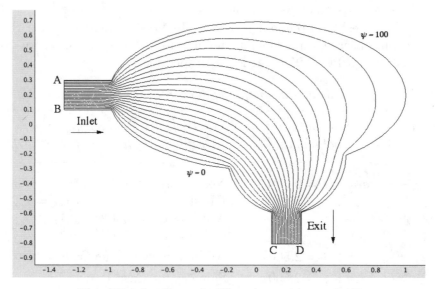

Fig. E7.4.3   Case 1.   The river enters at AB
and leaves at CD. *There is no wind blowing.*

Case 2. The streamlines are shown in Fig. E7.4.4. The lake is now completely closed, with no river flow, so the entire shoreline is the single streamline $\psi = 0$. However, there is now a nonuniform wind blowing, corresponding to Fig. E7.4.2, resulting in counterrotating circulatory patterns in the two halves of the lake. Arrows have been added to indicate the direction of motion. Recall that the value of the dependent variable $u\,(= \psi)$ can be obtained by clicking on any point (after double-clicking on "Snap," to disable the "snap to grid" feature). With this feature in mind, the COMSOL results show that the stream function is approximately $\psi = -68.9$ at the "eye" of the left-hand half, and $\psi = 73.7$ at the eye of the right-hand half.

Case 3. The streamlines are shown in Fig. E7.4.5. The river has been switched "on" again, and the nonuniform wind still blows. The results are essentially a combination of Cases 1 and 2. The river still enters at AB and leaves at CD, but has to follow a somewhat more tortuous path as it wends its way between the two circulating patterns, which are now diminished in size as a result. The stream function is approximately $\psi = -27.5$ at the left-hand eye and $\psi = 165.9$ at the right-hand eye.

Example 7.4—Flow Patterns in a Lake (COMSOL)   **377**

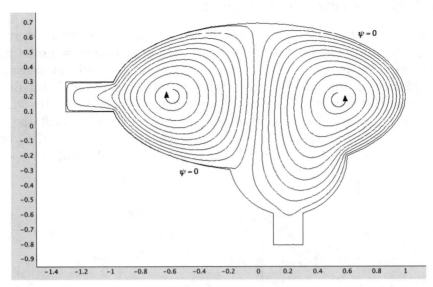

Fig. E7.4.4   *Case 2. The lake is closed, so the entire shoreline is the streamline $\psi = 0$. There is a nonuniform wind blowing, resulting in counterrotating streamlines with a negative vorticity in the left and a positive vorticity in the right.*

Fig. E7.4.5   *Case 3. The river enters at AB and leaves at CD. A nonuniform wind blows, but because of the river flow the vortices are pushed into smaller areas.*

7.6   Axially Symmetric Irrotational Flow[3]

Another important class of potential flows occurs when there is *rotational symmetry* about an axis, an example being the flow of an otherwise uniform stream past a sphere or a spherical bubble. For these flows, it is basically convenient to employ *spherical* coordinates r and θ; the symmetry condition then reduces any derivatives in the azimuthal (ϕ) direction to zero. The coordinates are shown in Fig. 7.10(b), in which the axial direction z is that from which the angle θ is measured. Thus, we shall find ourselves working with r, θ, and z, but it is important to realize that these are *not* the usual cylindrical coordinates.

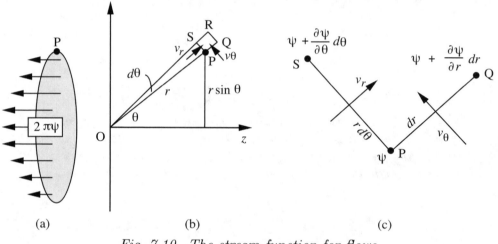

(a)   (b)   (c)

Fig. 7.10   The stream function for flows
that are axisymmetric about the z axis.

For this axisymmetric case, a new stream function ψ is now defined by reference to Fig. 7.10(a). Namely, the stream function at a point such as P is ψ if the flow rate is $2\pi\psi$, from *right to left*, through a circle formed by rotating P about the axis of symmetry. The factor of 2π is included because it will soon cancel with a similar quantity.

The two velocity components v_r and v_θ shown in Fig. 7.10(b) are now deduced by reference to Fig. 7.10(c). For example, the flow rate from right to left across the surface formed by rotating PS about the z axis is equal to 2π times the difference in stream functions at points S and P:

$$-v_r \underbrace{(2\pi r \sin\theta \, rd\theta)}_{\text{Area of PS, rotated}} = 2\pi \underbrace{\left(\psi + \frac{\partial\psi}{\partial\theta}d\theta - \psi\right)}_{\psi_S - \psi_P}, \qquad (7.40)$$

---

[3] The main goal of Sections 7.6–7.8 is to investigate flow around a sphere, needed in Chapter 10 when considering the rise of large bubbles in liquids and fluidized beds.

so that:

$$v_r = -\frac{1}{r^2 \sin\theta} \frac{\partial\psi}{\partial\theta}. \tag{7.41}$$

Similarly,

$$v_\theta \underbrace{(2\pi r \sin\theta \, dr)}_{\text{Area of PQ, rotated}} = 2\pi \underbrace{\left(\psi + \frac{\partial\psi}{\partial r} dr - \psi\right)}_{\psi_Q - \psi_P}, \tag{7.42}$$

giving:

$$v_\theta = \frac{1}{r \sin\theta} \frac{\partial\psi}{\partial r}. \tag{7.43}$$

Since $\mathbf{v} = \nabla\phi$, the velocities are given in terms of the potential ϕ by:

$$v_r = \frac{\partial\phi}{\partial r}, \qquad v_\theta = \frac{1}{r} \frac{\partial\phi}{\partial\theta}. \tag{7.44}$$

Continuity equation.   Under steady-state conditions, there is no accumulation in the element PQRS in Fig. 7.10(b), so that:

$$\frac{\partial}{\partial r}(v_r 2\pi r \sin\theta \, r d\theta) \, dr + \frac{\partial}{\partial\theta}(v_\theta 2\pi r \sin\theta \, dr) \, d\theta = 0, \tag{7.45}$$

or

$$\sin\theta \frac{\partial(r^2 v_r)}{\partial r} + r \frac{\partial(v_\theta \sin\theta)}{\partial\theta} = 0, \tag{7.46}$$

which is the *continuity equation* in axisymmetric coordinates, identical to Eqn. (5.50), multiplied through by $r^2 \sin\theta$. The reader may wish to check that the velocity components given in terms of the stream function in Eqns. (7.41) and (7.43) automatically satisfy Eqn. (7.46).

Irrotationality condition.   The flow will be irrotational if the ϕ component of the vorticity is zero. From the entry in Table 5.4 under spherical coordinates, the *irrotationality condition* is obtained by setting the ϕ component of $\nabla \times \mathbf{v}$, which is normal to the plane containing v_r and v_θ, to zero:

$$\frac{1}{r} \frac{\partial(r v_\theta)}{\partial r} - \frac{1}{r} \frac{\partial v_r}{\partial\theta} = 0. \tag{7.47}$$

Paralleling the x/y coordinate case, it may again be shown that the potential function in Eqn. (7.44) is compatible with this new irrotationality condition.

Laplace's equation.   Again paralleling the earlier x/y development, the velocity components derived from the velocity potential, which satisfies irrotationality, may be substituted into the continuity equation, (7.46), to give:

$$\sin\theta \frac{\partial}{\partial r}\left(r^2 \frac{\partial\phi}{\partial r}\right) + \frac{\partial}{\partial\theta}\left(\sin\theta \frac{\partial\phi}{\partial\theta}\right) = 0, \tag{7.48}$$

which is actually Laplace's equation in spherical coordinates (see Table 5.5), multiplied through by $r^2 \sin \theta$ (and with zero derivative in the ϕ direction).

Likewise, if the stream function, which satisfies continuity, is substituted into the irrotationality condition, we obtain:

$$\frac{1}{\sin \theta} \frac{\partial^2 \psi}{\partial r^2} + \frac{1}{r^2} \frac{\partial}{\partial \theta} \left(\frac{1}{\sin \theta} \frac{\partial \psi}{\partial \theta} \right) = 0. \tag{7.49}$$

Although Eqn. (7.49) is a second-order differential equation in the stream function, it is no longer Laplace's equation, nor a multiple of it.

7.7 Uniform Streams and Point Sources

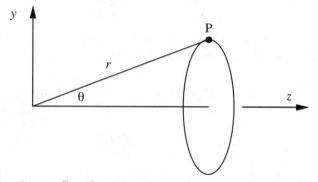

Fig. 7.11 *Coordinate systems for axisymmetric flows.*

We now examine a few cases of axisymmetric flows, for which it is convenient to use both the spherical and Cartesian coordinate frames, as shown in Fig. 7.11.

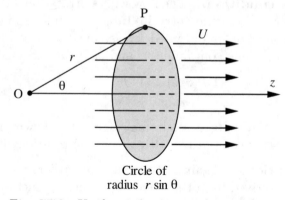

Circle of
radius $r \sin \theta$

Fig. 7.12 *Uniform stream in the z direction.*

Uniform stream.   A *uniform stream* flowing in the positive z direction with velocity U is shown in Fig. 7.12. By definition, the flow rate in the negative

z direction through the circle of radius $r \sin \theta$ passing through point P equals 2π times the stream function ψ at P:

$$-\pi (r \sin \theta)^2 U = 2\pi \psi, \tag{7.50}$$

so that:

$$\psi = -\frac{1}{2} U r^2 \sin^2 \theta. \tag{7.51}$$

The velocity potential is found by integrating the relationships given in Eqn. (7.44):

$$v_r = \frac{\partial \phi}{\partial r} = U \cos \theta, \qquad v_\theta = \frac{1}{r} \frac{\partial \phi}{\partial \theta} = -U \sin \theta, \tag{7.52}$$

which yield:

$$\phi = U r \cos \theta + f(\theta), \qquad \phi = U r \cos \theta + g(r). \tag{7.53}$$

These last two expressions are compatible if $f(\theta) = g(r) = c$, where c is a constant that is conveniently (but not necessarily) taken as zero, so that

$$\phi = U r \cos \theta = U z. \tag{7.54}$$

As a scalar, ϕ is invariant to the coordinate system used. However, when using partial derivatives in two different coordinate systems, we must pay attention to which coordinates are being kept constant during partial differentiation. Observe also that $\partial \phi / \partial z$ (with y constant) gives the z velocity component.

Fig. 7.13 *Flow emanating from a point source.*

Point source.   Fig. 7.13 shows a "point" source that emits a uniform flow in all radial directions. Its *strength* is defined as m if the total volumetric flow rate is $4\pi m$, which also equals at any radius r the radial velocity v_r times the surface area $4\pi r^2$ of the imaginary spherical shell through which the flow is passing:

$$4\pi m = v_r \, 4\pi r^2. \tag{7.55}$$

The radial velocity is therefore:

$$v_r = \frac{m}{r^2} = \frac{\partial \phi}{\partial r} = -\frac{1}{r^2 \sin \theta} \frac{\partial \psi}{\partial \theta}. \tag{7.56}$$

Observe that a point source by itself is not completely realistic, since the radial velocity is infinite at $r = 0$. Nevertheless, it is an important concept, as will be realized shortly. Integration of Eqn. (7.56) gives the velocity potential and stream function as:

$$\phi = -\frac{m}{r}, \qquad \psi = m\cos\theta = \frac{mz}{r}. \tag{7.57}$$

(Arbitrary functions of integration can be shown to be constants—which may be taken as zero—by integrating the corresponding relationships for v_θ, which is zero.)

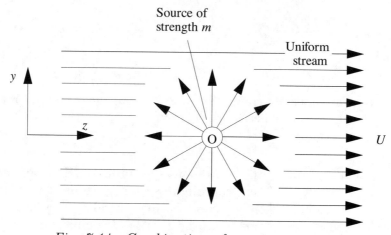

Source of
strength m

Uniform
stream

y

z

O

U

Fig. 7.14   Combination of a point source of strength m with a uniform stream of velocity U. The flow is axisymmetric about the z axis.

Point source in a uniform stream.   We now arrive at the intriguing possibility of *combining* the two previous concepts. As illustrated in Fig. 7.14, imagine a point source of strength m to be located at the origin O; superimposed on this flow is a uniform stream with velocity U flowing in the z direction. The previously derived velocity potentials and stream functions may be *added*, giving:

$$\phi = Ur\cos\theta - \frac{m}{r}, \qquad \psi = -\frac{1}{2}Ur^2\sin^2\theta + m\cos\theta. \tag{7.58}$$

Observe from the diagram that the part of the source flowing to the left opposes and hence tends to "neutralize" the oncoming stream from the left. In fact, a *stagnation point* S, with zero velocity components ($v_r = v_\theta = 0$), occurs when:

$$v_r = -\frac{1}{r^2\sin\theta}\frac{\partial\psi}{\partial\theta} = U\cos\theta + \frac{m}{r^2} = 0, \tag{7.59}$$

$$v_\theta = \frac{1}{r\sin\theta}\frac{\partial\psi}{\partial r} = -U\sin\theta = 0. \tag{7.60}$$

Therefore, as shown in Fig. 7.15, the stagnation point has coordinates:

$$\theta = \pi, \qquad r^2 = \frac{m}{U} \equiv a^2, \tag{7.61}$$

in which the ratio of the source strength to the stream velocity has been replaced by the square of a new quantity, a.

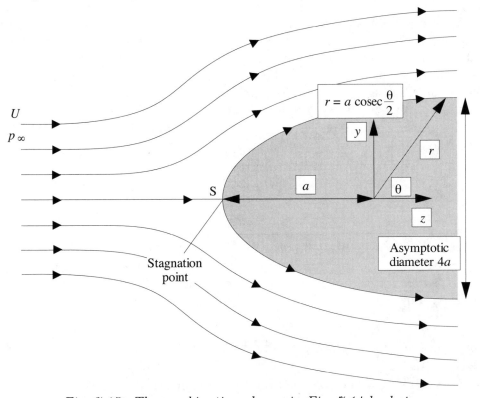

Fig. 7.15   The combination shown in Fig. 7.14 leads to axisymmetric irrotational flow around a blunt-nosed object.

From Eqns. (7.58) and (7.61), the value of the stream function at the stagnation point is $\psi = -m = -Ua^2$. Therefore, the streamline passing through the stagnation point has the equation[4]:

$$\psi = -\frac{1}{2} U r^2 \sin^2 \theta + m \cos \theta = -U a^2. \tag{7.62}$$

By using the identities:

$$y = r \sin \theta, \qquad 1 + \cos \theta = 2 \cos^2 \frac{\theta}{2}, \qquad \sin \theta = 2 \sin \frac{\theta}{2} \cos \frac{\theta}{2}, \tag{7.63}$$

---

[4] This does not imply that the fluid actually flows *through* the stagnation point. As an infinitesimally small element of fluid approaches the stagnation point S along the streamline from the left, it continuously decelerates, and it may be shown that it takes an infinite time to reach S.

the y and r coordinates of points on this streamline are found to be:

$$y^2 = 2a^2(1 + \cos\theta), \tag{7.64}$$

$$r = a\,\mathrm{cosec}\frac{\theta}{2}. \tag{7.65}$$

Equation (7.65) is drawn in Fig. 7.15 and is seen to be an axisymmetric curved surface with the stagnation point S at its "nose." It therefore follows that the combination of a source in a uniform stream represents potential flow past an axisymmetric blunt-nosed body, whose asymptotic diameter can readily be shown to equal $4a$. Of course, the equations also predict a flow pattern "inside" the body (what is it?), but we choose to ignore it here.

The *pressure* at any point can be obtained by first observing that the square of the velocity is:

$$\mathbf{v}^2 = v_r^2 + v_\theta^2 = \left(U\cos\theta + \frac{m}{r^2}\right)^2 + (-U\sin\theta)^2$$
$$= U^2\left(1 + 2\frac{a^2}{r^2}\cos\theta + \frac{a^4}{r^4}\right). \tag{7.66}$$

If the pressure in the undisturbed uniform stream (well away from the body) is p_∞, then—apart from any hydrostatic effects—Bernoulli's equation gives the local pressure at any point:

$$\frac{p}{\rho} = \frac{p_\infty}{\rho} - \frac{1}{2}U^2\left(2\frac{a^2}{r^2}\cos\theta + \frac{a^4}{r^4}\right). \tag{7.67}$$

Clearly, by taking other combinations, more sophisticated flow patterns can be generated, but our purpose is just to make the reader *aware* of such possibilities. However, another interesting combination *will* be considered in the next section.

7.8 Doublets and Flow Past a Sphere

A *doublet* is very much like a *dipole* in electricity and magnetism, and is introduced in Fig. 7.16(a).

Consider a point source and a point sink of strengths m and $-m$, respectively, located on the z axis at distances a to the right and left of the origin. Fluid emanating from the source will eventually flow back into the sink. For the combination, the velocity potential at a point P is:

$$-\phi = \frac{m}{r_2} - \frac{m}{r_1} = m\frac{r_1 - r_2}{r_1 r_2}. \tag{7.68}$$

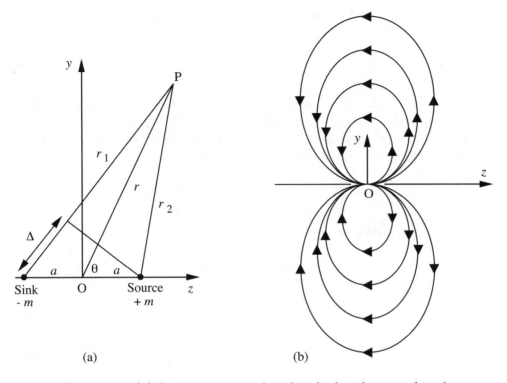

(a)   (b)

*Fig. 7.16   (a) Point source and sink, which, when combined
closely together lead to the flow from a doublet; (b) the resulting streamlines. The flow is axisymmetric about the z axis.*

Now let $a \rightarrow 0$ and $m \rightarrow \infty$, but in such a way that the product $2am \equiv s$
remains finite, where s is known as the *doublet strength*, which will be positive
when the direction from source to sink points in the negative z direction. The
velocity potential becomes:

$$-\phi = \frac{m\Delta}{r^2} = \frac{m(2a\cos\theta)}{r^2} = \frac{s\cos\theta}{r^2}. \tag{7.69}$$

Following a similar argument, the stream function for the doublet is:

$$\psi = -s\frac{\sin^2\theta}{r}. \tag{7.70}$$

The corresponding streamlines are shown in Fig. 7.16(b).

Combination of a doublet and a uniform stream.   In the same way
the combination of a *source* and a uniform stream was investigated in the previous
section, now consider the superposition of a *doublet* and a uniform stream.

Recall that for a *uniform stream* of velocity U in the z direction, the velocity potential and stream function are:

$$\phi = Uz = Ur\cos\theta, \qquad \psi = -\frac{1}{2}Ur^2\sin^2\theta. \tag{7.71}$$

Also consider a *doublet* of strength $s = -Ua^3/2$ at the origin, where U is again the velocity of the uniform stream and a is a variable whose significance is to be realized; the negative sign means that the orientation of the doublet is with the source to the left of the sink—*opposite* to that in Fig. 7.16:

$$\phi = \frac{1}{2}\frac{Ua^3\cos\theta}{r^2}, \qquad \psi = \frac{1}{2}Ua^3\frac{\sin^2\theta}{r}. \tag{7.72}$$

Thus, for the combination of the uniform stream and the doublet:

$$\phi = U\left(r + \frac{a^3}{2r^2}\right)\cos\theta, \tag{7.73}$$

$$\psi = -\frac{1}{2}Ur^2\left(1 - \frac{a^3}{r^3}\right)\sin^2\theta. \tag{7.74}$$

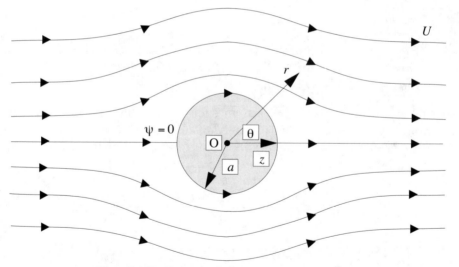

Fig. 7.17   *Potential flow around a sphere.*
The flow is axisymmetric about the z axis.

Investigation of Eqns. (7.73) and (7.74) leads to the following conclusions:

1. The streamline $\psi = 0$ occurs either for $\theta = 0$ or π, or for $r = a$. Therefore, the *sphere* $r = a$ may be taken as a solid boundary.
2. Both ϕ and ψ satisfy the general axisymmetric potential flow equations, (7.48) and (7.49).

3. Far away from the sphere, Eqns. (7.73) and (7.74) predict a uniform velocity U in the z direction.

4. These equations also predict a zero radial velocity v_r at the surface of the sphere.

The combination of a uniform stream and a doublet (with directions opposed) represents potential flow of a uniform stream with a solid sphere placed in it, as shown in Fig. 7.17. This result will help us in Chapter 10 to predict the motion of bubbles in liquids and fluidized beds. Also see Example 12.4 for an application to electroosmotic flow around a particle in a microchannel.

7.9 Single-Phase Flow in a Porous Medium

Laplace's equation also governs the flow of a viscous fluid, such as oil, in a porous medium, such as a permeable rock formation. Recall from d'Arcy's law in Section 4.4 that the superficial velocity v_x for one-dimensional flow is proportional to the pressure gradient:

$$v_x = -\frac{\kappa}{\mu}\frac{dp}{dx}, \tag{7.75}$$

where κ is the permeability of the medium and μ is the viscosity of the fluid. The minus sign occurs because the flow is in the direction of decreasing pressure. Equation (7.75) may be generalized to three-dimensional flow by using a vector velocity $\mathbf{v}$ and replacing dp/dx with the gradient of the pressure:

$$\mathbf{v} = -\frac{\kappa}{\mu}\nabla p. \tag{7.76}$$

But for an incompressible fluid the continuity equation is:

$$\nabla \cdot \mathbf{v} = 0. \tag{7.77}$$

The velocity may now be eliminated between these last two equations. If the permeability/viscosity ratio κ/μ is a *variable*, we obtain, after canceling the minus sign:

$$\nabla \cdot \frac{\kappa}{\mu}\nabla p = 0, \tag{7.78}$$

which simplifies in the case of *constant* κ/μ to:

$$\nabla^2 p = \frac{\partial^2 p}{\partial x^2} + \frac{\partial^2 p}{\partial y^2} = 0. \tag{7.79}$$

That is, the pressure distribution is *again* governed by Laplace's equation, which has been written out here for the case of two-dimensional rectangular coordinates, but is equally applicable in other coordinate systems. Thus, much of the theory in this chapter applies not only to flow of essentially *inviscid* fluids, but also to the flow of *viscous* fluids in porous media. Obvious important applications are to the recovery of oil from underground rock formations, and to the motion of groundwater in soil.

Example 7.5—Underground Flow of Water

A company has been disposing of some water containing a small amount of a chemical (1,4–dioxane) by injecting it into a permeable stratum of the ground via a vertical well, A, of radius a. Consequently, the groundwater has become slightly contaminated. As shown in Fig. E7.5.1(a), the company proposes to rectify the situation in a cleanup operation by drilling a second well, B, of the same radius, and separated by a distance L from A. Contaminated water will be withdrawn at a volumetric flow rate Q per unit depth from A, detoxified by irradiating it with ultraviolet light, and injecting the purified water back into the ground via well B. By modeling A as a line sink and B as a line source, derive an expression for the pressure difference $p_B - p_A$ between the two wells, in terms of Q, a, L, κ, and μ. Sketch several streamlines and isobars.

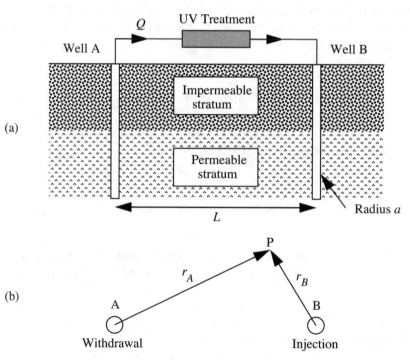

Fig. E7.5.1   *Withdrawal and injection wells A and B: (a) shows the side view of a vertical section through the formation; (b) shows the plan of a horizontal section of the permeable stratum, together with the radial coordinates of point P.*

Solution

For incompressible liquid flow in a porous medium, the superficial velocity vector (volumetric flow rate per unit area) $\mathbf{v}$ is given by d'Arcy's law:

$$\mathbf{v} = -\frac{\kappa}{\mu}\nabla p. \qquad (E7.5.1)$$

Example 7.5—Underground Flow of Water   **389**

Laplace's equation governs variations of pressure, which may be treated in virtually the same manner as the velocity potential in irrotational flow.

A plan of the permeable stratum is shown in Fig. E7.5.1(b), in which a general point such as P is seen to lie at radial distances r_A and r_B from the centers of wells A and B, respectively.

If B is considered to be a line source, with a total flow rate of Q per unit vertical depth (normal to the plane of Fig. E7.5.1(b)), the radially outward velocity v_{rB} at any distance r_B from it is obtained from continuity and d'Arcy's law:

$$v_{rB} = \frac{Q}{2\pi r_B} = -\frac{\kappa}{\mu}\frac{\partial p_B}{\partial r_B}. \tag{E7.5.2}$$

The corresponding pressure is obtained by integration:

$$p_B = -\frac{\mu Q}{2\pi\kappa}\ln r_B + f(z). \tag{E7.5.3}$$

For flow into the line sink at well A, the corresponding pressure is:

$$p_A = \frac{\mu Q}{2\pi\kappa}\ln r_A + f(z). \tag{E7.5.4}$$

In these last two equations, the function of integration $f(z)$ recognizes that the pressure also varies with the vertical distance z. Considering a fixed depth, the function effectively becomes a constant of integration, p_0.

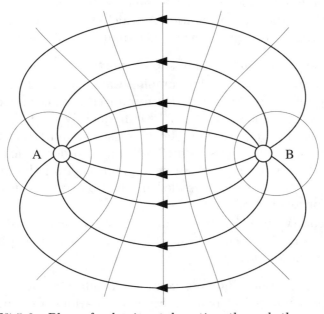

Fig. E7.5.2 Plan of a horizontal section through the permeable stratum, showing the streamlines (heavy) and isobars (light).

The combined effect of both injection and withdrawal wells leads to the pressure distribution:

$$p = p_0 + \frac{\mu Q}{2\pi\kappa} \ln \frac{r_A}{r_B}, \tag{E7.5.5}$$

in which the added pressure p_0 depends on the depth of the stratum and the overall level to which it is pressurized.

Equation (E7.5.5) is next applied at the outer radius of each well in turn, namely, for $r_B = a$, $r_A \doteq L$ (at well B) and $r_A = a$, $r_B \doteq L$ (at well A):

$$p_B = p_0 + \frac{\mu Q}{2\pi\kappa} \ln \frac{L}{a}, \qquad p_A = p_0 + \frac{\mu Q}{2\pi\kappa} \ln \frac{a}{L}. \tag{E7.5.6}$$

The required pressure *difference* between the wells is therefore:

$$p_B - p_A = \frac{\mu Q}{2\pi\kappa} \ln \frac{L^2}{a^2} = \frac{\mu Q}{\pi\kappa} \ln \frac{L}{a}. \tag{E7.5.7}$$

Finally, Fig. E7.5.2 shows several isobars (which are almost circular in the vicinity of the wells), together with the corresponding streamlines. □

7.10   Two-Phase Flow in Porous Media

An important application of fluid mechanics is the study of the flow of crude oil in the pores of the porous rock formation in which it is found. Increasing amounts of oil can be recovered in several stages, including the following:

1. *Primary* recovery, in which as much oil is pumped out as possible.
2. *Secondary* recovery, typically by pumping water down selected wells and displacing more of the oil, which is then produced at other selected wells. The "five-spot" pattern shown in Fig. 7.18 is frequently used. The duration of a single waterflood is many years, and it is usually important to simulate the results as accurately as possible beforehand, in order to maximize the production of oil. For example, if water is pumped down the wells too rapidly, it can "finger" towards the intended production wells, causing water to be "produced" instead of oil.
3. *Tertiary* recovery, in which surfactants are often used in order to reduce the capillary pressure between the oil and water, and to make it easier for the residual oil to be recovered.

The present discussion centers largely on secondary recovery, in which water and oil flow as two immiscible phases. Similar equations also govern the underground storage of natural gas, typically by the displacement of naturally occurring water. The notation to be used is given in Table 7.2.

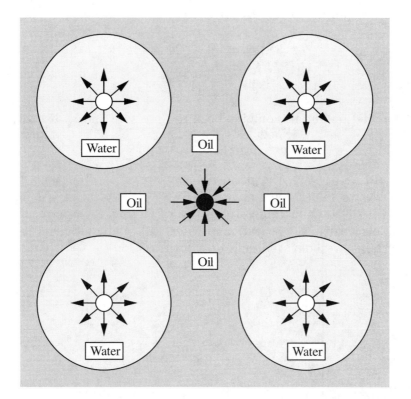

Fig. 7.18   Five-spot arrangement of wells for waterflooding.

Table 7.2   Variables for Two-Phase Immiscible Flow

Variable	Definition
p	Pressure
p_c	Capillary pressure, $p_o - p_w$
S	Saturation, fraction of pore space occupied by water
$\mathbf{v}$	Velocity vector
z	Elevation
o	Subscript for oil or nonwetting phase
w	Subscript for water or wetting phase
κ	Permeability
μ	Viscosity
ε	Porosity—fraction of total volume occupied by pores
Φ	Potential

d'Arcy's law can be applied to each of the two phases:

$$\mathbf{v}_o = -\frac{\kappa_o}{\mu_o}\nabla\Phi_o, \tag{7.80}$$

$$\mathbf{v}_w = -\frac{\kappa_w}{\mu_w}\nabla\Phi_w, \tag{7.81}$$

in which the fluid potentials combine both the pressure and hydrostatic effects:

$$\Phi_o = p_o + \rho_o gz, \qquad \Phi_w = p_w + \rho_w gz. \tag{7.82}$$

In Eqns. (7.80) and (7.81), the permeabilities are strong functions of the water saturation, relative values being shown in Fig. 7.19; typical maximum values are in the range of 10–100 md (millidarcies). Observe that each permeability is zero over a significant range of saturations; water will not flow unless its saturation reaches a certain threshold, and neither will the oil.

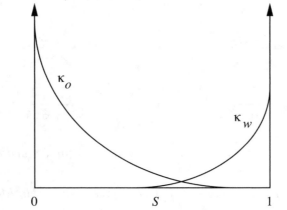

Fig. 7.19   Representative permeabilities for oil and water.

The capillary pressure is the excess pressure in the nonwetting phase over that in the wetting phase:

$$p_c = p_o - p_w, \tag{7.83}$$

and is again a strong function of the water saturation, as shown in Fig. 7.20(c). There, the *normalized* saturation $\bar{S}$ is shown, with the following extreme values:

1. For *connate* water (the water initially present in the oil-containing formation), $\bar{S} = 0$.
2. For *residual oil* (the residual oil when a porous medium containing oil is displaced by water), $\bar{S} = 1$.

Capillary pressure may be determined as shown in Fig. 7.20(a), by allowing a rock core, previously saturated with oil, to contact water at its base. Capillary attraction will cause the water to rise into the core, where its saturation will vary with elevation z.

If the path PQR is traversed, as shown in Fig. 7.20(b), the overall pressure change must be zero, since $p_R = p_P$:

$$p_R = p_P - \rho_w gz + p_c + \rho_o gz, \quad \text{or} \quad p_c = (\rho_w - \rho_g)gz. \tag{7.84}$$

If the water contains a small amount of an electrolyte such as sodium chloride, electrical conductivity measurements enable the saturation to be determined at various values of z, and hence p_c as a function of S or $\overline{S}$.

Fig. 7.20   *Capillary pressure for oil and water.*

Volumetric balances (the density is constant) on unit volumes give:

$$\nabla \cdot \mathbf{v}_o = -\varepsilon \frac{\partial(1 - S)}{\partial t} = \varepsilon \frac{\partial S}{\partial t},$$

$$\nabla \cdot \mathbf{v}_w = -\varepsilon \frac{\partial S}{\partial t}. \tag{7.85}$$

Substitution of the velocities from Eqns. (7.80) and (7.81) gives:

$$\nabla \cdot \left(\frac{\kappa_o}{\mu_o} \nabla \Phi_o \right) = -\varepsilon \frac{\partial S}{\partial t}, \tag{7.86}$$

$$\nabla \cdot \left(\frac{\kappa_w}{\mu_w} \nabla \Phi_w \right) = \varepsilon \frac{\partial S}{\partial t}. \tag{7.87}$$

Note that the time derivative of the saturation may be reexpressed in terms of the derivative of the potential difference:

$$\frac{\partial S}{\partial t} = \frac{\partial p_c}{\partial t} \underbrace{\frac{dS}{dp_c}}_{S'} = S' \frac{\partial (\Phi_o - \Phi_w)}{\partial t}, \tag{7.88}$$

in which $S' = dS/dp_c$ may be obtained by measuring the slope of the capillary-pressure curve at the appropriate value of the saturation.

With the above in mind, the transport equations become:

$$\nabla \cdot \left(\frac{\kappa_o}{\mu_o} \nabla \Phi_o \right) = -\varepsilon S' \frac{\partial (\Phi_o - \Phi_w)}{\partial t}, \tag{7.89}$$

$$\nabla \cdot \left(\frac{\kappa_w}{\mu_w} \nabla \Phi_w \right) = \varepsilon S' \frac{\partial (\Phi_o - \Phi_w)}{\partial t}. \tag{7.90}$$

Eqns. (7.89) and (7.90) must be solved numerically, and an appropriate rearrangement is to define sums and differences of the potentials and mobilities as:

$$P = \frac{1}{2}(\Phi_o + \Phi_w), \qquad R = \frac{1}{2}(\Phi_o - \Phi_w), \tag{7.91}$$

$$M = \frac{\kappa_o}{\mu_o} + \frac{\kappa_w}{\mu_w}, \qquad N = \frac{\kappa_o}{\mu_o} - \frac{\kappa_w}{\mu_w}, \tag{7.92}$$

so that Eqns. (7.89) and (7.90) can be rewritten as:

$$\nabla \cdot (M \nabla P) + \nabla \cdot (N \nabla R) = 0, \tag{7.93}$$

$$\nabla \cdot (N \nabla P) + \nabla \cdot (M \nabla R) = -4 \varepsilon S' \frac{\partial R}{\partial t}. \tag{7.94}$$

These last two equations are of *elliptic* and *parabolic* type, respectively, and may be solved by standard numerical techniques such as successive overrelaxation and the implicit alternating-direction method, as performed, for example, by Goddin et al.[5]

---

[5] C.S. Goddin, Jr., F.F. Craig, M.R. Tek, and J.O. Wilkes, "A numerical study of waterflood performance in a stratified system with crossflow," *Journal of Petroleum Technology*, Vol. 18, pp. 765–771 (1966). Also see Chapter 7 of B. Carnahan, H.A. Luther, and J.O. Wilkes, *Applied Numerical Methods*, Wiley & Sons, New York, 1969.

Underground storage of natural gas.   Several areas in the northern United States rely largely on the southern and southwestern part of the country for their supply of natural gas, which is typically conveyed by pipelines over distances sometimes amounting to 1,000–2,000 miles. Year-round operation of these pipelines is desirable in order to minimize costs. However, the consumer demand for the gas fluctuates substantially from a low point during the summer to a maximum during the winter. The excess gas pumped during the summer is most conveniently stored in underground porous-rock formations—preferably in depleted gas or oil fields, since these are known to be surmounted by an impermeable "caprock" formation that will prevent leakage.

The general scheme is shown in Fig. 7.21; the storage region is usually dome-shaped, with its highest elevation near the well or wells, thereby trapping the gas and preventing it from escaping laterally. During the summer, gas is pumped down one or more wells into the porous formation, whose storage region is called the gas "bubble," even though its horizontal extent may be thousands of feet. For the storage region, a typical vertical thickness is 100 ft, with a permeability in the range of 50–500 md. As the gas is injected, it displaces naturally occurring water laterally. During the winter, the gas is withdrawn, and the water moves back.

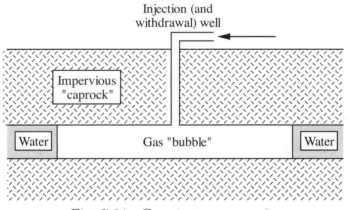

Fig. 7.21   Gas-storage reservoir.

The annual cycles of gas and water movement are governed by equations very similar to those already outlined for oil and water. However, significant additional simplifications may also be made because the gas and water regions can usually be treated realistically as single-phase regions. The following is a typical equation that governs pressure variations $p(x, y, t)$ of the gas, in which x and y are coordinates in the horizontal plane of the gas bubble and t is time:

$$\frac{\partial}{\partial x}\left(\frac{p}{zT}\frac{h\kappa}{\mu}\right)\frac{\partial p}{\partial x} + \frac{\partial}{\partial y}\left(\frac{p}{zT}\frac{h\kappa}{\mu}\right)\frac{\partial p}{\partial y} - \frac{MhR}{M_w} = \varepsilon h\frac{\partial(p/zT)}{\partial t}. \qquad (7.95)$$

Here, $M(x, y)$ is the local mass withdrawal rate of gas per unit volume (which will be zero if there is not a well in the vicinity), $h(x, y)$ is the local formation thickness,

$\kappa(x, y)$ is the local permeability, z is the compressibility factor of the gas, μ is the viscosity, T is the absolute temperature, M_w is the molecular weight of the gas, ε is the porosity or void fraction, and R is the gas constant.

7.11   Wave Motion in Deep Water

This chapter concludes with another, and quite different, application of potential flow theory—the motion of waves on an open body of water. We wish to find how the velocity of the water varies with time and position below the free surface.

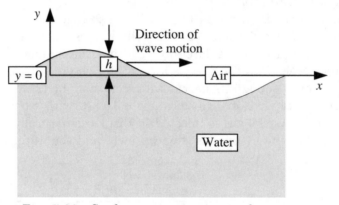

Fig. 7.22   Surface wave motion in deep water.

Fig. 7.22 shows a continuous wave that is moving to the right (in the x direction) on the surface of a body of very deep water of negligible viscosity. The equation of the free surface is:

$$h = a\sin(kx - \omega t) = a\sin\frac{2\pi}{\lambda}(x - ct), \qquad (7.96)$$

in which a is the amplitude, λ is the wavelength, $k = 2\pi/\lambda$ is the wave number, ω is the circular frequency, $c = \omega/k$ is the velocity of wave propagation, and the origin $y = 0$ corresponds to the level of the surface in the absence of waves.

The following potential function has been proposed for the motion of the water:

$$\phi = ace^{ky}\cos(kx - \omega t), \qquad (7.97)$$

which satisfies Laplace's equation, since:

$$\frac{\partial^2\phi}{\partial x^2} = -\frac{\partial^2\phi}{\partial y^2} = -k^2 ace^{ky}\cos(kx - \omega t). \qquad (7.98)$$

Note that in the very deep part of the water, y becomes highly negative, and the potential function and velocities approach zero.

Equation (7.8), for inviscid, irrotational, yet *transient* flow, with y now representing the vertical coordinate, leads to:

$$\frac{\partial \mathbf{v}}{\partial t} = -\nabla \left(\frac{p}{\rho} + \frac{1}{2} \mathbf{v}^2 + gy \right). \tag{7.99}$$

That is, any variation of the quantity in parentheses will generate a corresponding rate of change in the velocity. The substitution $\mathbf{v} = \nabla \phi$, coupled with the fact that the order of the gradient and partial differential operations can be interchanged, yields:

$$-\frac{\partial \phi}{\partial t} = \frac{p}{\rho} + \frac{1}{2} \mathbf{v}^2 + gy. \tag{7.100}$$

Eqn. (7.100) is now applied at the free surface, $y = h$. For waves of small amplitude, the kinetic energy term can be neglected, giving:

$$gh = - \left(\frac{\partial \phi}{\partial t} \right)_{y=h}. \tag{7.101}$$

Noting that the pressure is a constant at the free surface, differentiation of Eqn. (7.101) with respect to time gives:

$$g\frac{\partial h}{\partial t} = - \left(\frac{\partial^2 \phi}{\partial t^2} \right)_{y=h}. \tag{7.102}$$

The free surface is described by the equation $h = y$. Since a point on the surface moves with the liquid, we can employ differentiation following the liquid—that is, involve the *substantial derivative*, giving:

$$\frac{\mathcal{D}h}{\mathcal{D}t} = \frac{\mathcal{D}y}{\mathcal{D}t}, \tag{7.103}$$

or,

$$\frac{\partial h}{\partial t} + v_x \frac{\partial h}{\partial x} = v_y. \tag{7.104}$$

The terms on the left-hand side of Eqn. (7.104) have the following physical interpretations:

1. The first term, $\partial h/\partial t$, represents the rate at which the elevation of the free surface itself is increasing.
2. The second term, $v_x(\partial h/\partial x)$, represents the rate of increase of elevation of a fluid particle that is traveling with velocity v_x *and* is constrained to follow the contour of the free surface.

The combined effect then gives the actual y component of velocity, v_y. The situation is very much like the rate of increase of elevation of a person running up a hill, if the hill itself is moving upward, perhaps due to an earthquake.

For waves of small amplitude, the term involving the slope $\partial h/\partial x$ is small and can be neglected, resulting in:

$$\frac{\partial h}{\partial t} = (v_y)_{y=h} = \left(\frac{\partial \phi}{\partial y}\right)_{y=h}. \tag{7.105}$$

Elimination of $\partial h/\partial t$ between Eqns. (7.102) and (7.105) gives:

$$\left(\frac{\partial \phi}{\partial y}\right)_{y=h} = -\frac{1}{g}\left(\frac{\partial^2 \phi}{\partial t^2}\right)_{y=h}. \tag{7.106}$$

For the known potential function of Eqn. (7.97), these last two derivatives may be evaluated, leading to:

$$k = \frac{\omega^2}{g}. \tag{7.107}$$

But since the velocity of the waves is $c = \omega/k$ and $k = 2\pi/\lambda$, it follows that:

$$c = \sqrt{\frac{g\lambda}{2\pi}}, \tag{7.108}$$

giving not only the wave velocity but also demonstrating that waves of long wavelength travel faster than those of short wavelength. For a combination of waves of different wavelengths, there is a *dispersion* between the various wavelengths.

Finally, investigate the paths followed by individual liquid particles. By differentiation of the potential function, the velocity components—referred to the location $x = 0$ for simplicity—are:

$$v_x = -acke^{ky}\sin(kx - \omega t) = acke^{ky}\sin\omega t, \tag{7.109}$$

$$v_y = acke^{ky}\cos(kx - \omega t) = acke^{ky}\cos\omega t. \tag{7.110}$$

As ωt varies, the distances traveled in the x and y directions by a liquid particle can be obtained by integration of Eqns. (7.109) and (7.110) with respect to time:

$$x = \int v_x\,dt = -\frac{ack}{\omega}e^{ky}\cos\omega t + c_1, \qquad y = \int v_y\,dt = \frac{ack}{\omega}e^{ky}\sin\omega t + c_2. \tag{7.111}$$

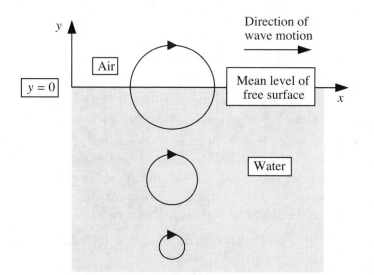

Fig. 7.23   Circular paths followed by liquid particles.

We can choose a starting point such that the constants of integration are both zero, and deduce from Eqn. (7.111) that:

$$x^2 + y^2 = \left(\frac{ack}{\omega}\right)^2 e^{2ky} = a^2 e^{2ky}. \tag{7.112}$$

That is, any liquid particle travels in a *circle* of radius ae^{ky}, which diminishes rapidly with depth away from the free surface, as shown in Fig. 7.23. In particular, note that the radius of motion at the free surface ($y = 0$) equals—as it must—the amplitude a of the passing wave.

PROBLEMS FOR CHAPTER 7

Unless otherwise stated, all flows are steady state, with constant density and viscosity (the latter for porous-medium flows).

1. *Flow past a sphere—M.* Consider the velocity potential and stream function for inviscid flow of an otherwise uniform stream U past a sphere of radius a, as given by Eqns. (7.73) and (7.74). Verify that these functions:

(a) Satisfy the appropriate axially symmetric equations, (7.48) and (7.49).
(b) Predict a uniform velocity U in the z direction far away from the sphere.
(c) Predict a zero radial velocity v_r at the surface of the sphere.
(d) Give a streamline $\psi = 0$ either for $\theta = 0$ or π, or for $r = a$, showing that the sphere $r = a$ may be taken as a solid boundary.

2. *Continuity and irrotationality—E.* Verify the following for irrotational flow in x/y coordinates:

(a) That the velocity components, defined in Eqn. (7.14) in terms of the velocity potential, automatically satisfy the irrotationality condition and—when substituted into the continuity equation—lead to Laplace's equation in ϕ.
(b) That the velocity components, defined in Eqn. (7.17) in terms of the stream function, automatically satisfy the continuity equation and—when substituted into the irrotationality condition—lead to Laplace's equation in ψ.

3. *Flow past a cylinder—E.* Consider the velocity potential and stream function for inviscid flow of an otherwise uniform stream U past a cylinder of radius a, as given by Eqns. (7.30) and (7.31). Verify that these functions:

(a) Satisfy Laplace's equations, (7.32).
(b) Predict a uniform velocity U in the x direction and zero velocity in the y direction far away from the cylinder.
(c) Give a streamline $\psi = 0$ either for $\theta = 0$ or π, or for $r = a$, showing that the cylinder $r = a$ may be taken as a solid boundary. Also prove that $v_r = 0$ at $r = a$.

4. *Stagnation and other flow—E.* Answer the following two parts:

(a) Derive the velocity potential function $\phi(x, y)$ for the stagnation flow whose stream function is given in Eqn. (E7.2.1) as $\psi = cxy$. Sketch several streamlines and equipotentials.
(b) For a certain flow, the velocity potential is given by:

$$\phi = ax + by.$$

Give physical interpretations of a and b, derive the corresponding stream function, $\psi(x, y)$, and sketch half a dozen streamlines and equipotentials.

5. *Axisymmetric continuity and irrotationality—E.* For axially symmetric irrotational flows, verify the following:

(a) That the velocity components, defined in Eqn. (7.44) in terms of the velocity potential, automatically satisfy the irrotationality condition and—when substituted into the continuity equation—lead to Eqn. (7.48) in ϕ.

(b) That the velocity components, defined in Eqns. (7.41) and (7.43) in terms of the stream function, automatically satisfy the continuity equation and—when substituted into the irrotationality condition—lead to Eqn. (7.49) in ψ.

6. *Uniform flow, point source, and line sink—M.* Fig. P7.6 shows a *line source*, in which the total strength m is distributed evenly along the z axis between the origin O and the point $z = a$. Thus, an element of differential length $d\zeta$ of this source will have a strength $m\,d\zeta/a$, and the corresponding stream function at a point P, whose coordinates are (r, θ), will be:

$$d\psi = \frac{m\,d\zeta}{a}\cos\alpha.$$

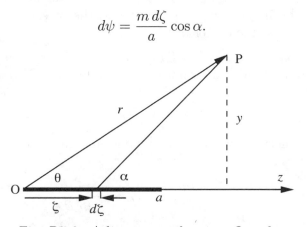

Fig. P7.6 *A line source between O and a.*

Prove by integration between $\zeta = 0$ and $\zeta = a$ for this axisymmetric flow problem that the stream function at P due to the entire line source is:

$$\psi = \frac{m}{a}\left(r - \sqrt{r^2 - 2az + a^2}\right).$$

Now consider the effect of a *combination* of the following three items:

(a) A uniform flow U in the z direction.

(b) A *point source* of strength m at the origin.

(c) A *line sink* of total strength m, extending between the origin O and a point $z = a$ on the axis of symmetry.

Derive an expression for the stream function ψ at any point, indicate what type of flow this represents, and sketch a representative sample of streamlines.

7. *Tornado flow—M.* This problem is one in axisymmetric *cylindrical* coordinates. Derive the velocity potential function for the following two flows:

(a) A free vortex, centered on the origin, in which the only nonzero velocity component is $v_\theta = \alpha/r$.
(b) A line sink along the z axis, in which the only nonzero velocity component is $v_r = -\beta/r$.

Hint: Remember to involve the components of $\nabla\phi$ in *cylindrical* coordinates.

Sketch a few streamlines for a tornado, which can be approximated by the following combination:

(a) A free vortex, centered on the origin at $r = 0$.
(b) A line sink flow, towards the origin.

At a distance of $r = 2,000$ m from the "eye" (center) of a tornado, the pressure is 1.0 bar and the two nonzero velocity components are $v_r = -0.25$ m/s and $v_\theta = 0.5$ m/s. The density of air is 1.2 kg/m^3. At a distance of $r = 10$ m from the eye, compute both velocity components and the pressure. What is likely to happen to a house near the center of the tornado?

8. *Spherical hole in a porous medium—M.* The velocity **v** of a liquid percolating through a porous medium of uniform permeability κ under a pressure gradient is given by d'Arcy's law:

$$\mathbf{v} = -\frac{\kappa}{\mu}\nabla p.$$

For an axially symmetric flow, you may assume that the corresponding superficial velocity components are:

$$v_r = -\frac{\kappa}{\mu}\frac{\partial p}{\partial r}, \qquad v_\theta = -\frac{\kappa}{\mu}\frac{1}{r}\frac{\partial p}{\partial \theta},$$

in which r and θ are spherical coordinates. Prove that the pressure p obeys the following equation:

$$\sin\theta\frac{\partial}{\partial r}\left(r^2\frac{\partial p}{\partial r}\right) + \frac{\partial}{\partial \theta}\left(\frac{\partial p}{\partial \theta}\sin\theta\right) = 0.$$

Water seeps from left to right through a porous medium that is bounded by two infinitely large planes AB and CD, separated by a distance $2L$, and whose pressures are P and $-P$, respectively, as shown in Fig. P7.8.

Halfway across, the medium contains a spherical hole of radius a, which offers no resistance to flow, and in which the pressure is $p = 0$. The dimensions are such that $a/L \ll 1$.

Show that the pressure distribution is given approximately by:

$$p = \left(\alpha r + \frac{\beta}{r^2}\right)\cos\theta,$$

and then determine the coefficients α and β in terms of L, P, and a. *Hint*: Check that this expression for p satisfies the governing differential equation and all boundary conditions.

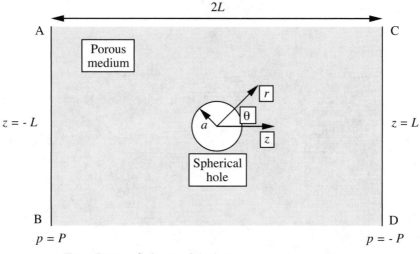

Fig. P7.8 *Spherical hole in a porous medium.*

Also, sketch a few streamlines and prove that the volumetric flow rate of water through the hole is:

$$Q = \frac{3\kappa P\pi a^2}{\mu L}.$$

If needed, $\sin 2\theta = 2\sin\theta\cos\theta$. (A very closely related problem occurs during the upwards motion of bubbles in a fluidized bed of catalyst particles, in which a significant amount of the fluidizing gas sometimes "channels" through the bubbles and hence has an adversely reduced contact with the catalyst.)

9. *Seepage under a dam—M.* The (vector) superficial velocity of an *incompressible* liquid of *constant* viscosity μ percolating through a porous medium of uniform permeability κ is given by d'Arcy's law:

$$\mathbf{v} = -\frac{\kappa}{\mu}\nabla p, \tag{P7.9.1}$$

where p is the pressure *beyond* that occurring due to usual hydrostatic effects.

(a) In one or two sentences, state why you would expect the flow to be irrotational.

(b) Use the vector form of the continuity equation to show that the excess pressure obeys Laplace's equation, $\nabla^2 p = 0$. (Note that in this case the excess pressure behaves very much like the velocity potential ϕ in irrotational flow, in which the Laplacian of ϕ is also zero.)

Fig. P7.9 shows seepage of water through the ground under a dam, caused by the excess pressure P that arises from the buildup of water behind the dam, which has (underground) a semicircular base of radius r_D. The following relation has been proposed for the excess pressure in the ground:

$$p = P\left(1 - \frac{\theta}{\pi}\right). \qquad (\text{P7.9.2})$$

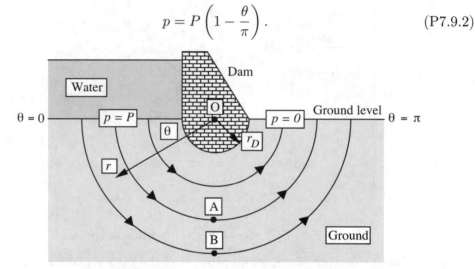

Fig. P7.9   Seepage of water under a dam.

(c) Prove that Eqn. (P7.9.2) is correct by verifying that it satisfies:

   (i) The conditions on pressure at the ground level.

   (ii) Laplace's equation, $\nabla^2 p = 0$, in cylindrical $(r/\theta/z)$ coordinates, in which all z derivatives are zero.

   (iii) Zero radial flow at the base of the dam.

(d) What are the r and θ components of ∇p in cylindrical $(r/\theta/z)$ coordinates? What then are the expressions for v_r and v_θ in terms of r and θ? If the corresponding expressions in terms of the stream function ψ are:

$$v_r = -\frac{1}{r}\frac{\partial \psi}{\partial \theta}, \qquad v_\theta = \frac{\partial \psi}{\partial r}, \qquad (\text{P7.9.3})$$

derive an expression for the stream function in terms of r and/or θ, and hence prove that the streamlines are indeed semicircles. Draw a sketch showing a few streamlines and isobars (lines of constant pressure—much like equipotentials).

(e) Between points A and B, some copper-impregnated soil has been detected, with the possibility that some of this toxic metal may leach out and have adverse effects downstream of the dam. To help assess the extent of this danger, derive an expression for the volumetric flow rate Q of water between A and B (per unit depth in the z direction, normal to the plane of the diagram). Your answer should give Q in terms of P, κ, μ, r_A, and r_B.

10. *Uniform stream and line sink—M.* In this two-dimensional problem, give your answers in terms of r and θ coordinates, except in one place where the y coordinate is specifically requested.

(a) A line sink at the origin O extends normal to the plane shown in Fig. P7.10. If the total volumetric flow rate withdrawn from it is $2\pi Q$ per unit length, prove that the velocity potential is:

$$\phi = -Q \ln r.$$

What is the corresponding equation for the stream function ψ?

(b) Now consider the addition of a stream of liquid, which far away from the sink has a uniform velocity U in the x direction. Show for the combination of uniform stream and sink that $\phi = Ur \cos \theta - Q \ln r$, and also obtain the corresponding expression for ψ.

(c) Give expressions for the velocity components v_r and v_θ, and from them demonstrate that the flow is irrotational.

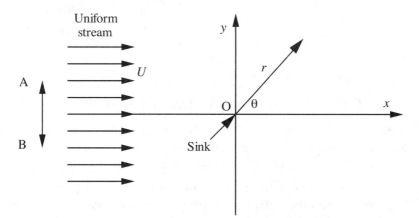

Fig. P7.10   Coordinate systems for uniform stream and sink.

(d) Where is the single stagnation point S located? What is the equation of the streamline that passes through S? Hence, by considering the y coordinate of this streamline far upstream of the sink, deduce, in terms of U and Q, the width AB of that portion of the uniform stream that flows into the sink.

(e) Sketch several equipotentials and streamlines for the combination of uniform stream and sink.

(f) For incompressible liquid flow in a porous medium, the superficial velocity vector (volumetric flow rate per unit area) $\mathbf{v}$ is given by d'Arcy's law:

$$\mathbf{v} = -\frac{\kappa}{\mu} \nabla p,$$

in which κ is the uniform permeability, μ is the constant viscosity, and p is the pressure. Prove that the pressure obeys Laplace's equation:

$$\nabla^2 p = 0,$$

so that it may be treated essentially as the velocity potential that we have been studying for irrotational flow, the only difference being the factor of $-\kappa/\mu$. Hence, if Fig. P7.10 now represents the horizontal cross section of a well in a porous soil formation, derive an expression showing how the pressure p deviates from its value p_S at the stagnation point.

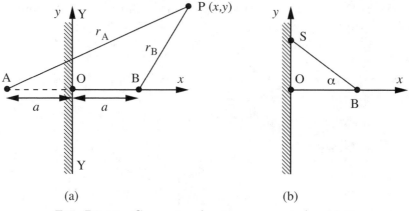

(a)   (b)

Fig. P7.11   *Geometry for two sources A and B.*

11. *Injection well near an impermeable barrier—M.* This problem involves potential flow in *two* dimensions—*not* in axisymmetric spherical coordinates. For convenience, both the x, y and r, θ coordinate systems will be needed.

(a) First consider a line source at the origin O, extending normal to the x, y plane. The flow is radially outwards everywhere (with $v_\theta = 0$), and the total volumetric flow rate issuing from the source is $2\pi Q$ per unit depth (corresponding to a strength Q). Deduce a simple expression for the radial velocity v_r as a function of radial distance r, and hence prove that the velocity potential is given by:

$$\phi = Q \ln r.$$

What is the corresponding equation for the stream function ψ?

(b) Derive an expression for the time t taken for a fluid particle to travel from the source to a radial location where $r = R$.

(c) Next, Fig. P7.11(a) shows a line source located at point B, which has coordinates $x = a$ and $y = 0$. There is an impermeable barrier to flow along the plane Y–Y, which extends indefinitely along the y axis. Explain why the potential ϕ and stream function ψ at a point P, with coordinates (x, y), may be modeled by *adding* to the source B an identical source of strength Q, but now located at a point A with coordinates $x = -a$ and $y = 0$.

(d) For the combination of the two sources, obtain an expression for the stream function at point P in terms of Q, x, and y. Assume the following identity if needed:

$$\tan^{-1}\left(\frac{y}{x+a}\right) + \tan^{-1}\left(\frac{y}{x-a}\right) = \tan^{-1}\left(\frac{2xy}{x^2 - y^2 - a^2}\right).$$

(e) Sketch several streamlines (for $x \geq 0$ only), clearly showing the direction of flow. If there is a stagnation point, identify it. If there isn't one, explain why not.

(f) Write down an expression for the potential function ϕ at point P in terms of Q, r_A, and r_B. Add several equipotentials to your diagram.

(g) Fig. P7.11(b) shows the same basic situation, but with several lines removed for clarity. Consider the velocity v_y (in the y direction) at a point S on the barrier. For what value of a would v_y be a maximum? Explain.

(h) Suppose now that there is a well W of radius R centered on B (where $R \ll a$), situated in a porous formation of permeability κ, and into which treated waste water of viscosity μ is being injected at a volumetric flow rate Q. Derive an expression for $p_W - p_O$, the pressure difference between well W and the origin O. *Hint*: recall that flow through a porous medium may be modeled as potential flow, provided that ϕ is replaced by $-\kappa p/\mu$.

12. *Circulation in vortices—E.* By examining the appropriate component of curl **v**, derive expressions for the *circulation*—defined in Eqn. (5.26)—in the forced- and free-vortex regions of Example 7.1, and comment on your findings.

13. *Motion of a particle in potential flow—D.* This problem has applications in the sampling of fluids for entrained particles. First, examine Example 7.3. A small spherical particle of diameter d_p and density ρ_p enters upstream of the slot at a location to be specified, with initial ($t = 0$) velocity components ($v_{x0}^p = 0$, $v_{y0}^p = -U$). The particle is sufficiently small so that Stokes' law gives the drag force on it:

$$F_x = 3\pi\mu_f d_p(v_x^f - v_x^p), \qquad F_y = 3\pi\mu_f d_p(v_y^f - v_y^p),$$

in which μ_f is the viscosity of the fluid, and superscripts f and p denote fluid and particle velocities, respectively.

Prove that the following dimensionless differential equations give the accelerations of the particle in the two principal directions:

$$\xi\frac{dV_X^p}{d\tau} = -\frac{X}{X^2 + Y^2} - V_X^p, \qquad \xi\frac{dV_Y^p}{d\tau} = -1 - \frac{Y}{X^2 + Y^2} - V_Y^p,$$

in which:

$$\tau = \frac{\pi U^2 t}{Q}, \quad \xi = \frac{\pi\rho_p d_p^2 U^2}{18\mu_f Q}, \quad X = \frac{\pi U x}{Q}, \quad Y = \frac{\pi U y}{Q}, \quad V_X^p = \frac{v_x^p}{U}, \quad V_Y^p = \frac{v_y^p}{U}.$$

Use a spreadsheet, MATLAB, or other computer software to determine the subsequent location of the particle (measured by X and Y) as a function of time, starting from $\tau = 0$ and proceeding downstream until either: (a) the particle enters the slot, (b) the particle hits the wall, or (c) the particle proceeds beyond the slot and has an insignificant value of V_X^p. Euler's method with a time step of $\Delta\tau$ is suggested for solving the differential equations (see Appendix A).

If you use a spreadsheet, the following column headings are suggested for tracking the various quantities over successive time steps: τ, X, Y, V_X^f, V_Y^f, V_X^p, V_Y^p, $dV_X^p/d\tau$, and $dV_Y^p/d\tau$.

Take Y_0 large enough so that its exact value is not critical. Plot four representative particle trajectories for various values of X_0 and ξ, and give comments with a few sentences on each trajectory.

14. *Packed-column flooding—D (C).* Consider the upward flow of an essentially inviscid gas of density ρ_G around a sphere of radius a. Well away from the sphere, the upward gas velocity is U, so the stream function is:

$$\psi = -\frac{1}{2}Ur^2\sin^2\theta\left(1 - \frac{a^3}{r^3}\right), \tag{P7.14.1}$$

in which r is the radial distance from the center of the sphere and the angle θ is measured *downward* from the vertical. Derive a relation for $v^2 = v_r^2 + v_\theta^2$ as a function of r and θ. Hence, prove that the pressure gradient in the θ direction at the surface of the sphere is:

$$\left(\frac{\partial p}{\partial\theta}\right)_{r=a} = -\frac{9}{4}\rho_G U^2\sin\theta\cos\theta. \tag{P7.14.2}$$

In an experiment to investigate the flooding of packed columns (the inability to accommodate ever-increasing simultaneous downward liquid flows and upward gas flows), an inviscid liquid of density ρ_L flows freely and symmetrically under gravity as a thin film covering the surface of a solid sphere. Gas of density ρ_G flows upward around the sphere, and its stream function is given by Eqn. (P7.14.1).

By considering the equilibrium of a liquid element that subtends an angle $d\theta$ at the center, show that the pressure gradient within the gas stream will prevent the liquid from running down when:

$$\frac{\partial p}{\partial\theta} = \rho_L ga\sin\theta, \tag{P7.14.3}$$

and that this will first be satisfied when:

$$U^2 = \frac{4\rho_L ga}{9\rho_G}. \tag{P7.14.4}$$

15. *Vortex lines and streamlines—E.* Consider the forced vortex region of a stirred liquid, as in Example 7.1. Ignoring gravitational effects, what is the shape of the surface on which $p/\rho + \mathbf{v}^2/2 + gz$ is constant? Illustrate with a sketch, showing the direction of the velocity and the vorticity at a representative point, together with the direction in which $p/\rho + \mathbf{v}^2/2 + gz$ is increasing.

16. *Velocity potential/stream function—M (C).* In a certain two-dimensional irrotational flow, the velocity potential is given by:

$$\phi = a(x^2 + by^2),$$

where $a = 0.1$ s^{-1}.

Evaluate the constant b, and derive an expression for the corresponding stream function. Considering only the region $x > 0$, $y > 0$, sketch a few typical streamlines and hence determine the flow pattern that ϕ represents.

If at time $t = 0$ an element of fluid is at the point P ($x = 10$ cm, $y = 20$ cm), find its position Q at a time $t = 2$ s, and illustrate its path on your diagram.

17. *Flow of water between two wells—M.* For the situation shown in Example 7.4, take the following values: $\mu = 1$ cP, $\kappa = 15$ darcies, $a = 0.05$ m, and $L = 1{,}000$ m. If the available pump limits the pressure drop between the two wells to 5.0 bar, what volumetric flow rate of water can be expected per unit vertical depth of the stratum? Give your answer in both cubic meters per second per meter depth, and in gpm per foot depth.

Under these conditions, what is the superficial velocity (m/s) of the water at a point exactly halfway between the two wells? With this velocity, how far would the water travel in a year if the stratum has a porosity of 0.30?

18. *Effect of waves below the surface—E.* Parallel waves whose peaks are spaced 5 m apart are traveling continuously on the surface of deep water. What is the velocity (m/s) of these waves? At what depth below the surface will the magnitude of the displacement be one-hundredth of its value at the surface?

19. *Sampling a fluid—M.* An *axisymmetric* inviscid incompressible flow consists of the combination of two basic elements:

(i) A uniform stream of fluid with velocity U in the positive z direction.

(ii) A point sink located at the origin, $r = 0$, that withdraws a sample of the fluid at a steady volumetric flow rate $4\pi Q$.

For the combined two flows:

(a) Give expressions in spherical coordinates (r and θ) for the stream function and potential function.

(b) Sketch several streamlines and equipotentials. Show all relevant details.

(c) From the stream function, deduce the corresponding velocity components, v_r and v_θ.

(d) Is there a stagnation point? If so, explain where it is located and why. If not, explain why not.

(e) Below what radius a, measured in the y direction, will fluid coming from a large distance in the negative z direction be withdrawn by the sampling point? The y and z coordinates are those shown in Fig. 7.11 of your notes.

(f) Consider a small element of the fluid starting at an upstream point ($z = -D$; $y = H$). Explain carefully how would you determine the time taken for it to reach the sink. You need not carry the derivation completely to its conclusion, but you should give sufficient detail so that—given enough time—it could be completed by somebody else.

20. *Storage of natural gas—E.* Prove Eqn. (7.95), for the pressure variations in the gas bubble during the underground storage of natural gas. A transient balance on an element of dimensions $dx \times dy \times h$ is recommended.

21. *Hele-Shaw flow—M* Flow of a liquid of viscosity μ occurs between two parallel horizontal transparent glass plates separated by a *small* distance H, as shown in Fig. P7.21. The plates are sealed around the edges to prevent leakage, except for small ports where the liquid may be introduced and withdrawn. By adding dye or small tracer particles, flow patterns may be observed by looking down through the top plate—including flows past obstacles that may be clamped between the plates. The object of this problem is to show that even though the flow is *viscous*, the mean velocity components $\bar{v}_x$ and $\bar{v}_y$ in the horizontal x/y plane are the gradients of a scalar ϕ—that is, the flow behaves as though it were *irrotational*, and the corresponding streamlines can be observed.

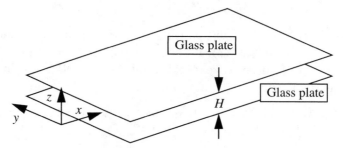

Fig. P7.21   Hele-Shaw "table."

The x velocity component obeys the simplified momentum equation:

$$\frac{\partial p}{\partial x} = \mu \frac{\partial^2 v_x}{\partial z^2} = \mu \frac{d^2 v_x}{dz^2}, \tag{P7.21}$$

where z is the vertical coordinate measured normal to the plates (with $z = 0$ at the midplane). A similar equation holds for the y velocity component.

(a) State the simplifying assumptions that Eqn. (P7.21) implies.

(b) Derive an expression for v_x in terms of H, z, μ, and $\partial p/\partial x$, and for $\bar{v}_x$ (the mean value of v_x between the two plates) in terms of H, μ, and $\partial p/\partial x$.

(c) Write down, without proof, the corresponding expression for the mean y velocity component $\bar{v}_y$.

(d) What is the potential function ϕ that gives $\bar{\mathbf{v}} = \nabla\phi$, the velocity vector whose components are $\bar{v}_x$ and $\bar{v}_y$?

22. *Flow pattern for wells—M.* The Hele-Shaw apparatus of Problem 7.21 is to be used for simulating the porous-medium flow in a horizontal plane (viewed from above in Fig. P7.22) between one injection well I and two withdrawal wells A and B (which share equally the flow rate injected, but which do not necessarily have the same pressure). The situation is very similar to that of Example 7.4, except that there are now *two* withdrawal wells and there is an *impermeable boundary* surrounding the region.

(a) Explain *briefly* why the Hele-Shaw apparatus can simulate flow in a porous medium.

(b) Sketch the flow patterns that are likely to result, as follows:

   (i) Eight streamlines (full curves with arrows) and six isobars or equipotentials (dotted curves) are expected, on a full-page diagram.

   (ii) Do *not* perform any calculations or algebra.

  (iii) Make sure you give reasonable coverage to the whole area—not just in the vicinity of the wells.

  (iv) Identical streamlines obtained from symmetry should be included, but will not contribute to the total of eight required.

   (v) Clearly indicate with the letter "S" (stagnation) any points where the velocity is zero.

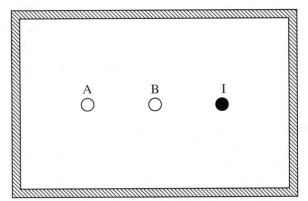

Fig. P7.22   Injection (I) and withdrawal (A and B) wells.

23. *Flow in wedge-shaped region—M.* Verify that the potential function:

$$\phi = -cr^3\cos 3\theta,$$

with c a constant, gives two-dimensional irrotational flow in the wedge-shaped region that is bounded by the planes $\theta = 0$ and $\theta = \pi/3$. Sketch a few streamlines, clearly showing the direction of flow. Is there a stagnation point? If so, where?

24. *Inviscid flow with rotation—E.* Referring to Example 7.5:

(a) Prove the validity of $-\nabla^2\psi = \zeta$, where $\zeta = (\partial v_y/\partial x - \partial v_x/\partial y)$ is the vorticity.

(b) Does the use of $-\nabla^2\psi = \zeta$ still imply that continuity is satisfied?

25. *True/false.* Check *true* or *false*, as appropriate:

    (a)   Euler's equation arises when inertial effects are ig-   T ☐ F ☐
nored, but viscous effects are included.

    (b)   In a forced vortex, the velocity component v_θ is uni-   T ☐ F ☐
form everywhere.

    (c)   It is possible to have an irrotational flow in which the   T ☐ F ☐
fluid is everywhere moving in concentric circles.

    (d)   Angular velocity is twice the vorticity.   T ☐ F ☐

    (e)   Velocities given in terms of the potential function by   T ☐ F ☐
$v_x = \partial\phi/\partial x$ and $v_y = \partial\phi/\partial y$ automatically satisfy
the continuity equation.

    (f)   Streamlines can cross one another.   T ☐ F ☐

    (g)   Velocity components may be deduced from expres-   T ☐ F ☐
sions for either the velocity potential *or* the stream
function.

    (h)   The stream function $\psi = Ux$ represents a uniform   T ☐ F ☐
flow in the x direction.

    (i)   The problem of irrotational flow past a cylinder is   T ☐ F ☐
best solved in axisymmetric spherical coordinates.

    (j)   In a potential flow solution, the velocity must always   T ☐ F ☐
be zero on a solid boundary.

    (k)   The stream function for axisymmetric irrotational   T ☐ F ☐
flows satisfies Laplace's equation.

    (l)   A point source of strength m emits a total volumetric   T ☐ F ☐
flow of m per unit time.

    (m)   Simple potential flows may be combined with one an-   T ☐ F ☐
other in order to simulate more complex flows.

    (n)   A stagnation point occurs when the pressure becomes   T ☐ F ☐
a minimum.

    (o)   A fluid doublet has equal effects in all directions.   T ☐ F ☐

(p)   The combination of a doublet and a uniform stream,   T □ F □
       with directions opposed to each other, represents ir-
       rotational flow past a solid sphere.

(q)   In the forced vortex of Fig. 7.1(a), the shear stress $\tau_{r\theta}$ T □ F □
       is zero.

(r)   For flow parallel to the x axis, as shown in Fig. 7.7,   T □ F □
       the stream function decreases in the y direction.

(s)   The radial velocity at a distance from a point source   T □ F □
       of strength m is given by $v_r = m/r$.

(t)   Tornado flow can be approximated by a combination   T □ F □
       of a line sink and a forced vortex.

(u)   When waves of a given amplitude travel on the surface   T □ F □
       of water, those having higher values of ω will have a
       larger effect on the velocities at greater depths than
       waves with lower values of ω.

(v)   In the theoretical treatment of potential flow past a   T □ F □
       cylinder, the solution must satisfy zero tangential ve-
       locity v_θ at the surface of the cylinder.

(w)   Laplace's equation governs the pressure distribution   T □ F □
       for steady single-phase flow in a porous medium.

Chapter 8

BOUNDARY-LAYER AND OTHER
NEARLY UNIDIRECTIONAL FLOWS

8.1 Introduction

ALTHOUGH the Navier-Stokes equations apply to the flow of a Newtonian fluid, they are—in general—very difficult to solve exactly. There is also the problem that an apparently correct solution may be physically unrealistic for high Reynolds numbers, because of turbulence or other instabilities.

Two particular types of simplifying approximations may be made, depending on the Reynolds number (the ratio of inertial to viscous effects), in order to make the situation more tractable:

1. Omit the inertial terms, as was done in the examples of Chapter 6. The resulting equations are typically appropriate for the flow near a fixed boundary, where viscous action is particularly important. (Another example, not given there, leads to the Stokes' solution for "creeping" flow round a sphere, but is valid only for Re < 1.)

2. Omit the viscous terms, as was done in Chapter 7. The resulting equations for inviscid flow are often fairly easy to handle, and are frequently adequate for applications away from boundaries, where viscous action is relatively unimportant.

Generally, however, there will be a region in which *both* viscous *and* inertial terms are of comparable magnitude; even for a low-viscosity fluid, such a region will occur near a solid boundary, where there is a high shear rate. These regions are called *boundary layers*, and are important in some chemical engineering operations because diffusion of heat or mass across them frequently controls the rate of some heat- and mass-transfer operations. Also, in aerospace engineering and naval architecture, a knowledge of boundary-layer theory is essential for determining the drag on airplane bodies and wings and on ship hulls.

A key feature of boundary-layer analysis is the existence of a primary flow direction, whose velocity component v_x is considerably larger than that in a transverse or secondary direction, such as v_y. The fact that $v_x \gg v_y$ typically implies essentially no pressure variation in the transverse direction—a substantial simplification. The same simplification also applies to lubrication problems, polymer calendering, paint spreading, and other "nearly unidirectional" flows.

8.2 Simplified Treatment of Laminar Flow Past a Flat Plate

The full analytical solution of the Navier-Stokes equations for boundary-layer flow is virtually impossible, so that in Sections 8.2–8.5 we are content to make simplifications and still obtain realistic solutions by the following two approaches:

1. Perform mass and momentum balances on an element of the boundary layer, then *assume* a reasonable shape for the velocity profile, and finally solve by the method outlined in this section for laminar flow and in Section 8.5 for turbulent flow.

2. For laminar boundary layers only, perform the classical *Blasius solution*, starting with the Navier-Stokes equations and dropping those terms that are expected to be small (see Section 8.3). As discussed in Section 8.4, a suitable change of variables leads to a third-order *ordinary* differential equation, which can be solved numerically in order to obtain the velocities.

Additionally, in Examples 8.3 and 8.5, we show how computational fluid dynamics can solve the full Navier-Stokes equations for boundary-layer and lubrication flows.

For the present, consider steady two-dimensional *laminar* flow past a flat plate, shown in Fig. 8.1, and assume the existence of a *boundary layer*, being a thin region in which the velocity changes from zero at the plate to practically its value $v_{x\infty}$ in the *mainstream*. The velocity $v_{x\infty}$ in the mainstream (where viscous effects are negligible) will be considered constant for the present; therefore, from Bernoulli's equation, the pressure in the mainstream does not vary. Further, since there is likely to be a negligible pressure variation *across* the boundary layer (see Section 8.3 for a rigorous proof), the pressure in the boundary layer is also constant.

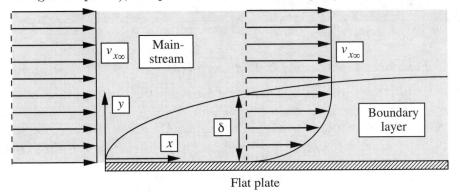

Flat plate

Fig. 8.1 Formation of a laminar boundary layer on a flat plate.

The almost exact Blasius solution, to be treated in Section 8.4, will show that the shear force τ_w exerted on the plate at any distance x from its leading edge is given in dimensionless form by the *drag coefficient*, c_f, as:

$$c_f = \frac{\tau_w}{\frac{1}{2}\rho v_{x\infty}^2} = \frac{0.664}{\sqrt{\mathrm{Re_x}}}, \quad \text{where} \quad \mathrm{Re}_x = \frac{\rho v_{x\infty} x}{\mu}. \tag{8.1}$$

In this section, we shall obtain a comparable result much more easily, by performing mass and momentum balances on an element of the boundary layer, as shown in Fig. 8.2.

Let the length of the element be dx, and consider unit width normal to the plane of the diagram. The y velocity component v_y arises because the plate retards the x velocity component; thus, $\partial v_x/\partial x$ is *negative*, and—from continuity—$\partial v_y/\partial y$ must be positive; since v_y is zero at the wall, it becomes increasingly positive away from the wall. By integration over the thickness of the boundary layer, the following total fluxes of mass (m) and momentum ($\dot{\mathcal{M}}$) entering through the left-hand face AD are obtained[1]:

$$m = \rho \int_0^\delta v_x \, dy,$$

$$\dot{\mathcal{M}} = \rho \int_0^\delta v_x^2 \, dy. \qquad (8.2)$$

Flow into ABCD from outside the boundary layer

A B

Mass m and momentum $\dot{M}$ flowing in the boundary layer

δ

dx

D C

Flat plate

τ_w

Fig. 8.2 Fluxes to and from an element of the boundary layer.

By the usual arguments, the mass flux leaving through the face BC will be $m + (dm/dx)dx$. That is, a mass flux of $(dm/dx)dx$ must be entering from the mainstream across the interface AB into the control volume ABCD. Since its velocity is $v_{x\infty}$, it will be bringing in with it an *x-momentum* flux of $(dm/dx)\,dx\,v_{x\infty}$. A *momentum* balance on ABCD in the x direction yields:

$$\dot{\mathcal{M}} + \frac{dm}{dx}\,dx\,v_{x\infty} - \tau_w dx - \left(\dot{\mathcal{M}} + \frac{d\dot{\mathcal{M}}}{dx}\,dx\right) = 0. \qquad (8.3)$$

Note that both *inertial* and *viscous* terms (the latter via the wall shear stress τ_w) are involved, consistent with the definition of a boundary layer expressed in Section

---

[1] There is no need for a dot over m, which has already been defined as the *rate* of transfer of mass, M.

8.1. There are no terms involving pressure, since this is constant everywhere. Substitution of m and $\dot{M}$ from Eqn. (8.2), and division by ρ, gives:

$$v_{x\infty}\frac{d}{dx}\int_0^\delta v_x\,dy = \frac{\tau_w}{\rho} + \frac{d}{dx}\int_0^\delta v_x^2\,dy. \qquad (8.4)$$

Equation (8.4) will enable the wall shear stress and the boundary-layer thickness to be predicted as a function of x if the velocity distribution $v_x = v_x(x, y)$ is known.

The simplified approach pursued here assumes that the velocity profile obeys the law:

$$\frac{v_x}{v_{x\infty}} = f(\zeta), \quad \text{where} \quad \zeta = \frac{y}{\delta(x)}. \qquad (8.5)$$

Equation (8.5) is assumed to hold at all distances x from the leading edge, even though the boundary-layer thickness δ depends on x; that is, the velocity profiles have similar shapes, but are "stretched" more in the y direction the further the distance from the leading edge, as shown in Fig. 8.3. Note that ζ is a *dimensionless distance* normal to the plate. The principle employed is that of "similarity of velocity profiles"; that is, at a particular distance from the plate—expressed as a *fraction* of the local boundary-layer thickness—the velocity will have built up to a certain fraction of the mainstream velocity, *independent* of the distance x from the leading edge.

Fig. 8.3 Similar velocity profiles.

The next step is to make a reasonable speculation about the form of $f(\zeta)$, which should conform to the following conditions at least:

1. $f(0) = 0$ (zero velocity at the wall, where $\zeta = 0$).
2. $f(1) = 1$ (the mainstream velocity $v_{x\infty}$ is regained at the edge of the boundary layer, where $\zeta = 1$).
3. $f'(1) = 0$ (the velocity profile has zero slope at the junction between the boundary layer and the mainstream).

If required, higher-order derivatives can be obtained at $y = 0$ and $y = \delta$ by successive differentiations of the simplified x momentum balance (see Section 8.3 for details):

$$v_x \frac{\partial v_x}{\partial x} + v_y \frac{\partial v_x}{\partial y} = -\frac{1}{\rho} \frac{\partial p}{\partial x} + \frac{\mu}{\rho} \frac{\partial^2 v_x}{\partial y^2} = v_{x\infty} \frac{dv_{x\infty}}{dx} + \frac{\mu}{\rho} \frac{\partial^2 v_x}{\partial y^2}$$

$$= \frac{\mu}{\rho} \frac{\partial^2 v_x}{\partial y^2} \qquad \text{for constant mainstream velocity.} \tag{8.6}$$

For example,

$$\text{At } y = 0: \quad v_x = v_y = 0; \quad \frac{\partial^2 v_x}{\partial y^2} = 0, \quad \frac{\partial^3 v_x}{\partial y^3} = 0, \text{ etc.} \tag{8.7}$$

$$\text{At } y = \delta: \quad \frac{\partial v_x}{\partial x} = \frac{\partial v_x}{\partial y} = 0; \quad \frac{\partial^2 v_x}{\partial y^2} = 0, \quad \frac{\partial^3 v_x}{\partial y^3} = 0, \text{ etc.} \tag{8.8}$$

The present choice will be:

$$\frac{v_x}{v_{x\infty}} = \sin\left(\frac{\pi y}{2\delta}\right), \tag{8.9}$$

which not only conforms to conditions 1, 2, and 3 above, but also yields the indicated values for the following integrals:

$$\frac{1}{v_{x\infty}} \int_0^\delta v_x \, dy = \frac{2\delta}{\pi}, \qquad \frac{1}{v_{x\infty}^2} \int_0^\delta v_x^2 \, dy = \frac{\delta}{2}. \tag{8.10}$$

Differentiation of both Eqns. (8.10) with respect to x, followed by multiplication by $v_{x\infty}^2$, gives:

$$v_{x\infty} \frac{d}{dx} \int_0^\delta v_x \, dy = v_{x\infty}^2 \frac{2}{\pi} \frac{d\delta}{dx}, \qquad \frac{d}{dx} \int_0^\delta v_x^2 \, dy = \frac{v_{x\infty}^2}{2} \frac{d\delta}{dx}, \tag{8.11}$$

which can then be substituted into the momentum balance of Eqn. (8.4):

$$v_{x\infty}^2 \frac{2}{\pi} \frac{d\delta}{dx} = \frac{\tau_w}{\rho} + \frac{v_{x\infty}^2}{2} \frac{d\delta}{dx}. \tag{8.12}$$

Equation (8.12) is an ordinary differential equation governing variations in the boundary-layer thickness with distance from the leading edge, and which can be integrated as soon as the wall shear stress is known. Differentiation of the velocity profile of Eqn. (8.9) leads to:

$$\tau_w = \mu \left(\frac{dv_x}{dy}\right)_{y=0} = \frac{\mu \pi v_{x\infty}}{2\delta}. \tag{8.13}$$

Equation (8.12) then becomes:

$$\frac{d\delta}{dx} c v_{x\infty}^2 = \frac{\mu \pi v_{x\infty}}{2 \rho \delta}, \quad \text{where} \quad c = \frac{2}{\pi} - \frac{1}{2} \doteq 0.137. \tag{8.14}$$

Separation of variables and integration gives:

$$\frac{2c}{\pi} \int_0^\delta \delta \, d\delta = \frac{\mu}{\rho v_{x\infty}} \int_0^x dx. \tag{8.15}$$

That is, the thickness of the boundary layer varies with distance from the leading edge as:

$$\frac{\delta}{x} = \sqrt{\frac{\pi}{c}} \sqrt{\frac{\mu}{\rho v_{x\infty} x}} = \frac{4.79}{\sqrt{\mathrm{Re_x}}}. \tag{8.16}$$

The wall shear stress can now be obtained from Eqns. (8.13) and (8.16). After some algebra, there results:

$$c_f = \frac{\tau_w}{\frac{1}{2} \rho v_{x\infty}^2} = \frac{\sqrt{c\pi}}{\sqrt{\mathrm{Re_x}}} = \frac{0.656}{\sqrt{\mathrm{Re_x}}}. \tag{8.17}$$

In view of the highly simplified approach taken, this last result compares very favorably with the Blasius solution of Section 8.4.

Of course, other approximations can be made for the velocity profile, such as $v_x/v_{x\infty} = a\zeta + b\zeta^2 + c\zeta^3 + \ldots$. Similar results follow, and Table 8.1 indicates the corresponding numerical coefficients that will appear in Eqn. (8.17).

Table 8.1 Solutions for Different Assumed Velocity Profiles

Approximation for $v_x/v_{x\infty}$	$c_f\sqrt{\mathrm{Re_x}}$
ζ	0.578
$2\zeta - \zeta^2$	0.730
$\frac{3}{2}\zeta - \frac{1}{2}\zeta^3$	0.646
$2\zeta - 2\zeta^3 + \zeta^4$	0.686
$\sin\left(\frac{\pi}{2}\zeta\right)$	0.656
Exact (Blasius)	0.664

Example 8.1—Flow in an Air Intake (C)

The device shown in Fig. E8.1 is used for measuring the flow of air of density ρ into the circular intake port of an engine. Basically, it uses Bernoulli's equation between two points: (a) far upstream of the intake, where the velocity is essentially zero and the pressure is p_1, and (b) the pressure tapping at a distance x from the inlet, where the pressure is measured to be p_2.

Fig. E8.1   *Device for measuring air flow rate into a port.*

The situation is complicated by the formation of a boundary layer as the air impinges against the inlet of the port, so the velocity profile at the location of the tapping consists of: (a) a central core, in which the velocity is uniformly u_2, and (b) a boundary layer in which the velocity declines from its mainstream value $v_{x\infty} = u_2$ to zero at the wall.

Bernoulli's equation can be applied, starting from the outside air (where $u_1 \doteq 0$) and continuing into the core (where viscosity is negligible), leading to:

$$u_2 = \sqrt{\frac{2(p_1 - p_2)}{\rho}}. \tag{E8.1.1}$$

However, the ultimate goal is to be able to determine the *mean* velocity u_m (defined as the total flow rate divided by the total area), which necessitates the introduction of a discharge coefficient C_D, so that:

$$u_m = C_D\sqrt{\frac{2(p_1 - p_2)}{\rho}}. \tag{E8.1.2}$$

The total mass flow rate of air is then:

$$m = \rho u_m A, \tag{E8.1.3}$$

where A is the cross-sectional area of the pipe, which has an internal diameter D. If $D = 2$ in., $x = 1$ ft, $\nu = 1.59 \times 10^{-4}$ ft^2/s, and $u_m = 20$ ft/s, estimate C_D.

Example 8.1—Flow in an Air Intake (C)   **421**

Solution

First, estimate the properties of the boundary layer. The largest value of the Reynolds number occurs opposite the pressure tapping, where $x = 1$ ft. Also, a rough estimate of the mainstream velocity is u_{m}, so that $v_{x\infty} \doteq 20$ ft/s. Thus, to a first approximation:

$$\mathrm{Re}_x \doteq \frac{v_{x\infty} x}{\nu} \doteq \frac{20 \times 1}{1.59 \times 10^{-4}} = 1.26 \times 10^5. \qquad (\mathrm{E}8.1.4)$$

From Section 8.5, the flow in the boundary layer is therefore laminar all the way from the inlet to the pressure tapping, and the result of Eqn. (8.16), which was based on the assumed velocity profile $v_x/v_{x\infty} = \sin(\pi y/2\delta)$, applies:

$$\frac{\delta}{x} = \frac{4.79}{\sqrt{\mathrm{Re}_x}} = \frac{4.79}{1.26 \times 10^5} = 0.0135, \qquad (\mathrm{E}8.1.5)$$

giving a first approximation of the thickness of the boundary layer:

$$\delta \doteq 0.0135 \times 12 = 0.162 \text{ in.} \qquad (\mathrm{E}8.1.6)$$

Thus, the boundary layer has an outer diameter of 2.000 in. and an inner diameter of $(2.000 - 2 \times 0.162) = 1.676$ in., for a mean diameter of $D_{\mathrm{m}} = (2.000 + 1.676)/2 = 1.838$ in.

Although the intake has circular geometry, the boundary layer can be approximated as that on a flat plate of width πD_{m}. The corresponding mass flow rate in the boundary layer is:

$$m \doteq \pi \rho D_{\mathrm{m}} \int_0^\delta v_x \, dy = \pi \rho D_{\mathrm{m}} \left(\frac{2\delta v_{x\infty}}{\pi} \right) = 2 \rho D_{\mathrm{m}} \delta v_{x\infty}. \qquad (\mathrm{E}8.1.7)$$

The total mass flow rate, which is based on the mean velocity, equals the sum of the mass flow rates in the core and in the boundary layer:

$$\underbrace{\left(\frac{\pi \times 2.000^2}{4} \right) \rho \times 20.0}_{\text{Total flow rate}} = \underbrace{\left(\frac{\pi \times 1.676^2}{4} \right) \rho v_{x\infty}}_{\text{Flow rate in core}} + \underbrace{2\rho \times 1.838 \times 0.162 \, v_{x\infty}}_{\text{Flow rate in boundary layer}}, \qquad (\mathrm{E}8.1.8)$$

which gives the improved estimate $v_{x\infty} \doteq 22.4$ ft/s. The calculations can be repeated iteratively; convergence essentially occurs after just one more iteration, giving a final value $v_{x\infty} \doteq 22.29$ ft/s. Since the pressure drop based on Bernoulli's equation gives $v_{x\infty} \doteq 22.29$, and the actual mean velocity is a lower value, $u_{\mathrm{m}} = 20.0$, the required discharge coefficient is:

$$C_D = \frac{20.0}{22.29} = 0.897. \qquad (\mathrm{E}8.1.9)$$

□

8.3   Simplification of the Equations of Motion

Here, we start with the Navier-Stokes and continuity equations, and inquire as to which terms are likely to be insignificant, so that they may be dropped and the equations thereby simplified. In two dimensions, assuming steady flow and the absence of gravitational effects, the starting equations are:

$$\frac{\mathcal{D}v_x}{\mathcal{D}t} = v_x\frac{\partial v_x}{\partial x} + v_y\frac{\partial v_x}{\partial y} = -\frac{1}{\rho}\frac{\partial p}{\partial x} + \nu\left(\frac{\partial^2 v_x}{\partial x^2} + \frac{\partial^2 v_x}{\partial y^2}\right), \tag{8.18}$$

$$\frac{\mathcal{D}v_y}{\mathcal{D}t} = v_x\frac{\partial v_y}{\partial x} + v_y\frac{\partial v_y}{\partial y} = -\frac{1}{\rho}\frac{\partial p}{\partial y} + \nu\left(\frac{\partial^2 v_y}{\partial x^2} + \frac{\partial^2 v_y}{\partial y^2}\right), \tag{8.19}$$

$$\frac{\partial v_x}{\partial x} + \frac{\partial v_y}{\partial y} = 0. \tag{8.20}$$

Now investigate the relative orders of magnitude of the various terms in Eqns. (8.18), (8.19), and (8.20). Let "O" denote "order of"; for example $v_x = \mathrm{O}(v_{x\infty})$, because for much of the boundary layer, the x velocity component v_x is a significant fraction (0.1 or more, for example) of the mainstream velocity $v_{x\infty}$; nowhere is v_x larger than $v_{x\infty}$. Also, make the following two key (and realistic) assumptions at any location x:

1. The derivative $\partial v_x/\partial x$ is of the same order of magnitude as the corresponding mean gradient from the origin to x. The situation is illustrated in Fig. 8.4(a), which shows for a given y how v_x declines from $v_{x\infty}$ at the leading edge ($x = 0$) to a much smaller value for significant values of x. The magnitude of the slope of the chord, which represents the mean velocity gradient, is approximately $v_{x\infty}/x$, and the diagram shows that the slope of the velocity profile at two representative points is not vastly different from $v_{x\infty}/x$. That is, $\partial v_x/\partial x = \mathrm{O}(v_{x\infty}/x)$.

2. The velocity gradient of v_x in the y direction is of the same order as the corresponding mean gradient from $y = 0$ to $y = \delta$. The situation is illustrated in Fig. 8.4(b), which shows for a given x how v_x increases from zero at the wall ($y = 0$) to $v_{x\infty}$ at the edge of the boundary layer ($y = \delta$). The magnitude of the slope of the chord, which represents the mean velocity gradient, is approximately $v_{x\infty}/\delta$. Further reflection shows that the slope of the velocity profile at a few representative locations between the wall and the edge of the boundary layer is nowhere vastly different from $v_{x\infty}/\delta$. That is, $\partial v_x/\partial y = \mathrm{O}(v_{x\infty}/\delta)$. Because δ is much smaller than x, the slope in Fig. 8.4(b) is much greater than that in Fig. 8.4(a), and representative intermediate slopes cannot be drawn clearly.

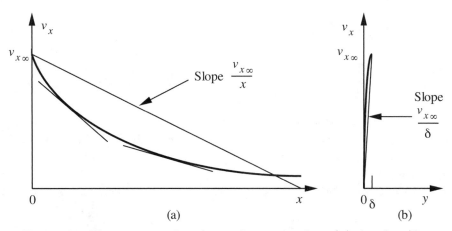

Fig. 8.4   *Illustration of orders of magnitude: (a) for $\partial v_x/\partial x$, with two representative intermediate tangents, and (b) for $\partial v_x/\partial y$. Note that the slope is much higher in (b) than in (a), because v_x changes from zero to $v_{x\infty}$ over a very short distance, δ.*

Orders of magnitude of individual derivatives.   Based on the above two assumptions, and by following similar arguments and also invoking the continuity equation in Eqn. (8.20), we obtain the following orders of magnitude:

$$\frac{\partial v_x}{\partial x} = \mathrm{O}\left(\frac{v_{x\infty}}{x}\right), \qquad \frac{\partial^2 v_x}{\partial x^2} = \mathrm{O}\left(\frac{v_{x\infty}}{x^2}\right); \tag{8.21}$$

$$\frac{\partial v_x}{\partial y} = \mathrm{O}\left(\frac{v_{x\infty}}{\delta}\right), \qquad \frac{\partial^2 v_x}{\partial y^2} = \mathrm{O}\left(\frac{v_{x\infty}}{\delta^2}\right); \tag{8.22}$$

$$\frac{\partial v_y}{\partial y} = -\frac{\partial v_x}{\partial x} = \mathrm{O}\left(\frac{v_{x\infty}}{x}\right). \tag{8.23}$$

Since $v_y = 0$ at $y = 0$, Eqn. (8.23) leads to:

$$v_y = \mathrm{O}\left(v_{x\infty}\frac{\delta}{x}\right). \tag{8.24}$$

The various derivatives of v_y therefore have the following orders of magnitude:

$$\frac{\partial^2 v_y}{\partial y^2} = \mathrm{O}\left(\frac{v_{x\infty}}{x\delta}\right), \qquad \frac{\partial v_y}{\partial x} = \mathrm{O}\left(v_{x\infty}\frac{\delta}{x^2}\right), \qquad \frac{\partial^2 v_y}{\partial x^2} = \mathrm{O}\left(v_{x\infty}\frac{\delta}{x^3}\right). \tag{8.25}$$

Examination of the x momentum balance.   Based on the above, all of the terms in Eqn. (8.18), except the pressure gradient, may be assigned their appropriate orders of magnitude:

$$v_x \frac{\partial v_x}{\partial x} \quad + \quad v_y \frac{\partial v_x}{\partial y} \quad = \quad -\frac{1}{\rho}\frac{\partial p}{\partial x} \quad + \quad \nu \left(\frac{\partial^2 v_x}{\partial x^2} + \frac{\partial^2 v_x}{\partial y^2} \right),$$

Order : $\qquad \dfrac{v_{x\infty}^2}{x} \qquad\qquad \dfrac{v_{x\infty}^2}{x} \qquad\qquad ? \qquad\qquad \nu \left(\dfrac{v_{x\infty}}{x^2} \quad \dfrac{v_{x\infty}}{\delta^2} \right).$

$$\text{(8.26)}$$

By hypothesis, the thickness of the boundary layer is much less than the distance from the leading edge. Hence, $\delta \ll x$, and on the right-hand side of Eqn. (8.26) the second-order derivative of v_x with respect to x is negligible compared to the derivative with respect to y. Also, because inertial and viscous terms are both important in the boundary layer, their orders of magnitude can be equated, giving:

$$\frac{v_{x\infty}^2}{x} = \mathrm{O}\left(\frac{\nu v_{x\infty}}{\delta^2} \right) \quad \text{or} \quad \left(\frac{\delta}{x} \right)^2 = \mathrm{O}\left(\frac{\nu}{v_{x\infty}x} \right). \tag{8.27}$$

Hence, since $\delta \ll x$, $\mathrm{Re}_x = v_{x\infty}x/\nu$ must be large, which excludes the analysis from applying very close to the leading edge. Also note that:

$$\nu = \mathrm{O}\left(\frac{\delta^2 v_{x\infty}}{x} \right), \quad \frac{1}{\rho}\frac{\partial p}{\partial x} = \mathrm{O}\left(\frac{v_{x\infty}^2}{x} \right). \tag{8.28}$$

This last equation gives the order of the pressure gradient $\partial p/\partial x$ if it is important; however, in certain cases—notably when the mainstream velocity is constant—this pressure gradient may be zero.

Examination of the y momentum balance. Similar considerations apply to the terms in Eqn. (8.19), except for the pressure gradient, whose order of magnitude is again unknown:

$$v_x \frac{\partial v_y}{\partial x} \quad + \quad v_y \frac{\partial v_y}{\partial y} \quad = \quad -\frac{1}{\rho}\frac{\partial p}{\partial y} \quad + \quad \nu \left(\frac{\partial^2 v_y}{\partial x^2} + \frac{\partial^2 v_y}{\partial y^2} \right),$$

Order : $\quad \dfrac{v_{x\infty}^2\delta}{x^2} \qquad \dfrac{v_{x\infty}^2\delta}{x^2} \qquad ? \qquad \dfrac{\delta^2 v_{x\infty}}{x}\left(\dfrac{v_{x\infty}\delta}{x^3} \quad \dfrac{v_{x\infty}}{x\delta} \right).$

$$\text{(8.29)}$$

Note that the $\partial^2 v_y/\partial x^2$ term is of a lower order than the others and can be neglected. It therefore follows that:

$$\frac{1}{\rho}\frac{\partial p}{\partial y} = \mathrm{O}\left(\frac{v_{x\infty}^2\delta}{x^2} \right) \ll \mathrm{O}\left(\frac{v_{x\infty}^2}{x} \right). \tag{8.30}$$

That is, the variation of pressure p across the boundary layer in the y direction is small in comparison to that in the x direction [see Eqn. (8.28)], so that p is (at most) a function of x only:

$$p = f(x). \tag{8.31}$$

The simplified equations of motion are therefore:

$$v_x \frac{\partial v_x}{\partial x} + v_y \frac{\partial v_x}{\partial y} = -\frac{1}{\rho} \frac{\partial p}{\partial x} + \nu \frac{\partial^2 v_x}{\partial y^2}, \tag{8.32}$$

$$p = f(x), \tag{8.33}$$

$$\frac{\partial v_x}{\partial x} + \frac{\partial v_y}{\partial y} = 0. \tag{8.34}$$

Pressure variation in the mainstream.   In the mainstream, where by hypothesis the effect of viscosity is negligible:

$$v_x \frac{\partial v_x}{\partial x} + v_y \frac{\partial v_x}{\partial y} = -\frac{1}{\rho} \frac{\partial p}{\partial x}. \tag{8.35}$$

Thus, if appreciable velocity gradients exist in the mainstream, it follows that $(\partial p/\partial x)/\rho$ is of the same order as $v_x(\partial v_x/\partial x)$, namely $O(v_{x\infty}^2/x)$, and must therefore be retained in the x momentum balance, Eqn. (8.18). An alternative viewpoint in the mainstream, which is removed from the plate and hence has negligible viscous effects, is to apply Bernoulli's equation and then differentiate it:

$$\frac{p}{\rho} + \frac{v_{x\infty}^2}{2} = \text{ constant}, \tag{8.36}$$

$$v_{x\infty} \frac{\partial v_{x\infty}}{\partial x} = -\frac{1}{\rho} \frac{\partial p}{\partial x}, \tag{8.37}$$

which is the same result as obtained from Eqn. (8.35) if the term involving v_y is neglected.

8.4 Blasius Solution for Boundary-Layer Flow

In Section 8.3, by considering the relative magnitudes of the various derivatives, we had greatly simplified the equations of motion. If the additional assumption is made that the mainstream velocity is constant, the pressure is also everywhere constant, leaving only the simplified x momentum balance and the continuity equation for consideration:

$$v_x \frac{\partial v_x}{\partial x} + v_y \frac{\partial v_x}{\partial y} = \nu \frac{\partial^2 v_x}{\partial y^2}, \tag{8.38}$$

$$\frac{\partial v_x}{\partial x} + \frac{\partial v_y}{\partial y} = 0, \tag{8.39}$$

In addition, the boundary conditions are that both velocity components are zero at the plate, and that the mainstream velocity is attained well away from the plate:

$$y = 0: \quad v_x = v_y = 0, \tag{8.40}$$

$$y = \infty: \quad v_x = v_{x\infty}. \tag{8.41}$$

The key to the solution of this simplified problem is to introduce a *new* space variable, ζ, which combines both x and y, and is motivated by observing that the simplified integral momentum balance led to δ being proportional to $\sqrt{\nu x / v_{x\infty}}$:

$$\zeta = y\sqrt{\frac{v_{x\infty}}{\nu x}}. \tag{8.42}$$

Also introduce a *stream function* (see also Section 7.3) ψ, such that:

$$v_x = \frac{\partial \psi}{\partial y}, \qquad v_y = -\frac{\partial \psi}{\partial x}, \tag{8.43}$$

which *automatically* satisfies the continuity equation, (8.39). Also try a separation of variables of the form:

$$\psi = g(x)f(\zeta). \tag{8.44}$$

$$v_x = \frac{\partial \psi}{\partial y} = g(x)f'(\zeta)\sqrt{\frac{v_{x\infty}}{\nu x}}. \tag{8.45}$$

Next, try to make $v_x/v_{x\infty}$ a function of ζ only, by choosing:

$$g(x) = \sqrt{\nu x v_{x\infty}}, \tag{8.46}$$

which gives:

$$\frac{v_x}{v_{x\infty}} = \frac{df(\zeta)}{d\zeta} \equiv f'(\zeta), \tag{8.47}$$

$$v_y = -\frac{\partial \psi}{\partial x} = \frac{1}{2}\sqrt{\frac{\nu v_{x\infty}}{x}}\left(\zeta f' - f\right). \tag{8.48}$$

Substitution of v_x and v_y and the derivatives of v_x from Eqns. (8.47) and (8.48) into Eqn. (8.38) leads to the following *ordinary* differential equation:

$$f\frac{d^2 f}{d\zeta^2} + 2\frac{d^3 f}{d\zeta^3} = 0, \tag{8.49}$$

subject to the boundary conditions:

$$\zeta = 0: \quad f = f' = 0, \tag{8.50}$$

$$\zeta = \infty: \quad f' = 1. \tag{8.51}$$

The original Blasius solution divided $\zeta = 0$ to ∞ into two regions and obtained approximate solutions (involving an infinite series and a double integral) for each such region. Three arbitrary constants appeared in the solution, and these were evaluated by requiring continuity of f, f', and f'' at the junction between the two regions.

With computer software readily available for solving simultaneous ordinary differential equations, we have opted for a numerical solution of Eqns. (8.49)–(8.51), first giving f as a function of ζ, and then the velocity components from Eqns. (8.47) and (8.48). The values shown in Fig. 8.5 agree completely with the original Blasius solution. The reader who wishes to check these results can do so fairly readily (see Problem 8.14).[2]

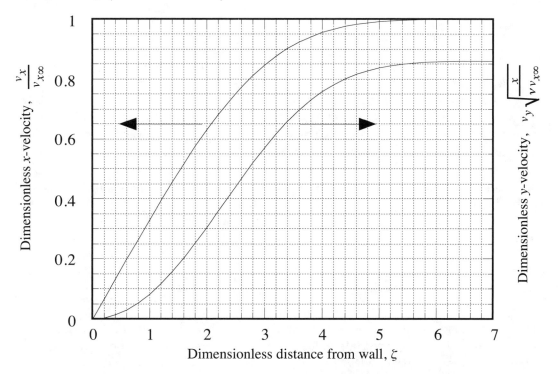

Fig. 8.5 *Computed velocities for the Blasius problem.*

As a by-product, the solution yields the following value for the second derivative of f at the wall:

$$\zeta = 0 : \quad f'' = 0.332, \tag{8.52}$$

which then enables the wall shear stress to be deduced:

$$\tau_w = \mu \left(\frac{\partial v_x}{\partial y} \right)_{y=0} = \mu v_{x\infty} \sqrt{\frac{v_{x\infty}}{\nu x}} \, f''(0) = 0.332 \, \mu v_{x\infty} \sqrt{\frac{v_{x\infty}}{\nu x}}. \tag{8.53}$$

[2] The values shown in Fig. 8.5 were computed by a *numerical* solution of Eqn. (8.49), by B. Carnahan, H.A. Luther, and J.O. Wilkes, as shown on page 414 of their book, *Applied Numerical Methods*, Wiley & Sons, New York, 1969.

The drag coefficient, which is a dimensionless shear stress, is then:

$$c_f = \frac{\tau_w}{\frac{1}{2}\rho v_{x\infty}^2} = \frac{0.664}{\sqrt{\mathrm{Re_x}}}. \tag{8.54}$$

Experiments of Nikuradse (Re_x from 1.08×10^5 to 7.28×10^5) verify the predicted velocity profile. There is a transition from laminar into turbulent flow at higher Reynolds numbers.

8.5   Turbulent Boundary Layers

The preceding analysis in Sections 8.2–8.4 was for *laminar* flow in the boundary layer. However, it is found experimentally that the boundary layer on a flat plate becomes *turbulent* if the Reynolds number, Re_x, exceeds approximately 3.2×10^5, which is next investigated. Assume for simplicity that the flow is everywhere turbulent—from the leading edge of the plate ($x = 0$) onward. In reality, the boundary layer will be laminar in the vicinity of the leading edge and would, for large enough x, undergo a transition to turbulent flow; thus, if the accuracy of the situation warranted it, the separate portions of the layer (laminar and turbulent) could be analyzed individually and then combined.

The analysis begins exactly as previously, in which a momentum balance on an element of the boundary layer leads to Eqn. (8.4):

$$v_{x\infty}\frac{d}{dx}\int_0^\delta v_x\,dy = \frac{\tau_w}{\rho} + \frac{d}{dx}\int_0^\delta v_x^2\,dy. \tag{8.55}$$

From experience with turbulent velocity profiles in pipe flow, assume (reasonably) that the one-seventh power law holds for the time-averaged x velocity in the turbulent boundary layer:

$$\frac{v_x}{v_{x\infty}} = \left(\frac{y}{\delta}\right)^{1/7}. \tag{8.56}$$

This last relation will be adequate for substitution into the two integrals of Eqn. (8.55) but it cannot be used for obtaining the shear stress at the wall—as was done previously in Eqn. (8.13)—since it predicts an infinite velocity gradient at the wall ($y = 0$).

For the shear stress we again have to draw an analogy with the shear stress for turbulent pipe flow. Recall the Blasius equation, (3.40), which is based on experiment and holds for smooth pipe up to a Reynolds number of about 10^5:

$$f_\mathrm{F} = 0.079\,\mathrm{Re}^{-1/4} = 0.079\left(\frac{\rho v_m D}{\mu}\right)^{-1/4}, \tag{8.57}$$

Example 8.2—Laminar and Turbulent Boundary Layers Compared   **429**

which, by substituting for the mean velocity v_m in terms of the centerline or maximum velocity v_{max}, may be rephrased as:

$$\frac{\tau_w}{\frac{1}{2}\rho v_{max}^2} = 0.0450 \left(\frac{\nu}{v_{max}a}\right)^{1/4}, \tag{8.58}$$

where ν is the kinematic viscosity and a is the pipe radius. Therefore, for the wall shear stress for the boundary-layer case, assume essentially the same law, namely:

$$\frac{\tau_w}{\frac{1}{2}\rho v_{x\infty}^2} = 0.0450 \left(\frac{\nu}{v_{x\infty}\delta}\right)^{1/4}. \tag{8.59}$$

From Eqns. (8.55), (8.56), and (8.59), there results:

$$0.0225 \left(\frac{\nu}{v_{x\infty}\delta}\right)^{1/4} = \frac{7}{72}\frac{d\delta}{dx}, \tag{8.60}$$

which, upon integration, gives:

$$\frac{\delta}{x} = \frac{0.376}{\text{Re}_x^{1/5}}, \qquad \text{where} \qquad \text{Re}_x = \frac{v_{x\infty}x}{\nu}, \tag{8.61}$$

with a corresponding local skin friction coefficient of:

$$c_f = \frac{\tau_w}{\frac{1}{2}\rho v_{x\infty}^2} = \frac{0.0576}{\text{Re}_x^{1/5}}. \tag{8.62}$$

Example 8.2—Laminar and Turbulent Boundary Layers Compared

Compare the relative thicknesses and drag coefficients of laminar and turbulent boundary layers at the transition Reynolds number, $\text{Re}_x = 3.2 \times 10^5$.

Solution

1. *Thicknesses.* The relevant equations for the ratio δ/x and their values at the indicated Reynolds number are:

$$\text{Laminar}: \quad \frac{4.79}{\sqrt{\text{Re}_x}} = 0.00847; \qquad \text{Turbulent}: \quad \frac{0.376}{\text{Re}_x^{1/5}} = 0.02980, \qquad \text{(E8.2.1)}$$

so that the turbulent boundary layer is about 3.52 times thicker than the laminar boundary layer at the transition.

2. *Drag coefficient.* The relevant equations for c_f and their values at the indicated Reynolds number are:

$$\text{Laminar}: \quad \frac{0.656}{\sqrt{\text{Re}_x}} = 0.00116; \qquad \text{Turbulent}: \quad \frac{0.0576}{\text{Re}_x^{1/5}} = 0.00456, \qquad \text{(E8.2.2)}$$

so that the turbulent boundary layer has a drag coefficient about 3.93 times that of the laminar boundary layer at the transition.

However, note that turbulence in the wake of a *nonstreamlined* object (see Section 8.7) can postpone *separation* of the boundary layer and lead to a *reduced* drag.   □

8.6   Dimensional Analysis of the Boundary-Layer Problem

Much information relating to the Blasius solution can be obtained by an intriguing dimensional analysis of the governing differential equations and boundary conditions.[3] From Eqns. (8.38) and (8.39):

x Momentum:

$$v_x \frac{\partial v_x}{\partial x} + v_y \frac{\partial v_x}{\partial y} = \nu \frac{\partial^2 v_x}{\partial y^2}. \tag{8.63}$$

Continuity:

$$\frac{\partial v_x}{\partial x} + \frac{\partial v_y}{\partial y} = 0. \tag{8.64}$$

Boundary conditions:

$$y = 0: \qquad v_x = v_y = 0, \tag{8.65}$$

$$y = \infty: \qquad v_x = v_{x\infty}. \tag{8.66}$$

First, introduce dimensionless coordinates and velocities as follows:

$$X = \frac{x}{x_r}, \quad Y = \frac{y}{y_r}, \quad U = \frac{v_x}{v_{xr}}, \quad V = \frac{v_y}{v_{yr}}. \tag{8.67}$$

Here, x_r and y_r are reference distances, and v_{xr} and v_{yr} are reference velocities. At this stage it is important *not* to equate x_r and y_r, nor v_{xr} and v_{yr}, as this may limit our options by being too restrictive. Note that we also resist for the present the temptation to equate either v_{xr} or v_{yr} to the mainstream velocity, $v_{x\infty}$.

In terms of the dimensionless variables, the governing equations become:

x Momentum:

$$U \frac{\partial U}{\partial X} + \frac{v_{yr} x_r}{v_{xr} y_r} V \frac{\partial U}{\partial Y} = \frac{\nu x_r}{v_{xr} y_r^2} \frac{\partial^2 U}{\partial Y^2}. \tag{8.68}$$

Continuity:

$$\frac{\partial U}{\partial X} + \frac{v_{yr} x_r}{v_{xr} y_r} \frac{\partial V}{\partial Y} = 0. \tag{8.69}$$

Boundary conditions:

$$Y = 0: \qquad U = V = 0, \tag{8.70}$$

$$Y = \infty: \qquad U = \frac{v_{x\infty}}{v_{xr}}. \tag{8.71}$$

---

[3]   A good summary of the approach is given by J.D. Hellums and S.W. Churchill, "Simplification of the mathematical description of boundary and initial value problems," *American Institute of Chemical Engineers Journal*, Vol. 10, No. 1, pp. 110–114 (1964).

Since these equations completely specify the problem, the dependent variables U and V must be functions of the independent variables X and Y *and* of the dimensionless groups that appear in the equations:

$$U = f\left(X,\ Y,\ \frac{v_{yr}x_r}{v_{xr}y_r},\ \frac{\nu x_r}{v_{xr}y_r^2},\ \frac{v_{x\infty}}{v_{xr}}\right),\tag{8.72}$$

$$V = g\left(X,\ Y,\ \frac{v_{yr}x_r}{v_{xr}y_r},\ \frac{\nu x_r}{v_{xr}y_r^2},\ \frac{v_{x\infty}}{v_{xr}}\right).\tag{8.73}$$

Here, f and g are two functions, unknown as yet.

Since the reference quantities are arbitrary, we may equate each dimensionless group to a constant, taken as unity for simplicity:

$$\frac{v_{x\infty}}{v_{xr}} = 1,\quad \frac{\nu x_r}{v_{xr}y_r^2} = 1,\quad \frac{v_{yr}x_r}{v_{xr}y_r} = 1.\tag{8.74}$$

Equation (8.74) gives the following expressions for the *three* reference quantities v_{xr}, y_r, and v_{yr}, in terms of the *fourth* reference quantity, x_r:

$$v_{xr} = v_{x\infty},\quad y_r = \sqrt{\frac{\nu x_r}{v_{x\infty}}},\quad v_{yr} = \sqrt{\frac{\nu v_{x\infty}}{x_r}}.\tag{8.75}$$

The functional relationship for the x-velocity component now becomes:

$$U = \frac{v_x}{v_{x\infty}} = f\left(\frac{x}{x_r},\ \frac{y}{y_r}\right) = f\left(\frac{x}{x_r},\ y\sqrt{\frac{v_{x\infty}}{\nu x_r}}\right).\tag{8.76}$$

Since x_r did not appear in the problem statement, it must disappear from Eqn. (8.76). Hence, f *must* be some function of the first group divided by the square of the second group, resulting in:

$$\frac{v_x}{v_{x\infty}} = f\left(\frac{\nu x}{y^2 v_{x\infty}}\right) = f\left(y\sqrt{\frac{v_{x\infty}}{\nu x}}\right).\tag{8.77}$$

Strictly speaking, the second function in Eqn. (8.77) should be distinguished from the first, but since both of them are unknown, we have used the common designation of "f." In proceeding from Eqn. (8.76) to (8.77), it looks as though we have merely set $x_r = x$, but this is *not* the case.

Similarly, the functional relationship for the y velocity component is:

$$V = \frac{v_y}{v_{yr}} = g\left(\frac{x}{x_r},\ \frac{y}{y_r}\right),\tag{8.78}$$

$$v_y\sqrt{\frac{x_r}{\nu v_{x\infty}}} = g\left(\frac{x}{x_r},\ y\sqrt{\frac{v_{x\infty}}{\nu x_r}}\right) = g\left(\frac{x}{x_r},\ y\sqrt{\frac{v_{x\infty}}{\nu x}}\right).\tag{8.79}$$

Again, since x_r did not appear in the problem definition, it *must* cancel out from both sides of Eqn. (8.79), whose form is therefore:

$$v_y \sqrt{\frac{x_r}{\nu v_{x\infty}}} = \sqrt{\frac{x_r}{x}} \, g\left(y\sqrt{\frac{v_{x\infty}}{\nu x}}\right), \tag{8.80}$$

so that:

$$v_y \sqrt{\frac{x}{\nu v_{x\infty}}} = g\left(y\sqrt{\frac{v_{x\infty}}{\nu x}}\right). \tag{8.81}$$

Note the following important points from the above development:

1. Equations (8.77) and (8.81) indicate how experimental data may be correlated most economically—that is, by constructing plots of:

$$\frac{v_x}{v_{x\infty}} \quad \text{vs.} \quad y\sqrt{\frac{v_{x\infty}}{\nu x}}, \quad \text{and} \quad v_y\sqrt{\frac{x}{\nu v_{x\infty}}} \quad \text{vs.} \quad y\sqrt{\frac{v_{x\infty}}{\nu x}}. \tag{8.82}$$

2. The reference distances and velocities were *not* the same, leading to a more precise result than if we had insisted on $x_r = y_r = \nu/v_{x\infty}$ and $v_{xr} = v_{yr} = v_{x\infty}$ at the outset.

3. The analysis clearly suggests the introduction of a new space variable,

$$\zeta = y\sqrt{\frac{v_{x\infty}}{\nu x}}, \tag{8.83}$$

and indeed this was employed in Blasius's solution.

The expression for the wall shear stress is:

$$\tau_w = \mu\left(\frac{\partial v_x}{\partial y}\right)_{y=0} = \mu\frac{\partial}{\partial y}\left[v_{x\infty} f\left(y\sqrt{\frac{v_{x\infty}}{\nu x}}\right)\right]_{y=0}$$

$$= \mu v_{x\infty}\sqrt{\frac{v_{x\infty}}{\nu x}} f'(0),$$

which gives, in dimensionless form:

$$\frac{\tau_w}{\rho v_{x\infty}^2} = \sqrt{\frac{\mu}{\rho v_{x\infty} x}} f'(0) = \mathrm{Re}_x^{-1/2} f'(0) = c\mathrm{Re}_x^{-1/2}. \tag{8.84}$$

Observe that apart from the single unknown coefficient c, dimensional analysis has predicted a form that is in complete agreement with Eqn. (8.54) from the Blasius solution. In principle, a single precise experiment then gives the solution, $c \doteq 0.332$.

8.7 Boundary-Layer Separation

Previous sections have discussed both laminar and turbulent boundary layers on a flat plate, in which there was a constant mainstream velocity and hence a constant pressure. Now consider Eqn. (8.32) applied close to the wall, where both v_x and v_y are small, so that:

$$\frac{\partial p}{\partial x} \doteq \mu \frac{\partial^2 v_x}{\partial y^2} = \mu \frac{\partial}{\partial y}\left(\frac{\partial v_x}{\partial y}\right). \tag{8.85}$$

Since $\partial p/\partial x = 0$, the second derivative of the velocity is also zero, so the velocity gradient $\partial v_x/\partial y$ is a constant near the wall, as in Fig. 8.6(a).

Fig. 8.6 *Effect of various pressure gradients on the velocity profile in the boundary layer.*

For more complex situations, the flow in the mainstream may not be uniform, leading to the following possibilities:

(a) A negative pressure gradient, $\partial p/\partial x < 0$, in which case the velocity gradient near the wall *decreases* as we proceed out from the wall. Since the velocity must match the mainstream value at the edge of the boundary layer, this implies that the velocity gradient is *steeper* at the wall, as in Fig. 8.6(b).

(b) A positive pressure gradient, $\partial p/\partial x > 0$, in which case the velocity gradient near the wall *increases* as we proceed out from the wall. Since the velocity must match the mainstream value at the edge of the boundary layer, this implies that the velocity gradient is *less steep* at the wall, as in Fig. 8.6(c). In a more extreme case, in which $\partial p/\partial x$ is more highly positive, the gradient can actually become zero (or even slightly negative) at the wall, as in Fig. 8.6(d).

The modification of the velocity profile by the pressure gradient is similar to that shown in Fig. 8.11 for the flow of lubricant in a bearing.

Consider now the flow past an object that is not particularly well streamlined, such as the sphere shown in Fig. 8.7(a). On the upstream hemisphere, the streamlines are pushed closer together, so in the direction of flow the velocity increases and the pressure decreases—a relatively stable situation. On the downstream hemisphere, the opposite occurs, and the pressure increases in the direction of flow, leading to the situations shown in Fig. 8.6(c) and (d).

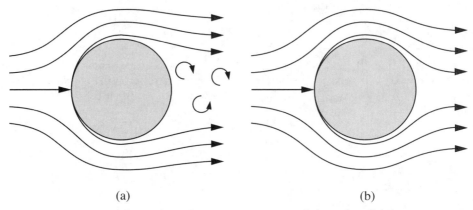

(a)   (b)

Fig. 8.7 Boundary-layer separation: (a) early; (b) late.

Under such circumstances, the boundary layer no longer "hugs" the surface but *separates* from it, leading to considerable turbulence and loss of pressure in the wake of the sphere, causing a substantial increase in the drag. It is clearly desirable in most cases to delay the onset of separation, and indeed this can be done and often totally prevented for flow past a well streamlined surface such as an airplane wing. Boundary-layer separation was first examined extensively by Ludwig Prandtl.

Another way of delaying the separation is somewhat paradoxical—to make sure that the boundary layer is *turbulent*, which can often be achieved by roughening part or all of the surface, as in the dimples on a golf ball.[4] Under these turbulent circumstances, momentum in the form of velocity is transmitted much more rapidly from the mainstream to the boundary layer, thus helping to insure that the velocity near the boundary does not stagnate, but continues with a positive value. The onset of separation is therefore postponed, as in Fig. 8.7(b).

The sudden "dip" in the drag coefficient for a sphere at a high Reynolds number, as in Fig. 4.7, can now be explained. It is because the flow in the boundary layer has become turbulent and the previous high degree of separation of the laminar boundary layer no longer exists. Such events—in conjunction with spin and seam orientation—are also important in determining the trajectory of a baseball.[5]

---

[4] Do you know how many dimples there are on a golf ball? According to Ian Stewart in "Mathematical Recreations" on page 96 of *Scientific American*, February 1997, considerations of symmetry dictate that the most frequently occurring numbers of dimples are 252, 286, 332, 336, 360, 384, 392, 410, 416, 420, 422, 432, 440, 480, 492, and 500.

[5] A.T. Bahill, D.G. Baldwin, and J. Venkateswaran, "Predicting a baseball's path," *Scientific American*, pp. 218–225 (May/June 2005).

Example 8.3—Boundary-Layer Flow Between Parallel Plates (COMSOL)   **435**

Prandtl, Ludwig, born 1875 in Freising, Germany, died 1953 in Göttingen, Germany. From 1894–1900, Prandtl studied engineering at the Technische Hochschule at Munich, where he wrote his doctoral dissertation on the lateral instability of beams in bending. He became a professor at the Hannover Technische Hochschule in 1901. His interests soon turned to fluid mechanics. When asked to improve a suction device for removing shavings in a factory, he soon found that a diverging pipe led to a separation of the mainstream from the wall. Thus, in 1904, Prandtl's famous boundary-layer theory was enunciated. He soon received a chair at Göttingen University, where he did research into aerodynamics and supersonic flow and developed the first German wind tunnel. Prandtl excelled at working individually with students but was a relatively poor lecturer. His most famous student was Theodore von Kármán, who discovered the importance of the "Kármán street" of alternating vortices shed from airfoils, increasing the drag beyond what could be accounted for solely from boundary-layer theory. Experiments by Prandtl and theory by Kármán led to further insights into turbulent mixing-length theory, published by Prandtl in 1933. Prandtl was largely bypassed during the German scientific effort of World War II, much of which was devoted to rocketry.

Source: *Dictionary of Scientific Biography*, Charles Scribner's Sons, New York, 1975.

Example 8.3—Boundary-Layer Flow Between Parallel Plates (COMSOL Library)

A fluid of density 1,000 and viscosity 0.01 flows freely in the y direction with an initially uniform velocity $v_0 = 0.02$. After a distance of $y = 0.01$ it encounters flat plates of length 0.04 located at $x = \pm 0.01$. The exit pressure from the plates is $p = 0$. The corresponding boundary conditions are shown in Fig. E8.3.1. All units are mutually consistent.

A. Initial check. This problem is entitled "Parallel Plates" in the COMSOL library. Open it, and first check:

1. The values for the viscosity and the density.
2. The six boundary conditions.

B. Overall plots. Print the following plots for the whole region in Fig. E8.3.1:

3. The streamlines.
4. Arrows showing the velocity vectors.
5. A two-dimensional contour plot of the velocity.

C. Cross-section plots. For $y = 0.015$, plot and print the following at section A–A between the two plates, and comment briefly on each:

6. The y velocity.

7. The x velocity.

8. The vorticity.

D. Values at point P. Finally, comment on the relative values of the pressure and y velocity v_y at $x = 0, y = 0.01$, marked as P in Fig. E8.3.1., and at the exit $(x = 0, y = 0.05)$.

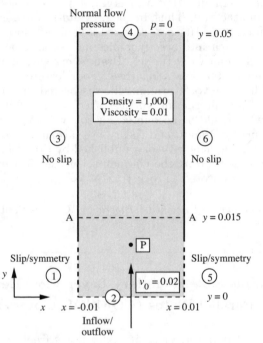

Fig. E8.3.1   Boundary conditions and other information for flow between two flat plates.

Solution

A. Initial check. In COMSOL, choose the sequence **Model Library**, **Chemical Engineering Module**, **Momentum Transport**, and **parallel plates**, which will show the velocity field in color.

Start by selecting **Constants** under the **Options** pulldown menu, and observe that `rho`, `eta`, and `v0` have been assigned the values 1,000, 0.01, and 0.02, respectively. Then choose **Subdomain Settings** under the **Physics** pulldown menu and highlight the solitary subdomain 1. The density and viscosity are clearly set to the constants `rho` and `eta`.

Likewise, choose **Boundary Settings** under the **Physics** pulldown menu and highlight each of the six boundary segments in turn. You will see that each segment corresponds to one the types indicated in Fig. E8.3.1, and that the corresponding boundary appears in red.

Example 8.3—Boundary-Layer Flow Between Parallel Plates (COMSOL)   **437**

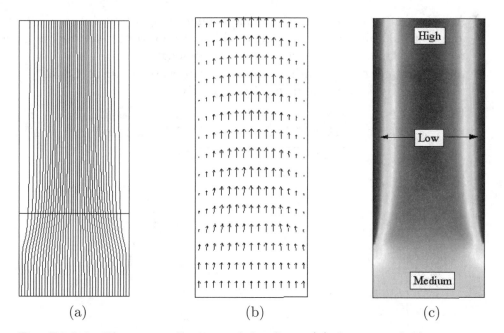

(a)   (b)   (c)

Fig. E8.3.2   Three overall views of the flow: (a) the upward moving stream-lines, for which 40 streamline "start points" were specified under plot param-eters; the horizontal line is the section A–A *in Fig. E8.3.1 and appears au-tomatically only after the cross-section plots have been done; (b) the velocity vectors; and (c) a two-dimensional plot (the original being in color) showing the velocities, the approximate magnitudes of which have been added.*

B. Overall plots. By selecting appropriate choices in **Plot Parameters** under **Postprocessing** (Streamline, Arrow, and Surface), the overall picture of the flow is displayed in Fig. E8.3.2. After the initial small "free-flow" entrance region, the formation of boundary layers on the two plates is very evident, either by examining the movement of the streamlines away from the plates or by noting the reduction in velocity adjacent to the plates. From continuity, this velocity reduction also causes an acceleration in the middle of the stream.

C. Cross-section plots. The three requested plots at $y = 0.015$ are obtained by selecting **Line/extrusion** under **Cross-Section Plot Parameters** and entering x0 = -0.01, x1 = 0.01, y0 = 0.015, y1 = 0.015 in the **Cross-section line data** window, then asking in sequence under **y-axis data** for plots of the y velocity, x velocity, and vorticity. The results are shown in Figs. E8.3.3–E8.3.5, for which we make the following comments:

1. The velocity v_y in the primary flow direction is still essentially flat in the central region, falling to zero as the wall is reached.

(a) At values of x that are clearly in the *boundary layer*, such as $x = -0.008$ and $x = 0.008$ in Fig. E8.3.3, v_y would continue to decrease in the y direction, so that $\partial v_y / \partial y$ is negative at these locations.

(b) In the *central core* v_y continues to accelerate as seen in Fig. E8.3.2(c), so that the reverse occurs and $\partial v_y / \partial y$ is positive.

Fig. E8.3.3   *The upwards y velocity between the plates.*

2. Following on Item 1(a), the continuity equation, $\partial v_x / \partial x + \partial v_y / \partial y = 0$, mandates that $\partial v_x / \partial x$ must be positive in the boundary layer, and this is evidenced in Fig. E8.3.4 at, for example, $x = -0.008$ and $x = 0.008$. Conversely, from Item 1(b), $\partial v_x / \partial x$ must be negative in the central core, and this is also clearly seen.

3. The vorticity is $\partial v_y / \partial x - \partial v_x / \partial y \doteq \partial v_y / \partial x$ (because $v_y \gg v_x$), leading to a positive vorticity (counterclockwise rotation) near the left-hand plate and a negative vorticity (clockwise rotation) near the right-hand plate. In the central region the vorticity is essentially zero because v_y is virtually constant there.

D. Values at point P. Finally, the pressure p and velocity v_y at the entrance point P are obtained by first choosing the sequence **Plot Parameters**, **Surface**, and **Predefined quantities**, and then requesting pressure and y velocity plots. Then, clicking on the point $(x = 0, y = 0.01)$ shows the values $p = 0.334018$ and $v_y = 0.022147$ in the lower left-hand corner. Similarly, the exit centerline values are $p = 0$ (given) and $v_y = 0.029838$. As expected at the centerline, there is an acceleration from P to the exit and a corresponding reduction in pressure.

Example 8.3—Boundary-Layer Flow Between Parallel Plates (COMSOL)   **439**

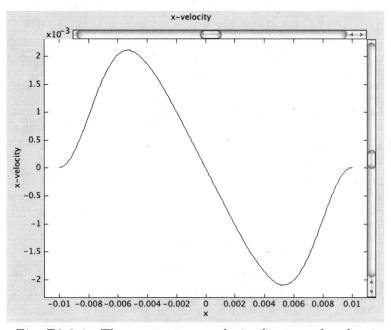

Fig. E8.3.4   The transverse x velocity between the plates.

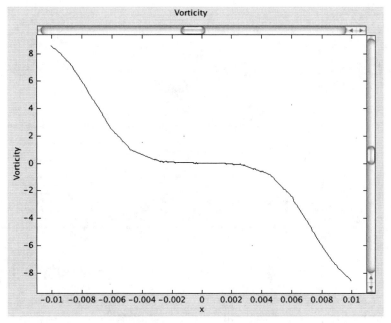

Fig. E8.3.5   The variation of vorticity between the plates.

Example 8.4—Entrance Region for
Laminar Flow Between Flat Plates

In Example 6.1, involving laminar flow between parallel plates, only the velocity in the axial direction, v_x, was considered to be nonzero. In reality, a flat profile at the entrance will gradually be transformed into a final parabolic shape by the viscous action of the walls. Fig. E8.4.1 shows three representative stages of the progression, in which a boundary layer builds up from a thickness of $\delta = 0$ at the entrance to a final value of $y = d$; the problem is to determine the necessary distance $x = L$. Note that in order to satisfy a mass balance, the "mainstream" velocity V is *not* constant, but continues to increase downstream along the path A–B–C. The solution should be in the form $L/d = c\rho \overline{V} d/\mu$, where the value of c is to be determined.

Solution

The analysis concentrates on half of the region, between the lower wall and the centerline, since the other half can be obtained from symmetry. All quantities are on the basis of unit depth normal to the plane of the diagram. First, assume a reasonable velocity profile in the boundary layer—one that gives zero velocity at the wall and matches both the mainstream velocity (V) and its slope (zero) at $y = \delta$:

$$v_x = V\left(2\frac{y}{\delta} - \frac{y^2}{\delta^2}\right). \tag{E8.4.1}$$

This particular choice has the further advantage that it exactly reproduces the well-known parabola when point C is reached.

Fig. E8.4.1   Development of velocity profile between flat plates.

On the basis of this velocity profile, the following useful quantities can be determined, in which it is convenient to define a fraction α:

$$\alpha = \frac{\delta}{d}. \tag{E8.4.2}$$

Example 8.4—Entrance Region for Laminar Flow Between Flat Plates   **441**

Mass flux

$$m = \underbrace{\rho V (d - \delta)}_{\text{Mainstream}} + \underbrace{\rho \int_0^\delta v_x \, dy}_{\text{Boundary layer}} = \rho V (d - \delta) + \frac{2}{3} \rho V \delta = \rho V d \left(1 - \frac{\alpha}{3}\right). \quad \text{(E8.4.3)}$$

Momentum flux

$$\dot{\mathcal{M}} = \underbrace{\rho V^2 (d - \delta)}_{\text{Mainstream}} + \underbrace{\rho \int_0^\delta v_x^2 \, dy}_{\text{Boundary layer}} = \rho V^2 (d - \delta) + \frac{8}{15} \rho V^2 \delta = \rho V^2 d \left(1 - \frac{7}{15} \alpha\right).$$

$$\text{(E8.4.4)}$$

Wall shear stress

$$\tau_w = \mu \left(\frac{dv_x}{dy}\right)_{y=0} = \frac{2\mu V}{\delta} = \frac{2\mu V}{\alpha d}. \quad \text{(E8.4.5)}$$

Mean/mainstream velocity relation

$$m = \rho \overline{V} d = \rho V d \left(1 - \frac{\alpha}{3}\right), \quad \text{or} \quad \overline{V} = V \left(1 - \frac{\alpha}{3}\right) \quad \text{or} \quad V = \frac{3\overline{V}}{3 - \alpha}. \quad \text{(E8.4.6)}$$

Bernoulli's equation in the mainstream

$$\frac{V^2}{2} + \frac{p}{\rho} = c, \quad \text{or} \quad \frac{dp}{dx} = -\rho V \frac{dV}{dx}. \quad \text{(E8.4.7)}$$

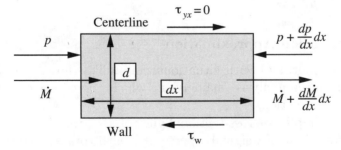

Fig. E8.4.2   Momentum balance on rectangular element of fluid.

Now perform a momentum balance in the x direction on the rectangular element shown in Fig. E8.4.2, situated between the wall and centerline, noting that at the centerline there is zero velocity gradient and hence zero shear stress:

$$\dot{\mathcal{M}} + pd - \tau_w dx - \left(\dot{\mathcal{M}} + \frac{d\dot{\mathcal{M}}}{dx} dx\right) - \left(p + \frac{dp}{dx} dx\right) d = 0, \quad \text{(E8.4.8)}$$

or:

$$\tau_w + \frac{d\dot{M}}{dx} + \frac{dp}{dx}d = 0. \tag{E8.4.9}$$

Differentiation of Eqn. (E8.4.4) with respect to x gives:

$$\frac{d\dot{M}}{dx} = \rho d \left[2V \frac{dV}{dx} \left(1 - \frac{7}{15}\alpha \right) - \frac{7}{15}V^2 \frac{d\alpha}{dx} \right]. \tag{E8.4.10}$$

After making the appropriate substitutions, Eqn. (E8.4.9) yields an ordinary differential equation for α as a function of x:

$$\frac{10\mu}{\rho \overline{V} d^2} = \frac{\alpha(6 + 7\alpha)}{(3 - \alpha)^2} \frac{d\alpha}{dx}. \tag{E8.4.11}$$

Separation of variables and integration between the channel inlet and the location at which the boundary layer reaches out to the centerline gives:

$$\frac{10\mu}{\rho \overline{V} d^2} \int_0^L dx = \int_0^1 \frac{\alpha(6 + 7\alpha)\, d\alpha}{(3 - \alpha)^2} \doteq 1.048. \tag{E8.4.12}$$

That is, the dimensionless entrance length is directly proportional to the Reynolds number (based in this case on half the separation between the plates):

$$\frac{L}{d} = 0.1048 \frac{\rho \overline{V} d}{\mu}. \tag{E8.4.13}$$

For flow in a pipe of *diameter* D, the corresponding result is $L/D \doteq 0.061\,\mathrm{Re}$, in which $\mathrm{Re} = \rho \overline{V} D/\mu$. □

8.8 The Lubrication Approximation

This chapter continues with situations not unlike those encountered in the earlier boundary-layer analysis—namely, in which there is a dominant velocity component in one direction.

Consider the simplified cross-sectional view of a *journal bearing*, shown in Fig. 8.8(a). The heavy shaft of weight W per unit length rotates counterclockwise inside the housing, the two being separated by a thin lubricant film of viscous oil. Although the pressure at the two extremities of the oil film is atmospheric ($p = 0$, say), the following analysis will show that the rotation of the shaft induces a significant pressure increase between these two points—enough to support the weight of the shaft. Observe that when equilibrium is reached, the shaft and housing are *not* concentric, so that the oil film, whose thickness is greatly exaggerated in Fig. 8.8(a), becomes thinner in the direction of rotation. The shaft becomes off-center automatically and to a very small degree.

Because the oil film is very thin, the analysis can justifiably be expedited by "unwinding" the film and considering it to be wedge-shaped, contained between an upper surface (the shaft) moving with velocity V and a lower stationary surface (the housing), an enlarged portion of which is shown in Fig. 8.8(b). Observe that the oil tends to be squeezed into an ever-narrowing space, and that this is the basic cause of the increased pressure.

(a)   (b)

Fig. 8.8   (a) Journal bearing, with exaggerated thickness of the oil film, and (b) magnified view of film.

A similar situation—that of a *thrust* bearing—is shown in Fig. 8.9. In this case, the *horizontal* thrust W of an axle is resisted by a series of wedge-shaped segments, arranged in a circle on a fixed end plate, adjacent to which is a disk attached to the rotating axle. The fluid between the disk and segments does not have to be oil—it will probably be air in the majority of cases.

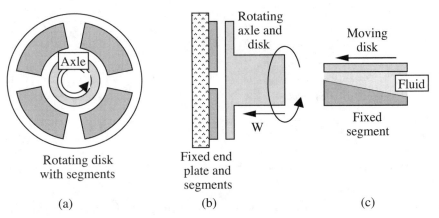

(a)   (b)   (c)

Fig. 8.9   Details of thrust bearing: (a) head-on view of end plate with segments; (b) side view of axial section; (c) the motion of the disk relative to one of the fixed segments on the end plate.

To proceed with the analysis of the journal bearing, make the following simplifying assumptions, all of which can be justified by arguments similar to those made when analyzing the formation of a boundary layer on a flat plate:

1. There is a dominant velocity component in one direction—x for example.
2. Pressure is only a function of the x coordinate—that in the primary flow direction, and not of the transverse coordinate y. That is, $p = p(x)$ and $\partial p/\partial y = 0$.
3. Inertial and gravitational terms are negligible in comparison with the two dominant effects of pressure and viscous forces.
4. There is no flow in the z direction, normal to the plane of the diagram. This is a poor assumption in some cases, not considered here, in which there is a continuous leakage of makeup lubricant along the direction of the shaft (with a corresponding injection of lubricant in the vicinity of the bearing surface).

Basic equations for flow in a bearing.   Item 1 above indicates that only the x momentum balance is needed from the full Navier-Stokes equations:

$$\rho \left(\frac{\partial v_x}{\partial t} + v_x \frac{\partial v_x}{\partial x} + v_y \frac{\partial v_x}{\partial y} + v_z \frac{\partial v_x}{\partial z} \right) = -\frac{\partial p}{\partial x} + \mu \left(\frac{\partial^2 v_x}{\partial x^2} + \frac{\partial^2 v_x}{\partial y^2} + \frac{\partial^2 v_x}{\partial z^2} \right) + \rho g_x.$$

With the simplifications just listed—known as the *lubrication approximation*—the momentum balance reduces to:

$$\frac{dp}{dx} = \mu \frac{\partial^2 v_x}{\partial y^2}. \tag{8.86}$$

At the boundaries, the fluid has zero velocity at the fixed lower surface, and moves with velocity $V \cos \theta \doteq V$ (θ is quite small) at the moving upper surface:

$$\begin{aligned} y = 0 : \quad & v_x = 0, \\ y = h : \quad & v_x = V. \end{aligned} \tag{8.87}$$

Double integration of Eqn. (8.86), incorporating these two boundary conditions, leads to the following expression for the velocity profile at any section in the x direction:

$$v_x = \frac{yV}{h} + \frac{1}{2\mu} \frac{dp}{dx} y(y - h). \tag{8.88}$$

Observe the interpretations of the two terms appearing on the right-hand side of Eqn. (8.88):

1. The first term represents *Couette* flow, in which there is a linear increase in v_x, from zero at the stationary lower surface to V at the moving upper surface.
2. The second term corresponds to pressure-driven *Poiseuille* flow, in which the velocity is zero at the two surfaces, varying parabolically between them.

The problem is not yet solved, because dp/dx is unknown. Therefore, now focus attention on determining how the pressure—and hence its gradient—varies over the length of the bearing. To start, the volumetric flow rate of lubricant along the bearing is obtained by integrating the velocity across the film:

$$Q = \int_0^h v_x \, dy = \int_0^h \left[\frac{yV}{h} + \frac{1}{2\mu} \frac{dp}{dx} y(y - h) \right] dy = \frac{Vh}{2} - \frac{h^3}{12\mu} \frac{dp}{dx}. \qquad (8.89)$$

Observe again the Couette and pressure-induced contributions to the flow rate. Unless there is leakage of lubricant normal to the plane of Fig. 8.8 (as could occur in a truly two-dimensional bearing), Q must remain constant. Thus, setting $dQ/dx = 0$ immediately leads to the differential equation that governs the variation of pressure along the bearing:

$$\frac{d}{dx} \frac{h^3}{\mu} \frac{dp}{dx} = 6V \frac{dh}{dx}, \qquad (8.90)$$

which is an example of *Poisson's equation*, and in which the expression on the right-hand side is effectively a "generation" term, which serves to boost the pressure. Note that the relative velocity of the two surfaces (V) and the steepness of their inclination (dh/dx) are important factors in causing the pressure increase.

Pressure variation in the bearing.   In a bearing, viscous friction causes temperature variations, so the viscosity depends on position. In such cases, Eqn. (8.90) can always be solved numerically for the pressure. However, to proceed further analytically at an introductory level it is necessary to make the following simplifying assumptions:

1. The fluid viscosity μ is constant.
2. The inclination dh/dx is also constant.
3. Eventually, a mean film thickness, h_m, may be introduced for further simplification.

After defining a positive quantity:

$$\alpha = -6\mu V \frac{dh}{dx}, \qquad (8.91)$$

(note that dh/dx is negative), the pressure equation becomes:

$$\frac{d}{dx} h^3 \frac{dp}{dx} = -\alpha, \qquad (8.92)$$

which, when integrated once, gives:

$$\frac{dp}{dx} = -\frac{\alpha x}{h^3} + \frac{c_1}{h^3}, \qquad (8.93)$$

in which c_1 is an integration constant.

A precise integration of Eqn. (8.93)—see Problem 8.10—demands that the exact expression for the film thickness, $h = h(x)$, be used. An approximate answer may be obtained by assuming that h maintains its mean value, h_m, in which case we obtain:

$$p = -\frac{\alpha x^2}{2h_m^3} + \frac{c_1 x}{h_m^3} + c_2. \tag{8.94}$$

The boundary conditions on the pressure are that it is zero at the beginning $(x = 0)$ and end $(x = L)$ of the film, giving:

$$c_2 = 0, \qquad c_1 = \frac{1}{2}\alpha L. \tag{8.95}$$

From Eqns. (8.94) and (8.95), the pressure obeys a *parabolic* distribution between the two end points:

$$p = \frac{\alpha}{2h_m^3}x(L - x), \tag{8.96}$$

which is also sketched in Fig. 8.10 as curve A, with a maximum p at the center.

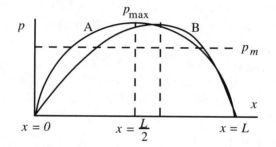

Fig. 8.10 Pressure variation along the lubricant film. Curves A and B are for the simplified (Eqn. (8.96)) and exact (Problem 8.10) methods, respectively.

The *mean* pressure in the lubricant film is readily obtained by integration as:

$$p_m = \frac{1}{L}\int_0^L p\,dx = \frac{\alpha L^2}{12h_m^3} = \frac{2}{3}p_{max}, \tag{8.97}$$

in which p_{max} is the maximum pressure, which occurs (in our simplified treatment) at $x = L/2$. In reality, if the assumption of $h \doteq h_m$ is *not* made, the point of maximum pressure will be shifted somewhat to the right of the midpoint.

Recalling the original curved nature of the surfaces, the total weight per unit length of the shaft that can be supported by the pressure generated in the lubricant film is:

$$W = \int_0^L p(x)\cos\phi\,dx, \tag{8.98}$$

in which ϕ is the angle between the vertical and the line drawn from the center of the bearing to the point distant by x from the thicker end of the film. Insertion of representative values (including a thin film thickness) will show that W can become quite appreciable.

Since the distribution of pressure is now known approximately, its gradient is:

$$\frac{dp}{dx} = \frac{\alpha}{h_m^3}\left(\frac{L}{2} - x\right), \tag{8.99}$$

which can be substituted into the original expression, Eqn. (8.88), for the velocity profile, giving:

$$v_x = \frac{yV}{h} + \underbrace{\frac{1}{2\mu}\frac{\alpha}{h_m^3}\left(\frac{L}{2} - x\right)}_{\text{Positive}} \underbrace{y(y-h)}_{\text{Negative}}. \tag{8.100}$$

Eqn. (8.100) is plotted in Fig. 8.11 at three representative locations along the bearing.

Fig. 8.11   *Lubricant velocity profile along the bearing.*

Along the moving surface the velocity is, of course, V. Observe that at the midpoint the velocity profile is purely Couette in nature, since the pressure gradient dp/dx is zero at this location. However, Fig. 8.10 shows that there are significant pressure gradients at $x = 0$ and $x = L$, and these create additional Poiseuille-type components at these locations. At the "exit" ($x = L$), the velocity profile is enhanced beyond its Couette value, whereas at the "inlet" ($x = 0$) it is reduced, and there is indeed some backflow occurring. In practice, there will be lubricant injected into the film in order to make up for the amounts leaking out at the end locations. The modification of the velocity profile by the pressure gradient is similar to that shown in Fig. 8.6 for the separation of a boundary layer.

Example 8.5—Flow in a Lubricated Bearing (COMSOL)

Consider the situation in Fig. E8.5.1, in which one inclined plate (surface 3) moves at a velocity V relative to a stationary plate at $y = 0$, the intervening space being occupied by a Newtonian fluid of density 1,000 and viscosity 1.0. The velocity components of the moving plate are $u_0 = 0.05$ in the x direction and $v_0 = -0.01$ in the y direction. The other three boundary conditions (on surfaces 1, 2, and 4) are shown in Fig. E8.5.1. The two "ends," at $x = -1.4$ and $x = 1.4$, are exposed to a pressure of $p = 0$. All units are mutually consistent. In reality, a bearing would have a much smaller mean thickness/length ratio, but the present choice allows the main features to be displayed without "stretching" the y coordinate.

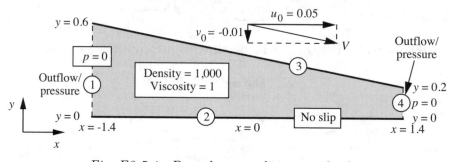

Fig. E8.5.1   *Boundary conditions and other*
information for flow in a lubricated bearing.

Solve for the streamlines, the pressure distribution, and the variation of pressure along the straight line between midpoints of the opposite "ends"—between points $(-1.4, 0.3)$ and $(1.4, 0.1)$.

Solution

1. In COMSOL, choose the sequence **New**, **Chemical Engineering Module**, **Momentum Balance**, and **Incompressible Navier-Stokes**.

2. Use the line tool to draw the trapezoid outlining the shaded region in Fig. E8.5.1, noting that the coordinates of the vertices are $(-1.4, 0)$, $(1.4, 0)$, $(1.4, 0.2)$, and $(-1.4, 0.6)$. Remember to terminate at the last point either by right-button click anywhere (PC) or by Control-click (Macintosh).

3. Under **Physics/Boundary Settings**, set the indicated boundary conditions on boundaries 1, 2, 3, and 4.

4. Under **Physics/Subdomain Settings**, indicate that $\rho = 1{,}000$ and $\eta = 1$.

5. Initialize and refine the mesh, and solve the problem.

6. Under **Postprocessing**, make appropriate choices in the **Plot Parameters** pulldown menu to generate the following:

(a) A picture that shows 20 streamlines.

Example 8.5—Flow in a Lubricated Bearing (COMSOL)   **449**

(b) A surface plot of the pressure distribution.

(c) A cross-section plot of the pressure between the points $(-1.4, 0.3)$ and $(1.4, 0.1)$. Here, you will need to choose the **Line/Extrusion** option in the **Cross-Section Plot Parameters** window.

Discussion of Results

Fig. E8.5.2   Streamlines: in the left-hand half the flow is clockwise, and in the right-hand half it is from left to right. The thin straight line joining points $(-1.4, 0.3)$ and $(1.4, 0.1)$ was added automatically when the pressure cross-plot requested in Item 6(c) above (see Fig. E8.5.4) was made.

Fig. E8.5.3   A two-dimensional representation (the original being in color) showing the pressure distribution, the approximate magnitudes of which have been added.

The COMSOL results are shown in Figs. E8.5.2–E8.5.4. The first of these displays the streamlines. In the upper part of the left-hand side, and for almost all of the right-hand side the motion is from left to right, driven by the moving upper plate. But much of the left-hand region is occupied by a clockwise-rotating vortex-type flow; in the bottom half of this vortex, the reason for the flow in the negative x direction is readily seen by examining Fig. E8.5.3, which shows the following important features:

(a) A pressure that essentially depends only on x and not on y—a fact that is at the heart of simplified treatments of boundary-layer and lubrication-type flows.

(b) A high-pressure zone centered at about $x = 0.75$. The maximum pressure is approximately $p = 2.2$.

Fig. E8.5.4 clearly shows the pressure variation between the midpoints of the two ends—between points $(-1.4, 0.3)$ and $(1.4, 0.1)$. The variation is broadly parabolic but with a peak that is noticeably skewed towards the right of the center.

Fig. E8.5.4   Cross-plot of the pressure between mid-points of the two ends, (−1.4, 0.3) and (1.4, 0.1).

8.9 Polymer Processing by Calendering

Calendering is the process of forming a continuous sheet by forcing a semi-molten material through one or more pairs of closely spaced counterrotating rolls. Calendering was first used—and is still used extensively—in the rubber industry. In the plastics industry, the most commonly used plastic that is calendered (about 80–85% of the total) is PVC, both soft (flexible PVC) and hard (rigid PVC). Other plastics that are calendered are mainly LDPE, HDPE, and PP.

Two main benefits can accrue to the calendered material:

1. It is formed to an accurate final thickness.
2. Depending on the nature of the surface of the rolls, a smooth, rough, or patterned surface is imparted to the finished sheet.

Referring to Fig. 8.12, heat-softened, but not fully molten, material is constantly fed into the *nip* of rolls 1 and 2. Complete melting and some mixing of additives (such as plasticizers, stabilizers, fillers, lubricants, and colorants) occurs in the first nip, where the stock also warms up. Shear effects also occur because roll 2 normally runs somewhat faster than roll 1—to encourage the band (the film produced) to stick to roll 2 so that it progresses down the rolls. The roll temperatures increase slightly from 1 to 3, but roll 4 is slightly cooler than roll 3 to encourage the film to strip easily off roll 4. The nip gaps get progressively smaller and that between rolls 3 and 4 should be almost the same as the final film thickness.

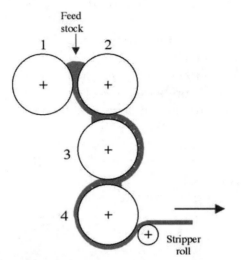

Fig. 8.12   An inverted "L" four-roll calender, courtesy John W. Ellis.[6]

The functions of the three nips are:

1. First nip (between rolls 1 and 2). The feed rate through the calender is determined here. This is also related to the material residence time, which is relatively short (good for PVC, which is easily degraded). Roll speeds obviously influence these factors.

2. The second nip (between rolls 2 and 3) functions as a kind of metering device. The melt gets spread along the rolls axially to its final width here, although most axial spreading should occur in the first nip.

3. The third nip (between rolls 3 and 4) is the final shaping step, analogous to a sheet extrusion die.

With its highly polished precision rolls and ancillary equipment, calendering is a very capital-intensive operation. Typical operating conditions are:

1. Roll diameters are generally in the range of 40–90 cm. Roll lengths are typically from 1–2.5 m, corresponding to a length-to-diameter ratio between 2 and 3.

2. Roll separation ("nip gap size") varies from one pair of rolls to the next but is adjustable up to about 5 mm. Production calenders turn out sheet with a thickness tolerance of ±0.005 mm over the width of the final sheet, whose thickness is usually 1 mm at most.

3. Roll speeds. Linear line speeds are generally in the range 30–120 m/min (100–400 ft/min) for production machines. The actual roll speeds are fairly low

---

[6] Mr. Ellis is a polymer testing and processing specialist at The Petroleum and Petrochemical College, Chulalongkorn University, Bangkok. We thank him for supplying all of the practical information on calendering in this section.

(e.g., 20 rpm). If, for example, the roll diameter is 70 cm and the rotational speed is 20 rpm, this works out to be 44 m/min linear line speed. Calendering typically involves shear rates from 1–100 s^{-1}.

During calendering, the melt needs to be fairly viscous, otherwise it can run all over the calender bowls. But conversely, too high a viscosity will create excessive roll bending due to the high pressure created between the rolls (up to 800 bar)! Typical viscosities would correspond to a MFI of 1 or 2, equivalent to a melt viscosity of between 10^3 and 10^4 N s/m^2 (10^4 to 10^5 poise).

An idealized representation of calendering is shown in Fig. 8.13, in which two counterrotating rollers with angular velocities ω, radii R, and closest separation $2H$ are employed for reducing the thickness of a sheet of material, ideally assumed to have a Newtonian viscosity μ, from an initial value of $2H_1$ to a final value $2H_2$.

Fig. 8.13 *Calendering operation (the sheet thickness and roller separation have been exaggerated).*

A key quantity in the analysis is the distance $h = h(x)$ of either roller from the centerline $y = 0$. To make reasonable headway without resorting entirely to numerical methods, it is necessary to assume that x/R is small, which is approximately true in the location of greatest interest (the "nip" region), where the separation of the rollers is small and the pressure gradients are the highest. Application of Pythagoras's theorem leads to:

$$h = R + H - \sqrt{R^2 - x^2} \doteq H + \frac{x^2}{2R} = H(1 + \alpha x^2), \quad \text{with } \alpha = \frac{1}{2HR}, \quad (8.101)$$

in which the approximation $\sqrt{1 - 2\varepsilon} \doteq 1 - \varepsilon$ has been made for small values of $\varepsilon = x/R$.

The situation and assumptions are very similar to those studied in the previous section, and the lubrication approximation again leads to pressure only being a function of x, together with the governing equation:

$$\frac{dp}{dx} = \mu \frac{\partial^2 v_x}{\partial y^2}. \tag{8.102}$$

The boundary conditions on the velocity v_x are that at the extremities of the gap it is approximately the same as the peripheral velocity of the rollers (at $y = h$), and that there is symmetry about the centerline of the sheet ($y = 0$):

$$y = h : v_x = R\omega, \qquad y = 0 : \frac{\partial v_x}{\partial y} = 0. \tag{8.103}$$

It is then easy to show that:

$$v_x = \frac{1}{2\mu} \frac{dp}{dx}(y^2 - h^2) + R\omega, \tag{8.104}$$

in which the pressure gradient is, for the present, unknown, and that the flow rate per unit depth, normal to the plane of Fig. 8.12, is:

$$Q = 2 \int_0^h v_x \, dy = 2h \left(R\omega - \frac{h^2}{3\mu} \frac{dp}{dx} \right), \tag{8.105}$$

which can be rearranged to give the pressure gradient as:

$$\frac{dp}{dx} = \frac{3\mu}{2h^3}(2R\omega h - Q). \tag{8.106}$$

Now focus on the point, $x = x_2$, where the calendered sheet leaves the rollers with a half thickness of $y = H_2$. From then on, the pressure is everywhere constant, since the sheet is exposed to the atmosphere. Thus, it is reasonable to suppose a flat velocity profile at the point of departure, the same as in the sheet after it has left the rollers. Thus, the flow rate is the peripheral velocity times the thickness:

$$Q = 2R\omega H_2, \tag{8.107}$$

which amounts to saying that the pressure gradient dp/dx is zero at this stage. Eqn. (8.106) then becomes, after invoking Eqn. (8.101):

$$\frac{dp}{dx} = \frac{3\mu R\omega(h - H_2)}{h^3} = \frac{3\mu\omega}{2H^3} \frac{x^2 - x_2^2}{(1 + \alpha x^2)^3}. \tag{8.108}$$

In order to proceed further, the following integrals will be needed, which in this case have conveniently been obtained from the computer program "Maple":

$$\int \frac{x^2}{(1+\alpha x^2)^3}\, dx = \frac{1}{8\alpha}\left[-\frac{2x}{(1+\alpha x^2)^2} + \frac{x}{1+\alpha x^2} + \frac{1}{\sqrt{\alpha}}\tan^{-1}(x\sqrt{\alpha})\right] \tag{8.109}$$

$$\int \frac{1}{(1+\alpha x^2)^3}\, dx = \frac{1}{8}\left[\frac{2x}{(1+\alpha x^2)^2} + \frac{3x}{1+\alpha x^2} + \frac{3}{\sqrt{\alpha}}\tan^{-1}(x\sqrt{\alpha})\right]. \tag{8.110}$$

Noting that the pressure has fallen to zero at the point of departure from the rolls, Eqn. (8.108) integrates to:

$$p = \int_0^p dp = \int_{x_2}^x \frac{3\mu\omega}{2H^3}\frac{x^2 - x_2^2}{(1+\alpha x^2)^3}\, dx$$

$$= \frac{3\mu\omega}{2H^3}\frac{1}{8\alpha^{3/2}}\left[\frac{x\sqrt{\alpha}[\alpha x^2(1-3\alpha x_2^2) - 1 - 5\alpha x_2^2]}{(1+\alpha x^2)^2} + \frac{x_2\sqrt{\alpha}(1+3\alpha x_2^2)}{1+\alpha x_2^2}\right.$$

$$\left. + (1 - 3\alpha x_2^2)[\tan^{-1}(x\sqrt{\alpha}) - \tan^{-1}(x_2\sqrt{\alpha})]\right]. \tag{8.111}$$

For a specified inlet sheet thickness $2H_1$, the corresponding value of x_1 will be known from Eqn. (8.101). The value of x_2 (the point at which the sheet leaves the rolls) can then be determined by using Eqn. (8.111) in conjunction with the upstream boundary condition, $p(x_1) = 0$ (atmospheric pressure), by solving the following equation for x_2:

$$\frac{x_1\sqrt{\alpha}[\alpha x_1^2(1-3\alpha x_2^2) - 1 - 5\alpha x_2^2]}{(1+\alpha x_1^2)^2} + \frac{x_2\sqrt{\alpha}(1+3\alpha x_2^2)}{1+\alpha x_2^2}$$

$$+ (1 - 3\alpha x_2^2)[\tan^{-1}(x_1\sqrt{\alpha}) - \tan^{-1}(x_2\sqrt{\alpha})] = 0. \tag{8.112}$$

Example 8.6—Pressure Distribution in a Calendered Sheet

To illustrate the preceding analysis, evaluate the point of departure, and the pressure distribution, for a calendering operation in which $R = 0.35$ m, $H = 0.00025$ m, and $x_1 = -0.01$ m. The rolls rotate at 20 rpm, and the viscosity of the melt is 0.5×10^4 P.

Solution

First, note that $\alpha = 1/2HR = 1/(2 \times 0.00025 \times 0.35) = 5,714$ m^{-2}. The solution of Eqn. (8.112) can then be effected by any standard numerical root-finding technique. The "Goal-Seek" feature of the Excel spreadsheeting application gives the downstream distance before separation as $x_2 = 0.00403$ m.

Example 8.6—Pressure Distribution in a Calendered Sheet **455**

When x_2 has thus been found, Eqn. (8.111) then gives the complete pressure distribution, which can again be evaluated with the aid of a spreadsheet. The results are shown in Fig. E8.6, which actually shows a closely related value as the ordinate (namely, the expression in the large brackets in Eqn. (8.111)):

$$p^* = \frac{16H^3\alpha^{3/2}}{3\mu\omega}p. \tag{E8.6.1}$$

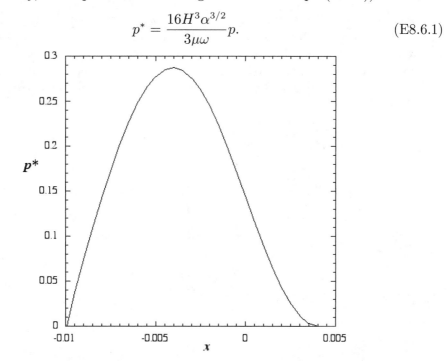

Fig. E8.6 *Pressure distribution along the calendered sheet.*

Observe that the pressure rises from zero at the inlet, to a maximum *upstream* of the "nip," and then declines to zero as the sheet leaves the calender. In accordance with an earlier observation, note that the pressure gradient is zero at the exit.

The angular velocity of the rolls is $\omega = 2\pi \times 20/60 = 2.09$ rad/s. Since the maximum value of p^* is about 0.288, the maximum pressure between the rolls is:

$$p_{\max} = \frac{3\mu\omega p^*}{16H^3\alpha^{3/2}} = \frac{3 \times 0.5 \times 10^4 \times 2.09 \times 0.288}{16 \times 0.00025^3 \times 5,714^{3/2}} = 8.38 \times 10^7 \text{ Pa} = 838 \text{ bar.}$$

As mentioned previously, the maximum pressure can be simply enormous. □

8.10   Thin Films and Surface Tension

This chapter continues with situations not unlike those encountered in the earlier boundary-layer analysis—namely, in which there is a dominant velocity component in one direction, and also in which surface tension may be important. Fig. 8.14 shows a thin liquid film, of density ρ and viscosity μ, after it leaves the distributor at the top of a wetted-wall column, which can be used for gas-absorption and reaction studies in relatively simple geometries. Of interest here is the shape of the free surface once the liquid is free of the influence of the distributor, a situation that has been investigated both theoretically and experimentally by Nedderman and Wilkes.[7]

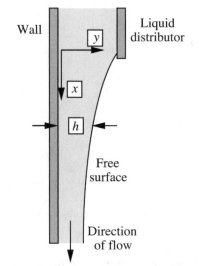

Fig. 8.14   Liquid film on a wetted-wall column.

Since the film is thin, curvature effects of the cylindrical tube wall can be neglected, and the problem can be solved in x/y coordinates, leading to the following parabolic velocity profile:

$$v_x = \frac{3Q}{2h^3}\left(2hy - y^2\right). \tag{8.113}$$

Here, Q is the total volumetric flow rate (per unit depth, normal to the plane of the figure), h is the local film thickness, and y is the coordinate measured from the wall into the film.

Now perform a momentum balance on an element of the film, shown in Fig. 8.15, analogous to a similar treatment in Section 8.2 for the analysis of boundary-layer flow. The surface BC is taken to be *just* on the *liquid* side of the liquid/air

[7]   R.M. Nedderman and J.O. Wilkes, "The measurement of velocities in thin films of liquid," *Chemical Engineering Science*, Vol. 17, pp. 177–187 (1962). This investigation *was* started by Prof. Terence Fox, to whom this book is dedicated, and represents the only case known to the author in which the legendary Prof. Fox initiated any doctoral research (see the Preface).

interface, and will have a pressure acting normally on it, whose mean value is $p + (dp/dx)(dx/2)$.

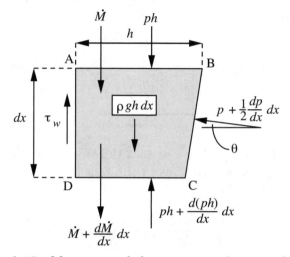

Fig. 8.15 *Momentum balance on an element of film.*

If $\dot{M}$ is the x convected momentum flux in the film, a downward steady-state momentum balance gives:

$$\dot{M} + ph + \rho gh\,dx - \tau_w\,dx - \left(\dot{M} + \frac{d\dot{M}}{dx}\,dx\right) - \left(ph + \frac{d(ph)}{dx}\,dx\right)$$

$$- \left(p + \frac{1}{2}\frac{dp}{dx}\,dx\right)dx\,\sin\theta = 0. \qquad (8.114)$$

Here, the $\sin\theta$ term recognizes that the normal force on the surface BC must be resolved in the vertical direction. After substituting for the wall shear stress, noting that $\sin\theta \doteq -dh/dx$, and neglecting a second-order differential quantity $(dx)^2$, Eqn. (8.114) simplifies to:

$$\frac{d\dot{M}}{dx} = \rho gh - \mu\left(\frac{\partial v_x}{\partial y}\right)_{y=0} - h\frac{dp}{dx}. \qquad (8.115)$$

Now observe that the momentum flux per unit depth in the film is:

$$\dot{M} = \rho\int_0^h v_x^2\,dy = \frac{6\rho Q^2}{5h}. \qquad (8.116)$$

Also note that the pressure p in the film will in general differ from the pressure $p_0 = 0$ in the gas outside the liquid film because of surface tension σ. The pressure *decreases* as we proceed across the free surface into the film, by an amount that equals the product of the surface tension and the curvature of the film. Provided that $|dh/dx| \ll 1$, and observing the expression for curvature in Appendix A, the result may be simplified:

$$p = p_0 - \sigma \underbrace{\frac{\dfrac{d^2h}{dx^2}}{\left[1 + \left(\dfrac{dh}{dx}\right)^2\right]^{3/2}}}_{\text{Surface curvature}} \doteq -\sigma \frac{d^2h}{dx^2}. \tag{8.117}$$

That is, Eqns. (8.115)–(8.117) lead to a third-order nonlinear ordinary differential equation that governs variations in the film thickness, in which $\nu = \mu/\rho$ is the kinematic viscosity:

$$\sigma \frac{d^3h}{dx^3} + \frac{6\rho Q^2}{5h^3} \frac{dh}{dx} - \frac{3\mu Q}{h^3} = -\rho g. \tag{8.118}$$

In general, Eqn. (8.118) cannot be solved analytically. However, Nedderman and Wilkes (*loc. cit.*) do give solutions for the following limiting cases:

1. For negligible surface tension, the analytical solution shows that the film thickness steadily approaches its final steady asymptotic value, $h_\infty = (3Q\mu/\rho g)^{1/3}$.
2. For finite surface tension, and with the film thickness near h_∞, an approximate solution for the film thickness shows that the expression for h consists of two parts:

   (a) A disturbance that decays downstream.
   (b) A series of standing waves that become amplified downstream. That is, the film tends to become unstable. If the wave amplitude is significant, however, the assumption of a parabolic velocity profile becomes questionable— in other words, the above simplified theory cannot be "pushed" too far.

Surface-tension forces can play vital roles in stabilizing or destabilizing coating operations and paint leveling, as suggested by Problems 8.6 and 8.13.

PROBLEMS FOR CHAPTER 8

Unless otherwise stated, all flows are steady
state, with constant density and viscosity.

1. *Boundary layer with linear velocity profile—E.* Review the simplified treatment of Section 8.2 for the boundary layer on a flat plate. Repeat the development with the following expression for the velocity profile in the boundary layer, instead of Eqn. (8.9):

$$\frac{v_x}{v_{x\infty}} = \zeta = \frac{y}{\delta}.$$

You should obtain expressions for both δ/x and c_f, similar to Eqns. (8.16) and (8.17). Also see Table 8.1. What are the merits and disadvantages of this linear velocity profile?

2. *Boundary layer with suction—M.* Fig. P8.2 shows the development of a boundary layer on a *porous* plate, in which fluid is sucked from the boundary layer with a superficial velocity v_s normal to the plate. (This arrangement has been proposed for airplane wings, in order to reduce the drag by preventing the boundary layer from becoming turbulent at high Reynolds numbers.)

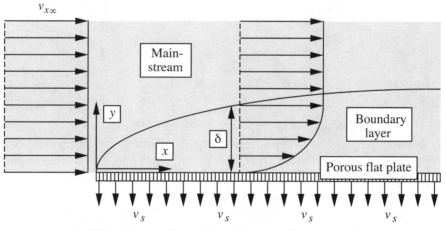

Fig. P8.2   *Boundary layer with suction.*

Use the simplified treatment to prove that the momentum balance of Eqn. (8.4) becomes:

$$v_{x\infty} \frac{d}{dx} \int_0^\delta v_x \, dy = \frac{\tau_w}{\rho} - v_{x\infty} v_s + \frac{d}{dx} \int_0^\delta v_x^2 \, dy.$$

Assuming a velocity profile of the form $v_x/v_{x\infty} = 2\zeta - \zeta^2$, where $\zeta = y/\delta$, derive an expression for δ as a function of x. Hence, show that the boundary layer will eventually reach a *constant* thickness—which no longer changes with distance x—and derive an expression for this limiting thickness.

3. *Tank draining with boundary layer—M.* For a laminar boundary layer on a flat plate, the solution in Section 8.2 gave the following expression for the wall shear stress at a point distant x from the leading edge:

$$\frac{\tau_w}{\frac{1}{2}\rho v_{x\infty}^2} = \frac{0.656}{\sqrt{\text{Re}_x}}. \tag{P8.3}$$

Derive an expression for the total drag force F exerted by a plate of length L on the fluid, per unit distance normal to the plane of Fig. 8.3. Your answer should give F in terms of $v_{x\infty}$, L, ρ, and μ or ν.

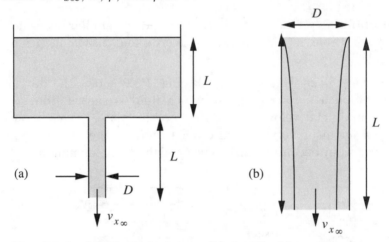

Fig. P8.3 (a) Tank draining; (b) detail of boundary layer.

As shown in Fig. P8.3(a), a viscous liquid is draining steadily from a large-diameter tank. As it enters the pipe of diameter D and length L, it has a flat velocity profile, and a thin laminar boundary layer subsequently forms, as in Fig. P8.3(b). Derive an expression for the velocity of the liquid $v_{x\infty}$ in the mainstream, in terms of any or all of D, L, g, ρ, and μ. Assume that the boundary layer is sufficiently thin so that $v_{x\infty}$ is essentially constant; also, since the velocities are small, neglect any inertial terms in a momentum balance. Evaluate $v_{x\infty}$ if $L = 1$ m, $D = 0.1$ m, $\rho = 1,000$ kg/m³, $g = 9.81$ m/s², and $\mu = 1$ kg/m s.

4. *Power for the "Queen Mary"—M.* The RMS *Queen Mary* embarked on her maiden transatlantic voyage from Southampton, England, on May 27, 1936. On the crossing to New York, she achieved an average speed of 29.13 knots (1 knot = 1.15 miles/hour). Her length at the waterline was 1,004 ft, with a draft of 38.75 ft and a beam (width) of 118 ft. Anybody who has been *underneath* the *Queen Mary*, when in dry dock, can attest that the bottom of the ship is almost completely flat.

Estimate the horsepower needed to overcome skin friction on the *Queen Mary's* maiden voyage, and compare with the total of 160,000 HP that was nominally delivered to the quadruple screws of the ship. Assume a density of 64.0 lb$_m$/ft³ and a viscosity of 3.45 lb_m/ft hr for seawater. Clearly state any reasonable simplifying assumptions.

5. *Pressure in a bearing—E.* The shaft in a journal bearing has a diameter of 2 ft and rotates at 60 rpm. The lubricant film extends all the way around the lower half of the shaft, with maximum and minimum thicknesses of 0.003 and 0.001 ft, respectively. If the viscosity of the lubricant is $\mu = 1,000$ cP, estimate the mean pressure (psi) in the film.

6. *Coating a moving substrate—M.* A continuous strip of a flexible metal substrate is being pulled to the right with a steady velocity V as it leaves a reservoir of a protective liquid polymer coating of viscosity μ and surface tension σ, as shown in Fig. P8.6.

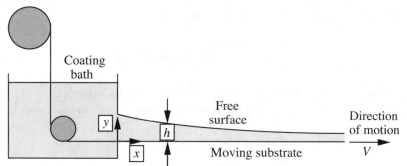

Fig. P8.6 *Coating a moving substrate.*

Assuming that pressure and viscous forces dominate, what is the simplified x momentum balance? If the pressure in the liquid film only depends on x, show that the velocity profile is given by:

$$v_x = V - \frac{1}{2\mu}\left(\frac{dp}{dx}\right)y(2h-y),$$

where $h = h(x)$ is the local film thickness. If surface tension is important, why is dp/dx in the film positive for the free-surface shape as drawn in Fig. P8.6? Sketch the resulting velocity profile.

If the final thickness of the coating is h_∞, prove that the intermediate film thicknesses obey:

$$\frac{\sigma}{3\mu V h_\infty}h^3\frac{d^3h}{dx^3} + \frac{h}{h_\infty} = 1.$$

7. *Wetted-wall column without surface tension—M.* For the film on the wetted-wall column discussed in Section 8.10, variations of film thickness are governed by the differential equation:

$$\sigma\frac{d^3h}{dx^3} + \frac{6\rho Q^2}{5h^3}\frac{dh}{dx} - \frac{3\mu Q}{h^3} = -\rho g. \qquad (P8.7.1)$$

Consider the situation in which surface tension is negligible.

The analytical solution of Eqn. (P8.7.1) is somewhat complicated. However, show that if the film thickness is not too far from its asymptotic value ($h/h_\infty \doteq 1$), the solution may be approximated by:

$$\frac{h}{h_\infty} \doteq 1 + ce^{-15\mu x/2\rho Qh_\infty}. \tag{P8.7.2}$$

in which $h_\infty = (3Q\mu/\rho g)^{1/3}$ and c is a constant of integration. Give a physical interpretation of Eqn. (P8.7.2).

8. *Wetted-wall column with surface tension—D.* Consider the film on the wetted-wall column discussed in Section 8.10, in which the variations of film thickness were governed by the differential equation:

$$\sigma \frac{d^3 h}{dx^3} + \frac{6\rho Q^2}{5h^3}\frac{dh}{dx} - \frac{3\mu Q}{h^3} = -\rho g. \tag{P8.8.1}$$

Supposing that the film thickness is close to its steady value (that is, $h = h_\infty + \varepsilon$, where $\varepsilon \ll h_\infty$, and $h_\infty = (3Q\mu/\rho g)^{1/3}$), prove that the governing equation becomes:

$$\sigma \frac{d^3 \varepsilon}{dx^3} + \frac{6\rho Q^2}{5h_\infty^3}\frac{d\varepsilon}{dx} + \frac{3\rho g\varepsilon}{h_\infty} = 0. \tag{P8.8.2}$$

Now attempt a solution of the form:

$$\varepsilon = ae^{\alpha x} + be^{\beta x} + ce^{\gamma x}, \tag{P8.8.3}$$

for which a representative term is:

$$\varepsilon = de^{\delta x}, \tag{P8.8.4}$$

and derive the corresponding cubic equation in δ. Now form $(\delta - \alpha)(\delta - \beta)(\delta - \gamma) = 0$, in which:

$$\alpha = -\eta, \qquad \beta = \frac{\eta}{2} + i\zeta, \qquad \gamma = \frac{\eta}{2} - i\zeta, \tag{P8.8.5}$$

and compare it with the cubic equation. Hence, verify that the three roots for δ are those given in Eqn. (P8.8.5), where $i = \sqrt{-1}$, both η and ζ are real, and η is positive. Hence, show, by noting that b and c must be complex if the final solution is to be real, that:

$$h = h_\infty + ae^{-\eta x} + \kappa e^{\eta x/2}\sin(\zeta x + \lambda), \tag{P8.8.6}$$

in which a, κ, and λ are unknown real constants. Give a physical interpretation of each of the terms in Eqn. (P8.8.6).

9. *Alternative thin-film momentum balance—E.* Perform a *careful* momentum balance on an element of a thin liquid film, as in Fig. 8.14. However, now take the surface BC to be *just* on the air side of the liquid/air interface, so the pressure acting on it is $p_0 = 0$. Make sure that you obtain the same final result, namely, Eqn. (8.115). There will now be surface tension forces σ (per unit depth normal to the plane of Fig. 8.14) acting tangentially along the interface, and these will have to be resolved in the vertical direction—through angles that are marginally different at B and C.

10. *Exact pressure distribution in a lubricant film—D.* In Section 8.8, the final expression for the pressure distribution in the lubricant film, Eqn. (8.96), assumed that the film thickness h maintained its mean value h_m. Refine the solution by conducting an exact integration of Eqn. (8.93), now with the film thickness $h(x)$ varying linearly between h_0 (at $x = 0$) and h_L (at $x = L$) according to:

$$h = h_0 - \beta x, \qquad \text{where} \qquad \beta = \frac{h_0 - h_L}{L}. \tag{P8.10.1}$$

Prove that the pressure distribution is given by:

$$p = \frac{\alpha x (L - x)}{(h_0 - \beta x)^2 (2h_0 - \beta L)}. \tag{P8.10.2}$$

Use a standard computer application such as Excel to plot $ph_m^3/\alpha L^2$ against x/L, for $\beta L/h_m = 0$, 0.1, and 0.5. Explain your observations qualitatively.

11. *Trying to keep warm—M.* You are walking along the street in Chicago when the $0\,°F$ air is moving down the street with a "mainstream" velocity of $v_{x\infty} = 30$ mph. At each block, as the air impacts the corner of a building, a boundary layer builds up along the side of the building, just as it does on a flat plate.

To avoid getting too cold, you walk very close to the building, hoping to stay inside the boundary layer. At what minimum distance from the corner can you be guaranteed that the boundary layer will be at least 2 ft wide? For air, $\rho = 0.086$ lb_m/ft^3, $\mu = 0.0393$ lb_m/ft hr.

12. *Velocity profiles in calendering—M.* For the calendering problem for which the pressure distribution is shown in Fig. E8.4, use a spreadsheet application to plot the velocity profiles at $x = -0.05$, -0.025, 0, and 0.0225. A plot of v_x/ω versus y/h is recommended. Comment briefly on your results.

13. *Leveling of a paint film—M.* The problem of the leveling of paint films after they have been sprayed on automobile body panels is becoming quite important as increased-solids paints of lower polymeric molecular weight are being mandated by the government.[8] An idealized situation is shown in Fig. P8.13, in which a paint film, of density ρ and Newtonian viscosity μ, has at an initial time $t = 0$ a nonuniform thickness $h = h_0(x)$. You are called in as a consultant to help derive a differential equation, whose solution would then indicate how h subsequently varies with time t and position x.

Fig. P8.13   *Leveling of a paint film.*

You may make the following assumptions:

(a) Because the paint film is baked in an oven, there will be known temperature variations and hence known changes in the surface tension σ along the surface of the film; these give rise to a *known* shear stress $(\tau_{yx})_{y=h} = \mu(\partial v_x/\partial y)_{y=h} = \partial\sigma/\partial x$ at every point on the surface.

(b) Surface tension causes the pressure to change from atmospheric pressure (zero) just outside the film to a different value just inside the film, in exactly the same manner as in Eqn. (8.117). In addition, *gravity is now important,* and the pressure increases hydrostatically from just inside the free surface to the body panel.

(c) The usual lubrication approximation/boundary layer theory still holds; that is, inertial effects can be neglected, and the flow in the film is mainly in one direction, so that you need only be concerned with the velocity component v_x.

Now answer the following:

(a) Derive an expression for the pressure p at any point in the film, as a function of ρ, σ, g, h, x, and y. Then form $\partial p/\partial x$, and show that this is not a function of y.

(b) To what two terms does the x momentum balance simplify? Integrate it twice with respect to y, apply the boundary conditions, and show that the velocity profile is given by:

$$v_x = \frac{1}{\mu}\left[\frac{\partial p}{\partial x}\left(\frac{y^2}{2} - yh\right) + y\frac{\partial\sigma}{\partial x}\right]. \tag{P8.13.1}$$

[8] This problem follows part of the doctoral research of Richard Blunk of the General Motors Corporation; his help and insight are gratefully acknowledged.

(c) Integrate the velocity to obtain an expression for the flow rate Q in the paint film.

(d) By means of a mass balance on an element of height h and width dx, prove that variations of the film thickness h with time are related to variations of the flow rate Q with distance by:

$$\frac{\partial h}{\partial t} = -\frac{\partial Q}{\partial x}. \tag{P8.13.2}$$

(e) Indicate briefly what you would next do in order to obtain a differential equation governing variations of h with x and t. (However, do not perform these steps.)

14. *A check on the Blasius problem—M.* Concerning the mathematics leading up to the solution of the Blasius problem, first check Eqn. (8.48). Then verify that substitution of v_x and v_y and the derivatives of v_x from Eqns. (8.47) and (8.48) into Eqn. (8.38) leads to the following *ordinary* differential equation:

$$f \frac{d^2 f}{d\zeta^2} + 2\frac{d^3 f}{d\zeta^3} = 0, \tag{P8.14.1}$$

subject to the boundary conditions:

$$\zeta = 0: \quad f = f' = 0, \tag{P8.14.2}$$

$$\zeta = \infty: \quad f' = 1. \tag{P8.14.3}$$

15. *Numerical solution of the Blasius problem—D.* This problem concerns the numerical solution of the third-order ordinary differential equation and accompanying boundary conditions of Eqns. (8.49)–(8.51), also restated in Problem P8.14 above.

(a) Verify that by defining new variables $z = \zeta$, $f_0 = f$, $f_1 = df/dz$, and $f_2 = d^2 f/dz^2$, Eqn. (8.49) can be expressed as *three* first-order ordinary differential equations:

$$\frac{df_0}{dz} = f_1, \quad \frac{df_1}{dz} = f_2, \quad \frac{df_2}{dz} = -\frac{1}{2}f_0 f_2, \tag{P8.15.1}$$

subject to the boundary conditions:

$$z = 0: \quad f_0 = f_1 = 0, \qquad z = \infty: \quad f_1 = 1. \tag{P8.15.2}$$

(b) As discussed in Appendix A, use a spreadsheet to implement Euler's method to solve Eqns. (P8.15.1) subject to boundary conditions (P8.15.2), from $z = 0$ to $z = 10$ (which latter value is effectively infinity). (Any other standard software for solving differential equations may be used instead.) A step-size of $\Delta z = 0.1$ or even smaller is recommended, otherwise Euler's method is not very accurate. Try different values for Δz in order to get a feel for what is happening.

(c) You will need to supply the "missing" boundary condition for f_2 at $z = 0$, so that f_1 is very close to the actual boundary condition $f_1 = 1$ at $z = 10$. If you are using Excel, you will find the "Goal-Seek" feature under the "Formula" pull-down menu to be very helpful.

(d) When all is working satisfactorily, add another column for the dimensionless y velocity, and then plot the two dimensionless velocities against the dimensionless distance from the wall. Clearly label everything, and check your results against those given in Fig. 8.5.

16. *Developing boundary layer—M.* Consider a cylinder of radius a that starts to rotate at $t = 0$ with an angular velocity of ω in an otherwise stagnant liquid of kinematic viscosity ν. We wish to calculate the thickness $\delta = \delta(t)$ of the boundary layer that starts to build up around the cylinder and grows with time. Since $\delta \ll a$, we may approximate the situation in x/y coordinates, as shown in Fig. P8.16, in which $U = a\omega$ is the velocity of the surface of the cylinder, and y is the radially outward direction.

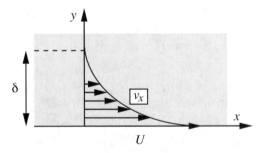

Fig. P8.16   Developing boundary layer.

If the pressure is uniform everywhere, what justification is there for assuming that v_x obeys the equation:

$$\frac{\partial v_x}{\partial t} = \nu \frac{\partial^2 v_x}{\partial y^2} ? \tag{P8.16.1}$$

To solve Eqn. (P8.16.1) approximately, use the similarity approach, in which the velocity profile is assumed to be of the following form for all values of δ:

$$v_x = U \left(1 - 2\frac{y}{\delta} + \frac{y^2}{\delta^2} \right). \tag{P8.16.2}$$

(a) Explain why the form of v_x given by Eqn. (P8.16.2) is reasonable.

(b) By substituting for v_x from Eqn. (P8.16.2) into Eqn. (P8.16.1) and integrating both sides from $y = 0$ to $y = \delta$, and then integrating the resulting differential equation in δ and t, prove that the boundary-layer thickness varies with time according to:

$$\delta = \sqrt{c\nu t}, \tag{P8.16.3}$$

and determine the value of the constant c.

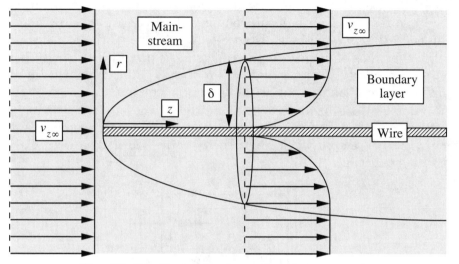

Fig. P8.17   *Boundary layer forming on a wire.*

17. *Boundary layer on a wire—M (C).* As shown in Fig. P8.17, a stream of fluid flows with mainstream velocity $v_{z\infty}$ parallel to the axis of a thin cylindrical wire of radius a. Within the laminar boundary layer, which builds up near the surface of the wire, the velocity profile in the fluid is given by $v_z = v_{z\infty} \sin(\pi r/2\delta)$, where δ is the boundary-layer thickness, r is the radial distance from the surface of the wire, and z is the axial coordinate. Perform a simplified momentum balance on an element of the boundary layer of axial extent dz.

Assume that the wire is very thin, so that in any integrals for mass and momentum fluxes, you may take the lower limit as $r = 0$. The only time you will require the wire radius a is when you need an area on which the wall shear stress acts.

Hence, prove that the thickness of the boundary layer varies with axial distance as:

$$\delta^3 = \frac{3\pi^3}{(12 - \pi^2)} \frac{a\nu z}{v_{z\infty}}. \tag{P8.17.1}$$

Rearrange this result in terms of three dimensionless groups, δ/z, a/z, and $v_{z\infty}z/\nu$.

The following integrals may be quoted without proof:

$$\int r \sin \alpha r \, dr = \frac{1}{\alpha^2} \sin \alpha r - \frac{r}{\alpha} \cos \alpha r,$$

$$\int r \sin^2 \alpha r \, dr = \frac{r^2}{4} - \frac{r}{4\alpha} \sin 2\alpha r - \frac{1}{8\alpha^2} \cos 2\alpha r.$$

18. *Wetted-wall column with boundary layer—D (C).* Fig. P8.18 shows idealized flow conditions in a wetted-wall column, which consists of two wide parallel plates of length L separated by a distance h. The insides of the plates are covered by uniform thin films of liquid, whose surfaces move down with velocity W. Air of density ρ and viscosity μ enters the bottom of the column from a large reservoir at a small pressure P above that at the top of the column.

The value of W is unaffected by the airflow, whose pattern at any section consists of a central core of uniform upward velocity U, and a boundary layer of thickness $\delta = \delta(x)$, across which the velocity varies linearly between the central core and the liquid film.

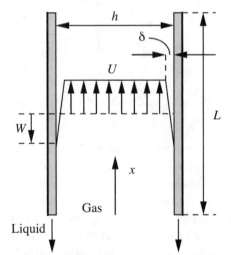

Fig. P8.18   Parallel-plate wetted-wall column.

Show that $\delta = h(U - U_0)/(U + W)$, where U_0 is the value of U at the bottom of the column. Also prove that the momentum passing across any section is, per unit depth, normal to the plane of the diagram:

$$\rho \left[(h - 2\delta)U^2 + \frac{2\delta(U^3 - W^3)}{3(U + W)} \right]. \tag{P8.18.1}$$

Outline—but do not attempt to carry to completion—the steps that are needed to determine U_0 as a function of L, h, W, ρ, and μ. *Hint:* Consider the development of Example 8.1. Ignore any gravitational effects.

19. *Isokinetic sampling of a dusty gas—D (C).* Consider the boundary layer of thickness δ that develops along a flat plate for laminar flow with mainstream velocity $v_{x\infty}$ in a fluid of density ρ and viscosity μ.

(a) If the velocity profile in the boundary layer is of the form:

$$\frac{v_x}{v_{x\infty}} = 2\frac{y}{\delta} - \left(\frac{y}{\delta}\right)^2, \qquad (P8.19.1)$$

derive an expression that gives the ratio δ/x as a function of the Reynolds number $Re_x = \rho v_{x\infty} x/\mu$, based on the distance x from the leading edge. If needed, you may quote an integral momentum balance without proof.

(b) Hence, prove that the corresponding expression for the *local* wall shear stress is given by:

$$\frac{\tau_w}{\frac{1}{2}\rho v_{x\infty}^2} = \frac{0.730}{\sqrt{Re_x}}. \qquad (P8.19.2)$$

(c) If the mean wall shear stress between $x = 0$ and $x = L$ is defined by:

$$\overline{\tau}_w = \frac{1}{L}\int_0^L \tau_w\,dx, \qquad (P8.19.3)$$

derive an expression for $\overline{\tau}_w$ that is very similar to Eqn. (P8.19.2), except that the Reynolds number $Re_L = \rho v_{x\infty} L/\mu$, based on the total length, L, now appears.

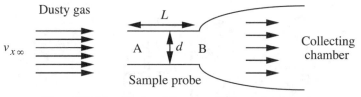

Fig. P8.19 Probe for sampling a dusty gas.

(d) Fig. P8.19 shows a probe of length L and internal diameter d for sampling a gas of density ρ and kinematic viscosity ν that contains fine dust particles. Upstream of the probe, the mainstream velocity is $v_{x\infty}$, and it is essential that the velocity at the inlet A of the probe must be the same, otherwise a biased sample will occur. (For example, if the velocity at A exceeds $v_{x\infty}$, some dust particles, because of their inertia, will deviate little from their path and will not be drawn into the sampling tube.) To achieve this condition of "isokinetic" sampling, the pressure at B must be reduced below that at A, to compensate for the wall shear stress that builds up in the boundary layer that coats the inside wall of the sample probe. Derive an approximate expression that gives the required pressure difference $(p_A - p_B)$ for isokinetic sampling.

20. *Boundary layer in stagnation flow—M.* Consider the upper half plane ($y \geq 0$) of the potential stagnation flow of Example 7.2, and take the x axis to be the surface of the ground, with a vertically downward wind impinging on it. Since viscosity is now important adjacent to the solid surface at $y = 0$, there will be a developing boundary layer. Noting from symmetry that only $x > 0$ needs to be studied, the thickness $\delta = \delta(x)$ of the boundary layer will be a function of the distance from the origin.

Since for the stagnation flow $v_x = v_{x\infty} = cx$, the mainstream velocity is *not* constant; thus, the pressure $p = p(x)$ will also be a function of location. The y component of the mainstream velocity can be ignored, since $v_y = -cy$, and y is very small.

If the flow is laminar with $v_x/v_{x\infty} = \sin[\pi y/(2\delta)]$, perform a simplified momentum balance on an element of the boundary layer and obtain a differential equation that governs δ as a function of x. If the boundary-layer thickness reaches a steady asymptotic value (in which case assume $d\delta/dx = 0$ without solving the differential equation), prove that this value equals $0.5\pi\sqrt{\nu/c}$.

Fig. P8.21   Coating a continuous paper sheet.

21. *Film-thickness variations—M.* In the coating experiment shown in Fig. P8.21, a flexible paper sheet is being withdrawn vertically upward at a velocity V from a bath containing a Newtonian polymeric liquid of viscosity μ, density ρ, surface tension σ, and local thickness h. Assume that a momentum balance gives the following velocity profile:

$$v_x = V - \frac{1}{2\mu}\left(\rho g + \frac{dp}{dx}\right)y(2h - y). \tag{P8.21}$$

If the net upward liquid flow rate Q (per unit depth normal to the plane of Fig. P8.21) is zero (as much is being dragged upwards as falls under gravity), derive a differential equation that governs the variation of film thickness h with distance x. Assume steady flow—that is, there are no time derivatives.

If, for large x, the film were to come to a constant thickness h_∞, express h_∞ in terms of μ, V, ρ, and g.

Suppose that the film thickness deviates by a small amount ε from h_∞, so that $h = h_\infty + \varepsilon$. Derive a differential equation for ε as a function of x. Try a solution of the form $\varepsilon = de^{\delta x}$, and obtain an equation from which δ could be calculated (but do *not* attempt to obtain these values). If needed, $1/[1 + (\varepsilon/h_\infty)] \doteq 1 - (\varepsilon/h_\infty)$.

22. *True/false.* Check *true* or *false*, as appropriate:

(a) A boundary layer is a region in which neither viscous nor inertial terms can be neglected.   T ☐ F ☐

(b) A drag coefficient is a measure of the ratio of viscous to inertial forces.   T ☐ F ☐

(c) In boundary layers, "similarity of velocity profiles" means that the profiles of v_x are "stretched" in both the x and y directions.   T ☐ F ☐

(d) In the simplified analysis of a boundary layer on a flat plate, the assumption that $v_x = y/\delta$ leads to a drag coefficient that is within 5% of that predicted by the more exact Blasius solution.   T ☐ F ☐

(e) In the analysis of boundary layers, Bernoulli's equation can be used to predict $\partial p/\partial x$ if the mainstream velocity $v_{x\infty}$ is known as a function of the distance x.   T ☐ F ☐

(f) The Blasius analysis of the boundary-layer problem converts the equations of motion, which contain partial derivatives, into a second-order ordinary differential equation.   T ☐ F ☐

(g) At the transition from one to the other, a laminar boundary layer is significantly thinner than a turbulent one.   T ☐ F ☐

(h) The relative motion of two surfaces, slightly inclined to each other, can cause appreciable pressure rises in a fluid that is located in the narrow gap between the two surfaces.   T ☐ F ☐

(i) In lubrication analysis, the Couette flow component indicates a parabolic velocity profile.   T ☐ F ☐

(j)   In calendering, a knowledge of the entering sheet   T ☐   F ☐
        thickness, together with the geometry of the rolls and
        their speed, will enable the exiting sheet thickness to
        be determined.

(k)   For the Blasius problem, dimensional analysis can   T ☐   F ☐
        lead to the exact form of the relation between the
        drag coefficient and the Reynolds number, apart from
        a single unknown constant.

(l)   In a lubricant film, the maximum pressure occurs   T ☐   F ☐
        somewhat closer to the end of the film that has the
        greater thickness.

(m)   In the boundary layer on a flat plate, the flow is en-   T ☐   F ☐
        tirely parallel to the plate, with zero velocity in the
        direction normal to the plate.

(n)   In the drag coefficient plot for flow past a sphere, the   T ☐   F ☐
        sudden "dip" at high Reynolds number is because the
        boundary layer suddenly becomes turbulent.

(o)   In the simplified analysis of a boundary layer on a flat   T ☐   F ☐
        plate, the value of $\partial^2 v_x/\partial y^2$ at the plate is zero when
        $v_{x\infty}$ is constant.

(p)   Boundary-layer separation is likely to occur in flow in   T ☐   F ☐
        a diverging channel.

(q)   Boundary-layer separation is likely to occur in flow of   T ☐   F ☐
        a circular jet of air normal to a flat plate ("stagnation
        flow").

(r)   The total shaft weight that can be supported by a   T ☐   F ☐
        lubricant film in a journal bearing increases with de-
        creasing lubricant viscosity.

(s)   The addition of a surface-active agent to decrease the   T ☐   F ☐
        surface tension of a liquid is likely to increase the
        instabilities in thin-film flow of the liquid.

Chapter 9

TURBULENT FLOW

9.1 Introduction

TURBULENT fluid motion has been aptly described as "an irregular condition of flow in which the various quantities show a random variation with time and space coordinates, so that statistically distinct average values can be discerned."[1]

A thorough understanding of turbulence has not been achieved to date, and may never be attained in the foreseeable future. For this reason, the subject is sometimes neglected in introductory fluid mechanics courses. But this should not be the case, and we should not be discouraged from trying to learn something about it, because—as nice as the solutions for laminar flow are—turbulence is the more normal state of affairs in fluid flow. To be sure, there are important cases in which the flow is indeed laminar, and at least two fields come readily to mind: polymer processing (typically with high viscosities) and microfluidics (with very tiny channel dimensions). However, there are countless other situations involving turbulence, such as gas flow in long-distance pipelines, liquid flow in heat exchangers, blood flow in the heart, combustion of gases, many industrial reactors, and so on.

We cannot resist the temptation of reproducing the famous quotation recalled by Parviz Moin and John Kim[2]: "The difficulty with turbulence was wittily expressed in 1932 by the British physicist Horace Lamb, who, in an address to the British Association for the Advancement of Science, reportedly said, 'I am an old man now, and when I die and go to heaven there are two matters on which I hope for enlightenment. One is quantum electrodynamics, and the other is the turbulent motion of fluids. And about the former I am rather optimistic.'"

In this chapter, our goals will be:

1. To gain some understanding of turbulent flow in pipes, channels, and jets.
2. To investigate the analogies between the turbulent transport of momentum and thermal energy.
3. To learn more about how computational fluid dynamics (CFD) is starting to allow us to tackle turbulence in more complicated situations by the numerical

---

[1] J.O. Hinze, *Turbulence—an Introduction to Its Mechanism and Theory*, McGraw-Hill, New York, 1959.

[2] *Tackling Turbulence with Supercomputers*, http://turb.seas.ucla.edu/~jkim/sciam/turbulence.html

solution of certain equations of motion. In particular, we shall study the k/ε method, which deals with the production and dissipation of turbulent kinetic energy. Although the method has some limitations, its relatively easy implementation (by FlowLab and COMSOL, for example) does at least lead to plausible and useful results in many instances.

The reader will already know that the Reynolds number, being a measure of the relative importance of inertial effects to viscous effects, is an important factor in determining whether turbulence is likely to occur or not. At low Reynolds numbers, viscous effects dominate, and will *dampen* out any small perturbations in the flow. But at higher Reynolds numbers this is no longer the case, and the predominance of the nonlinear inertial terms in the Navier-Stokes equations will lead to an *amplification* of small perturbations. The consequence is that in addition to the general trend of flow there will be superimposed *eddies*, being regions that are typically rotating and hence have vorticity, behaving as little whirlpools. For pipe flow, the eddies that are initially formed at the expense of pressure energy are of a size that is the same order of magnitude as the pipe diameter. But these larger eddies are themselves unstable and break down into progressively smaller eddies, the general scheme being shown in simplified form in Fig. 9.1.

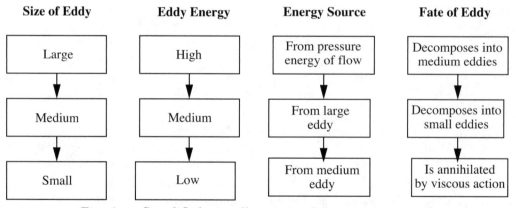

Fig. 9.1   Simplified overall picture of the cascade process.

Although the energy per eddy is highest for the large eddies, there are so many more of the smaller eddies that the total amount of mechanical energy passed down the chain or *cascade* remains essentially constant. But the process eventually terminates, when the eddies become so small that their velocity variations occur over such short distances (known as the Kolmorogov limit) that the resulting high-velocity gradients cause viscosity to kick in and convert the mechanical energy into heat.

Mathieu and Scott, authors of an excellent and comprehensive introduction to turbulence, state that the energy cascade from large to small turbulent eddies was first postulated by Lewis F. Richardson.[3] And it was in Richardson's publication that he summarized his beliefs in the famous lines:

> Big whorls have little whorls,
> Which feed on their velocity;
> And little whorls have lesser whorls,
> And so on to viscosity.

Although turbulence is more likely to occur at the higher Reynolds numbers, it is also promoted by the shearing action of a solid surface, and tends to be enhanced in the wake downstream of an object such as an airplane wing or golf ball. But note that at moderate Reynolds numbers it is possible to have a *laminar* flow that is *unstable* but cannot be classified as turbulent. The last example in Chapter 13 shows how CFD can simulate (with considerable accuracy) the flow past a cylinder at a moderate Reynolds number of 100. In that event, the instability of the laminar flow causes vortices to be shed alternately from opposite sides of the cylinder, forming a *Kármán vortex street*. However, the vortices are generated in a regular pattern and therefore do not satisfy the criterion of randomness or chaotic behavior that characterizes turbulence.

Generally, turbulent flow occurs at sufficiently high values of the Reynolds number, at conditions under which laminar flow becomes unstable and the velocities and pressure no longer have constant or smoothly varying values. For pipe flow, the general picture is that relatively large rotational eddies are formed in the region of high shear near the wall, and that these degenerate into progressively smaller eddies, in which the energy is dissipated into heat by the action of viscosity.

Fig. 9.2 Fluctuating velocity components.

To proceed, we recognize that turbulence is inherently a three-dimensional phenomenon, and at any point P there will be fluctuations in time of the three velocity components shown in Fig. 9.2.[4] It is first necessary to introduce the concept of a *time-averaged* quantity, denoted by an overbar, defined as the mean

---

[3] J. Mathieu and J. Scott, *An Introduction to Turbulent Flow*, Cambridge University Press, 2000, p. 61. The following reference is given: L.F. Richardson, *Weather Prediction by Numerical Process*, Cambridge University Press, 1922.

[4] Nevertheless, useful information about turbulence can still be obtained in situations that have just two coordinates of primary interest, such as r and z that occur in pipe flow.

value of that quantity over a time period T that is very large in comparison with the time scale of the individual fluctuations. Thus, $\bar{v}_x$ is the time-averaged value of the x velocity component, v_x.

Then, at any point in space, we can think of the three individual velocity components and the pressure at any *instant of time* as being given by:

$$v_x = \bar{v}_x + v_x', \qquad v_y = \bar{v}_y + v_y',$$
$$v_z = \bar{v}_z + v_z', \qquad p = \bar{p} + p'. \tag{9.1}$$

Here, v_x' etc. denote fluctuations from the mean values, and which may change very rapidly—in a matter of milliseconds, as shown in Fig. 9.3; in such event, a value of T equal to a few seconds would be appropriate for the time averaging just described. This chapter generally deals with flows such as those in pipes and jets that are *steady in the mean*—that is, these time-averaged quantities do not vary with time; in other situations, they may fluctuate slowly with time.

Fig. 9.3   Variation of instantaneous velocity v_x with time t. The fluctuations typically occur over a period of a few milliseconds.

If the fluctuations shown in Fig. 9.3 could be examined under a succession of increasingly powerful magnifying glasses, the fluctuations would not be smooth and a certain amount of "fuzziness" would be seen, corresponding to the continuously smaller eddies in the overall energy cascade. However, a limit would eventually be reached (the Kolmorogov limit) when the length scale is sufficiently small that the fluctuations *do* finally look smooth.

By definition, the time averages of the fluctuations are zero, since—on the average—they are equally likely to be above the mean or to fall below it:

$$\overline{v_x'} = \overline{v_y'} = \overline{v_z'} = \overline{p'} = 0. \tag{9.2}$$

Therefore, the time averages of the mean quantities are the mean quantities themselves; for example:

$$\overline{\bar{v}_x} = \bar{v}_x. \tag{9.3}$$

The *intensity* of the turbulence is typically in the range given by:

$$\frac{\widetilde{v'}}{\bar{v}} \sim 0.01 \text{ to } 0.1, \tag{9.4}$$

in which the root-mean-square of the fluctuating velocity component is defined as:

$$\widetilde{v'} = \sqrt{\overline{v'^2}}. \tag{9.5}$$

Time-averaged continuity equation.   From Eqn. (5.52), the continuity equation with constant density is:

$$\frac{\partial v_x}{\partial x} + \frac{\partial v_y}{\partial y} + \frac{\partial v_z}{\partial z} = 0. \tag{9.6}$$

Now substitute the instantaneous velocity components from Eqn. (9.1) and time average the entire equation. A representative term, and its subsequent rearrangement, is:

$$\overline{\frac{\partial(\bar{v}_x + v'_x)}{\partial x}} = \frac{\partial\overline{(\bar{v}_x + v'_x)}}{\partial x} = \frac{\partial(\overline{\bar{v}_x} + \overline{v'_x})}{\partial x} = \frac{\partial\bar{v}_x}{\partial x}. \tag{9.7}$$

Observe the four intermediate steps involved:

1. The time average of the partial derivative of a quantity equals the partial derivative of the time average of that quantity.
2. The time average of the sum of two quantities equals the sum of their individual time averages.
3. From Eqn. (9.3), the time average of a mean quantity, $(\bar{v}_x)$ in this case, is that mean quantity itself.
4. From Eqn. (9.2), the time average of a fluctuating component is zero.

Time averaging the other derivatives as well yields the *time-averaged* continuity equation:

$$\frac{\partial\bar{v}_x}{\partial x} + \frac{\partial\bar{v}_y}{\partial y} + \frac{\partial\bar{v}_z}{\partial z} = 0, \tag{9.8}$$

which is identical with the original continuity equation *except* that time-averaged velocities have replaced the instantaneous velocities.

Time-averaged momentum balances.   (The reader who is not interested in tedious derivations, or doesn't have time for them, can skip immediately from Eqn. (9.9) to the final result, Eqn. (9.13).) The x momentum balance (for example) in terms of the *viscous* stress components is, from Eqn. (5.72):

$$\rho\left(\frac{\partial v_x}{\partial t} + v_x\frac{\partial v_x}{\partial x} + v_y\frac{\partial v_x}{\partial y} + v_z\frac{\partial v_x}{\partial z}\right) = -\frac{\partial p}{\partial x} + \underbrace{\frac{\partial\tau_{xx}}{\partial x} + \frac{\partial\tau_{yx}}{\partial y} + \frac{\partial\tau_{zx}}{\partial z}}_{\text{Viscous stress terms}} + \rho g_x. \tag{9.9}$$

We again substitute the instantaneous velocity components from Eqn. (9.1) and time average the resulting equation to give the *time-averaged* x momentum balance. However, the following intermediate steps, relating to the nonlinear terms on the left-hand side, are useful. First note that these terms can be rewritten, with the aid of the continuity equation, (9.6), as:

$$v_x \frac{\partial v_x}{\partial x} + v_y \frac{\partial v_x}{\partial y} + v_z \frac{\partial v_x}{\partial z} = \frac{\partial(v_x v_x)}{\partial x} + \frac{\partial(v_x v_y)}{\partial y} + \frac{\partial(v_x v_z)}{\partial z}. \tag{9.10}$$

Eqn. (9.10) may also be proved by noting that the left-hand side can be written as $\mathbf{v} \cdot \nabla v_x$. The vector identity $\nabla \cdot v_x \mathbf{v} = v_x \nabla \cdot \mathbf{v} + \mathbf{v} \cdot \nabla v_x = \mathbf{v} \cdot \nabla v_x$ for an incompressible fluid [see Eqn. (5.24)] leads directly to the right-hand side of Eqn. (9.10).

Now substitute the instantaneous velocity components from Eqn. (9.1) into the right-hand side of Eqn. (9.10) and time-average the result. We now have to deal with the time averages of quantities such as $v_x v_y = (\overline{v}_x + v_x')(\overline{v}_y + v_y') = (\overline{v}_x \overline{v}_y + \overline{v}_x v_y' + v_x' \overline{v}_y + v_x' v_y')$. Time-averaged in the usual way, these terms become:

$$\frac{\partial(\overline{v}_x \overline{v}_x + \overline{v}_x v_x' + v_x' \overline{v}_x + v_x' v_x')}{\partial x} + \frac{\partial(\overline{v}_x \overline{v}_y + \overline{v}_x v_y' + v_x' \overline{v}_y + v_x' v_y')}{\partial y}$$

$$+ \frac{\partial(\overline{v}_x \overline{v}_z + \overline{v}_x v_z' + v_x' \overline{v}_z + v_x' v_z')}{\partial z}$$

$$= \frac{\partial(\overline{v}_x \overline{v}_x)}{\partial x} + \frac{\partial(\overline{v}_x \overline{v}_y)}{\partial y} + \frac{\partial(\overline{v}_x \overline{v}_z)}{\partial z} + \frac{\partial(\overline{v_x' v_x'})}{\partial x} + \frac{\partial(\overline{v_x' v_y'})}{\partial y} + \frac{\partial(\overline{v_x' v_z'})}{\partial z}, \tag{9.11}$$

which, with the aid of time-averaged continuity equation, (9.8), may be rephrased as:

$$\overline{v}_x \frac{\partial \overline{v}_x}{\partial x} + \overline{v}_y \frac{\partial \overline{v}_x}{\partial y} + \overline{v}_z \frac{\partial \overline{v}_x}{\partial z} + \frac{\partial \overline{(v_x' v_x')}}{\partial x} + \frac{\partial \overline{(v_x' v_y')}}{\partial y} + \frac{\partial \overline{(v_x' v_z')}}{\partial z}. \tag{9.12}$$

By replacing the time-averaged nonlinear terms on the left-hand side of Eqn. (9.9) with those from Eqn. (9.12), the complete time-averaged x momentum balance becomes:

$$\rho \left(\frac{\partial \overline{v}_x}{\partial t} + \overline{v}_x \frac{\partial \overline{v}_x}{\partial x} + \overline{v}_y \frac{\partial \overline{v}_x}{\partial y} + \overline{v}_z \frac{\partial \overline{v}_x}{\partial z} \right) = -\frac{\partial \overline{p}}{\partial x}$$

$$+ \frac{\partial(\overline{\tau}_{xx} - \rho \overline{v_x' v_x'})}{\partial x} + \frac{\partial(\overline{\tau}_{yx} - \rho \overline{v_x' v_y'})}{\partial y} + \frac{\partial(\overline{\tau}_{zx} - \rho \overline{v_x' v_z'})}{\partial z} + \rho g_x. \tag{9.13}$$

Here, the time-averaged viscous shear stresses, such as $\overline{\tau}_{xx}$, are based in the usual way by Newton's law on the time-averaged velocity profiles. Observe that the time-averaged momentum balance is the same as the original (instantaneous) momentum balance, except that:

Example 9.1—Numerical Illustration of a Reynolds Stress Term   **479**

1. Time-averaged values have replaced the original instantaneous values.
2. Additional stresses, known as the *Reynolds stresses*, have now appeared on the right-hand side, and represent the transport of momentum due to turbulent velocity fluctuations. A physical interpretation of these stresses will be given in the next section. Calculations for all but the lowest intensities of turbulence show that a Reynolds stress such as $\tau_{yx}^t = -\rho \overline{v_x' v_y'}$ (the supercript t denotes turbulence) is very much larger than its viscous or molecular counterpart $\overline{\tau}_{yx}$ (except very close to a wall), so the latter can be ignored in most situations.

Time-averaged y and z momentum balances give similar results.

Example 9.1—Numerical Illustration of a Reynolds Stress Term

A short example will give a better understanding of a term such as $\overline{v_x' v_y'}$ that appears in a typical Reynolds stress. For simplicity, consider a hypothetical situation in which only "quantum" values of the fluctuating components are permitted, and only allow each of v_x' and v_y' to have values of either -1 (low) or 1 (high). Investigate the possible values for $\overline{v_x' v_y'}$. (A more elaborate investigation would involve a statistical analysis of a *continuous distribution* of velocities, but the main thrust would not differ significantly from that in the present, more elementary, treatment.)

Table 9.1 Possible Values for Velocity Fluctuations

Correlated			*Correlated*			*Independent*		
v_x'	v_y'	$v_x' v_y'$	v_x'	v_y'	$v_x' v_y'$	v_x'	v_y'	$v_x' v_y'$
-1	-1	1	-1	1	-1	-1	-1	1
-1	-1	1	-1	1	-1	-1	1	-1
1	1	1	1	-1	-1	1	-1	-1
1	1	1	1	-1	-1	1	1	1
Average:		1	Average:		-1	Average:		0

Solution

Table 9.1 shows three possibilities, two in which the fluctuations v_x' and v_y' are correlated, and the other in which they are independent of each other.

Observe that in the first two cases, the time-averaged values $\overline{v_x' v_y'}$ of the product are *nonzero*, being $(1+1+1+1)/4 = 1$ when the individual fluctuations move in concert with each other (high values occur together, and so do low values), and -1 when they are opposed to each other (a high value of one occurs with a low value of the other). In the third example, when the fluctuations are quite independent of each other, the time-averaged value $\overline{v_x' v_y'}$ is zero. The Reynolds stresses are therefore only nonzero when there is a certain degree of correlation between the fluctuations in the different coordinate directions. Such correlations *do* tend to occur in turbulent flow.   □

9.2   Physical Interpretation of the Reynolds Stresses

The Reynolds stresses may also be understood by considering the fragments of fluid, or "eddies," that suddenly jump across a unit area of the plane X–X due to turbulent motion, as shown in Fig. 9.4. Here, we are considering two-dimensional flow with $v_x = \overline{v}_x + v_x'$ and $v_y = 0 + v_y'$ (no overall motion in the y direction).

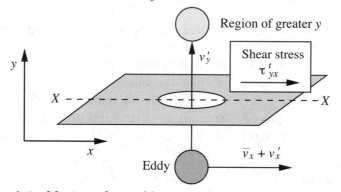

Fig. 9.4   Motion of an eddy upward across the plane X–X.

Observe that the *instantaneous* rate of transfer of x momentum in the y direction, from below X–X to above it, is, per unit area:

$$\rho v_y'(\overline{v}_x + v_x').   \tag{9.14}$$

This result is obtained by noting that the x momentum of the eddy is $\rho(\overline{v}_x + v_x')$ per unit volume, and that the volume flux in the y direction is simply the velocity v_y'. Hence, considering fluid crossing X–X from *either* above or below, at all possible velocities, the *time-averaged* rate of transfer of x momentum in the y direction, is, per unit area:

$$\overline{\rho v_y'(\overline{v}_x + v_x')} = \overline{\rho v_y' \overline{v}_x} + \overline{\rho v_y' v_x'} = 0 + \rho \overline{v_y' v_x'}.   \tag{9.15}$$

The conventional direction for shear stress is expressed in Fig. 5.12 and also by the box in Fig. 9.4, so we need the rate of transfer of x momentum from *above* X–X to *below* it—that is, in the negative y direction; the resulting shear stress due to *turbulent fluctuations* must therefore include a negative sign:

$$\tau_{yx}^t = -\rho \overline{v_y' v_x'}.   \tag{9.16}$$

9.3   Mixing-Length Theory

Unfortunately, there is no universal law whereby the turbulent shear stress of Eqn. (9.16) can be predicted, and it is necessary to resort to a semitheoretical approach in conjunction with experimental observations.

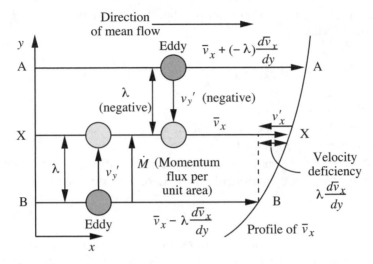

Fig. 9.5   Transport of x momentum across the plane X–X.

Consider the rate of transport $\dot{\mathcal{M}}$ of x momentum, whose time-averaged value is $\rho \bar{v}_x$ per unit volume, across the representative plane X–X shown in Fig. 9.5. Suppose that this momentum is transferred from one plane to another by a series of eddies that—on the average—move a distance λ in the y direction and then suddenly lose their identity and mix completely with the surrounding fluid.

In more detail, there will be x momentum transport across unit area of X–X by elements of fluid coming with velocities v'_y from planes A–A and B–B at distances λ from *above* and *below* X–X. Since v'_y and λ have the same sign for the same eddy (both are positive for the eddy from B–B and both are negative for the eddy from A–A), the same equation holds in either direction. Thus, we only need to consider the eddy traveling upward from B–B.

Also note from Fig. 9.5 that if the value of momentum is $\rho \bar{v}_x$ at the midplane X–X, it will be less than this value at the lower plane B–B; assuming that the gradient $d(\rho \bar{v}_x)/dy$ is approximately constant, the deficiency is $\lambda d(\rho \bar{v}_x)/dy$. A total derivative is used, it being assumed that the time-averaged velocity profile is fully established.

Per unit area, the rate of turbulent transfer of x momentum from B–B across X–X is:

$$\dot{\mathcal{M}} = v'_y \left(\rho \bar{v}_x - \lambda \frac{d(\rho \bar{v}_x)}{dy} \right) \quad \text{(upward)}. \tag{9.17}$$

As noted, Eqn. (9.17) holds equally well for the transfer in *either* direction.

The time-averaged net upward rate of transport, in the y direction, is therefore:

$$\overline{\dot{\mathcal{M}}} = \overline{v'_y \left(\rho \bar{v}_x - \lambda \frac{d(\rho \bar{v}_x)}{dy} \right)} = \overline{v'_y \rho \bar{v}_x} - \overline{v'_y \lambda \frac{d(\rho \bar{v}_x)}{dy}}. \tag{9.18}$$

But, by definition, $\overline{v'_y} = 0$. Also assume that the gradient $d(\rho \bar{v}_x)/dy$ is constant over the region, giving:

$$\overline{\dot{\mathcal{M}}} = -\overline{v'_y \lambda} \frac{d(\rho \bar{v}_x)}{dy} = -\nu_{\mathrm{T}} \frac{d(\rho \bar{v}_x)}{dy}, \tag{9.19}$$

in which the time-averaged quantity $\nu_{\mathrm{T}} = \overline{v'_y \lambda}$ is called the *eddy* kinematic viscosity; the subscript T emphasizes a turbulent quantity.

To proceed further, we must transform $\overline{v'_y \lambda}$ into something more tractable. To start, it can be reexpressed as the product of the RMS (root-mean-square) values of its individual components, multiplied by a correlation-coefficient R:

$$\overline{v'_y \lambda} = R \widetilde{v'_y} \widetilde{\lambda}. \tag{9.20}$$

Also consider the mechanism that causes a turbulent fluctuation v'_x in the x velocity component at the central plane X–X, noting that the curve in Fig. 9.5 represents the time-averaged x velocity, $\bar{v}_x$. Clearly, the x velocity is lower at the plane B–B than it is at the midplane X–X, by an amount $\lambda d\bar{v}_x/dy$, also known as the velocity "deficiency." Thus, an eddy suddenly moving from B–B to X–X will also have an x momentum deficiency, and will impart a negative "kick" to the x velocity at the plane X–X. The resulting fluctuation will be proportional to the velocity deficiency. Approximately, if α is a constant:

$$v'_x = -\alpha \lambda \frac{d\bar{v}_x}{dy}. \tag{9.21}$$

The same result will hold for an eddy traveling downward from A–A to X–X, for in that case λ is negative and the fluctuation is positive. In terms of RMS values:

$$\widetilde{v'_x} = \alpha \widetilde{\lambda} \left| \frac{d\bar{v}_x}{dy} \right|. \tag{9.22}$$

Experimentally, $\widetilde{v'_y}$ tends to be proportional to $\widetilde{v'_x}$—that is, the velocity fluctuations are generally correlated with each other:

$$\widetilde{v'_y} = \beta \widetilde{v'_x} = \alpha \beta \widetilde{\lambda} \left| \frac{d\bar{v}_x}{dy} \right|. \tag{9.23}$$

Here, the coefficient β would be unity for *isotropic turbulence*, in which the intensity of turbulence is uniform in all directions.

From Eqns. (9.19), (9.20), and (9.23):

$$\overline{\mathcal{M}} = -R\widetilde{v_y'}\widetilde{\lambda}\frac{d(\rho\overline{v}_x)}{dy} = -R\alpha\beta\widetilde{\lambda}\widetilde{\lambda}\frac{d(\rho\overline{v}_x)}{dy}\left|\frac{d\overline{v}_x}{dy}\right|$$

$$= -\ell^2\left|\frac{d\overline{v}_x}{dy}\right|\frac{d(\rho\overline{v}_x)}{dy}, \tag{9.24}$$

a result obtained by Prandtl in 1925, in which:

$$\ell = \widetilde{\lambda}\sqrt{R\alpha\beta} \tag{9.25}$$

is the *Prandtl mixing length*.

Also note that the eddy kinematic viscosity can be expressed in terms of the Prandtl mixing length and the velocity gradient:

$$\nu_{\mathrm{T}} = \ell^2\left|\frac{d\overline{v}_x}{dy}\right|. \tag{9.26}$$

Similar analyses can be performed for the rates of turbulent transport of heat and mass (the last typically for a single chemical species when others are present). All results are summarized in Table 9.2, in which c_p denotes specific heat, and $\overline{T}$ and $\overline{c}$ are the time-averaged temperature and species concentration, respectively. Since the shear stress conventionally denotes a transfer of momentum in the *negative y* direction (see Section 9.2), there is a minus sign preceding τ_{yx}^t; however, heat and mass transfer are conventionally taken in the *positive y* direction so no minus sign is needed in these cases. Partial derivatives are used for the temperature and concentration gradients because in a heat exchanger or tubular reactor $\overline{T}$ and $\overline{c}$ would almost certainly vary both in the transverse (y) direction *and* in the direction of the main flow.

Table 9.2 Summary of Relations for Turbulent Transport

Property Being Transported		Rates of Turbulent Transport $\overline{N}^t$				
Momentum	$\rho\overline{v}_x$	$-\rho\nu_{\mathrm{T}}\dfrac{d\overline{v}_x}{dy}$	$=$	$-\rho\ell^2\dfrac{d\overline{v}_x}{dy}\left\|\dfrac{d\overline{v}_x}{dy}\right\|$	$=$	$-\tau_{yx}^t$
Heat	$\rho c_p\overline{T}$	$-\rho c_p\nu_{\mathrm{T}}\dfrac{\partial\overline{T}}{\partial y}$	$=$	$-\rho c_p\ell^2\dfrac{\partial\overline{T}}{\partial y}\left\|\dfrac{d\overline{v}_x}{dy}\right\|$	$=$	q^t
Mass	$\overline{c}$	$-\nu_{\mathrm{T}}\dfrac{\partial\overline{c}}{\partial y}$	$=$	$-\ell^2\dfrac{\partial\overline{c}}{\partial y}\left\|\dfrac{d\overline{v}_x}{dy}\right\|$	$=$	N^t

Table 9.2 assumes that the eddy thermal diffusivity α_T and the eddy diffusivity $\mathcal{D}_T$ are the same as the eddy kinematic viscosity ν_{T}, which is true only as a first approximation. Section 9.13 explores the relation between turbulent heat and momentum transport in more detail.

9.4 Determination of Eddy Kinematic Viscosity and Mixing Length

Fig. 9.6 *Shear stresses for (a) parallel plates, (b) pipe.*

Note that the eddy kinematic viscosity can be determined fairly readily as a function of position. For flow between parallel plates and for pipe flow, momentum balances on the elements shown in Fig. 9.6 give:

$$\text{Parallel plates}:\ \tau_{yx}^{\ell}+\tau_{yx}^{t} \doteq \tau_{yx}^{t} = y\left(\frac{dp}{dx}\right),\quad \text{Pipe flow}:\ \tau_{rz}^{\ell}+\tau_{rz}^{t} \doteq \tau_{rz}^{t} = \frac{r}{2}\left(\frac{dp}{dz}\right),$$

(9.27)

where y is here measured from the centerline. Here, the typically very small laminar or viscous shear stresses $\tau_{yx}^{\ell} = \mu d\bar{v}_x/dx$ and $\tau_{rz}^{\ell} = \mu d\bar{v}_z/dr$ have been ignored, but could easily be retained if high accuracy is needed, especially near the walls. Since pressure drops and hence pressure gradients can easily be measured, the turbulent shear stress τ_{yx}^{t} or τ_{rz}^{t} can be deduced at any location. Velocity profiles and hence velocity gradients such as $d\bar{v}_x/dy$ can also be measured by means of a Pitot tube or laser-Doppler velocimetry. The mixing length ℓ and the eddy kinematic viscosity ν_{T} are then obtained from:

$$\text{Parallel plates}:\ \tau_{yx}^{t} = \rho\,\underbrace{\ell^2\left|\frac{d\bar{v}_x}{dy}\right|}_{\nu_{\mathrm{T}}}\frac{d\bar{v}_x}{dy},\quad \text{Pipe flow}:\ \tau_{rz}^{t} = \rho\,\underbrace{\ell^2\left|\frac{d\bar{v}_z}{dr}\right|}_{\nu_{\mathrm{T}}}\frac{d\bar{v}_z}{dr}.$$

(9.28)

For both smooth *and* rough pipes, experiments show[5] that the ratio of the mixing length to the pipe radius is given very accurately by:

$$\frac{l}{a} = 0.14 - 0.08\left(1-\frac{y}{a}\right)^2 - 0.06\left(1-\frac{y}{a}\right)^4,$$

(9.29)

[5] See Eqn. (20.18) in H. Schlichting, *Boundary Layer Theory*, Pergamon Press, New York, 1955.

which is plotted in Fig. 9.7. Note that as the wall is approached, $l = 0.4y$, which is the simplified result used in the Prandtl hypothesis. However, at greater distances from the wall, the eddies have more freedom to travel and a maximum value is reached at the centerline.

Fig. 9.7 *Variation of dimensionless mixing length with location.*

Further, the experiments of Nikuradse in *smooth* pipes show the variation of the dimensionless eddy kinematic viscosity is given approximately by:

$$\frac{\nu_\text{T}}{a\sqrt{\tau_\text{w}/\rho}} \doteq 0.32\,\frac{y}{a}\left(1 - \frac{y}{a}\right), \tag{9.30}$$

and very accurately by:

$$\frac{\nu_\text{T}}{a\sqrt{\tau_\text{w}/\rho}} = 0.4070\,\frac{y}{a} - 0.7012\left(\frac{y}{a}\right)^2 + 0.4505\left(\frac{y}{a}\right)^3 - 0.1437\left(\frac{y}{a}\right)^5, \tag{9.31}$$

in which there is no term in $(y/a)^4$. The accurate correlation of Eqn. (9.31) is plotted in Fig. 9.8.

Recall from Eqn. (9.28) that the eddy viscosity is given by the product of the square of the mixing length and the magnitude of the velocity gradient:

$$\nu_\text{T} = l^2\left|\frac{d\bar{v}_z}{dr}\right| = l^2\left|\frac{d\bar{v}_z}{dy}\right|. \tag{9.32}$$

The shape of the curve in Fig. 9.8 is now easily explained. Moving away from the wall, as l increases, the eddy viscosity also increases and indeed reaches a maximum halfway to the centerline. Thereafter, the declining velocity gradient is the dominant factor, and the eddy viscosity falls significantly (but not quite to zero) as the centerline is reached.

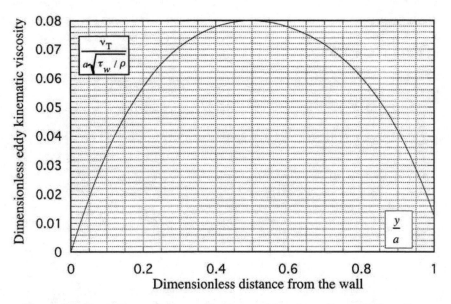

Fig. 9.8   *Variation of dimensionless eddy viscosity with location.*

9.5   Velocity Profiles Based on Mixing-Length Theory

The Prandtl hypothesis.   Consider pipe flow, with axial coordinate z. The simplest model—that of Prandtl—assumes a direct proportionality between the mixing length ℓ and the distance y *from the wall*, the argument being that the eddies have more freedom of motion the further they are away from the wall. Also assume a constant shear stress τ_{yz}^t, equal to its value τ_w at the wall, which is strictly only true for a small interval near the wall:

$$\ell = ky, \tag{9.33}$$

$$\tau_{yz}^t \doteq \tau_w. \tag{9.34}$$

Actually, Eqns. (9.33) and (9.34) are overestimates for both ℓ and τ_{yz}^t; fortuitously, both overestimates will tend to cancel each other and give an excellent result for the turbulent velocity profile! Note that for the moment it is more expedient to work with y than r as the transverse coordinate, hence the notation τ_{yz}^t.

Mixing-length theory also gave the formula:

$$\tau_{yz}^t = \rho \ell^2 \left(\frac{d\overline{v}_z}{dy} \right)^2, \tag{9.35}$$

Example 9.2—Investigation of the von Kármán Hypothesis   **487**

in which $d\bar{v}_z/dy$ is recognized as positive, since the time-averaged velocity increases as the distance from the wall increases.[6] From the above:

$$\tau_w = \rho k^2 y^2 \left(\frac{d\bar{v}_z}{dy} \right)^2, \tag{9.36}$$

which integrates to:

$$\bar{v}_z = \frac{1}{k} \sqrt{\frac{\tau_w}{\rho}} \ln y + c, \tag{9.37}$$

in which c is a constant of integration. Experimentally, it is found, perhaps rather surprisingly, that the velocity profile of Eqn. (9.37) *also* holds in the central region of a pipe, with $k \doteq 0.4$. Since $\ln y$ is fairly insensitive to changes in y, Eqn. (9.34) predicts a fairly "flat" turbulent velocity profile, in accordance with experimental observation.

Note that the logarithmic law cannot possibly hold *very* close to the wall, because as y tends to zero, it gives an ever-increasing negative velocity, and an ever-increasing velocity gradient.

Example 9.2—Investigation of the von Kármán Hypothesis

Von Kármán proposed that the mixing length should be proportional to the ratio of the first two derivatives of the velocity:

$$\ell = -k \frac{d\bar{v}_z/dy}{d^2\bar{v}_z/dy^2}, \tag{E9.2.1}$$

in which k is a dimensionless constant. Investigate the merits of this hypothesis and compare its predictions for the velocity profile with that of Prandtl.

Solution

The Prandtl hypothesis expressed the mixing length in terms of the distance from the wall. But turbulence can occur in a region well away from a wall (in the upper atmosphere, to take an extreme example), where the impact of the wall is small—that is, where there is no obvious length scale on which to base ℓ.

First, observe that the ratio of the first and second derivatives of the velocity in Eqn. (E9.2.1) has indeed the desired dimensions of length. Thus, an obvious advantage of the von Kármán hypothesis is that it gives the mixing length in terms of the *local* behavior of the velocity profile, rather than relying on the presence of a wall for the length scale.

---

[6] For the rest of the chapter, the designation "time-averaged" will be dropped, it being assumed that any quantity with an overline is so denoted.

From Eqns. (9.32) and (E9.2.1), again assuming near the wall that $\tau_{yz}^t \doteq \tau_w$:

$$\sqrt{\frac{\tau_w}{\rho}} = \ell \frac{d\bar{v}_z}{dy} = -k \frac{(d\bar{v}_z/dy)^2}{d^2\bar{v}_z/dy^2}, \tag{E9.2.2}$$

which integrates to:

$$\int \frac{d\left(\dfrac{d\bar{v}_z}{dy}\right)}{\left(\dfrac{d\bar{v}_z}{dy}\right)^2} = -\frac{k}{\sqrt{\tau_w/\rho}} \int dy, \quad \text{or} \quad \frac{1}{\dfrac{d\bar{v}_z}{dy}} = \frac{ky}{\sqrt{\tau_w/\rho}} + c_1. \tag{E9.2.3}$$

The constant of integration c_1 disappears if we assume that the velocity gradient $d\bar{v}_z/dy$ becomes very steep and approaches infinity at the wall, where $y = 0$. A second integration gives:

$$\bar{v}_z = \frac{1}{k}\sqrt{\frac{\tau_w}{\rho}} \ln y + c, \tag{E9.2.4}$$

which is identical with the result from Prandtl's hypothesis.

A disadvantage of the von Kármán hypothesis is that it fails if $d^2\bar{v}_z/dy^2$ is zero, a situation that *can* quite easily occur—for example, at the inflection points in the velocity profiles of the turbulent jets discussed in Section 9.13.   □

Suppose that the more realistic shear-stress distribution for pipe flow, $\tau_{rz}^t = -\tau_w(r/a)$, where a is the pipe radius and r is the distance from the centerline, is used in conjunction with the mixing-length theory of Eqn. (9.29) and the von Kármán hypothesis of Example 9.2. A double integration, noting an infinite velocity gradient at the wall to determine one constant of integration, gives:

$$\bar{v}_z = (\bar{v}_z)_{r=0} + \frac{1}{k}\sqrt{\frac{\tau_w}{\rho}}\left[\sqrt{\frac{r}{a}} + \ln\left(1 - \sqrt{\frac{r}{a}}\right)\right], \tag{9.35}$$

in which $(\bar{v}_z)_{r=0}$ is the centerline velocity. For reasons already given, this seemingly more sophisticated result is not as accurate as the simpler law of Eqn. (9.34).

9.6 The Universal Velocity Profile for Smooth Pipes

The logarithmic law of Eqn. (9.37) is now rephrased in terms of dimensionless variables. First, define a dimensionless distance from the wall, y^+, and a dimensionless velocity, v_z^+, by:

$$y^+ = \frac{y\sqrt{\tau_w/\rho}}{\nu}, \tag{9.38}$$

$$v_z^+ = \frac{\overline{v}_z}{\sqrt{\tau_w/\rho}}, \tag{9.39}$$

in which ν is the kinematic viscosity. Equation (9.37) can then be rewritten as:

$$\frac{\overline{v}_z}{\sqrt{\tau_w/\rho}} = A + B \ln \frac{y\sqrt{\tau_w/\rho}}{\nu}, \tag{9.40}$$

or:

$$v_z^+ = A + B \ln y^+, \tag{9.41}$$

in which A is a constant of integration and $B = 1/k$. The quantity $\sqrt{\tau_w/\rho}$ is known as the *friction velocity* and is given the symbol $\overline{v}_z^*$.

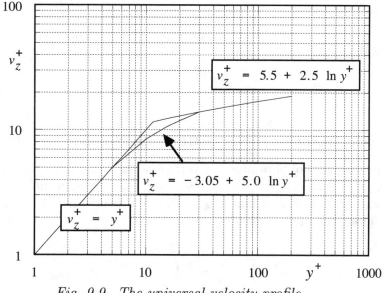

Fig. 9.9   The universal velocity profile.

Equation (9.41) is known as the *universal velocity profile*, because it gives exceptionally good agreement with experimental velocities over a very wide range of Reynolds number for *smooth pipes*.[7] Experimentally, the constants are $A = 5.5$ and $B = 2.5$, giving:

$$v_z^+ = 5.5 + 2.5 \ln y^+. \tag{9.42}$$

Of course, Eqn. (9.42) does not hold very close to the wall, where, by integration of $\tau_{yz} = \mu \, d\overline{v}_z/dy$ in the laminar sublayer:

$$v_z^+ = y^+. \tag{9.43}$$

As shown in Fig. 9.9, the turbulent and laminar velocity profiles of Eqns. (9.42) and (9.43) intersect at $y^+ \doteq 11.63$.

---

[7] For example, see the velocities displayed on page 405 of Schlichting, *op. cit.*

To represent the experimental data of Nikuradse even better, particularly in a *buffer region* between the laminar sublayer and the fully developed turbulent core, von Kármán divided the complete velocity profile into three regions:

Laminar sublayer:

$$0 < y^+ < 5: \qquad v_z^+ = y^+. \tag{9.44}$$

Buffer region:

$$5 < y^+ < 30: \qquad v_z^+ = -3.05 + 5 \ln y^+. \tag{9.45}$$

Turbulent core:

$$30 < y^+: \qquad v_z^+ = 5.5 + 2.5 \ln y^+. \tag{9.46}$$

The following velocity profile, based on a theory of Sleicher and discussed further in Problem 9.10, also gives an excellent fit of turbulent velocities in the laminar sublayer and buffer regions:

$$v_z^+ = \frac{1}{0.088} \tan^{-1}(0.088 y^+). \tag{9.47}$$

9.7 Friction Factor in Terms of Reynolds Number for Smooth Pipes

For the purposes of obtaining the total flow rate, and hence the mean velocity, the logarithmic profile $v_z^+ = A + B \ln y^+$ can be assumed to hold over the whole pipe, because the laminar sublayer is so thin and contributes virtually nothing to the flow. Thus, the mean z velocity is:

$$\bar{v}_{zm} = \frac{1}{\pi a^2} \int_0^a 2\pi r \bar{v}_z \, dr, \tag{9.48}$$

which can be shown by integration (see Example 9.3) to equal:

$$\bar{v}_{zm} = \sqrt{\frac{\tau_w}{\rho}} \left[B \ln \frac{a\sqrt{\tau_w/\rho}}{\nu} + A - \frac{3}{2} B \right]. \tag{9.49}$$

Thus, for the profile with $A = 5.5$ and $B = 2.5$ in Eqn. (9.46):

$$v_z^+ = 5.5 + 2.5 \ln y^+, \tag{9.46}$$

the integration gives:

$$\bar{v}_{zm} = \sqrt{\frac{\tau_w}{\rho}} \left[2.5 \ln \left(\frac{D\bar{v}_{zm}}{2\nu} \sqrt{\frac{\tau_w}{\rho \bar{v}_{zm}^2}} \right) + 1.75 \right]. \tag{9.50}$$

Example 9.3—Expression for the Mean Velocity   **491**

Recall the Fanning friction factor, defined as:

$$f_F = \frac{\tau_w}{\frac{1}{2}\rho\bar{v}_{zm}^2},$$

(9.51)

in which $\bar{v}_{zm}$ has the same meaning as u_m used in Chapter 3. It follows from Eqns. (9.50) and (9.51) that—based on the universal velocity profile—the friction factor and Reynolds number are related by:

$$\sqrt{\frac{1}{f_F}} = 1.77\ln\left(\text{Re}\sqrt{f_F}\right) - 0.601.$$

(9.52)

Experimentally, a law similar to Eqn. (9.52),

$$\sqrt{\frac{1}{f_F}} = 1.74\ln\left(\text{Re}\sqrt{f_F}\right) - 0.391.$$

(9.53)

agrees with experiment for smooth pipes over a very wide range of Re (from 5×10^3 up to at least 3.4×10^6 and probably beyond).

Example 9.3—Expression for the Mean Velocity

For the velocity profile given by $v_z^+ = A + B\ln y^+$, prove the formula for the mean velocity given in Eqn. (9.49).

Solution

Noting the definitions of the dimensionless velocity and distance from the wall, the given velocity profile can be rewritten as:

$$\bar{v}_z = \sqrt{\frac{\tau_w}{\rho}}\left[A + B\ln\frac{y\sqrt{\tau_w/\rho}}{\nu}\right].$$

(E9.3.1)

The mean velocity $\bar{v}_{zm}$ is therefore:

$$\frac{1}{\pi a^2}\int_{r=0}^{r=a} 2\pi r\bar{v}_z\,dr = \frac{1}{\pi a^2}\int_{y=0}^{y=a} 2\pi\bar{v}_z(a-y)\,dy \;=$$

(E9.3.2)

$$\frac{2}{a^2}\sqrt{\frac{\tau_w}{\rho}}\left[\int_0^a A(a-y)\,dy + \int_0^a B(a-y)\ln y\,dy + \int_0^a B(a-y)\ln\frac{\sqrt{\tau_w/\rho}}{\nu}\,dy\right] \;=$$

$$\frac{2}{a^2}\sqrt{\frac{\tau_w}{\rho}}\left[\int_0^a\left(A + B\ln\frac{\sqrt{\tau_w/\rho}}{\nu}\right)(a-y)\,dy + \int_0^a Ba\ln y\,dy - \int_0^a By\ln y\,dy\right].$$

Noting that $\int\ln y\,dy = y\ln y - y$ and $\int y\ln y\,dy = (y^2/2)\ln y - (y^2/4)$, integration, insertion of limits, and collection of terms gives:

$$\bar{v}_{zm} = \sqrt{\frac{\tau_w}{\rho}}\left[B\ln\frac{a\sqrt{\tau_w/\rho}}{\nu} + A - \frac{3}{2}B\right],$$

(9.49)

which is the desired result.   □

9.8 Thickness of the Laminar Sublayer

The thickness of the laminar sublayer will be diminished as the Reynolds number increases, because the more vigorous turbulent eddies approach the wall more closely. For simplicity in the following analysis, ignore the presence of the buffer region, and assume just a "two-piece" model, which is needed for the Prandtl-Taylor analogy of Section 9.12.

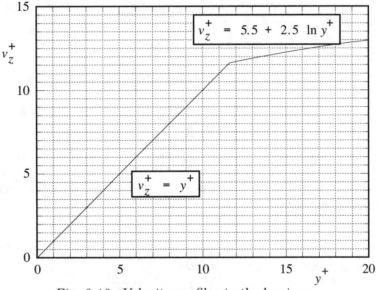

Fig. 9.10   *Velocity profiles in the laminar sublayer and part of the turbulent core.*

Start with the known velocity profiles in the two regions, which are also shown in Fig. 9.10:

Laminar sublayer:

$$0 < y^+ < 11.6: \qquad v_z^+ = y^+. \tag{9.54}$$

Turbulent core:

$$11.6 < y^+: \qquad v_z^+ = 5.5 + 2.5 \ln y^+. \tag{9.55}$$

The thickness δ of the laminar sublayer is now determined as a function of the Reynolds number. At the junction between the two regions, the velocities must match each other:

$$v_z^+ = y^+ = 5.5 + 2.5 \ln y^+, \qquad \text{giving} \qquad y^+ = 11.6. \tag{9.56}$$

That is, the thickness δ of the laminar sublayer and the velocity $\overline{v}_{z\delta}$ at the laminar sublayer/turbulent core junction are:

$$\delta = 11.6 \frac{\nu}{\sqrt{\tau_w/\rho}}, \qquad \text{and} \qquad \overline{v}_{z\delta} = 11.6 \sqrt{\frac{\tau_w}{\rho}}. \tag{9.57}$$

That is,

$$\frac{\delta}{D} = 11.6\,\frac{\mu}{\rho\bar{v}_{zm}D}\sqrt{\frac{\rho\bar{v}_{zm}^2}{\tau_w}} = \frac{16.4}{\text{Re}\sqrt{f_{\text{F}}}}. \tag{9.58}$$

Although f_{F} is given by Eqn. (9.53), a somewhat simpler relationship—of admittedly more limited range ($\text{Re} < 10^5$)—is the *Blasius* law:

$$f_{\text{F}} = 0.0790\,\text{Re}^{-1/4}. \tag{9.59}$$

From Eqns. (9.56) and (9.57), the ratio of the laminar sublayer thickness to the pipe diameter is:

$$\frac{\delta}{D} = \frac{58.3}{\text{Re}^{7/8}}. \tag{9.60}$$

For a slightly different velocity-profile equation, the *one-seventh* power law,

$$\frac{\bar{v}_z}{\bar{v}_{zc}} = \left(\frac{y}{a}\right)^{1/7}, \tag{9.61}$$

in which $\bar{v}_{zc}$ is the centerline velocity, the result is only marginally different:

$$\frac{\delta}{D} = \frac{61.7}{\text{Re}^{7/8}}. \tag{9.62}$$

If representative values for the Reynolds numbers are substituted into the above equations, they will show that:

1. The laminar sublayer is an *extremely* thin region, across which the velocity builds up to a significant portion of its centerline value.
2. The thickness of the laminar sublayer decreases almost in direct proportion to the value of the Reynolds number.

If a *three-piece* velocity profile is considered, as in Fig. 9.9, then the outer edge of the laminar sublayer occurs at $y^+ = 5$, and the ratio δ/D becomes:

$$\frac{\delta}{D} = \frac{25.2}{\text{Re}^{7/8}}. \tag{9.63}$$

9.9 Velocity Profiles and Friction Factor for Rough Pipe

In turbulent flow, the wall irregularities may or may not project through the laminar sublayer. Depending on the extent of the wall roughness ε as discussed in Section 3.4, three regimes may be delineated—each supported by experiment—with differing dependencies for the friction factor:

Hydraulically smooth pipe:

$$0 \; < \; \frac{\varepsilon \overline{v}_z^*}{\nu} \; < \; 5: \qquad f_F = f_F(\text{Re}). \tag{9.64}$$

Transition state:

$$5 \; < \; \frac{\varepsilon \overline{v}_z^*}{\nu} \; < \; 70: \qquad f_F = f_F\left(\text{Re}, \; \frac{\varepsilon}{a}\right). \tag{9.65}$$

Completely rough pipe:

$$70 \; < \; \frac{\varepsilon \overline{v}_z^*}{\nu} : \qquad f_F = f_F\left(\frac{\varepsilon}{a}\right). \tag{9.66}$$

The turbulent velocity profiles in all three regions can be represented by:

$$\frac{\overline{v}_z}{\overline{v}_z^*} = 2.5 \ln \frac{y}{\varepsilon} + B, \tag{9.67}$$

in which the values of B and the corresponding velocity profiles, except for the transition region, are:

Hydraulically smooth pipe:

$$B = 5.5 + 2.5 \ln \frac{\varepsilon \overline{v}_z^*}{\nu}, \qquad \frac{\overline{v}_z}{\overline{v}_z^*} = 5.5 + 2.5 \ln \frac{y\sqrt{\tau_w/\rho}}{\nu}. \tag{9.68}$$

Completely rough pipe:

$$B = 8.5, \qquad \frac{\overline{v}_z}{\overline{v}_z^*} = 8.5 + 2.5 \ln \frac{y}{\varepsilon}. \tag{9.69}$$

For smooth pipe, we have already noted that Eqn. (9.68) is the universal velocity profile. For completely rough pipe, Eqn. (9.69) is independent of the pipe radius, and its validity is completely substantiated by experiments over a very wide range of roughness ratios ε/D.[8]

The *Colebrook and White* equation is consistent with the above observations and represents the variation of the Fanning friction factor with the relative roughness and the Reynolds number over a very wide range of roughnesses and (turbulent) Reynolds numbers:

$$\frac{1}{\sqrt{f_F}} = -1.737 \ln \left(0.269 \frac{\varepsilon}{D} + \frac{1.257}{\text{Re}\sqrt{f_F}} \right). \tag{9.70}$$

---

[8] See page 420 of Schlichting, *op. cit.*, for experimental evidence.

9.10 Blasius-Type Law and the Power-Law Velocity Profile

Experimentally, for smooth pipe, provided that the Reynolds number is no larger than 10^5:

$$f_{\mathrm{F}} = 0.0790 \left(\frac{2a\bar{v}_{zm}}{\nu} \right)^{-1/4}. \tag{9.71}$$

More generally, for other values of the Reynolds number:

$$f_{\mathrm{F}} = A \left(\frac{a\bar{v}_{zm}}{\nu} \right)^{-1/n}. \tag{9.72}$$

But, by definition of the friction factor:

$$\tau_w = \frac{1}{2} f_{\mathrm{F}} \rho \bar{v}_{zm}^2 = \rho (\bar{v}_z^*)^2. \tag{9.73}$$

From Eqns. (9.71) and (9.72), elimination of the friction factor gives:

$$\frac{\bar{v}_{zm}}{\bar{v}_z^*} = \left(\frac{2}{A} \right)^{n/(2n-1)} \left(\frac{\bar{v}_z^* a}{\nu} \right)^{1/(2n-1)}. \tag{9.74}$$

Now take a power-law velocity profile of the general form:

$$\frac{\bar{v}_z}{\bar{v}_z^*} = c \left(\frac{y}{y^*} \right)^m, \tag{9.75}$$

which gives, after suitable integration:

$$\frac{\bar{v}_{zm}}{\bar{v}_z^*} = \left(\frac{\bar{v}_z^* a}{\nu} \right)^m \frac{2c}{(m+1)(m+2)}. \tag{9.76}$$

By equating exponents on the Reynolds number in Eqns. (9.74) and (9.76):

$$m = \frac{1}{2n-1}. \tag{9.77}$$

For example, if $n = 4$, then $m = 1/7$, which is a familiar result, and is shown in Fig. 9.11.

There is also another relation between c in the velocity profile and A in the expression for f_{F}. Determination of either of these experimentally shows that $c = 8.74$, that is:

$$\frac{\bar{v}_z}{\bar{v}_z^*} = 8.74 \left(\frac{y\bar{v}_z^*}{\nu} \right)^{1/7}. \tag{9.78}$$

Fig. 9.11   One-seventh power-law velocity profile.

Solving for $\overline{v}_z^*$:

$$\overline{v}_z^* = 0.150\,\overline{v}_z^{7/8}\left(\frac{\nu}{y}\right)^{1/8}, \tag{9.79}$$

which can now be back-substituted into the expression for the wall shear stress, Eqn. (9.73):

$$\tau_w = \rho(\overline{v}_z^*)^2 = 0.0225\,\rho\overline{v}_z^2\left(\frac{\nu}{\overline{v}_z y}\right)^{1/4}. \tag{9.80}$$

This last expression is analogous to the Blasius law, but now contains the velocity $\overline{v}_z$ at a definite distance y from the wall. Note that it was used for finding the thickness of the turbulent boundary layer in Section 8.5.

9.11   A Correlation for the Reynolds Stresses

Fig. 9.12 shows how the total shear stress for flow in a pipe is the sum of the viscous and turbulent contributions. Note that the turbulent or Reynolds stress term dominates over most of the region, but declines sharply to zero at the wall, whereas that the viscous term is the main effect very close to the wall.

In recent years, S.W. Churchill and S.C. Zajic have proposed a valuable alternative to mixing length and eddy viscosity correlations for flow in simple geometries, and their approach concentrates instead on the Reynolds stress term.[9]

---

[9] S.W. Churchill and S.C. Zajic, "Prediction of fully developed turbulent convection with minimal explicit empiricism," *American Institute of Chemical Engineers Journal*, Vol. 48, No. 5 (May 2002), p. 927.

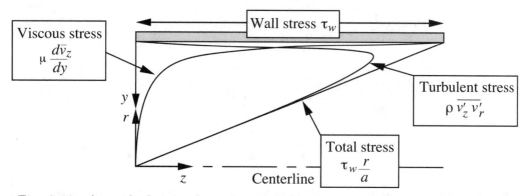

Fig. 9.12 A graph showing how the total shear stress is the sum of the viscous and turbulent stresses. All quantities shown are positive. (Not to scale.)

For pipe flow the starting point is the time-averaged z momentum equation, similar to Eqn. (9.13). With the usual simplifications, there results:

$$\frac{d\overline{p}}{dz} = \frac{1}{r}\frac{d[r(\overline{\tau}_{rz} - \rho\overline{v_z'v_r'})]}{dr}. \tag{9.81}$$

The wall shear stress τ_w, which is traditionally taken to be a positive quantity, is related to the pressure gradient by Eqn. (9.27):

$$-\tau_w = \frac{a}{2}\frac{d\overline{p}}{dz}. \tag{9.82}$$

Thus, separation of variables and integration of Eqn. (9.81) gives:

$$-2\frac{\tau_w}{a}\int_0^r r\,dr = \int_0^r d[r(\overline{\tau}_{rz} - \rho\overline{v_z'v_r'})], \tag{9.83}$$

Since both integrands are zero at the lower limit, the result is:

$$\tau_w\frac{r}{a} = -\overline{\tau}_{rz} + \rho\overline{v_z'v_r'}. \tag{9.84}$$

But the viscous stress is given by $\overline{\tau}_{rz} = \mu\,d\overline{v}_z/dr = -\mu\,d\overline{v}_z/dy$, in which $y = a - r$ is the distance from the wall, giving:

$$\underbrace{\tau_w\frac{r}{a}}_{\text{Total}} = \underbrace{\mu\frac{d\overline{v}_z}{dy}}_{\text{Viscous}} + \underbrace{\rho\overline{v_z'v_r'}}_{\text{Reynolds}}. \tag{9.85}$$

Fig. 9.12 illustrates Eqn. (9.85) by showing approximately how the viscous and Reynolds stresses vary, such that their sum equals the total stress.

Churchill and coworkers have developed an accurate correlation for a quantity that they designate $(\overline{v'_z v'_r})^{++}$, namely $\rho \overline{v'_z v'_r}/[\tau_w (r/a)]$, the fraction of the total stress that is attributable to the Reynolds stress:

$$(\overline{v'_z v'_r})^{++} = \frac{\rho \overline{v'_z v'_r}}{\tau_w (r/a)} = f(y^+, a^+) = \left(\left[0.7 \left(\frac{y^+}{10} \right)^3 \right]^{-8/7} \right.$$

$$\left. + \left| \exp \left(\frac{-1}{0.436 y^+} \right) - \frac{1}{0.436 a^+} \left(1 + \frac{6.95 y^+}{a^+} \right) \right|^{-8/7} \right)^{-7/8}. \quad (9.86)$$

Here, the form and values of the coefficients have been chosen by a variety of methods to give the best fit to available evidence.

The seemingly curious arrangements of the exponents comes directly from an ingenious way of interpolating between two asymptotes or limiting values, documented by Churchill and Usagi[10]:

$$[y(x)]^p \doteq [y_0(x)]^p + [y_\infty(x)]^p. \quad (9.87)$$

The functions $y_0(x)$ and $y_\infty(x)$ are arranged so that they dominate, respectively, at the opposite extremes of the range of values of x, so that $y(x)$ is correctly represented at these extremes. The exponent p is chosen to represent the data most accurately between these limits.

Churchill[11] points out that the primary purpose of the correlation of Eqn. (9.86) is to provide predictions for $\overline{v}_z$ and $\overline{v}_{zm}$ that are almost exact. Starting from Eqn. (9.85), substituting for $\rho \overline{v'_z v'_r}$ from Eqn. (9.86), noting that $y = a - r$, and rearranging, we have:

$$\tau_w \left(1 - \frac{y}{a} \right) \left(1 - f(y^+, a^+) \right) = \mu \frac{d\overline{v}_z}{dy}. \quad (9.88)$$

After expressing $\overline{v}_z$ and y in dimensionless form, this differential equation may be integrated successively to give the velocity profile and the mean velocity, and hence—by a development similar to that in Section 9.7—the friction factor.

An additional article by Churchill, Shinoda, and Arai gives an excellent historical review of several correlative strategies for turbulent flow in pipes.[12]

---

[10] S.W. Churchill and R. Usagi, "A general expression for the correlation of rates of transfer and other phenomena," *American Institute of Chemical Engineers Journal*, Vol. 18, p. 1121 (1972).

[11] Private communication, April 2005.

[12] S.W. Churchill, M. Shinoda, and N. Arai, "An appraisal of experimental, predictive, and correlative contributions to fully developed turbulent flow in a round tube," *Thermal Science & Engineering*, Vol. 10, No. 2 (2002).

9.12 Computation of Turbulence by the k/ϵ Method

Mixing-length theory as presented in the first part of this chapter provides a good explanation for turbulent velocities in a pipe, but suffers from two insurmountable drawbacks as the basis for turbulent flow in more general situations:

1. It essentially describes a *local* phenomenon. But turbulence at one location could easily be influenced by convection currents from turbulence upstream, or even downstream in the case of a recirculating flow—as in the eddies for the orifice plate in Fig. 2.19 or the sudden expansion in Fig. 2.20.
2. It cannot predict turbulent flow patterns for any new situation that involves complex geometry.

Hitherto, mixing-length theory gave a turbulent velocity fluctuation, such as v'_x, that was proportional to the mean velocity gradient, such as $d\bar{v}_x/dy$ in Eqn. (9.21). But Prandtl and Kolmorogov reasoned that it would be more logical to relate quantities such as v'_x and the Reynolds stresses to a property of the *turbulence itself*. Two such properties are:

1. The turbulent *kinetic energy* per unit mass of the fluctuating components: $k = (v'^2_x + v'^2_y + v'^2_z)/2$, with dimensions L^2/T^2. In *index notation*, where summation is implied for repeated subscripts, the kinetic energy can also be written as $k = v'_i v'_i/2$.
2. The turbulent *dissipation rate* ε of this kinetic energy, with dimensions L^2/T^3.

Recall from Example 5.9 the vector form of the three momentum balances for a fluid of constant density but *variable* viscosity:

$$\rho\frac{\partial \mathbf{v}}{\partial t} + \rho\mathbf{v} \cdot \nabla\mathbf{v} = -\nabla p + \nabla \cdot \mu(\nabla\mathbf{v} + (\nabla\mathbf{v})^T) + \mathbf{F}. \tag{9.89}$$

For *turbulent* flow, the Reynolds stresses can be included by enhancing the molecular viscosity μ with the turbulent eddy viscosity $\rho\nu_T$. Bearing in mind that the time-averaged velocity $\bar{\mathbf{v}}$ and pressure $\bar{p}$ are now involved, we have:

$$\rho\frac{\partial \bar{\mathbf{v}}}{\partial t} + \rho\bar{\mathbf{v}} \cdot \nabla\bar{\mathbf{v}} = -\nabla\bar{p} + \nabla \cdot (\mu + \rho\nu_T)(\nabla\bar{\mathbf{v}} + (\nabla\bar{\mathbf{v}})^T) + \mathbf{F}, \tag{9.90a}$$

$$\nabla \cdot \bar{\mathbf{v}} = 0. \tag{9.90b}$$

in which the continuity equation has been included for completeness.

The k/ε method asserts that the eddy kinematic viscosity at any point should depend only on k and ε at that point. Since ν_T has dimensions L^2/T, it follows that:

$$\nu_T = C_\mu \frac{k^2}{\varepsilon}, \tag{9.91}$$

in which C_μ is the first of five adjustable coefficients.

We are then faced with the problem of how to compute k and ε as functions of time and position. Starting from the time-averaged equations of motion, and also considering the rate of dissipation of energy by viscous action, a long and involved derivation, with a small amount of empiricism, leads to the transport equation for the kinetic energy. The reader who wishes to learn the details of this and/or related matters, is referred to a number of sources.[13, 14, 15, 16, 17] Here, we only really need the final transport equation for the turbulent kinetic energy:

$$\rho\frac{\partial k}{\partial t} + \rho\overline{\mathbf{v}}\cdot\nabla k = \nabla\cdot\left[\left(\mu + \rho\frac{C_\mu k^2}{\sigma_k\varepsilon}\right)\nabla k\right] + \rho C_\mu\frac{k^2}{\varepsilon}(\nabla\overline{\mathbf{v}} + (\nabla\overline{\mathbf{v}})^T)^2 - \rho\varepsilon. \quad (9.92)$$

Largely by analogy—that is, with less rigor—a similar equation governs variations of the dissipation rate:

$$\rho\frac{\partial \varepsilon}{\partial t} + \rho\overline{\mathbf{v}}\cdot\nabla \varepsilon = \nabla\cdot\left[\left(\mu + \rho\frac{C_\mu k^2}{\sigma_\varepsilon\varepsilon}\right)\nabla \varepsilon\right] + \rho C_{\varepsilon 1}C_\mu k(\nabla\overline{\mathbf{v}} + (\nabla\overline{\mathbf{v}})^T)^2 - \rho C_{\varepsilon 2}\frac{\varepsilon^2}{k}. \quad (9.93)$$

The five model constants have been chosen so that the k/ε method gives predictions that conform reasonably well with experiments:

$$C_\mu = 0.09, \quad C_{\varepsilon 1} = 0.1256, \quad C_{\varepsilon 2} = 1.92, \quad \sigma_k = 1.0, \quad \sigma_\varepsilon = 1.3. \quad (9.94)$$

The k/ε method is implemented by solving Eqns. (9.90a), (9.90b), (9.92), and (9.93) simultaneously—a daunting task indeed, and one that must be performed by the digital computer. An additional problem occurs in that the equations either start to break down in the viscous layer next to a boundary wall, or would need an impossibly fine mesh in the wall region to achieve accurate results. One approach, implemented by COMSOL (and slightly differently by FlowLab), is to assume (quite reasonably) that the velocity component parallel to the wall is a logarithmic function of the distance from the wall, as in the universal velocity profile. If needed, complete details can be found in one of the COMSOL manuals.[18]

Two examples of software for implementing the method are presented in this book—COMSOL (based on the finite-element method) in the present chapter and FlowLab (based on the finite-volume method) in Chapter 13. As with all simulations, the user should at least ponder—do the results appear reasonable or not? Certainly, even though it gives reasonably good results in many circumstances, we should have no illusion that the k/ε method can completely solve all turbulent-flow problems.

[13] B.E. Launder and D.B. Spalding, *Lectures in Mathematical Models of Turbulence*, Academic Press, London, 1972.

[14] H. Tennekes and J.L. Lumley, *A First Course in Turbulence*, The MIT Press, Cambridge, MA, 1972.

[15] P.A. Libby, *Introduction to Turbulence*, Taylor & Francis, Washington, D.C., 1996.

[16] P.A. Durbin and B.A. Petterson Reif, *Statistical Theory and Modeling for Turbulent Flows*, Wiley & Sons, Ltd., Chichester, 2001.

[17] J. Mathieu and J. Scott, *An Introduction to Turbulent Flow*, Cambridge University Press, Cambridge, 2000.

[18] For example, see p. 93 of *FEMLAB 3 Chemical Engineering Module User's Guide*, COMSOL AB, 2003.

Example 9.4—Flow Through an Orifice Plate (COMSOL)   **501**

Example 9.4—Flow Through an Orifice Plate (COMSOL)

Use COMSOL to investigate the flow pattern in a short segment of pipe that contains an orifice plate, as presented in Fig. 2.19. The fluid is water, with density $\rho = 62.4 \ \text{lb}_\text{m}/\text{ft}^3$ and kinematic viscosity $\nu = 1.077 \times 10^{-5} \ \text{lb}_\text{m}/\text{ft s}$. The dimensions in feet represented in Fig. E9.4.1 are $L = 0.5$, $D = 0.1$, and $b = 0.1$. The dimensions of the intrusion representing the orifice plate are given below. The axial inlet velocity is uniformly $v_\text{inlet} = 1.0$ ft/s (with zero radial velocity u_0), corresponding to a Reynolds number of 9,285, well in the turbulent-flow region even upstream of the orifice plate. The exit pressure is zero.

Fig. E9.4.1   Geometry of an orifice plate in a pipe.

Solution

1. In COMSOL, choose the sequence **New**, **Chemical Engineering Module**, **Momentum Balance**, **Incompressible Navier-Stokes**, and **K-Epsilon Turbulence Model**. Also set **Space dimension** to **Axial symmetry 2D**, followed by **OK**.

2. Because of symmetry, only the "bottom" half of the pipe needs to be drawn. Draw *any* rectangle R1 and double-click on it. In the window that opens, enter width = 0.05, height = 0.3, base = corner, $r = 0$, $z = 0$, followed by **OK**. In the **Options** pulldown menu, choose **Zoom** and **Zoom Extents**—or just click on the **Zoom Extents** button—so that the rectangle now occupies most of the vertical extent of the window and the r and z axes labels are adjusted accordingly. Note that COMSOL orients the z axis in the "vertical" direction.

3. Similarly, draw a second rectangle R2 of width = 0.02, height = 0.01, base = corner, $r = 0.03$, and $z = 0.1$.

4. Form the composite object R1-R2, amounting to cutting out the smaller rectangle from the larger one and giving the composite object CO1.

5. Also draw a circle C1 of radius 0.005 and center (0.03, 0.105). Then form the final composite object CO1-C1.

6. Under **Options/Constants**, set the values of rho, nu, and Vinlet as indicated above.

7. Under **Physics/Subdomain Settings**, indicate that the density and kinematic viscosity are `rho` and `nu`, respectively.

Fig. E9.4.2   (a) Finite-element mesh, (b) streamlines (with arrow showing the general direction of flow), and (c) a surface plot showing the magnitudes of the velocity.

8. Under **Physics/Boundary Settings**, establish the following boundary conditions:

   (a) Along the centerline $r = 0$: Axial symmetry.
   (b) At the inlet $z = 0$: Inflow, with $u0 = 0$, $v0 =$ `Vinlet`.
   (c) At the exit $z = 0.5$: Normal flow/pressure, with $p0 = 0$.
   (d) On the remaining six boundary segments: Logarithmic wall function.

9. Initialize the mesh, without refinement, as shown in Fig. E9.4.2(a).

10. To assist convergence in this very nonlinear problem, open **Solver/Solver Parameters**. Under the **Nonlinear** tab, select the **Parametric nonlinear** option,

Example 9.4—Flow Through an Orifice Plate (COMSOL)   **503**

and note that the **General** becomes highlighted. Under **Parameter**, enter nu for the name of the parameter, and 1.077e-5*[100 10 1] in the list of parameter values. As a result, COMSOL will start with an artificially high viscosity (100 times its true value) in order to dampen out any instabilities, and then approach the solution by gradually decreasing the viscosity. The intermediate value "10" means that the current solution at that stage (when the viscosity is 10 times its true value) will be saved and will be available for eventual plotting if needed.

11. Solve the problem. As the solution progresses, note in the **Progress** window that the kinematic viscosity nu is progressively decreased from 1.083e-3 to the final true value of 1.08e-5.

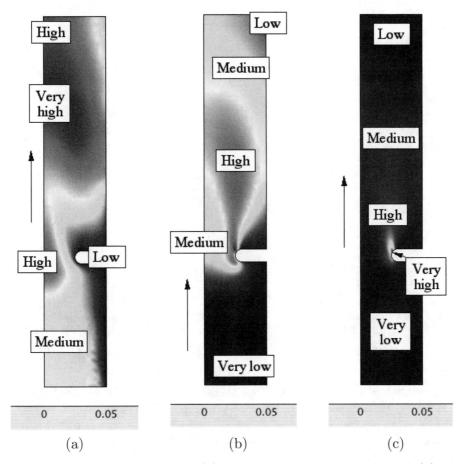

Fig. E9.4.3 Surface plots of (a) turbulent kinematic viscosity, (b) turbulent kinetic energy, and (c) turbulent dissipation. Arrows show the general direction of flow.

Display of Results and Comments

Make diagrams of the following:

1. The finite-element mesh, readily obtained from the **Mesh/Mesh Mode** pull-down menu, as in Fig. E9.4.2(a). For this problem, good results can be obtained with no mesh refinement.

2. Streamlines, for example by using **Postprocessing/Quick Plots/Streamline Plot**, as in Fig. E9.4.2(b). Recall that these streamlines represent the *time-averaged* motion, and *not* the paths taken by the water. Observe the contraction through the orifice and the substantial recirculating eddy in the "backwater" just downstream of the orifice plate.

Fig. E9.4.4   Variation of pressure from inlet to exit.

3. Starting with **Postprocessing/Plot Parameters**, make a series of four colored **Surface** plots, for the following **Predefined quantities** in succession. Since some of the information is lost in black-and-white, we have added boxes that show general trends of values. The results are shown in Figs. E9.4.2(c) and E9.4.3(a), (b), and (c).

   (i) Velocity field. Either by observing the original color bar at the right, or clicking on various parts of the diagram (with "Snap" turned off by double-clicking on it), the highest velocity is found to occur on the centerline, immediately downstream of the orifice, with a value of about 4.4 ft/s.

Example 9.5—Turbulent Jet Flow (COMSOL)   **505**

(ii) Turbulent (kinematic) viscosity, whose maximum value is about 5.9×10^{-3}, almost 550 times the molecular kinematic viscosity. The values can be placed in perspective by noting that $\nu_T = C_\mu k^2/\varepsilon$, and comparing ν_T with k and ε from the turbulent kinetic energy (k) and dissipation (ε) figures.

(iii) Turbulent kinetic energy. Note that there is a "plume" of high values emanating from the rounded edge of the orifice plate.

(iv) Turbulent dissipation. Note that the general distribution is *roughly* the same as that of the turbulent kinetic energy, but that it decays somewhat faster. Here, the color bar at the right of the original can be somewhat misleading because a small area of extremely high values (again at the edge of the orifice plate) can make the rest appear unduly small in comparison.

4. A cross plot of the pressure along the centerline, from inlet to exit, as in Fig. E9.4.4. Start with **Postprocessing/Cross-Section Plot Parameters**. Under **Solutions to use**, select under **General** the smallest viscosity (recall that COMSOL automatically reduced it during the solution). Then, under the **Line/Extrusion tab**, select `Pressure` as the **Predefined quantity** and for the **Cross-section line data** enter the coordinates `r0 = 0`, `r1 = 0`, `z0 = 0`, and `z1 = 0.3`, followed by **OK**. As the velocity increases from 1.0 to about 4.4 ft/s, there is—from Bernoulli's equation—a sharp but smooth reduction in pressure as the *vena contracta* is reached at $z \doteq 0.13$. Thereafter, as some kinetic energy is converted back to pressure energy, there is some recovery of pressure as the exit is approached; the curve in this last section is somewhat "rough," but remember that it's turbulent flow that we're solving!   □

Example 9.5—Turbulent Jet Flow (COMSOL)

Fig. E9.5.1 shows a rectangular channel ABCD with indicated x and y dimensions in meters, with three inlet ports at the left and one exit at the right. A jet of air (density $\rho = 1.18$ kg/m³ and kinematic viscosity $\nu = 1.83 \times 10^{-5}$ m²/s) flows with velocity $V_{\text{jet}} = 11$ m/s into the central port and entrains two other streams whose entering velocities are both $V = 2$ m/s. Use COMSOL to conduct a detailed investigation of the resulting turbulent flow. The exit pressure along BC is zero.

Fig. E9.5.1   Flow of jets into a channel.

Solution

1. In COMSOL, choose the sequence **New, Chemical Engineering Module, Momentum Balance, Incompressible Navier-Stokes,** and **K-Epsilon Turbulence Model**, followed by **OK**.

2. Draw *any* rectangle R1 and double-click on it. In the window that opens, enter width = 4, height = 1, and base = corner, $x = 0$, $y = 0$, followed by **OK**. In the **Options** pulldown menu, choose **Zoom** and **Zoom Extents**—or just click on the **Zoom Extents** button—so that the rectangle now occupies most of the width of the window.

3. Similarly, draw two more rectangles R2 and R3 each of width = 0.5 and height = 0.1, base = corner. The coordinates (x, y) of the corners are (0, 0.3) for R2 and (0, 0.6) for R3.

4. Form the composite object R1-R2-R3, cutting out the two smaller rectangles from the larger one.

5. Under **Options/Constants**, set the values of rho, nu, Vjet, and V as indicated above.

6. Under **Physics/Subdomain Settings**, indicate that ρ and ν are designated rho and nu, respectively.

7. Under **Physics/Boundary Settings**, set all 12 boundary segments to "Logarithmic wall function" except for the following four segments:

(a) At the central jet: Inflow, with $u0$ and $v0$ equal to Vjet and zero.

(b) At the two other inlets: Inflow, with $u0$ and $v0$ equal to V and zero.

(c) At the exit : Normal flow/pressure, with $p0 = 0$.

8. Initialize and refine the mesh just once, and solve the problem.

Display of Results and Comments

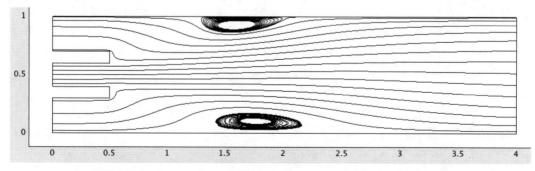

Fig. E9.5.2 Streamlines (from left to right), showing the time-averaged motion. The faster-moving central jet entrains the two side jets and there are two small vortices next to the walls.

Example 9.5—Turbulent Jet Flow (COMSOL)   **507**

1. To visualize the (time-averaged) flow pattern, generate first a *streamline plot*, shown, with comments, in Fig. E9.5.2. To obtain full control, begin with **Plot Parameters/Streamlines** and **Specify start point coordinates**, and to ensure 20 streamlines, all passing through points with $x = 1.9$ and y-coordinates equally spaced between 0 and 1, specify x: `linspace(1.9,1.9,20)` and y: `linspace(0,1,20)`, Also make an arrow plot, shown in Fig. E9.5.3.

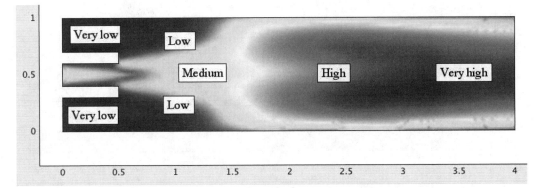

Fig. E9.5.3 Arrow plot. The fast-moving central jet spreads out and decelerates as its momentum is transferred in the transverse direction by turbulence.

2. Now make a series of colored **Surface** plots, for the following **Surface data/Predefined quantities** in succession: turbulent viscosity, turbulent kinetic energy, dissipation, and pressure. Since some of the information is lost in black-and-white, we have added boxes in Figs. E9.5.4–E9.5.7 that show general trends of values.

Fig. E9.5.4 Turbulent kinematic viscosity ν_T. Successively higher values are attained as full mixing occurs downstream. The highest value of the turbulent kinematic viscosity is about 0.09, some 4,900 times the molecular kinematic viscosity. The values can be placed in perspective by noting that $\nu_T = C_\mu k^2/\varepsilon$, and comparing ν_T with k and ε from the next two figures.

Fig. E9.5.5 Turbulent kinetic energy k, showing a gradual decrease from its highest value at the jet entrance to a low value at the exit.

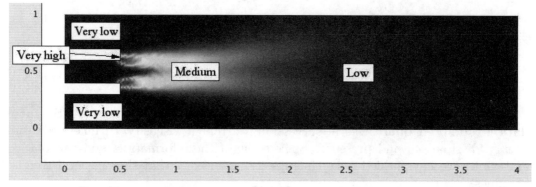

Fig. E9.5.6 Turbulent dissipation ε, again very high at the jet entrance and declining very significantly towards the exit.

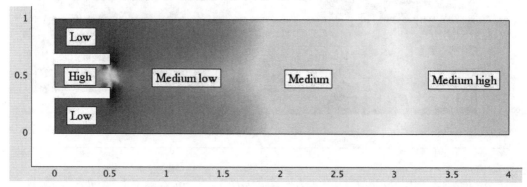

Fig. E9.5.7 Pressure distribution. The fast-moving central jet needs a high pressure to be injected into the channel, and there is a loss of pressure as it interacts with the side jets. As the central jet decelerates, some of its kinetic energy is recovered as pressure energy, so there is a partial pressure recovery near the exit.

9.13   Analogies Between Momentum and Heat Transfer

The fact that turbulent eddies can transport heat and mass as well as momentum has already been mentioned in Section 9.3. The question then arises—for example—can an experimental observation on *momentum* transport be used to make a prediction about *heat* transfer? The answer, as we shall see, is yes, provided that a realistic model is used.

The Reynolds analogy.   Fig. 9.13 shows the simplest model, used in the *Reynolds analogy,* in which idealized turbulent eddies are constantly moving in both directions between the turbulent "core," or mainstream, and the walls of the containing duct or pipe. Let m denote the mass flux (mass per unit time per unit area) of such motion in either direction. The rate of transport of z momentum from the mainstream to the wall is $m\bar{v}_{zm}$, where $\bar{v}_{zm}$ is the mean velocity in the mainstream; in the reverse direction it is $m\bar{v}_{zw} = 0$, since the velocity $\bar{v}_{zw}$ at the wall is zero. The difference between these two rates of transport corresponds to a net shear force exerted by the fluid on the wall:

Net rate of momentum transport:

$$\tau_w = m\bar{v}_{zm} - m\bar{v}_{zw} = m\bar{v}_{zm}. \tag{9.95}$$

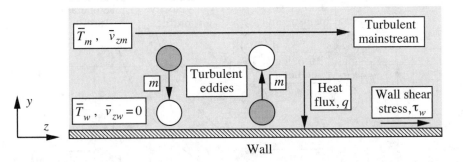

Fig. 9.13   *Momentum and heat transport by turbulent eddies.*

By the same token, the rate of heat transport from the mainstream to the wall is $mc_p\bar{T}_m$, where $\bar{T}_m$ is the mean temperature in the mainstream and c_p is the specific heat; in the reverse direction it is $mc_p\bar{T}_w$. The difference between these two rates of transport corresponds to a net heat flux from the mainstream to the wall:

Net rate of heat transport:

$$q = mc_p\bar{T}_m - mc_p\bar{T}_w = mc_p(\bar{T}_m - \bar{T}_w). \tag{9.96}$$

The unknown mass flux m is conveniently eliminated from Eqns. (9.95) and (9.96):

$$\frac{\tau_w}{\bar{v}_{zm}} \; [= m] = \frac{q}{c_p(\bar{T}_m - \bar{T}_w)}. \tag{9.97}$$

Division by $\rho\bar{v}_{zm}$ gives:

$$\frac{\tau_w}{\rho\bar{v}_{zm}^2} = \frac{q}{\rho\bar{v}_{zm}c_p(\overline{T}_m - \overline{T}_w)}. \tag{9.98}$$

But the first group is just half the Fanning friction factor, $f_F/2$; from the definition of the heat-transfer coefficient, h, we can also substitute $q = h(\overline{T}_m - \overline{T}_w)$, giving:

$$\frac{1}{2}f_F = \frac{h}{\rho\bar{v}_{zm}c_p} = \text{St}, \tag{9.99}$$

in which St is a dimensionless group known as the *Stanton number*.

Note that each of the dimensionless groups in Eqn. (9.99) has a ready physical interpretation. The friction factor measures the ratio of the overall momentum transport (to the wall) to inertial effects in the mainstream; and the Stanton number indicates the relative importance of the overall heat transport (to the wall) to convective effects in the mainstream. In effect, the Reynolds analogy simply states that these two ratios are identical, because the same basic model has been assumed for both momentum and heat transfer.

Fig. 9.14   Addition of a laminar sublayer to the turbulent mainstream.

Prandtl-Taylor analogy.   An obvious refinement to the Reynolds analogy, which is employed in the *Prandtl-Taylor analogy*, is to insert between the wall and the eddies a *laminar sublayer*, of thickness δ, in which turbulent effects are negligible, momentum transport being determined by viscous action and heat transport being controlled by thermal conduction. Referring to the general scheme and notation shown in Fig. 9.14, and assuming linear velocity and temperature variations in the very thin laminar sublayer, the two transport rates are now:

Example 9.6—Evaluation of the Momentum/Heat-Transfer Analogies   **511**

Net rate of momentum transport:

$$\tau_w = \mu \frac{\overline{v}_{z\delta} - 0}{\delta} = m(\overline{v}_{zm} - \overline{v}_{z\delta}). \qquad (9.100)$$

Net rate of heat transport:

$$q = k \frac{\overline{T}_\delta - \overline{T}_w}{\delta} = mc_p(\overline{T}_m - \overline{T}_\delta). \qquad (9.101)$$

Either of these last two pairs of equations can be divided by the other in order to eliminate both δ and m. Either $\overline{T}_\delta$ or $\overline{v}_{z\delta}$ can then be eliminated from the remaining two equations, the former being preferred since little is known about it. Several lines of algebra, including the manipulations already seen in the Reynolds analogy, lead to:

$$\text{St} = \frac{f_\text{F}/2}{1 + \dfrac{\overline{v}_{z\delta}}{\overline{v}_{zm}}(\text{Pr} - 1)}. \qquad (9.102)$$

Observe that for $\text{Pr} = 1$, the Prandtl-Taylor analogy simplifies to the Reynolds analogy. Further, from Eqns. (9.61) and (9.62), the required velocity ratio can be shown to be:

$$\frac{\overline{v}_{z\delta}}{\overline{v}_{zm}} = \frac{2.44}{\text{Re}^{1/8}}. \qquad (9.103)$$

Further refinements may be made by inserting a buffer region between the turbulent mainstream and the laminar sublayer. Mass transfer may also be included in the analogies.

Example 9.6—Evaluation of the Momentum/Heat-Transfer Analogies

Investigate the merits of the Reynolds and Prandtl-Taylor analogies by comparing their predictions for the Nusselt number $\text{Nu} = hD/k$ against those obtained from the well-known Dittus-Boelter equation, a correlation that is based on *experimentally determined* heat-transfer coefficients in smooth pipe:

$$\text{Nu} = 0.023\,\text{Re}^{0.8}\,\text{Pr}^{0.4}. \qquad (\text{E9.6.1})$$

Solution

1. *Reynolds analogy.* For simplicity, use the predictions of the Blasius equation for the friction factor, which holds over a fairly wide range of Reynolds numbers:

$$f_\text{F} = 0.0790\,\text{Re}^{-1/4}. \qquad (\text{E9.6.2})$$

A combination of Eqns. (9.99) and (E9.6.2) gives:

$$\frac{h}{\rho \bar{v}_{zm} c_p} = \frac{1}{2} f_F = 0.0395 \left(\frac{\rho \bar{v}_{zm} D}{\mu} \right)^{-1/4}. \tag{E9.6.3}$$

Multiplication of both sides by $\rho \bar{v}_{zm} c_p D / k$ and rearrangement leads to:

$$\frac{hD}{k} = 0.0395 \left(\frac{\rho \bar{v}_{zm} D}{\mu} \right)^{3/4} \frac{\mu c_p}{k}, \tag{E9.6.4}$$

or:

$$\mathrm{Nu} = 0.0395 \, \mathrm{Re}^{3/4} \mathrm{Pr}, \tag{E9.6.5}$$

in which the definitions of the *Nusselt* and *Prandtl* numbers have been recognized. Thus, we now have a basis for predicting the heat-transfer coefficient h via the Nusselt number. For a representative Reynolds number of $\mathrm{Re} = 50,000$, Table E9.6 gives in the column "RA" (Reynolds analogy) values of Nu from Eqn. (E9.6.5) for three different Prandtl numbers. Note also the column "DB", which gives values for Nu from the well-known Dittus-Boelter equation, (E9.6.1), representative of experimental values:

$$\mathrm{Nu} = 0.023 \, \mathrm{Re}^{0.8} \, \mathrm{Pr}^{0.4}. \tag{E9.6.6}$$

Table E9.6   Nusselt Numbers at $\mathrm{Re} = 50,000$

Pr	RA	DB	PTA
0.5	67	100	95
1.0	130	130	130
2.0	264	173	162

Observe the outstanding *agreement* at $\mathrm{Pr} = 1$ between the prediction of the Reynolds analogy and the essentially experimental value given by the Dittus-Boelter equation. Note also the significant *differences* between the two for $\mathrm{Pr} = 0.5$ and 2. The essence of this discrepancy may be found by reexpressing the Prandtl number as:

$$\mathrm{Pr} = \frac{\mu c_p}{k} = \frac{\frac{\mu}{\rho}}{\frac{k}{\rho c_p}} = \frac{\nu}{\alpha}, \tag{E9.6.7}$$

in which ν and α are the kinematic viscosity and thermal diffusivity, respectively. These last two represent the rates at which momentum and heat diffuse through a fluid by molecular action—factors that are completely ignored in the Reynolds analogy.

2. *Prandtl-Taylor analogy.* By similar arguments, if the Blasius equation is again assumed for the friction factor, the Prandtl-Taylor analogy of Eqn. (9.102) yields:

$$\mathrm{Nu} = \frac{0.0395\,\mathrm{Re}^{3/4}\mathrm{Pr}}{1 + \dfrac{\overline{v}_{z\delta}}{\overline{v}_{zm}}\,(\mathrm{Pr} - 1)}, \tag{E9.6.8}$$

in which the velocity ratio is given in Eqn. (9.103).

The Nusselt numbers predicted from Eqn. (E9.6.8) can also be evaluated, three representative values being given in the last column of Table E9.4. Note that the agreement with the Dittus-Boelter values is now much improved for both $\mathrm{Pr} = 0.5$ and 2, thus illustrating the success of the more realistic Prandtl-Taylor analogy. □

9.14 Turbulent Jets

The previous sections in this chapter have discussed turbulence in which the presence of a wall exerted a profound effect on the flow pattern. It is possible to extend the ideas developed so far to the situation of *free turbulence*, essentially unaffected by such confining boundaries. Two situations are of general interest:

1. The turbulent *jet*, in which fluid issues from a narrow constriction, to form a turbulent plume of ever-increasing breadth and decreasing velocity.
2. The turbulent *wake* behind a stationary nonstreamlined object situated in a fluid stream (or an object that is moving in an otherwise stagnant fluid). Again, the wake gradually broadens out downstream.

To a first approximation, both the jet and the wake share five important features in common with the boundary layers discussed in Chapter 8:

1. The flow is primarily in one principal direction, corresponding to the axis of the jet.
2. There is a region of turbulence whose extent gradually increases in the down-stream direction.
3. The derivatives of the velocities are considerably larger in the direction trans-verse to the jet or wake than they are in the principal direction of flow.
4. An order-of-magnitude analysis can be used to simplify the equations of motion.
5. In some cases, the solution is facilitated by introducing a stream function.

Under these circumstances, the arguments extended in Section 8.3 for the simplification of the equations of motion still apply, except of course that time-averaged velocities must now be used. There are two further modifications: (a) a turbulent shear stress must be used instead of the viscous stress, and (b) the pressure is essentially constant, apart from a region immediately behind the object in the case of the wake. Only the turbulent jet will be studied here, because it is of greater interest to chemical engineers who are concerned with the rate at which two reactants (represented by the jet and its environment) interact with each other. Two types of jets will be considered:

1. A jet issuing from a long slot—the plane turbulent jet.
2. A jet issuing from a small circular orifice—the axisymmetric turbulent jet.

Fig. 9.15   Plane turbulent jet (the broadening effect is exaggerated).

The plane turbulent jet.   First, consider flow in the x/y plane, in which Fig. 9.15 shows fluid issuing to the right from a narrow slotlike opening into a "sea" of fluid in which the pressure is constant. The flow rate is sufficiently high so that the jet is turbulent, and its half-width b increases in the x direction, linearly as it transpires, although this is not assumed *a priori*. The more usual, but theoretically somewhat more complicated case, of axisymmetric flow will be discussed later. The simplified equations of motion are:

$$\bar{v}_x \frac{\partial \bar{v}_x}{\partial x} + \bar{v}_y \frac{\partial \bar{v}_x}{\partial y} = \frac{1}{\rho} \frac{\partial \bar{\tau}_{yx}^t}{\partial y}, \tag{9.104}$$

$$\frac{\partial \bar{v}_x}{\partial x} + \frac{\partial \bar{v}_y}{\partial y} = 0. \tag{9.105}$$

Note the absence of any time derivative—that is, the flow is "steady in the mean." Also, because the pressure is constant and there is no external force to modify it, the rate of momentum transfer per unit depth in the x direction, $\dot{\mathcal{M}}$, must be constant:

$$\dot{\mathcal{M}} = \rho \int_{-\infty}^{\infty} \bar{v}_x^2 \, dy. \tag{9.106}$$

Next, consider the turbulent shear stress, for which two treatments are possible, depending on the representation of the eddy kinematic viscosity:

1. From the Prandtl mixing-length theory (this subsection).
2. From a simplified eddy-viscosity theory (the next subsection).

Now proceed with the first of these alternatives. According to the mixing-length theory, which still holds in the absence of a wall, the turbulent shear stress is given by:

$$\overline{\tau}^t_{yx} = \rho \nu_{\mathrm{T}} \frac{\partial \overline{v}_x}{\partial y} = \rho \underbrace{\ell^2 \left| \frac{\partial \overline{v}_x}{\partial y} \right| \frac{\partial \overline{v}_x}{\partial y}}_{\nu_{\mathrm{T}}}, \tag{9.107}$$

in which the underbrace emphasizes the terms that comprise the eddy kinematic viscosity.

As in the case of the boundary-layer problem, we hope that similarity of velocity profiles can be achieved, that this will be expedited by the following assumptions:

1. Paralleling the Blasius solution, a new dimensionless space coordinate ζ can be introduced.
2. The mixing length ℓ depends only on x, and is directly proportional to b.
3. The half-width of the turbulent zone is proportional to some power p of the x coordinate.
4. The velocity is proportional to a function of ζ and inversely proportional to another power q of x:

$$\ell = c_1 b, \qquad b = c_2 x^p, \qquad \zeta = c_3 \frac{y}{b} = c_4 \frac{y}{x^p}, \qquad \overline{v}_x = \frac{g(\zeta)}{x^q}. \tag{9.108}$$

The values of p and q can be found by investigating the dependency of each term of Eqns. (9.104) and (9.106) on x. Note that $\overline{v}_x$ is of order x^{-q}, $\overline{v}_x^2$ is of order x^{-2q}, and the integral of Eqn. (9.106) is of order x^{p-2q}. But since $\dot{M}/\rho$ is constant, it can have no dependency on x, so that:

$$p = 2q. \tag{9.109}$$

By similar arguments, the reader can discover the dependencies on x shown in Table 9.4, in which the continuity equation, (9.108), has been invoked to determine the order of $\overline{v}_y$.

For Eqn. (9.104) to balance, we must have $1 + 2q = p + 2q$, so that:

$$p = 1, \qquad q = \frac{1}{2}. \tag{9.110}$$

Physically, the fact that $p = 1$ confirms the earlier supposition that the jet spreads linearly with distance x, and $q = \frac{1}{2}$ also indicates that the velocities decline proportionally to $x^{-1/2}$.

Table 9.4 Order of Terms

Term	Order
$\bar{v}_x \dfrac{\partial \bar{v}_x}{\partial x}$	$x^{-(1+2q)}$
$\bar{v}_y \dfrac{\partial \bar{v}_x}{\partial y}$	$x^{-(1+2q)}$
$\dfrac{1}{\rho} \dfrac{\partial \bar{\tau}_{yx}^{t}}{\partial y}$	$x^{-(p+2q)}$

The solution is expedited by introducing a stream function ψ such that:

$$\bar{v}_x = \frac{\partial \psi}{\partial y}, \qquad \bar{v}_y = -\frac{\partial \psi}{\partial x}. \tag{9.111}$$

Bearing in mind the preceding development, we can now set:

$$\ell = \alpha x, \qquad \zeta = \frac{\beta y}{x}, \qquad \psi = \gamma \sqrt{x} f(\zeta), \tag{9.112}$$

from which the velocities are:

$$\bar{v}_x = \frac{\beta \gamma f'}{\sqrt{x}}, \qquad \bar{v}_y = \frac{\gamma}{\sqrt{x}} \left(\zeta f' - \frac{f}{2} \right), \tag{9.113}$$

in which the prime denotes differentiation with respect to ζ.

Substitute these expressions for the velocities into the momentum balance, and choose $\beta = (2\alpha^2)^{-1/3}$ so that α is absorbed into the definition of ζ. After some algebra, Eqn. (9.104) yields an ordinary differential equation in the unknown function $f(\zeta)$:

$$\text{With } \beta = \frac{1}{(2\alpha^2)^{1/3}}, \qquad \left(\frac{df}{d\zeta} \right)^2 + f \frac{d^2 f}{d\zeta^2} = \frac{d}{d\zeta} \left(\frac{d^2 f}{d\zeta^2} \right)^2. \tag{9.114}$$

This last equation integrates directly to:

$$f \frac{df}{d\zeta} = \left(\frac{d^2 f}{d\zeta^2} \right)^2, \tag{9.115}$$

the constant of integration being zero because at the centerline ($\zeta = 0$), considerations of symmetry ($\bar{v}_y = d\bar{v}_x/dy = 0$) give $f = f'' = 0$. A numerical integration of Eqn. (9.115), subject to the initial conditions at $\zeta = 0$ of $f = 0$ and $f' = 1$ (an arbitrary choice—the value of γ can then be selected so that the integrated velocity matches any specified flow rate from the jet), yields the plots of f' (needed for $\overline{v}_x$) and $\zeta f' - f/2$ (needed for $\overline{v}_y$) versus ζ, shown in Fig. 9.16.

The two time-averaged turbulent velocities $\overline{v}_x$ and $\overline{v}_y$ can then be obtained for any values of γ (determined by the strength of the jet) and the downstream distance x. Finally, note that the numerical results are in complete agreement with the experimental values determined by Förthmann.[19] Experimental evidence of Reichardt gives $\beta = 7.67$, corresponding to $\alpha = 0.0333$.[20]

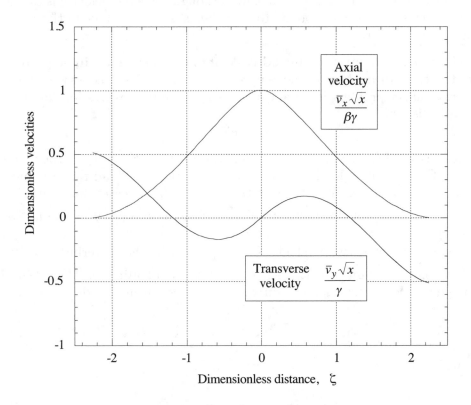

Fig. 9.16 Axial and transverse velocities for plane turbulent jet.

As expected, the axial velocity follows a bell-shaped curve, symmetrical about the centerline, with the following characteristics:

1. Since $f' = \overline{v}_x \sqrt{x}/\beta\gamma$ is plotted, the actual velocities $\overline{v}_x$ will *decrease* as the distance x from the jet increases.

---

[19] As reported on p. 594 of S. Goldstein's *Modern Developments in Fluid Dynamics*, Clarendon Press, Oxford, 1957.

[20] As reported on p. 500 of Schlichting, *op. cit.*

2. Because the dimensionless distance $\zeta = \beta y/x$ is plotted, the region in which $\overline{v}_x$ is significant will *increase* as x increases.

That is, the jet simultaneously spreads out and gets slower.

The transverse velocity also shows some interesting features. Consider, for example, positive values of the transverse dimensionless coordinate:

1. For $0 < \zeta < 1.25$ approximately, $\overline{v}_y$ is positive, corresponding to the spreading of the jet.
2. For the region $\zeta > 1.25$, $\overline{v}_y$ becomes negative and eventually levels out at a significant nonzero negative value, even though the axial velocity is essentially zero. This region corresponds to *entrainment* of fluid outside the turbulent zone into the jet.

Alternative form for the eddy kinematic viscosity in a plane turbulent jet.   An analytical and quicker, but slightly less accurate, solution to the turbulent jet problem is obtained by using a simpler expression for the eddy kinematic viscosity:

$$\nu_{\mathrm{T}} = cb\overline{v}_{xc}, \tag{9.116}$$

in which c is another constant, b is the half-width of the jet, and $\overline{v}_{xc}$ is the centerline velocity. Observe that the dimensions of ν_{T} are still L^2/T, and that the model is still physically quite realistic, because the value of ν_{T} increases both with:

1. The space available for turbulent fluctuations, as expressed by the half-width b of the jet.
2. The general intensity of velocities, as expressed by the centerline velocity $\overline{v}_{xc}$.

The solution again uses the stream function ψ of Eqn. (9.111). Bearing in mind the preceding development, in which b varied linearly with x, we can now set:

$$\overline{v}_{xc} = \frac{V}{\sqrt{x}}, \quad \nu_{\mathrm{T}} = \alpha V \sqrt{x}, \quad \zeta = \frac{\beta y}{x}, \quad \psi = \gamma V \sqrt{x} f(\zeta), \tag{9.117}$$

from which the velocities are:

$$\overline{v}_x = \frac{\beta \gamma V f'}{\sqrt{x}}, \qquad \overline{v}_y = \frac{\gamma V}{\sqrt{x}} \left(\zeta f' - \frac{f}{2} \right). \tag{9.118}$$

Substitution into the momentum balance and simplification gives:

$$\frac{d}{d\zeta} \left(f \frac{df}{d\zeta} \right) = -\frac{2\alpha\beta}{\gamma} \frac{d^3 f}{d\zeta^3}, \tag{9.119}$$

which integrates once to:

$$f \frac{df}{d\zeta} = -\frac{2\alpha\beta}{\gamma} \frac{d^2 f}{d\zeta^2}, \tag{9.120}$$

the constant of integration being zero because at the centerline symmetry again gives $f = f'' = 0$. A second integration yields:

$$f^2 = -\frac{4\alpha\beta}{\gamma}\frac{df}{d\zeta} + c, \tag{9.121}$$

the value of the constant being obtained by noting that at the centerline, $f = 0$ and $f' = 1$. Thus, with an appropriate choice for β, we have:

$$\beta = \frac{\gamma}{4\alpha}, \qquad f^2 + \frac{df}{d\zeta} = 1. \tag{9.122}$$

The solution of Eqn. (9.122) and the expression for f', from which the axial velocity profile can be obtained, are:

$$f = \tanh\zeta, \qquad \frac{df}{d\zeta} = 1 - \tanh^2\zeta. \tag{9.123}$$

This analytical solution agrees quite well with the somewhat more accurate numerical solution displayed in Fig. 9.16, except that it slightly overpredicts the velocities for $\zeta > 1.5$.

The axisymmetric turbulent jet. For a jet issuing from a circular orifice, cylindrical coordinates with axisymmetry are used, as in Fig. 9.17.

Fig. 9.17 Axisymmetric turbulent jet.

The simplified equations of motion are:

$$\bar{v}_r\frac{\partial\bar{v}_z}{\partial r} + \bar{v}_z\frac{\partial\bar{v}_z}{\partial z} = \frac{1}{\rho r}\frac{\partial(r\bar{\tau}_{rz}^t)}{\partial r}, \tag{9.124}$$

$$\frac{1}{r}\frac{\partial(r\bar{v}_r)}{\partial r} + \frac{\partial\bar{v}_z}{\partial z} = 0. \tag{9.125}$$

Also, the rate of momentum transfer in the z-direction, $\dot{\mathcal{M}}$, must be constant:

$$\dot{\mathcal{M}} = \rho \int_0^\infty 2\pi r \bar{v}_z^2 \, dr. \tag{9.126}$$

The turbulent shear stress and the eddy kinematic viscosity are also given by:

$$\bar{\tau}_{rz}^t = \rho \nu_\text{T} \frac{\partial \bar{v}_z}{\partial r}, \qquad \nu_\text{T} = c_1 b \bar{v}_{zc}, \tag{9.127}$$

in which c_1 is a constant, b is the half-width of the jet, and $\bar{v}_{zc}$ is the centerline velocity. Similar to the earlier development of Eqn. (9.108), we can introduce:

$$b = c_2 z^p, \qquad \zeta = c_3 \frac{r}{z^p}, \qquad \bar{v}_z = \frac{g(\zeta)}{z^q}. \tag{9.128}$$

From arguments similar to those centered on Eqns. (9.109) and (9.110), and Table 9.4:

$$p = 1, \qquad q = 1. \tag{9.129}$$

The eddy kinematic viscosity ν_T is therefore constant everywhere, so the equations reduce to:

$$\bar{v}_r \frac{\partial \bar{v}_z}{\partial r} + \bar{v}_z \frac{\partial \bar{v}_z}{\partial z} = \nu_\text{T} \frac{1}{r} \frac{\partial}{\partial r} \left(r \frac{\partial \bar{v}_z}{\partial r} \right) \tag{9.130}$$

$$\frac{1}{r} \frac{\partial (r \bar{v}_r)}{\partial r} + \frac{\partial \bar{v}_z}{\partial z} = 0. \tag{9.131}$$

Although these equations appear formidable, they nevertheless have the following analytical solution, with $\bar{v}_z$ again appearing as a bell-shaped curve[21]:

$$\bar{v}_z = \frac{3\dot{\mathcal{M}}}{8\pi \nu_\text{T} \rho z \left(1 + \frac{\zeta^2}{4} \right)^2}, \tag{9.132}$$

$$\bar{v}_r = \frac{1}{4z} \sqrt{\frac{3\dot{\mathcal{M}}}{\pi \rho}} \left[\frac{\zeta - \frac{\zeta^3}{4}}{\left(1 + \frac{\zeta^2}{4} \right)^2} \right], \tag{9.133}$$

in which:

$$\zeta = \frac{1}{4\nu_\text{T}} \sqrt{\frac{3\dot{\mathcal{M}}}{\pi \rho}} \frac{r}{z}. \tag{9.134}$$

---

[21] See, for example, pp. 161 and 500 of Schlichting, *op. cit.*

The above theoretical axial velocity profile is well substantiated by the experimental evidence of Reichardt, although the use of mixing-length theory—as opposed to the above assumption of a constant eddy viscosity—gives somewhat better agreement for large values of ζ.[22] Reichardt's experiments also gave $\nu_T \rho / \dot{\mathcal{M}} = 0.0161$.

PROBLEMS FOR CHAPTER 9

Unless otherwise stated, all flows are steady
state, with constant density and viscosity.

1. *Agitation of particles—M (C).* A liquid containing a suspension of particles is flowing down a channel inclined at an angle θ to the horizontal. The particles are kept in suspension by the turbulent motion of the liquid, which has a uniform depth λ.

Assuming that the Prandtl mixing length has the form $ky\sqrt{1 - (y/\lambda)}$, where k is a constant, show that the velocity $\bar{v}_x$ at a point distant y from the base of the channel may be expressed in the following form, in which c is a constant:

$$\bar{v}_x = \frac{1}{k}\sqrt{\lambda g \sin\theta}\,\ln y + c.$$

Assuming that all the particles settle with the same vertical velocity v_g relative to the liquid, and that the mixing length for the particles is the same as for the turbulent motion of the liquid, prove that (except near the lower surface) the number of particles n per unit volume varies with depth according to:

$$\frac{n}{n_0} = \left[\frac{y_0(\lambda - y)}{y(\lambda - y_0)}\right]^\beta,$$

where n_0 is the value of n at a distance y_0 above the base of the channel, and in which:

$$\beta = \frac{v_g}{k\sqrt{\lambda g \sin\theta}}.$$

If $y_0 = \lambda/2$, plot n/n_0 against y/λ for $\beta = 0.1$ and 1. Comment briefly on your results.

2. *Logarithmic velocity profile from dimensional analysis—M.* For fully developed turbulent flow in a smooth pipe, the following suggestions have been made concerning the distribution of the velocity $\bar{v}_z$:

---

[22] See, for example, p. 501 of Schlichting, *op. cit.*

(a) Near the wall, $\bar{v}_z$ depends only on ρ, μ, τ_w, and y, but *not* on the pipe radius a; that is, $\bar{v}_z = f(\rho, \mu, \tau_w, y)$. Perform a dimensional analysis along the lines of Section 4.6, and show that $\bar{v}_z/\bar{v}_z^* = f_1(y/y^*)$, where $\bar{v}_z^* = \sqrt{\tau_w/\rho}$ and $y^* = \nu/\sqrt{\tau_w/\rho}$.

(b) Near the center, the deviation of the velocity from its centerline value $\bar{v}_{zc}$ is determined purely by turbulence. That is, $\bar{v}_z - \bar{v}_{zc} = f(\rho, a, \tau_w, y)$, *independent* of the viscosity μ. Use dimensional analysis to obtain a functional form for the velocity deviation in this turbulent region.

(c) If, generally, $\bar{v}_z = f(\rho, \mu, \tau_w, y, a)$, use dimensional analysis to obtain the general functional form for $\bar{v}_z/\bar{v}_z^*$.

Hence, if there exists a buffer region in which both (a) and (b) are applicable, demonstrate on dimensional grounds that the velocity profile *must* be logarithmic in form, both in this buffer region and in the fully turbulent core.

3. *Blasius and power-law relationship—E.* In Section 9.10 the following statement was made: "There is also another relation between c in the velocity profile and A in the expression for f_F." What is that relation?

4. *A novel turbulent velocity profile—M.* The suggestion has been made for turbulent flow in a pipe of diameter $D = 2a$ that the eddy kinematic viscosity at any location is of the form $\nu_T = cD\bar{v}_z$, where $\bar{v}_z$ is the time-averaged velocity at that location. Based on this model:

(a) What are the dimensions of the constant c?

(b) Give a brief possible explanation for the model.

(c) What is a realistic expression for the turbulent shear stress τ_{rz}^t at any radius r in terms of the wall shear stress τ_w?

(d) Obtain an expression for the velocity profile $\bar{v}_z$ in terms of r, c, a, ρ, τ_w, and the centerline velocity $\bar{v}_{zc}$.

(e) Assuming that this velocity profile holds all the way up to the wall, obtain an expression for $\bar{v}_z$ that depends only on r, a, and $\bar{v}_{zc}$.

(f) Evaluate $\bar{v}_z/\bar{v}_{zc}$ for $r/a = 0$, 0.25, 0.5, 0.75, and 1, and comment briefly whether or not this velocity profile is realistic.

5. *Turbulent condensate film—M.* Consider the flow of a film of liquid condensate of thickness λ on the outside of a vertical tube, approximated in Fig. P9.5 as a vertical plate.

You are to model the flow in two regimes, without resorting to the universal velocity profile as discussed earlier:

1. A very thin laminar sublayer (LSL) of thickness δ, in which the shear stress is virtually the same as its value τ_w at the wall.

2. A much thicker turbulent mainstream (TMS), in which the velocity can be assumed to be of the form:

$$\bar{v}_x = \frac{1}{k}\sqrt{\frac{\tau_w}{\rho}} \ln y + c.$$

The condensate density is ρ and its viscosity is μ, both assumed constant. Also assume continuity of both the velocity *and* the velocity gradient at the LSL/TMS junction.

(a) What is the reason for supposing that $\sqrt{\tau_w/\rho} = \sqrt{\lambda g}$?

(b) If the liquid flowing down the plate is iso-octane, show that the wall shear stress equals $\tau_w \doteq 81.5$ dynes/cm^2, and then obtain numerical values for the following:

   (i) The thickness δ (cm) of the laminar sublayer.
   (ii) The velocity (cm/s) at the LSL/TMS junction.
   (iii) The velocity at the free surface of the condensate.

Data (for iso-octane, at 99.3 °C): $\rho = 0.692$ g/cm^3, $g = 981$ cm/s^2, $\mu = 0.266$ cP $= 0.00266$ g/cm s, $\lambda = 0.12$ cm, $k = 0.40$.

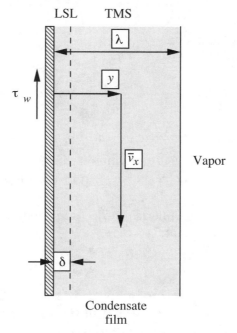

Fig. P9.5   Turbulent condensate film with laminar sublayer.

6. *Turbulent mixing of reactants—M (C).* A constant stream of gas A flows turbulently along a smooth pipe. A relatively small flow rate of a second reacting gas B is to be injected axially by a ring of jets concentric with the axis of the pipe.

At what radius R, expressed as a fraction of the pipe radius a, would you place the ring of jets to give the most rapid initial mixing between A and B? You may assume that:

(a) The most rapid mixing occurs where the eddy diffusivity (taken to be identical with the eddy kinematic viscosity ν_T) is a maximum.

(b) The velocity profile obeys the one-seventh power law:

$$\frac{\overline{v}_z}{\overline{v}_{zc}} = \left(\frac{y}{a}\right)^{1/7},$$

in which $\overline{v}_{zc}$ is the centerline velocity.

7. *Pneumatic particle transport—M.* A gas of density ρ_g and viscosity μ_g flows turbulently with mean velocity $\overline{v}_{zm}$ in a horizontal tube of diameter D that is partly filled with solid spherical particles of diameter d and density ρ_p. At sufficiently high gas velocities, the particles will be agitated by the turbulence and conveyed along the tube by the gas—a phenomenon known as *pneumatic transport*. We wish to investigate the principal dimensionless group that could be used for correlating experimental data on variables such as flow regimes and the fraction ν_T of the total volume occupied by the particles.

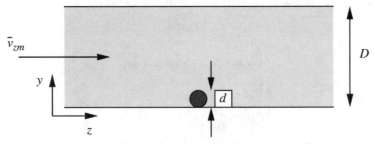

Fig. P9.7   *Particle at wall in turbulent flow.*

As shown in Fig. P9.7, conduct a preliminary investigation into the gas velocity that is needed to pick up a *single* particle from its initial position on the wall. Assume the following:

(a) Analogous to the drag on spherical objects in fluid streams, we can define a dimensionless drag coefficient:

$$C_D = \frac{F_y}{\rho_g d^2 (v'_y)^2},$$

which can reasonably be expected to be constant at sufficiently high Reynolds numbers. Here, F_y is the force exerted in the y direction on the particle due to a sudden transverse velocity fluctuation v'_y.

(b) The fluctuation v'_y is proportional to the magnitude of a similar fluctuation v'_z in the axial velocity, which is given by a standard expression from mixing-length theory.

(c) The mixing length ℓ' is proportional to the pipe diameter.

Demonstrate from the above that the value of the following dimensionless group is likely to determine whether or not the particle will be picked up from the wall:

$$\frac{\rho_g}{\rho_p}\left(\frac{\bar{v}_{zm}^2}{gd}\right).$$

8. *Velocity at the outer edge of the laminar sublayer—M.* Starting from Eqns. (9.61) (the one-seventh power law) and (9.62) (for the thickness of the laminar sublayer), prove Eqn. (9.103), which gives the velocity at the outer edge of the laminar sublayer:

$$\frac{\bar{v}_{z\delta}}{\bar{v}_{zm}} = \frac{2.44}{Re^{1/8}}. \tag{9.103}$$

Evaluate this velocity ratio for Re $= 10^4$, 10^5, and 10^6, and comment on the results.

9. *Various quantities in turbulent flow—M.* Quoting *any* formula(s) for the universal velocity profile, prove the following relationship between the dimensionless mean velocity and the dimensionless pipe radius for flow in a smooth pipe:

$$v_{zm}^+ = 2.5 \ln a^+ + 1.75.$$

Comment briefly on any assumptions that have been made in the derivation of this equation.

Water ($\rho = 1,000$ kg/m^3, $\mu = 0.001$ kg/m s) flows through a hydraulically smooth 5.0-cm (0.05 m) diameter pipe under a pressure gradient of $-dp/dz = 14,000$ N/m^3. Find the wall shear stress (N/m^2), the mean velocity (m/s), the Reynolds number, the thickness of the laminar sublayer (m), and the velocity (m/s) at the junction between the laminar sublayer and the buffer region.

Estimate the heat-transfer coefficient (W/m^2 °C) between the water and the pipe wall if water has a specific heat of $c_p = 4,184$ J/kg °C.

10. *An alternative to the Prandtl hypothesis—M.* Sleicher proposed that in a region close to the wall, the eddy kinematic viscosity obeys[23]:

$$\nu_T = \nu(cy^+)^2.$$

By considering *both* turbulent *and* viscous contributions to the shear stress, which still essentially equals its wall value, prove that the corresponding velocity profile is:

$$v_z^+ = \frac{1}{c}\tan^{-1} cy^+.$$

If $c \doteq 0.088$, show that the above velocity profile merges smoothly—both in magnitude and slope—with $v_z^+ = 2.5 \ln y^+ + 5.5$, and discover the value of y^+ at which the transition occurs.

---

[23] C.A. Sleicher, Jr., "Experimental velocity and temperature profiles for air in turbulent pipe flow," *Transactions of the American Society of Mechanical Engineers*, Vol. 80, pp. 693–704 (1958).

11. *Turbulent mass transfer—D.* Considering gas flow of molecular weight M_w in a pipe as represented by a central turbulent core with a laminar sublayer next to the wall, derive the following analogy between mass and momentum transfer when the partial pressure of A is p_{Am} in the mainstream and zero at the wall:

$$N_A = \frac{\tau p_{Am}}{P M_w [u_m + u_L(\text{Sc} - 1)]}.$$

Here, N_A is the molal flux, P is the total pressure, and $\text{Sc} = \nu/\mathcal{D}$ is the Schmidt number, where $\mathcal{D}$ is the diffusion coefficient across the laminar sublayer.

The catalytic isomerization of butene–1 to butene–2,

$$\text{CH}_3\text{CH}_2\text{CH}=\text{CH}_2 \longrightarrow \text{CH}_3\text{CH}=\text{CHCH}_3$$

is carried out continuously and isothermally at 800 °F and 1 atm in a lime-coated porcelain tube of 2-in. I.D., and with the tube wall serving as the catalyst. Assuming that the rate of reaction is essentially the rate of which butene–1 diffuses to the wall, estimate the length of tube required to give 80% conversion of a pure butene–1 feed. Assume that the laminar sublayer is thin and contributes little to the overall flow, and that the concentration and velocity profiles are flat over most of the turbulent region.

Data: at 800 °F, $\mathcal{D} = 0.59$ ft^2/hr, $\mu = 0.044$ lb$_m$/ft hr, $c_f = 0.0086$, and $u_L/u_m = 0.72$.

12. *Thickness of the laminar sublayer—M.* In Section 9.8, the thickness of the laminar sublayer was shown to be $\delta/D = 58.3/\text{Re}^{7/8}$ for a logarithmic velocity profile in the mainstream.

Conduct a similar derivation to verify the corresponding result for a velocity profile in the turbulent region given by the one-seventh power law, $\bar{v}_z/\bar{v}_{zc} = (y/a)^{1/7}$. Evaluate this thickness ratio for $\text{Re} = 10^4$, 10^5, and 10^6. Comment on your results.

13. *Turbulent heat transfer—M (C).* In a gas-cooled nuclear reactor, the fissile material is made into straight tubular fuel elements of length L and internal diameter D. Coolant passes through the tubes, which are embedded in a block of moderating material of thickness L, shown in Fig. P9.13. At a distance x from the inlet, the rate of heat production per unit length is $q\sin(\pi x/L)$.

Fig. P9.13   Representative coolant tube in a nuclear reactor.

By using the Reynolds analogy, prove that the position x_m at which the temperature at the inner surface of the fuel element is a maximum is given by:

$$x_m = \frac{L}{\pi} \tan^{-1} \left(\frac{-\pi D}{2 f_{\mathrm{F}} L} \right),$$

in which f_{F} is the Fanning friction factor.

Assume that an energy balance leads to the following equation for the variation of coolant temperature T with distance x:

$$m c_p \frac{dT}{dx} = \pi D h (T_w - T) = q \sin \frac{\pi x}{L},$$

where m is the mass flow rate of coolant and c_p is its specific heat, h is the heat-transfer coefficient, and T_w is the local wall temperature.

14. *Turbulent velocity profile—M (C).* A fluid in turbulent flow passes through a smooth circular pipe, and the velocity distribution is given by the equation:

$$\overline{v}_z^+ = 5.5 + 2.5 \ln y^+, \quad \text{where} \quad \overline{v}_z^+ = \frac{\overline{v}_z}{\sqrt{\tau_w/\rho}}, \quad \text{and} \quad y^+ = \frac{y\sqrt{\tau_w/\rho}}{\nu}.$$

If this equation holds across the whole pipe, calculate the ratio of the maximum velocity to the mean velocity at $\mathrm{Re} = 10^5$. At what position should a Pitot tube be placed to measure the mean velocity at this Reynolds number?

15. *Reynolds analogy for minimum pumping power—M (C).* A gas of molecular weight M_w is pumped through a cylindrical coolant duct in order to remove heat from part of a nuclear reactor. The temperature rise ΔT, the logarithmic-mean temperature difference $\Delta T_{log\ mean}$, the transfer area A, and heat load q are *specified*, so in the heat-balance equation:

$$q = m c_p \Delta T = A h \Delta T_{log\ mean}, \tag{4}$$

the *only* variables are the mass flow rate m and the specific heat c_p of the gas. Thus, $m c_p$ is a constant, c_1 for example.

Use the Reynolds analogy to prove that the pumping power $P = Q \Delta p$ (where Q is the volumetric flow rate and Δp is the pressure drop in the duct) is lowest if a gas is selected with the largest possible value of $M_w^2 c_p^3$. Assume $\rho = M_w p / RT$ for the gas, with p, R, and T effectively constant, so that $\rho = c_2 M_w$, where c_2 is another constant.

16. *Turbulent mass flux—E.* Consider flow in a pipe of diameter D and length L with a mean axial velocity $\overline{v}_{zm}$. Note that the Reynolds analogy gives the turbulent mass flux per unit area of the wall as $m = \tau_w / \overline{v}_{zm}$. Let m_t be the total such mass flux based on the *total wall area* of the pipe. Also define m_c as the convective mass flow rate along the pipe. For flow in a smooth pipe of diameter $D = 0.05$ m at a Reynolds number $\mathrm{Re} = 10^4$, how long would the pipe have to be so that $m_t = m_c$?

17. *Theory for a plane turbulent jet—D.* Prove Eqn. (9.114) from the development that precedes it.

18. *Calculation of velocities for a plane turbulent jet—D.* (a) Verify that by defining new variables $f_0 = f$ and $f_1 = df/d\zeta$, Eqn. (9.115) can be expressed as *two* first-order ordinary differential equations:

$$\frac{df_0}{d\zeta} = f_1, \qquad \frac{df_1}{d\zeta} = \sqrt{f_0 f_1}, \qquad\qquad (\text{P9.18.1})$$

subject to the boundary conditions:

$$\zeta = 0: \ \ f_0 = 0, \ \ f_1 = 1, \qquad\qquad (\text{P9.18.2})$$

(b) As discussed in Appendix A, use a spreadsheet to implement Euler's method (or similar) to solve Eqn. (P9.18.1) subject to the boundary conditions (P9.18.2), from $\zeta = 0$ to a value where the velocities are negligible. (Any other standard software for solving differential equations may be used instead.) Try different step-sizes $\Delta\zeta$ to make sure that you have used one that is sufficiently small for accurate results. Plot f' and $\zeta f' - f/2$ against both negative and positive values of ζ, one-half of which can be obtained from symmetry.

Hint: if your computed values appear improbable, did you make the right choice when faced with an alternative?

19. *Streamlines and velocities for a plane turbulent jet—M.* Sketch the streamlines for the turbulent jet, for which the velocities are shown in Fig. 9.16. Include both the jet *and* the region outside it. Also show the axial velocity profiles at three stages in the development of the jet.

20. *Analytical solution for the plane turbulent jet—M.* Verify Eqn. (9.123) from the development that precedes it, and compare the numerical values for f' with those deduced from Fig. 9.16.

21. *Entrainment by a plane turbulent jet—M.* For the plane turbulent jet discussed in Section 9.14, what is the dependency of the total mass flow rate per unit depth in the jet on the axial distance x? Comment critically on your answer. Assume the "simplified" analytical solution given by Eqn. (9.123).

22. *Axisymmetric turbulent jet—D.* For the axisymmetric turbulent jet, verify that:

(a) The powers of z appearing in Eqn. (9.128) for ζ and $\bar{v}_z$ are $p = q = 1$.
(b) The velocities given in Eqns. (9.132) and (9.133) satisfy the appropriate differential momentum and mass balances and that the total rate of momentum transfer is $\dot{M}$.

23. *Entrainment by an axisymmetric turbulent jet—M.* For the axisymmetric turbulent jet discussed in Section 9.14, what is the dependency of the total mass flow rate in the jet on the axial distance z? Comment critically on your answer.

24. *Streamlines and velocities in an axisymmetric turbulent jet—M.* (a) Verify that the velocities in Eqns. (9.132) and (9.133) are consistent with the stream function:

$$\psi = \nu_T z f(\zeta), \quad \text{where} \quad \zeta = \frac{\beta r}{z}, \quad f = \frac{\zeta^2}{1 + \dfrac{\zeta^2}{4}}.$$

(b) Sketch several streamlines, both within the jet and outside it. Also show the axial velocity profiles at three stages in the development of the jet.

(c) Plot dimensionless axial and radial velocity profiles in the manner of Fig. 9.16.

25. *The Prandtl number and the Reynolds analogy—E.* Table E9.4 showed that the Reynolds analogy underestimated/overestimated the heat-transfer coefficient for Prandtl numbers lower/higher than one. With reference to the definition of the Prandtl number as the ratio of the kinematic viscosity to the thermal diffusivity, explain on physical grounds why this is so.

26. *Order of terms for turbulent jet—E.* Verify the orders of the three terms appearing in Table 9.4 relating to a plane turbulent jet.

27. *True/false.* Check *true* or *false*, as appropriate:

(a) Turbulent fluctuations typically occur in a matter of microseconds.   T ☐ F ☐

(b) The time-averaged continuity equation contains the time-averaged velocities instead of the instantaneous velocities.   T ☐ F ☐

(c) Extra terms, known as the Reynolds stresses, occur when the momentum balances are time-averaged.   T ☐ F ☐

(d) The Reynolds stresses can be explained by considering the transport of momentum by turbulent eddies.   T ☐ F ☐

(e) Mixing-length theory attempts to unify the turbulent transport of momentum, heat, and mass.   T ☐ F ☐

(f) The Prandtl hypothesis assumes that the mixing length ℓ does not depend on the distance from the wall.   T ☐ F ☐

(g) The Prandtl hypothesis, in conjunction with the assumption of an approximately constant shear stress, leads to a logarithmic-type velocity profile.   T ☐ F ☐

(h)   The laminar sublayer is a thin region next to the wall,   T □ F □
in which turbulent fluctuations are essentially negligible.

(i)   A friction factor is the ratio of the wall shear-stress   T □ F □
effects to inertial effects.

(j)   In the universal velocity profile, the junction between   T □ F □
the laminar sublayer and the turbulent core occurs at
approximately $y^+ = 25$.

(k)   The thickness of the laminar sublayer increases as the   T □ F □
Reynolds number increases.

(l)   The Colebrook and White equation gives the friction   T □ F □
factor for both laminar and turbulent flow in a pipe.

(m)   The Blasius law, $f_F = 0.079\mathrm{Re}^{-1/4}$, is consistent with   T □ F □
the one-seventh power law turbulent velocity profile.

(n)   The Prandtl hypothesis, $\ell = ky$, somewhat underestimates the mixing length for general values of y.   T □ F □

(o)   The Prandtl number is the ratio of the kinematic viscosity to the thermal diffusivity.   T □ F □

(p)   If the random fluctuations v'_x and v'_y are correlated in   T □ F □
some way, then $\overline{v'_x v'_y}$ will not be zero.

(q)   In the simple model we discussed for turbulence, the   T □ F □
same eddies can transport mass, momentum, and
thermal energy.

(r)   A hydraulically smooth pipe is one in which the   T □ F □
"hills" on the wall do not extend beyond the laminar sublayer.

(s)   When the general momentum balances are time-   T □ F □
averaged, the only significant changes are that time-
averaged values (such as $\bar{p}$) have replaced the instantaneous values (such as p).

(t)   For flow in a smooth pipe of diameter $D = 12$ in with   T □ F □
Re $= 10^5$, the thickness of the laminar sublayer is
0.138 ± 0.02 in.

(u)   For a turbulent eddy, a high Prandtl number means   T □ F □
that its momentum content is likely to be dissipated
more quickly than its thermal energy content before
it reaches its destination.

Chapter 10

BUBBLE MOTION, TWO-PHASE
FLOW, AND FLUIDIZATION

10.1 Introduction

THE simultaneous motion of two or more immiscible fluids is of considerable importance in chemical engineering, and there is an abundance of examples, such as:

1. Mixing of coal particles with water and pumping the resulting slurry through a pipeline.
2. Simultaneous production of oil and natural gas upward through a vertical well.
3. Vaporization of water or other liquids inside the heated tubes of a vertical-tube evaporator.
4. Contacting a reacting fluid with catalyst particles in a fluidized bed.
5. Absorption of a component from a gas stream that is rising through a packed bed in a tower, down through which an absorbing liquid is flowing.
6. Aerated bioreactors, with air bubbles affording a large surface area for reaction.

The complete study of any one of these topics requires much more space than is available here. Indeed, we shall see later that an entire book has been written on one-dimensional two-phase flow, and that another one has been devoted exclusively just to the annular regime in two-phase flow. Nevertheless, the reader will be introduced in this chapter to the important concepts and should gain a substantial insight into the key issues.

10.2 Rise of Bubbles in Unconfined Liquids

In order to predict the performance of two-phase flow in vertical tubes (Section 10.4) and fluidized beds (Sections 10.7 and 10.8), it is first necessary to study the rate of rise of single bubbles.

Small bubbles. The results of Peebles and Garber[1] are summarized in Table 10.1 for bubbles with volumes in the range equivalent to spheres with radii between

---

[1] F.N. Peebles and H.J. Garber, "Studies on the motion of gas bubbles in liquids," *Chemical Engineering Progress*, Vol. 49, No. 2, pp. 88–97 (1953).

0.024 and 0.3 in. There are four distinct rise modes, generally corresponding to increasing bubble size (Re_b is the Reynolds number of the bubble, based on its rise velocity u_b, and R_b is the radius of a sphere having the same volume as the bubble, and is half the equivalent diameter D_e introduced later):

1. For $Re_b < 2$, the bubbles behave as buoyant solid spheres, rising vertically, with their motion determined by Stokes' law.
2. For larger values of Re_b (up to a limit determined by properties of the liquid) the bubbles also rise vertically as spheres, but with a drag coefficient slightly less than that of a solid sphere of the same volume.
3. For a range of Re_b determined by the liquid properties, the bubbles are flattened and rise in a zigzag pattern, with significantly increased drag coefficients.
4. The largest bubbles rise almost vertically with significantly increasing drag coefficients, and assume irregular mushroom-cap shapes. For this region, the constant 1.53 recommended by Harmathy has been used instead of the value of 1.18 from Peebles and Garber, because it correlates better with experiment.[2]

Table 10.1   Terminal Velocities for Bubbles.
Values for Regions 1–4 are from Peebles and Garber.

Region	Range of Applicability	Terminal Velocity, u_b
1	$Re_b \leq 2$	$\dfrac{2R_b^2(\rho_l - \rho_g)g}{9\mu_l}$
2	$2 < Re_b \leq 4.02G_1^{-0.214}$	$0.33g^{0.76}\left(\dfrac{\rho_l}{\mu_l}\right)^{0.52} R_b^{1.28}$
3	$4.02G_1^{-0.214} < Re_b \leq 3.10G_1^{-0.25}$	$1.35\left(\dfrac{\sigma}{\rho_l R_b}\right)^{0.5}$
4	$3.10G_1^{-0.25} < Re_b$	$1.53\left(\dfrac{\sigma g}{\rho_l}\right)^{0.25}$
5	$R_b \geq 2.3\sqrt{\dfrac{\sigma}{g\rho_l}}$	$1.00\sqrt{gR_b}$

$$\text{In the above:} \quad Re_b = \frac{2\rho_l u_b R_b}{\mu_l}, \quad G_1 = \frac{g\mu_l^4}{\rho_l \sigma^3},$$

---

[2] T.Z. Harmathy, "Velocity of large drops and bubbles in media of infinite or restricted extent," *American Institute of Chemical Engineers Journal*, Vol. 6, No. 2, pp. 281–288 (1960).

Peebles and Garber also reported the results of Davies and Taylor for larger bubbles (see below), but did not include them in their table. For completeness, however, these have been included under Region 5 in Table 10.1. The limit of bubble radius R_b beyond which Region 5 applies has been obtained by equating the velocities given by Region 4 and Eqn. (10.8) for spherical-cap bubbles.

Large "spherical-cap" bubbles. For otherwise stationary liquids that are not very viscous, the shape of a "large" bubble (whose volume equals that of a sphere whose diameter is several mm) is roughly *lenticular* or that of a *spherical cap*, as shown in the upper part of Fig. 10.1. A somewhat similar shape occurs for a large gas bubble rising in a fluidized-bed reactor, although in this case the bubble is more nearly a complete sphere.

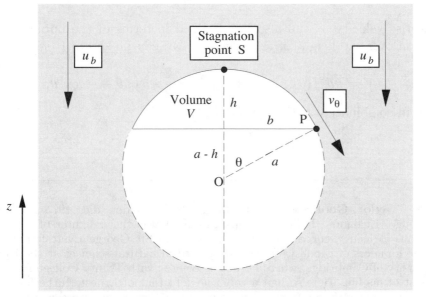

Fig. 10.1   Flow around a spherical cap bubble.

The rise velocity u_b of such bubbles is analyzed by first superimposing a *downward* velocity u_b on the whole system, so that we have a downward streaming of the liquid—in potential flow—past a *stationary* spherical-cap bubble. Because the bubble is large, surface-tension effects are negligible.

Noting that the pressure is constant because of the gas in the bubble, application of Bernoulli's equation between S and P gives the tangential velocity v_θ at point P as:

$$v_\theta = \sqrt{2gh} = \sqrt{2ga(1 - \cos\theta)}. \tag{10.1}$$

But, from Eqn. (7.73), for potential flow around a sphere, with $U = -u_b$:

$$v_\theta = \frac{1}{r}\frac{\partial \phi}{\partial \theta} = u_b\left(1 + \frac{a^3}{2r^3}\right)\sin\theta, \tag{10.2}$$

which gives, at the bubble surface $r = a$:

$$v_\theta = \frac{3}{2} u_b \sin \theta. \tag{10.3}$$

If the two effects give the same result, (10.1) and (10.3) may be equated, giving:

$$2ga(1 - \cos \theta) = \frac{9}{4} u_b^2 (1 - \cos \theta)(1 + \cos \theta), \tag{10.4}$$

which can be satisfied for small values of θ (that is, near the bubble "nose" S, where $1 + \cos \theta \doteq 2$), giving a bubble rise velocity of:

$$u_b \doteq \frac{2}{3} \sqrt{ga}. \tag{10.5}$$

The rise velocity u_b can also be obtained in terms of the bubble volume V. Integration over the height of the spherical cap can be shown to give its volume as:

$$V = \pi a^3 f(\theta), \quad \text{where} \quad f(\theta) = \frac{2}{3} - \cos \theta + \frac{1}{3} \cos^3 \theta, \tag{10.6}$$

yielding, in combination with Eqn. (10.5):

$$u_b = \frac{2}{3\pi^{1/6}} V^{1/6} \sqrt{g} \, [f(\theta)]^{-1/6}. \tag{10.7}$$

Taylor, Geoffrey Ingram, born 1886 in London; died 1975 in Cambridge, England. Taylor's mother Margaret was the daughter of the famous logician George Boole and the niece of Sir George Everest, one of the founders of geodesy. Taylor graduated in natural sciences at the University of Cambridge, where he was associated with Trinity College for the rest of his life. He also had a room next to that of Ernest Rutherford at the Cavendish Laboratory. During a six-month voyage on a scientific expedition in the North Atlantic in 1912, Taylor studied mixing in the lower atmosphere, and this led to his concept of a mixing length for turbulent processes. During World War I, he did research on the strength of propeller shafts, leading to a theory on dislocations in metal crystals, published in 1934. From 1923 to 1952 he held a Royal Society research professorship, allowing him to concentrate on research and relieving him from teaching— invoking a comment from Rutherford that he was "paid provided he does no work." During this period his main work was on turbulent diffusion and the statistical properties of turbulence. He was an expert on blasts and shock waves during World War II and was involved with the Manhattan Project. Taylor and his wife Stephanie were avid sailors.

Source: *Dictionary of Scientific Biography*, Charles Scribner's Sons, New York, 1975.

The experiments of Davies and Taylor gave a similar result:[3,4]

$$u_b = 0.792\, V^{1/6}\sqrt{g} = 0.711\sqrt{gD_e}, \tag{10.8}$$

in which the *equivalent diameter* D_e of the bubble is defined as the diameter of a sphere that has the same volume as the bubble:

$$V = \frac{1}{6}\,\pi D_e^3. \tag{10.9}$$

We now need to reexpress $\sqrt{gD_e}$ in terms of $\sqrt{ga}$, so that the Davies and Taylor result of Eqn. (10.8) can be compared directly with the theoretical prediction of Eqn. (10.5). From Eqns. (10.6) and (10.9) and the geometry of Fig. 10.1, a few lines of algebra give two relations for the bubble volume:

$$V = \frac{2}{3}\,\pi a^2 h - \frac{1}{3}\,\pi b^2 (a - h) = \frac{1}{6}\,\pi D_e^3, \tag{10.10}$$

from which the values given in Table 10.2 can be determined.

Table 10.2   Numerical Values for Spherical-Cap Bubbles

h/a	b/a	$V/\pi a^3$	D_e/a	c
0.100	0.436	0.0097	0.387	0.442
0.250	0.661	0.0573	0.701	0.595
0.359	0.767	0.1133	0.879	0.667
0.500	0.866	0.2084	1.077	0.738

If the Davies and Taylor experimental correlation of Eqn. (10.8) is translated into the form:

$$u_b = c\sqrt{ga}, \tag{10.11}$$

then Table 10.2 gives the appropriate values for the coefficient c. The bubble angle that corresponds to the approximate theoretical value $c = 2/3$ is $\theta = \cos^{-1}(1 - 0.359) = 50.1°$. A more accurate treatment would require that the shape of the bubble be adjusted, and the flow pattern recomputed, so that the Bernoulli condition, Eqn. (10.1), is satisfied *everywhere*—not just in the vicinity of the nose.

---

[3] R.M. Davies and G.I. Taylor, "The mechanics of large bubbles rising through extended liquids and through liquids in tubes," *Proceedings of the Royal Society*, Vol. 200A, pp. 375–390 (1950).

[4] The author attended a G.I. Taylor seminar during the sesquicentennial celebrations of the University of Michigan in 1967. Taylor related the following technique for locating the position of a sailboat lost in a deep fog near a coastline (whose general direction was known). Sail due west (for example) and record the distance traveled until the coast is reached. Return to the original position by sailing due east for the same distance. Then sail north for a known number of miles. Repeat the entire process a few times. On a suitable scale, the result can be transcribed to resemble a comb with a few teeth of unequal length, which can then be adjusted until it fits a map of the coastline, thereby determining the ship's bearings!

Example 10.1—Rise Velocity of Single Bubbles

Plot the rise velocity u_b, cm/s, for single bubbles of air in water against the equivalent bubble diameter D_e, from $D_e = 0.01$ to 10 cm.

Solution

From Chapter 1, the physical properties of water at 20 °C are: $\rho = 0.998$ g/cm^3, $\sigma = 72.75$ dynes/cm ($=$ g/s^2), $\mu = 1.21$ cP or 0.0121 g/cm s, and $g = 981$ cm/s^2. A spreadsheet is then used to implement the correlations for the five regions of Table 10.1. The results for u_b vs. D_e are shown in Fig. E10.1.

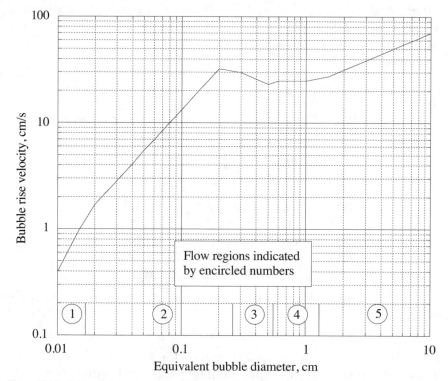

Fig. E10.1 *Bubble rise velocity as a function of equivalent diameter.*

After the peak in u_b at the higher end of Region 2, the moderate decline in Region 3 *is* observed experimentally. The slight jump in passing from Region 3 to Region 4 is a consequence of the correlations used. The Reynolds number range was 0.5–70,420.   ☐

10.3 Pressure Drop and Void Fraction in Horizontal Pipes

In this chapter, the discussion of two-phase flow in pipes will center largely on *vertical* flow because of its importance in gas and oil wells, evaporators, bioreactors, and boilers. However, the topic is introduced by considering two-phase flow in *horizontal* pipes. Although there are several different regimes in horizontal flow in which the gas and liquid interact with each other in different ways (somewhat similar to those in Fig. 10.3 for vertical flow), a simpler approach is taken in the following model. See Table 10.3 for the notation for this and the next section.

The *void fraction* ε is the fraction of the total volume in the pipe that is occupied by the gas phase, and is related to the individual volumetric flow rates G and L and the mean velocities v_g and v_l by:

$$G = \varepsilon v_g A, \qquad L = (1 - \varepsilon)v_l A. \qquad (10.12)$$

The *quality* x of the flowing stream (a term used when analyzing the performance of boilers) is the fraction of the total mass flow rate that is in the gas phase:

$$x = \frac{\rho_g G}{\rho_g G + \rho_l L}. \qquad (10.13)$$

Table 10.3   Variables for Two-Phase Gas/Liquid Flow

Variable	Definition
A	Cross-sectional area of pipe
D	Pipe diameter
f_F	Fanning friction factor
G	Gas volumetric flow rate
j	Superficial velocity (volumetric flow rate per unit area)
L	Liquid volumetric flow rate
p	Pressure
v	Mean velocity
x	Quality—fraction of flowing mass that is gas
z	Axial coordinate
ε	Void fraction
ρ	Density
ϕ^2	Multiplier used for obtaining two-phase pressure gradient
X^2	Ratio ϕ_g^2/ϕ_l^2
Subscripts	
g, l	Gas, liquid
tp	Two-phase
go	Gas only
lo	Liquid only

As a *first approximation*, the frictional pressure drop for two-phase flow in the horizontal tube of diameter D in Fig. 10.2(a) may be deduced by supposing that the gas and liquid flow individually in cylinders of diameters D_g and D_l, respectively, as in (b) and (c). The void fraction is then the ratio of two areas:

$$\varepsilon = \frac{\pi D_g^2/4}{\pi D^2/4} = \frac{D_g^2}{D^2}. \qquad \text{Likewise}: \quad 1 - \varepsilon = \frac{D_l^2}{D^2}. \qquad (10.14)$$

As another preliminary, observe the definitions of two important types of velocities:

1. The *superficial* velocities are those that would occur if each phase flowed by itself in a pipe of cross-sectional area A, the other phase being absent. Thus, if the volumetric gas and liquid flow rates are G and L, respectively:

$$j_g = \frac{G}{A,} \qquad j_l = \frac{L}{A}. \qquad (10.15)$$

2. The mean velocities are the superficial velocities divided by the gas and liquid fractions:

$$v_g = \frac{j_g}{\varepsilon}, \qquad v_l = \frac{j_l}{1 - \varepsilon}. \qquad (10.16)$$

Fig. 10.2 Separated horizontal flow model. Simultaneous gas/liquid flow in (a) is considered as the combination of gas and liquid flows, as in (b) and (c).

First consider the pressure gradient in the gas, for the following two cases:

1. Gas only flowing by itself and occupying the entire cross section of the pipe:

$$\left(\frac{dp}{dz}\right)_{go} = -\frac{2f_F \rho_g j_g^2}{D}. \qquad (10.17)$$

2. Gas flowing by itself in the hypothetical cylinder of diameter D_g. Note that this must also equal the pressure gradient for the combined two-phase flow:

$$\left(\frac{dp}{dz}\right)_{tp} = \left(\frac{dp}{dz}\right)_g = -\frac{2f_F \rho_g v_g^2}{D_g}. \tag{10.18}$$

The ratio of these last two equations yields:

$$\frac{(dp/dz)_{tp}}{(dp/dz)_{go}} = \left(\frac{v_g}{j_g}\right)^2 \frac{D}{D_g} = \frac{1}{\varepsilon^{5/2}}. \tag{10.19}$$

in which Eqns. (10.14) and (10.16) have been used to eliminate the ratios v_g/j_g and D/D_g in terms of the void fraction.

Equation (10.19) can be rewritten in the form:

$$\left(\frac{dp}{dz}\right)_{tp} = \phi_g^2 \left(\frac{dp}{dz}\right)_{go}, \quad \text{in which} \quad \phi_g^2 = \frac{1}{\varepsilon^{5/2}}. \tag{10.20}$$

Similarly, by considering the pressure drop in the liquid phase:

$$\left(\frac{dp}{dz}\right)_{tp} = \phi_l^2 \left(\frac{dp}{dz}\right)_{lo}, \quad \text{in which} \quad \phi_l^2 = \frac{1}{(1-\varepsilon)^{5/2}}. \tag{10.21}$$

The use of ϕ^2 instead of ϕ perpetuates the notation used by early researchers in the field and has no particular significance.

Elimination of the void fraction from the second relations in Eqns. (10.20) and (10.21) gives:

$$\frac{1}{(\phi_g^2)^{2/5}} + \frac{1}{(\phi_l^2)^{2/5}} = 1. \tag{10.22}$$

Note that these results are based on a *simplified* model, and that the same value of the friction factor is also assumed for both phases. Nevertheless, the early studies of Lockhart and Martinelli correlate remarkably well with modest modifications to Eqn. (10.22).[5] These researchers investigated a wide variety of liquids flowing with air at roughly atmospheric pressure in small-diameter (up to 1.0–in. I.D.) tubes, and their results are summarized in Table 10.4. They recognized four regimes, comprising all possible combinations of viscous (Re < 1,000) or turbulent (Re > 2,000) flow for each phase, calculated as though it flowed by itself and occupied the whole cross section.

The square of the parameter X is defined as the ratio of the liquid-only and gas-only pressure gradients:

$$X^2 = \frac{(dp/dz)_{lo}}{(dp/dz)_{go}} = \frac{\phi_g^2}{\phi_l^2}. \tag{10.23}$$

---

[5] See R.W. Lockhart and R.C. Martinelli, "Proposed correlation of data for isothermal two-phase, two-component flow in pipes," *Chemical Engineering Progress*, Vol. 45, No. 1, pp. 39–48 (1949).

Large or small values of X correspond to a flow regime that is primarily liquid or primarily gas, respectively.

Equation (10.22) is a special case of the more general form:

$$\frac{1}{(\phi_g^2)^{1/n}} + \frac{1}{(\phi_l^2)^{1/n}} = 1, \tag{10.24}$$

in which the value of n depends on the flow regime as indicated in Table 10.5.

Table 10.4   Values from Lockhart and Martinelli

X	Void Fraction ε	Liquid: Gas:	Turb Turb ϕ_g	Visc Turb ϕ_g	Turb Visc ϕ_g	Visc Visc ϕ_g
0.01			1.28	1.20	1.12	1.05
0.02			1.37	1.28	1.16	1.07
0.04			1.54	1.36	1.24	1.12
0.07	0.96		1.71	1.45	1.35	1.19
0.1	0.95		1.85	1.52	1.45	1.24
0.2	0.91		2.23	1.78	1.74	1.40
0.4	0.86		2.83	2.25	2.20	1.70
0.7	0.81		3.53	2.85	2.85	2.16
1.0	0.77		4.20	3.48	3.48	2.61
2.0	0.69		6.20	5.25	5.24	4.12
4.0	0.60		9.50	8.20	8.60	7.00
7.0	0.52		13.7	12.1	12.8	11.2
10.0	0.47		17.5	15.9	16.6	15.0
20.0	0.34		29.5	28.0	28.8	27.3
40.0	0.24		51.5	50.0	50.0	50.0
70.0	0.16		82.0	82.0	82.0	82.0
100.0	0.10		111.0	111.0	111.0	111.0

The tabulated values of ε are *empirically* correlated (with an error no larger than 0.02) by:

$$\varepsilon = \frac{1}{(1 + 0.0904\, X^{0.548})^{2.82}}. \tag{10.25}$$

From Eqns. (10.23) and (10.24):

$$\phi_g = (1 + X^{2/n})^{n/2}, \tag{10.26}$$

Example 10.2—Two-Phase Flow in a Horizontal Pipe   **541**

and it is easy to show from a spreadsheet that the values of Table 10.4 are well represented by Eqn. (10.26) with the exponents given in Table 10.5.

Table 10.5 Exponents for Two-Phase Correlation

	n	Liquid Flow	Gas Flow
	4.12	Turbulent	Turbulent
	3.61	Viscous	Turbulent
	3.56	Turbulent	Viscous
$X < 1$:	2.68	Viscous	Viscous
$X > 1$:	3.27	Viscous	Viscous

Refinements of the above procedure are available from Baroczy, who investigated a wide variety of fluids, including steam and water up to pressures of 1,400 psia.[6] Chisholm has also adapted Baroczy's work to the prediction of pressure drops in evaporating flows.[7]

Example 10.2—Two-Phase Flow in a Horizontal Pipe

Gas and oil are flowing at volumetric flow rates $G = 10.0$, $L = 0.5$ ft^3/s in a horizontal pipeline of internal diameter $D = 0.5$ ft. Estimate the fraction of gas, the pressure gradient, and the mean velocities of each phase. The physical properties are $\rho_g = 0.15$, $\rho_l = 50.0$ lb$_\text{m}$/ft^3, $\mu_g = 0.025$, $\mu_l = 20.0$ lb$_\text{m}$/ft hr; the Fanning friction factor is $f_\text{F} = 0.0045$.

Solution

The following quantities are first calculated:

Cross-sectional area of pipe

$$A = \frac{\pi D^2}{4} = \frac{\pi \times 0.5^2}{4} = 0.196 \text{ ft}^2.$$

Superficial velocities

$$j_g = \frac{G}{A} = \frac{10.0}{0.196} = 51.0, \qquad j_l = \frac{L}{A} = \frac{0.5}{0.196} = 2.55 \text{ ft/s}.$$

[6] C.J. Baroczy, "A systematic correlation for two-phase pressure drop," *Chemical Engineering Progress Symposium Series* Vol. 62, No. 64, pp. 232–249 (1966).

[7] D. Chisholm, "Pressure gradients due to friction during the flow of evaporating two-phase mixtures in smooth tubes and channels, *International Journal of Heat and Mass Transfer*, Vol. 16, pp. 347–358 (1973).

Reynolds numbers

$$\text{Re}_g = \frac{0.15 \times 51.0 \times 0.5 \times 3{,}600}{0.025} = 5.51 \times 10^5,$$

$$\text{Re}_l = \frac{50.0 \times 2.55 \times 0.5 \times 3{,}600}{20.0} = 11{,}475.$$

Thus, the flow is turbulent for the liquid and turbulent for the gas, with $n = 4.12$ from Table 10.5.

The quantity X^2, being the ratio of the liquid-only and gas-only pressure gradients is next calculated from Eqns. (10.23), (10.17), and a similar one for the liquid pressure gradient. The friction factor and the diameters are the same in each case and cancel, giving:

$$X^2 = \frac{(dp/dz)_{lo}}{(dp/dz)_{go}} = \frac{\rho_l \, j_l^2}{\rho_g \, j_g^2} = \frac{50 \times 2.55^2}{0.15 \times 51.0^2} = 0.833, \quad \text{or} \quad X = 0.913.$$

From Eqn. (10.26) (or Table 10.4), the value of ϕ_g is:

$$\phi_g = (1 + X^{2/n})^{n/2} = (1 + 0.913^{2/4.12})^{4.12/2} = 3.99.$$

From Eqn. (10.25) (or Table 10.4) the fraction of gas is:

$$\varepsilon = \frac{1}{(1 + 0.0904 \, X^{0.548})^{2.82}} = \frac{1}{(1 + 0.0904 \times 0.913^{0.548})^{2.82}} = 0.792.$$

The relevant pressure drops are obtained from Eqns. (10.17) and (10.20):

Gas only

$$\left(\frac{dp}{dz}\right)_{go} = -\frac{2 f_F \rho_g j_g^2}{D} = -\frac{2 \times 0.0045 \times 0.15 \times 51.0^2}{0.5 \times 32.2 \times 144} = -1.51 \times 10^{-3} \text{ psi/ft.}$$

Two-phase flow

$$\left(\frac{dp}{dz}\right)_{tp} = \phi_g^2 \left(\frac{dp}{dz}\right)_{go} = 3.99^2 \times (-1.51 \times 10^{-3}) = -0.0241 \text{ psi/ft.}$$

Finally, the individual mean velocities are obtained:

Gas

$$v_g = \frac{G}{\varepsilon A} = \frac{10.0}{0.792 \times 0.196} = 64.4 \text{ ft/s.}$$

Liquid

$$v_l = \frac{L}{(1 - \varepsilon)A} = \frac{0.5}{(1 - 0.792) \times 0.196} = 12.3 \text{ ft/s.} \qquad \square$$

10.4 Two-Phase Flow in Vertical Pipes

Two-phase gas/liquid flow in *vertical* pipes will be treated in more detail because of its importance in evaporators and in the simultaneous transport of oil and gas in wells. Several flow *regimes* can occur, depending on many factors, including the individual magnitudes of the liquid and gas flow rates, and physical properties such as density, surface tension, and viscosity.

There are four principal flow regimes, shown in Fig. 10.3, which occur successively at ever-increasing gas flow rates:

(a) *Bubble* flow, in which the gas is dispersed as small bubbles throughout the liquid, which is the continuous phase.

(b) *Slug* flow, in which the individual small bubbles have started to coalesce together in the form of gas *slugs*. The liquid phase is still continuous. For high gas flows, the liquid "pistons" between the slugs start to narrow down and form irregularly shaped "bridges," a situation that is known as *churn* flow.

(c) *Annular* flow, in which the fast-moving gas stream is now a continuous phase that encompasses the central portion of the pipe, with the liquid forming a relatively thin film on the pipe wall. Partial entrainment of the liquid into the gas stream may occur.

(d) *Mist* flow, in which the velocity of the continuous gas phase is so high that it reaches as far as the pipe wall and entrains all of the liquid in the form of droplets.

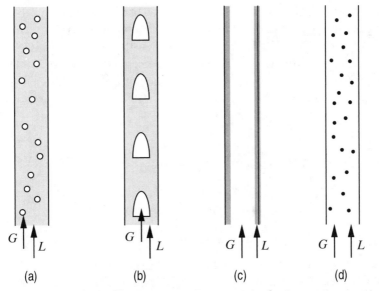

Fig. 10.3 *Two-phase flow regimes in a vertical pipe at increasing gas flow rates: (a) bubble, (b) slug, (c) annular, and (d) mist flow. In each case, the gas is shown in white, and the liquid is shaded or black.*

Determination of flow regime. A typical situation occurs when the gas and liquid volumetric flow rates G and L are specified, and the pressure gradient dp/dz and void fraction ε are to be calculated. Correlations for these last two variables are more likely to be successful if they can recognize the particular flow regime and develop relationships specifically for it. The following approximate demarcations are recognized:

1. *Bubble/slug flow transition.* Small bubbles introduced at the base of a column of liquid will usually eventually coalesce into slugs. The transition depends very much on the size of the bubbles, how they were introduced, the distance from the inlet, and on surface-tension effects, so there is no simple criterion for the transition. Also, for practical purposes of determining pressure gradients and void fraction, an exact knowledge of the transition point is unlikely to be needed. Experiments of Nicklin indicate that the correlation for the void fraction based on slug flow (see below) extends accurately into the bubble region for void fractions as low as a few percent.[8] It appears that if a void fraction of $\varepsilon = 0.1$ is somewhat arbitrarily taken for the transition, little significant error will arise in pressure-gradient and void-fraction calculations.

2. *Slug/annular flow transition.* In the narrow gap between the gas and the tube wall at the base of the slugs, there is a significant *downward* flow of liquid, and hence a fairly strong relative velocity between gas and liquid in this point. For increasing gas flow rates, the result is an instability of the liquid film, which can start bridging the whole cross section of the tube. These bridges can in turn be broken up by the gas and the flow becomes chaotic. Eventually, the bridges disappear and the flow consists of a gas core with a liquid film lining the wall of the tube. Wallis reports the results of several investigations into the slug/annular transition.[9] One of these used a probe located in the center of the tube to detect the presence or absence of liquid bridges, resulting in the following correlation for the transition:

$$j_g^* = 0.5 + 0.6j_l^*, \tag{10.27}$$

where the dimensionless superficial gas and liquid velocities are defined by:

$$j_g^* = j_g\sqrt{\frac{\rho_g}{gD(\rho_l - \rho_g)}}, \qquad j_l^* = j_l\sqrt{\frac{\rho_l}{gD(\rho_l - \rho_g)}}, \tag{10.28}$$

in which $(\rho_l - \rho_g)$ can usually be simplified to ρ_l.

3. *Annular/mist flow transition.* The transition is ill-defined because most annular flow entrains some droplets. In their very comprehensive book, Hewitt

---

[8] D.J. Nicklin, *Two-Phase Flow in Vertical Tubes* (PhD dissertation), University of Cambridge, Cambridge, 1961.

[9] G.B. Wallis, *One-Dimensional Two-Phase Flow*, McGraw-Hill, New York, 1969.

and Hall-Taylor devote a detailed 37-page chapter to the formation and entrainment of droplets, and strongly caution against the use of any correlation beyond the immediate experiments on which it was based.[10] Therefore, we are unable to present any general correlations regarding mist flow.

Fig. 10.4   Bubble flow.

Bubble flow.   Fig. 10.4 shows gas bubbles and liquid in upward cocurrent flow. Consider the plane A–A, drawn so that it lies entirely in the liquid. If $\overline{u}_l$ is the mean upward liquid velocity across A–A, continuity requires that:

$$A\overline{u}_l = G + L, \qquad \text{or} \qquad \overline{u}_l = \frac{G + L}{A}. \tag{10.29}$$

(Over a relatively short vertical span, the pressure varies little and the gas bubbles have an essentially constant volume.) Thus, the gas bubbles just below A–A are rising relative to a liquid that is already moving at a velocity $\overline{u}_l$, so that the velocity of the gas bubbles is:

$$v_g = \overline{u}_l + u_b = \frac{G + L}{A} + u_b, \tag{10.30}$$

in which u_b is the bubble velocity rising into a stagnant liquid. But the total volumetric flow rate of gas is:

$$G = \varepsilon A v_g, \tag{10.31}$$

so that the void fraction is given by:

$$\frac{G}{\varepsilon A} = \frac{G + L}{A} + u_b \qquad \text{or} \qquad \varepsilon = \frac{G}{G + L + u_b A}. \tag{10.32}$$

---

[10] G.F. Hewitt and N.S. Hall-Taylor, *Annular Two-Phase Flow*, Pergamon Press, Oxford, 1970.

This relation for the void fraction holds within certain limits for G and/or L *negative*—that is, for *downflow* of one or both phases, as discussed in Example 10.3.

Concerning the pressure gradient in the upwards vertical direction, note that the density of the liquid, which occupies a fraction $(1 - \varepsilon)$ of the total volume, is much greater than that of the gas. Also, for the relatively low liquid velocities likely to be encountered in the bubble-flow regime, friction is negligible. Therefore, the pressure gradient is given fairly accurately by considering only the hydrostatic effect:

$$\frac{dp}{dz} = -\rho_l g(1 - \varepsilon). \tag{10.33}$$

Example 10.3—Limits of Bubble Flow

For specified gas and liquid volumetric flow rates G and L, investigate the feasibility of bubble-flow operation of a column. Examine all four combinations of gas and liquid in upflow and downflow.

Solution

Start by rearranging Eqn. (10.32), the relation for void fraction in terms of the gas and liquid flow rates, into:

$$G\left(\frac{1 - \varepsilon}{\varepsilon}\right) = L + u_b A. \tag{E10.3.1}$$

Now examine Eqn. (E10.3.1) to see if it can realistically be balanced and hence be solved for ε. Then accept the situation as bubble flow only if $\varepsilon < 0.1$.

1. *Gas upflow, liquid upflow.* Both sides of Eqn. (E10.3.1) are positive, and a value of ε (between 0 and 1) can always be found to balance the equation.

2. *Gas upflow, liquid downflow.* The liquid flow rate L is now negative. The left-hand side of Eqn. (E10.3.1) is always positive, but the right-hand side will be negative for sufficiently large negative L. Thus, a solution for ε is still possible only for modest downflows of liquid.

3. *Gas downflow, liquid upflow.* Since G is negative and L is positive, Eqn. (E10.3.1) cannot be balanced and this type of operation is impossible.

4. *Gas downflow, liquid downflow.* The left-hand side of Eqn. (E10.3.1) is negative, and so will the right-hand side be, provided L is sufficiently negative. Thus, a solution for ε is possible for strong downflows of liquid. □

Slug flow.   The general situation is shown in Fig. 10.5(a), in which the gas and liquid travel upward together at individual volumetric flow rates G and L, respectively, in a pipe of internal diameter D. In general, there will be an upward liquid velocity $\bar{u}_l$ across a plane A–A just ahead of a gas slug. By applying continuity and considering the gas to be incompressible over short distances, the total upward volumetric flow rate of liquid across A–A must be the *combined* gas and liquid flow rates entering at the bottom, namely, $G + L$. The *mean* liquid velocity at A–A is therefore $\bar{u}_l = (G + L)/A$, where A is the cross-sectional area of the pipe.

Next, consider Fig. 10.5(b), which shows a somewhat different situation—that of a single bubble, which is moving steadily upward with a rise velocity u_b in an otherwise *stagnant* liquid. For liquids such as water and light oils that are not very viscous, the situation is one of *potential* flow in the liquid. Under these circumstances, Davies and Taylor used an *approximate* analytical solution (see Problem 10.8 to retrace their analysis) that gave[11]:

$$u_b = c\sqrt{gD}, \tag{10.34}$$

in which $c \doteq 0.33$ and g is the gravitational acceleration. Experimental evidence shows that the constant should be $c \doteq 0.35$.

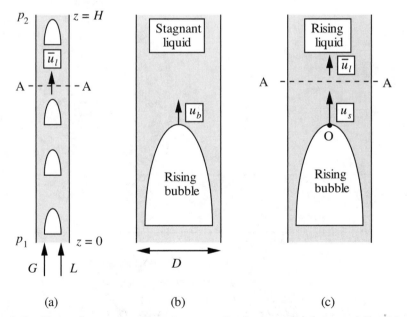

(a)   (b)   (c)

Fig. 10.5   Two-phase slug flow in a vertical pipe: (a) gas and liquid ascending; (b) bubble rising in stagnant liquid; (c) bubble rising in moving liquid.

---

[11] Davies and Taylor, *op. cit.*

The situation of Fig. 10.5(a) is now shown enlarged, in Fig. 10.5(c). The slug is no longer rising in a stagnant liquid, as in Fig. 10.5(b), but in a liquid whose mean velocity just ahead of it is $\bar{u}_l$. Further, near the "nose" O of the slug—at the center of the pipe, where the velocity is the highest—the liquid velocity will be somewhat larger, namely, about $1.2\bar{u}_l$, as shown by Nicklin, Wilkes, and Davidson, provided that the Reynolds number between slugs exceeds 8,000.[12] Therefore, the actual rise velocity of the slug is:

$$u_s = 1.2\frac{G+L}{A} + u_b = 1.2\frac{G+L}{A} + 0.35\sqrt{gD}. \tag{10.35}$$

By conservation of the gas:

$$G = u_s A \varepsilon, \tag{10.36}$$

in which ε is the *void* fraction (the fraction of the total volume that is occupied by the gas). Hence, eliminating u_s between Eqns. (10.35) and (10.36):

$$\frac{G}{\varepsilon A} = 1.2\frac{G+L}{A} + 0.35\sqrt{gD}. \tag{10.37}$$

Fig. 10.6 Experimental liquid void fractions for two-phase air/water flow in a vertical 1.02-in. diameter tube, as a function of the gas superficial velocity, with a liquid superficial velocity of 3.0 ft/s. From D.J. Nicklin, "Two-Phase Flow in Vertical Tubes," Ph.D. dissertation, University of Cambridge, 1961.

[12] D.J. Nicklin, J.O. Wilkes, and J.F. Davidson, "Two-phase flow in vertical tubes," *Transactions of the Institution of Chemical Engineers*, Vol. 40, No. 1, pp. 61–68 (1962).

If G and L are known, Eqn. (10.37) gives the void fraction, which is the dominant factor in determining the pressure gradient in the vertical direction:

$$\varepsilon = \frac{G}{1.2(G+L) + 0.35A\sqrt{gD}} = \frac{j_g^* \sqrt{\frac{\rho_l}{\rho_g}}}{1.2 \left(j_g^* \sqrt{\frac{\rho_l}{\rho_g}} + j_l^* \right) + 0.35}, \qquad (10.38)$$

in which ρ_g has been neglected in comparison with ρ_l in the definitions of the dimensionless superficial velocities from Eqn. (10.28). Fig. 10.6 shows that the above theory is completely substantiated by experiments reported by Nicklin, Wilkes, and Davidson, not only for the complete slug-flow regime, but (somewhat unexpectedly) for the adjoining parts of the bubble- and semiannular-flow regimes as well.

Concerning the pressure gradient, note that the density of the liquid, which occupies a fraction $(1-\varepsilon)$ of the total volume, is much greater than that of the gas. Therefore, the pressure gradient is given to a good approximation by considering only the hydrostatic effect:

$$\frac{dp}{dz} = -\rho_l g(1-\varepsilon). \qquad (10.39)$$

A secondary correction to (10.39) would include the wall friction on the liquid "pistons" between successive gas slugs. Thus, if $(dp/dz)_{sp}$ is the single-phase frictional pressure gradient for *liquid only*, flowing at a mean velocity $\bar{u}_l$, a more accurate expression for the pressure gradient is:

$$\frac{dp}{dz} = (1-\varepsilon) \left[-\rho_l g + \left(\frac{dp}{dz} \right)_{sp} \right]. \qquad (10.40)$$

The above analysis is for an ideal liquid of negligible viscosity, in which inertial and gravitational effects are dominant. White and Beardmore have investigated slug rise velocities for a wide variety of liquids and have plotted the results as Fr (the Froude number) versus Eo (the Eötvös number), with a property group Y as a parameter, where[13]:

$$\text{Fr} = \frac{u}{\sqrt{gD}}, \qquad \text{Eo} = \frac{\rho g D^2}{\sigma}, \qquad Y = \frac{g\mu^4}{\rho\sigma^3}. \qquad (10.41)$$

For all liquids, Fr rises from zero to an asymptotic value of about 0.345 as Eo increases, but rises more slowly for larger values of Y. The authors conclude that distilled water ($\rho = 0.997$ g/ml, $\mu = 0.87$ cP, $\sigma = 71.5$ g/s^2, $Y = 1.54 \times 10^{-11}$) behaves ideally, with Fr essentially at its full value of 0.345, for Eo > 70, which

---

[13] E.T. White and R.H. Beardmore, "The velocity of rise of single cylindrical air bubbles through liquids contained in vertical tubes," *Chemical Engineering Science*, Vol. 17, pp. 351–361 (1962).

occurs provided that the pipe diameter $D > 2.26$ cm. Basically, viscous effects are relatively unimportant for most low-viscosity liquids in pipes likely to be used in practice. A representative lubricant, Voluta oil ($\rho = 0.902$ g/ml, $\mu = 294$ cP, $\sigma = 30.8$ g/s^2, $Y = 2.8$) would need Eo $> 3,000$ to be considered ideal, which would occur for pipes with $D > 10.22$ cm.

Example 10.4—Performance of a Gas-Lift Pump

Fig. E10.4 shows a gas-lift "pump," in which the buoyant action of a volumetric flow rate G of gas serves to lift a volumetric flow rate L of liquid from a height H_0 in a reservoir to a height H in a vertical pipe of diameter D and cross-sectional area A, in which slug flow may be assumed. Neglect liquid friction in both the supply pipe and in the vertical pipe.

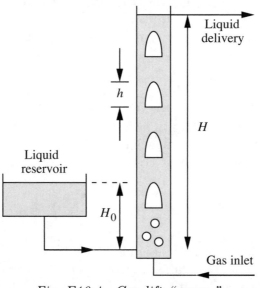

Fig. E10.4   Gas-lift "pump."

(a) Derive an expression for the void fraction ε in the column in terms of H and H_0.

(b) If $L = 0$ (that is, the liquid is just on the verge of reaching the top), prove that the height H is given by:

$$\frac{H}{H_0} = \frac{1.2G/A + u_b}{0.2G/A + u_b},$$

where $u_b = 0.35\sqrt{gD}$.

(c) If the value of G/A is much larger than u_b, what is the maximum height H_{max} to which the liquid can be raised (still with $L = 0$)? Give your answer as a multiple of H_0.

Example 10.4—Performance of a Gas-Lift Pump   **551**

(d) If the column has a diameter $D = 0.1$ m and operates with a positive liquid flow rate such that $\varepsilon = 0.5$ and $G/A = 2L/A$, what is the value of G/A (m/s)?

Solution

(a) From hydrostatics, the pressure at the base of the column is obtained in two ways:

$$\rho_l g H_0 = \rho_l g H(1 - \varepsilon), \qquad (\text{E}10.4.1)$$

which gives:

$$\varepsilon = 1 - \frac{H_0}{H}. \qquad (\text{E}10.4.2)$$

(b) From Eqn. (10.38), with $L = 0$ and $u_b = 0.35\sqrt{gD}$:

$$\varepsilon = \frac{G/A}{1.2G/A + u_b} = 1 - \frac{H_0}{H}. \qquad (\text{E}10.4.3)$$

Rearrangement and solution for the height ratio yields the desired result:

$$\frac{H}{H_0} = \frac{1.2G/A + u_b}{0.2G/A + u_b}. \qquad (\text{E}10.4.4)$$

(c) For $G/A \gg u_b$, the velocity u_b in the numerator and denominator of Eqn. (E10.4.4) can be neglected, so that:

$$\frac{H}{H_0} = \frac{1.2G/A}{0.2G/A} = 6. \qquad (\text{E}10.4.5)$$

Thus, the maximum height attainable is six times the reservoir height H_0.

(d) Inserting numerical values:

$$u_b = 0.35\sqrt{gD} = 0.35\sqrt{9.81 \times 0.1} = 0.347 \text{ m/s}. \qquad (\text{E}10.4.6)$$

$$\frac{G}{\varepsilon A} = 1.2\left(\frac{G}{A} + \frac{L}{A}\right) + u_b \quad \text{or} \quad 2\frac{G}{A} = 1.2 \times 1.5\frac{G}{A} + 0.347. \qquad (\text{E}10.4.7)$$

The superficial velocity of the gas is therefore:

$$\frac{G}{A} = 1.733 \text{ m/s}, \qquad (\text{E}10.4.8)$$

which then suffices to achieve the desired liquid flow rate. ☐

Annular flow. Consider the simultaneous flow of gas and liquid in a *vertical* tube as in Fig. 10.7. One approach, recommended by Wallis (*op. cit.*), is to use the Lockhart/Martinelli correlation for the frictional part of the pressure drop, and supplement this with appropriate gravitational terms.

First, consider just the flow of gas in the inner core. Since the gas velocity v_g is typically much higher than that of the liquid, the pressure gradient may be approximated as if the gas were flowing with velocity v_g in a pipe of diameter D_g, giving:

$$\left(\frac{dp}{dz}\right)_{\text{tp}} = \left(\frac{dp}{dz}\right)_g = -\frac{2f_{\text{F}}\rho_g v_g^2}{D_g} - \rho_g g \tag{10.42}$$

$$= -\underbrace{\left(\frac{v_g^2 D}{j_g^2 D_g}\right)}_{\phi_g^2}\underbrace{\left(\frac{2f_{\text{F}}\rho_g j_g^2}{D}\right)}_{(dp/dz)_{go}} - \rho_g g = \underbrace{\phi_g^2\left(\frac{dp}{dz}\right)_{go}}_{\text{Friction}} - \underbrace{\rho_g g}_{\text{Gravity}}, \tag{10.43}$$

in which the substitution $\phi_g^2 = v_g^2 D/j_g^2 D_g$ comes from Eqns. (10.19) and (10.20).

Second, consider the entire flow, obtaining the frictional contribution from the viewpoint of the liquid:

$$\left(\frac{dp}{dz}\right)_{\text{tp}} = \underbrace{\phi_l^2\left(\frac{dp}{dz}\right)_{lo}}_{\text{Friction}} - \underbrace{[\varepsilon\rho_g + (1-\varepsilon)\rho_l]g}_{\text{Gravity}}. \tag{10.44}$$

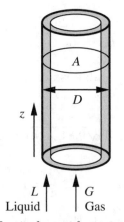

Fig. 10.7 Vertical annular two-phase flow.

For specified gas and liquid flow rates, the derivatives on the right-hand sides of Eqns. (10.43) and (10.44) will be known. These equations can then be solved simultaneously for the pressure gradient $(dp/dz)_{\text{tp}}$ and the void fraction ε. One suitable procedure involves:

Example 10.5—Two-Phase Flow in a Vertical Pipe   **553**

(a) Equating (10.43) and (10.44), thereby eliminating $(dp/dz)_{tp}$ and giving a single equation with unknowns ϕ_g^2, ϕ_l^2, and ε.

(b) Using Eqn. (10.23) to eliminate ϕ_l^2 in favor of X and ϕ_g^2, giving a single equation with unknowns ϕ_g^2, X, and ε.

(c) Solving for ϕ_g^2, X, and ε by using the information in Table 10.4.

Example 10.5—Two-Phase Flow in a Vertical Pipe

Gas and oil are flowing upwards at volumetric flow rates $G = 10.0$, $L = 0.1$ ft^3/s in a vertical pipeline of internal diameter $D = 0.25$ ft. Estimate the fraction of gas, the pressure gradient, and the mean velocities of each phase. The physical properties are $\rho_g = 0.15$, $\rho_l = 50.0$ lb$_m$/ft^3, $\mu_g = 0.025$, $\mu_l = 20.0$ lb$_m$/ft hr; the Fanning friction factor is $f_F = 0.0045$.

Solution

The following quantities are first calculated:

Cross-sectional area of pipe

$$A = \frac{\pi D^2}{4} = \frac{\pi \times 0.25^2}{4} = 0.0491 \text{ ft}^2.$$

Superficial velocities

$$j_g = \frac{G}{A} = \frac{10.0}{0.0491} = 204, \qquad j_l = \frac{L}{A} = \frac{0.1}{0.0491} = 2.04 \text{ ft/s}.$$

Dimensionless superficial velocities from Eqn. (10.28)

$$j_g^* = j_g \sqrt{\frac{\rho_g}{gD(\rho_l - \rho_g)}} = 204 \sqrt{\frac{0.15}{32.2 \times 0.25 \times (50.0 - 0.15)}} = 3.94,$$

$$j_l^* = j_l \sqrt{\frac{\rho_l}{gD(\rho_l - \rho_g)}} = 2.04 \sqrt{\frac{50.0}{32.2 \times 0.25 \times (50.0 - 0.15)}} = 0.720.$$

Recall that from Eqn. (10.27) that the approximate value of j_g^* at which slug flow becomes annular flow is given by $0.5 + 0.6 j_l^*$. This transition value is clearly exceeded, and the flow is taken to be annular. The appropriate pressure gradients for inclusion in Eqns. (10.43) and (10.44) are:

Gas only

$$\left(\frac{dp}{dz}\right)_{go} = -\frac{2 f_F \rho_g j_g^2}{D} = -\frac{2 \times 0.0045 \times 0.15 \times 204^2}{0.25 \times 32.2 \times 144} = -0.0485 \text{ psi/ft}.$$

Liquid only

$$\left(\frac{dp}{dz}\right)_{lo} = -\frac{2 f_F \rho_l j_l^2}{D} = -\frac{2 \times 0.0045 \times 50.0 \times 2.04^2}{0.25 \times 32.2 \times 144} = -1.62 \times 10^{-3} \text{ psi/ft}.$$

Reynolds numbers

$$\text{Re}_g = \frac{0.15 \times 204 \times 0.25 \times 3{,}600}{0.025} = 1.10 \times 10^6,$$

$$\text{Re}_l = \frac{50.0 \times 2.04 \times 0.25 \times 3{,}600}{20.0} = 4{,}590.$$

Thus, the flow is turbulent for the liquid and turbulent for the gas, with $n = 4.12$ (from Table 10.5) for use in Eqn. (10.26). Equations (10.43) and (10.44) now become:

$$\frac{dp}{dz} = -0.0485\phi_g^2 - \frac{0.15 \times 32.2}{32.2 \times 144} = -0.0485\phi_g^2 - 0.00104,$$

$$\frac{dp}{dz} = -1.62 \times 10^{-3}\phi_l^2 - \frac{32.2[0.15\varepsilon + 50.0(1 - \varepsilon)]}{32.2 \times 144} = -1.62 \times 10^{-3}\phi_l^2 - 0.347 + 0.346\varepsilon.$$

Now equate the two pressure gradients, substitute $\phi_l^2 = \phi_g^2/X^2$ from Eqn. (10.23), and collect terms:

$$\phi_g^2\left(29.9 - \frac{1}{X^2}\right) = 214(1 - \varepsilon),$$

which can be solved in conjunction with Eqns. (10.25) and (10.26) to give:

$$X = 0.198, \qquad \varepsilon = 0.902, \qquad \phi_g = 2.17, \qquad \phi_g^2 = 4.70.$$

Thus, the fraction of gas is 0.902, so the oil occupies a fairly thin film on the inside of the tube. The pressure gradient is next calculated.

Two-phase flow pressure gradient

$$\left(\frac{dp}{dz}\right)_{tp} = \phi_g^2\left(\frac{dp}{dz}\right)_{go} - \rho_g g = -4.70 \times 0.0485 - \frac{0.15 \times 32.2}{32.2 \times 144} = -0.229 \text{ psi/ft.}$$

Finally, the individual mean velocities are obtained:

Gas

$$v_g = \frac{G}{\varepsilon A} = \frac{10.0}{0.902 \times 0.0491} = 226 \text{ ft/s.}$$

Liquid

$$v_l = \frac{L}{(1 - \varepsilon)A} = \frac{0.1}{(1 - 0.902) \times 0.0491} = 20.8 \text{ ft/s.} \qquad \square$$

10.5 Flooding

Consider two-phase flow in a vertical tube with the gas flowing upward and the liquid downward. For sufficiently high relative velocities, unstable waves will form at the gas/liquid interface, eventually resulting in fragments of liquid that break away, to be carried upward by the gas, at which stage normal countercurrent operation ceases and *flooding* occurs. For small liquid flow rates, Wallis (*op. cit.*) gives the condition under which no liquid will flow downward as:

$$j_g^* = j_g \sqrt{\frac{\rho_g}{gD(\rho_l - \rho_g)}} \doteq 0.9, \tag{10.45}$$

(in which j_g is the same as the gas velocity v_g in Fig. 10.8), and also presents the following limiting condition for flooding in tubes with sharp-edged flanges:

$$\sqrt{j_g^*} + \sqrt{j_l^*} = 0.725, \tag{10.46}$$

with the constant rising to the range 0.88–1.00 if end effects are minimized.

Fig. 10.8   Model for flooding correlation.

A condition remarkably close to that of Eqn. (10.45) can be derived by an elegant method suggested by Whalley and also paralleled by Problem 2.44 of the present book.[14] The general idea is shown in Fig. 10.8, in which an annular-shaped "wave" with a semicircular cross section of small radius δ is supposed to have formed on the thin film lining the inner surface of the tube. The gas accelerates between locations 1 and 2, with an attendant pressure decrease from the Bernoulli effect. There is some recouping of the pressure between locations 2 and 3, but only a partial amount because of turbulence in the expanding flow.

---

[14] P.B. Whalley, *Boiling, Condensation, and Gas-Liquid Flow*, Clarendon Press, Oxford, 1987.

From Eqn. (2.65), the overall pressure drop is:

$$p_1 - p_3 = \rho_g \frac{v_g^2}{2} \left(\frac{A_1}{A_2} - 1 \right)^2, \tag{10.47}$$

in which the areas at the indicated sections are (for $\delta \ll D$):

$$A_1 = \frac{\pi D^2}{4}, \qquad A_2 \doteq \frac{\pi (D - 2\delta)^2}{4} \doteq \frac{\pi D (D - 4\delta)}{4}. \tag{10.48}$$

The area ratio and pressure drop then become:

$$\frac{A_1}{A_2} = \frac{D^2}{D(D - 4\delta)} \doteq 1 + \frac{4\delta}{D}, \qquad p_1 - p_3 = \rho_g \frac{v_g^2}{2} \frac{16\delta^2}{D^2}. \tag{10.49}$$

At the onset of flooding, the pressure drop just suffices to counterbalance the weight of the wave, which has a circumference πD and a cross-sectional area $\pi \delta^2 / 2$, so that:

$$(p_1 - p_3) \frac{\pi D^2}{4} \doteq \pi D \frac{\pi \delta^2}{2} \rho_l g. \tag{10.50}$$

Elimination of $(p_1 - p_3)$ from the last two equations gives, upon rearrangement:

$$j_g^* = v_g \sqrt{\frac{\rho_g}{g D \rho_l}} = \sqrt{\frac{\pi}{4}} \doteq 0.886, \tag{10.51}$$

which is in essential agreement with Eqn. (10.45). Note that a different assumed shape of the wave cross section would give a somewhat different numerical value for the constant in Eqn. (10.51). However, even with some uncertainty in the value, the analysis demonstrates that j_g^* is an important dimensionless group to include in any attempted correlation for flooding. Also see Problem 7.14.

Packed columns. Packed columns are used in chemical engineering for absorbing a component from a gas stream into a liquid stream, and also for conducting certain distillation operations. As shown in Fig. 10.9, the liquid is sprayed through a distributor at the top of the column and flows countercurrent to the gas, which is admitted at the bottom.

In order to increase the surface area for mass transfer, the column is filled with inert "packing," of which the ceramic Raschig ring is an early example. Other types of packing are Pall rings and Burl saddles, and often contain more convoluted internal surfaces.

If attempts are made to increase the gas and liquid throughputs, a stage will be reached at which the pressure drop in the vertical direction suffices to arrest the downflow of the liquid, at which stage the column fills up or "floods" with liquid, and normal operation ceases. The flooding limits are shown in Fig. 10.10, which is derived from information in an article by Eckert.[15] Note that the ordinate is not dimensionless, and that the variables *must* have the units indicated in Table 10.6. The reader may wish to check the extent to which the ordinate is related to the dimensionless velocity in Eqn. (10.51).

Table 10.6 Variables for Packed-Column Flooding Correlation

Variable	Definition
F	Packing factor (see Table 10.7)
g_c	Conversion factor, 32.2 $\mathrm{lb_m}$ ft/$\mathrm{lb_f}$ $\mathrm{s^2}$
G_g	Gas mass velocity, $\mathrm{lb_m}/\mathrm{ft^2}$ s
G_l	Liquid mass velocity, $\mathrm{lb_m}/\mathrm{ft^2}$ s
μ_l	Liquid viscosity, cP
ρ_g	Gas density, $\mathrm{lb_m}/\mathrm{ft^3}$
ρ_l	Liquid density, $\mathrm{lb_m}/\mathrm{ft^3}$
Ψ	Density of water divided by density of liquid

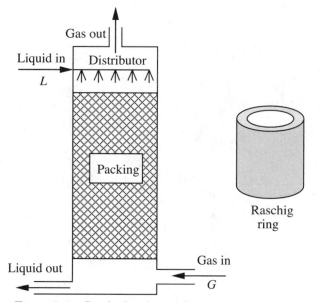

Fig. 10.9 Packed column for gas absorption.

Representative packing factors are given in Table 10.7, and the capacity of the column is inversely proportional to the square root $\sqrt{F}$ of the packing factor. Eckert discusses advantages and disadvantages of several packing materials. Although

---

[15] J.S. Eckert, "Selecting the proper distillation column packing," *Chemical Engineering Progress*, Vol. 66, No. 3, pp. 39–44 (1970).

the smaller size packings enhance mass transfer, they have lower capacities and cost more. Nominal 2-in. packing appears optimal in most circumstances.

Table 10.7   Representative Values of the Packing Factor F

Type	Material	Nominal Packing Size, in.			
		1/2	1	2	3
Intalox saddles	Ceramic	200	98	40	22
Raschig rings	Ceramic	580	155	65	37
Pall rings	Plastic	—	52	25	—
Pall rings	Metal	—	48	20	—
Raschig rings (1/16-in. wall)	Metal	410	137	57	32

Fig. 10.10   Correlation for flooding in packed columns.

10.6 Introduction to Fluidization

(a)   (b)

Fig. 10.11   Fluidization: (a) upward flow through a packed bed;
(b) variation of bed height and pressure drop with superficial velocity.

Consider the upward flow of a fluid through a bed packed with small particles. As the superficial fluid velocity u (defined as Q/A, the volumetric flow rate divided by the cross-sectional area) is increased, the behavior of the bed height h and the pressure drop $\Delta p = p_1 - p_2$ is shown in Fig. 10.11. A point is reached— that of *incipient fluidization*—when the pressure drop just suffices to support the downward weight per unit area of the bed. At this point, the superficial velocity is denoted by u_0, and the void fraction by ε_0 (typically between 0.4 and 0.5).

For $u > u_0$, the bed starts to expand and behaves essentially as a *fluid*; there is little subsequent increase in the pressure drop. There are two distinct *modes* of fluidization, depending mainly on the relative values of ρ_s, the density of the solid particles, and ρ_f, the density of the fluid.

Particulate fluidization. If ρ_s and ρ_f are of the same order of magnitude, the bed is usually a homogeneous mixture of fluid and particles. This situation occurs mainly for *liquid*-fluidized beds, of secondary practical importance. For $u > u_0$, the bed expands uniformly and the void fraction obeys the *Richardson-Zaki* correlation[16]:

$$\frac{u}{u_t} = \varepsilon^n, \tag{10.52}$$

in which u_t is the terminal velocity of a single particle, and the exponent n varies with the particle Reynolds number, $\text{Re} = \rho_f u d_p / \mu_f$ according to Table 10.8. Here, d_p is the effective particle diameter and μ_f is the viscosity of the fluid.

---

[16] J.F. Richardson and W.N. Zaki, "Sedimentation and fluidisation," *Transactions of the Institution of Chemical Engineers*, Vol. 32, p. 35 (1954).

Table 10.8 Values of the
Richardson-Zaki Exponent

Re	n
$Re < 0.2$	4.65
$0.2 < Re < 1$	$4.35Re^{-0.03}$
$1 < Re < 500$	$4.45Re^{-0.1}$
$Re > 500$	2.39

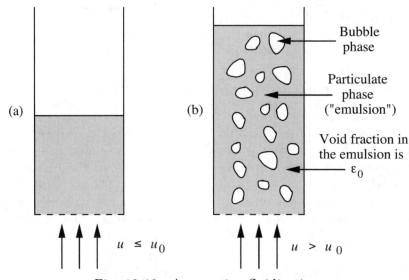

(a)

(b)

Bubble phase

Particulate phase ("emulsion")

Void fraction in the emulsion is ε_0

$u \leq u_0$

$u > u_0$

Fig. 10.12 Aggregative fluidization:
(a) compacted bed; (b) fluidized bed.

Aggregative fluidization.   When the density of the fluid is considerably less than that of the particles, the bed is found to consist of two distinct phases:

1. A *particulate phase* or *emulsion*, in which the void fraction remains essentially at the incipient value ε_0, even though the total fluid flow rate may considerably exceed the incipient value.

2. A *bubble phase*, which contains the extra fluid beyond that needed for incipient fluidization.

Aggregative fluidization occurs mainly in *gas*-fluidized beds, of considerable importance in industrial catalytic reactors. The general scheme is shown in Fig. 10.12. The interactive motion of the particles and bubbles will be considered in detail in Section 10.8.

10.7 Bubble Mechanics

Rise velocity of a continuous swarm of small bubbles in a liquid.
Paralleling the earlier discussion of gas slugs in Section 10.2, Nicklin considered
the bubble regime in two-phase gas/liquid flow, again distinguishing, as shown in
Fig. 10.13, between the motion of a single swarm of bubbles and of continuously
generated bubbles.[17] In both cases, the bubbles are produced by the passage of
gas at a flow rate G through small holes in a distributor at the bottom of a column
of cross-sectional area A containing a liquid.

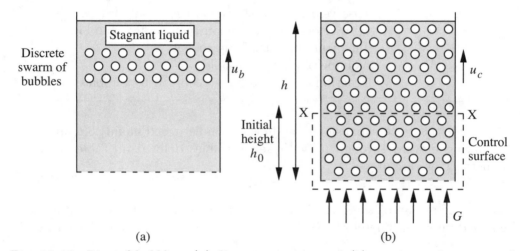

Fig. 10.13 Rise of bubbles: (a) discrete swarm, and (b) continuously generated.

Let a *single cluster* of bubbles rise with a velocity u_b relative to the stagnant
liquid above. Then, from continuity (a volume balance) on the control surface in
Fig. 10.13(b), the mean upwards velocity of the liquid across X–X between two
"layers" of bubbles is G/A. Hence, by the same argument used for gas slugs rising
in pipes, the rise velocity u_c of *continuously generated* bubbles is:

$$u_c = \frac{G}{A} + u_b. \tag{10.53}$$

Note that relative to the bubbles the motion is (fairly) steady, and that liquid
streamlines can be visualized. Such is not the case for a fixed observer, who will
see an *upflow* of liquid across X–X at one instant and a *downflow* of liquid across
X–X a moment later when a layer of bubbles has advanced to that location. If
there is no net flow of liquid (it is not overflowing the column), then these two
effects must cancel each other on the average.

---

[17] D.J. Nicklin, "Two-phase bubble flow," *Chemical Engineering Science*, Vol. 17, pp. 693–702 (1962).

Table 10.9   Variables for Rising Groups of Bubbles

Variable	Definition
A	Cross-sectional area of column
D_e	Effective bubble diameter
G	Gas volumetric flow rate
h_0	Depth of undisturbed liquid
h	Depth of liquid with bubbles rising through it
u_b	Rise velocity of a single swarm of bubbles
u_c	Rise velocity of continuously generated bubbles
u_0	Incipient superficial gas velocity
V	Volume of a single bubble
ε	Fraction of total volume occupied by bubbles

Consider next the expansion of an initially undisturbed liquid column by bubbles flowing upward through it, with key variables defined in Table 10.9. The volumetric flow rate of gas is:

$$G = \varepsilon A u_c. \tag{10.54}$$

Also, the expansion of the initially undisturbed column is due to the presence of the added bubbles. Therefore, we can equate the extra volume to that of the gas:

$$A(h - h_0) = \varepsilon h A. \tag{10.55}$$

From Eqns. (10.53)–(10.55):

$$\frac{u_b}{G/A} = \frac{h_0}{h - h_0}, \tag{10.56}$$

which enables u_b to be deduced from experimental observations of the bed height h at different values of G/A.

Extension to a fluidized bed.   The above treatment still holds for an aggregatively fluidized bed if G/A is replaced by $(G/A - u_0)$, which simply recognizes the fact that all gas above the incipient rate passes through the bed as bubbles, so the rise velocity of *continuously* generated bubbles is:

$$u_c = \frac{G}{A} - u_0 + u_b. \tag{10.57}$$

As an *approximation* (there may be a certain amount of interference between the bubbles), the rise velocity u_b can be *estimated* from Eqn. (10.8) at the beginning of the chapter for inviscid liquids as a function of the equivalent bubble diameter D_e, or its equivalent in terms of the bubble volume $V = \pi D_e^3/6$:

$$\text{Inviscid liquids:} \qquad u_b \doteq 0.711\sqrt{gD_e} = 0.792g^{1/2}V^{1/6}. \qquad (10.8)$$

However, for bubbles rising in fluidized-bed columns, a somewhat reduced value of the constant has been obtained instead of 0.792. The following two investigations also made the necessary corrections in order to give the rise velocities for bubbles unimpeded by walls, which can retard the bubbles if their diameter is not small compared to that of the column:

(a) Davidson et al. performed experiments in 3-in. diameter fluidized beds of glass beads and sand, with bubble volumes ranging from 5.80–15.0 ml; their results translate into a value of 0.728 for the constant.[18]

(b) Harrison and Leung investigated the rise of bubbles with volumes from 25–10,000 ml in fluidized sand particles contained in a bed with a 61-cm square cross section; their results give the value of the constant as 0.712.[19]

Taking an average value of 0.720 for the constant, we shall use the following expressions for the bubble rise velocity:

$$\text{Fluidized beds:} \qquad u_b \doteq 0.646\sqrt{gD_e} = 0.720g^{1/2}V^{1/6}. \qquad (10.58)$$

Thus, the analog of Eqn. (10.56) for the expansion of fluidized beds is:

$$\frac{0.720g^{1/2}V^{1/6}}{G/A - u_0} = \frac{h_0}{h - h_0}, \qquad (10.59)$$

which can be used to estimate the bubble volume (and its equivalent diameter) by measuring h as a function of G/A.

Volume of a bubble formed at an orifice.   Potential flow theory may also be used to determine the volume of a bubble growing and finally detaching from an orifice submerged either in a column of liquid *or* in an aggregatively fluidized bed.

Fig. 10.14 shows three stages in the development of an essentially spherical bubble growing at such an orifice. The volume V and radius r of the bubble keep increasing, as does the orifice-to-bubble-center-distance $s = \text{OC}$ (this last because of buoyancy). Ultimately, s will exceed r and the bubble detaches from the orifice. The cycle then repeats, and may be analyzed as follows, after certain assumptions are made:

---

[18] J.F. Davidson, R.C. Paul, M.J.S. Smith, and H.A. Duxbury, "The rise of bubbles in a fluidised bed," *Transactions of the Institution of Chemical Engineers*, Vol. 37, pp. 323–328 (1959).

[19] D. Harrison and L.S. Leung, "The rate of rise of bubbles in fluidised beds," *Transactions of the Institution of Chemical Engineers*, Vol. 40, pp. 146–151 (1962).

1. The gas volumetric flow rate G is constant. (A more complicated alternative is to consider gas supplied from a reservoir at a constant pressure, in which the flow rate fluctuates as the bubble grows and detaches.)
2. The gas is supplied by a small bore tube, and may therefore be considered as issuing from a small point. (An alternative is gas rising from a small hole in a horizontal plate, in which case none of the gas can flow downward.)
3. The gas-flow rate is neither so small that surface tension is important, nor so large that the inertia of the gas is significant (in which case a jet might form); also, viscous effects are unimportant. That is, the dominant effects are those of gravity and the acceleration of the liquid surrounding the bubble.

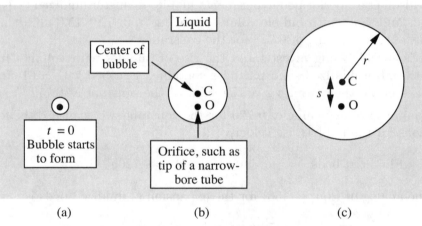

(a)   (b)   (c)

Fig. 10.14 Formation of a bubble at an orifice.

Since the gas flow rate is constant, the bubble volume increases linearly with time:

$$V = \frac{4}{3}\pi r^3 = Gt. \tag{10.60}$$

As the bubble rises, it accelerates the surrounding fluid. Although the bubble has essentially zero density, it has an *effective* mass because any bubble motion also causes the surrounding fluid (of significant density ρ_l) to move. The reader who solves Problem 10.4 will discover that this effective mass equals one-half of the mass of fluid displaced by the spherical bubble.

A momentum balance, equating the buoyant force to the rate of increase of effective momentum of the bubble, gives:

$$\rho_l V g = \frac{d}{dt}\left(\frac{1}{2}\rho_l V \frac{ds}{dt}\right). \tag{10.61}$$

Elimination of V between Eqns. (10.60) and (10.61), followed by separation of variables and integration twice, yields a surprisingly simple result:

$$\rho_l G g \int_0^t t\,dt = \int_0^{\frac{1}{2}\rho_l V ds/dt} d\left(\frac{1}{2}\rho_l V \frac{ds}{dt}\right), \tag{10.62}$$

$$\rho_l G g \frac{1}{2} t^2 = \frac{1}{2} \rho_l G t \frac{ds}{dt}, \tag{10.63}$$

$$\frac{ds}{dt} = gt, \tag{10.64}$$

$$s = \frac{1}{2} g t^2. \tag{10.65}$$

Fig. 10.15   *Volumes of bubbles formed at orifices in incipiently fluidized beds. Reproduced with permission of IChem^E from Harrison and Leung[20].*

From Eqns. (10.60) and (10.65), setting $r = s$, the *departure time* is:

$$t_D = \left(\frac{6G}{\pi g^3} \right)^{1/5}, \tag{10.66}$$

so that the bubble volume at detachment is:

$$V = G t_D = 1.138 \frac{G^{6/5}}{g^{3/5}}. \tag{10.67}$$

The experiments of Harrison and Leung[20] and others, as reported by Davidson and Harrison[21] agree well with Eqn. (10.67) under an *exceptionally* wide range of gas flow rates, for orifices of varying sizes situated in both liquids *and* fluidized beds, as shown in Fig. 10.15 for the latter. Davidson and Harrison offer several explanations for the modest deviations from theory, including: (a) bubbles not completely spherical, (b) circulating currents induced near the orifice by the train of rising bubbles, and (c) coalescence of bubbles at high gas flow rates.

The treatment for an orifice located in a horizontal plate is similar, and has been studied by Davidson and Schüler.[22] The effective mass of the bubble is now 11/16 times the mass of the displaced liquid, and the final result is $V = 1.378\, G^{6/5}/g^{3/5}$.

10.8   Bubbles in Aggregatively Fluidized Beds

The ultimate goal of this section is to examine the interaction between two gas regimes—that in the rising bubbles and that in the particulate phase—for aggregatively fluidized beds. The treatment is that of Harrison and Davidson in their 1963 book (*op. cit.*), which is a model of clarity and conciseness. The fluid mechanics principles involved afford a nice culmination to the previous work, especially that in Chapter 7.

Consider first the definitions shown in Table 10.10. Since both the particles and the fluidizing fluid are effectively incompressible, the continuity equation can be applied to both phases:

Fluid:

$$\nabla \cdot \mathbf{v}^f = 0, \tag{10.68}$$

Particles:

$$\nabla \cdot \mathbf{v}^p = 0. \tag{10.69}$$

The bed also behaves as a porous medium, so d'Arcy's law applies:

$$\mathbf{v}^f - \mathbf{v}^p = -\frac{\kappa}{\mu}\nabla p. \tag{10.70}$$

Although the *superficial* velocity was given by d'Arcy's law as discussed in previous sections, $\mathbf{v}^f$ in Eqn. (10.70) is the mean *interstitial* fluid velocity in the interstices or spaces between the particles.

[20] D. Harrison and L.S. Leung, "Bubble formation at an orifice in a fluidised bed," *Transactions of the Institution of Chemical Engineers*, Vol. 39, pp. 409–414 (1961).

[21] J.F. Davidson, and D. Harrison, *Fluidised Particles*, Cambridge University Press, Cambridge, 1963, pp. 53, 59.

[22] J.F. Davidson and B.O.G. Schüler, "Bubble formation at an orifice in an inviscid liquid," *Transactions of the Institution of Chemical Engineers*, Vol. 38, pp. 335–342 (1960).

Table 10.10   Variables for a Bubble Rising in a Fluidized Bed

Variable	Definition
a	Bubble radius
κ	Bed permeability
p	Pressure
r, θ	Spherical coordinates, as shown in Fig. 7.11
u_b	Absolute bubble rise velocity
u_i	Interstitial fluid velocity relative to particles, far away from the bubble
u_0	Incipient superficial fluid velocity
$\mathbf{v}^f$	Fluidizing fluid velocity vector
$\mathbf{v}^p$	Particle velocity vector
v_r^f, v_θ^f	Components of $\mathbf{v}^f$
v_r^p, v_θ^p	Components of $\mathbf{v}^p$
ε_0	Void fraction under incipient conditions (*not* the fraction of the total volume occupied by bubbles)
μ	Fluid viscosity
ξ	Vertical pressure gradient far away from rising bubble

The fluid and particle velocity vectors may be eliminated from Eqns. (10.68)–(10.70), yielding a simple and familiar result:

$$\nabla^2 p = 0. \tag{10.71}$$

That is, the pressure distribution obeys Laplace's equation, and is *independent* of the particle motion. Representative isobars are shown in Fig. 10.16, where, for simplicity, a rising spherical bubble is assumed to be at the midpoint between high- and low-pressure regions of pressure $p = \pm P$, separated by a distance 2h. The coordinates are such that $r = 0$ is at the center of the bubble, and θ is measured downward from the vertical. The reader may check that the following pressure distribution satisfies Eqn. (10.71) and the boundary conditions at the top and bottom of the bed, and on the surface of the bubble:

$$p = -\xi \left(r - \frac{a^3}{r^2} \right) \cos \theta, \tag{10.72}$$

in which the overall pressure gradient well away from the bubble is:

$$\xi = \frac{P}{h}. \tag{10.73}$$

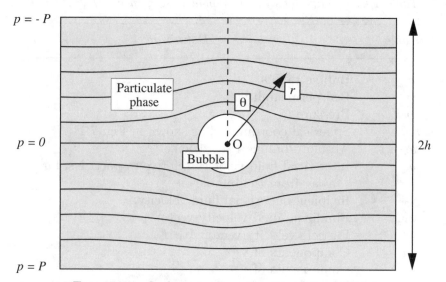

Fig. 10.16   *Isobars in the vicinity of the bubble.*

In catalytic reactors, it may be important to know the volumetric flow rate Q of the fluidizing fluid that is entering the lower hemisphere of the bubble and departing from its upper hemisphere. The development outlined in Problem 10.5 shows that by integrating over one of these hemispheres,

$$Q = 3u_0\pi a^2. \tag{10.74}$$

Since the upward superficial fluid velocity well away from the bubble is u_0, Eqn. (10.74) shows that the cavity—and reduced resistance—caused by the bubble attracts three times the normal fluid flow rate.

Well away from the bubble, the interstitial fluid velocity u_i *relative to the particles* is in the vertically upward direction and obeys d'Arcy's law:

$$u_i = -\frac{\kappa}{\mu}\frac{\partial p}{\partial z} = \frac{\kappa\xi}{\mu}. \tag{10.75}$$

The *superficial* velocity at incipient fluidization, which is based on the total cross-sectional area—not the restricted area between the particles—will be lower, equaling the void fraction times the relative interstitial velocity:

$$u_0 = \varepsilon_0 u_i = \frac{\varepsilon_0\kappa\xi}{\mu}. \tag{10.76}$$

Motion of the particles.   For an observer moving upwards with a bubble, the velocity potential for the streaming motion of the particles is obtained from Eqn. (7.73):

$$\phi = -u_b\left(r + \frac{a^3}{2r^2}\right)\cos\theta. \tag{10.77}$$

The corresponding streamlines are shown in Fig. 10.17.

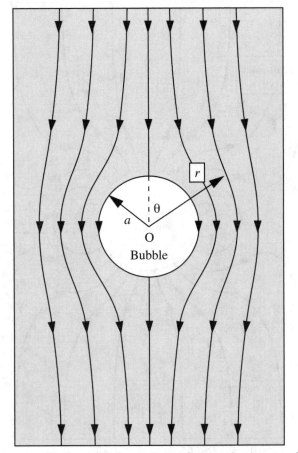

Fig. 10.17 Downward streaming of particles past a bubble.

Motion of the fluidizing fluid.   The individual velocity components for the fluidizing fluid are again obtained from d'Arcy's law, Eqn. (10.70):

$$v_r^f = v_r^p - \frac{\kappa}{\mu}\frac{\partial p}{\partial r}, \tag{10.78}$$

$$v_\theta^f = v_\theta^p - \frac{1}{r}\frac{\kappa}{\mu}\frac{\partial p}{\partial \theta}, \tag{10.79}$$

in which the *particle* velocity components are given by:

$$v_r^p = \frac{\partial \phi}{\partial r}, \qquad v_\theta^p = \frac{1}{r}\frac{\partial \phi}{\partial \theta}, \tag{10.80}$$

where the velocity potential ϕ has already been given in Eqn. (10.77).

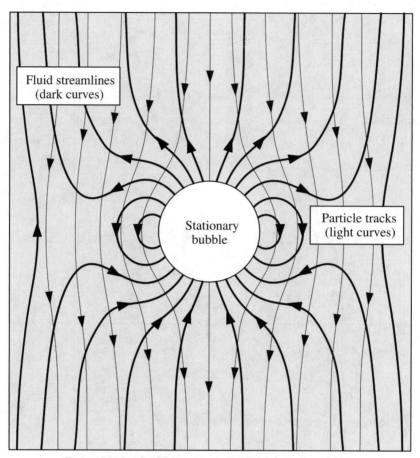

Fig. 10.18   Bubble rise velocity u_b is less than the incipient interstitial fluidizing velocity u_i.

From Eqns. (10.72), (10.75), and (10.77)–(10.80), the *fluid* velocities are therefore:

$$v_r^f = \left[\frac{a^3}{r^3}(u_b + 2u_i) - (u_b - u_i) \right] \cos\theta, \tag{10.81}$$

$$v_\theta^f = \left[\frac{a^3}{r^3}\left(\frac{1}{2}u_b + u_i \right) + (u_b - u_i) \right] \sin\theta. \tag{10.82}$$

The flow pattern of the fluid may be traced by examining its stream function:

$$v_r^f = -\frac{1}{r^2 \sin\theta}\frac{\partial\psi}{\partial\theta}, \tag{10.83}$$

$$v_\theta^f = \frac{1}{r \sin\theta}\frac{\partial\psi}{\partial r}. \tag{10.84}$$

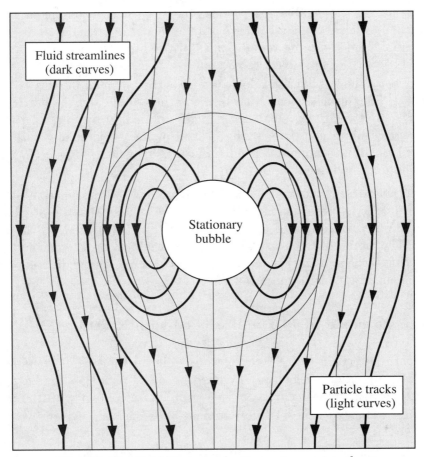

*Fig. 10.19   Bubble rise velocity u_b is greater than
the incipient interstitial fluidizing velocity u_i.*

The following solution is consistent with Eqns. (10.81)–(10.84):

$$\psi = \frac{1}{2}(u_b - u_i)\left(1 - \frac{\alpha^3}{r^3}\right)r^2 \sin^2\theta, \qquad (10.85)$$

in which

$$\alpha^3 = a^3\left(\frac{u_b + 2u_i}{u_b - u_i}\right). \qquad (10.86)$$

The corresponding streamlines for both the fluid and the particles are presented in Figs. 10.18 and 10.19, relative to an observer traveling upward with the bubble. Although in both cases the particles "rain down" around the bubble identically, there are two quite distinct flow regimes as far as the fluidizing fluid is concerned:

1. For $u_b < u_i$, the bubble rise velocity is less than the incipient interstitial fluidizing velocity. Generally, there is an excellent interchange of fluid between the bubble and the more remote parts of the fluidized bed, with a relatively small "recycle" of fluid back to the bubble.

2. For $u_b > u_i$, the bubble rise velocity exceeds the incipient interstitial fluidizing velocity. The interchange of fluid between the bubble and the particles is now confined to a relatively small zone in the immediate vicinity of the bubble. All fluid within a sphere of radius α continuously recirculates in and out of the bubble; this is apparent from Eqn. (10.85), which, for $r > \alpha$, represents streaming around a sphere of radius α.

When the gas stream contains reactive components whose reaction is catalyzed by the particles, the performance of the bed as a reactor may be predicted from relatively simple extensions of the above theory, as illustrated by the following example. The works of Orcutt, Davidson, and Pigford, on the decomposition of ozone, and Harrison and Davidson (*op. cit.*), who discuss several different reactions, are recommended for further information.[23]

Example 10.6—Fluidized Bed with Reaction (C)

The particle bed in Fig. E10.6 contains uniform bubbles of volume V when fluidized at a superficial velocity U by gas that contains a small entering concentration c_0 of a reactant, which decomposes rapidly on contact with the particles.

Show that the exit concentration at the top of the bed is given by:

$$\frac{c_e}{c_0} = \left(1 - \frac{u_0}{U}\right) e^{-X},$$

where $X = Qh/u_cV$, u_0 is the incipient fluidizing velocity, Q is the volumetric cross-flow rate between a bubble and the particulate phase, h is the bed height, and u_c is the absolute velocity of the continuously generated bubbles.

Use these results to estimate c_e/c_0 for a bed of large diameter containing bubbles of equivalent diameter $D_e = 0.15$ m, where $\pi D_e^3/6 = V$. Take $u_0 = 0.2$ m/s, $U = G/A = 1.5$ m/s, and h_0 (the height of the bed at incipient fluidization) $= 0.5$ m. The bubble-rise velocity is $u_c = U - u_0 + 0.646\sqrt{gD_e}$.

Solution

Consider a single bubble rising a distance dz—which it will traverse in a time dz/u_c—with the concentration of reactant changing from c_b to $c_b + (dc_b/dz)dz$. The rate of loss of reactant due to recirculation *out* of the bubble is Qc_b; since the

---

[23] J.C. Orcutt, J.F. Davidson, and R.L. Pigford, "Reaction time distributions in fluidized catalytic reactors," *Chemical Engineering Progress Symposium Series*, Vol. 58, No. 38, pp. 1–15 (1962).

Example 10.6—Fluidized Bed with Reaction (C)   **573**

reaction is virtually instantaneous in the emulsion phase, no reactant *enters* the bubble. A mass balance on the bubble gives:

$$V\left[c_b - \left(c_b + \frac{dc_b}{dz}dz\right)\right] = Qc_b\frac{dz}{u_c}. \qquad \text{(E10.6.1)}$$

$$\underbrace{\phantom{V\left[c_b - \left(c_b + \frac{dc_b}{dz}dz\right)\right]}}_{\text{Rate of loss of reactant}}$$

Separation of variables and integration between the inlet and exit, where the concentration in the bubbles is c_{bh}, gives:

$$\int_{c_0}^{c_{bh}} \frac{dc_b}{c_b} = -\frac{Q}{u_c V}\int_0^h dz, \qquad \text{or} \qquad \frac{c_{bh}}{c_0} = e^{-Qh/u_c V}. \qquad \text{(E10.6.2)}$$

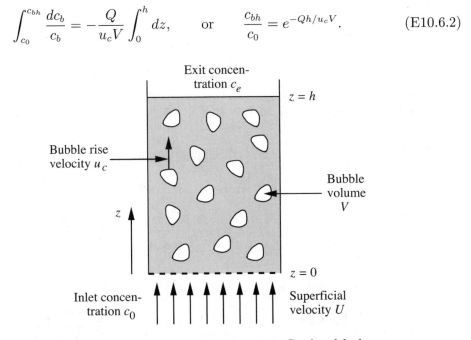

Fig. E10.6 *Reacting gas rising in a fluidized bed.*

Since the entering superficial velocity is U and that needed for incipient fluidization is only u_0, the fraction of entering gas passing through as bubbles is $(U - u_0)/U = 1 - u_0/U$. Also, the gas passing through in the emulsion phase reacts quickly and contributes nothing to the reactant concentration at the exit. Thus, the net reactant concentration at the exit is given by:

$$\frac{c_e}{c_0} = \left(1 - \frac{u_0}{U}\right)e^{-Qh/u_c V}. \qquad \text{(E10.6.3)}$$

The cross-flow rate in and out of the bubbles for a spherical bubble of radius a is given by Eqn. (10.74):

$$Q = 3u_0\pi a^2. \qquad \text{(10.74)}$$

(In some situations, this flow rate may be enhanced by diffusion of gas from the bubble to the surrounding particles, but this effect is neglected here.)

The height h of the bed after it has expanded from its incipient value h_0 can be obtained from Eqn. (10.59), recognizing that $G/A = U$:

$$\frac{0.646\sqrt{gD_e}}{U - u_0} = \frac{h_0}{h - h_0}, \quad \text{or} \quad h = h_0 \left(1 + \frac{U - u_0}{0.646\sqrt{gD_e}}\right). \quad \text{(E10.6.4)}$$

All necessary quantities can now be calculated:

Bed height

$$h = 0.5 \left(1 + \frac{1.5 - 0.2}{0.646\sqrt{9.81 \times 0.15}}\right) = 1.33 \text{ m.}$$

Bubble rise velocity

$$u_c = U - u_0 + 0.646\sqrt{gD_e} = 1.5 - 0.2 + 0.646\sqrt{9.81 \times 0.15} = 2.08 \ \frac{\text{m}}{\text{s}}.$$

Bubble volume

$$V = \frac{\pi D_e^3}{6} = \frac{\pi \times 0.15^3}{6} = 0.00177 \text{ m}^3.$$

Cross flow rate

$$Q = 3u_0\pi a^2 = 3 \times 0.2 \times \pi \times (0.15/2)^2 = 0.0106 \ \frac{\text{m}^3}{\text{s}}.$$

Fraction of reactant in the exiting gas stream

$$\frac{c_e}{c_0} = \left(1 - \frac{u_0}{U}\right)e^{-Qh/u_cV} = \left(1 - \frac{0.2}{1.5}\right)\exp\left(-\frac{0.0106 \times 1.33}{2.08 \times 0.00177}\right) = 0.0188.$$

For finite reaction rates in the bed, two fruitful avenues are available, as discussed by Orcutt et al. (*op. cit.*) and Davidson and Harrison (*op. cit.*):

1. Treat the particulate phase as well-mixed, with a single concentration throughout, in which case an overall balance for the reacting species can be performed on the bed.

2. Consider the concentration in the particulate phase as a function of elevation z, in which case a *differential* balance must be performed for the reacting species in the bed. $\square$

PROBLEMS FOR CHAPTER 10

*Unless otherwise stated, all flows are steady
state, with constant density and viscosity.*

1. *Gas slug between parallel plates—D (C).* Fig. P10.1 shows the cross section through a large bubble OPQ between two parallel vertical plates AA and BB that are separated by a distance $2a$. For purposes of experimental observation, the bubble is held stationary by a downward flow of an inviscid liquid whose velocity is U far upstream of the bubble. The motion is two-dimensional, so that the flow pattern is the same in all planes parallel to that of the diagram.

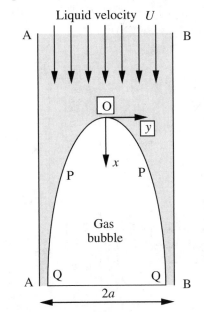

Fig. P10.1   *Flow downward past a stationary gas bubble.*

The following approximate relation has been proposed for the stream function in the liquid:

$$\psi = Uy - \frac{Ua}{\pi}e^{\pi x/a}\sin\left(\frac{\pi y}{a}\right).$$

Verify that this stream function satisfies:

(a) Laplace's equation.
(b) The boundary condition far above the bubble.
(c) The boundary conditions along the walls AA and BB.

Prove that for small values of y, the equation for the free surface OPQ is:

$$x = \frac{\pi y^2}{6a}.$$

What boundary condition must be satisfied along this free surface? Show that if this boundary condition is satisfied for small values of y, then the bubble-rise velocity in an otherwise *stagnant* liquid is:

$$U = \sqrt{\frac{ga}{3\pi}}.$$

2. *Pressure drop in slug flow—M.* A light oil and a gas are flowing upward in a vertical tube of 4.0-in. I.D. at superficial velocities of $L/A = 1.2$ ft/sec and $G/A = 4.8$ ft/sec, respectively. The viscosities of the oil and gas are 15.6 and 0.038 lb_m/ft hr, respectively; their densities are 48 and 0.42 lb_m/ft^3.

Assuming commercial steel pipe ($f_F = 0.00675$) with slug flow prevailing, calculate the pressure drop (psi) in the pipe for every 10 ft of vertical rise.

Fig. P10.3   Gas-lift "pump."

3. *Gas-lift pump—M.* Fig. P10.3 shows a gas-lift "pump," in which the buoyant action of a volumetric flow rate G of gas serves to lift a liquid from a height H_0 to a height H above the bottom of a vertical pipe of diameter D and cross-sectional area A. Analyze the most favorable case with $L = 0$—that is, with *no* net liquid flow:

(a) Assuming slug flow of the gas, what superficial gas velocity G/A is then needed to "expand" the liquid column from an initial height H_0 to a final height H? Take $H_0 = 20$ ft, $H = 60$ ft, and $D = 3$ in.
(b) What is the maximum *upward* liquid velocity, and where does it occur?
(c) What is then the maximum *downward* liquid velocity, and where does it occur? (On the average, the slug length is 2 ft.)

(d) To a first approximation, what gas inlet pressure (psig) is needed if the liquid is water?

4. *Potential flow around a bubble—M.* This problem is closely related to Section 10.7, concerning the volume of bubbles produced at an orifice. Consider a spherical bubble of radius a moving upward with a velocity u_b in an inviscid liquid. Relative to a stationary observer, what is the justification for supposing that the potential function for the flow of the liquid is:

$$\phi = \frac{u_b a^3}{2r^2} \cos \theta?$$

As long as the circumstances are explained properly, you may quote any equation in the chapter in order to arrive at your conclusion.

Next, prove that the magnitude u of the velocity at any point (r, θ) in the liquid is given by:

$$u^2 = \frac{u_b^2 a^6 \left(1 + 3\cos^2 \theta\right)}{4r^6}.$$

Finally, consider the total kinetic energy of the liquid surrounding the bubble:

$$\int_{r=a}^{\infty} \int_{\theta=0}^{\pi} \frac{1}{2} u^2 \, dm,$$

where dm is a differential element of mass, and prove that this equals half the kinetic energy of a sphere of radius a and density equal to that of the liquid. In other words, prove that the effective mass of the surrounding fluid is one half of the mass displaced by the spherical bubble.

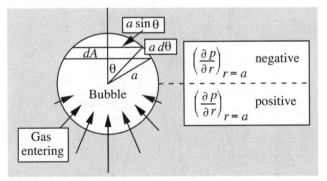

Fig. P10.5   Geometry for gas passing through a bubble.

5. *Gas percolation through a bubble—M.* Fig. P10.5 shows a detailed view of a single bubble at the midpoint of an aggregatively fluidized bed, as in Fig. 10.16. Arrows represent the gas flow entering the lower hemisphere; a similar flow rate of gas departs from the upper hemisphere, although (to avoid confusion) the arrows are omitted in this case.

Following the notation of Section 10.8, consider the radial pressure gradient at the surface of the bubble, $r = a$. By integration over one of the hemispheres, verify Eqn. (10.74) for the flow rate of gas from the bed through the bubble:

$$Q = 3u_0\pi a^2. \tag{10.74}$$

6. *Pressure distribution in a fluidized bed—M.* For a fluidized bed, check that the continuity equations for the fluid and particles, when combined with d'Arcy's law, lead to Laplace's equation, (10.71), for the pressure. Then verify that the proposed pressure distribution from Eqn. (10.72) for the situation shown in Fig. 10.16 satisfies:

(a) Laplace's equation.
(b) The condition that $p = 0$ at the bubble surface.
(c) The conditions that $p = \pm P$ at the bottom and top of the bed, respectively.

7. *Stream function in a fluidized bed—M.* Consider the motion of the *fluid* in a fluidized bed. Starting from Eqns. (10.78) and (10.79), verify the relations given in Eqns. (10.81) and (10.82) for the velocity components v_r^f and v_θ^f. Then prove the relations given in Eqns. (10.85) and (10.86) for the fluid stream function.

8. *Slug flow in a tube—D.* This problem parallels the work of Davies and Taylor for determining the rise velocity u_b of a gas slug or bubble ascending in an otherwise stagnant inviscid liquid in a vertical tube of radius a.[24] By superimposing a downward velocity u_b on the system, the analysis is performed for liquid streaming downward past a *stationary* bubble, as shown in Fig. P10.8.

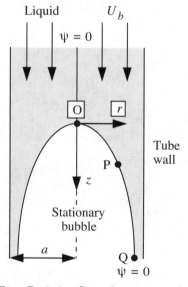

Fig. P10.8   Gas slug in a tube.

[24]   R.M. Davies and G.I. Taylor, "The mechanics of large bubbles rising through extended liquids and through liquids in tubes," *Proceedings of the Royal Society*, Vol. 200A, pp. 375–390 (1950).

The following relationships may be assumed for the velocities, potential function, and stream function in the coordinate system shown, with symmetry about the z axis:

$$v_r = \frac{\partial \phi}{\partial r} = \frac{1}{r}\frac{\partial \psi}{\partial z}, \qquad v_z = \frac{\partial \phi}{\partial z} = -\frac{1}{r}\frac{\partial \psi}{\partial r}. \tag{P10.8.1}$$

$$\frac{\partial^2 \phi}{\partial r^2} + \frac{1}{r}\frac{\partial \phi}{\partial r} + \frac{\partial^2 \phi}{\partial z^2} = 0, \qquad \frac{\partial^2 \psi}{\partial r^2} - \frac{1}{r}\frac{\partial \psi}{\partial r} + \frac{\partial^2 \psi}{\partial z^2} = 0. \tag{P10.8.2}$$

The following forms are assumed for the potential and stream functions:

$$\phi = u_b z - A e^{kz/a} f(r), \qquad \psi = -\frac{1}{2} u_b r^2 + A r e^{kz/a} g(r). \tag{P10.8.3}$$

Also, note that the solution of the differential equation,

$$\frac{d^2 y}{dx^2} + \frac{1}{x}\frac{dy}{dx} + \left(1 - \frac{n^2}{x^2}\right) y = 0, \tag{P10.8.4}$$

is given by $y = J_n(x)$, the nth-order Bessel function of the first kind. The following relations may be needed:

$$\frac{d}{dx}J_0(x) = -J_1(x), \quad \frac{d}{dx}[xJ_1(x)] = xJ_0(x), \quad J_0(0) = 1, \quad J_1(0) = 0. \tag{P10.8.5}$$

Development.   With the above in mind, verify the following:

(a) By substituting for ϕ and ψ from Eqns. (P10.8.3) into Eqns. (P10.8.2), check that:

$$f(r) = J_0\left(\frac{kr}{a}\right), \qquad g(r) = J_1\left(\frac{kr}{a}\right). \tag{P10.8.6}$$

Thus, the potential and stream functions are now:

$$\phi = u_b z - A e^{kz/a} J_0\left(\frac{kr}{a}\right), \qquad \psi = -\frac{1}{2} u_b r^2 + A r e^{kz/a} J_1\left(\frac{kr}{a}\right). \tag{P10.8.7}$$

(b) Derive expressions for v_r and v_z from both ϕ and ψ, and make sure that both approaches agree with each other. Far above the nose of the bubble, show that the velocity is uniformly u_b downward.

(c) Since there is no radial velocity component at the tube wall, check that:

$$J_1(k) = 0, \tag{P10.8.8}$$

an equation whose lowest root you can assume is $k = 3.832$. (There are additional roots, such as the nth-order root k_n, so more terms such as $A_n J_0(k_n r/a)$ could be included for the potential function, but we are only considering the simplest case here.)

(d) Along $r = 0$, the centerline above the nose of the bubble at the origin O, verify that the stream function is zero (and continues this value along the surface of the bubble to points such as P and Q). Prove that the wall ($r = a$) is also a streamline, whose value of ψ corresponds to a flow rate through the tube of $\pi a^2 u_b$.

(e) At the stagnation point O, check that $v_r = 0$ is immediately satisfied. Also prove that the requirement that v_z is also zero there leads to:

$$A = \frac{a u_b}{k}. \tag{P10.8.9}$$

(f) Note that because of the gas in the bubble, the pressure is uniform along OPQ, so that the Bernoulli condition along the bubble surface is:

$$v_r^2 + v_z^2 = 2gz. \tag{P10.8.10}$$

With only a single term in the approximation for ϕ, this condition can only be observed at a *single* point, which we shall take at $r = a/2$. By considering the streamline $\psi = 0$, and assuming that $J_1(3.832/2) = 0.580$, show that this occurs when $z/a = 0.131$. Also assuming that $J_0(3.832/2) = 0.273$, verify that the Bernoulli condition leads to the relation:

$$u_b = c_1\sqrt{ga} = c_2\sqrt{gD}, \tag{P10.8.11}$$

where D is the tube diameter. What are the values for the constants c_1 and c_2?

9. *Performance of a bubble reactor—M.* Fig. P10.9 shows a *bubble reactor*, which uses air bubbles both for removing a trace of bacteria from water and for generating a circulation of water down a central tube of length H and up the surrounding annulus, each of which has the same cross-sectional area A. The air is injected as bubbles at a volumetric flow rate G at a height h above the bottom of the tube, and eventually disengages at the top of the annulus.

The diameter of the bubbles, which may be assumed constant, is such that a discrete swarm of the bubbles (not continuously generated) would rise with velocity u_b in *stagnant* water. Answer the following, using any or all of the variables A, g, G, h, H, L, and u_b:

(a) In terms of A, G, L, and u_b, give expressions for the absolute velocities of the bubbles that are *continuously* generated: u_{ct} downward in the tube and u_{ca} upward in the annulus. Bear in mind that there are net flows of both the gas (G) and liquid (L).

(b) Still bearing in mind that the total air flow rate is G, derive expressions for the void fractions—ε_t in the tube and ε_a in the annulus, in terms of A, G, L, and u_b.

(c) If the total head loss due to the fluid circulation is $c(L/A)^2$, where c is a constant, prove that:

$$\frac{L}{A} = \sqrt{\frac{1}{c}(\varepsilon_a H - \varepsilon_t h)}.$$

Hint: Cycle around from any starting point in the loop, such as the gas injection nozzles, and note that the net pressure change must equal zero when the starting point is again reached. Pressure and head changes are related by $\Delta p = \rho g \Delta h$.

(d) If $G/A = 0.1$ m/s, $H = 10$ m, $h = 5$ m, $L/A = 0.9$ m/s, and $u_b = 0.25$ m/s, calculate the values of ε_t, ε_a, and c, and give the units for c.

(e) How would you recommend starting up the reactor?

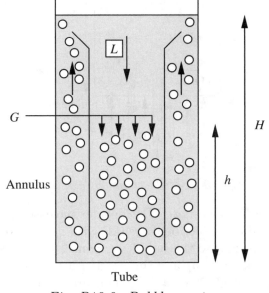

Tube

Fig. P10.9 Bubble reactor.

10. *Bubble motion in a fluidized bed—M.* An experimental aggregatively fluidized bed is operated with an incipient superficial velocity of $u_0 = 10$ cm/s, and the corresponding void fraction is $\varepsilon_0 = 0.45$. As shown in Fig. P10.10, a *single* spherical bubble is generated from the orifice by a gas flow rate of $G = 100$ cm³/s, after which the gas flow is abruptly terminated. If needed, $g = 981$ cm/s².

Now answer the following questions:

(a) Verify that the volume of the bubble is approximately 4.58 cm³. What is its diameter (cm)?

(b) Assuming that it is quickly reached, what is the steady upward rise velocity of the bubble (cm/s)? How long (s) does it take for the bubble to reach the top of the bed after leaving the orifice?

(c) During this period, how much gas enters and leaves the bubble (expressed as a multiple of the bubble volume?

(d) Is the entering gas in (c) essentially fresh gas from the rest of the bed, or is it mainly gas that is recycled from the bubble itself?

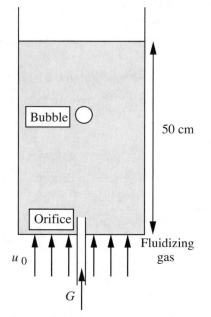

Fig. P10.10   Bubble rising in a fluidized bed.

11. *Flow of expanding gas slugs—D.* Consider the case of slug flow in a vertical pipe of diameter D and cross-sectional area A, with no net liquid flow. The height H of the pipe is substantial, so the bubbles expand as they rise toward the top of the pipe, where the pressure is p_T and the gas density is ρ_{gT}. Assuming isothermal ideal-gas behavior with negligible wall friction, prove that the necessary pressure at the bottom of the pipe is:

$$p_B = p_T + \rho_l g H - \frac{m p_T}{\rho_{gT} A u_b} \ln \left[\frac{0.2 m p_T + \rho_{gT} A u_b p_B}{0.2 m p_T + \rho_{gT} A u_b p_T} \right].$$

Here, ρ_l is the liquid density, m is the mass flow rate of gas, and $u_b = 0.35\sqrt{gD}$ is the rise velocity of a single slug in a stagnant liquid.

12. *Bubble rise in a fluidized bed—M (C).* The superficial velocity of air needed for incipient fluidization of a bed of catalyst particles is 5.4 cm/s, and the mean bed depth is then 22.2 cm. At higher air velocities, the bed consists of a dense phase, together with bubbles that all have the same equivalent diameter, D_e.

Estimate D_e when the superficial air velocity is 28.4 cm/s and the mean bed height is 41.5 cm. How long do the bubbles take to rise through the bed, and what proportion of the air passing through the bed do they represent? What is the mean residence time for the air in the dense phase?

Assume that the void fraction of the dense phase remains constant at 0.4, the value at incipient fluidization.

13. *Fluidized-bed residence time—D (C).* Explain the importance of the incipient fluidizing velocity u_0, and the bubble-rise velocity u_b, in understanding the behavior of an aggregatively fluidized bed. Outline the steps in the derivation of the following results:

$$u_0 = \frac{\varepsilon_0^3(\rho_S - \rho_F)gD^2}{200(1-\varepsilon_0)\mu}, \qquad u_b = 0.646\sqrt{gD_e}.$$

A shot of gaseous tracer is suddenly injected into the base of an air fluidized bed of spherical particles. Calculate the mean residence time of the tracer when the superficial velocity $U = u_0$. When $U = 10$ cm/s, the mean bubble diameter D_e may be taken as 2 cm; estimate the time for the first tracer material to appear at the top of the bed. Show on a sketch how the outlet concentration of tracer varies with time after injection. Use the values given in Table P10.13.

Table P10.13   Data and Notation

Quantity	Symbol	Value	Units
Density of particles	ρ_S	2.5	g/cm^3
Density of fluid (air)	ρ_F	0.0012	g/cm^3
Fluid viscosity	μ	0.00018	poise
Particle diameter	D	0.01	cm
Void fraction at incipient fluidization	ε_0	0.4	—
Bed height at incipient fluidization	H_0	30.0	cm

14. *U-tube bubble reactor—M.* Fig. P10.14 shows a bubble reactor of height H, in which volumetric flow rates G_1 and G_2 of gas are injected at the top of the downcomer and at the base of the riser, respectively. The cross-sectional areas, void fractions, and absolute gas-bubble velocities are denoted by A, ε, and u_c, with subscripts 1 and 2 referring to the downcomer and riser, respectively. The volumetric liquid circulation rate is L. All injected bubbles leave at the top of the riser.

Prove that the void fraction in the downcomer is:

$$\varepsilon_1 = \frac{G_1}{G_1 + L - u_b A_1},$$

in which u_b is the rise velocity of a discrete swarm of bubbles relative to *stagnant* liquid. Obtain the corresponding expression for ε_2 in the riser.

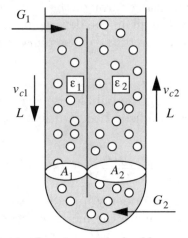

Fig. P10.14   Reactor with double gas injection.

The frictional head loss per unit depth is found experimentally to be c times the square of the corresponding superficial liquid velocity, where $c = 0.0452$ s^2/m^2. If $A_1 = 0.02$, $A_2 = 0.1$ m^2, $G_1 = 0.01$, $G_2 = 0.05$ m^3/s, and $u_b = 0.25$ m/s, calculate ε_1, ε_2, and L (m^3/s).

15.   *Two-phase horizontal flow—M.* Consider Example 10.2, with the sole change that the liquid volumetric flow rate is reduced to $L = 0.025$ ft^3/s. Calculate the new values of X, ϕ_g, ε, and $(dp/dz)_{tp}$. Any values reused from Example 10.2 may be quoted without recalculating them. *Note:* for laminar pipe flow, $f_F = 16/\text{Re}$.

16. *Bubble diameter in a fluidized bed—D(C).* Define the two-phase theory of gas fluidization, and describe briefly how it is applied in practice.

Use the theory to show that:

$$U - u_0 = 0.646\sqrt{gD_e}\left(\frac{h - h_0}{h_0}\right)$$

for a gas-fluidized bed, where u_0 and h_0 are the superficial gas velocity and bed height at incipient fluidization, h is the bed height at a gas velocity U, and D_e is the equivalent bubble diameter in the bed. State any assumptions that you make.

For a bed of spherical catalyst particles of diameter $d_p = 0.03$ cm and density $\rho_p = 2.6$ g/cm^3 fluidized by air, the bed height is 24.2 cm at incipient fluidization. At $U = 1.5u_0$ the mean bed height is observed to be 30.6 cm. For air: $\rho_f = 0.00121$ g/cm^3, $\mu_f = 0.0183$ cP. Estimate the equivalent bubble diameter at this higher gas flow rate.

Use the Ergun equation for packed beds:

$$\frac{-\Delta p}{\rho_f u_0^2} \frac{d_p}{h_0} \frac{\varepsilon_0^3}{(1 - \varepsilon_0)} = 1.75 + \frac{150}{\text{Re}}, \quad \text{where} \quad \text{Re} = \frac{\rho_f u_0 d_p}{\mu_f (1 - \varepsilon_0)},$$

in which ε_0 is the void fraction at the point of incipient fluidization, as distinct from the fraction of the expanded bed that is occupied by bubbles. To determine ε_0, assume that the spherical particles are touching one another in a cubic manner.

17. *Particulate fluidization—M (C).* What is the difference between particulate and aggregative fluidization? The following results were obtained for the expansion of a bed of ball bearings by a fluid:

Fluidizing velocity (U cm/s):	18	28	43	56
Void fraction (ε):	0.5	0.6	0.7	0.8

The free-fall velocity of a single ball bearing through the fluid is 96.0 cm/s.

Show that these results obey the Richardson and Zaki correlation, and evaluate the corresponding index. Find also the minimum fluidizing velocity, U_{mf}, if the void fraction is 0.4 at minimum fluidization.

Indicate how Stokes' law, $F = 3\pi\mu Ud$, can be used in conjunction with these results to show that:

$$U_{mf} = \kappa(\rho_s - \rho_f)\frac{gd^2}{\mu},$$

and evaluate the constant κ. In these expressions, F is the drag force on a single particle, ρ_s and ρ_f are the densities of the solid and fluid, respectively, μ is the viscosity of the fluid, and d is the diameter of the particles.

18. *Aerated treatment of sewage—D (C).* Fig. P10.18 shows a "deep-shaft" unit for treating sewage. The shaft (much deeper than indicated here) is partitioned into two sections D and R of equal area. Air is injected into the downcomer at a depth F. The bubbles are carried down to the bottom, up the riser R, and are then separated at S so that clear liquid returns to D. This direction of circulation is established by a start-up air supply (not shown) in R, which is turned off during normal operation.

Assume: (a) void fractions $\varepsilon_D, \varepsilon_R \ll 1$; (b) constant bubble-rise velocity u_b relative to the liquid; (c) negligible bubble absorption; (d) in calculating the change of bubble volume, the pressure at position x is $p_0 + \rho_l gx$, in which p_0 is the absolute pressure at $x = 0$, and ρ_l is the liquid density.

Show that at a given x, $\varepsilon_R/\varepsilon_D = (w - u_b)/(w + u_b)$, where w is the liquid velocity, and that the liquid head ΔH available to overcome friction and other hydraulic losses in the circuit is given by:

$$\frac{\Delta H}{\varepsilon_{D0} H_0} = \left(\frac{w - u_b}{w + u_b}\right) \ln\left(\frac{H_0 + H - F}{H_0 - F}\right) - \ln\left(\frac{H_0 + H - F}{H_0}\right),$$

where $\rho_l gH_0 = p_0$, and $\varepsilon_D = \varepsilon_{D0}$ at $x = 0$.

Fig. P10.18   Deep-shaft aeration of sewage.

19. *Slug flow fluidized-bed reactor—M (C).* Consider a fluidized-bed reactor identical to that in Example 10.6, except that it now occurs in a tube of diameter D operating in the slug flow mode. The slugs have a uniform volume V and rise with an absolute velocity u_b. The superficial velocity is again U and the incipient fluidizing velocity is u_0; the bed height is h_0 when compacted and h when fluidized.

If diffusion is negligible, outline the arguments leading to a volumetric cross flow rate $Q = \pi D^2 u_0/4$.

Estimate c_e/c_0 for a bed in a tube of diameter $D = 0.07$ m, with $D_e = 0.15$ m. $h_0 = 0.5$ m, $u_0 = 0.2$ m/s, and $U = 1.5$ m/s. Take $u_b = U - u_0 + 0.35\sqrt{gD}$. The result of Eqn. (E10.6.3) may be assumed.

20. *Bubble formation in a viscous liquid—M.* This problem is based on investigations of Davidson and Schüler.[25] Consider the growth of a single spherical bubble at an orifice in a liquid of significant kinematic viscosity ν. If the gas flow rate G is sufficiently small so that: (a) inertial effects in the liquid can be ignored, and (b) the velocity of the bubble is always at its Stokes value, prove that the bubble volume on detachment is:

$$V = \left(\frac{4\pi}{3}\right)^{1/4} \left(\frac{15\nu G}{2g}\right)^{3/4}.$$

21. *Group used in flooding correlation—E.* Investigate the extent to which the ordinate $G_g^2 F\psi\mu_l^{0.2}/\rho_g\rho_l g_c$ of Fig. 10.10 is related to the dimensionless velocity $v_g\sqrt{\rho_g/gD\rho_l}$ in Eqn. (10.51).

[25] J.F. Davidson and B.O.G. Schüler, "Bubble formation at an orifice in a viscous liquid," *Transactions of the Institution of Chemical Engineers*, Vol. 38, pp. 144–154 (1960).

22. *Well-mixed fluidized-bed reactor—D.* Consider the situation of Example 10.6, but now with a first-order reaction with rate constant k in the particulate phase, which is assumed to be well mixed, with a uniform reactant concentration c_p.

(a) Using the same notation as in Example 10.6, prove that the reactant concentration c_b in the bubble at height z above the inlet is:

$$c_b = c_p + (c_0 - c_p)e^{-Qz/u_c V}.$$

(b) Explain briefly why the total rate of addition of reactant from the bubbles to the particulate phase over the entire bed equals:

$$r_{b \to p} = \int_0^h Q c_b N A \, dz,$$

where N is the number of bubbles per unit volume and A is the bed cross-sectional area.

(c) Also perform a material balance for the reactant in the particulate phase of the entire bed, and prove that:

$$(c_0 - c_p)\left[u_0 + \varepsilon u_c\left(1 - e^{-Qh/u_c V}\right)\right] = h(1 - \varepsilon)kc_p,$$

in which the void fraction is given by $\varepsilon = NV$, and u_0 is the incipient fluidizing velocity. Hence, determine an expression for the fraction of the reactant leaving the bed unconverted.

23. *Cylindrical bubble in fluidized bed—M.* Experimental fluidized beds are sometimes operated as shown in Fig. P10.23, with a *cylindrical* bubble rising between two transparent parallel plates, because the bubble can be photographed easily. Analyze the fluid mechanics surrounding such a bubble that is halfway up the bed, as in Fig. 10.16.

(a) Prove that the following pressure distribution in the particulate phase satisfies the appropriate Laplace's equation:

$$p = -\xi\left(r - \frac{a^2}{r}\right)\cos\theta.$$

(b) If the velocity potential for the streaming of the particles around the bubble is:

$$\phi = -u_b\left(r + \frac{a^2}{r}\right)\cos\theta,$$

derive expressions for the particle velocity components v_r^p and v_θ^p.

(c) Finally, obtain expressions for the velocity components v_r^f and v_θ^f of the fluidizing fluid, and the stream function ψ, in terms of r, θ, a, u_b, and u_i.

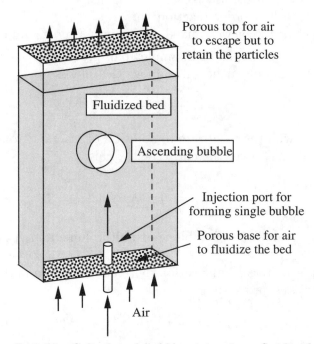

Fig. P10.23   Cylindrical bubble rising in a fluidized bed.

24. *Two-phase bubble flow—E.* Consider two-phase gas/liquid flow in a vertical column of cross-sectional area 100 cm². The rise velocity of the bubbles relative to a stagnant liquid above them is 25 cm/s. The regime is thought to be that of bubble flow.

(a) If the upward gas and liquid volumetric flow rates are $G = 250$ cm³/s and $L = 0$, respectively, what is the void fraction, and is this likely to correspond to bubble flow? If the liquid is water, and the gas density is negligible, and the height of the two-phase mixture is 100 cm, what is the pressure drop in Pa across the column? If the gas is turned off, to what height does the liquid settle?

(b) For $G = L = 250$ cm³/s, both upward, what is the void fraction?

(c) For $G = 200$ cm³/s upward, is it possible to adjust the liquid flow rate L to achieve a void fraction of $\varepsilon = 0.08$? If so, what is the value of L? If not, explain why such a void fraction is impossible.

25. *Two-phase vertical annular flow—M.* Consider Example 10.5, with the sole change that the gas volumetric flow rate is reduced to $G = 5.0$ ft³/s.

Calculate the new values for X, ε, ϕ_g, and $(dp/dz)_{\mathrm{tp}}$. Any values reused from Example 10.5 may be quoted without recalculating them.

26. *True/false.* Check *true* or *false*, as appropriate:

(a) The rise velocity of a spherical-cap bubble can be determined fairly accurately from potential-flow theory. T ☐ F ☐

(b) The equivalent diameter of a spherical-cap bubble is defined as the diameter of a sphere that has the same surface area as the bubble. T ☐ F ☐

(c) For two-phase gas/liquid flow in vertical tubes, the bubble-flow regime occurs at low gas flow rates. T ☐ F ☐

(d) The void fraction in the slug flow regime of two-phase flow in vertical tubes is important because it is a key factor in determining the overall pressure drop. T ☐ F ☐

(e) The rise velocity of a single gas slug in a tube containing a stagnant liquid above it is approximately $0.711\sqrt{gD}$, where D is the internal tube diameter. T ☐ F ☐

(f) Particulate fluidization occurs when excess gas—beyond that needed to effect the fluidization—travels as discrete bubbles. T ☐ F ☐

(g) In a liquid, the rise velocity of continuously generated bubbles is less than that of a single swarm of bubbles. T ☐ F ☐

(h) A fluidized bed of particles behaves almost as though it were an inviscid liquid. T ☐ F ☐

(i) A bubble, formed at an orifice in a liquid, detaches when the distance from its center to the orifice just exceeds the radius of the bubble. T ☐ F ☐

(j) The pressure in a fluidized bed obeys Laplace's equation. T ☐ F ☐

(k) A spherical bubble cavity in a fluidized bed causes three times the normal gas flow rate to flow through it. T ☐ F ☐

(l) For a single bubble in an aggregatively fluidized bed, the nature of the motion of the particles around the bubble depends on whether or not the bubble rise velocity exceeds the incipient interstitial fluidizing velocity. T ☐ F ☐

(m) In vertical two-phase slug flow, the highest downward liquid velocity occurs in the liquid film at the base of each slug. T ☐ F ☐

(n)   In the bubble reactor problem—P10.9—the reactor   T ☐ F ☐
      can be started with a fast initial air flow rate; throt-
      tling it back then gives the desired circulation.

(o)   The volume of a bubble generated at an orifice in   T ☐ F ☐
      a low-viscosity liquid depends almost entirely on the
      gas flow rate and the acceleration of gravity.

(p)   A single bubble rising in a fluidized bed will have more   T ☐ F ☐
      gas recycled back to itself the slower it goes.

(q)   For vertical two-phase gas/liquid slug flow in a pipe,   T ☐ F ☐
      the maximum upward liquid velocity occurs about
      halfway between successive gas slugs.

(r)   For vertical two-phase gas/liquid slug flow in a pipe   T ☐ F ☐
      of known diameter, a knowledge of the gas and liquid
      flow rates will enable the void fraction to be predicted.

(s)   For vertical two-phase gas/liquid slug flow in a pipe   T ☐ F ☐
      with zero net liquid flow, a succession of continuously
      generated gas slugs will rise more slowly than a single
      slug would in an otherwise stagnant liquid.

(t)   The rise velocities of single bubbles in a liquid increase   T ☐ F ☐
      monotonically as the bubble volume increases.

(u)   The Lockhart and Martinelli two-phase pressure-   T ☐ F ☐
      gradient correlation explicitly recognizes flow regimes
      such as bubble flow, slug flow, and annular flow.

(v)   For two-phase gas/liquid flow in vertical tubes, the   T ☐ F ☐
      transition between slug and annular flow is well de-
      fined.

(w)   One model for predicting the onset of flooding de-   T ☐ F ☐
      pends on Bernoulli's equation for determining the
      pressure drop as rising gas accelerates between liquid
      waves.

(x)   For two-phase gas/liquid bubble flow in a vertical   T ☐ F ☐
      tube, a solution cannot exist for gas downflow and
      liquid upflow.

(y)   The Eötvös number is a measure of the ratio of grav-   T ☐ F ☐
      itational forces to viscous forces.

(z)   Although the Lockhart/Martinelli correlation was   T ☐ F ☐
      originally developed for two-phase flow in horizontal
      tubes, it can be used to predict the frictional part of
      the pressure drop for two-phase flow in vertical tubes.

Chapter 11

NON-NEWTONIAN FLUIDS

11.1 Introduction

$\mathbf{T}$HE discussion thus far has been mainly for viscous *fluids*, all of which share the characteristic that they tend to deform *continuously* under the influence of an applied stress. However, fluids represent just one end of a wide spectrum of materials, for which a very simplified overview is given in Fig. 11.1. *Solids* lie at the other extreme; they, too, will deform under an applied stress, but will reach a position of equilibrium, in which further deformation ceases; and—if the stress is removed—will recover their original shape. Thus, fluids generally exhibit *viscosity*, and solids typically show *elasticity*. In between these two extremes lie materials that, depending on the particular circumstances, can exhibit both viscosity *and* elasticity. Such materials are called *viscoelastic fluids*, and are typified by polymer melts and polymer solutions.

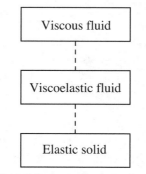

Fig. 11.1 Simplified spectrum of material types.

To this point, the treatment has also been further restricted, in that we have largely considered only incompressible fluids that exhibit a *linear* response to the applied strain. These fluids are called *Newtonian*, and the constitutive equation, which describes the stress/strain rate relationship, is given by Newton's law of viscosity, typified by Eqn. (5.59):

$$\tau_{xy} \left(= \tau_{yx}\right) = \mu \left(\frac{\partial v_x}{\partial y} + \frac{\partial v_y}{\partial x} \right). \tag{11.1}$$

However, many incompressible fluids do not exhibit such a simple linear relationship between the stress and strain rate, and this broad class of fluids is called *non-Newtonian*.

This chapter discusses the flow of incompressible non-Newtonian fluids. The analysis of such flows is identical to that of Newtonian fluids, except that a different *constitutive* equation is used to relate the shear and normal stresses to the velocity gradients. Unfortunately, no single constitutive equation can describe all non-Newtonian fluids. In fact, the development of constitutive equations for non-Newtonian fluids, especially polymeric fluids, is an active—and often frustrating—area of research.

11.2 Classification of Non-Newtonian Fluids

Non-Newtonian fluids encompass all viscous fluids that do not obey Newton's law of viscosity, regardless of whether or not they exhibit elastic behavior. Since the study of non-Newtonian fluids can become quite complicated, we start at an introductory level, discussing fluids that are essentially inelastic.

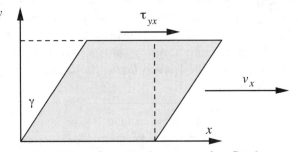

Fig. 11.2   Simple shearing of a fluid.
The strain angle γ is exaggerated.

Consider the situation of Fig. 11.2, in which a fluid element is subjected to a simple shearing action; the size of the strain or angle of deformation, γ, is exaggerated. Based on their responses under these circumstances, different types of non-Newtonian fluids can be defined. In each case, $d\gamma/dt = \dot{\gamma}$, the *rate* of increase of the angle, known variously as the rate of deformation, the rate of shear, or simply the *strain rate*, is important. Fig. 11.3 shows the stress/strain rate behavior of four such types of fluids.[1]

1. *Newtonian fluids.* As already seen, *Newtonian* fluids exhibit a linear response to an applied strain rate. The slope of the shear stress/strain rate line is the

---

[1] For a "popular-science" demonstration of the various types of fluids, see "Serious fun with Polyox, Silly Putty, Slime and other non-Newtonian fluids" in Jearl Walker's "The Amateur Scientist" column in *The Scientific American*, pp. 186–196 (November 1978). With great effect and to the amusement of his audiences, the late Prof. Karl Weissenberg used to give similar demonstrations in some of his lectures on non-Newtonian fluids.

Newtonian viscosity, η, which is constant regardless of the strain rate. When discussing non-Newtonian fluids, the symbol η is often employed to represent a viscosity or effective viscosity, to distinguish it from a purely Newtonian viscosity, μ. Examples of Newtonian fluids are gases, water, liquid hydrocarbons, and most organic liquids of relatively low molecular weight.

2. *Pseudoplastic* or shear-thinning fluids exhibit a viscosity that decreases with increasing strain rate. Most polymer solutions and polymer melts, such as polystyrene and nylon, behave as shear-thinning fluids; shearing tends to cause the entangled long-chain molecules to straighten out and become aligned with the flow, thus reducing the effective viscosity. Many paints, being essentially dispersions of pigment particles in polymer solutions, also exhibit pseudoplastic behavior. When the paint is applied to a surface by brushing or spraying, which shears the paint, its viscosity decreases. When the paint is on the surface, no longer subjected to brushing, its viscosity increases and prevents it from "sagging" or flowing under the action of gravity.

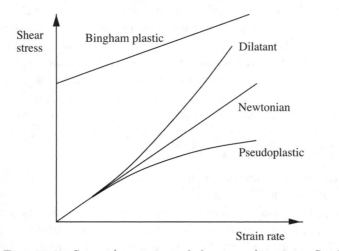

Fig. 11.3 Stress/strain rate behavior of various fluids.[2]

3. *Dilatant* or shear-thickening fluids exhibit a viscosity that *increases* with increasing strain rate. Although dilatant fluids are less common than pseudoplastic fluids, some particulate suspensions such as sand or cornstarch in water are dilatant, in which the microstructure rearranges under shear so that the fluid provides *more* resistance to flow. For example, if a sand/water suspension has settled for some time, particularly if has been subject to a *small* shear, the sand grains will have assumed an orientation of the closest possible packing to one another; that is, the void fraction occupied by the water is a *minimum*. Any significant shear will disturb the close packing and the void fraction will

---

[2] The curves for the dilatant, Newtonian, and pseudoplastic fluids have been drawn with a common low strain-rate slope only for the purpose of demonstrating the relative behavior at higher strain rates.

increase slightly. Consequently, the water will no longer fill the entire space between many of the sand grains, and the lack of lubrication will cause an *increased* resistance to flow.

This dilatant phenomenon can be demonstrated by filling a plastic bottle (or rubber balloon) with sand or glass beads and water—enough of the latter is added so that its level is near the top of a glass tube passing through a rubber stopper in the neck of the bottle. Close packing is obtained by tapping or *gently* massaging the sides of the bottle. If the bottle is then squeezed, it will offer considerable resistance and the water level in the tube will *drop*, much to the relief of members of the audience who think that they are about to be squirted with water. Note that dilatancy usually occurs only for a limited range of strain rates; in fact, dilatant fluids often exhibit a region of shear thinning at low shear rates.

4. *Bingham plastics.* The fourth type of fluid illustrated in Fig. 11.3 is a *Bingham plastic.* A finite stress, or *yield stress,* is required before the fluid flows. Once the fluid flows, the viscosity of a Bingham plastic fluid can be constant, as shown in Fig. 11.3, or it can either increase or decrease with increasing rate of strain. The identification of whether or not a fluid has a yield stress depends on the time scale of the experiment. For example, concentrated particulate systems may exhibit very high viscosities in the limit of very small strain rates, and consequently may falsely appear to have a yield stress if observed over a short time period. Examples of Bingham-plastic fluids are ketchup, mayonnaise, toothpaste, blood, many paints, some printers' inks, and slurries of particles (such as coal) in liquids.[3]

The above discussion focused on fluids whose viscosities were functions only of strain rate. Other non-Newtonian fluids exhibit viscosities that depend not only on strain rate but also on *time*. A *thixotropic* fluid exhibits a viscosity that decreases with time at a constant strain rate. Some particulate suspensions that contain dissolved polymer molecules are thixotropic. The decrease in the viscosity with time results from the gradual breakdown of the polymer network in the fluid. A *rheopectic* fluid, on the other hand, is characterized by a viscosity that increases with time at a constant strain rate.

Viscoelastic fluids exhibit characteristics common to both viscous liquids *and* elastic solids. The stress is a function of the strain, similar to a solid, and is *also* a function of the strain rate, similar to a liquid. The elastic component of a viscoelastic fluid provides memory to the fluid; that is, the fluid "remembers" its state prior to a deformation and attempts to return to this state when the applied strain is removed. The phenomenon of the fluid returning to its previous state, at

---

[3] Values for the yield stress and effective viscosity of several Bingham plastics are given by R. Darby and J. Melson, in "How to predict the friction factor for flow of Bingham plastics," *Chemical Engineering* (December 28, 1981).

least partially, is called *recoil*—a term that brings to mind the "snap" of a rubber band. The degree to which the fluid can return to its previous state depends on the relative elastic and viscous natures of the fluid. Viscoelasticity is discussed in more detail in Section 11.4.

11.3 Constitutive Equations for Inelastic Viscous Fluids

A constitutive equation, or *rheological* equation of state, typically relates the stress tensor τ in a fluid to either or both the strain tensor γ and the rate-of-strain tensor $\dot{\gamma}$.[4] The stress, strain, and rate of strain are second-rank *tensors*, each composed of nine components; *vectors*, on the other hand, are composed of three components. As noted in Chapter 5, second-order tensors such as the stress tensor can be written in matrix form as:

$$\tau = \begin{pmatrix} \tau_{11} & \tau_{12} & \tau_{13} \\ \tau_{21} & \tau_{22} & \tau_{23} \\ \tau_{31} & \tau_{32} & \tau_{33} \end{pmatrix}. \tag{11.2}$$

In the rectangular Cartesian coordinate system, for example, the subscripts 1, 2, and 3 become x, y, and z, respectively, and the stress tensor is then:

$$\tau = \begin{pmatrix} \tau_{xx} & \tau_{xy} & \tau_{xz} \\ \tau_{yx} & \tau_{yy} & \tau_{yz} \\ \tau_{zx} & \tau_{zy} & \tau_{zz} \end{pmatrix}. \tag{11.3}$$

Likewise, the subscripts correspond to r, θ, z in cylindrical coordinates, and to r, θ, and ϕ in spherical coordinates.

A constitutive equation can be derived from a *microscopic* viewpoint, in which the motions of the individual molecules are considered, or from a *macroscopic* viewpoint, in which the fluid is considered as a homogeneous system. While one constitutive equation describes all Newtonian fluids, no single constitutive equation describes all non-Newtonian fluids. In fact, constitutive equations have not been developed for all non-Newtonian fluids. This section reviews the most common constitutive equations for incompressible non-Newtonian fluids based on the macroscopic viewpoint. Larson has written an excellent book on constitutive equations for a wide variety of important fluid types.[5]

Newtonian fluids.   Newton's law of viscosity is written in the general form:

$$\tau = \mu\,\dot{\gamma}. \tag{11.4}$$

---

[4] *Rheology* is the science of the deformation and flow of matter.

[5] R.G. Larson, *Constitutive Equations for Polymer Melts and Solutions*, Butterworths, Boston, 1988. The book includes a wealth of experimental evidence.

Thus, the stress tensor is directly proportional to the strain-rate tensor, given in rectangular coordinates by Eqn. (5.70):

$$
\dot{\gamma} =
\begin{pmatrix}
2\dfrac{\partial v_x}{\partial x} & \left(\dfrac{\partial v_x}{\partial y} + \dfrac{\partial v_y}{\partial x}\right) & \left(\dfrac{\partial v_x}{\partial z} + \dfrac{\partial v_z}{\partial x}\right) \\[3mm]
\left(\dfrac{\partial v_y}{\partial x} + \dfrac{\partial v_x}{\partial y}\right) & 2\dfrac{\partial v_y}{\partial y} & \left(\dfrac{\partial v_y}{\partial z} + \dfrac{\partial v_z}{\partial y}\right) \\[3mm]
\left(\dfrac{\partial v_z}{\partial x} + \dfrac{\partial v_x}{\partial z}\right) & \left(\dfrac{\partial v_z}{\partial y} + \dfrac{\partial v_y}{\partial z}\right) & 2\dfrac{\partial v_z}{\partial z}
\end{pmatrix}
= \nabla\mathbf{v} + (\nabla\mathbf{v})^T, \quad (11.5)
$$

in which the *dyadic product* $\nabla\mathbf{v}$ and its transpose are defined in Eqn. (11.37). The constant of proportionality in Eqn. (11.4) is equal to the fluid viscosity η.

As shown in Fig. 11.3, a plot of shear stress versus strain rate for a Newtonian fluid yields a straight line with a slope of η. A common unit of viscosity is the *poise*, abbreviated P, which is equivalent to one g/cm s. A centipoise, abbreviated cP, is equal to 0.01 P. In SI, the unit of viscosity is the Pascal second, abbreviated Pa·s or Pa s. One Pa s equals 10 P. Examples of Newtonian fluids include water, glycerol, and honey, with viscosities of approximately 10^{-2}, 10, and 10^2 P, respectively. Even glass is considered to be Newtonian, with a viscosity of 10^{40} P!

This realization illustrates that all materials can be considered to be fluids, if the time scale of the "experiment" is long enough. Consider the following excerpt from the book of Judges in the Old Testament, which is attributed to Deborah[6]: "The mountains melted [flowed] from before the Lord . . . . " This statement led to the definition of the dimensionless Deborah number:

$$
\mathrm{De} = \frac{\lambda}{T}, \tag{11.6}
$$

where λ is a characteristic time of the *material* and T is a characteristic time of the deformation *process*. In broad terms, λ for the material is the time it takes to respond when subjected to some external influence such as an applied stress or strain. In contrast to solids, it is relatively easy for liquids to respond because of the greater ease with which the liquid molecules can rearrange themselves.

The Deborah number is important when analyzing viscoelastic materials. If the relaxation time of the fluid is less than the characteristic deformation time ($De < 1$), the fluid appears more viscous than elastic (more fluid than solid). Conversely, if the relaxation time of the fluid is greater than the characteristic deformation time ($De > 1$), it appears more elastic than fluid. The classification of a viscoelastic fluid thus depends on the specific deformation process. The phenomenon of viscoelasticity is considered further in Section 11.4.

---

[6] Judges, chapter 5, verse 5.

Invariants of the strain-rate tensor.   To proceed further, it is necessary to examine the strain-rate tensor $\dot{\boldsymbol{\gamma}}$ of Eqn. (11.5) in more detail. The goal is to extract from it a *single* scalar quantity, known as the *strain rate* $\dot{\gamma}$, which broadly characterizes the rate of deformation of the fluid, and which will find immediate use in the definition of a generalized Newtonian fluid (see below).

The quantity that we seek must possess the property of *independence* of the particular orientation of axes employed in the coordinate system being used. Stated without proof, there are three such scalars, known as the *invariants* of the strain-rate tensor, which remain unchanged when this tensor is transformed through a rotation of axes. For an incompressible fluid, the invariants are:

$$I_1 = \sum_{i=1}^{3} \dot{\gamma}_{ii} = 2\nabla \cdot \mathbf{v} = 0. \tag{11.7}$$

$$I_2 = \dot{\boldsymbol{\gamma}} : \dot{\boldsymbol{\gamma}} = \sum_{i=1}^{3}\sum_{j=1}^{3} \dot{\gamma}_{ij}\dot{\gamma}_{ji} = 2\Phi. \tag{11.8}$$

$$I_3 = \det(\dot{\boldsymbol{\gamma}}) \doteq 0 \ \ (\text{see text}). \tag{11.9}$$

In Eqn. (11.8), the double-dot product of two tensors is defined, and by performing the ensuing double summation on the elements of the strain-rate tensor, Φ is given (for an incompressible fluid) by:

$$\Phi = 2\left[\left(\frac{\partial v_x}{\partial x}\right)^2 + \left(\frac{\partial v_y}{\partial y}\right)^2 + \left(\frac{\partial v_z}{\partial z}\right)^2\right]$$
$$+ \left(\frac{\partial v_y}{\partial x} + \frac{\partial v_x}{\partial y}\right)^2 + \left(\frac{\partial v_z}{\partial y} + \frac{\partial v_y}{\partial z}\right)^2 + \left(\frac{\partial v_x}{\partial z} + \frac{\partial v_z}{\partial x}\right)^2. \tag{11.10}$$

Consider again the case of simple shear, shown in Fig. 11.4, in which the only nonzero velocity component is v_x, and which varies linearly by an amount V over a distance h in the y direction.

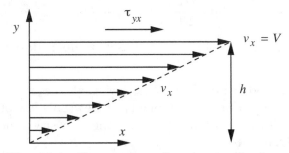

Fig. 11.4   Couette-type flow in simple shear.

The only nonzero velocity gradient in the strain-rate tensor is:

$$\frac{\partial v_x}{\partial y} = \frac{V}{h} = s, \tag{11.11}$$

which has also been designated by the symbol s. The strain-rate tensor then simplifies to:

$$\dot{\gamma} = \begin{pmatrix} 0 & \dfrac{\partial v_x}{\partial y} & 0 \\ \dfrac{\partial v_x}{\partial y} & 0 & 0 \\ 0 & 0 & 0 \end{pmatrix} = s \begin{pmatrix} 0 & 1 & 0 \\ 1 & 0 & 0 \\ 0 & 0 & 0 \end{pmatrix}. \tag{11.12}$$

Now, reexamine the three invariants. The first, being the divergence of the velocity, is always zero for an incompressible fluid. The third, for the case of simple shear just examined, is also found by evaluating the determinant to be zero; it is also known to be zero or approximately so for a number of other commonly occurring cases. Thus, the only invariant that shows any promise for our purpose is the second one, which for the case of simple shear becomes:

$$I_2 = 2 \left(\frac{\partial v_x}{\partial y} \right)^2 = 2s^2. \tag{11.13}$$

With the above in mind, now define the *strain rate* as:

$$\dot{\gamma} = \sqrt{\frac{1}{2} I_2} = \sqrt{\Phi} = s \text{ (for simple shear)}, \tag{11.14}$$

it being understood that the *positive* root is always taken.

In terms of obtaining a physical "feel" for the situation, note that Φ is also the *viscous dissipation function* that appears in the general energy balance, used when studying heat transfer:

$$\rho c_p \frac{DT}{Dt} = k \nabla^2 T + \mu \Phi, \tag{11.15}$$

in which c_p is the specific heat of the fluid, T is its temperature, and k is its thermal conductivity.

Generalized Newtonian fluids.   Now that the strain rate $\dot{\gamma}$ has been defined for a general situation, the simplicity of Newton's law of viscosity leads to a variation for non-Newtonian fluids in which the viscosity is not constant, but is a function of the strain rate:

$$\boldsymbol{\tau} = \eta(\dot{\gamma})\dot{\gamma}, \tag{11.16}$$

where $\eta(\dot\gamma)$ is the generalized viscosity.[7] Fluids whose behavior can be described by this form of constitutive equation are called *generalized Newtonian* fluids.

Many empirical models exist for $\eta(\dot\gamma)$. The well-known *power law*, or Ostwald-de-Waele, model expresses the viscosity as a function of a power of the strain rate $\dot\gamma$:

$$\eta = \kappa\dot\gamma^{n-1}, \tag{11.17}$$

where κ is the consistency index (with units of $P\cdot s^{n-1}$) and n is the power-law index. Concerning the exponent, $n < 1$ describes a *pseudoplastic* or shear-thinning fluid, in which the viscosity decreases as the strain rate increases. On the other hand, $n > 1$ describes a *dilatant* fluid, in which the viscosity increases as the magnitude of the strain rate increases. And $n = 1$ corresponds to a *Newtonian* fluid, for which κ is just the Newtonian viscosity, μ. A logarithmic plot of viscosity versus strain rate ($\ln\eta$ vs. $\ln\dot\gamma$) yields a straight line with a slope of $n - 1$ and an intercept of $\ln\kappa$.

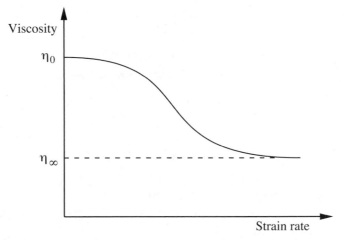

Fig. 11.5   Characteristic viscosity/strain-rate behavior of shear-thinning fluids.

In practice, shear-thinning fluids do not exhibit viscosities that continuously decrease over a large strain-rate range, but instead exhibit the behavior shown in Fig. 11.5. A plateau in the viscosity exists at both low and high strain rates, separated by a shear-thinning region of decreasing viscosity with increasing strain rate. The constant viscosity at low strain rates is called the zero-shear viscosity η_0 and at high strain rates is called the infinite-shear viscosity η_∞. For polymer melts, η_∞ is zero, and for polymer solutions it is the viscosity of the solvent.

Since the power-law model can only approximate the shear-thinning behavior, a *four* parameter model such as that of *Carreau* is needed to approximate the flow behavior over the entire shear-rate range:

---

[7] The symbol μ is reserved exclusively for the viscosity of a *Newtonian* fluid.

$$\frac{\eta - \eta_\infty}{\eta_0 - \eta_\infty} = \frac{1}{(1 + \lambda^2 \dot\gamma^2)^{(1-n)/2}}. \tag{11.18}$$

Here, λ is a relaxation parameter of the fluid and for a wide variety of polymer melts n is a dimensionless constant approximately in the range 0.2–0.5.[8] Thus, the right-hand side of Eqn. (11.18) generally *decreases* as the strain rate *increases*. At low or high strain rates, the viscosity approaches η_0 or η_∞, respectively.

For $\eta_0 \gg \eta_\infty$ and high strain rates, the Carreau model gives:

$$\eta = \eta_\infty + \frac{\eta_0}{(\lambda \dot\gamma)^{1-n}}. \tag{11.19}$$

For $n = 0$, the viscosity decreases with increasing strain rate:

$$\eta = \eta_\infty + \frac{\eta_0}{\lambda \dot\gamma}, \tag{11.20}$$

which, when substituted into Eqn. (11.16), with η_0/λ interpreted as a yield stress τ_0, gives the high shear part of the *Bingham* model, whose complete form is:

$$
\begin{aligned}
\text{Low shears:} \quad & \frac{1}{2}(\boldsymbol{\tau} : \boldsymbol{\tau}) \leq \tau_0^2, \quad \dot{\boldsymbol\gamma} = 0, \\
\text{High shears:} \quad & \frac{1}{2}(\boldsymbol{\tau} : \boldsymbol{\tau}) > \tau_0^2, \quad \boldsymbol{\tau} = \left(\eta + \frac{\tau_0}{\dot\gamma}\right)\dot{\boldsymbol\gamma},
\end{aligned}
\tag{11.21}
$$

for which the double-dot product is similar to that already defined for I_2. For the situation in which the only nonzero shear stresses are the equivalent pair τ_{yx} and τ_{xy}, there results $(\boldsymbol{\tau} : \boldsymbol{\tau})/2 = \tau_{yx}^2$. Note that a Bingham fluid will flow only when the stress in the fluid is greater than the yield stress. The Bingham model, as written above, assumes Newtonian flow behavior for stresses greater than the yield stress.

Example 11.1—Pipe Flow of a Power-Law Fluid

Consider the steady pressure-driven flow of an incompressible non-Newtonian fluid in a horizontal circular pipe of radius a and length L, shown in Fig. E11.1.1. Derive the resulting velocity profile and volumetric flow rate if the fluid viscosity is given by the power-law model. Sketch the velocity profiles for $n < 1$, $n = 1$, and $n > 1$.

---

[8] See, for example, Z. Tadmor and C.G. Gogos, *Principles of Polymer Melt Processing*, Wiley & Sons, New York, 1979, p. 694.

Example 11.1—Pipe Flow of a Power-Law Fluid   **601**

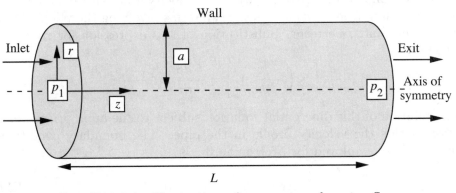

Wall

Inlet

r

a

Exit

p_1

z

p_2

Axis of
symmetry

L

Fig. E11.1.1 Illustration of parameters for pipe flow.

Solution

The equation of continuity for an incompressible fluid, with v_z as the only nonzero velocity component, reduces to:

$$\frac{\partial v_z}{\partial z} = 0, \tag{E11.1.1}$$

so that the axial velocity depends only on the radial location, $v_z = v_z(r)$. Note that the continuity equation in this form applies to *all* incompressible fluids, regardless of whether they are Newtonian or non-Newtonian.

The velocity profile is found by using the z component of the equations of motion in cylindrical coordinates, written in terms of the shear stress—*not* the Navier-Stokes equations, which apply only to incompressible *Newtonian* fluids. Noting that $v_r = v_\theta = 0$ and using Eqn. (E11.1.1), the z component, given in Eqn. (5.74), becomes:

$$0 = -\frac{dp}{dz} + \frac{1}{r}\frac{\partial}{\partial r}\left(r\tau_{rz}\right), \tag{E11.1.2}$$

which is completely general for laminar, steady flow of an incompressible fluid.

By arguments similar to the development leading to Eqn. (11.14), the reader can first check that the strain rate is:

$$\dot{\gamma} = \left|\frac{\partial v_z}{\partial r}\right| = \left|\frac{dv_z}{dr}\right| = -\frac{dv_z}{dr}. \tag{E11.1.3}$$

Note that $\partial v_z/\partial r < 0$ for pipe flow, and that the total derivative is used because v_z depends only on the radial location, r.

By examining the appropriate constitutive equation relating the shear stress to the strain rate, the shear stress τ_{rz} can be replaced by a function of the velocity. For a power-law fluid, the stress τ_{rz} is given by:

$$\tau_{rz} = \eta\dot{\gamma}_{rz} = \kappa\dot{\gamma}^{n-1}\dot{\gamma}_{rz} = -\kappa\left(-\frac{dv_z}{dr}\right)^{n-1}\left(-\frac{dv_z}{dr}\right) = -\kappa\left(-\frac{dv_z}{dr}\right)^n. \tag{E11.1.4}$$

The incorporation of a negative sign with the velocity gradient (and a corresponding negative sign in front of κ) allows the two terms involving the velocity gradient to be combined into one term. Substitution of this expression for τ_{rz} into Eqn. (E11.1.2) gives:

$$\frac{1}{r}\frac{d}{dr}\left[r\kappa\left(-\frac{dv_z}{dr}\right)^n\right] = -\frac{dp}{dz} = \frac{p_1 - p_2}{L}. \tag{E11.1.5}$$

The solution of this differential equation subject to the appropriate boundary conditions yields the velocity profile in the pipe. The boundary conditions are identical to those employed for Newtonian fluids:

$$\text{At } r = 0: \qquad \tau_{rz} = 0. \tag{E11.1.6}$$

$$\text{At } r = a: \qquad v_z = 0. \tag{E11.1.7}$$

Using Eqn. (E11.1.4) for τ_{rz}, the symmetry boundary condition at $r = 0$ becomes:

$$\text{At } r = 0: \qquad \frac{dv_z}{dr} = 0. \tag{E11.1.8}$$

Integration of Eqn. (E11.1.5) once yields:

$$\int d\left[r\kappa\left(-\frac{dv_z}{dr}\right)^n\right] = -\frac{dp}{dz}\int r\,dr$$

or:

$$r\kappa\left(-\frac{dv_z}{dr}\right)^n = -\frac{dp}{dz}\frac{r^2}{2} + c_1, \tag{E11.1.9}$$

where c_1 is an integration constant that is found to be zero by applying the boundary condition at $r = 0$. Rearrangement of Eqn. (E11.1.9) to isolate the velocity derivative gives:

$$-\frac{dv_z}{dr} = \left(-\frac{dp}{dz}\frac{r}{2\kappa}\right)^{1/n}. \tag{E11.1.10}$$

A second integration leads to an expression for the velocity profile:

$$\int dv_z = -\int\left(-\frac{dp}{dz}\frac{r}{2\kappa}\right)^{1/n}dr$$

or:

$$v_z = -\left(-\frac{dp}{dz}\frac{1}{2\kappa}\right)^{1/n}\frac{r^{1+1/n}}{\dfrac{1}{n}+1} + c_2. \tag{E11.1.11}$$

Example 11.1—Pipe Flow of a Power-Law Fluid   **603**

The constant c_2 is found by applying the boundary condition at $r = a$:

$$c_2 = \left(-\frac{dp}{dz}\frac{1}{2\kappa}\right)^{1/n}\frac{a^{1+1/n}}{\frac{1}{n}+1}. \tag{E11.1.12}$$

From the last two equations, the velocity profile is then:

$$v_z(r) = \left(-\frac{dp}{dz}\frac{1}{2\kappa}\right)^{1/n}\frac{a^{1+1/n}-r^{1+1/n}}{\frac{1}{n}+1}. \tag{E11.1.13}$$

The volumetric flow rate is:

$$Q = \int_0^a v_z 2\pi r\, dr = \left(-\frac{dp}{dz}\frac{1}{2\kappa}\right)^{1/n}\frac{\pi n a^{3+1/n}}{1+3n}. \tag{E11.1.14}$$

If $n = 1$ and $\kappa = \eta$, the expression for the velocity profile reduces to that for a Newtonian fluid:

$$v_z(r) = \frac{1}{4\eta}\left(-\frac{dp}{dz}\right)(a^2 - r^2), \tag{E11.1.15}$$

which is the expected Poiseuille flow velocity profile (middle curve in Fig. E11.1.2).

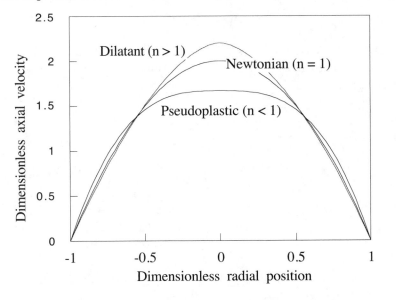

Fig. E11.1.2 Dimensionless velocity profiles (ratio of actual velocity to mean velocity) for dilatant, Newtonian, and pseudoplastic fluids in pressure-driven pipe flow.

Under conditions of equal volumetric flow rates, the velocity profiles for $n = 0.5$, $n = 1$, and $n = 1.5$ are sketched in Fig. E11.1.2. Note that for a shear-thinning or pseudoplastic fluid, the profile is flatter in the center, where it resembles plug flow, and is steeper near the wall, where it has a higher velocity than either the Newtonian or dilatant fluids. Consequently, shear-thinning fluids experience *increased* heat-transfer rates when they flow in the tubes of a heat exchanger because of the enhanced convective energy transport near the wall. Conversely, shear-thickening or dilatant fluids experience *lower* heat-transfer rates.   □

As shown in the above example, the analysis of the flow of *generalized Newtonian* fluids is nearly identical to that of *Newtonian* fluids. However, it is important to remember to start from the equation of motion written in terms of the *shear stress*, not from the Navier-Stokes equations. The appropriate constitutive equation is then used to relate the shear stress to the velocity.

Example 11.2—Pipe Flow of a Bingham Plastic

Investigate the nature of the shear stress and velocity profile for flow of a Bingham plastic, with yield stress τ_0 and effective viscosity η, in a cylindrical pipe of radius a and length L. The inlet and outlet pressures are p_1 and p_2, respectively.

Solution

The shear-stress distribution is independent of the nature of the fluid, and a steady-state momentum balance in the z direction on a cylinder of fluid of radius r and length L gives:

$$\underbrace{2\pi r L \tau_{rz}}_{\text{Shear}} + \underbrace{(p_1 - p_2)\pi r^2}_{\text{Pressure}} = 0, \qquad (\text{E11.2.1})$$

or:

$$\tau_{rz} = -\frac{r}{2}\left(\frac{p_1 - p_2}{L}\right) = -\frac{r}{2}\underbrace{\left(-\frac{dp}{dz}\right)}_{\underbrace{\text{Positive}}_{\text{Negative}}}, \qquad (\text{E11.2.2})$$

which can also be derived by integration of the last of Eqn. (5.74). Fig. E11.2 shows the resulting linear variation of the shear stress, and the arrows pointing to the left reflect that τ_{rz} (by convention, the stress exerted by the region of greater r on the region of lesser r) is physically in the negative-z direction. Let τ_w denote the value of τ_{rz} at the wall.

Example 11.2—Pipe Flow of a Bingham Plastic   **605**

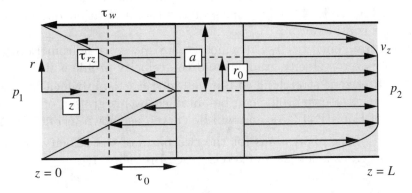

Fig. E11.2 *Shear stress and velocity profile for Bingham-plastic pipe flow, for $|\tau_w| > \tau_0$. At $r = r_0$, $|\tau_{rz}| = \tau_0$.*

Next consider the stress and strain-rate tensors for the case of $v_r = v_\theta = 0$:

$$\boldsymbol{\tau} = \begin{pmatrix} \tau_{rr} & \tau_{rz} & \tau_{r\theta} \\ \tau_{zr} & \tau_{zz} & \tau_{z\theta} \\ \tau_{\theta r} & \tau_{\theta z} & \tau_{\theta\theta} \end{pmatrix} = \begin{pmatrix} 0 & \tau_{rz} & 0 \\ \tau_{zr} & 0 & 0 \\ 0 & 0 & 0 \end{pmatrix}, \quad \dot{\boldsymbol{\gamma}} = \begin{pmatrix} 0 & \dfrac{dv_z}{dr} & 0 \\ \dfrac{dv_z}{dr} & 0 & 0 \\ 0 & 0 & 0 \end{pmatrix}, \quad (E11.2.3)$$

leading to:

$$\boldsymbol{\tau} : \boldsymbol{\tau} = 2\tau_{rz}^2, \qquad \dot{\gamma} = \sqrt{\frac{1}{2} I_2} = \sqrt{\frac{1}{2} \dot{\boldsymbol{\gamma}} : \dot{\boldsymbol{\gamma}}} = \left| \frac{dv_z}{dr} \right|. \qquad (E11.2.4)$$

The Bingham model of Eqn. (11.21) then reduces to:

Low shears: $|\tau_{rz}| \le \tau_0, \quad \dfrac{dv_z}{dr} = 0,$

High shears: $|\tau_{rz}| > \tau_0, \quad \tau_{rz} = \left(\eta + \dfrac{\tau_0}{\dot{\gamma}} \right) \dfrac{dv_z}{dr} = -\tau_0 + \eta \dfrac{dv_z}{dr}.$

$(E11.2.5)$

There are two cases to be considered, depending on the relative values of $|\tau_w|$ and τ_0:

1. If the maximum shear stress never exceeds the threshold value, $|\tau_w| \le \tau_0$, Eqn. (11.2.5) indicates a zero velocity gradient everywhere. Since the velocity is zero at the wall, the velocity is also zero at all radial locations. That is, the applied pressure gradient is insufficient to move the fluid, which effectively remains as a rigid body.

2. If the maximum shear stress exceeds the threshold value, $|\tau_w| > \tau_0$ (the case shown in Fig. E11.2), let $|\tau_{rz}| = \tau_0$ at the radial location r_0. There are two regions for consideration:

   (a) A central zone, $r \leq r_0$, in which $|\tau_{rz}| \leq \tau_0$, so that from Eqn. (E11.2.5) there is zero velocity gradient and the velocity profile is flat.

   (b) An outer zone stretching to the wall, $a \geq r > r_0$, in which $|\tau_{rz}| > \tau_0$. The resulting velocity profile can be obtained by integration of the high-shear part of Eqn. (E11.2.5), and will be curved as shown in Fig. E11.2.

Derivation of the expressions for the two parts of the velocity profile for the second of these cases is left as an exercise—see Problem 11.1. □

Example 11.3—Non-Newtonian Flow in a Die (COMSOL Library)

Fig. E11.3.1   (a) Cross section of axisymmetric extrusion die; (b) the region in which COMSOL solves the problem—rotated clockwise from (a) by 90° because in cylindrical coordinates COMSOL always has the $r = 0$ symmetry axis running "vertically."

This example involves taking one of the sample problems that is part of the COMSOL **Model Library**, so we will not have to set up the problem, but can concentrate on interpreting the results. The problem is that of non-Newtonian flow of a solution of polystyrene in 1-chloronaphthalene, which is being extruded through the axisymmetric die shown in Fig. E11.3.1(a); because of symmetry about the centerline, COMSOL only needs to solve the problem in the region shown in Fig. E11.3.1(b)—rotated through 90° to coincide with the direction given in the COMSOL library example. The dimensions of the die in meters are given above

Example 11.3—Non-Newtonian Flow in a Die (COMSOL Library)   **607**

and in Fig. 14.20, which also shows how it is drawn. Later, variations of properties over the narrow cross section AB will be of special interest. The problem will be solved for six different values of the inlet pressure p_{in} (Pa).

The viscosity of the polystyrene solution is predicted by the Carreau model of Eqn. (11.18):

$$\eta = \eta_\infty + (\eta_0 - \eta_\infty) \frac{1}{[1 + (\lambda\dot\gamma)^2]^{(1-n)/2}}, \tag{E11.3.1}$$

in which the strain rate is given by Eqn. (11.14), which in axisymmetric cylindrical coordinates amounts to:

$$\dot\gamma = \sqrt{\frac{1}{2}\left[\left(2\frac{\partial v_r}{\partial r}\right)^2 + 2\left(\frac{\partial v_z}{\partial r} + \frac{\partial v_r}{\partial z}\right)^2 + \left(2\frac{\partial v_z}{\partial z}\right)^2\right]}. \tag{E11.3.2}$$

The polystyrene solution is typical of *shear-thinning* polymers, which have a value $n < 1$. Thus, as the strain rate $\dot\gamma$ *increases*, the effective viscosity η *decreases*. Values of the density and the Carreau model parameters appear in Table E11.3.

Table E11.3   Parameter Values

Parameter	Value	Units
η_∞	0	kg/m s
η_0	166	kg/m s
λ	0.0173	s
n	0.538	—
ρ	450	kg/m^3

If **u** is the velocity vector, then the continuity and two momentum balances being solved are:

$$\nabla \cdot \mathbf{u} = 0, \tag{E11.3.3}$$

$$\rho\frac{\partial \mathbf{u}}{\partial t} + \rho(\mathbf{u} \cdot \nabla)\mathbf{u} = -\nabla p + \nabla \cdot \eta\left(\nabla\mathbf{u} + (\nabla\mathbf{u})^\mathrm{T}\right). \tag{E11.3.4}$$

Preliminary Investigation

(a) Open COMSOL and in the **Model Navigator** select the **Model Library** tab. Double click on the following in sequence: **Chemical Engineering Module**, **Momentum Transport**, and **non-Newtonian flow**.

(b) Examine **Physics/Boundary Settings**, which will show "slip/symmetry" for boundary 1 (the axis of symmetry), "outflow/pressure" with pressure **pin** for boundary 2 (the inlet), "outflow/pressure" with zero pressure for boundary 4 (the exit), and no slip for boundaries 3, 5, and 6. The problem is actually being solved for *six* different values of the variable inlet pressure, which may be found in **Solve/Solver Parameters/General**. For the parameter `pin`, the entry "10000:40000:210000" means that the first value of `pin` is 10,000, the second is 10,000 + 40,000 = 50,000, etc., until 210,000 is reached.

(c) In **Physics/Subdomain Settings**:

   (i) Observe (for Subdomain 1) the equations that are being solved (steady-state momentum balance and continuity), with $\rho = 450$. (COMSOL assumes that the user is working in a consistent set of known units; therefore, it does not show units for variables or parameters.)

   (ii) Then select the **Carreau** tab, and check that the values for the four parameters are the same as in Table E11.3.

Fig. E11.3.2   Two-dimensional representation of the magnitudes of the velocities for $p_{inlet} = 210{,}000$ Pa. The colors in the original ranged from red (high velocity) to blue (low velocity).

(d) In **Mesh Mode**, look at the FEM mesh (you can zoom in to see the fine detail

Example 11.3—Non-Newtonian Flow in a Die (COMSOL Library)   **609**

near the die walls at section A–B), and check **Mesh Statistics** to see how many simultaneous equations have to be solved.

(e) Then use **Solve Problem** to obtain the solution, to give (selecting **Velocity field**) the **2D Surface** color plot of Fig. E11.3.2, which should of course be the same as that already seen when the problem was opened. The reason for solving the problem again is to get some idea of the time that the FEM solution takes—probably a couple of minutes or so. During the solution, COMSOL has to cope with:

(i) Many thousands of simultaneous equations.

(ii) A nonlinear problem, with much iteration needed. The computed velocities depend on the non-Newtonian viscosity η that appears in Eqn. (E11.3.4), and η in turn depends on the velocities according to Eqns. (E11.3.1) and (E11.3.2).

pin(6)=2.1e5   Arrow: Velocity field

5   0   0.005   0.01   0.015   0.02

Fig. E11.3.3   Arrow plot for $p_{inlet} = 210{,}000$ Pa.

Required Plots

(a) Under **Postprocessing** and **Quick Plots**, make an arrow plot showing the velocity vectors, as in Fig. E11.3.3. Observe that both this plot and the 2D color plot of Fig. E11.3.2 show the following velocities:

(i) At the inlet, a progression of velocities from high at the centerline to low near the wall.

(ii) At the constriction AB, a generally high velocity.

(iii) At the exit, where there is a relatively large area for flow, a generally low velocity.

(b) Under **Postprocessing/Cross-Section Plot Parameters**, under the **General** tab, observe that **Solutions to use** has the following six values of the inlet pressure: 10,000, 50,000, 90,000, 130,000, 170,000, and 210,000 Pa. All six values will probably be highlighted. If not, select the first (10000), hold down the **Shift** key, and select the last (2.1e5).

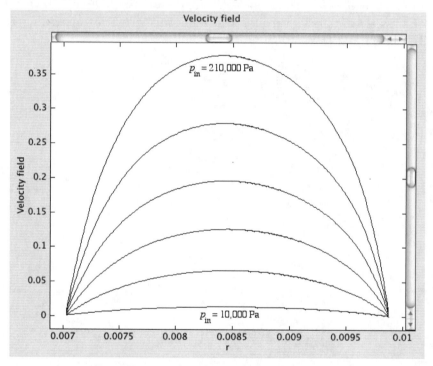

Fig. E11.3.4   Velocity profiles from A to B for the six levels of inlet pressure.

(c) Under the **Line/Extrusion** tab, note the following coordinates have been entered for the points A and B: $r0 = 0.007$, $r1 = 0.01$, $z0 = 0.0007$, and $z0 = 0.004$ (these values are somewhat different from those in FEMLAB 2.3, which has $r0 = 0.00798$, $r1 = 0.00989$, $z0 = 0.0008738$, and $z0 = 0.003616$). Set the selection in the **Predefined quantities** window to **Velocity field**. Then click **OK**. The resulting plot contains six curves, one for each of the inlet pressures (Fig. E11.3.4). The velocities, of course, increase as the inlet pressure rises. But the profiles are only *approximately* parabolic, and are steeper than

Example 11.3—Non-Newtonian Flow in a Die (COMSOL Library)   **611**

parabolas near the walls A and B because shear thinning reduces the viscosity in those regions.

(d) Make three more plots, corresponding to **Predefined quantities** entries of **dynamic viscosity, pressure**, and **vorticity**.

In Fig. E11.3.5, the shear thinning is seen to cause a drastic reduction in the viscosity as the walls are approached because of the higher velocity gradients and hence strain rates in those regions. The effect is enhanced as the inlet pressure is increased because it leads to generally higher velocities and velocity gradients.

Fig. E11.3.6 shows that the pressure is nearly constant across the section A–B. This observation lends strong support to a key conclusion in boundary-layer and other nearly unidirectional flows, namely, that the pressure is essentially constant in the direction transverse to the primary flow direction.

For axisymmetric cylindrical coordinates, the θ-component of the vorticity (see Table 5.4) is $(\partial v_r / \partial z - \partial v_z / \partial r)$. A careful examination of the velocities across AB from Fig. E11.3.5, bearing in mind the coordinate directions, will indeed verify the change of sign of the vorticity evidenced in Fig. E11.3.7.

Fig. E11.3.5 Non-Newtonian viscosity profiles from A to B for the six levels of inlet pressure.

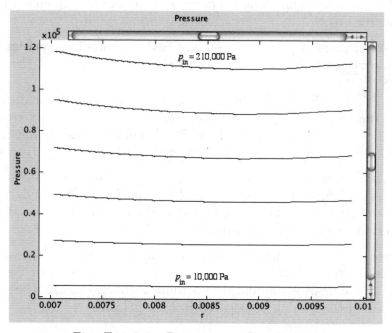

*Fig. E11.3.6   Pressure profiles from A
to B for the six levels of inlet pressure.*

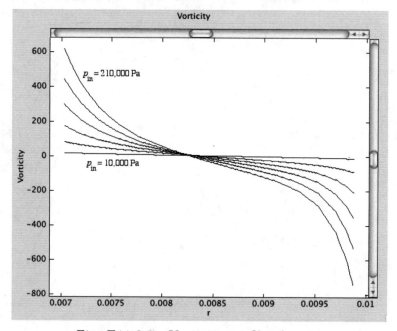

*Fig. E11.3.7   Vorticity profiles from A
to B for the six levels of inlet pressure.*

(e) In FEMLAB 2.3, running in conjunction with MATLAB, a strain-rate plot can also be made from A to B, as shown in Fig. E11.3.8. The increased strain rates as the walls are approached account for the shear-thinning behavior.

Fig. E11.3.8   Shear or strain rate profiles from A to B for the six levels of inlet pressure.

11.4   Constitutive Equations for Viscoelastic Fluids

The discussion now turns to visco*elastic* fluids, in which several additional phenomena are now present, some of which are illustrated in Fig. 11.6. The *primary normal stress difference*, $N_1 = \tau_{11} - \tau_{22}$, is a key quantity, where 1 denotes the coordinate in the primary direction of flow, and 2 refers to the direction in which the velocity changes. (Recall from Chapter 5 that in our sign convention, positive normal stresses are tensile, negative normal stresses are compressive.) For inelastic fluids, $N_1 = 0$; for elastic fluids, N_1 is usually positive and roughly proportional to $\dot{\gamma}^2$, the square of the strain rate.

Fig. 11.6 illustrates the following phenomena:

(a) Consider a viscoelastic fluid in simple shear between two plates, so that $N_1 = \tau_{xx} - \tau_{yy}$. But τ_{xx} is essentially zero, and since N_1 is positive, τ_{yy} must be negative, indicating a compressive stress—like pressure—that tends to push the two plates apart.

(b) A viscoelastic fluid such as "silly putty" leaving the confines of a tube increases its diameter by exhibiting the *die-swell* effect. A plausible explanation is that the tube flow tends to straighten out the polymer molecules in the direction of flow, parallel to the z axis. However, after these molecules are released from the confines of the tube, they attempt to return to their original more tangled configurations, thus "expanding" (the fluid is incompressible and its total volume is unchanged) in the radial direction—with, of course, a consequent reduction in the axial velocity.

(c) Frederickson reports an experiment by Phillippoff, in which "a glass sphere was dropped into a graduate filled with a solution of an aluminum soap in a hydrocarbon.[9] The ball penetrated the fluid for a few centimeters and then bounced up and down, with decreasing amplitude of oscillation, before settling slowly and steadily to the bottom of the graduate." This phenomenon—of elastic recoil—is shown in Fig. 11.6(c), in which the sphere has been given some artificial horizontal motion so that its up-and-down movement can readily be visualized.

(a) Normal-stress effect

(b) Die swell

(c) Elastic recoil

(d) Rod climbing

Fig. 11.6 Characteristics of an elastic fluid.

(d) If a viscoelastic fluid is stirred by a rotating rod, it shows the rod-climbing or *Weissenberg* effect.[10] The stirring tends to straighten out the polymer molecules in the direction of rotation; however, these molecules attempt to

---

[9] A.G. Frederickson, *Principles and Applications of Rheology*, Prentice-Hall, Englewood Cliffs, NJ, 1964, p. 121.

[10] Several examples are reported by L. Chan, *Experimental Observations and Numerical Simulation of the Weissenberg Climbing Effect*, Ph.D. dissertation, p. 34, University of Michigan, Ann Arbor, 1972. He studied Separan AP30 (polyacrylamide, Dow Chemical Co.) in water/glycerol solutions, and carboxymethyl cellulose and hydroxymethyl cellulose (both from Hercules, Inc.) in water.

return to their original configuration, thereby creating a tensile hoop stress. The fluid amounts to a series of stretched rubber bands, which can minimize their lengths by climbing up the rod. Another viewpoint is that $N_1 = \tau_{\theta\theta} - \tau_{rr}$ is positive; since τ_{rr} is essentially zero, $\tau_{\theta\theta}$ is positive and a tension exists in the direction of rotation.

In addition to these elastic effects, polymer melts usually exhibit shear thinning. Clearly, more elaborate constitutive equations than those discussed so far are needed in an attempt to describe these phenomena.

General linear viscoelastic fluids.   Viscoelastic fluids are often described by equations based on the following or similar forms:

$$\boldsymbol{\tau} = f(\boldsymbol{\gamma}, \dot{\boldsymbol{\gamma}}, t), \tag{11.22}$$

where $\boldsymbol{\tau}$, $\boldsymbol{\gamma}$, and $\dot{\boldsymbol{\gamma}}$ are the stress, strain, and strain-rate tensors, respectively. According to Eqn. (11.22), the shear stress in a viscoelastic fluid may be a function of the strain (characteristic of solids), the *rate* of strain (characteristic of liquids), and time t. The following discussion is restricted to materials that are subject to small deformations and small deformation rates (linear viscoelastic regime).

The Maxwell model.   Polymeric fluids are viscoelastic in nature; that is, they exhibit characteristics common to both viscous liquids *and* elastic solids. The *Maxwell model*, which serves mainly as an *introduction* to viscoelasticity, incorporates both of these characteristics, and is best introduced by considering again the case of simple shear, shown in Fig. 11.7, but this time for either a fluid *or* a solid. For simplicity, assume that $\tau = \tau_{yx}$ is constant.

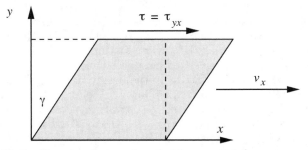

Fig. 11.7 Simple shear of a fluid or solid (γ exaggerated).

Assuming that the deformation γ is small, the two extremes of behavior are represented by a *Newtonian* fluid, of viscosity $\eta \ (= \mu)$, and a *Hookean* elastic solid, whose rigidity modulus is G. We then have:

Newtonian fluid

$$\tau = \eta \frac{dv_x}{dy} = \eta \frac{d\gamma}{dt} = \eta \dot{\gamma} \qquad \text{or} \qquad \dot{\gamma} = \frac{\tau}{\eta}. \tag{11.23}$$

Hookean elastic solid

$$\tau = \gamma G \qquad \text{or} \qquad \dot{\gamma} = \frac{d\gamma}{dt} = \frac{1}{G}\frac{d\tau}{dt}. \tag{11.24}$$

The Maxwell model then views the total strain rate as the sum of the strain rates for the Newtonian fluid and Hookean solid, resulting in the following constitutive equation:

$$\frac{\tau}{\eta} + \frac{1}{G}\frac{d\tau}{dt} = \dot{\gamma}. \tag{11.25}$$

Although η and G both increase in progressing from liquids to solids, η increases faster than G, so the overall effect is an increase in the *ratio* $\lambda = \eta/G$, which is known as the *relaxation time*, and is low for fluids and high for solids. The basic reason is that stresses in liquids can relax more quickly than those in solids because of the higher mobility of the liquid molecules. In terms of λ, the Maxwell model becomes:

$$\tau + \lambda\frac{d\tau}{dt} = \eta\frac{d\gamma}{dt} = \eta\dot{\gamma}. \tag{11.26}$$

Maxwell, James Clerk, born 1831 in Edinburgh, Scotland; died 1879 in Cambridge, buried in Parton village, Scotland. Maxwell obtained his degree in mathematics from the University of Cambridge in 1854 and subsequently held chairs of Natural Philosophy at Marischal College, Aberdeen, and in Physics and Astronomy at King's College, London. In 1871 he was the first professor of Experimental Physics at the University of Cambridge and played a pivotal role in establishing the Cavendish Laboratory there. Maxwell was a genius in both theory and experiment, and his research was devoted to a wide diversity of topics, including elasticity, capillary action, double refraction in viscous liquids subject to shear stresses, the stability of Saturn's rings, thermodynamics, and color photography. He was a pioneer in the kinetic theory of gases. However, starting even in his undergraduate days, Maxwell's *magnum opus* was his unified theory of electromagnetism, depending on field theory rather than the interaction at a distance of point sources of charge and magnetism. This work culminated in 1873 in the publication of his treatise on *Electricity and Magnetism*. In 1931, Albert Einstein claimed that Maxwell's work was "the most profound and the most fruitful that physics has experienced since the time of Newton."

Source: *The Encyclopædia Britannica*, 11th ed., Cambridge University Press (1910–1911), and also later editions.

The principle of the Maxwell model is shown schematically in Fig. 11.8. The Newtonian behavior of the fluid is represented by a "dashpot," in which the force is proportional to the rate of extension. The dashpot is in series with an elastic spring that represents the Hookean behavior of the fluid—in which the force is proportional to the extension. Note that the Maxwell model is restricted to deformations that are small, so that the response of the fluid is linear—a phenomenon known as *linear viscoelasticity*.

Fig. 11.8   The Maxwell model of linear viscoelasticity.

Fig. 11.9 shows a representative situation, in which the strain is first steadily increased, and then held constant. The Maxwell model predicts a stress that also first increases (but whose rate of increase starts to diminish), and that subsequently decays with time.

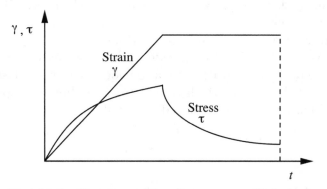

Fig. 11.9   Stress resulting from an applied strain.

The Maxwell model makes the following realistic predictions:

1. Steady shear causes flow.
2. If the strain no longer changes (as for the larger values of time in Fig. 11.9), $d\gamma/dt = 0$ and the stress *relaxes* from its maximum value τ_0 according to:

$$\tau = \tau_0 e^{-Gt/\eta} = \tau_0 e^{-t/\lambda}, \tag{11.27}$$

from which the relaxation time λ is clearly the time taken for the stress to fall to $1/e = 36.8\%$ of its initial value.
3. If the stress is removed, so that $\tau = 0$, there will be some recoil or *elasticity*.

Eqn. (11.26) is a first-order linear differential equation that can be solved to yield an expression for the shear-stress τ[11]:

$$\tau(t) = \int_{-\infty}^{t} \underbrace{\left[\frac{\eta}{\lambda} e^{-(t-t')/\lambda}\right]}_{\substack{\text{Relaxation modulus} \\ G(t-t')}} \dot{\gamma}(t')\, dt'. \qquad (11.28)$$

The stress in the fluid is a function not only of the strain rate at the current time t but *also* of the strain rate at previous times, designated by t'. The term in brackets is the *relaxation modulus* G, which weights the effect of the strain rate at previous times less than t. Note that the relaxation modulus equals unity when $t = t'$ and is equal to zero when $t = \infty$. Consequently, the fluid has a "memory" of past deformations, which decays exponentially with time.

The shear-stress can also be written in terms of the difference $[\gamma(t) - \gamma(t')]$ between the strains at times t and t':

$$\tau(t) = \int_{-\infty}^{t} \underbrace{\left[\frac{\eta}{\lambda^2} e^{-(t-t')/\lambda}\right]}_{\substack{\text{Memory function} \\ M(t-t')}} [\gamma(t) - \gamma(t')]\, dt'. \qquad (11.29)$$

The quantity in brackets is the *memory function*, which describes the effect of strain prehistory on the stress.

Unfortunately, with a single relaxation time λ and a single viscosity η, the two-parameter Maxwell model does not describe linear viscoelastic behavior particularly well. Another constitutive equation for linear viscoelastic fluids is the *generalized Maxwell model*, in which a moderate number n of springs and dashpots of varying characteristics are placed in parallel:

$$\tau = \sum_{k=1}^{n} \tau_k(t), \qquad (11.30)$$

with an individual Maxwell-type equation for every value of k:

$$\tau_k + \lambda_k \frac{\partial \tau_k}{\partial t} = \eta_k \dot{\gamma}. \qquad (11.31)$$

A *generalized linear viscoelastic* model is one in which the relaxation and memory moduli are not constrained to be of the exponential form appearing in Eqn. (11.28), but are properties of the fluid:

$$\tau(t) = \int_{-\infty}^{t} G(t - t')\dot{\gamma}(t')\, dt', \qquad (11.32)$$

---

[11] Equation (11.28) may be verified by forming $\tau + \lambda \partial \tau / \partial t$, then using Leibnitz's rule (see Appendix A), and finally checking that the result equals $\eta \dot{\gamma}$.

$$\tau(t) = \int_{-\infty}^{t} M(t - t')[\gamma(t) - \gamma(t')] \, dt'. \tag{11.33}$$

The White-Metzner model.   For *any* number of space dimensions, the Maxwell model can supposedly be generalized to:

$$\tau + \lambda \frac{\partial \tau}{\partial t} = \eta \dot{\gamma}. \tag{11.34}$$

but Middleman and others point out some basic drawbacks and inconsistencies of it—notably the fact that $\partial \tau / \partial t$ is *not* a tensor and hence is incompatible with the other two terms of the equation.[12] Also, even the simple model of Eqn. (11.26) is not necessarily true if the fluid is flowing.

These last two objections may be overcome by using a more sophisticated time derivative and rewriting the Maxwell model as:

$$\tau + \lambda \frac{\delta \tau}{\delta t} = \eta \dot{\gamma}, \tag{11.35}$$

in which the contravariant form of the *Oldroyd* derivative gives in a fixed coordinate system the rate of change with time in a coordinate system that moves and deforms with the fluid[13]:

$$\frac{\delta \tau}{\delta t} = \frac{\mathcal{D} \tau}{\mathcal{D} t} - \tau \cdot (\nabla \mathbf{v}) - (\nabla \mathbf{v})^T \cdot \tau. \tag{11.36}$$

The first term on the right-hand side will be remembered as the *convective* derivative, which accounts for the time variation of a scalar (such as ρ or v_x) following the path taken by the fluid. However, when dealing with a tensor, two additional terms are required, in which the dyadic product $\nabla \mathbf{v}$ and its transpose are defined by:

$$\nabla \mathbf{v} = \begin{pmatrix} \dfrac{\partial v_x}{\partial x} & \dfrac{\partial v_y}{\partial x} & \dfrac{\partial v_z}{\partial x} \\[2mm] \dfrac{\partial v_x}{\partial y} & \dfrac{\partial v_y}{\partial y} & \dfrac{\partial v_z}{\partial y} \\[2mm] \dfrac{\partial v_x}{\partial z} & \dfrac{\partial v_y}{\partial z} & \dfrac{\partial v_z}{\partial z} \end{pmatrix}, \qquad (\nabla \mathbf{v})^T = \begin{pmatrix} \dfrac{\partial v_x}{\partial x} & \dfrac{\partial v_x}{\partial y} & \dfrac{\partial v_x}{\partial z} \\[2mm] \dfrac{\partial v_y}{\partial x} & \dfrac{\partial v_y}{\partial y} & \dfrac{\partial v_y}{\partial z} \\[2mm] \dfrac{\partial v_z}{\partial x} & \dfrac{\partial v_z}{\partial y} & \dfrac{\partial v_z}{\partial z} \end{pmatrix}. \tag{11.37}$$

Unfortunately, the development rapidly becomes quite complicated. The dot product of two tensors $\mathbf{a}$ and $\mathbf{b}$ is also a tensor, such that the i, j component of $\mathbf{a} \cdot \mathbf{b}$ is $\sum_k a_{ik} b_{kj}$. Clearly, $\tau \cdot (\nabla \mathbf{v})$ and $(\nabla \mathbf{v})^T \cdot \tau$ give lengthy terms when written out fully. The reader may wish to check the following:

---

[12] S. Middleman, *Fundamentals of Polymer Processing*, McGraw-Hill, New York, 1977, pp. 54–55.

[13] J.G. Oldroyd, *Proceedings of the Royal Society (London)*, Vol. 200A, pp. 523–541 (1950).

1. The Oldroyd derivative of a component τ_{ij} of the stress tensor is:

$$\frac{\delta \tau_{ij}}{\delta t} = \frac{\partial \tau_{ij}}{\partial t} + \sum_k v_k \frac{\partial \tau_{ij}}{\partial x_k} - \sum_k \tau_{ik} \frac{\partial v_j}{\partial x_k} - \sum_k \tau_{kj} \frac{\partial v_i}{\partial x_k}, \tag{11.38}$$

in which the summation for k is over the three coordinate directions x, y, and z, such that, for example:

$$\sum_k v_k \frac{\partial \tau_{ij}}{\partial x_k} = v_x \frac{\partial \tau_{ij}}{\partial x} + v_y \frac{\partial \tau_{ij}}{\partial y} + v_z \frac{\partial \tau_{ij}}{\partial z}. \tag{11.39}$$

2. Even for the simplified case of two-dimensional flow, the relation for just the component τ_{xx} derived from Eqn. (11.38) is:

$$\tau_{xx} + \lambda \left(\frac{\partial \tau_{xx}}{\partial t} + v_x \frac{\partial \tau_{xx}}{\partial x} + v_y \frac{\partial \tau_{xx}}{\partial y} - 2\tau_{xx} \frac{\partial v_x}{\partial x} - 2\tau_{xy} \frac{\partial v_x}{\partial y} \right) = 2\eta \frac{\partial v_x}{\partial x}. \tag{11.40}$$

The following variation is known as the White-Metzner equation, in which λ and η are functions of the second invariant of the rate-of-strain tensor:

$$\boldsymbol{\tau} + \lambda(I_2) \frac{\delta \boldsymbol{\tau}}{\delta t} = \eta(I_2) \dot{\boldsymbol{\gamma}}, \qquad \text{with} \qquad \lambda = \frac{\eta}{G}, \tag{11.41}$$

in which G is the rigidity modulus as previously defined. Middleman points out that this model gives both realistic shear-thinning and primary normal-stress behavior.[14] Numerical solutions are clearly needed in all but the simplest cases.

11.5 Response to Oscillatory Shear

Linear viscoelastic fluids are usually characterized by their response to small-amplitude oscillatory shear, which can be achieved by rotational rheometers, as outlined in Section 11.6. Consider an imposed strain of the following form:

$$\gamma(t) = \gamma_0 \cos \omega t, \tag{11.42}$$

where γ_0 is the strain amplitude and ω is the frequency. The strain amplitude is small enough so that the fluid responds linearly to the applied strain. The strain *rate* is given by:

$$\dot{\gamma} = \frac{d\gamma}{dt} = -\gamma_0 \omega \sin \omega t. \tag{11.43}$$

---

[14] See S. Middleman, *op. cit.*, p. 60.

The corresponding stress, which can be checked by substitution into the Maxwell model of Eqn. (11.26), is:

$$\tau = \underbrace{\frac{\eta \omega^2 \lambda}{1 + \omega^2 \lambda^2} \gamma_0 \cos \omega t}_{\text{In phase}} - \underbrace{\frac{\eta \omega}{1 + \omega^2 \lambda^2} \gamma_0 \sin \omega t,}_{\text{Out of phase}} \qquad (11.44)$$

in which first term is in-phase, and the second term out-of-phase, with the applied strain. The stress can also be written as:

$$\tau = G' \gamma_0 \cos \omega t - G'' \gamma_0 \sin \omega t, \qquad (11.45)$$

in which G' is the *elastic* or *storage* modulus, and G'' is the *viscous* or *loss* modulus. A comparison of Eqns. (11.44) and (11.45) shows that:

$$G' = \frac{\eta \omega^2 \lambda}{1 + \omega^2 \lambda^2}, \qquad G'' = \frac{\eta \omega}{1 + \omega^2 \lambda^2}. \qquad (11.46)$$

An analysis of the behavior of τ will give G' and G'' and hence the material properties λ and η.

By using the identity $\cos(\omega t + \delta) = \cos \omega t \cos \delta - \sin \omega t \sin \delta$, the stress of Eqn. (11.45) can be rephrased as:

$$\tau = \gamma_0 \sqrt{(G')^2 + (G'')^2} \, \cos(\omega t + \delta), \qquad (11.47)$$

in which the *loss angle* δ describes the phase shift between the applied strain and the stress response and is given by:

$$\tan \delta = \frac{G''}{G'}. \qquad (11.48)$$

Note that the ratio of G'' to G' specifies the relative viscous-to-elastic nature of the fluid, and equals the tangent of the loss angle δ. For Maxwell fluids:

$$\tan \delta = \frac{1}{\omega \lambda}. \qquad (11.49)$$

A purely viscous fluid (zero relaxation time) is characterized by $\tan \delta = \infty$ ($\delta = \pi/2 = 90°$), while a purely elastic fluid (infinite relaxation time) is characterized by $\tan \delta = 0$ ($\delta = 0°$). If a fluid is equally viscous and elastic, $\tan \delta$ equals unity ($\delta = \pi/4 = 45°$). The wave forms of an applied oscillatory strain and the resulting stress response, together with loss angle δ, are shown in Fig. 11.10 for an elastic fluid, a viscoelastic fluid, and a viscous fluid. The dimensionless time is $\omega t/2\pi$.

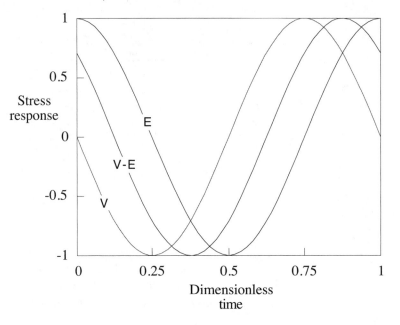

Fig. 11.10   Waves (scaled to maximum amplitude) for an applied oscillatory strain and the stress response. E: applied strain and stress response for an elastic fluid. V–E: stress response for a viscoelastic fluid with a loss angle of $\delta = 45°$. V: stress response for a viscous fluid; the stress leads the strain by $\delta = 90°$.

As shown in Eqns. (11.46) and (11.48), G', G'', and δ are also functions of the frequency ω. At high frequencies, in which the characteristic time of the deformation is much shorter than the relaxation time of the fluid, $\tan\delta$ approaches zero and the response of the fluid is that of an elastic solid. In this case, the stress is in phase with the applied strain. At low frequencies, in which the characteristic time of the deformation is much greater than the relaxation time of the fluid, $\tan\delta$ approaches infinity and the response of the fluid is that of a viscous liquid. Then, the stress is out of phase with the applied strain, leading the strain by 90°. The next section discusses dynamic methods of measuring G' and G''.

Physical explanation of phase relation between stress and strain. Fig. 11.11 illustrates how stress and strain can be out-of-phase for a viscous fluid. Consider a fluid layer, such as might occur in a viscometer, the upper surface being subjected to an oscillating strain, and the resulting stress being measured at the fixed lower surface. Assume that initial transients have died out, so that the indicated pattern is repeated from one cycle to the next.

At stage (a), the strain is one unit to the right (+1), and subsequently cycles through the values 0, −1, 0, and back to +1 in stages (b)–(e). The stress at the lower surface is measured by the slope of the displacement. Since shearing motions are not transmitted instantaneously through viscous fluids, it takes time for the strain at the top surface to manifest itself at the lower surface. Note that the stress at the lower surface becomes one unit to the right only at stage (d), which may be viewed as *lagging* the strain by three-quarters of a cycle or *leading* it by one-quarter of a cycle.

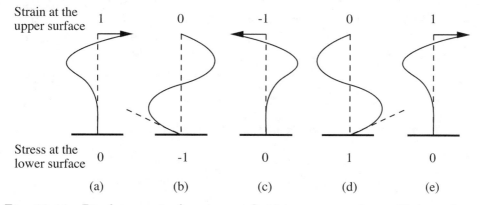

Fig. 11.11 *Displacement of a viscous fluid in response to oscillatory shear.*

11.6 Characterization of the Rheological Properties of Fluids

It is critical to characterize the rheological properties of a fluid accurately. For example, a shear-thickening fluid presents special problems with respect to pumping requirements. Since the fluid becomes more viscous as the strain rate increases, a shear-thickening fluid may "gel" in piping and pumping equipment. Shear-thinning fluids, on the other hand, become less resistant to flow at higher strain rates. It is then necessary to quantify the decrease in the viscosity—as the driving force and thus the strain rate increases—to control the fluid flow rate precisely.

Viscometers are devices used for measuring the viscosities of fluids. In these instruments, a precisely known fluid velocity profile is established, and either the resistance to flow or the fluid flow rate is measured. Shear-thinning and shear-thickening fluids can be differentiated from Newtonian fluids by studying the resistance of the fluid to flow over a range of strain rates.

Viscometers belong to a larger class of devices called *rheometers.* In addition to measuring the viscosity of the fluid, rheometers also characterize the viscoelastic properties such as the elastic and viscous moduli. Common viscometers and rheometers are described below.

Capillary viscometer.   A capillary viscometer is based on the principle developed in Example 11.1: For a given pressure drop, the volumetric flow rate of a fluid through a circular tube is a function of the fluid viscosity. In practice, the time is measured for a given volume of fluid to travel a known distance through the capillary, typically under the action of gravity. Since the flow rate depends on the capillary radius, which is difficult to measure, the capillary viscometer is calibrated with Newtonian oils of known viscosities. The kinematic viscosity of a fluid with unknown viscosity can then be calculated by multiplying the time required for flow by the calibration constant. Commercial capillary viscometers include the Cannon-Fenske and the Ubbelohde instruments.

For a Newtonian fluid, the strain rate and the shear stress vary linearly from zero at the center of the capillary ($r = 0$) to a maximum at the wall ($r = a$). The strain rate $\dot{\gamma}_w$ at the wall for a Newtonian fluid is given by:

$$\dot{\gamma}_w = \frac{4Q}{\pi a^3}. \tag{11.50}$$

A similar expression can be written for a non-Newtonian fluid

$$\dot{\gamma}_w = \frac{1}{\pi a^3 \tau_w^2} \frac{d(\tau_w^3 Q)}{d\tau_w} = \frac{4Q}{\pi a^3} \left[\frac{3}{4} + \frac{1}{4} \frac{d(\ln Q)}{d(\ln \tau_w)} \right], \tag{11.51}$$

where τ_w is the shear stress at the wall. The term in brackets is the *Rabinowitsch correction*. For power-law fluids, $d(\ln Q)/d(\ln \tau_w) = 1/n$. The contribution of this term to the viscosity can thus be significant for $n \ll 1$.

Example 11.4—Proof of the Rabinowitsch Equation

Prove the Rabinowitsch equation, (11.51), which enables the stress/strain relation to be established from data on steady fluid flow in a pipe, as shown in Fig. E11.4.1, and is therefore useful for developing a model for the behavior of the fluid.

Fig. E11.4.1   Flow of fluid through a pipe.

Solution

The flow rate is given by the integral of the velocity profile:

$$Q = \int_0^a 2\pi r v_z \, dr = \pi \int_0^a v_z \frac{d}{dr}(r^2) \, dr. \tag{E11.4.1}$$

Example 11.4—Proof of the Rabinowitsch Equation   **625**

Integration by parts yields:

$$Q = \left[\pi v_z r^2\right]_0^a - \pi \int_0^a r^2 \frac{dv_z}{dr}\, dr = -\pi \int_0^a r^2 \frac{dv_z}{dr}\, dr. \qquad (E11.4.2)$$

Now change the variable of integration from r to τ_{rz} (which will be denoted as τ for simplicity). The shear stress is known to vary linearly with the radial coordinate, so that:

$$\frac{\tau}{\tau_w} = \frac{r}{a}, \qquad \text{or} \qquad dr = \frac{a}{\tau_w}\, d\tau. \qquad (E11.4.3)$$

In terms of this new variable, the flow rate becomes, from Eqns. (E11.4.2) and (E11.4.3):

$$Q = -\pi \int_0^{\tau_w} \frac{a^2 \tau^2}{\tau_w^2} \frac{dv_z}{dr} \frac{a}{\tau_w}\, d\tau = -\pi a^3 \int_0^{\tau_w} \frac{\tau^2}{\tau_w^3} \frac{dv_z}{dr}\, d\tau. \qquad (E11.4.4)$$

Leibnitz's rule (given in Appendix A) is next used for differentiating the flow rate with respect to the wall shear stress:

$$\frac{dQ}{d\tau_w} = -\frac{\pi a^3}{\tau_w}\left(\frac{dv_z}{dr}\right)_{r=a} - \pi a^3 \int_0^{\tau_w} -\frac{3\tau^2}{\tau_w^4} \frac{dv_z}{dr}\, d\tau$$

$$= -\frac{\pi a^3}{\tau_w}\left(\frac{dv_z}{dr}\right)_{r=a} - \frac{3Q}{\tau_w}, \qquad (E11.4.5)$$

in which the last term is developed with the aid of Eqn. (E11.4.4).

The strain rate at the wall is therefore:

$$\dot{\gamma}_w = \left(-\frac{dv_z}{dr}\right)_{r=a} = \frac{1}{\pi a^3}\left(\tau_w \frac{dQ}{d\tau_w} + 3Q\right) = \frac{1}{\pi a^3 \tau_w^2} \frac{d(\tau_w^3 Q)}{d\tau_w}, \qquad (E11.4.6)$$

which is known as the *Rabinowitsch equation*.

Since $\tau_w = a(-dp/dz)/2$ (in which dp/dz is negative) is readily obtainable from the pressure drop, and Q is easily measured, Eqn. (E11.4.6) permits $\dot{\gamma}_w$ to be obtained as a function of τ_w, thus enabling the fluid to be characterized. $\square$

Rotational rheometers. Numerous rheometers, suitable for characterizing both viscous and elastic fluids, are *rotational-type* instruments; examples, which are discussed below, include the cone-and-plate, parallel-plate, and concentric-cylinder rheometers. One component of the rheometer is rotated at a constant angular velocity or is oscillated at a known frequency, and the other component is held stationary. Either the torque required to rotate the component or the torque required to hold the other component stationary is measured. (Note that these torques are equal in magnitude but opposite in direction.) Alternatively, a constant torque can be applied to one component and the resulting rotation rate or frequency measured. In all of these instruments, it is essential to establish a well-defined velocity profile in the fluid. As a result, most rotational rheometers are designed so that the fluid experiences a constant strain rate.

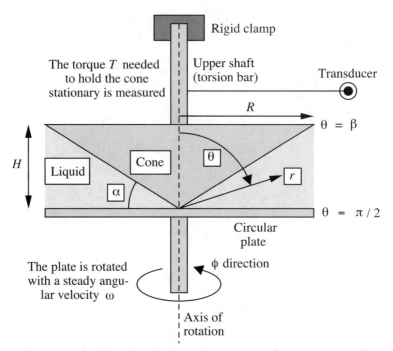

Fig. 11.12 *Cone-and-plate viscometer (α exaggerated).*

1. Cone-and-plate rheometer. A cone-and-plate rheometer consists of a cone of prescribed angle α, positioned above a flat disk, as shown in Fig. 11.12.

A sample of fluid is placed in the gap between the cone and the plate. When the plate is rotated at a constant angular velocity ω, for example, the fluid velocity in the gap is approximated closely by the following for small cone angles:

$$v_\phi = \omega r \left[1 - \frac{z}{h(r)}\right], \qquad (11.52)$$

where $h(r)$ is the gap distance between the plate and the cone and z is measured from the plate. The strain rate is *constant* across the gap, independent of r and θ:

$$\dot{\gamma} = \frac{\omega}{\alpha}, \qquad (11.53)$$

where $\alpha \doteq H/R$ is the small cone angle in radians. As shown in Problem 6.15, the viscosity can be calculated from the torque T required to rotate the plate at the angular velocity ω:

$$\eta = \frac{3T\alpha}{2\pi\omega R^3}. \qquad (11.54)$$

In practice, the apex of the cone is slightly truncated so that the cone does not touch the plate during measurements. This feature also allows the cone-and-plate rheometer to be used to measure the rheological properties of suspensions. The fluid can then be considered to be a continuum everywhere throughout the gap.

A considerably more detailed and somewhat more exact treatment of a cone-and-plate rheometer—the Weissenberg rheogoniometer—is given in Example 6.7.

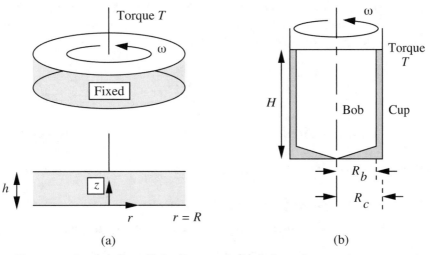

Fig. 11.13   (a) Parallel-plate and (b) bob-and-cup viscometers.

2. *Parallel-plate rheometer.* The parallel-plate rheometer, illustrated in Fig. 11.13(a), resembles the cone-and-plate rheometer, except that the cone is replaced by a flat circular plate of radius R, held at a known distance $h \ll R$ above the lower plate. A sample of fluid is placed in the gap between the two plates, and one plate either rotates at a known speed or oscillates at a known frequency. For steady rotation of the upper plate, for example, the velocity profile in the gap is given by:

$$v_\theta = \frac{rz}{h}\omega, \qquad (11.55)$$

with a corresponding strain rate:

$$\dot{\gamma} = \frac{r}{h}\omega. \qquad (11.56)$$

The viscosity is then:

$$\eta = \frac{3Th}{2\pi R^4 \omega \left[1 + \frac{1}{3}\frac{d(\ln T)}{d(\ln \omega)}\right]}, \qquad (11.57)$$

in which a Rabinowitsch-type approach has been used, and where T is the measured torque or couple. If the constitutive equation for the fluid is known, the torque can be related to the strain rate to simplify the derivative in Eqn. (11.57).

3. Coaxial-cylinder rheometer. Fig. 11.13(b) shows a coaxial-cylinder rheometer, which consists of two concentric cylinders, the outer of which is called the *cup* and the inner of which is called the *bob*. Either the cup or the bob is rotated at a known speed or frequency, and the torque required to sustain the rotation or the torque required to hold the other component stationary is measured. If the gap between the two cylinders is small with respect to the radii of both cylinders (the bob radius is at least 97% of the cup radius), the strain rate in the fluid is very nearly constant across the gap:

$$\dot{\gamma} = \frac{\omega}{1 - \dfrac{R_b}{R_c}},\tag{11.58}$$

where R_b and R_c are the radii of the bob and cup, respectively. Generally, the length of the bob that is immersed in the fluid is much greater than the gap distance, so that end effects are negligible.

Example 11.5—Working Equation for a
Coaxial-Cylinder Rheometer: Newtonian Fluid

Consider a coaxial cylinder rheometer in which the bob of radius R_b and height H rotates at an angular velocity ω in the θ direction, and the torque T required to hold the cup of radius R_c stationary is measured. Derive a relationship between the torque and the viscosity η if the fluid contained in the rheometer is Newtonian.

Solution

In the absence of secondary flows, the fluid flow in the gap between the cylinders will be in the θ direction only. The Navier-Stokes equation in the θ direction reduces to [see Eqn. (5.75)]:

$$\eta \frac{\partial}{\partial r}\left[\frac{1}{r}\frac{\partial(rv_\theta)}{\partial r}\right] = 0,\tag{E11.5.1}$$

subject to the boundary conditions:

$$\text{At } r = R_c: \qquad v_\theta = 0.\tag{E11.5.2}$$

$$\text{At } r = R_b: \qquad v_\theta = R_b\omega.\tag{E11.5.3}$$

(Note that if the fluid is non-Newtonian, the appropriate governing equation is the equation of motion in the θ direction:

$$\frac{1}{r^2}\frac{\partial(r^2\tau_{r\theta})}{\partial r} = 0,\tag{E11.5.4}$$

where the shear stress $\tau_{r\theta}$ is related to the velocity using a constitutive equation.)

Example 11.5—Working Equation for a Coaxial-Cylinder Rheometer   **629**

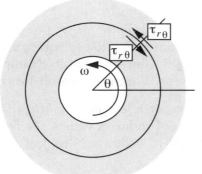

Fig. E11.5.1   Sign convention for shear stresses.

Integration of Eqn. (E11.5.1) and application of boundary conditions in the usual way gives the velocity profile:

$$v_\theta = \frac{\omega R_b^2}{R_c^2 - R_b^2} \left(\frac{R_c^2 - r^2}{r} \right). \tag{E11.5.5}$$

Referring to Fig. E11.5.1, the shear stress is:

$$\tau_{r\theta} = \eta r \frac{d}{dr} \left(\frac{v_\theta}{r} \right), \quad \text{leading to} \quad (\tau_{r\theta})_{r=R_c} = -\frac{2\eta\omega R_b^2}{R_c^2 - R_b^2}. \tag{E11.5.6}$$

The torque T (measured as positive in the direction of increasing θ), required to keep the cup stationary, equals the product of the force and the lever arm:

$$T = [2\pi R_c H(-\tau_{r\theta})|_{r=R_c}]R_c, \tag{E11.5.7}$$

which becomes (still for a Newtonian fluid):

$$T = -4\pi\eta H\omega \frac{R_b^2 R_c^2}{R_c^2 - R_b^2}. \tag{E11.5.8}$$

Note that the first minus sign arises because the torque required to hold the cup stationary is in the negative θ direction. From Eqn. (E11.5.8), this torque is directly proportional to the viscosity, which can then be calculated.   □

PROBLEMS FOR CHAPTER 11

1. *Bingham-plastic velocity profile—M.* A pressure gradient of $-dp/dz$ in a horizontal tube of radius a causes the flow of a Bingham plastic of yield stress τ_0 and "viscosity" η. Following the development in Example 11.2, derive the velocity profile. Assume that $|\tau_w| > \tau_0$. What is the result for $|\tau_w| \leq \tau_0$?

2. *Volumetric flow rate of a Bingham plastic—M.* A pressure gradient of $-dp/dz$ in a horizontal tube of radius a causes the flow of a Bingham plastic of yield stress τ_0 and "viscosity" η. Following Problem 11.1, prove for $|\tau_w| > \tau_0$ that the volumetric flow rate is:

$$Q = \frac{\pi a^4}{2\eta}\left(-\frac{dp}{dz}\right)\left(\frac{1}{4} - \frac{\beta}{3} + \frac{\beta^4}{12}\right), \quad \text{where} \quad \beta = \frac{2\tau_0}{a(-dp/dz)}.$$

3. *Power-law film flow—M.* Consider the film flow of a power-law fluid with parameters κ and n and density ρ down a plate inclined at angle θ with respect to the horizontal. If the film thickness is H:

(a) Derive an expression for the resulting velocity profile v_x as a function of y (distance from the plate), H, κ, n, g, ρ, and θ.

(b) Sketch three representative velocity profiles (each having, for simplicity, the same maximum velocity, which in practice would mean different values of κ), for $n < 1$, $n = 1$, and $n > 1$, and comment briefly on the important features.

(c) Derive an expression for the volumetric flow rate of the liquid per unit width of the plate. Check your answer against the known expression for a Newtonian fluid:

$$Q = \frac{H^3 \rho g \sin\theta}{3\eta}.$$

4. *Bingham-plastic film flow—M.* Repeat Problem 11.3 for a Bingham-plastic fluid.

5. *Rotating rod—M.* What is the strain-rate tensor in cylindrical coordinates? *Hint:* First examine the stress components given in Table 5.8 for a Newtonian fluid. Also write down the corresponding viscous-stress tensor.

A circular cylindrical rod of length L and radius R rotates steadily about its axis at an angular velocity ω in an otherwise stagnant power-law fluid that obeys the model $\eta = \kappa \dot\gamma^{n-1}$, in which $n < 2$. The length of the rod is much larger than the radius, so that end effects are negligible.

(a) Prove that the strain rate is given by:

$$\dot\gamma = -r\frac{d}{dr}\left(\frac{v_\theta}{r}\right) = \left| r\frac{d}{dr}\left(\frac{v_\theta}{r}\right)\right|.$$

(b) Check that the θ momentum balance simplifies to:

$$\frac{d}{dr}(r^2 \tau_{r\theta}) = 0,$$

and derive the expression for the velocity profile in the fluid, $v_\theta = v_\theta(r)$.

(c) Sketch the velocity profile for two values of n, taking care to show which profile belongs to the larger value of n.

(d) As n goes to infinity prove that the fluid is effectively in solid-body rotation.

(e) Derive an expression for the torque required to rotate the rod.

6. *Shear-thinning wire coating*—*M.* In a certain constant-pressure process, a wire is to be coated with PVC (polyvinylchloride) supplied at a flow rate Q, as illustrated in Fig. P11.6. The radius of the wire is r_1 and the radius of the die is r_2. It is critical to estimate the coating thickness δ from a knowledge of the velocity U at which the wire is drawn through the die.

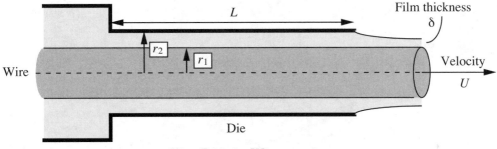

Fig. P11.6   Wire-coating.

(a) Derive an equation relating the coating thickness δ to Q, U, and r_1.

(b) PVC is a shear-thinning power-law fluid with an index n. If $q = 1 - (1/n)$, show that the velocity profile in the die is given by:

$$v_z = U \left(\frac{r^q - r_2^q}{r_1^q - r_2^q} \right).$$

(c) Hence, derive an expression for Q in terms of U, r_1, r_2, and q.

7. *Rabinowitsch correction for power-law fluids*—*E.* Concerning Eqn. (11.51), prove that $d(\ln Q)/d(\ln \tau_w)$ is equal to $1/n$ for power-law fluids. Any formulas in the text may be used directly without further proof.

8. *Fluid characterization*—*M.* You have been given a fluid of unknown rheological type. Design one or more experiments to characterize the fluid.

9. *Viscoelastic-fluid response*—*E.* For an applied strain of $\gamma = \gamma_0 \cos(\omega t)$, sketch the response of a viscoelastic fluid characterized by $\delta = 60°$.

10. *Bingham plastic in a vertical tube—M.* A long vertical pipe of radius a and length L, open to the atmosphere at both ends, is filled with a Bingham plastic of density ρ and rheological parameters τ_0 and η_0. If the z axis points vertically downwards, prove that the shear-stress distribution is given by:

$$\tau_{rz} = -\frac{r\rho g}{2}.$$

Derive an expression for the minimum pipe radius, a_{min}, that will just cause flow of the Bingham plastic under gravity. Your answer should be in terms of any or all of the parameters: ρ, τ_0, η_0, L, and g. You may assume that:

$$\frac{1}{2}\boldsymbol{\tau} : \boldsymbol{\tau} = \tau_{rz}^2, \qquad \dot\gamma = \left|\frac{\partial v_z}{\partial r}\right|.$$

If a pipe of radius $2a_{min}$ is now filled with the same material, and both ends are allowed to communicate with the atmosphere, derive the velocity profile v_z as a function of r and any or all of the parameters previously listed.

11. *Bingham-plastic paint film—E.* Many paints behave like Bingham-plastic fluids. As shown in Fig. P11.10, consider a paint film of thickness δ spread uniformly on a flat wall inclined at 30° to the horizontal.

Fig. P11.10   *Paint film on inclined wall.*

What is the maximum film thickness (cm) for which the paint film will *not* flow under the action of gravity?

Physical properties of the paint:	*Useful information:*
Yield stress (τ_0) = 70 dynes/cm^2.	$g = 981$ cm/s^2.
Viscosity (η) = 10 cP (constant).	1 dyne = 1 g cm/s^2.
Density (ρ) = 1.4 g/cm^3.	1 P = 1 g/(cm s).

12. *Oscillatory shear of a Maxwell fluid—E.* For small amplitude oscillatory shear of a Maxwell fluid with an imposed strain $\gamma(t) = \gamma_0 \cos \omega t$, the corresponding stress is postulated:

$$\tau = \frac{\eta\omega^2\lambda}{1 + \omega^2\lambda^2}\gamma_0 \cos\omega t \; - \; \frac{\eta}{1 + \omega^2\lambda^2}\gamma_0\omega \sin\omega t.$$

Check that this expression for the shear stress satisfies the Maxwell model.

13. *Parameters for a Bingham plastic—D (C).* A polymer solution of density ρ behaves as a Bingham plastic and for simple shear as in Fig. 11.7 has the stress/rate-of-strain relationships:

$$\tau_{yx} \leq \tau_0 : \quad s = 0,$$

$$\tau_{yx} > \tau_0 : \quad \eta s = (\tau_{yx} - \tau_0),$$

where τ_{yx} is the shear stress, $s = dv_x/dy$ is the rate of strain, and both τ_0 and η are constants.

This solution flows down a flat plate of breadth b inclined at an angle θ to the horizontal, and forms a film of thickness h measured normal to the plate. Show that the volumetric flow rate of liquid is given by:

$$Q = \frac{1}{6\eta}\,(1 - \beta)^2(2 + \beta)\rho g b h^3 \sin\theta,$$

where $\beta = \tau_0/(\rho g h \sin\theta)$. State all necessary assumptions.

Suppose 500 g of the polymer solution are placed uniformly on a plate of length 50 cm and breadth 15 cm. When the plate is tilted to an angle of 34° to the horizontal, the solution just begins to flow. If—at this same inclination—the solution is now supplied at a rate of 2.0 g/s uniformly to the top end of the plate, the mass of solution on the plate increases to 750 g. Given that $\rho = 1.29$ g/cm³, deduce values of the parameters τ_0 and η for the polymer solution.

14. *Dynamic testing—D (C).* The dynamic behavior of a certain polymer has been investigated by means of an apparatus that applies a sinusoidally varying shear stress to one face of a block of the polymer. In the context of these experiments, indicate what is meant by the storage modulus and the loss modulus.

T (K)	298	298	298	293	293	293	288	288	288	283
Storage modulus (MN/m²):	0.759	1.26	1.66	2.51	3.72	6.02	15.8	38.0	79.4	126.0
Frequency of testing (Hz):	70.8	398	1260	79.5	317	1260	31.6	159	1590	39.8

Experimental data (above) were obtained for the storage modulus at a number of temperatures and at frequencies limited by the apparatus. Prepare a master curve for the logarithm of the storage modulus at 298 K. The shift function—the ratio of the relaxation time at temperature T to its value at 298 K—has the empirical equation:

$$\log_{10} a_T = -\frac{17.4(T - 298)}{51.6 + (T - 298)}.$$

15. *Determination of constitutive equation—M.* The flow rate Q (m³/s) of a non-Newtonian fluid was measured under different applied pressure gradients $-dp/dz$ (Pa/m) in a horizontal tube of radius $a = 0.01$ m, resulting in the following data:

$-dp/dz$	Q	$-dp/dz$	Q
10,000	5.37×10^{-5}	60,000	3.36×10^{-3}
20,000	2.64×10^{-4}	70,000	4.87×10^{-3}
30,000	6.89×10^{-4}	80,000	7.13×10^{-3}
40,000	1.29×10^{-3}	90,000	9.12×10^{-3}
50,000	2.35×10^{-3}	100,000	1.11×10^{-2}

Find an appropriate constitutive equation for the fluid, which is thought to conform to either a Bingham plastic (Problem 11.2) or a shear-thinning power-law fluid (Example 11.1). *Hint:* use a spreadsheet to find the pair of parameters (τ_0, η) or (n, κ) that minimizes the sum of squares of differences between observed and predicted values for Q.

16. *Non-Newtonian fluid characterization—M.* A non-Newtonian liquid is tested by placing it between the two concentric cylinders of a viscometer. Since the gap h between the two surfaces is very small, they may be approximated by two planes as shown in Fig. P11.16, one surface being stationary and the other moving. The instrument essentially measures the shear stresses $\tau = (\tau_{yx})_{y=h}$ needed to move the upper plate at a variety of steady velocities V.

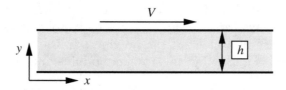

Fig. P11.16   Opposed surfaces of a viscometer.

Explain in detail how you would discover the model to which the liquid conforms—it may be either a power-law fluid or a Bingham plastic—and how you could determine from the data the two parameters (such as κ and n, or τ_0 and η) for either model. Use the symbol s for the rate of strain dv_x/dy.

17. *Tank draining with a power-law fluid—M.* A very viscous non-Newtonian liquid in simple shear obeys the relation:

$$\tau = c\sqrt{s},$$

where τ is the shear stress, s is the strain rate (of a typical form $\partial v_x/\partial y$), and c is a constant.

A uniform thin film of the liquid, of thickness h, flows steadily under gravity down a vertical plate. Show that the mass flow rate m per unit plate width is:

$$m = \frac{\rho^3 g^2 h^4}{4c^2}.$$

For a vertical plate from which the liquid is now *draining*, h will be a function of time *and* of the vertically downwards distance x. Prove, by means of a suitable transient mass balance on a differential element, and using the above equation for m, that:

$$\frac{\partial h}{\partial t} = -\frac{\rho^2 g^2 h^3}{c^2} \frac{\partial h}{\partial x}.$$

A tank with vertical sides is initially full, and at $t = 0$ is rapidly drained of the liquid. If x is measured from the top of the tank, verify that the film thickness on the sides is given by:

$$h = \left(\frac{c^2 x}{\rho^2 g^2 t} \right)^{1/3}.$$

18. *The Voigt model—M.* The *Voigt* model for a viscoelastic fluid is shown in Fig. P11.18, and consists of a dashpot and spring in *parallel*, with constants η and G, respectively.

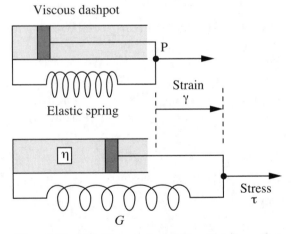

Viscous dashpot

Elastic spring

Strain γ

Stress τ

Fig. P11.18   Elements of the Voigt model.

When a stress τ is applied at point P, the resulting strain is γ.

(a) By noting the resisting stresses offered by the dashpot and spring, derive the differential equation that relates τ, γ, η, and G.

(b) If a constant stress τ_0 is applied, with an initial strain of zero, solve this differential equation for the strain as a function of time. Introduce the parameter $\lambda = \eta/G$ if possible. *Hint*: if needed, try $\gamma = c_1 + c_2 e^{-c_3 t}$ as a solution.

(c) If, after a long time, the stress is completely removed, what happens to the strain?

(d) Draw a diagram that shows both the stress and the strain as functions of time for both (b) and (c) above.

19. *Strain-rate solution of the Maxwell model—E.* Verify that the stress given by Eqn. (11.28) satisfies the Maxwell model of Eqn. (11.26).

20. *Strain solution of the Maxwell model—M.* Verify that the stress given by Eqn. (11.29) satisfies the Maxwell model of Eqn. (11.26).

21. *Strain decay for Maxwell fluid—E.* For a suddenly imposed strain γ_0, what is the stress τ in a Maxwell fluid of parameters λ and G as a function of time? Sketch τ as a function of time.

22. *Oldroyd derivative—E.* Verify Eqn. (11.38), which gives the Oldroyd derivative of a component τ_{ij} of the stress tensor.

23. *Maxwell model with Oldroyd derivative—E.* Verify Eqn. (11.40), the relation for the stress component τ_{xx}.

24. *Tests on a Maxwell fluid—M.* A viscoelastic Maxwell fluid is subjected to an oscillating strain $\gamma_0 \cos \omega t$ in a rheometer, and loss angles of $\delta = 0.375$ and 0.158 radians are found for angular frequencies $\omega = 2$ and 5 s^{-1}, respectively.

(a) Show that these two data points are consistent with each other, and evaluate the relaxation time λ (s).

(b) At $\omega = 2$ s^{-1}, the tests show that the amplitude of the oscillations of τ/γ_0 is 2.32 kg/m s^2. Evaluate the viscosity η (kg/m s) of the sample.

(c) If, in another experiment, starting from zero strain and zero shear, the sample is strained by a steady amount $\dot{\gamma} = c$, verify that the corresponding shear stress is:

$$\tau = \eta c \left(1 - e^{-t/\lambda}\right).$$

After what time will the stress have risen to half of its final asymptotic value?

25. *True/false.* Check *true* or *false*, as appropriate:

(a)	The term "non-Newtonian" applies only to those fluids that exhibit elasticity.	T ☐ F ☐
(b)	In simple shear, the strain rate refers to the rate of angular deformation.	T ☐ F ☐
(c)	The effective viscosity of pseudoplastic fluids decreases as the strain rate increases.	T ☐ F ☐
(d)	At a constant strain rate, a thixotropic fluid exhibits a viscosity that increases with time.	T ☐ F ☐

(e)   The second invariant of the strain-rate tensor is important because it enables a generalized strain rate $\dot{\gamma}$ to be defined.   T ☐   F ☐

(f)   The Carreau model needs just three independent parameters in its formulation.   T ☐   F ☐

(g)   A Bingham plastic will not flow unless the strain rate exceeds a certain minimum threshold.   T ☐   F ☐

(h)   For flow in a horizontal pipe under a specified pressure gradient, the shear-stress distribution, $\tau_{rz} = \tau_{rz}(r)$, differs between a Newtonian and a non-Newtonian fluid.   T ☐   F ☐

(i)   The physical interpretation of the Maxwell model is that of a spring and a dashpot in series.   T ☐   F ☐

(j)   The relaxation time of a viscoelastic fluid is the ratio of its elastic or rigidity modulus to its viscosity.   T ☐   F ☐

(k)   If a Maxwell model fluid is stretched to a certain point, but no more, the stress will decay exponentially.   T ☐   F ☐

(l)   For a generalized viscoelastic fluid, the memory function weights the strain rate in the integral expression for the stress tensor.   T ☐   F ☐

(m)   If a viscoelastic fluid is tested in oscillatory shear, the part of the shear that is in-phase with the strain depends on the elasticity of the fluid, rather than on its viscosity.   T ☐   F ☐

(n)   Elastic solids, and viscous fluids, have high and low relaxation times, respectively.   T ☐   F ☐

(o)   A parallel-plate rheometer has the advantage that the strain rate is constant throughout the fluid being tested.   T ☐   F ☐

(p)   In a coaxial-cylinder rheometer, the torque required to prevent the cup from rotation depends very much on the separation between the bob and cup.   T ☐   F ☐

(q)   The strain rate and the viscous dissipation function are closely related to each other.   T ☐   F ☐

(r)   For viscoelastic fluids, the primary normal-stress difference N_1 is approximately proportional to the strain rate $\dot{\gamma}$.   T ☐   F ☐

(s)   The Rabinowitsch equation enables stress/strain rate   T ☐   F ☐
        relationships to be determined from flow rate and
        pressure drop measurements.

(t)   The relaxation time of a solid is very small.   T ☐   F ☐

(u)   The Maxwell model for certain viscoelastic fluids can   T ☐   F ☐
        be reformulated as an equivalent "fading-memory"
        model.

(v)   Key steps in the formulation of the Rabinowitsch   T ☐   F ☐
        equation are integration by parts and the application
        of Leibnitz's rule.

(w)   The first and third invariants of the strain-rate tensor   T ☐   F ☐
        are frequently zero, and are therefore of little use for
        formulating a generalized Newtonian viscosity based
        on strain rate.

(x)   When shearing a viscoelastic liquid between two   T ☐   F ☐
        parallel plates, the primary normal-stress difference
        tends to push the plates apart.

(y)   There is more than one plausible explanation for the   T ☐   F ☐
        Weissenberg rod-climbing effect.

(z)   The Oldroyd derivative is useful in extending the   T ☐   F ☐
        Maxwell model to flowing situations.

(A)   Pseudoplasticity is a universal feature of most partic-   T ☐   F ☐
        ulate suspensions.

(B)   A more solid-like viscoelastic fluid is characterized by   T ☐   F ☐
        $G' > G''$.

(C)   The apparent viscosity of a non-Newtonian fluid is es-   T ☐   F ☐
        sentially constant everywhere in the fluid in a narrow-
        gap coaxial-cylinder rheometer operating at a fixed
        rotation rate.

Chapter 12

MICROFLUIDICS AND ELECTROKINETIC FLOW EFFECTS

12.1 Introduction

MICROFLUIDICS, as described here, refers to a fluidic regime (characterized by geometric length scales usually between 100 nm and 100 μm) in which the dominant transport physics change as compared to macroscopic fluid mechanics, due either to changes in Reynolds number, the relative importance of surface effects, or relative changes in the importance or character of different mixing and reaction processes. The tools used currently to understand microfluidic systems have historically been used to study flow in capillaries, both natural (in our circulatory system) and artificial (most commonly in glass capillaries), as well as flow through porous media such as soil. While the fluid mechanics tools have retained a similar structure, the applications have evolved, becoming more wide-ranging and more interesting with the development and ubiquity of microfabrication techniques.

Modern microfluidic systems are typically fabricated from glass, silicon, or a variety of ceramics and polymers. Initial systems were first developed in silicon and silica, owing to the prevalence of silicon micromachining expertise and the long history of using glass capillaries. More recently, a variety of polymeric substrates, including poly(dimethylsiloxane), polycarbonate, and poly(methylmethacrylate), have become more and more important due to their relatively low cost or their unique and useful properties.

Most microfluidic devices have been motivated by the desire to optimize fluidic handling for chemical, biological, and medical applications. Optimization can be achieved via massive parallelization or simply because the different fluidic regimes (e.g., surface-tension-dominated, electrokinetic) afford new techniques for manipulating and controlling fluids. Applications include protein crystallization as well as protein and DNA separation and identification, for which pictures of example devices are shown in Figure 12.1. Because of our interests in chemical/biological/medical applications, we will focus on primarily liquid-phase applications in this chapter.

(a)   (b)

Fig. 12.1   Two examples of microfluidic chips. (a) Protein-sizing chip for use on Agilent 2100 Bioanalyzer. The 16 fluid reservoirs in the plastic caddy mate with the glass chip underneath. The microfluidic chip itself occupies just the area under the reservoirs. From A. Kopf-Sill Lab Chip, Vol. 2, pp. 42N–47N, 2002. Reproduced by permission of The Royal Society of Chemistry; (b) The microfluidic chip and fluidic compression manifold of the Sandia National Labs μChemLab^{TM} microfluidic bioanalysis device. Reprinted with permission from R.F. Renzi et al., Analytical Chemistry, Vol. 77, pp. 435–441, 2005. Copyright 2005, American Chemical Society.

12.2   Physics of Microscale Fluid Mechanics

Dimensional analysis (Chapter 4) shows us that the character of flow through a given geometry depends only on the Reynolds number and not on the length scale *per se*. Similarly, scalar transport depends only on the Péclet number (defined using thermal diffusivity for heat-transfer problems or mass diffusivity for mass-transfer problems). Why, then, discuss microfluidics separately from Stokes flow? The answer is that, despite the fact that many of the distinguishing aspects of microfluidic flows follow directly from their laminar, low-Re character, many others do not. Key differences in the relevant physics of microscale fluid transport include:

- Flows are typically laminar and have a low Reynolds number.
- Both pressure and electric fields are used to actuate flow, so the transport equations must be modified to include electroosmotic flow and electrophoresis of molecules or particles.
- Boundary conditions are often more difficult to define, since slip velocities need be included, depending on the length scale, and these slip velocities are often not well known or understood.
- If multiple phases exist, surface tension can dominate over pressure forces.

After a discussion specific to dispersion in pressure-driven flow in microscale tubes, this chapter will focus primarily on electrokinetic phenomena.

Example 12.1—Calculation of Reynolds Numbers   **641**

12.3 Pressure-Driven Flow Through Microscale Tubes

Recall the flow inside a circular pipe induced by a pressure gradient (Chapter 3). For a given pressure gradient and fluid, the character of the flow changes from laminar to transitional to turbulent as the diameter of the pipe is increased. Microfluidic channels have extraordinarily small diameters and therefore flow in microchannels typically occurs at low Re.

Example 12.1—Calculation of Reynolds Numbers

Calculate the Reynolds number Re for the following. Assume that the fluid has properties of water: density $\rho = 1{,}000$ kg/m^3, viscosity $\mu = 10^{-3}$ kg/m s, and kinematic viscosity $\nu = 10^{-6}$ m^2/s.

- Flow of drugs injected using thumb pressure (about 5 atm) on a 5-ml syringe through a $\frac{1}{2}$-inch long, 27-gauge (internal diameter d approximately 300 μm) needle.
- Flow of cell culture media at a mean velocity of 10 μm/s through a $d = 20$-μm diameter tube machined in a glass microchip, used to culture cells in what is called "continuous perfusion cell culture."
- Flow of water at a mean velocity of 80 μm/s through $d = 50$-nm diameter tubes machined in silicon, used to sort and measure individual DNA molecules.

Solution

Syringe: Because of the high pressure drop, we anticipate that the flow is probably turbulent, in which case we can use Eqns. (3.33) for the pressure drop and (3.40) for the friction factor in a smooth tube:

$$-\Delta p = 2 f_{\mathrm{F}} \rho u_{\mathrm{m}}^2 \frac{L}{d}, \qquad f_{\mathrm{F}} = 0.0790 \, \mathrm{Re}^{-1/4} = 0.0790 \left(\frac{\mu}{\rho u_{\mathrm{m}} d} \right)^{1/4}.$$

Elimination of the friction factor and rearrangement gives the mean velocity:

$$u_{\mathrm{m}} = \frac{(-6.33 \, \Delta p)^{4/7} \, d^{5/7}}{\mu^{1/7} \, \rho^{3/7} \, L^{4/7}}$$

$$= \frac{(6.33 \times 5 \times 10^5 \ \mathrm{Pa})^{4/7} \, (300 \times 10^{-6} \ \mathrm{m})^{5/7}}{(10^{-3} \ \mathrm{kg/m \ s})^{1/7} \, (10^3 \ \mathrm{kg/m}^3)^{3/7} (0.5 \times 0.0254)^{4/7}} = 26.6 \ \mathrm{m/s}$$

$$\mathrm{Re} = \frac{u_{\mathrm{m}} d}{\nu} = \frac{26.6 \times 300 \times 10^{-6}}{10^{-6}} = 7{,}974,$$

thus verifying the assumption of turbulent flow.

Cell culture media:

$$Re = \frac{u_\mathrm{m}d}{\nu} = \frac{10 \times 10^{-6} \times 20 \times 10^{-6}}{10^{-6}} = 2 \times 10^{-4}.$$

DNA measurement:

$$Re = \frac{u_\mathrm{m}d}{\nu} = \frac{80 \times 10^{-6} \times 50 \times 10^{-9}}{10^{-6}} = 4 \times 10^{-6}. \qquad \square$$

12.4   Mixing, Transport, and Dispersion

Mixing is critical for countless chemical problems. Often, complete mixing is required to achieve a chemical reaction. In other cases, such as chemical separations, it is critical to minimize the extent to which different sections of fluid mix. Because microfluidic systems are used in both these situations, and because geometries of practical interest involve known solutions of the laminar Navier-Stokes equations, microfluidic systems serve as excellent model systems for discussing mixing and dispersion.

Taylor dispersion in a capillary tube.   Consider the flow of two miscible liquids (for example, clear water and water with food dye mixed in). Specifically, consider a band of water with width w containing a species at concentration c_0 amidst a flow of clear water ($c = 0$). As will be seen later, this flow is relevant to a number of chemical separations.

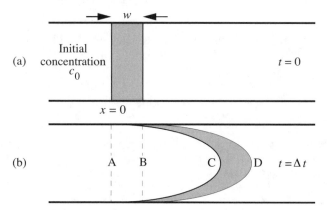

Fig. 12.2   Taylor dispersion caused by parabolic pipe flow: (a) initially; (b) at a later time Δt. See text for explanation.

First we consider one-dimensional *diffusion alone* with no flow, and consider the radially averaged concentration $\bar{c}$. The equation governing the diffusion of the band is:

$$\frac{\partial \bar{c}}{\partial t} = D \frac{\partial^2 \bar{c}}{\partial x^2}. \qquad (12.1)$$

And it can be shown that, for pure diffusion, the width of the concentration profile (which, for long times, is Gaussian in shape) grows as $w \propto \sqrt{Dt}$. Thus, as time proceeds, the width of the band increases proportional to the square root of time.

Consider now the opposite case. Ignoring diffusion and for the moment considering only convection, the parabolic velocity distribution distorts the interface between the two zones, as in Fig. 12.2.[1] Given a parabolic velocity distribution with mean velocity U_0 in a circular pipe, first note that the centerline velocity is $2U_0$; thus, a key quantity $w/2U_0$ is the time taken for the liquid at the center to traverse a distance equal to the initial width w. It can be shown that for $\Delta t > w/2U_0$, the *radially averaged* concentration $\bar{c}$ takes on a trapezoidal shape:

$$
\begin{aligned}
\bar{c} &= 0, & x &< 0, \\
\bar{c} &= c_0 \left(\frac{x}{2U_0\Delta t} \right), & 0 &< x < w, \\
\bar{c} &= c_0 \left(\frac{w}{2U_0\Delta t} \right), & w &< x < 2U_0\Delta t, \\
\bar{c} &= c_0 \left(\frac{w + 2U_0\Delta t - x}{2U_0\Delta t} \right), & 2U_0\Delta t &< x < 2U_0\Delta t + w, \\
\bar{c} &= 0, & x &> 2U_0\Delta t + w.
\end{aligned}
\tag{12.2}
$$

From Eqns. (12.2) it is clear that, for this *diffusionless flow*, the maximum concentration c_0 has been attenuated as the species are spread out by the flow; furthermore, the maximum *height* of the average concentration profile ($c_0 w/2U_0\Delta t$) decreases linearly with time and the *width* of the profile increases linearly with time. The five segments of Eqn. (12.2) correspond to regions to the left of A, between A and B, between B and C, between C and D, and to the right of D.

Unsurprisingly, the two models (first, diffusion only; second, convection only) discussed above do not satisfactorily describe the observed sample band growth, since both omit a key aspect of the transport. Experimentally, it is observed that the growth of sample band width convected by parabolic pipe flow profiles increase proportional to the square root of time, but with a coefficient of proportionality larger than $D^{1/2}$:

$$
\frac{\partial \bar{c}}{\partial t} = D \left(1 + \frac{\text{Pe}^2}{48} \right) \frac{\partial^2 \bar{c}}{\partial x^2}.
\tag{12.3}
$$

Here, the species diffusion Péclet number is defined as $\text{Pe} = U_0 a/D$, where a is the radius of the pipe. The Péclet number is analogous to the Reynolds number; whereas the Reynolds number expresses the ratio of advective fluxes of momentum to diffusive fluxes of momentum, the Péclet number expresses the ratio of advective fluxes of species concentration to the diffusive fluxes of species concentration.

---

[1] A short animation is available at http://www.fiu.edu/~thorned/LB/html/ani03_poisseuille.html (yes, the website has the spelling "poisseuille").

As Pe becomes small (diffusion dominates), Eqn. (12.3) becomes equal to Eqn. (12.1). As Pe becomes large (diffusion can be neglected), the term $1 + Pe^2/48$ becomes large, and diffusion is rapid. Note that the applicability of Eqn. (12.3) does not extend to Pe $= \infty$, so the result in Eqn. (12.2) cannot be recovered by simple substitution in Eqn. (12.3). Since we expect to apply Eqn. (12.3) to microscale flows, Re is small. Note, however, that we often deal with samples with very large Schmidt numbers (Sc $=$ Pe/Re) since their diffusivities are very low. Thus, Pe values as high as 1,000 are common, even in microfluidic applications.

12.5 Species, Energy, and Charge Transport

Before addressing the double layer *per se*, we must expand the transport equations from Chapter 5 to include chemical species, thermal energy, and charge. The discussion is by necessity brief, and much of the derivation is left for the reader.

Constitutive relations. Recall from Chapter 5 that the heat flux **q** defines the rate at which thermal energy is conducted due to temperature gradients and that Fourier's law states:

$$\mathbf{q} = -k\nabla T. \tag{5.23}$$

This constitutive relation is the necessary link between the temperature field and the governing transport equation that expresses conservation of thermal energy fluxes, just as Newton's law of viscosity is the necessary link between the velocity field and the governing transport equation for conservation of fluid momentum. Fick's law of diffusion:

$$\mathbf{J}_i = -D\nabla\rho_i, \tag{12.4}$$

similarly relates the species diffusive mass flux $\mathbf{J}_i$ to the gradient of the species mass density ρ_i.

While the constitutive relations that govern diffusion for species and temperature are quite similar, we must consider different effects when we treat charged species in an electric field. If a gradient in the electrostatic potential is present, a force is exerted on a charged molecule. The electric field **E** is defined as the negative gradient of the electrostatic potential:

$$\mathbf{E} = -\nabla\phi, \tag{12.5}$$

and the force exerted on a molecule is given by

$$\mathbf{F}_\mathrm{E} = qe\mathbf{E}, \tag{12.6}$$

where q is the charge number (the electrical charge normalized by the elementary charge, 1.602×10^{-19} coulombs) and e is the elementary charge (1.602×10^{-19} C). An electric field is thus seen to cause a force that is proportional to that field. The motion of ions is retarded, though, by collisions with the surrounding molecules. In the steady state, the ion motion (termed *electrophoresis*) takes on a velocity given by:

$$\mathbf{u}_{ep} = \mu_{ep}\mathbf{E}, \tag{12.7}$$

and the molar charge flux is

$$\mathbf{j} = \mu_{ep}zc\mathbf{E}, \tag{12.8}$$

where c is the species concentration. The electrophoretic mobility μ_{ep} is typically determined experimentally; values of μ_{ep} for some small ions are given in Table 12.1.

Table 12.1 Electrophoretic mobilities of some ions at infinite dilution in water, 298 K.†

Ion	μ_{ep} [m^2/V s]	Ion	μ_{ep} [m^2/V s]
H$^+$	3.63×10^{-4}	OH$^-$	-2.05×10^{-4}
Li$^+$	4.01×10^{-5}	Cl$^-$	-7.91×10^{-5}
Na$^+$	5.19×10^{-5}	Br$^-$	-8.11×10^{-5}
Ca^{2+}	3.08×10^{-5}	NO$_3^-$	-7.40×10^{-5}
Cu^{2+}	2.80×10^{-5}	HCO$_3^-$	-4.61×10^{-5}
La^{3+}	2.30×10^{-5}	SO$_4^{2-}$	-4.15×10^{-5}

† Values calculated from G. Atkinson, 1972: "Electrochemical information," in *American Institute of Physics Handbook*, 3rd ed., (ed. D.E. Gray), pp. 5-249–5-263, McGraw-Hill, New York.

Transport equations.   In Chapter 5, the transport equations for momentum (i.e., the Navier-Stokes equations) were derived. The transport equations for temperature, species, and charge are similar. We will treat mass and thermal energy as conserved scalars, i.e., we will assume that they are transported by the fluid flow but are neither created nor destroyed. To make this approximation, our assumptions include: (a) that no chemical reaction occurs, and (b) that the heat generated by viscous dissipation can be assumed small. With these assumptions, and using the convective derivative $\mathcal{D}/\mathcal{D}t$ we can write a transport equation for a species concentration:

$$\frac{\mathcal{D}c}{\mathcal{D}t} = D\nabla^2 c, \tag{12.9}$$

and temperature:

$$\frac{\mathcal{D}T}{\mathcal{D}t} = \alpha\nabla^2 T. \tag{12.10}$$

Note that Eqns. (12.9) and (12.10) are similar to the Navier-Stokes equations (Eqn. (5.68)) except that: (a) there is no pressure or gravity forcing term involved,

(b) species molar diffusivity D or thermal diffusivity α plays the same role as the kinematic viscosity μ/ρ, and (c) the equations are linear (the nonlinearity in the Navier-Stokes equations stems from the nonlinear convective term). Also note that the species and temperature transport equations, as written, do not couple back to the Navier-Stokes equations—that is, the fluid flow affects species and temperature transport but the species and temperature transport does not affect the flow. The same cannot be said for electrical charge, for which the laws of electrodynamics must be coupled to the flow equations.

If we assume that magnetic fields may be ignored (i.e., externally applied magnetic fields are small, current-induced magnetic fields are small, and the electric field is quasi-steady), then the system can be assumed electrostatic and the electrical potential ϕ is related to the local charge density via Poisson's equation:

$$\nabla^2\phi = -\frac{\rho_e}{\varepsilon}, \tag{12.11}$$

where ρ_e is the local net charge density, and ε is the electrical permittivity of the liquid. The electrical permittivity of free space, ε_0, is equal to 8.85×10^{-12} C/V m; the permittivity of various materials are typically described through their dielectric constant, $\varepsilon/\varepsilon_0$. The dielectric constant of water at room temperature is approximately 80.

Just as a gravitational body force term $\rho\mathbf{g}$ must be included in the Navier-Stokes equations when gravity plays a role (e.g., Eqn. (5.68)), so must the *Lorentz* body force term be included when there is a net charge density and an electric field. This term can be written as:

$$\mathbf{f}_{\text{Lorentz}} = \rho_e\mathbf{E}, \tag{12.12}$$

where $\mathbf{f}_{\text{Lorentz}}$ is the force (per unit volume) caused by the electrostatic force felt by the charge density ρ_e in the electric field. With this force, the Navier-Stokes equations relevant for microfluidic applications become:

$$\rho\frac{\mathcal{D}\mathbf{u}}{\mathcal{D}t} = -\nabla p + \mu\nabla^2\mathbf{u} + \rho_e\mathbf{E}. \tag{12.13}$$

Note that gravity has been ignored and incompressibility has been assumed. The effect of the Lorentz force will be of immediate impact when we discuss the electrical double layer and the microscopic derivation of the electroosmotic flow velocity.

12.6   The Electrical Double Layer and Electrokinetic Phenomena

If one drinks water from a glass, there is hardly any reason to think that electric fields play a prominent role in the motion as the water pours out into one's mouth. However, we can infer from some simple experiments that electric fields *do* play a significant role in fluid transport at small scales. For example, if a pressure gradient of 5 atm is applied to water to induce Poiseuille flow through a glass capillary of, say, 100 μm diameter, a voltage difference of approximately 1 volt can be measured between the two ends of the capillary. Similarly, if the pressure is removed and a voltage gradient (say 10 kV/m) is applied across the same capillary, the water inside will be observed to move uniformly at approximately 1 mm/s. These phenomena stem from an electrochemical phenomenon known as the *electrical double layer*, discussed in this section.

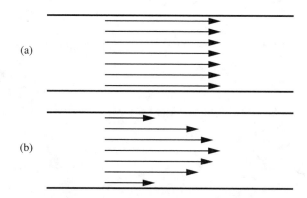

(a)

(b)

Fig. 12.3   (a) Electroosmotic flow in a tube can be approximated as uniform throughout; (b) laminar pressure-driven flow in a tube takes on the well-known parabolic form.

Electroosmosis.   Before discussing the detailed root cause of the electrical double layer, we will begin by describing its most obvious effect—*electroosmotic flow*. Experiments show that if an extrinsic electric field **E** is applied across a tube containing an electrolyte solution, a uniform flow will result, with a velocity given by:

$$\mathbf{u} = -\mu_{eo}\mathbf{E} = -\frac{\varepsilon\zeta}{\mu}\mathbf{E}. \tag{12.14}$$

Here, ζ is called the zeta potential, and $\mu_{eo} \equiv \varepsilon\zeta/\mu$ is termed the *electroosmotic mobility*. Both of these values are functions of both the wall material and the nature of the liquid. Unlike Poiseuille flow through a tube, the velocity of this flow is not a function of the size of the tube, within certain limits (roughly speaking, for tube diameters above 100 nm or so). Figure 12.3 shows the velocity distribution for electroosmotic flow as compared to pressure-driven flow.

Example 12.2—Relative Magnitudes of
Electroosmotic and Pressure-Driven Flows

Consider a circular tube of diameter $d = 5$ μm ($r = 2.5$ μm) and length $L = 10$ cm filled with an aqueous solution. If $\zeta = -100$ mV, calculate the pressure required to give the same total flow that would be generated by the application of 1 kV.

Solution

Electroosmotic flow:

$$Q = uA = -\frac{\varepsilon \zeta}{\mu} \frac{V}{L} A$$

$$= -\frac{80 \times 8.85 \times 10^{-12} \text{ C/V m} \times (-0.1) \text{ V}}{10^{-3} \text{ Pa s}} \times \frac{10^3 \text{ V}}{0.1 \text{ m}} \times \pi (2.5 \times 10^{-6} \text{ m})^2$$

$$= 1.39 \times 10^{-14} \text{ m}^3/\text{s} = 1.39 \times 10^{-11} \text{ l/s} = 13.9 \text{ p(pico)l/s}.$$

Pressure-driven flow:

$$Q = \frac{\pi r^4}{8\mu} \left(-\frac{dp}{dx} \right) = \frac{\pi r^4}{8\mu} \frac{(-\Delta p)}{L};$$

$$-\Delta p = \frac{8 Q \mu L}{\pi r^4}$$

$$= \frac{8 \times 1.39 \times 10^{-14} \text{ m}^3/\text{s} \times 10^{-3} \text{ Pa s} \times 0.1 \text{ m}}{\pi (2.5 \times 10^{-6} \text{ m})^4}$$

$$= 9.06 \times 10^4 \text{ Pa (about 0.9 atmospheres).} \qquad \square$$

Detailed structure of the electrical double layer. The simple preceding discussion is an accurate description of the majority of the flow field and predicts velocities and flow rates quite well; however, it is certainly dissatisfying from the standpoint of our intuition regarding the boundary conditions that apply to fluid flow. By describing electroosmotic flow as uniform, we are implying that the velocity at the wall is equal to the bulk velocity (i.e., full slip and a completely inviscid interface)! In the following paragraphs we will show that indeed the no-slip condition applies, but the region in which the flow is retarded is remarkably thin and, for the purposes of mass flow rate, can be ignored.

From a fluid mechanics point of view, we consider the wall to be simply the location of the no-slip boundary condition; chemically, however, things are much less simple. The surface chemical groups on the wall undergo reactions with the liquid and achieve an equilibrium chemical state. The surface charge at the wall is in general nonzero since these reactions typically involve adsorption of charged ions or creation of charged groups by addition or removal of protons from the surface material. The details of the surface chemistry are beyond the scope of this discussion; however, in general we can assume that the surface has a nonzero charge. Further, we can assume that the aqueous solution has some concentration of electrolytes (even pure water has 10^{-7} M H_3O^+ and 10^{-7} M OH^-).

When an electrolyte solution comes into contact with a charged surface, the charged surface locally changes the ion distribution (which would otherwise be uniform). Ions with an opposite charge to the surface (termed *counterions*) are attracted to the surface, while ions with similar charge (termed *coions*) are repelled. An equilibrium is achieved between: (a) electrostatic attraction/repulsion and (b) Brownian (thermal) motion, which tends to distribute ions randomly in an exponential (Boltzmann) distribution, as in Fig. 12.4.

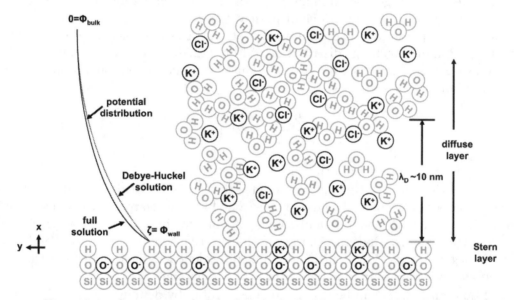

Fig. 12.4   Structure of the electrical double layer.   The situation shown is for a negatively charged surface (O^-), with a negative zeta potential.   The monolayer with the occasional K^+ ion is the Stern layer.   Note the preponderance of positive ions in the double layer.

The Boltzmann distribution can be derived from equilibrium arguments that say the likelihood of a system being in a state with molar potential energy E is proportional to $e^{-E/RT}$. The potential energy of a mole of charge held at a potential ϕ is $E = qF\phi$, where q is the ion charge (for example, $q^+ = 1$ for Na^+, $q^- = -1$ for Cl^-, $q^+ = 2$ for Mg^{2+}, etc.) and F is the Faraday constant, 9.65×10^4 C/mol. If the molar concentration of ions in the electrolyte solution far from the wall (where the potential is zero) is c_∞, the concentration of ions at a location with a potential ϕ conforms to *Boltzmann's equation*:

$$\frac{c}{c_\infty} = e^{-E/RT} = e^{-qF\phi/RT}. \tag{12.15}$$

As the potential becomes more positive, the concentration of positive ions will decrease and that of negative ions will increase.

If the counterions and coions have charges with the same magnitude (e.g., $+1$ and -1 or $+2$ and -2), the net charge density (see Problem 12.9) is given by:

$$\rho_e = F(q^+ c^+ + q^- c^-) = -2zFc_\infty \sinh\left(\frac{zF\phi}{RT}\right), \tag{12.16}$$

where $z = |q|$ is the valence of the ions (the magnitude of the charge normalized by the electron charge, for example, 1 for Na^+ and Cl^- and 2 for Mg^{2+} and SO_4^{2-}), c^+ and c^- are the concentrations of positive and negative ions, respectively, and ϕ is the local potential. The combination of the Poisson equation (Eqn. (12.11)) with the Boltzmann charge distribution of Eqn. (12.16) gives a relation termed the nonlinear Poisson-Boltzmann equation, which in a one-dimensional system (such as an infinite flat plate) is:

$$\frac{d^2\phi}{dx^2} = \frac{2zFc_\infty}{\varepsilon} \sinh\left(\frac{zF\phi}{RT}\right), \tag{12.17}$$

where x is the distance from the edge of the Stern layer.

Recall that the Poisson equation relates the Laplacian of the potential field to the charge density—here the charge density is directly related to the exponential distribution of both counter- and coions. Again, we have assumed a symmetric electrolyte, that is, that the charge of the positive ions is equal to the charge of the negative ions. Generalizing to multiply-valent ions is straightforward (but algebraically complex) and adds little to this discussion.

Equation (12.17) can be simplified considerably by defining the *Debye length*, followed by a suitable normalization of the potential. The Debye length λ_D is defined as $\sqrt{\varepsilon RT/2F^2 z^2 c_\infty}$; this length is approximately equal to the thickness of the double layer and will later be shown to be the characteristic length of the exponential decay of the potential in certain situations. We can also normalize ϕ by RT/zF ($\phi^* = \phi zF/RT$); RT/zF is the potential at which a coion of valence z has potential energy equal to RT. With these definitions, Eqn. (12.17) can be rearranged to give:

$$\frac{d^2\phi^*}{dx^2} = \frac{1}{\lambda_D^2} \sinh\phi^*. \tag{12.18}$$

If $\phi^* \ll 1$ (the so-called Debye-Huckel limit), the one-dimensional nonlinear Poisson-Boltzmann equation can be simplified (by using the approximation $\sinh x \doteq x$) to give (the asterisks can be removed from both sides):

$$\frac{d^2\phi}{dx^2} = \frac{\phi}{\lambda_D^2}, \tag{12.19}$$

which, for a flat plate in an infinite reservoir, leads to exponential solutions:

$$\phi = \zeta e^{-x/\lambda_D}. \tag{12.20}$$

The zeta potential, ζ, is the potential at the wall ($x = 0$), and the Debye length is the $1/e$ decay distance of the potential as the distance from the wall increases. For univalent electrolytes, λ_D is approximately 10 nm at 1 mM concentration.

We now consider electroosmosis in detail. Modeling the flow in the double layer as unidirectional and one-dimensional, and combining the Navier-Stokes equations (with the Lorentz source term, Eqn. (12.13)) with the Poisson equation (Eqn. (12.11)), a relation can be derived (see Problem 12.10) between the flow and the potential in the double layer:

$$\mu\frac{d^2u}{dx^2} = \varepsilon\frac{d^2\phi}{dx^2}E_y. \tag{12.21}$$

This relation has separated the potential field into two components: (1) the intrinsic potential due to the surface charge, denoted by ϕ, and (2) the extrinsic potential applied to drive the electroosmotic flow, denoted by the externally applied electrostatic field E_y.

Equation (12.21) may be integrated. For a flat wall in an infinite reservoir of fluid, this is straightforward analytically; requiring that the velocity is bounded at $x = \infty$ and zero at the wall, we can show that

$$u = -\frac{\varepsilon E_y}{\mu}(\zeta - \phi) = -\frac{\varepsilon\zeta E_y}{\mu}(1 - e^{-x/\lambda_D}), \tag{12.22}$$

which is valid for all x. From Eqn. (12.22) we note that, for a flat surface, the fluid velocity and the intrinsic potential in the double layer are similar, i.e., they have the same functional form. Thus the velocity and potential can be graphed simultaneously, as done in Fig. 12.5.

One particularly illustrative solution of Eqn. (12.19) (or, more generally, Eqn. (12.18)) is the solution between two infinite parallel plates. This solution (Fig. 12.5) shows the impact of wall spacing on the character of the potential distribution and, as we will see in the following paragraph, electroosmotic flow. For large wall spacings (as compared to the double layer thickness, which can be characterized using the Debye length), the potential is uniform throughout most of the fluid, decaying rapidly to the zeta potential in thin layers near the walls. Such a profile is typified by wall spacings roughly 100 times the double layer thickness, or $>$ 1 μm at 1 mM concentration. This is the regime described by the introductory discussion. When the wall spacing is no longer much greater than the double layer, the potential can no longer be described as largely uniform, and the resulting average electroosmotic velocity is no longer given by $u = -\varepsilon\zeta E/\mu$.

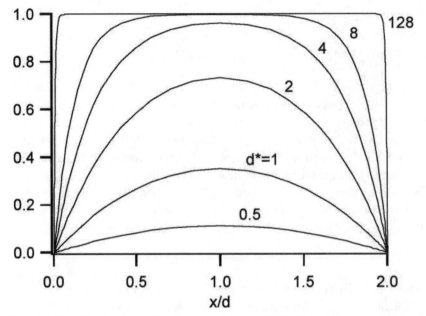

Fig. 12.5   Electroosmotic flow and potential distributions between two infinite parallel plates separated by a distance 2d. Spatial profiles of nondimensionalized velocity $u^* = u(\varepsilon\zeta E/\mu)^{-1}$ are shown as a function of $d^* = d/\lambda_D$, the ratio of the half width to the Debye length. Calculations are shown in the Debye-Huckel limit (i.e., solutions to Eqn. (12.19)).

Similitude between the velocity field and the extrinsic electric field. Recall from Chapter 7 that, absent electrokinetic effects, vorticity is generated at fluid-wall interfaces and propagates from the interfaces into the flow. The vorticity transport equation, though, has no source term, i.e., vorticity is not generated within the flow field. In Chapter 7 it was noted that, far from solid boundaries, the vorticity is often small and can be neglected. One of the more interesting results of electroosmotic flow is that the flow is irrotational outside the electrical double layer, and thus the flow field is given by a solution of Laplace's equation. In fact, it can be shown that if ζ is constant, double layers are thin compared to the geometry or the radius of curvature of any geometric features, and if electrokinetic forces are the only flow effects (i.e., no pressure gradient exists) the simple relation in Eqn. (12.14) applies to the entire flow field (except the double layer) *regardless of the complexity of the geometry*.

Thus, the electrical potential and the velocity potential differ only by the multiplier $\varepsilon\zeta/\mu$. The formal proof of this requires substantial vector operations, and will not be covered here. Equation (12.14) thus shows that, outside the double layer, the extrinsic electric field and the velocity are *similar* (i.e., proportional to

Example 12.4—Electroosmosis in a Microchannel (COMSOL)   **653**

each other). This *similitude* and the ability of Laplace's equation to produce the flow solution has several results. First of all, Laplace's equation is much more easily solved than the Navier-Stokes equations, both analytically (for which the Navier-Stokes equations typically have no solution) and numerically (for which the Navier-Stokes equations require more computing power). Second, the uniform flow allows fluid samples to be moved throughout a microchannel without dispersing the sample.

Example 12.3—Electroosmotic Flow Around a Particle

Assume a spherical particle of diameter $a \gg \lambda_D$ is suspended and held stationary inside a microchannel with diameter $d \gg a$ and that the particle and channel are made from the same material and have the same surface potential. An electric field E_∞ is applied to the system. Calculate the flow solution (in terms of velocity potential ϕ) around this particle.

Solution

Recall from Eqns. (7.71)–(7.74) and Fig. 7.17 that potential flow around a sphere can be modeled as the superposition of a uniform stream ($\phi = Ur\cos\theta$) with a doublet of strength $s = -Ua^3/2$, for which $\phi = Ua^3\cos\theta/2r^2$. This solution identically solves the electroosmotic flow around a particle, with U given by $\varepsilon\zeta E_\infty/\mu$. Since $d \gg a$, it is correct to model the electroosmotic flow as a uniform stream. $\qquad\square$

Example 12.4—Electroosmosis in a Microchannel (COMSOL)

The microchannel ABCD in Fig. E12.4.1, with length $L = 5 \times 10^{-4}$ m and width $H = 5 \times 10^{-5}$ m, contains an electrically conducting liquid whose physical properties are given in Table E12.4.1, in which "S" denotes units of *Siemens*, also equivalent to ohm^{-1}. Note also that a coulomb (C) equals an ampere-second (A s), and that the dielectric constant is also known as the relative permittivity. Electric potentials of zero and 1 V are applied at the left and right ends AB and CD, respectively, and we wish to find the resulting liquid velocities and electric potential distribution.

Fig. E12.4.1 Cross section of microchannel (not to scale).

Table E12.4.1   Physical Properties of Liquid

Property	Value	Units	COMSOL Name
Density, ρ	1,000	kg/m^3	rho
Viscosity, η	0.001	kg/m s	eta
Electric conductance, k_1	0.11845	S/m	k1
Dielectric constant	80.2	—	—
Permittivity of free space, ε_0	8.85×10^{-12}	C/V m	
Liquid permittivity, $\varepsilon_w = 80.2\varepsilon_0$	7.097×10^{-10}	C/V m	epsw
Wall zeta potential, ζ_0	-0.1	V	zet0

Table E12.4.2   Boundary Conditions

Boundary	Navier-Stokes Mode	Cond. Media Mode
1	Outflow/pressure $p_0 = 0$	Electric potential $V_0 = 0$
2	Inflow/outflow velocity $u_0 = v_x = (\varepsilon_w \zeta_0/\eta)\partial V/\partial x$ $v_0 = v_y = (\varepsilon_w \zeta_0/\eta)\partial V/\partial y$	Electric insulation
3	Inflow/outflow velocity $u_0 = v_x = (\varepsilon_w \zeta_0/\eta)\partial V/\partial x$ $v_0 = v_y = (\varepsilon_w \zeta_0/\eta)\partial V/\partial y$	Electric insulation
4	Neutral (no normal applied stress) $\mathbf{n} \cdot (-p\mathbf{I} + \eta(\nabla\mathbf{u} + (\nabla\mathbf{u})^{\mathrm{T}})) = 0$	Electric potential $V_0 = 1.0$

Solution

This problem falls into the *multiphysics* category, in that it involves both simultaneous motion of fluid and electric charge. In the **Model Navigator**, make sure the **New** tab is highlighted. Then, under **Fluid Dynamics**, select **Incompressible Navier-Stokes**, followed on the right-hand side by **Multiphysics**, and **Add**. Repeat the procedure for **Electromagnetics**, **Conductive Media DC**, **Add**, followed by **OK**, so the stage is all set in the graphical user interface for multiphysics—which you can check by pulling down the **Multiphysics** menu, which will display both modes:

1. Incompressible Navier-Stokes (ns).
2. Conductive Media DC (dc).

Example 12.4—Electroosmosis in a Microchannel (COMSOL)   **655**

Later, when media properties and boundary conditions are being set, we can toggle between the two modes.

Now execute the following steps:

1. Draw the rectangle ABCD.

2. Under **Options/Constants**, store the values from Table 12.4.1 for rho, eta, k1, epsw, and zet0, followed by **OK**. As usual, you make the appropriate entries in the "Name" and "Expression" columns (the latter may contain arithmetic operators), and the "Value" column is automatically computed.

3. Under **Physics/Subdomain Settings**, for the two modes, indicate that rho and eta are the expressions for the density and dynamic viscosity, and that k1 is the conductivity. To switch from one mode to the other, use the **Multiphysics** pulldown menu.

4. Under **Physics/Boundary Settings**, for the two modes, set the boundary conditions given in Table E12.4.2, noting for example that Vx is the COMSOL designation for $\partial V/\partial x$, the x-derivative of the electric potential. Thus, the entry for u_0 will be (epsw*zet0/eta)*Vx. Cutting, pasting, and editing expressions will save time.

5. Initialize and refine the mesh *twice*, as in Fig. E12.4.2, noting the number of elements and degrees of freedom.

6. Under **Solve/Solver Parameters**, click on the **Advanced** tab, and change **Scaling of variables** from **Automatic** to **None**—an *essential* step for the following none-too-obvious reason. Namely, for accuracy in solving simultaneous equations, COMSOL tries to *scale* the dependent variables so that they are all of order 1. The scaling is accomplished by dividing each variable by a reference quantity. But in our present problem, because there is no applied potential difference in the y direction, there is no flow in that direction, so that $v_y = 0$ and the scaling is impossible. If this precaution were not taken, the attempted solution would not converge.

Fig. E12.4.2   Mesh after two refinements.

7. Solve the problem.

Results and Discussion

1. Generate and print an arrow plot of the velocities, shown in Fig. E12.4.3. Note that in line with electroosmotic theory, the velocity is constant everywhere, and that the liquid, which has a negative zeta-potential ζ_0 (and hence an

excess of positive ions next to the walls), flows uniformly in the direction of decreasing electric potential—from right to left.

Fig. E12.4.3   Arrow plot. The ordinate scale also has a multiplier of 10^{-4}.

2. Generate and print a streamline plot—Fig. E12.4.4 shows the streamlines as parallel straight lines pointing in the negative x-direction. To obtain full control, begin with **Plot Parameters/Streamline/Specify start point coordinates**, and specify x: `linspace(0,0,10)` and y: `linspace(0,5e-5,10)`, which will ensure ten streamlines, all starting along AB ($x = 0$) and then being equally spaced along AB (between $y = 0$ and $y = 0.00005$).

Fig. E12.4.4   Streamlines, with arrows added to show the flow direction.

3. Generate a surface plot of the velocity distribution (not reproduced here) and note—by both observing the color bar at the right and also clicking on various points and looking at the lower left-hand corner—that the velocity is everywhere essentially uniform, with a value of 1.42×10^{-4} m/s (in the negative x direction, as evidenced by the arrow plot).

Fig. E12.4.5   Electric potential distribution.

4. Generate and print a contour plot of the electric potential—Fig. E12.4.5 shows a series of equally spaced equipotentials, varying from 1 V at the right-hand boundary CD to zero at the left-hand boundary AB.   □

Example 12.5—Electroosmotic Switching (COMSOL)   **657**

Example 12.5—Electroosmotic Switching
in a Branched Microchannel (COMSOL)

This example investigates the possibility of using electroosmosis as a "switching" mechanism for controlling the flow of liquids in microchannels.

Modify Example 12.4 so that it now includes a "vertical" branch of width 5×10^{-5} m, extending in "height" as far as $y = 2.5 \times 10^{-4}$ m, as shown below. Investigate the streamlines and equipotentials for the two sets of applied voltages shown in Figs. E12.5.1 and E12.5.2. All physical properties are identical with those given in Example 12.4.

Solution

The solution is very similar to that of Example 12.4, and most of the basics will not be repeated. The easiest approach is to make of copy of the `*.fl` file from the previous example and modify it. The additional rectangular channel extending in the y direction is easily added to the existing rectangle of Example 12.4 by forming a composite object, *unchecking* **Keep interior boundaries**. To smooth the corners where the two rectangles join, observe the following steps:

1. Select the **Fillet/Chamfer** drawing tool (the bottom tool at the left of the screen) and highlight the **Fillet** option.
2. Under **Vertex selection**, click on the composite object icon, and note that eight vertices are listed.
3. Find the vertex that corresponds to the coordinates $(2.25 \times 10^{-4}, 0.5 \times 10^{-4})$, enter a radius of 0.25×10^{-4}, and click **OK**.
4. Repeat for the vertex a little further to the right.

No changes are needed to the **Physics/Subdomain** settings. For boundary conditions, under the **Physics/Boundary Settings**, first enter the three electric potentials, -1, 0, and 1 V, shown in Fig. E12.5.1, and, for the Navier-Stokes boundaries, note that the right and upper channel entrances will be "neutral." Otherwise, the boundary conditions are very similar to those of Example E12.4 shown in Table E12.4.2, namely, "electric insulation" for the conductive media mode and "inflow/outflow velocity" for the Navier-Stokes mode.

Initialize the mesh (no refinement is needed) and solve the problem. Because there is now a flow in the y-direction, under **Solve/Solver Parameters** the scaling of variables can be left as **Automatic** under the **Advanced** tab.

Finally, display the results as a superposition of streamlines and equipotentials, as in Fig. E12.5.1. To accomplish this, under **Postprocessing/Plot Parameters**:

1. Under **General**, check both **Contour** and **Streamline**.
2. Under **Contour/Predefined quantities**, select `Electric potential (dc)`.

Repeat the process for the three electric potentials shown in Fig. E12.5.2.

Fig. E12.5.1   Streamlines and equipotentials for applied potentials of 0, 1, and −1 V at the left, right, and upper ends of the branch, respectively.

Fig. E12.5.2   Streamlines and equipotentials for applied potentials of −1, 1, and 0 V at the left, right, and upper ends of the branch, respectively.

3. We wish to ensure ten streamlines, all starting along the right-hand entrance ($x = 5 \times 10^{-4}$) equally spaced between $y = 0$ and $y = 0.5 \times 10^{-4}$. Under **Streamline/Specify start point coordinates**, specify x: `linspace(5e-4,5e-4,10)` and y: `linspace(0,5e-5,10)`.

Discussion of Results

From the two diagrams, it is very apparent that the direction of liquid flow can be controlled by suitable adjustment of the applied voltages. Observe that in both cases the entrance to the system that has no flow has an electric potential (of zero) that is midway between those of the entrances between which there *is* a flow. Note also that the streamlines are everywhere orthogonal to the electric equipotentials.

By a similar adjustment of electric potentials, liquid flows can be controlled in a much more complex arrangement of microchannels. □

12.7 Measuring the Zeta Potential

Two major techniques exist to determine the zeta potential experimentally:

1. Measuring electroosmotic velocities.
2. Measuring the streaming potential.

Electroosmotic velocities. The measurement of electroosmotic velocities is the most straightforward and intuitive technique for measuring the zeta potential. Recall from Eqn. (12.14) that the electroosmotic velocity is proportional to the applied electric field. If the electroosmotic velocity is measured under a known applied electric field, the zeta potential may be calculated directly. Electroosmotic velocities can be measured by optically tracking uncharged fluorescent tracers, weighing a volume of electroosmotically driven fluid to determine the mass flow rate and therefore the average velocity, or by tracking changes in electrical current as fluids with different conductivity are moved through a channel.

Streaming potential. Streaming-potential techniques are a less obvious but effective experimental means for measuring the zeta potential. Consider pressure-driven flow through a capillary with radius a and zeta potential ζ. Assume that the double-layer thickness is small compared to a and that the Debye-Huckel approximation may be made. In this case, the charge density near the wall is given by:

$$\rho_e = -\frac{\varepsilon \zeta}{\lambda_D^2} e^{-x/\lambda_D}. \tag{12.23}$$

The Hagen-Poiseuille flow through this capillary will tend to displace this charge density and transport it along the capillary. This net charge transport (i.e., current) is given by the integral of the velocity/charge-density product:

$$I = \int_{r=0}^{a} u\rho_e 2\pi r \, dr. \tag{12.24}$$

This current leads to a potential buildup at the downstream end of the capillary. At equilibrium, the current induced by the fluid flow is balanced by the current due to this potential. It can be shown (see Problem 12.15) that for a fluid with conductivity σ, the equilibrium potential difference is given by:

$$\Delta V = \frac{\varepsilon\zeta}{\sigma\mu}\Delta p, \tag{12.25}$$

in which $\Delta V = V_{\text{outlet}} - V_{\text{inlet}}$ and $\Delta p = p_{\text{outlet}} - p_{\text{inlet}}$. Eqn. (12.25) is a rather remarkable result. The streaming potential, in the thin-double-layer, Debye-Huckel limit, is *not* a function of the geometry of the capillary, nor is it a function of the double layer thickness!

Example 12.6—Magnitude of Typical Streaming Potentials

Above, we asserted that streaming potential can be used to measure the zeta potential. Of interest is the magnitude of these streaming potentials. Consider a capillary subjected to a pressure difference of 5 atmospheres (this is easy to generate with a handheld syringe), inducing flow of an aqueous solution with a conductivity of 180 μS/cm. If the surface zeta potential is -20 mV, what will the measured streaming potential be? Will these voltages be easy to measure?

Solution

The Helmholtz-Smoluchowski equation gives:

$$\Delta\phi = \frac{\varepsilon\zeta}{\sigma\mu}\Delta p,$$

$$\varepsilon = 80 \times 8.85 \times 10^{-12} = 7.1 \times 10^{-10} \text{ C/V m},$$

$$\sigma = 180 \times 10^{-6} \text{ S/cm} = 180 \times 10^{-4} \text{ S/m},$$

$$\mu = 1 \text{ mPa s} = 10^{-3} \text{ Pa s},$$

$$\Delta p = -5 \text{ atm} = -5 \times 0.101 \text{ MPa} \doteq -0.5 \times 10^6 \text{ Pa},$$

$$\zeta = -20 \text{ mV} = -0.02 \text{ V},$$

$$\Delta\phi = \frac{(7.1 \times 10^{-10})(-0.02)}{(180 \times 10^{-4})(10^{-3})}(-0.5 \times 10^6) = 0.39 \text{ V},$$

in which the outlet (low pressure) will have the more positive voltage. Voltages on the order of 1 volt are quite easy to measure! □

12.8 Electroviscosity

An additional consequence of the streaming potential described in the previous section is a phenomenon known as *electroviscosity*. When pressure is used to induce flow through small capillaries, the observed flow rate is at times below that value calculated using the Hagen-Poiseuille solution for flow in a circular tube. The reason for this is that the streaming potential generated by a pressure-driven flow induces a counterpropagating electroosmotic flow that tends to reduce the net flow rate. If one does not account for electrokinetic effects, this reduction in flow rate appears to result from an increase in viscosity in the fluid, hence the term *electroviscosity*; however, this phenomenon has nothing to do with changes in viscosity.

A detailed treatment of electroviscosity can involve quite detailed and extensive integration; however, we will consider electroviscosity in a simple manner only. In general, pressure applied to a circular pipe or capillary will induce a flow rate from the Hagen-Poiseuille solution:

$$Q = \frac{\pi a^4}{8\mu}\left(-\frac{dp}{dx}\right).$$

(12.26)

Eqns. (12.14) and (12.25) can be combined to give the counterpropagating flow due to streaming potential-induced electroosmosis:

$$Q = \frac{\varepsilon^2\zeta^2\pi a^2}{\sigma\mu^2}\left(-\frac{dp}{dx}\right).$$

(12.27)

For low-conductivity solutions (e.g., deionized water) and small channels ($< 1~\mu$m), the counterpropagating flow can become significant. For electrolyte solutions at 0.1 mM or higher concentration and channels larger than 1 μm, electroviscosity can safely be neglected.

12.9   Particle and Macromolecule Motion in Microfluidic Channels

In this section we are concerned with the motion of particles and macromolecules in microfluidic channels. The particles of interest include:

- Solid microspheres, typically made from latex or silica, often used for a variety of biochemical assays.
- Insoluble colloids.
- Cells.

The macromolecules of interest are typically polymers, and include:

- Synthetic polymers (e.g., polyacrylamide, polyethylene glycol).
- Proteins (polymers of amino acids).
- Ribonucleic acids (e.g., DNA and RNA, polymers of ribonucleotides).
- Carbohydrates (polymers of sugars).

We will start by considering particles. The particles of interest in microscale flows typically have sizes ranging from 10 nm to 10 μm; if we assume typical velocities of 10^{-5} to 10^{-3} m/s, typical Reynolds numbers (in water) will range from 10^{-2} to 10^{-7}. Thus, the flow around these particles is well within the low Reynolds number Stokes limit.

First, consider the Stokes drag on a particle. Recall the following from Chapter 4:

1. The flow regime (laminar vs. turbulent) is based on the particle Reynolds number $\text{Re} = \rho u d/\mu$, where d is the particle diameter.
2. The transition from laminar to turbulent occurs gradually, and only flows with $\text{Re} < 1$ are completely laminar.
3. The drag coefficient $C_D = F_D A_p/(\rho u_\infty^2/2)$ is used for correlating the drag with the Reynolds number.
4. For $\text{Re} < 1$, $C_D = 24/\text{Re}$ and Stokes law gives $F_D = 3\pi\mu u_\infty D$.
5. Particles settle when a body force such as gravity is applied. The terminal velocity is reached when the drag force equals the body force.

Example 12.7—Gravitational and Magnetic Settling of Assay Beads

Problem Statement (Part 1)

Numerous biochemical assays use solutions of magnetic beads. Typically, a solution of beads is flushed into a system where a chemical reaction takes place at the surface of the beads (perhaps chemical binding due to antibodies). Magnets are used to control whether beads flow freely or are held stationary, as in Fig. E12.5. Two characteristic settling times are important:

1. The gravitational settling time (i.e., how long can a solution of beads be stored before the beads fall to the bottom of the container?).
2. The magnetic trapping time (i.e., how quickly will a magnet pull the particles into a wall?).

Given that $\text{Re} < 1$ for a microsphere, derive the terminal velocity for a microsphere. Estimate the settling time in hours for 1-μm diameter magnetic microspheres in water in a container of height $L = 1$ cm by calculating the time required for particles at the terminal velocity to travel 0.5 cm. Assume that the magnetic microspheres have a density of 4 g/ml.

Solution

For $\text{Re} < 1$, $F_D = 3\pi\mu u_\infty D$. Since the drag force must balance the gravitational body force $(\pi D^3/6)(\rho_s - \rho_f)g$:

$$3\pi\mu u_\infty D = \frac{\pi D^3}{6}(\rho_s - \rho_f)g.$$

Example 12.7—Gravitational and Magnetic Settling of Assay Beads   **663**

First, solve for u_∞:

$$u_\infty = u_t = \frac{D^2(\rho_s - \rho_f)g}{18\mu}.$$

Then evaluate:

$$u_\infty = u_t = \frac{(10^{-6}\text{m})^2 \left((4-1) \times 10^3 \text{ kg/m}^3\right)(9.8 \text{ m/s}^2)}{18 \times (10^{-3} \text{ Pa s})} = 1.6 \ \mu\text{m/s},$$

$$t = \frac{L}{u_t} = \frac{0.5 \times 10^{-2} \text{ m}}{0.16 \times 10^{-6} \text{ m/s}} = 3{,}125 \text{ s} = 0.87 \text{ hours}.$$

Problem Statement (Part 2)

Assume that a magnet is used to apply a magnetic force F_M of approximately 1 pN to these microspheres, which is a typical value for polystyrene beads with ~15% ferrite and typical rare-earth magnets. How many seconds would it take for these microspheres to move from one end of a 20-μm diameter channel to the other? As before, use the terminal velocity to estimate the elapsed time.

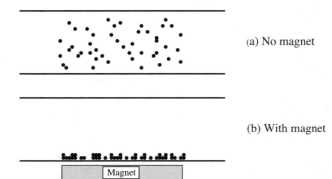

(a) No magnet

(b) With magnet

Magnet

Fig. E12.5   Magnets can be used to control the movement of particles. This is typically used to facilitate flushing steps for chemical assays.

Solution

$$3\pi\mu u_\infty D = F_M, \qquad u_\infty = u_t = \frac{F_M}{3\pi\mu D},$$

$$u_\infty = u_t = \frac{10^{-12} \text{ N}}{3\pi(10^{-3} \text{ Pa s})(10^{-6} \text{ m})} = 106 \ \mu\text{m/s},$$

$$t = \frac{L}{u_t} = \frac{20 \ \mu\text{m}}{106 \ \mu\text{m/s}} = 0.19 \text{ s}.$$

$\square$

Electrokinetic forces.   Having discussed the impact of Stokes flow solutions on particle motion in a flow with body forces, let us now turn to particle migration with surface forces, namely *electrokinetic* forces. As was indicated earlier, voltage is often used in microfluidic systems to actuate ions, particles, or fluids. *Electrophoresis* is the term used to denote the motion of an electrically charged molecule or particle in response to an electric field.

The electrophoretic velocity of a particle can be predicted using arguments similar to those used to derive the electroosmotic velocity of flow past an immobile surface. Presuming the double layer is thin compared to the diameter of the particle, the one-dimensional relations derived earlier hold, and the electrophoretic velocity of the particle—*relative to any other liquid motion*—is again given by:

$$\mathbf{u}_{ep} = \frac{\varepsilon \zeta_p}{\mu} \mathbf{E}, \tag{12.28}$$

where ζ_p is the zeta potential of the *particle surface*. Note the absence of a minus sign in Eqn. (12.28)—in *electrophoresis* a negative particle (for example) is attracted to a positive electrode, while in *electroosmosis* the positive ions near a negative wall are attracted to a negative electrode. Eqn. (12.28) further assumes that the electric field is not perturbed by the presence of the double layer nor is the double layer perturbed by the particle motion. Both of these assumptions often do not hold rigorously.

In the limit where the double layer is very thick compared to the particle size, a relation similar to Eqn. (12.28) can be derived, which differs only by a factor of two-thirds:

$$\mathbf{u}_{ep} = \frac{2}{3} \frac{\varepsilon \zeta_p}{\mu} \mathbf{E}. \tag{12.29}$$

Henry's equation (not shown here) can be used to derive the electrophoretic velocity for double layers that are of similar order to the particle diameter; however, since Eqns. (12.28) and (12.29) are only approximately correct, the detailed solution of Henry's equation typically does not greatly improve the accuracy of these relations.

Electrophoretic separations.   One common application of microfluidic channels, particularly commercially, is for electrophoresis separations of chemical species or particles. Recall from earlier in the chapter that chemical species each have an electrophoretic mobility that, when multiplied by the electric field, gives the motion of the species. Similarly, we can define an electrophoretic mobility of a particle from Eqns. (12.28) and (12.29):

$$\mu_{ep} = A \frac{\varepsilon \zeta_p}{\mu}, \tag{12.30}$$

where the factor A has unity value for thin double layers, two-thirds for thick double layers, and is otherwise given by Henry's equation. For ions, μ_{ep} is typically defined phenomenologically by $\mathbf{u}_{ep} = \mu_{ep}\mathbf{E}$ and cannot be related directly to fundamental parameters.

A capillary electrophoresis separation entails applying an electric field to a microchannel containing a short bolus of fluid (e.g., Fig. 12.2) containing multiple chemical species. The electric field in general induces two phenomena, namely electroosmotic fluid flow and electrophoretic molecular migration. Since the flow can be assumed uniform if the channel is large compared to the double layer thickness, electroosmosis transports all chemical species equivalently down the microchannel. Electrophoresis induces electromigration of each chemical species j such that its velocity is the vector sum of the electroosmotic flow and the electrophoretic migration of that species:

$$\mu_{net,j} = \mu_{eo} + \mu_{ep,j}. \tag{12.31}$$

Further, assuming the electrophoretic mobilities of the different species are different, species will in general move away from each other with a velocity equal to the electric field times the difference in electrophoretic mobilities.

Band broadening and resolution. The resolution of a capillary electrophoresis system is a quantitative measure of the ability of the system to resolve different chemical species. Typically, electrophoresis separations are monitored via an *electropherogram*, which is a measurement of some phenomenon related to the passage of each chemical species past a detector near the end of the channel (e.g., laser-induced fluorescence detector, electrochemical detector, absorbance detector). Each chemical species thus appears as a peak in the signal level. The resolution R of a chemical separation is defined for two peaks, and is given (for peaks m and n) by the peak separation Δt_{mn} normalized by the width w of the peaks:

$$R_{mn} = \frac{\Delta t_{mn}}{w}. \tag{12.32}$$

If electroosmotic flow is used and the chemical bands widen only due to diffusion, w is proportional to $\sqrt{t}$ while Δt_{mn} is proportional to t, so the resolution of a capillary electrophoresis system is proportional to $\sqrt{t}$. If there is any pressure-driven flow, though (recall Taylor-Aris dispersion from earlier in this chapter), the resolution may not improve with time.

PROBLEMS FOR CHAPTER 12

1. *Taylor-Aris dispersion—M.* Derive Eqns. (12.2).

2. *Chromatographic separations—D.* Chromatographic separations take advantage of the fact that different chemical species will travel at different speeds through a tube if they have a varying chemical affinity for a wall. In general, if the average flow velocity is U_0, then the velocity of species i is given by:

$$u_i = \frac{U_0}{k_i},$$

where k_i is the retardation coefficient for species i. For species that do not stick to the wall, $k_i = 1$. For species that do stick to the wall, $k_i > 1$.

Chemical separations require that two or more species be separated by a distance d that is greater than the width of each sample band, where the width of each sample band is dictated by the effective diffusion coefficient.

Assume that a band of two species with $k_1 = 1$ and $k_2 = 2$ is forced through a capillary tube of diameter 5 μm and length $L = 30$ cm with a pressure of 3 MPa. Assume species 1 has a diffusivity of 5×10^{-11} m^2/s and species 2 has a diffusivity of 2×10^{-11} m^2/s.

(a) Calculate the diffusion Péclet number for species 1 and 2.

(b) Calculate the effective diffusion coefficient for species 1 and 2.

(c) Using the relation $w = \sqrt{Dt}$—that is, assuming the injected bands are infinitesimally thin at time zero—derive relations for the widths of each band as a function of time.

(d) Define the time at which the two species are separated as the time when the band separation due to differential retardation is equal to the sum of the two band widths. Calculate this time.

(e) Rework the above calculations for an unknown pressure (or average velocity). What velocity is optimal to minimize the separation time?

Hint: Work mainly in terms of symbols and insert numbers when required.

3. *Derivation of the species transport equation—M.* Taking Fick's law as given, derive Eqn. (12.9) in rectangular coordinates. Either consider the fluxes across the faces of a rectangular control volume (e.g., Fig. 5.5) or invoke the properties of the divergence of a vector. What additional assumptions are required?

4. *Derivation of the heat transport equation—M.* Taking Fourier's law as given, derive Eqn. (12.10) in rectangular coordinates. Either consider the fluxes across the faces of a rectangular control volume (e.g., Fig. 5.5) or invoke the properties of the divergence of a vector. What additional assumptions are required?

5. *Response time of electrophoresis—E.* Assume a hydrogen ion is floating in an aqueous solution, and an electric field is applied. When the field is first applied, the velocity of the ion relative to the surrounding fluid is zero, and the ion begins to accelerate due to the force given by Eqn. (12.6). The ion's averaged final velocity can be calculated from the ion mobility. Given these two relations, and neglecting fluid drag, estimate how long it takes for an ion to accelerate to steady-state under application of an electric field.

6. *Electrokinetic pressure generation I—M.* Stokes-flow solutions are *linear* and *superposable*, meaning that solutions satisfying various boundary conditions can be solved separately and added. For flows through tubes in the presence of both electric fields and pressure gradients, this means that the total flow rate Q_{tot} can be calculated by adding the flow rate caused by the electric field Q_{eo} and the flow rate caused by the pressure gradient Q_p. Consider an aqueous solution in a closed tube ($Q_{tot} = 0$) with an electric field applied (Fig. P12.6). Example 12.2 effectively shows how the pressure generated by electroosmotic flow in a closed tube can be calculated.

Fig. P12.6 *Electroosmotic pressure generation in a closed tube. Directions are shown for a negative wall charge. Left: uniform electroosmotic flow is caused by the electric field. Middle: since the closed tube forces a net mass flow of zero, a pressure is generated at right that causes a parabolic flow from right to left. Right: the net flow is a linear sum of the two flows, which integrates to zero but shows net rightward flow near the wall and net leftward flow at center.*

(a) Rework Example 12.2 in the general case. Work in terms of symbols and determine the generated pressure p_L in terms of any or all of ζ, ε, μ, V_0, L, and d.

(b) If our goal is to generate the largest pressure possible, how should the tube size be selected?

(c) What key parameters are not in this relation?

7. *Electrokinetic pressure generation II—D.* Expand the analysis in the previous problem to allow the net flow rate ($Q_{tot} = Q_{eo} - Q_p$) to be a variable (the tube is no longer closed). Assume the solution has a conductivity σ, i.e., that the electrical current I through the flow field is given by $I = V\sigma A$. Derive expressions for the following:

(a) The pressure p_L at the outlet as a function of Q_{tot}.
(b) The mechanical output work $p_L Q_{tot}$.
(c) The electrical input work $V_0 I$.
(d) The thermodynamic efficiency $p_L Q_{tot} / V_0 I$.
(e) Evaluate the thermodynamic efficiency for the conditions in Example 12.2, assuming a conductivity of 100 μS (S = Siemens = ohm^{-1}). Take Q_{tot} to be half of the forward electroosmotic flow in Example 12.2.

8. *Ion fluxes and current—E.* Consider a 10-cm long channel of diameter $d = 50$ μm fabricated in glass with a 1-mM solution (M = mol/l) of sodium nitrate ($\mu = 0.001$ kg/m s) at pH = 7 ([H$^+$] = 10^{-7}, [OH$^-$] = 10^{-7}). Assume the glass surface has a zeta potential of -60 mV at this condition. Assume a voltage of $V_L = 1$ kV is applied at $x = 10$ cm and the location $x = 0$ is grounded ($V = 0$). Calculate:

(a) The steady-state velocity of all four ions if the fluid velocity is assumed small compared to the ion velocities.
(b) The velocity of the bulk fluid motion—is the assumption made in (a) a good one?
(c) The net charge flux.
(d) The current I and power dissipation rate $V_L I$.

9. *Charge density in the electrical double layer—E.* The Boltzmann distribution of concentration of an ion in a potential field is given by $c = c_\infty e^{-qF\phi/RT}$, where q is the ion charge (the valence $z = |q|$). Assuming that the ions come from NaCl, write the concentration distributions of Na$^+$ and Cl$^-$ ions as a function of the potential ϕ. Given that the local charge density ρ_e is defined as the sum of the charge density of all species i ($\rho_e = \sum_i q_i F c_i$), show that the charge density is:

$$\rho_e = -2zFc_\infty \sinh\left(\frac{zF\phi}{RT}\right).$$

10. *Derivation of electroosmotic flow in the double layer—M.* Derive the one-dimensional steady, uniform pressure flow equation in the double layer:

$$\mu \frac{d^2 u}{dx^2} = \varepsilon \frac{d^2 \phi}{dx^2} E_y$$

from the relevant full momentum equations (i.e., Navier-Stokes with the Lorentz body force term) and the Poisson equation.

11. *Debye lengths—E.* Calculate λ_D for the following conditions ($T = 300$ K):
(a) Deionized water ($z = 1$, $c = 10^{-7}$ M).
(b) Water at equilibrium with air, which has dissolved CO$_2$ and HCO$_3^-$ at equilibrium ($z = 1$, $c = 2 \times 10^{-6}$ M).
(c) 1 mM Epsom salts ($z = 2$, $c = 2 \times 10^{-3}$ M).
(d) Phosphate-buffered saline, which is often used for biochemical analysis or temporary cell storage ($z = 1$, $c = 0.18$ M).

12. *Derivation of electroosmotic flow outside the double layer—M.* Integrate the one-dimensional boundary-layer equation derived in Problem 12.10 to derive the relation in Eqn. (12.14) for the electroosmotic flow velocity.

13. *Nonlinear Poisson-Boltzmann solutions: infinite reservoir—D.* Using software such as MATLAB, Excel, or Mathematica, integrate the nonlinear Poisson-Boltzmann equation at $c_\infty = 10^{-3}$ M with boundary conditions: (a) $\phi_{x=0} = \zeta = -100$ mV, (b) $\phi_{x=\infty} = 0$, corresponding to a surface in an infinite reservoir of fluid. Take $z = 1$ and $T = 300$ K, and plot ϕ against the distance x from the wall.

14. *Nonlinear Poisson-Boltzmann solutions: parallel plates—D.* Repeat the previous problem, but use the boundary conditions: (a) $\phi_{x=0} = \zeta = -100$ mV, (b) $(d\phi/dx)_{x=10 \text{ nm}} = 0$, corresponding to two parallel plates separated by 20 nm.

15. *Streaming potential/streaming current—D.* Assume pressure is used to induce laminar flow in a cylindrical tube with diameter $2d$ and surface potential ζ. Assume $d \gg \lambda_D$. Since there is a net excess of counterions in the double layer, the net charge flux $\int u \rho_e \, dA$ caused by the flow is nonzero, i.e., a net current is induced by this flow. Make the Debye-Huckel approximation and perform the following calculations:

(a) Calculate the net charge flux (streaming current) caused by the flow as a function of the applied pressure, surface potential, viscosity, and permittivity.

(b) Assume the system reaches equilibrium, i.e., that a potential is generated by the net advective charge flux so that the total flux (advective plus conductive) is zero. Assuming that the conductive charge flux is $-\sigma AV/L$ (i.e., that excess conductivity in the double layer can be ignored), show that the resulting potential is given by the Helmholtz-Smoluchowski equation:

$$\Delta V = \frac{\varepsilon \zeta}{\sigma \mu} \Delta p.$$

16. *Electroviscosity I—M.* Using Eqns. (12.26) and (12.27), derive a relation for the net flow in a capillary as a function of a, σ, ζ, ε, μ, and dp/dx (assuming thin Debye layers and using the Debye-Huckel approximation). Use this relation to derive a relation for the ratio of the apparent fluid viscosity (i.e., what fluid viscosity would be needed to generate the same net flow rate, ignoring electrokinetic effects) to the true viscosity. Evaluate this ratio for deionized water ($\sigma = 0.1$ μS/cm) and a surface potential of -20 mV.

17. *Electroviscosity II—E.* A simple way to measure the internal diameter of a capillary is to measure the flow rate of fluid with a known viscosity upon application of a known pressure gradient. Suppose you were using a microcapillary system and water to make such a measurement. What would be the simplest and most effective way to ensure that electroviscous effects did not affect your experiment?

18. *Scaling laws for electrophoretic separations—E.* Assume that two chemical species with electrophoretic mobilities μ_{ep1} and μ_{ep2} are separated in a channel of length L using an applied field E. How is the resolution affected by the following:

(a) Doubling the electric field at constant L.

(b) Doubling the voltage at constant E.

(c) Doubling the length at constant V.

19. *True/false.* Check *true* or *false*, as appropriate:

(a)	Electroosmosis is the motion of fluid caused by an electric field.	T ☐	F ☐
(b)	Electroosmotic flow is irrotational everywhere in the flow field.	T ☐	F ☐
(c)	Electrophoretic separations can be used to separate uncharged molecules.	T ☐	F ☐
(d)	Most of the ions in the electrical double layer are counterions.	T ☐	F ☐
(e)	The Debye-Huckel approximation may be made only when the potential is small.	T ☐	F ☐
(f)	The electroosmotic velocity is not a function of the size of the channel.	T ☐	F ☐
(g)	In a channel with $d \gg \lambda_D$, there is an electroosmotic velocity at the center of the channel, but there is no electrostatic force there.	T ☐	F ☐
(h)	Pressure-driven flows in microchannels are more dispersive than electroosmotic flows.	T ☐	F ☐
(i)	Streaming potential measurements require sensitive voltmeters that can measure tiny potentials.	T ☐	F ☐
(j)	In an electrolyte solution with a Debye length of 100 nm, particles of 10 nm and particles of 1000 nm will have the same electrophoretic velocity.	T ☐	F ☐
(k)	Péclet numbers for microscale flows are always < 1.	T ☐	F ☐
(l)	The width of the electrical double layer is independent of the charge of the ions.	T ☐	F ☐

Chapter 13

AN INTRODUCTION TO COMPUTATIONAL FLUID DYNAMICS AND FLOWLAB

13.1 Introduction and Motivation

T HE elegant mathematical formulation of the laws governing fluid dynamics has been complete since the late 19th century. In the ensuing half a century, a tremendous amount of progress has been made by famous scientists and fluid dynamicists to provide insights, clarity, and understanding about the solutions to the core equation system, which consists of the continuity, momentum, and energy equations, together with other supporting equations such as the equation of state for gases and rate equations for chemical reactions, etc. For example, Ludwig Prandtl (1875–1953) developed the boundary-layer theory, which is most crucial to advance the state of the art in modern fluid dynamics; G.I. Taylor (1886–1975) contributed enormously to wide-ranging topics such as hydrodynamic instabilities, statistical theory of turbulence, drops and bubbles, and many other areas; and Theodore von Kármán (1881–1963) laid down the foundations of high-speed aerodynamics and aerothermodynamics, in addition to advancing Prandtl's boundary-layer theory and making numerous other contributions.

But a ponderous fact about fluid dynamics remains unchanged: we are still not able to solve its core second-order nonlinear partial differential equation system analytically, except for some very special cases such as Couette or Poiseuille flow.[1]

In order to obtain approximate solutions to the Navier-Stokes or Euler equations, fluid dynamicists and applied mathematicians had been earnestly developing new theories and tools in numerical analysis along the way. Yet the need for more detailed solutions to the general fluid dynamics equations far exceeded what could be offered by theoretical and numerical analysis alone. Hence the only way for engineers and scientists to patch things up was to rely heavily on experiments in fluids and thermal design—until the arrival of computational fluid dynamics (CFD) in the second half of the 20th century.

CFD combines the fruits of numerical analysis in partial differential equations (PDEs) and linear algebra with the ever-increasing number-crunching power of modern digital computers to solve the governing PDEs approximately. By itself,

---

[1] Another major obstacle is that we still have very limited knowledge in turbulent flows. The cost of the state of our ignorance cannot be easily measured, but it is definitely a huge number.

CFD is not the answer to all questions regarding complex thermal-fluid phenomena, but it has established firmly as an indispensable tool, together with the traditional theoretical and experimental methods, in the analysis, design, optimization, and trouble-shooting of thermal-fluid systems.[2]

In this chapter we can only scratch the surface of the wide-ranging topics in CFD, but we hope the reader will appreciate the importance of the tool so that a further study in CFD is taken as a high priority in preparing and advancing one's own professional career. Thus, we wish to promote the use of CFD generally, and not any particular software. Of necessity, however, we have had to choose two examples—FlowLab and COMSOL in this and the next chapter. These are but two of a multitude of CFD "packages," a fairly complete list of which is available at the website `http://www.cfd-online.com/Resources/soft.html`

CFD applications in chemical engineering.   CFD has been extensively used in the chemical, oil and gas, pharmaceutical, and process industries. In part it is due to advances made in the complex physical models (e.g., multiphase flows, chemically reacting flows, and advanced turbulence models, etc.) that have become available in the general-purpose commercial CFD codes.

In comparison with the "traditional" approach of relying on empiricism and experimental correlations for nonuniform or nonequilibrium conditions in the design, scale-up, and operation of processes and unit operations in the chemical industry, CFD analysis provides a more fundamental view of the picture. The numerical computation contains full-field data of velocity, temperature, and species concentrations, etc., which engineers can use to understand better the pattern and visualize the physics of the flow. Furthermore, many "what-if" studies can be performed by CFD to examine quickly the influences of various parameters (physical or geometrical) on flow patterns and system performance.

Chemical engineers have been successfully utilizing CFD as design tool in a wide range of unit operations. A small sampling is listed below:

1. Chemical reactions: fluidized beds, bubble columns, and packed beds.
2. Heat transfer processes: heat exchangers, boilers, and evaporators.
3. Fluid transport processes: pumps, compressors, manifolds, headers, pipes, and valves.
4. Mixing processes: stirred tank reactors, static mixers, in-line mixers, and jet-mixed systems.
5. Separation processes: cyclones, scrubbers, precipitators, and filtration systems.

For each of the diverse group of operations, CFD simulation provides detailed field data (and the derived fluxes, stresses, and forces, etc. of interest) in the solu-

---

[2] From its brief historical background, it should be clear that CFD is more a branch of numerical analysis applied to fluid dynamics than a computer-aided design tool. Hence, the emphasis for successful CFD practitioners lies primarily on their competence in the knowledge of physics of fluids, heat transfer and other associated phenomena, and numerical methods.

tion domain. In many cases this valuable insight enables engineers to accomplish a better design, which results in improved performance and efficiency, minimization of power consumption and waste, optimization of the process, and better scale-up of the system, etc.

13.2 Numerical Methods

Governing equations.   For a general single-species flow, the mass balance equation and convection-diffusion equation for a scalar ϕ can be written in vector form as follows:

$$\frac{\partial \rho}{\partial t} + \nabla \cdot (\rho \mathbf{v}) = 0, \tag{13.1}$$

$$\rho \frac{\partial \phi}{\partial t} + \rho \mathbf{v} \cdot \nabla \phi = \nabla \cdot (\kappa_\phi \nabla \phi) + S_\phi, \tag{13.2}$$

where ϕ is the scalar variable (e.g., v_x, v_y, v_z, and enthalpy h, etc.), κ_ϕ is the "diffusion" coefficient associated with ϕ, and S_ϕ is usually called the source term in the ϕ equation, but actually it contains all other terms that do not fit into the given format.

Following the chain rule of differentiation, we have:

$$\nabla \cdot (\rho \mathbf{v} \phi) = \rho \mathbf{v} \cdot \nabla \phi + \phi \nabla \cdot (\rho \mathbf{v}), \tag{13.3}$$

$$\frac{\partial (\rho \phi)}{\partial t} = \rho \frac{\partial \phi}{\partial t} + \phi \frac{\partial \rho}{\partial t}. \tag{13.4}$$

Thus, Eqn. (13.2) for ϕ can be rewritten as:

$$\frac{\partial (\rho \phi)}{\partial t} + \nabla \cdot (\rho \mathbf{v} \phi) - \underbrace{\phi \frac{\partial \rho}{\partial t} - \phi \nabla \cdot (\rho \mathbf{v})}_{=0} = \nabla \cdot (\kappa_\phi \nabla \phi) + S_\phi. \tag{13.5}$$

Note that the last two terms on the left-hand side are zero due to the mass conservation equation, Eqn. (13.1). It is customary to join the diffusion term with the convection term together on the left-hand side of the equation as follows:

$$\frac{\partial (\rho \phi)}{\partial t} + \nabla \cdot (\rho \mathbf{v} \phi - \kappa_\phi \nabla \phi) = S_\phi. \tag{13.6}$$

The general convection/diffusion equation is now written in the *strong conservation form* (in which all spatial derivatives are expressed as divergence terms). We will see the benefit of using the conservation form shortly in the next few sections.

Discretization.   The objective of any numerical methods is to reduce a given continuum problem (an infinite number of degrees of freedom (DOFs)) to a discrete problem (a finite number of DOFs). Generally, CFD methods use two levels of discretization or approximation, as follows:

1. *Domain discretization (or grid generation)*: in this step, the spatial domain is subdivided into a number of smaller, regular, and connected "subregions" called *cells* or *elements* (collectively, the cells/elements are called a *grid* or a *mesh*). The distribution of the density of the cells should follow the physics of the flow closely, namely, the grid density should be finer where flow conditions change rapidly (large gradients), and it can be coarser where flow conditions are constant or change very gradually.

2. *Equation discretization*: the governing PDEs are converted into algebraic forms for each and every cell or element by various numerical methods. Details will be shown in the following sections.

The resulting system of algebraic equations is then solved numerically after the application of appropriate boundary and initial conditions (for steady-state problems, initial conditions are simply a set of guessed values to initiate the computation).

There exist many well-established ways for discretizing the PDEs, such as: finite-difference, finite-volume, finite-element, spectral-element, boundary-element methods, and so on, and there are still more variants even within each "family." Because of its applicability to FlowLab we will focus on finite-volume methods, with briefer mentions of finite-difference and finite-element methods.

Finite-difference methods.   Finite-difference methods (FDMs) first discretize the domain with a grid, then approximate each derivative in the governing PDE by using a corresponding truncated Taylor's series expansion. We illustrate with the following two-dimensional diffusion equation (Poisson's equation):

$$\kappa \left(\frac{\partial^2 \phi}{\partial x^2} + \frac{\partial^2 \phi}{\partial y^2} \right) + S(x, y) = 0, \tag{13.7}$$

in which κ is a conductivity or diffusion coefficient and S is a source or generation term. The corresponding grid is shown in Fig. 13.1(a), with Fig. 13.1(b) showing a typical grid point P and its four neighbors, in which subscripts i and j denote positions in the x and y directions, and Δx and Δy are the corresponding grid spacings.

The Taylor's series expansion for ϕ at the right-hand point $(i+1, j)$ about its central value $\phi_{i,j}$ is:

$$\phi_{i+1,j} = \phi_{i,j} + \left(\frac{\partial \phi}{\partial x} \right)_{i,j} \Delta x + \left(\frac{\partial^2 \phi}{\partial x^2} \right)_{i,j} \frac{(\Delta x)^2}{2} + \left(\frac{\partial^3 \phi}{\partial x^3} \right)_{i,j} \frac{(\Delta x)^3}{6} + O((\Delta x)^4),$$

$$\tag{13.8}$$

where we have used $\phi_{i,j}$ to represent $\phi(x_i, y_j)$, and $O((\Delta x)^4)$ indicates that the remaining terms in the expansion have a truncation error of order $(\Delta x)^4$. Similarly:

$$\phi_{i-1,j} = \phi_{i,j} - \left(\frac{\partial \phi}{\partial x}\right)_{i,j} \Delta x + \left(\frac{\partial^2 \phi}{\partial x^2}\right)_{i,j} \frac{(\Delta x)^2}{2} - \left(\frac{\partial^3 \phi}{\partial x^3}\right)_{i,j} \frac{(\Delta x)^3}{6} + O((\Delta x)^4).$$

$$(13.9)$$

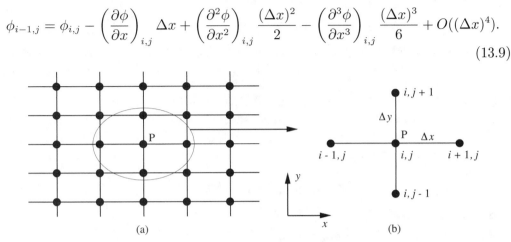

Fig. 13.1   (a) A simple two-dimensional finite-difference grid; (b) a representative point P and its four neighbors.

Addition of the last two equations and rearrangement gives us the *central-difference* approximation for $\partial^2 \phi / \partial x^2$ at (x_i, y_j):

$$\left(\frac{\partial^2 \phi}{\partial x^2}\right)_{i,j} = \frac{\phi_{i-1,j} - 2\phi_{i,j} + \phi_{i+1,j}}{(\Delta x)^2} + O((\Delta x)^2), \qquad (13.10)$$

with a similar approximation for $\partial^2 \phi / \partial y^2$. Substitution of the difference approximations for the derivatives in Eqn. (13.7), and approximation of the source term $S(x_i, y_j)$ by $\overline{S}_{i,j}$ (an averaged value in the vicinity of point P) gives:

$$\kappa \left(\frac{-\phi_{i-1,j} + 2\phi_{i,j} - \phi_{i+1,j}}{(\Delta x)^2} + \frac{-\phi_{i,j-1} + 2\phi_{i,j} - \phi_{i,j+1}}{(\Delta y)^2} \right) = \overline{S}_{i,j}, \qquad (13.11)$$

which is a discretized finite-difference form of the original differential equation.

Repetition of the procedure for each point in the two-dimensional grid yields a system of algebraic equations, which—after incorporating known values from the boundary conditions—has the same number of equations as the number of unknowns (the values of $\phi_{i,j}$ at the grid points, $i = 1, 2, \ldots, j = 1, 2, \ldots$). This set of equations can be solved for the unknowns $\phi_{i,j}$ by numerical methods developed for linear algebra.

The big advantage of the FDM is its simplicity. A major disadvantage is that for nonrectangular regions, especially with non-Dirichlet boundary conditions, special approximations have to be used to accommodate curved boundaries.

Finite-volume methods. In order to demonstrate the salient features of the *finite-volume methods* (FVM), we use a steady-state, convection-diffusion PDE in strong conservation form for illustration:

$$\nabla \cdot (\rho \mathbf{v} \phi - \kappa_\phi \nabla \phi) = S_\phi. \tag{13.12}$$

Without losing generality of the algorithm development, we will discretize the equation for a general two-dimensional mesh as shown in Fig. 13.2, in which the solution domain is subdivided into a finite number of nonoverlapping small control volumes, called cells. The computational nodes are placed at the *centroid* of each cell, in other words, all unknowns are stored in the cell centers. The starting point of the finite-volume methods is to use the integral form of the PDE with respect to each cell's domain (Ω):

$$\int_\Omega \nabla \cdot (\rho \mathbf{v} \phi - \kappa_\phi \nabla \phi) \, d\Omega = \int_\Omega S_\phi \, d\Omega. \tag{13.13}$$

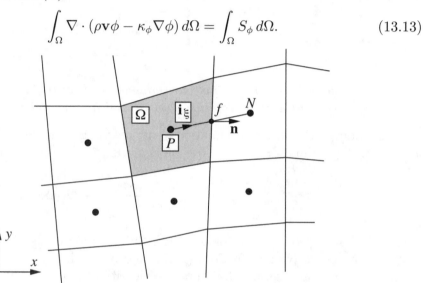

Fig. 13.2   *A typical nonorthogonal two-dimensional finite-volume grid around a cell P and the notation used in the equation discretization.*

Applying the Gauss divergence theorem in vector calculus, the volume integral on the left-hand side of Eqn. (13.13) becomes a surface integral over the boundary surface of the cell:

$$\int_S (\rho \mathbf{v} \phi - \kappa_\phi \nabla \phi) \cdot \mathbf{n} \, dS = \int_\Omega S_\phi \, d\Omega, \tag{13.14}$$

where $\mathbf{n}$ is the outward unit normal vector of the boundary surface S. For the sake of clarity, we separate the left-hand side integral of Eqn. (13.14) into two parts:

$$\underbrace{\int_S (\rho \mathbf{v} \phi) \cdot \mathbf{n} \, dS}_{\text{Convective term}} - \underbrace{\int_S (\kappa_\phi \nabla \phi) \cdot \mathbf{n} \, dS}_{\text{Diffusive term}} = \int_\Omega S_\phi \, d\Omega. \tag{13.15}$$

At this point, the advantage of the finite-volume formulation should become clear: it is inherently conservative because the strong conservation form provides for each cell or control volume a balance of fluxes across the cell's boundary surface, just like we apply conservation laws to an open system.[3]

Each finite-volume cell is bounded by a group of flat boundary surfaces, hence the surface integral can be expressed as the sum of integrals over all faces of the cell. Now the convective flux in Eqn. (13.15) can be treated as follows:

$$F^C \equiv \int_S (\rho \mathbf{v} \phi) \cdot \mathbf{n} \, dS = \sum_f \int_f (\rho \mathbf{v} \phi) \cdot \mathbf{n} \, dS \approx \sum_f (\rho \mathbf{v} \phi)_f \cdot \mathbf{S}_f, \qquad (13.16)$$

where $\mathbf{S}_f = \mathbf{n} S_f$ is the surface normal vector of face f. In Eqn. (13.16) we adopt the "midpoint rule" to calculate approximately the surface integral on a face by the inner product of the integrand evaluated at the centroid of the face and the face vector $\mathbf{S}_f$. The midpoint rule can be shown to be second-order spatially accurate. The diffusive flux in Eqn. (13.15) can be treated similarly:

$$F^D \equiv \int_S (\kappa_\phi \nabla \phi) \cdot \mathbf{n} \, dS \approx \sum_f (\kappa_\phi \nabla \phi)_f \cdot \mathbf{S}_f. \qquad (13.17)$$

Both the convective and diffusive fluxes in Eqns. (13.16) and (13.17) require knowledge of solution variables and properties on the cell's boundary faces, but in the cell-centered finite-volume formulation (see Fig. 13.2), all unknowns are stored in the cell centroids, e.g., ϕ, velocity components (v_x, v_y, v_z), pressure (p) and physical properties (μ, ρ, k, . . .), etc. Hence we need to utilize some kind of interpolation or approximation techniques to evaluate the integrands on the cell faces in Eqns. (13.16) and (13.17) from the values at the cell centers. Many methods have been developed for this purpose, and here we illustrate the procedure with a general technique that can be used in any kind of mesh topology. First, we turn to the Taylor's series expansion of ϕ in multiple dimensions:

$$\phi(\mathbf{r}) = \phi_P + (\mathbf{r} - \mathbf{r}_P) \cdot (\nabla \phi)_P + O(|(\mathbf{r} - \mathbf{r}_P)^2|), \qquad (13.18)$$

where P denotes the centroid of a cell, and $\mathbf{r}$ is the position vector. In other words, it states that the following linear distribution of ϕ about P:

$$\phi(\mathbf{r}) \approx \phi_P + (\mathbf{r} - \mathbf{r}_P) \cdot (\nabla \phi)_P, \qquad (13.19)$$

is second-order spatially accurate because the truncation error term scales with $O(|(\mathbf{r} - \mathbf{r}_P)^2|)$. In order to use Eqn. (13.19), we need to know the gradient $(\nabla \phi)_P$

---

[3] In contrast, a finite-difference formulation based on PDEs in nonconservation form may lead to numerical difficulties in situations where the dependent variables may be discontinuous, as in flows containing shocks.

at the centroid. The following vector calculus identity derived from the Gauss divergence theorem is very handy for that purpose:

$$\int_{\Omega} \nabla \phi \, d\Omega = \int_{S} \phi \, \mathbf{n} \, dS. \tag{13.20}$$

By using the midpoint rule in Eqn. (13.20), the cell-centroid gradient can be calculated as

$$(\nabla \phi)_P \approx \frac{1}{V_P} \sum_f \phi_f \mathbf{S}_f, \tag{13.21}$$

where V_P is the cell's volume. The value of ϕ_f at the boundary-face center can be linearly reconstructed by using Eqn. 13.19:

$$\phi_f \approx \phi_P + (\mathbf{r}_f - \mathbf{r}_P) \cdot (\nabla \phi)_P. \tag{13.22}$$

Now that the variables on each face of cell P are obtained, the convective flux F^C in Eqn. (13.16) can be calculated (simple initializations and the iterative nature of the CFD solver will take care of the "circular relationship" between $(\nabla \phi)_P$ and ϕ_f after just a few iterations). However, we need to point out that the linear reconstruction scheme outlined here belongs to the family of central-differencing scheme (CDS), which is known to cause the problem of spurious oscillations, or unboundedness, of the solution when the solution field changes rapidly in a region with a relatively coarse mesh. An alternative approach that guarantees boundedness is to determine ϕ_f according to direction of the flow approaching the face as follows:

$$\phi_f = \begin{cases} \phi_P, & \text{when } F^C \geq 0, \\ \phi_N, & \text{when } F^C < 0. \end{cases} \tag{13.23}$$

This is called the *upwind differencing scheme* (UDS). Unfortunately the boundedness of the UDS is secured at the expense of accuracy, since it is only first-order spatially accurate. In fact, the UDS scheme effectively introduces an excessive amount of "numerical diffusion" to the discretization of the convective terms. One trick to preserve the boundedness and accuracy simultaneously is to use a deferred correction scheme:

$$F^C = (F^C)_{UDS} + \underbrace{\left[(F^C)_{CDS} - (F^C)_{UDS} \right]^{old}}_{\text{From the previous iteration}}. \tag{13.24}$$

Another way is to adopt one of the bounded second-order upwind schemes, which will not be discussed here. Interested readers may refer to the references for details [3, 4, 11].

In order to calculate the diffusive flux F^D, we need to obtain the gradient of ϕ at the cell face centers in Eqn. (13.17). For the face f of control volume P in

Fig. 13.2, notice that the only contribution to the diffusive flux through f is from the derivative in the surface-normal direction (n):

$$F_f^D = (\kappa_\phi)_f \underbrace{(\nabla\phi)_f \cdot \mathbf{n}}_{=(\partial\phi/\partial n)_f} S_f = (\kappa_\phi)_f \left(\frac{\partial\phi}{\partial n}\right)_f S_f. \tag{13.25}$$

To represent $(\partial\phi/\partial n)_f$ implicitly (i.e., by using the unknowns stored at the neighboring centroids), the derivative in Eqn. (13.25) is approximated by a central differencing scheme in the ξ direction (the direction of the line connecting centroids P and N), then correcting by the difference between the gradients in the ξ and n directions obtained explicitly from the previous ("old") iteration:

$$F_f^D = (\kappa_\phi)_f S_f \frac{\phi_N - \phi_P}{L_{\overline{PN}}} + (\kappa_\phi)_f S_f \overline{(\nabla\phi)}_f^{\,\text{old}} \cdot (\mathbf{n} - \mathbf{i}_\xi), \tag{13.26}$$

where $L_{\overline{PN}}$ is the distance between P and N, and $\mathbf{i}_\xi$ is the unit vector along line $\overline{PN}$. To obtain $\overline{(\nabla\phi)}_f^{\,\text{old}}$ we may first use Eqn. (13.21) to calculate the cell-centered gradients at P and N, then interpolate to face f:

$$\overline{(\nabla\phi)}_f^{\,\text{old}} = r_d (\nabla\phi)_P + (1 - r_d) (\nabla\phi)_N, \qquad \text{where} \quad r_d = \frac{L_{\overline{fN}}}{L_{\overline{PN}}}. \tag{13.27}$$

The approximation of the diffusive flux F_f^D in Eqn. (13.26) is second-order accurate on uniform grids, free of potential un-physically oscillatory solutions, and applicable to arbitrary, nonorthogonal finite-volume meshes.

Finally, the discretization of the source term on the right-hand side of Eqn. (13.15) is carried out by first "linearizing" S_ϕ as:

$$S_\phi = S_U + S_P\phi, \tag{13.28}$$

where S_U and S_P are expressed in terms of constants and/or known values of field variables (possibly ϕ itself) based on the previous iteration. Then the volume integral of the source term is calculated as:

$$\int_\Omega S_\phi \, d\Omega \approx (S_U + S_P\phi_P)V_P. \tag{13.29}$$

By substituting the discretized convection, diffusion, and source terms back into Eqn. (13.15), we can rewrite the discretization equation in the following format:

$$a_P\phi_P + \sum_K^{\text{nb}} a_K\phi_K = b, \tag{13.30}$$

where the summation is over all neighboring cells (nb) sharing faces with cell P. In summary, the discretized equation quantitatively relates the unknown variables (ϕ_K) stored in neighboring cell centroids and sources to ϕ_P. When this procedure is repeated for each cell throughout the domain, and after appropriate boundary conditions are applied, we obtain a large system of algebraic equations, expressed as:

$$\mathbf{A}\phi = \mathbf{b}, \tag{13.31}$$

where $\mathbf{A}$ is the system matrix, $\mathbf{b}$ is the right-hand-side vector, and ϕ is the solution vector. This large, sparse system must be handled by iterative solution techniques. First, the so-called *inner iterations* start solving the linearized system by some iterative methods (for instance, Gauss-Seidel or the more advanced Krylov subspace methods) [4]. Then in the *outer iterations* the entries in $\mathbf{A}$ and $\mathbf{b}$ are updated so that the next round of inner iterations can work on an improved linearized system. This cycle is repeated until a converged result is obtained.

Within the limitations of space and scope in this chapter, we have managed to cover a central part of the finite-volume algorithm. For a complete treatment of the FVM, many topics—such as how to calculate the pressure and treat the pressure-velocity link (i.e., how to treat the continuity equation), and methods for unsteady problems, etc., need to be covered. But we will leave those topics for interested readers to pursue in the references [4, 12].

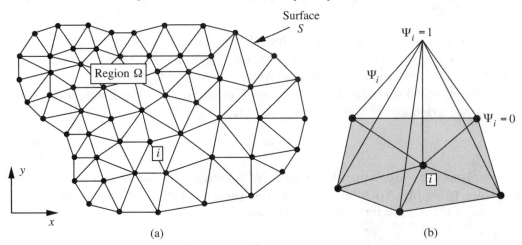

Fig. 13.3   (a) Subdivision of a region into triangular finite elements, (b) a typical node i and neighboring triangular elements, with a linear shape function Ψ_i.

Finite-element methods.   We again use the diffusion equation for a brief discussion of the finite-element method (FEM). At its simplest, as in Fig. 13.3(a), the region Ω is subdivided into a number of triangular elements interconnected at a total of n nodes. Note that by using varying sizes and orientations of elements, a curved boundary can be approximated fairly easily. As in Fig. 13.3(b), each node i has associated with it a *shape* or *basis* function Ψ_i that equals one at node i and declines linearly to zero at all the immediately adjacent nodes (and is zero beyond them).

Let $\overline{\phi}$ represent an *approximate* solution to the PDE of Eqn. (13.7). If ϕ_i denotes the value of $\overline{\phi}$ at node i, the dependent variable $\overline{\phi}$ at any point is expressed by:

$$\overline{\phi} = \sum_{i=1}^{n} \phi_i \Psi_i(x, y), \qquad (13.32)$$

which amounts to representing $\overline{\phi}$ over the whole region by linear interpolation of the nodal values ϕ_i.

Since $\overline{\phi}$ is an approximate solution, it does *not* satisfy the differential equation exactly, there being a non-zero *residual* R when $\overline{\phi}$ is substituted for ϕ:

$$\nabla \cdot \kappa \nabla \overline{\phi} + S(x, y) = \frac{\partial}{\partial x} \kappa \frac{\partial \overline{\phi}}{\partial x} + \frac{\partial}{\partial y} \kappa \frac{\partial \overline{\phi}}{\partial y} + S(x, y) = R. \qquad (13.33)$$

Note that the PDE of Eqn. (13.33) is more general than Eqn. (13.7) and allows for a spatial variation of κ—a feature that is easily accommodated by the FEM.

The next step employs the method of *weighted residuals*, in which the weighted integrals of R over the entire region Ω are set to zero, as expressed in the following n equations:

$$\int_{\Omega} w_j R \, d\Omega = \int_{\Omega} \Psi_j \left[\nabla \cdot \kappa \nabla \overline{\phi} + S(x, y) \right] d\Omega = 0, \quad j = 1, 2, \ldots, n. \qquad (13.34)$$

Here, the w_j are n different *weight functions*, which—in line with the popular *Galerkin method*—are simply chosen to be the same as the shape functions, so that $w_j = \Psi_j$. The formidable-looking integral over the whole region Ω is greatly simplified because Ψ_j is non-zero only in the immediate vicinity of node j, and the integral therefore reduces to summation of contributions from a mere handful of elements. The method essentially amounts to satisfying the PDE in an average way in n small subregions centered on each of the nodes.

The discretized equations for the Galerkin method are obtained from Eqn. (13.34) by substituting the shape function representation for $\overline{\phi}$ from Eqn. (13.32). The integral of $\Psi_j \nabla \cdot \kappa \nabla \overline{\phi}$ is further simplified by applying Green's formula, which amounts to integration by parts in any number of space dimensions:

$$\int_{\Omega} \Psi_i \nabla \cdot \kappa \nabla \overline{\phi} \, d\Omega = -\int_{\Omega} \kappa \nabla \Psi_i \cdot \nabla \overline{\phi} \, d\Omega + \int_{S} \Psi_i \kappa \frac{\partial \overline{\phi}}{\partial n} \, dS. \qquad (13.35)$$

Eqn. (13.35) has two important merits:

1. It replaces a second-order derivative ($\Psi_j \nabla \cdot \kappa \nabla \overline{\phi}$) by the product of two first-order derivatives ($\nabla \Psi_i$ and $\nabla \overline{\phi}$), each of which just amounts to a constant because the shape functions are linear.

2. It introduces the normal derivative of the dependent variable at the surface, which often appears in boundary conditions that involve a conductive or diffusive flux $\kappa \, \partial \overline{\phi} / \partial n$ of mass, energy, or momentum.

Again we will get a large system of algebraic equations, $\mathbf{A}\phi = \mathbf{b}$, for which the coefficient matrix $\mathbf{A}$ is typically "banded," and for which special solution techniques are available for the solution vector ϕ of nodal values $\phi_1, \phi_2, \ldots \phi_n$. Improved accuracy is obtained by using more elements and/or more sophisticated ones that involve quadratic or even higher-order shape functions. The above method is *vertex-based*, but *element-based* methods can also be applied, in which the weighted residual is set to zero for every element.

A major advantage of the FEM is that the mathematical properties of its algorithms are developed more rigorously than those of the FVM and FDM. For example, the convergence analysis and error estimates of the FEM algorithms can be shown to fit snugly into the framework of modern theory of differential equations [7]. A major disadvantage of the FEM for CFD is its substantially higher cost in CPU time and computer memory in comparison with the other two methods.

13.3 Learning CFD by Using FlowLab

The study of CFD for engineers should not end with theories only, it also needs to include practical examples in order to gradually build up the necessary problem-solving capabilities. In the remaining sections of this chapter, we will utilize FlowLab from Fluent Inc. to facilitate this process.

FlowLab is an educational CFD software package (based on the FVM) designed to be a *virtual fluid laboratory* using computers to conduct numerical experiments, in order to teach and enhance physical concepts in fluid mechanics and heat transfer. FlowLab provides an easy and friendly user interface so that users can start using CFD before acquiring more extensive knowledge in its theories and methodologies. It is available on the current Linux and Microsoft Windows platforms. Information regarding FlowLab's software license, installation, documentation and support, etc. can be found at the website: `http://flowlab.fluent.com`

FlowLab structure. FlowLab is designed to integrate seamlessly a CFD preprocessor (to create the domain and generate a suitable mesh), a solver, and postprocessor (to visualize the solution and extract useful information from the vast amount of data) in a single graphical user interface (GUI). This is managed by using problem-specific template files. From the user's viewpoint, FlowLab is built on top of a number of ready-to-use exercises,[4] which are called templates created

---

[4] New exercises are continuously created by FlowLab developers and users in academia. Available FlowLab

from the general-purpose CFD software FLUENT (solver and postprocessor) and its preprocessor called GAMBIT.[5] A sample screen shot of the GUI is shown in Fig. 13.4.

Fig. 13.4 FlowLab's graphical user interface (GUI).

How to use FlowLab.   The following general steps will give you the general idea of using FlowLab to solve a problem in fluid mechanics.

1. When you start FlowLab, a launcher window will appear. Select Start a new session for a new problem, and highlight the template problem you want to work on, then click Start.

Fig. 13.5   FlowLab's Operation Toolpad buttons.

---

exercises include flow over a Clark-Y airfoil, flow in an orifice meter, flow over a heated plate, flow over a cylinder, and a sudden expansion in a pipe, etc. A complete and up-to-date list is available at http://flowlab.fluent.com

[5] FLUENT is a comprehensive finite-volume solver (for both structured and unstructured grids) for a wide-range of flow modeling applications. See the website http://www.fluent.com for additional information about FLUENT.

2. On the upper right portion of the FlowLab GUI, shown in Fig. 13.4, lies the Operation Toolpad, shown in Fig. 13.5. It consists of a series of command buttons, each of which is connected to the corresponding graphical panel. The layout of the buttons from left to right reflects the general order of operation when setting up a CFD problem: Geometry to define the problem domain, Physics to specify physical models, boundary conditions, and material properties, Mesh to create/refine mesh, Solve to start a CFD computation, Reports to analyze the computational results and generate x/y plots, and Postprocessing to create plots of contours, vectors, iso-surfaces, and particle tracks, etc.

Fig. 13.6 The Geometry *panel.*

3. A series of screen shots of the Operation Toolpad panels are presented below in order give you a visual overview of a typical FlowLab session (involving a developing pipe-flow problem):

   (a) Fig. 13.6 gives the Geometry panel for specifying the size and other geometrical features of the problem.

Fig. 13.7 The Physics, Boundary Condition, *and* Materials *panels.*

   (b) Fig. 13.7 shows the panels for Physics, Boundary Condition, and Material properties. Users can freely specify the physical model (laminar or turbulent, with or without heat transfer), boundary condition, and material properties for each simulation so that it is a well-posed problem—in other words, all the model conditions should be physically consistent and correct. For example, one cannot expect FlowLab to generate correct results if the flow is specified as laminar while the Reynolds number is 10^6!

Fig. 13.8   The Mesh *and* Solve *panels.*

(c) Fig. 13.8 shows the Mesh and Solve panels. In the Mesh panel, you can select a mesh from various levels of fineness from the drop-down menu, click the generate button, and a grid will be generated and shown in the graphics window. You can use the mouse to examine the mesh more closely (drag the right button to zoom in or zoom out). In the Solve panel, you can set up the number of iterations, convergence limit (for the scaled residuals of the discretized equations), and also select the locations where you will be able to view profiles of solution variables after the computation is converged. Finally, click the iterate button, the solver will start the calculation, and an x/y plot of the residuals' history will pop up and show the progress of the iteration process (the magnitudes of residuals should drop substantially before the calculation can be declared as a converged solution). After a while it will prompt you when the solution has converged. Now you can move on to the Postprocessing phase to view the results of the numerical model.

(d) Fig. 13.9 shows the **Reports** and **Postprocessing** panels, from where you can generate plots and visualize field data in many different ways, including animations.

Fig. 13.9   The **Reports** *and* **Postprocessing** *panels.*

4. Views and display attributes of the model can be easily manipulated by using **Global Control** buttons that are located in the lower right portion of the FlowLab GUI. A brief description of the function of each button is provided in the **Description** window (located in the lower center of the GUI) when the mouse cursor is placed on top of that button. Detailed description of the use of **Global Control** buttons can be found in the *FlowLab User's Guide*.

5. Before studying the CFD examples in the next section, you should read the first three chapters of the *FlowLab User's Guide* for an overview of the software package and its graphical user interface. A sample CFD session is also included in the documentation to guide you step by step from problem setup to examination of CFD results. You will quickly find out the FlowLab software is very easy to learn and use, so you can concentrate on the physics of the fluid dynamics problems.

13.4 Practical CFD Examples

In the remaining part of this chapter, we solve four example problems to introduce the basic features and limitations of CFD software. For a beginner, it is usually best to learn from simpler problems before trying to attack a more complicated one. In the following examples, we emphasize the proper understanding of the physics of each problem when setting up the case. After we have obtained a converged solution, we need to check the results very carefully in order to make sure the solution is either validated by experimental results or is physically reasonable. You are also encouraged to modify the setup of each case, including the sizes of the flow domain, mesh density, boundary conditions (BCs), and physical properties, etc. in order to build up your own hands-on experience.

Example 13.1—Developing Flow in a Pipe Entrance Region (FlowLab)   **687**

Example 13.1—Developing Flow in a Pipe Entrance Region (FlowLab)

Fig. E13.1.1   *Geometry of a developing pipe flow problem.*

Problem Setup

We consider a steady-state, laminar developing flow (Re_D=300) in the pipe entrance region shown in Fig. E13.1.1.

The flow is steady, incompressible, and axisymmetric. The domain's axial length is at least $50D$ (50 pipe diameters). The following key points should be taken into consideration while setting up the model:

- The flow is laminar when $\text{Re}_D < 2300$.

- Because the gradient of velocity in the near-wall region is high, the mesh should be finer in the radial direction near the wall. We first select the "medium" mesh density and click on the **Create** button to establish a mesh for the computational domain.

Fig. E13.1.2   *An enlarged view of the finite-volume mesh used in the developing flow calculation in FlowLab Example 13.1. It is only shown partially due to the high aspect ratio of the domain.*

- *Inlet* boundary condition: constant velocity at the inlet. (To make the setup process simple for FlowLab users, the following BCs are specified without requesting user's input: *Outlet*: constant static pressure. *Wall*: no-slip. *Axis of symmetry*: axisymmetry.)

- Given a pipe diameter, for example, $D = 0.1$ m (caution: in FlowLab you specify the radius, not the diameter, during the setup of the problem), and length L is 5 m so that $L/D = 50$. In order to ensure that the Reynolds number of the flow is 300, one can pick any fluid (so the kinematic viscosity ν is fixed), then calculate the corresponding inlet velocity. For example, a fluid with $\rho = 1$ kg/m³ and $\mu = 10^{-6}$ kg/m s, we need to set $\overline{V} = 0.003$ m/s for $\mathrm{Re}_D = 300$.

- The mesh ("medium" mesh density) created by FlowLab is shown in Fig. E13.1.2. In this particular mesh there are 6,678 two-dimensional rectangular cells. For calculating the flow, three equations (mass conservation, x- and y-momentum equations) are needed for each cell, so in total 20,034 simultaneous equations are solved numerically. If heat transfer is also calculated (using the energy equation), another 6,678 equations would be added to the system.

Fig. E13.1.3   *Axial velocity profiles for a develop-ing laminar pipe flow* ($Re_D = 300$, $D = 0.1$ m).

Results and Discussion

- The axial length of the pipe entrance region, called the hydrodynamic entrance length L_e, can be shown to scale with $D\,\mathrm{Re}_D$ by using the integral boundary-layer method [13]:

$$L_e \approx 0.058 D\,\mathrm{Re}_D. \tag{E13.1.1}$$

Example 13.1—Developing Flow in a Pipe Entrance Region (FlowLab)   **689**

For $\text{Re}_D = 300$, we get $L_e/D \approx 17.4$. In order to verify this result, we can use FlowLab's convenient x/y plot tool to plot the axial velocity profiles along several axial locations. As shown in Fig. E13.1.3, the velocity profiles (at $x/D = 0$, 5, and 10) are still developing before $x/D = 17.4$. Beyond this point, e.g., at $x/D = 20$, 25, and 30, all profiles become identical and collapse into a single curve in the figure.

• For a fully developed laminar pipe flow, the static pressure should vary linearly along the axial direction (in other words, dp/dx is constant). The plot of static pressure versus the axial distance (x) from our FlowLab calculation verifies this point. You can see pressure variation is closely linear in the fully developed region ($L_e > 0.174$ m) in Fig. E13.1.4.

Fig. E13.1.4 Static pressure distribution in the axial direction for a developing laminar pipe flow ($\text{Re}_D = 300$, $D = 0.1$ m).

• You can also plot the wall friction factor f along the axis. The hydrodynamic entrance length L_e is also defined as the distance required for the friction factor to reach within 5% of its fully developed value. For laminar flow, $f = 64/\text{Re}_D$. Use the plot to verify the CFD results.

• Thermal entrance length (L_{eth}): Perform the heat transfer calculation of this developing flow problem by enabling the energy equation for the model in the **Physics** panel and go through the iterations until convergence. Then plot the wall Nusselt number Nu along the axial direction. The thermal entrance length is defined as the distance from the pipe inlet required for the Nusselt number to reach within 5% of its fully developed value Nu_∞ (for instance, Nu = 4.364 for a constant heat-flux wall). Examine the relationship between the thermal entrance length and the hydrodynamic entrance length L_e for different Prandtl numbers (Pr).

• Grid independence (effect of the grid resolution): Compute the same flow with the coarse and fine mesh densities, then compare the predicted results of the entrance length among the different grids. Verify that the laminar flow entrance length predicted by FlowLab is not affected by the grid resolution.

Example 13.2—Pipe Flow Through a Sudden Expansion (FlowLab)

Problem Setup

Flow through a sudden expansion such as that in Fig. E13.2.1 occurs frequently in a piping system. Here we study an axisymmetric case of such a flow.

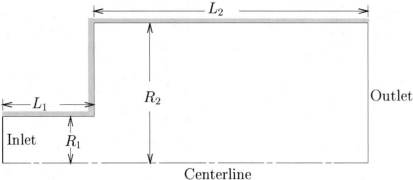

Fig. E13.2.1   *A sudden expansion in a pipe.*

The problem is set up according to the following considerations:

• The flow is assumed to be incompressible—an assumption that is valid for almost all liquids, and for gases when the fluid speed is less than 30% of the local speed of sound.

• The inlet pipe length L_1 should be at least ten times the inlet pipe radius R_1. The length L_2 should be 50 times the step size $(R_2 - R_1)$ or longer. A domain configured as such allows reasonably accurate specifications of the inlet and outlet boundary conditions detailed below. (These are very conservative ratios—you may shorten the outlet and inlet lengths without appreciably changing the numerical results.)

Example 13.2—Pipe Flow Through a Sudden Expansion (FlowLab)   **691**

• Based on the inlet pipe diameter, the flow is laminar when Re < 1,000. At a higher Reynolds number, the flow is in the turbulent regime and needs to be modeled by a suitable turbulence model. Here we use the standard k-ε two-equation model [8]. For an excellent review of eddy-viscosity turbulence models, see [2].

• *Inlet* boundary condition: constant velocity normal to the inlet. To make the problem setup process simple for FlowLab users, the following BCs are specified without requesting user's input. *Inlet*: k and ε are specified via the turbulence intensity and a turbulence length scale. *Outlet*: constant static pressure. *Axis*: axisymmetry. *Wall*: no-slip.

• In order to resolve steep gradients, the mesh needs to be very fine near the wall. We also need a fine mesh for the regions above and below the imaginary line extending downstream from the corner of the step, because there exists a boundary-free shear-layer structure (i.e., with large gradients) beyond the step's corner.[6]

• For a sample calculation, take $R_1 = 0.1$ m, $R_2 = 0.2$ m, $L_1 = 1$ m, $L_2 = 5$ m. Use standard air properties (the default in FlowLab) for the fluid. At the inlet, the velocity $\overline{V} = 1.34$ m/s is given, and the Reynolds number (based on the inlet pipe diameter) for this case is 20,000. Finally, use the "coarse" mesh density and turbulence model's enhanced wall treatment (designated *Enh-wall* in FlowLab) for the wall boundary.[7]

Taking all the points above into account, we can properly set up the governing equations, geometry, fluid properties, boundary conditions, and mesh for the problem in FlowLab.

Results and Discussion

• We can use the postprocessing tools of FlowLab to visualize the computed results. For example, the streamline plot of Fig. E13.2.2 provides us the visualization of the flow separation and reattachment zone behind the step. You can use the tools to plot velocity vectors, contours, and more.

Fig. E13.2.2 Streamline plot for the recirculation zone in a pipe's sudden expansion: Re = 20,000 (based on the inlet pipe diameter), expansion ratio $R_2/R_1 = 2$.

---

[6] This is a good example of applying basic understanding in fluid mechanics when running CFD. The need for a fine mesh in the shear-layer structure is not as obvious as the requirement in the near-wall region, but it is critical for obtaining accurate numerical results.

[7] Enhanced wall treatment is to use an advanced two-layer zonal model for the near-wall region combined with a *blending* wall function for the complete turbulent boundary layer including viscous sublayer, buffer layer, log-law layer and outer layer [5].

• For calculating the head loss generated by a pipe expansion, we can evaluate the discharge coefficient K of the sudden expansion, which is defined as:

$$K = \frac{2(p_1 - p_2)/\rho + (V_1^2 - V_2^2)}{V_1^2}, \qquad \text{(E13.2.1)}$$

where "1" is at the expansion point and "2" is at the reattachment point of the separation. For the current sample calculation ($R_2/R_1=2$), FlowLab reports $K = 0.549$. The discharge coefficient for other expansion ratios can be quickly obtained by changing the expansion ratio (we may still fix $R_1 = 0.1$ m, and Re = 20,000), re-generating the domain and the mesh, then running the CFD calculations. For example, using the coarse mesh option, a few computed versus theoretical values of K [13] are given in Table E13.2.1.

Table E13.2.1   Values of the Discharge Coefficient, K

Expansion Ratio, R2/R1	Computed	Theoretical
1.5	0.265	0.308
2.0	0.549	0.563
2.5	0.699	0.705
3.0	0.785	0.790
3.5	0.839	0.843

• Grid independence: You can run these same cases with different mesh densities and compare the results with theoretical values.

• Many other interesting features of this problem can be examined by using FlowLab. For example: predictions of the re-attachment point, the centerline velocity and pressure distributions, skin friction factor at the wall downstream of the step, and centerline total pressure distribution, etc.

Example 13.3—A Two-Dimensional Mixing Junction (FlowLab)

Problem Setup

This example studies two-dimensional turbulent flow in the channel junction of Fig. E13.3.1. The cold fluid coming into the main channel from the left (inlet 1) is mixed with the same fluid at a higher temperature from a smaller side channel at the corner (inlet 2). We will use this example to investigate the complex flow mixing patterns and the effect of velocity and temperature on the mixing results.

Example 13.3—A Two-Dimensional Mixing Junction (FlowLab)   **693**

• The flow is incompressible and turbulent (one should check the Reynolds number to confirm that this is the case). Since we are interested in the convection heat transfer of the mixing flow, we need to enable the heat transfer calculation in the **Physics** panel.

• The outlet section length L_2 is adjustable by the user. Note that when changing the inlet velocities (especially the velocity of the hot stream), it is possible to cause a flow recirculation zone in the main channel. If that happens, the outlet section length must be long enough in order to get an accurate numerical result. The point is to make sure that the outlet plane is not cutting through the recirculation zone. This requirement is quite common in all CFD simulation, it is due to the fact that general boundary conditions at outlets do not work well when significant backflow occurs.

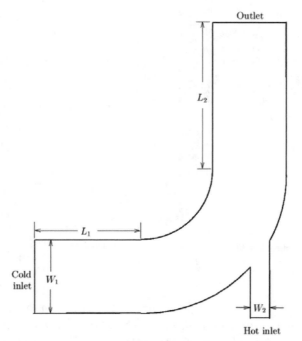

Fig. E13.3.1   Schematic of the two-dimensional mixing channel.

• *Inlet* boundary conditions: constant fluid velocity and temperature are specified at the respective inlets. (To make the setup simple for FlowLab users, the following BCs are specified without requesting user's input. *Wall*: no slip and adiabatic (zero heat flux). *Outlet*: constant pressure.)

• Users must exercise caution and make sure the Reynolds number based on the hydraulic diameter D_H of the main channel ($D_{H_1} = 2W_1$) is high enough for the flow to be turbulent. The standard two-equation k-ε turbulence model is used in this example, together with "Std-wall" (standard wall-function—*the logarithmic law of the wall*) as the wall BC.

• A case under investigation can be set up as follows: Geometry: outlet length $L_2 = 4$ m, cold-inlet width $W_1 = 0.4064$ m, hot-inlet width $W_2 = 0.1016$ m; fluid property $\nu = 10^{-6}$ m^2/s, mesh: medium density; BCs: $T_1 = 293$ K, $\overline{V}_1 = 0.492126$ m/s, $\overline{V}_2 = 1$ m/s, and $T_2 = 313$ K. The resulting main inlet Reynolds number Re$_1$ is 400,000. A second case with identical setup except for a different hot-inlet velocity ($\overline{V}_2$) will be used later for the study of flow separation pattern in the mixing junction.

Fig. E13.3.2   Streamline patterns of the two-dimensional mixing junction flows: all conditions are the same for both cases except that the hot-inlet velocity on the right is 2.5 times higher than that of the case on the left. The higher hot-inlet velocity gives rise to a sizable separation/recirculation zone downstream of the junction.

Results and Discussion

• Streamline and temperature contours for two different flow conditions are shown in Fig. E13.3.2 and Fig. E13.3.3, respectively. The two cases have the same cold-inlet velocity (hence the Reynolds number Re$_1$), but have different velocities at the hot-stream (small) inlet. The streamline contours in Fig. E13.3.2 indicate that the resulting flow patterns are very different between the two. When the ratio of hot-inlet to the cold-inlet velocities $\overline{V}_2/\overline{V}_1$ is 4, no separation occurs at

Example 13.3—A Two-Dimensional Mixing Junction (FlowLab)   **695**

the mixing junction; but when the ratio is 10 a sizable separation and recirculation zone appears. The flow pattern affects very strongly the mixing of the cold and hot streams, as illustrated by the temperature contours shown in Fig. E13.3.3.

Fig. E13.3.3   Temperature contours of the two-dimensional mixing junction flows: all conditions are the same for both cases except that the hot-inlet velocity on the right is 2.5 times higher than that of the case on the left. Due to the presence of a sizable separation/recirculation zone for the case on the right, the temperature distribution of the mixing junction is drastically different than that of the case on the left, which has no separation zone.

• This example serves to illustrate an advantage of CFD in many industrial applications: numerical simulations provide excellent quantitative results for engineering design and analysis. For example, it is almost impossible to predict the flow patterns (with or without separation) under different inlet velocities in the mixing junction by either physical intuition or by using basic principles in fluid mechanics without the knowledge of complete velocity and pressure fields of the flow.

• The standard k-ε turbulence model is known to be very *dissipative*; in other words, the high turbulent viscosity in the recirculation region tends to "damp out" vortices. Therefore when the standard k-ε model predicts separation, it is indeed likely to be the case. But additional validations should be conducted to support this conclusion.

• A cautionary statement should be given here about turbulence models and CFD: although numerous turbulence models have been developed over the last twenty some years, there is not yet a single turbulence model that can reliably predict all turbulent flows in industrial applications with sufficient accuracy.[8] You may find the previous statement very disappointing or even depressing, but the state of our lack of understanding of the essential physics of turbulence was succinctly summarized by Richard P. Feynman (Nobel Laureate in Physics, 1965)— turbulence is *"the most important unresolved problem of classical physics."* Surely we do need turbulence modeling for CFD (so it is very important to know the strengths and weaknesses of turbulence models), but we—as CFD practitioners— also need to examine very carefully the results and conduct further validations, for instance, by comparing CFD predictions with experimental data, or with other similar flow data published in the literature, etc. before jumping to a conclusion.

Example 13.4—Flow Over a Cylinder (FlowLab)

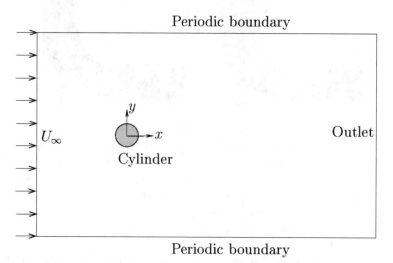

Fig. E13.4.1   *Uniform cross flow over a circular cylinder.*

Problem Setup

Consider the flow in Fig. E13.4.1 about a circular cylinder of diameter D immersed in a steady cross flow of velocity U_∞ (the uniform undisturbed stream velocity). One might guess the flow to be perfectly steady, but, depending on the Reynolds number, it is actually oscillatory. When $\mathrm{Re}_D = \rho U_\infty D/\mu > 40$, the flow

---

[8] Yet proponents of a particular turbulence model might want you to think otherwise.

Example 13.4—Flow Over a Cylinder (FlowLab)   **697**

becomes unsteady and regularly sheds an array of alternating vortices from the cylinder. The flow pattern is called a *Kármán vortex street*, named after Theodore von Kármán, who provided a theoretical analysis of this phenomenon in 1912.

This is a fundamental fluid mechanics problem of practical importance. We will use this example to examine the predicted drag coefficient, the vortex-shedding frequency of the flow, and the beautiful flow patterns.

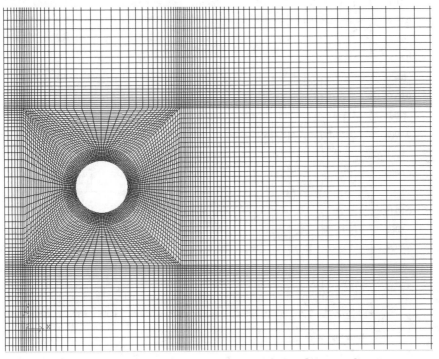

Fig. E13.4.2   An enlarged view of the finite-volume mesh in the vicinity of the cylinder in the cross flow.

- The flow is modeled as a two-dimensional flow over a circle. The flow domain is created as follows: diameter of the cylinder is 0.1 m, the upstream distance is 5 times the diameter, and downstream distance is 25 times the diameter. Width of the domain is 25 times the diameter.

- The flow is laminar if $\mathrm{Re}_D < 1{,}000$. The FlowLab template for flow over a cylinder is designed to work well in the range $1 < \mathrm{Re}_D < 1{,}000$. Here we set $\mathrm{Re}_D = 100$ (say, $\rho = 1$ kg/m^3, $\mu = 10^{-5}$ kg/m s with $U_\infty = 0.01$ m/s), so that the case lies in the range for the laminar, incompressible, and vortex-shedding flow.

- Boundary condition: constant velocity $U_\infty = 0.01$ m/s at the inlet (on the left). To make the setup simple for FlowLab users, the following BCs are specified without requesting user input. Cylinder wall: no-slip. Outlet (on the right): constant pressure. Top and bottom boundaries: *periodic* (which means that solutions along the pair of boundaries are completely identical).

● Initial condition: the following initial condition is given without requesting user input: uniform $u = v_x = 0.01$ m/s; in order to break the symmetry of the flow and "trigger" the vortex shedding, a very small *perturbation* is applied to the flow by initially setting $v = v_y = 0.00005$ m/s.

Fig. E13.4.3   Time history plot of y velocity at a point (x = 0.7 m) in the cylinder's wake.

● Select "medium" mesh and click **Generate** button in the **Mesh** panel. An enlarged view of the finite-volume mesh in the vicinity of the cylinder is shown in Fig. E13.4.2. The quality of the mesh near the cylinder's surface and in the wake of the cylinder is very critical. This mesh has a total of 16,176 quadrilateral cells in the solution domain.

● Starting from a uniform flow approaching the cylinder from the left-hand side, this unsteady-flow case requires a substantial amount of computing time in order to observe counterrotating vortices to develop and then move downstream. Make sure the time-step size (Δt) is not too large (we use a "conservative" $\Delta t = 1$ s to resolve the fluctuations in the time domain) and that the iterations have sufficiently converged during each time step before advancing to the next time step. You can also specify how often (in terms of the flow time) the field data set are saved for visualizing the flow at the postprocessing stage.

Example 13.4—Flow Over a Cylinder (FlowLab)   **699**

- For this current case (medium mesh density, $\Delta t = 1$ s), it takes about 80 minutes of *wall clock time* to simulate 1,000 time-steps (i.e., 1,000 seconds of flow time) on an IBM PC workstation (2.8 GHz Pentium 4 CPU, 2 GB of RAM, Windows XP operating system).

Results and Discussion

- The vortex-shedding frequency n is often expressed as a non-dimensional quantity called the Strouhal number S:

$$S = \frac{nD}{U_\infty}. \qquad (E13.4.1)$$

In the range $40 < S < 10^4$, the Strouhal number depends uniquely on the Reynolds number Re_D. We can use FlowLab's postprocessing feature (a point object) to monitor the time history, e.g., the y velocity history at $x = 0.7$ m in the wake region, shown in Fig. E13.4.3, then derive the frequency of the oscillations from it. The frequency n is simply the inverse of the period, which is 64 seconds measured in the figure between two adjacent peaks. With $U_\infty = 0.01$ m/s and $D = 0.1$ m, the Strouhal number S calculated by FlowLab is 0.156, which is within 5% of the experimental value of 0.165 ± 0.08 [10].

- The drag coefficient of the cylinder is defined as:

$$C_D = \frac{F_{\text{drag}}}{0.5\rho\, U_\infty^2 A}, \qquad (E13.4.2)$$

where A is the *frontal area* as seen from the approaching stream. For a two-dimensional cylinder in a cross flow, $A = D \times 1$ (a unit depth perpendicular to the plane). FlowLab calculations predict $C_D = 1.33$. The reported experimental values of C_D at $Re_D = 100$ have quite a wide variation in the literature, ranging from 1.26 [1], 1.41 [6], 1.46 [9] to 1.6 [10]. The discrepancy is possibly due to the three-dimensional effects or fluctuations of the free stream in the wind tunnel. The numerical prediction given by FlowLab falls right within the range of the data and is considered to be quite satisfactory.

- Figs. E13.4.4 and E13.4.5 show a series of velocity-magnitude contour plots of the cylinder flow, from the initial symmetric flow pattern to the final regular shedding of alternating vortices. The flow is quite symmetric initially ($t < 50$ s); at around 200 s, the tail end of the wake has started to flap. The pattern of alternating vortices shed from the cylinder is increasingly clear at 250 s, and the Kármán vortex street has reached its regular (periodic) stage at $t = 400$ s. In FlowLab you can easily generate *animations* to view such unsteady flow patterns.

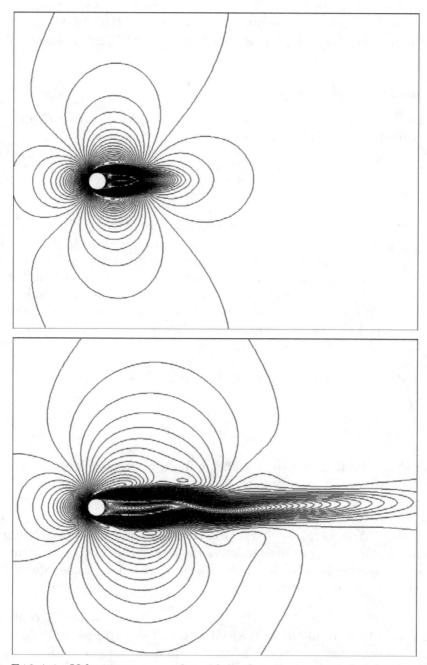

Fig. E13.4.4   Velocity contour plots of the flow over a cylinder ($Re_D=100$), at t = 50 s (top) and t = 200 s (bottom) from the beginning of the simulation. At t = 200 s, the tail end of the wake has started to flap and the initial symmetry (as seen at t = 50 s) of the flow has been destroyed.

Example 13.4—Flow Over a Cylinder (FlowLab)   **701**

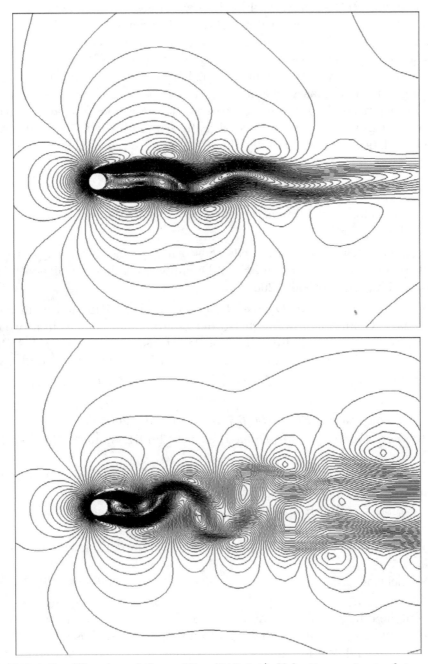

*Fig. E13.4.5   (Continued from Fig. E13.4.4) Velocity contour plots at t =
250 s (top) and t = 400 s (bottom). After being shed from the cylinder, the
vortices move alternately clockwise (top row) and counterclockwise (bottom
row). The Kármán vortex street is already regularly periodic at the later time.*

Summary. In the previous FlowLab examples we have demonstrated that CFD can do a very good job in flows with well-understood physics (e.g., incompressible, steady or unsteady, laminar flow), if and when users provide a *well-posed problem* with a sufficient mesh resolution and with proper boundary/initial conditions. When the flow is turbulent and/or is more involved in terms of complex physics (e.g., multiple phases, chemical reactions, etc., which are beyond the scope of this chapter), users should exercise extra caution in properly selecting turbulence and additional physical models, and critically examine the CFD results to avoid possible blunders in trusting blindly those "seemingly reasonable" predictions.

References for Chapter 13

1. G.K. Batchelor, *An Introduction to Fluid Dynamics*, Cambridge University Press, 1967, p. 261.

2. J. Bredberg, "On two-equation eddy-viscosity models," *Internal Report 01/8*, Dept. of Thermodynamics and Fluid Dynamics, Chalmers University of Technology, Göteborg, Sweden, 2001.

3. C.-Y. Cheng, E.A. Lim, D. Fee, T.M. Foster, H.S. Kunanayagam and P.R. Tuladhar, "Comparisons of higher-order differencing schemes in a two dimensional, incompressible finite-volume scalar transport code," *Joint ASME and JSME Fluids Engineering Conference*, Hilton Head, SC, August 13–18, 1995.

4. J.H. Ferziger and M. Perić, *Computational Methods for Fluid Dynamics*, Springer Verlag, 2nd ed., 1999.

5. Fluent Incorporated, *FLUENT 6.1 User's Guide*, 2003.

6. S. Hoerner, *Fluid-Dynamic Drag*, Hoerner Fluid Dynamics, Bakersfield, CA, 1965.

7. A. Iserles, *A First Course in the Numerical Analysis of Differential Equations*, Cambridge University Press, Cambridge, 1996.

8. B.E. Launder and D.B. Spalding, *Lectures in Mathematical Models of Turbulence*, Academic Press, London, 1972.

9. A.F. Mills, *Basic Heat and Mass Transfer*, Richard D. Irwin, Inc., Chicago, 1995, p. 251.

10. H. Schlichting, *Boundary-Layer Theory*, 7th ed., McGraw-Hill, New York, 1968.

11. J.C. Tannehill, D.A. Anderson, and R.H. Pletcher, *Computational Fluid Mechanics and Heat Transfer*, 2nd ed., Taylor & Francis, Washington DC, 1997.

12. H. Versteeg and W. Malalasekra, *Introduction to Computational Fluid Dynamics: The Finite-Volume Approach*, Prentice-Hall, Upper Saddle River, NJ, 1996.

13. F.M. White, *Fluid Mechanics*, 4th ed., McGraw-Hill, New York, 1999.

Chapter 14

COMSOL (FEMLAB) MULTIPHYSICS FOR SOLVING FLUID MECHANICS PROBLEMS

14.1 Introduction to COMSOL Multiphysics

Chapter 13 introduced the general concept and importance of computational fluid dynamics (CFD), together with the three main types of programs (based on the FDM, FEM, and FVM). The chapter concluded with four illustrations of the versatile capabilities of FlowLab, which is based on the finite-volume method.

The present chapter introduces COMSOL Multiphysics (which we shall usually abbreviate as "COMSOL"), another type of CFD software, which is based on the *finite-element method*. COMSOL (known as FEMLAB until September 2005, when COMSOL 3.2 was introduced) is developed by COMSOL Inc., and its broad capabilities are best introduced by quoting (with permission) several paragraphs (until the middle of the next page) from the earlier *FEMLAB 3 User's Guide*:

"FEMLAB is designed to make it as easy as possible to model and simulate physical phenomena. With FEMLAB, you can perform free-form entry of custom partial differential equations (PDEs) or use specialized physics application modes. These physics modes consist of predefined templates and user interfaces that are already set up with equations and variables for specific areas of physics. Further, by combining any number of these application modes into a single problem description, you can model a *multiphysics* problem—such as one involving *simultaneous* mass and momentum transfer.

The *Model Library* is an essential part of the package. It contains complete models taken from various engineering fields, including a special *Chemical Engineering Library Module*. Each model includes extensive documentation. Because the models already come with the finite-element mesh and the solution, you can immediately try out various postprocessing options.

Since its introduction in 1999, FEMLAB has been fully integrated with MATLAB. In fact, until the release of FEMLAB version 3.0, MATLAB was required, although this is no longer the case. Even so, those who wish to run FEMLAB along with MATLAB can do so through a new interface that make it possible to do FEMLAB script programming within the MATLAB language.

FEMLAB is a powerful, interactive environment for modeling and solving all kinds of scientific and engineering problems that are governed by partial differential equations (PDEs). By using the built-in *physics modes* it is possible to build models by defining the relevant physical quantities such as material properties, loads, constraints, sources, and fluxes, rather than by defining the underlying equations. FEMLAB then internally compiles a set of PDEs representing the entire model. You access FEMLAB as a stand-alone product through a flexible graphical user interface (GUI) or by script programming in the MATLAB language.

The underlying FEMLAB structure is a system of PDEs, three ways of describing these being:

- *Coefficient form*, suitable for linear or nearly linear models.
- *General form*, suitable for nonlinear models.
- *Weak form*, useful for models with PDEs on boundaries or those using terms with mixed space and time derivatives.

With these application modes, you can perform various types of analysis, including:

- Eigenfrequency and modal analysis.
- Stationary and time-dependent analysis.
- Linear and nonlinear analysis.

FEMLAB can solve a very wide range of problems, including those in fluid dynamics, chemical reactions, electromagnetics, fuel cells and electrochemistry, transport phenomena, polymer processing, and so on."

COMSOL, as discussed in the present chapter, affords an ideal entry point to the realm of CFD, for the following reasons:

- Its interface is easy to use.
- It is available on a number of computing platforms.
- It affords great flexibility.
- In common with some of the larger and more general-purpose simulators, COMSOL uses the relatively sophisticated finite-element method (FEM) for solving the governing partial differential equations.

Equations solvable by COMSOL. Section 14.5 gives full details of the various fluid-mechanics-related differential equations and accompanying boundary conditions that can be solved by COMSOL. For the present, it suffices to know that COMSOL can solve virtually all of the problems hitherto encountered in this book, provided their governing partial differential equation(s) and boundary conditions are known. Thus, COMSOL can solve problems in the following areas:

- Viscous flows, governed by the Navier-Stokes equations (Chapters 5, 6, and parts of 8).
- Turbulent flows, according to the k/ε method of Chapter 9.
- Irrotational flows, governed by the Laplace and Poisson equations (Chapter 7).

Example 14.1—Flow in a Porous Medium (COMSOL)   **705**

- Non-Newtonian flows, according to the power-law or Carreau models presented in Chapter 11.
- Porous-medium flows, governed by d'Arcy's law (Chapter 7).
- Compressible inviscid flows, of which a single one-dimensional example was given in Section 3.6, and which may even involve supersonic flow.

In this introductory chapter we shall confine the discussion to two-dimensional flow, although COMSOL also has three-dimensional capabilities.

Additional resources. If more complete details are needed, they can be found from the following manuals (or similar COMSOL versions), which are published by COMSOL AB (COMSOL Inc. in the United States):

- *FEMLAB 3 User's Guide.*
- *FEMLAB 3 User's Guide—Chemical Engineering Module.*
- *FEMLAB 3 Modeling Guide.*
- *FEMLAB 3 Model Library.*
- *FEMLAB 3 Installation Guide.*
- *FEMLAB 3 Quick Start.*
- *FEMLAB 3 MATLAB Interface Guide.*

Readers are encouraged to examine the website `www.comsol.com` for further information.

14.2   How to Run COMSOL

Our approach in learning COMSOL is to jump right in with an example, which—if repeated by the user—will enable him or her to become acquainted quickly with the most important features of the software. After that, we shall elaborate in greater detail on the various options in COMSOL and the problems that it can tackle. With this approach it is essential that the beginner repeat all the steps in the example, in order to learn several of the basic aspects of COMSOL— many of which will not be repeated later.

Example 14.1—Flow in a Porous Medium
with an Obstruction (COMSOL)

Solve the problem of two-dimensional flow in the rectangular region ABCD of the porous medium shown in Fig. 14.1, in which there is a central circular region E of essentially zero permeability, which allows no flow through it. The boundaries AB and CD are also impervious to flow. The pressure along the two ends, AD and BC, are $p = 10$ and $p = 0$, respectively. The permeability/viscosity ratio is constant, with $\kappa/\mu = 1$ throughout; all units are mutually consistent. Points A and C have x and y coordinates $(-1.25, 0.8)$ and $(1.25, -0.8)$, respectively; the circle is centered at the origin and has a radius of 0.5.

The pressure is governed by d'Arcy's equation:

$$\nabla \cdot \frac{\kappa}{\mu} \nabla p = \frac{\partial}{\partial x} \frac{\kappa}{\mu} \frac{\partial p}{\partial x} + \frac{\partial}{\partial y} \frac{\kappa}{\mu} \frac{\partial p}{\partial y} = 0. \qquad (\text{E}14.1.1)$$

Note that Problem 14.2 requests a modification, in which the circular region has a higher permeability, with $\kappa = 5$.

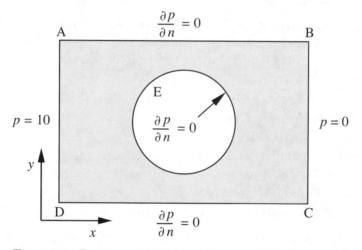

Fig. 14.1 *Region and its boundary conditions, extending any distance normal to the plane of the figure.*

Solution

1. Load COMSOL

The loading of COMSOL will vary from one installation to the next, so local procedures must be consulted. For the earlier FEMLAB 2.3 on the author's Macintosh G4 computer, MATLAB must first be loaded, and then `"femlab"` is typed at the prompt. For FEMLAB 3, used for all the examples in this book, only the FEMLAB icon needs to be clicked for both the Macintosh and PC versions. The program produces extremely similar results for both types of computers—compare, for example, the **Model Navigator** windows in Figs. 14.2(a) (PC) and 14.2(b) (Macintosh), whose formats are virtually identical, except for minor details such as the reversal of the order of the **OK** and **Cancel** buttons in the lower right-hand corners. This chapter is generally illustrated by Macintosh screen captures, except for the very similar PC windows shown in Figs. 14.2(a) and 14.16(b).

2. Model Navigator

After COMSOL is loaded, the foremost window is the **Model Navigator**. Fig. 14.2(b) shows its appearance on a Macintosh computer after the following steps:

Example 14.1—Flow in a Porous Medium (COMSOL)   707

Fig. 14.2(a)   The model navigator window—PC version, showing the options available after selecting "New," "Chemical Engineering Module," "Momentum balance" and "Darcy's law."

Fig. 14.2(b)   The model navigator window—Macintosh version, after the same selections as in Fig. 14.2(a).

(a) Make sure the space dimension is set to **2D**, but also note the other available options.

(b) Under the **New** tab:

- Double-click **Chemical Engineering Module**.
- Double-click **Momentum balance**.
- Double-click **Darcy's law**.

You will now see the COMSOL or FEMLAB graphical interface (Fig. 14.3) ready for you to enter the details of the problem to be solved.

Fig. 14.3   The graphical interface.

In the graphical interface, take a minute to examine the menu bar and the main toolbar buttons at the top of the screen (Fig. 14.4/a/b/c/d) and the draw toolbar buttons at the left of the screen (Fig. 14.7). The menu bar consists of 11 pulldown menus (ten for a PC, which lacks the COMSOL or FEMLAB pulldown), each of which has a few or several options. A typical FEMLAB solution involves most—but not usually all—of these pulldown menus, proceeding from left to right. The toolbar buttons at the top of the graphical interface offer a quick duplication of the more common pulldown menu items. Similarly, the draw toolbar buttons offer a convenient alternative to several of the items (drawing a circle, for example) that appear in the **Draw** pulldown menu.

Example 14.1—Flow in a Porous Medium (COMSOL)   **709**

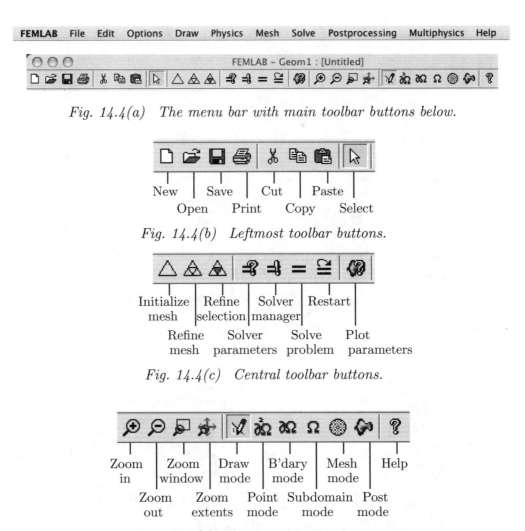

Fig. 14.4(a)   The menu bar with main toolbar buttons below.

Fig. 14.4(b)   Leftmost toolbar buttons.

Fig. 14.4(c)   Central toolbar buttons.

Fig. 14.4(d)   Rightmost toolbar buttons.

3. *Options*

(a) Pull down the **Options** menu to **Axes/Grid Settings**.

(b) Under the **Axis** option (Fig. 14.5):
- Set **x min** and **x max** to -1.5 and 1.5, respectively.
- Set **y min** and **y max** to -1 and 1, respectively.

(c) Under the **Grid** option (Fig. 14.6):
- Clear **Auto**.
- Set **x spacing** and **y spacing** to 0.25 and 0.1, respectively.
- Click **OK**.

Fig. 14.5 *The axes-settings dialog box. For the PC, the positions of the "Apply" and "OK" buttons are reversed.*

Fig. 14.6 *The grid-settings dialog box. For the PC, the positions of the "Apply" and "OK" buttons are reversed.*

4. Draw

(a) *Either* pull down the **Draw** menu to **Draw Objects** and thence to **Rectangle/Square** *or* click once on the corresponding button at the top of the column on the left of your screen. (Before you click, but with the cursor moved to this or any other button, you will soon see a notice telling you the function of the button.)

- Draw a rectangle with two of its opposite corners having coordinates $(-1.25, 0.8)$ and $(1.25, -0.8)$. Note that this rectangle will automatically be labeled **R1**. It also appears in pink, indicating it is the (only) currently selected object.

Example 14.1—Flow in a Porous Medium (COMSOL)　　**711**

(b) *Either* pull down the **Draw** menu to **Ellipse/Circle (Centered)** *or* click once on the corresponding button near the top of the column on the left of your screen.

- Hold down the **Control** key (this will insure you get a circle and not an ellipse) and draw a circle of radius 0.5 centered at (0,0). The x/y coordinates of the cursor position are shown in the window at the bottom left corner. As you draw the circle, this window will also indicate its radius. Note that this circle will automatically be labeled C1. It also appears in pink, whereas the rectangle R1 is now gray.

(c) Pull down **Edit** to **Select All**, and note that R1 and C1 are now both pink. The same result can be achieved by clicking the mouse with the cursor at the top left of the screen and dragging it to the lower right, thus encompassing both R1 and C1 with a dotted rectangle.

—Draw rectangle/square
—Draw rectangle/square (centered)
—Draw ellipse/circle (the latter with Control key)
—Draw centered ellipse or circle (the latter with Control key)
—Draw point
—Draw line
—Draw 2nd-degree Bézier curve
—Draw 3rd-degree Bézier curve
—Array
—Move
—Rotate
—Scale
—Union (combination of objects)
—Intersection (common region of objects)
—Difference (one region removed from another)
—Create composite object
—Coerce to solid
—Coerce to curve
—Split object
—Delete interior boundaries
—Fillet/chamfer

Fig. 14.7　The left column of draw toolbar buttons. The right column (not shown above) is not needed for introductory work.

(d) *Either* pull down the **Draw** menu to **Create Composite Object** or click once on the corresponding button, which is sixth from the bottom in the column on the left of your screen. See Fig. 14.8.

- In the **Set formula** window, type R1-C1, in which "-" is the set *difference* operator. Ignore **Keep interior boundaries** at present.
- Click **OK**, and note that the single region (a rectangle with a hole in it) is now labeled C01 (Fig. 14.9).

Drawn objects can be selected, moved, rotated, and deleted, etc., as with most other drawing programs. Note that each object drawn is automatically assigned a name and a label, such as R1 for the first rectangle or E2 for the second ellipse. To determine precise information about an object—such as the coordinates of the corners of a rectangle, or of the center of a circle, double-click on the object while in **Draw** mode; a dialog box will open with all pertinent information. If you wish, you can change any of the object's geometric parameters (independently of the prevailing grid) or even its name.

Fig. 14.8   The composite-object dialog box.

5. *Boundary*

Pull down the **Physics** menu to **Boundary Settings**. Highlight each of the eight boundary segments in turn in the **Boundary selection** box (see Fig. 14.11) and use either the **Pressure condition** or **Insulation/Symmetry** to establish the following boundary conditions:

- Pressure $p = 10$ on the left-hand (inlet) edge AD of the rectangle.
- Pressure $p = 0$ on the right-hand (exit) edge BC of the rectangle.
- Insulated boundary conditions $\mathbf{n} \cdot \mathbf{u} = 0$ on the remaining six boundary segments.

Example 14.1—Flow in a Porous Medium (COMSOL)   **713**

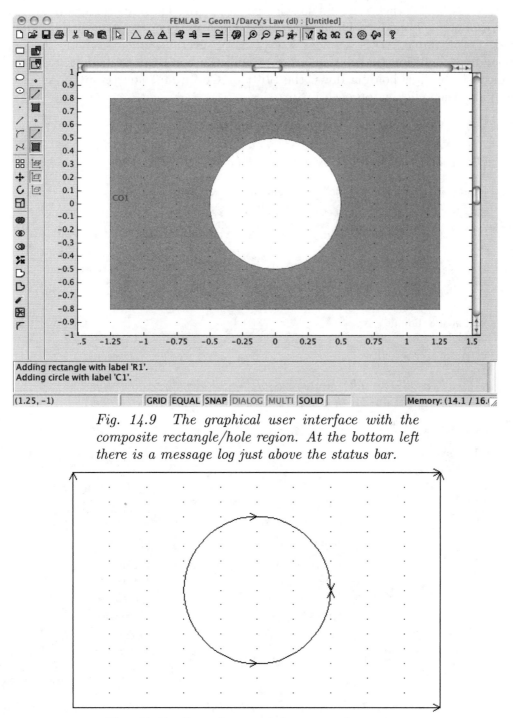

Fig. 14.9   The graphical user interface with the composite rectangle/hole region. At the bottom left there is a message log just above the status bar.

Fig. 14.10   Boundaries of the composite region.

Note that as each segment is selected inside the **Domain selection** box, the corresponding boundary in the drawing behind it (Fig. 14.10) is highlighted in red. For example, AD is segment 1, etc. Also, several segments having the same boundary condition (notably 5–6–7–8 around the hole) can be highlighted simultaneously by first holding down the **Shift** or **Command** key (**Control** key for a PC) and then clicking on each in turn.

Fig. 14.11   *The boundary-settings dialog box. For the PC, the positions of the "Apply" and "OK" buttons are reversed.*

Fig. 14.12   *The subdomain-settings dialog box. For the PC, the positions of the "Apply" and "OK" buttons are reversed.*

Example 14.1—Flow in a Porous Medium (COMSOL)   **715**

6. Subdomain

Pull down the **Physics** menu to **Subdomain Settings**, and observe that the region is currently transparent.

- In the subdomain-settings dialog box (Fig. 14.12), select subdomain 1 (the only one we have) and observe that the region becomes pink. The window also shows that the equation $\nabla \cdot (-\kappa/\eta \nabla p) = F$ is being solved.
- Set the permeability $\kappa = 1$ and the viscosity $\eta = 1$, with the generation term $F = 0$. Ignore the time-scaling coefficient.
- Click **OK**.

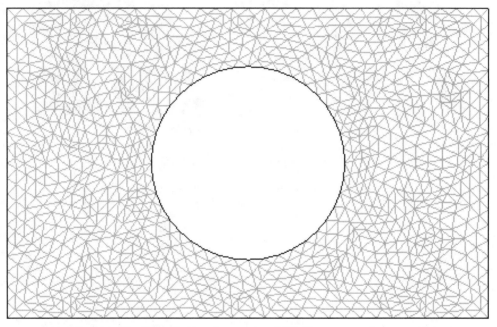

Fig. 14.13   The finite-element grid (after one refinement).

7. Mesh

- The mesh for the finite-element solution is generated automatically by pulling down the **Mesh** menu to **Initialize Mesh**. You will see many triangular elements and can find out exactly how many by pulling down to **Mesh Statistics**; the number of degrees of freedom tells you how many unknowns (both velocities and pressures in the case of the Navier-Stokes equations) exist in the simultaneous equations that are about to be solved.
- Also pull down the **Mesh** menu to **Refine Mesh**, as in Fig. 14.13, which would be particularly desirable for solving problems whose geometry is more complicated than in the present case.

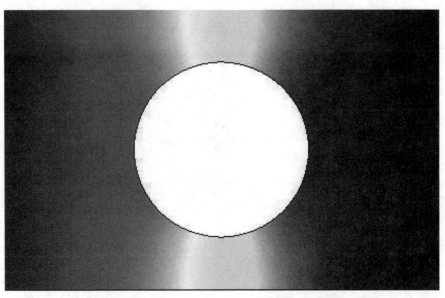

Fig. 14.14 Two-dimensional surface plot. In color, there is a progression from red (high pressure) at the left to blue (low pressure) at the right.

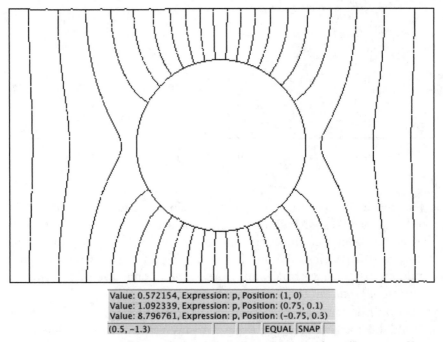

Value: 0.572154, Expression: p, Position: (1, 0)
Value: 1.092339, Expression: p, Position: (0.75, 0.1)
Value: 8.796761, Expression: p, Position: (−0.75, 0.3)

(0.5, −1.3)   EQUAL SNAP

Fig. 14.15 Isobars. A forthcoming version of COMSOL will allow labeling of the isobars with the values of the pressure.

Example 14.1—Flow in a Porous Medium (COMSOL) **717**

8. Solve

Pull down the **Solve** menu to **Solve Problem**, which will effect the solution of the simultaneous equations generated by the finite-element method. The results then appear in a **2D Surface** plot as the velocity field. However, we want the pressure field, so pull down the **Postprocessing** menu to **Plot Parameters**; select the **Surface** option, and adjust **Predefined quantities** to **Pressure**. After **OK**, you will obtain Fig. 14.14, in which the pressure varies from high (red) at the left to low (blue) at the right, with symmetry about the x axis ($y = 0$).

Fig. 14.16(a)   The Plot Parameters dialog box with the contour tab selected— Macintosh version. Note that 20 pressure contours (isobars) are requested. For the PC, the positions of the "Apply" and "OK" buttons are reversed. A click on the arrow at upper right leads to "Deform" and "Animate" options.

9. Postprocessing

Now that the finite-element method solution is complete, the results may be displayed in several different ways. **Quick Plots** gives rapid results, but without complete control, for example:

- Verify the nature of the plot just generated by pulling down the **Postprocessing** menu to **Quick Plots** and thence to **2D Surface Plot** and observe that nothing changes.

- **3D Surface** plot. By "grabbing" the base of the "box" you can rotate the plot to any desired position.

For complete control of the plots, select **Postprocessing**, **Plot Parameters**, and the **General** tab, as in Fig. 14.16(b), which allows you to check exactly those plot(s) you want, for example:

- After checking **Contour** under the **General** tab, select pressure under the **Contour** tab (Fig. 14.16(a)), click **OK** and observe the isobars (contours of equal pressure) in Fig. 14.15. Note that the default case is 20 contours.
- Again under the **Contour** tab, reduce the number of levels to 10 and change the color uniformly to black. Click **OK** and observe the results.
- **Arrow** plot gives arrows that indicate the flow direction, with the velocities proportional to the lengths of the arrows (Fig. 14.17, with contours).
- **Streamline** plot shows the paths taken by the fluid.
- Try another contour plot, but now change **Predefined quantities** to **Pressure gradient**, ∇p, click **OK** and observe the results.

Note that there are many other options—explore them!

Fig. 14.16(b)   The Plot Parameters dialog box with the general tab selected—PC version. Note that a surface plot is requested.

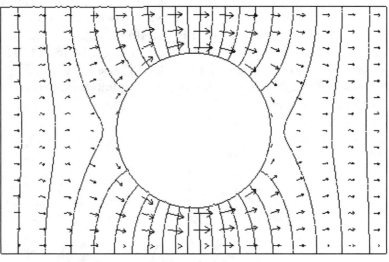

Fig. 14.17 Two plots have been requested—contours (without labels) and arrows (showing the streamlines).

10. Conclusion

- To save your work at any stage in a file, pull down the **File** menu to **Save** and enter a name for the file. There are two formats (in FEMLAB 3): FEMLAB Model File (`*.fl`), and Model M-file (`*.m`), the latter for use with MATLAB only. FEMLAB 2.3 uses Model MAT Files (`*.mat`).

- To leave COMSOL pull down **COMSOL** to **Quit COMSOL** (for a PC, **File** to **Exit** or click on the cross in the upper right-hand corner).

- Perform any log-out procedure, as appropriate. Next time, try the toolbar buttons and/or keystrokes instead of the pulldown menus.

14.3 Draw Mode

    ▭ —Draw rectangle/square

    ▣ —Draw rectangle/square (centered)

    ◯ —Draw ellipse/circle (the latter with Control key)

    ◉ —Draw centered ellipse/circle (the latter with Control key)

    · —Draw point

    ╱ —Draw line

    ⌐ —Draw 2nd-degree Bézier curve ("arc" for short)

    ∿ —Draw 3rd-degree Bézier curve

Fig. 14.18 Some draw toolbar buttons (direct drawing).

The various modes can be activated either by pulling down the appropriate menu from the menu bar at the top of the screen or by clicking on one of the main toolbar buttons immediately below (see Fig. 14.4(a)).

The draw-mode toolbar buttons are shown in Figs. 14.18, 14.21, 14.22, and 14.23, which are segmented repeats of Fig. 14.7. Those shown in Fig. 14.18—for drawing various shapes and lines—are almost self-explanatory, except that the last two need a little more explanation.

Fig. 14.19   Region to be drawn
for polymer extrusion example.

1. Draw 2nd-degree Bézier Curve ("Arc" for short)

We wish to draw the region depicted in Fig. 14.19, which is needed in Example 11.3 for solving non-Newtonian flow in an extrusion die. Referring to Fig. 14.20(a), start with the two rectangles R1 and R2.[1] Then, as in Fig. 14.20(b), select the **Draw "arc"** tool and click on points A, B, and C to draw an elliptical arc, tangential to AB and BC, that is a quarter of an ellipse with semiaxes equal to AB and BC. If AB = BC, the arc will be circular.

Continue by clicking on two additional points, D and E, still with the **Draw "arc"** tool selected. Select the **Draw line** tool before clicking on the last point F. To stop drawing and fill-in to give the pink-shaded composite object C01, click on any drawing tool (such as a rectangle) except a line or either of the Bézier curves (or right-click anywhere on a PC). Use a similar procedure to generate the second composite object C02, as in Fig. 14.20(c). Finally, as in Fig. 14.8, open the **Create Composite Object** window, and enter R1+R2-C01-C02 in the **Set formula** window to give the single desired final object, now labeled C01, shown in Fig. 14.20(d).

---

[1]   Fig. 14.20 was actually drawn in FEMLAB 2.3, but the same result is also achieved with FEMLAB 3.

Fig. 14.20   *Four stages in the development of a composite object using the "draw arc" (2nd-degree Bézier curve) tool; (a) and (b) are at top, (c) and (d) below.*

2. Draw Cubic Bézier Curve

A cubic or 3rd-degree Bézier curve is determined by clicking on *four* points in sequence, and permits more control of its shape than does an elliptical or circular arc. After such a curve has been generated, it may be reshaped either by grabbing and moving any of its four "vertex" points, or by pulling down the **Draw** menu, selecting **Object Properties**, and entering the desired new information. When filled in, the object may be moved either by grabbing it and relocating to the desired new position, or by selecting the object and using the **Move** tool to specify the extent of movement in the two coordinate directions.

3. Array/Move/Rotate/Scale—Fig. 14.21

After an object has been selected, **Array** enables it to be repeated and moved any number of times in the x and y directions. Selection of the **Move** tool opens a window in which the desired object displacements in the x and/or y directions can be entered. Similarly, the **Rotate** tool opens a window asking for the desired rotation angle in degrees and the x and y coordinates of the center of rotation.

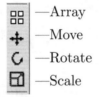

—Array

—Move

—Rotate

—Scale

Fig. 14.21   Some draw toolbar buttons (array/move/rotate/scale).

The **Scale** tool window asks for x and y scale factors and the corresponding base coordinates; the selected object will be expanded or contracted by the factors entered for the two coordinate directions, relative to the base coordinates.

4. Boolean Operations—Fig. 14.22

Consider two objects "A" and "B" that have been drawn and have a common region in which they overlap each other. Now select both of them, either by using the cursor to outline a rectangle that includes both of them, or by pulling down the **Edit** menu to **Select all**, or by using the **Control**–A combination.

—Union (combination of objects)

—Intersection (common region of objects)

—Difference (one region removed from another)

Fig. 14.22   Some draw toolbar buttons (Boolean operations).

The **Union** tool will give an object that includes both A and B. The **Intersection** tool will give the region that is common to both A and B. And the **Difference** tool will create a region in which A has deleted from it that region in which it is overlapped by B.

5. Miscellaneous Operations—Fig. 14.23

—Create composite object

—Coerce to solid

—Coerce to curve

—Split object

—Delete interior boundaries

—Fillet/chamfer

Fig. 14.23 Some draw toolbar buttons (special effects).

The **Create composite object** tool opens a window whose use has already been illustrated. It also displays two radio buttons that allow shortcuts for the **Union** and **Intersection** operations just described above.

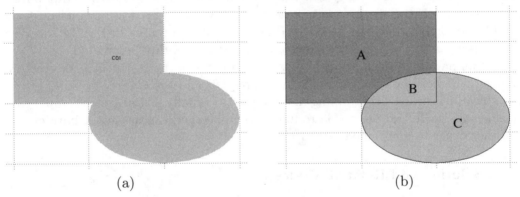

(a)   (b)

Fig. 14.24   (a) Composite object formed from the union
of a rectangle and an ellipse. In (b), the letters A, B, C
have been added later. See the text for an explanation.

The **Create composite object** window also displays another button—**Keep interior boundaries**. To explain, consider the composite object CO1 formed from a rectangle R1 and an ellipse E1, either by typing R1+E1 in the **Set formula** window or by using the **Union** shortcut. The result is shown in Fig. 14.24a, whether or not the **Keep interior boundaries** button is activated. However, when setting the differential equation coefficients later in the **Subdomain mode**, there will be a difference, as follows:

1. If the **Keep interior boundaries** button is not activated, the appearance will be similar to that in Fig. 14.24(a), except that there will be a single continuous border around the whole composite object.

2. If, however, the **Keep interior boundaries** is activated, the appearance will be that in Fig. 14.24(b). Although there is still one object, it is now subdivided in this example into three *subdomains*, which we have labeled as A, B, and C. (Region A has been selected and will appear in pink on the screen, with B and C in gray.) Such an arrangement may be advantageous because it allows different physical properties to be assigned to each of these subdomains.

Fig. 14.25 After the "split" operation, a composite object such as that in Fig. 14.24(a) can be dissected into its component parts, each of which can then be moved independently of the others.

The **Split** button can be used on a composite object. Consider for example the composite object CO1 in Fig. 14.24(a). When it is selected and the **Split** button activated, its three subregions (A, B, and C in Fig. 14.24(b)) will be labeled 01, 02, and 03, each of which can be selected and moved around, as in Fig. 14.25. With some imagination and practice, unusual shapes can readily be created.

If we have an object whose area is shaded (as is normal), the **Coerce to curve** button will yield just the boundary of the object, and the **Coerce to solid** button will have the reverse effect, filling in a boundary. Finally, **Fillet** (see Example 12.5) enables a selected corner to be rounded to a specified radius, and **Chamfer** cuts off the corner to a specified distance.

14.4 Solution and Related Modes

The following steps are needed to solve the problem at hand. Most of the basic ideas have been adequately covered in Example 14.1 and will not be repeated here. Only a few extra points will be mentioned. Most of the operations can be achieved either through the toolbar buttons or by the more comprehensive pulldown menus.

Setting the boundary conditions.   The boundary conditions will vary with the type of problem being solved. COMSOL can accommodate all boundary conditions of physical significance. Also, if the dependent variable were to vary along a boundary as, for example, $u = 10\sin[5(x - 0.5)]$, where x is the horizontal coordinate, then the arithmetic expression `10*sin(5*(x - 0.5))` would be entered into the appropriate window, the understanding being that arguments for the trigonometric functions are in radians.

Inserting equation coefficients.   As indicated in Fig. 14.12, the dialog box for the subdomain settings provides the opportunity for setting the values of the various coefficients in the PDE. Again, these values can be specified as functions of position if desired.

Establishing the finite-element mesh.   The **Initialize Mesh** option is usually activated to start the mesh generation, followed by as many applications of **Refine Mesh** as are needed to obtain the desired accuracy. In practice, just one application of **Refine Mesh** will suffice for most problems. Also, if extra definition is needed in one part of the region, activation of **Refine Selection** followed by dragging the cursor over a region will generate more and finer finite elements in that region.

Solving the simultaneous equations.   Activation of **Solve Problem** then solves the simultaneous equations generated by the finite-element method. Clearly, the more elements that are created, the longer is the time needed for the solution. The progress of the solution is automatically indicated by a window that displays how far the solution has progressed, with messages such as "LU factorization," "Assembling," and "Solving linear equations," which are mainly of significance to those interested in numerical methods.

Examining the results.   There is a multitude of ways in which the results can be displayed. The most important of these have already been introduced in Part 9, "Postprocessing," of Example 14.1, and will not be repeated here. The user should gain experience by experimenting with many of the various options. COMSOL also allows the opportunity for "zooming" in (and out) to vary the magnification of the displayed results. Also, the **Zoom Extents** button is very useful—it will give an image that automatically fills the screen as much as possible.

14.5   Fluid Mechanics Problems Solvable by COMSOL

This section lists seven general types of fluid-mechanics problems that can be solved by COMSOL. The software is exceptionally versatile and can also solve *multiphysics* problems with, e.g., combined heat and mass transfer, fluid mechanics and electromagnetics—as illustrated for electroosmosis in Examples 12.4 and 12.5.

In their appearance the following equations parallel those in the FEMLAB manuals cited earlier, although slight changes have occasionally been made to conform to the notation in this book. The equations have been left in vector form, the understanding being that COMSOL will automatically reformat them according to the number of dimensions and the coordinate system specified by the user.

1. Navier-Stokes Equations

The governing equations for incompressible flow are the momentum balance and continuity equation:

$$\rho \left(\frac{\partial \mathbf{v}}{\partial t} + \mathbf{v} \cdot \nabla \mathbf{v} \right) + \nabla p - \nabla \cdot \mu (\nabla \mathbf{v} + (\nabla \mathbf{v})^T) = \mathbf{F}, \qquad (14.1)$$

$$\nabla \cdot \mathbf{v} = 0. \tag{14.2}$$

Although the body-force vector $\mathbf{F}$ per unit volume usually accounts for gravitational attraction, it could also be the force on a charged particle in an electric field (see Examples 12.4 and 12.5). Further, if the energy equation is being solved simultaneously for temperature, $\mathbf{F}$ can account for free convection if temperature-dependent density variations are accommodated.

Boundary conditions. COMSOL can accommodate any or all of the following boundary conditions, each corresponding to a practical physical situation. As usual, $\mathbf{n}$ is the unit vector in the outwards normal direction.

No slip

$$\mathbf{v} = \mathbf{0}. \tag{14.3}$$

Specified velocity with normal component v_0

$$\mathbf{v} \cdot \mathbf{n} = v_0. \tag{14.4}$$

Slip or symmetry (no normal velocity component)

$$\mathbf{v} \cdot \mathbf{n} = 0. \tag{14.5}$$

Specified pressure p_0

$$p = p_0. \tag{14.6}$$

"Straight-out" condition—zero tangential velocity and zero pressure

$$\mathbf{v} \cdot \mathbf{t} = 0, \tag{14.7a}$$
$$p = 0. \tag{14.7b}$$

Neutral—no transport by shear stress

$$\mu(\nabla \mathbf{v} + (\nabla \mathbf{v})^T) \cdot \mathbf{n} = 0. \tag{14.8}$$

2. Turbulent-Flow Equations

Here, $\mathbf{v}$ is the time-averaged velocity, k is the turbulent kinetic energy per unit mass, and ε is the rate of dissipation of this energy. The first two equations are the momentum balance and continuity equation ($C_\mu k^2/\sigma_k \varepsilon$ is the same as ν_T in Eqn. (9.90a) since $\sigma_k = 1$), followed by equations for the rate of growth of k and ε:

$$\rho\left(\frac{\partial \mathbf{v}}{\partial t} + \mathbf{v} \cdot \nabla \mathbf{v}\right) + \nabla p - \nabla \cdot \left[\left(\mu + \rho\frac{C_\mu k^2}{\sigma_k \varepsilon}\right) \cdot (\nabla \mathbf{v} + (\nabla \mathbf{v})^T)\right] = 0 \tag{14.9}$$

$$\nabla \cdot \mathbf{v} = 0. \tag{14.10}$$

$$\rho\left(\frac{\partial k}{\partial t} + \mathbf{v} \cdot \nabla k\right) - \nabla \cdot \left[\left(\mu + \rho\frac{C_\mu k^2}{\sigma_k \varepsilon}\right)\nabla k\right] = \rho C_\mu \frac{k^2}{\varepsilon}(\nabla\mathbf{v} + (\nabla\mathbf{v})^T)^2 - \rho\varepsilon, \tag{14.11}$$

$$\rho\left(\frac{\partial \varepsilon}{\partial t} + \mathbf{v} \cdot \nabla\varepsilon\right) - \nabla \cdot \left[\left(\mu + \rho\frac{C_\mu k^2}{\sigma_\varepsilon \varepsilon}\right)\nabla\varepsilon\right] = \rho C_{\varepsilon 1}C_\mu k(\nabla\mathbf{v} + (\nabla\mathbf{v})^T)^2 - \rho C_{\varepsilon 2}\frac{\varepsilon^2}{k}. \tag{14.12}$$

In the above, the model constants are

$$C_\mu = 0.09, \quad C_{\varepsilon 1} = 0.1256, \quad C_{\varepsilon 2} = 1.92, \quad \sigma_k = 1.0, \quad \sigma_\varepsilon = 1.3 \tag{14.13}$$

Wall boundary conditions. COMSOL can accommodate most of the boundary conditions listed under the Navier-Stokes equations above—with one major exception. Namely, because there are extremely sharp velocity gradients in the very thin region next to a wall, it is quite impractical to use a small enough mesh to accommodate such gradients accurately. Therefore, we lean on the knowledge that the velocity profile is logarithmic near the wall (but not all the way up to the wall), and set the following conditions close to the boundary, in which v is the velocity component parallel to the wall, n is the normal distance from the wall, ν is the kinematic viscosity, $\kappa \doteq 0.42$, and $C \doteq 5$ for smooth walls.

Velocity parallel to a wall

$$v = \sqrt{\frac{\tau_w}{\rho}}\left[\frac{1}{\kappa}\ln\left(\frac{y\sqrt{\tau_w/\rho}}{\nu}\right) + C\right], \quad \text{in which} \quad \frac{\tau_w}{\rho} = \frac{C_\mu k^2}{\varepsilon}\frac{\partial v}{\partial n}. \tag{14.14}$$

Turbulent energy

$$k = \frac{\tau_w/\rho}{\sqrt{C_\mu}} = \frac{\sqrt{C_\mu}k^2}{\varepsilon}\frac{\partial v}{\partial n}. \tag{14.15}$$

Turbulent dissipation

$$\varepsilon = \frac{(\tau_w/\rho)^{3/2}}{\kappa y} = \frac{1}{\kappa y}\left(\frac{C_\mu k^2}{\varepsilon}\frac{\partial v}{\partial n}\right)^{3/2}. \tag{14.16}$$

3. Laplace and Poisson Equations

COMSOL is capable of solving many "classical" types of partial differential equations. Of these, the following two will be useful when solving irrotational (inviscid) flow problems.

Laplace's equation

$$\nabla \cdot \nabla u = \nabla^2 u = 0. \tag{14.17}$$

Poisson's equation

$$\nabla \cdot c\nabla u + f = 0. \tag{14.18}$$

In the last equation c is essentially a conductivity and f is a source or generation term; both c and f may be functions of position. The corresponding boundary conditions are either of the Dirichlet type (u specified) or of the Neumann type (involving the normal derivative $\mathbf{n} \cdot \nabla u = \partial u / \partial n$).

4. Non-Newtonian Flows

The momentum and mass balances are the same as for the Navier-Stokes equations for Newtonian flows except that η is used instead of μ for the viscosity.

$$\rho \left(\frac{\partial \mathbf{v}}{\partial t} + \mathbf{v} \cdot \nabla \mathbf{v} \right) + \nabla p - \nabla \cdot \eta (\nabla \mathbf{v} + (\nabla \mathbf{v})^T) = \mathbf{F}, \tag{14.19}$$

$$\nabla \cdot \mathbf{v} = 0. \tag{14.20}$$

The variety of available boundary conditions is the same as for the Navier-Stokes equations. Two models are available for the non-Newtonian viscosity η, in which the strain rate $\dot{\gamma}$ (also known as the shear rate) is automatically computed by COMSOL as a function of the several velocity gradients as in Eqn. (11.10):

Power-law model

$$\eta = \kappa \dot{\gamma}^{n-1}. \tag{14.21}$$

Here, κ is the consistency index and n is the power-law index.

Carreau model

$$\eta = \eta_\infty + \frac{\eta_0 - \eta_\infty}{(1 + \lambda^2 \dot{\gamma}^2)^{(1-n)/2}}. \tag{14.22}$$

Here, η_0 and η_∞ are the viscosities at zero and very high strain rates, respectively, λ is a relaxation time of the fluid, and n is a dimensionless constant approximately in the range 0.2–0.5 for many polymer melts.

5. Flow in a Porous Medium

Incompressible flow in a porous medium is governed by the continuity equation, $\nabla \cdot \mathbf{v} = \mathbf{0}$, which, when coupled with d'Arcy's law, $\mathbf{v} = -(\kappa/\mu)\nabla p$, gives:

$$\nabla \cdot \frac{\kappa}{\mu} \nabla p = 0. \tag{14.23}$$

For *compressible* flow, the continuity equation is $\nabla \cdot \rho \mathbf{v} = \mathbf{0}$, so the governing equation now includes the variable density:

$$\nabla \cdot \rho \frac{\kappa}{\mu} \nabla p = 0. \tag{14.24}$$

In the important case of underground storage of natural gas of molecular weight M_w and compressibility factor z at an absolute temperature T, Eqn. (14.24) becomes:

$$\nabla \cdot \left(\frac{M_w p}{zRT}\right) \frac{\kappa}{\mu} \nabla p = 0. \tag{14.25}$$

Boundary conditions. There are three relevant types of boundary conditions, in which $\mathbf{n}$ is the unit outward normal:

Specified pressure p_0

$$p = p_0, \tag{14.26}$$

Impervious (no flow) boundary or plane of symmetry

$$\nabla p \cdot \mathbf{n} = 0. \tag{14.27}$$

Specified superficial velocity u_0 in the outward normal direction

$$-\frac{\kappa}{\mu} \nabla p \cdot \mathbf{n} = u_0. \tag{14.28}$$

6. Induced Flow in a Porous Medium—the Brinkman Equation

Some problems involve a porous medium in contact with an "open" region in which flow is unimpeded by such a porous medium; such conditions prevail for flow in the vicinity of porous catalysts and in fuel cells. In this event, there is a shear-stress induced flow in the porous medium, where the situation is governed by the Brinkman equations:

$$\rho \frac{\partial \mathbf{v}}{\partial t} - \nabla \cdot \mu (\nabla \mathbf{v} + (\nabla \mathbf{v})^T) - \left(\frac{\mu}{\kappa} \mathbf{v} + \nabla p - \mathbf{F}\right) = 0, \tag{14.29a}$$

$$\nabla \cdot \mathbf{v} = 0. \tag{14.29b}$$

In the "open" region, the usual Navier-Stokes equations govern the viscous flow. By selecting the Brinkman option, such problems can be solved by COMSOL. The **F** term can accommodate mild compressibility effects.

7. Compressible Inviscid Flows

The continuity, momentum, and energy equations are:

$$\frac{\partial \rho}{\partial t} + \nabla \cdot \rho \mathbf{v} = 0, \tag{14.30}$$

$$\frac{\partial (\rho \mathbf{v})}{\partial t} + \nabla \cdot (\rho \mathbf{v} \mathbf{v}) + \nabla p = \mathbf{F}, \tag{14.31}$$

$$\frac{\partial e}{\partial t} + \nabla \cdot (e + p)\mathbf{v} = \mathbf{Q}, \tag{14.32}$$

in which the dyadic product **vv** is defined in Appendix C, e is the total (potential plus kinetic) energy per unit volume, **F** is a body force per unit volume, and **Q** is a heat generation per unit volume. Three additional equations are needed, assuming ideal gas behavior. First, the pressure is calculated from:

$$p = (\gamma - 1)\left(e - \frac{1}{2}\rho|\mathbf{v}|^2\right),$$
(14.33)

in which $\gamma = c_p/c_v$ is the ratio of specific heats. Second, the speed of sound is obtained from Eqn. (3.80):

$$c = \sqrt{\frac{\gamma p}{\rho}}.$$
(14.34)

Third, the Mach number is needed because it determines how the boundary conditions are handled:

$$M = \frac{|\mathbf{v}|}{c}.$$
(14.35)

For *subsonic* flow ($M < 1$), the density, velocity, and pressure are specified at both the inlet and outlet, and a slip condition is applied at solid boundaries.

For *supersonic* flow ($M > 1$) the same conditions will hold at the inlet and solid boundaries, but the outlet is governed by a *neutral* boundary condition, which does not add any constraints to the flow.

PROBLEMS FOR CHAPTER 14

There are only two problems here, but completion of both of them will reinforce many of the fundamentals of COMSOL. For additional practice the reader is encouraged to read the several COMSOL examples throughout the book, also perhaps duplicating or modifying them. Nothing succeeds like constant practice!

1. *Introductory check.* Repeat the steps in Example 14.1, and make five representative prints, corresponding to Figs. 14.9, 14.13, 14.14, 14.15, and 14.17.

2. *Central conductive hole.* Also solve the problem in Example 14.1, except that the central circular region now has a permeability of $\kappa = 5$. Note that you no longer need to consider the internal interface between the circle C1 and rectangle R1, which is automatically accommodated by COMSOL. When you open the **Create Composite Object** dialog box, you will now need to:

- In the **Set formula** window, type R1+C1, in which "+" is the set *union* operator. (If ever needed, "*" is the set *intersection* operator.)
- Make sure the **Keep interior boundaries** box is checked.

Make two additional prints for this second problem, analogous to Figs. 14.15 and 14.17.

Appendix A

USEFUL MATHEMATICAL RELATIONSHIPS

HERE, we present several commonly used mathematical formulas, the majority of which will be needed in this text. Although the reader should already have encountered these relationships in first- and second-year college mathematics courses, we believe that it is helpful to summarize them in one location.

Geometrical Shapes

In Table A.1, r is the radius of an object, D is its diameter, and L is the length of the cylinder.

Table A.1

	Circle	*Sphere*	*Cylinder*
Circumference:	$2\pi r = \pi D$	—	—
Surface area:	$\pi r^2 = \dfrac{\pi D^2}{4}$	$4\pi r^2 = \pi D^2$	$\underbrace{2\pi r^2}_{\text{Ends}} + \underbrace{2\pi r L}_{\substack{\text{Curved} \\ \text{surface}}}$
Volume:	—	$\dfrac{4\pi r^3}{3} = \dfrac{\pi D^3}{6}$	$\pi r^2 L = \dfrac{\pi D^2 L}{4}$

Additionally, the area of a triangle of base B and altitude H is $A = BH/2$; and the area of a rectangle of breadth B and width W is $A = BW$.

Derivatives

$$\frac{d(x^n)}{dx} = nx^{n-1}, \qquad \frac{d(\sin x)}{dx} = \cos x, \qquad \frac{d(\cos x)}{dx} = -\sin x, \qquad \frac{d(e^x)}{dx} = e^x,$$

$$\frac{d(\ln x)}{dx} = \frac{1}{x}, \qquad \frac{d(uv)}{dx} = u\frac{dv}{dx} + v\frac{du}{dx}, \qquad \frac{d(u/v)}{dx} = \frac{v(du/dx) - u(dv/dx)}{v^2},$$

$$\frac{da^x}{dx} = a^x \ln a, \qquad \frac{d\tan^{-1}x}{dx} = \frac{1}{1+x^2}.$$

731

Integrals

In the following, a constant of integration is omitted throughout:

$$\int x^n \, dx = \frac{x^{n+1}}{n+1} \quad (n \neq -1), \qquad \int \frac{dx}{x} = \ln x, \qquad \int \sin x \, dx = -\cos x,$$

$$\int \cos x \, dx = \sin x, \qquad \int e^x \, dx = e^x, \qquad \int \frac{dx}{a+bx} = \frac{1}{b}\ln(a+bx),$$

$$\int \cos^2 x \, dx = \frac{1}{2}\left(x + \frac{1}{2}\sin 2x\right), \qquad \int \sin x \cos x \, dx = -\frac{1}{4}\cos 2x,$$

$$\int \frac{dx}{\sqrt{a+bx}} = \frac{2}{b}\sqrt{a+bx}, \qquad \int \frac{dx}{a+b\sqrt{x}} = \frac{2}{b^2}\left(b\sqrt{x} - a\ln\left|a+b\sqrt{x}\right|\right),$$

$$\int \ln y \, dy = y\ln y - y, \qquad \int \frac{dx}{a - \sqrt{b+x}} = -2\left[\sqrt{(b+x)} + a\ln\left|a - \sqrt{(b+x)}\right|\right],$$

$$\int \frac{dx}{a - bx^2} = \frac{1}{2\sqrt{ab}}\ln\frac{\sqrt{a}+x\sqrt{b}}{\sqrt{a}-x\sqrt{b}}, \qquad \int r\ln r \, dr = \frac{1}{2}r^2\left(\ln r - \frac{1}{2}\right),$$

$$\int \cos^n(ax) \, dx = \frac{1}{na}\cos^{n-1}(ax)\sin(ax) + \frac{n-1}{n}\int \cos^{n-2}(ax) \, dx,$$

$$\int \frac{x\,dx}{a+bx} = \frac{x}{b} - \frac{a}{b^2}\ln(a+bx), \qquad \int \frac{dx}{a^2+x^2} = \frac{1}{a}\tan^{-1}\frac{x}{a}.$$

Trigonometric Identities

$$\sin 2\theta = 2\sin\theta\cos\theta, \qquad e^{i\theta} = \cos\theta + i\sin\theta,$$

$$\sin(a+b) = \sin a \cos b + \cos a \sin b, \qquad \cos(a+b) = \cos a \cos b - \sin a \sin b.$$

Hyperbolic Functions

$$\sinh x = \frac{1}{2}(e^x - e^{-x}), \qquad \cosh x = \frac{1}{2}(e^x + e^{-x}), \qquad \tanh x = \frac{\sinh x}{\cosh x},$$

$$\text{sech } x = \frac{1}{\cosh x}, \qquad \frac{d\tanh x}{dx} = \text{sech}^2 x, \qquad \tanh^2 x + \text{sech}^2 x = 1.$$

Taylor's Expansion

$$f(x_0 + h) = f(x_0) + h \left(\frac{df}{dx}\right)_{x=x_0} + \frac{h^2}{2!} \left(\frac{d^2 f}{dx^2}\right)_{x=x_0} + \frac{h^3}{3!} \left(\frac{d^3 f}{dx^3}\right)_{x=x_0} + \cdots$$

If h is taken to be a differential increment dx, and $x_0 = x$, all the derivatives of second order and higher have vanishingly small coefficients and can be neglected, leading to:

$$f(x + dx) = f(x) + \frac{df}{dx}\, dx,$$

which has important applications throughout the book when performing mass, energy, and momentum balances. It is, of course, completely compatible with the definition of a derivative:

$$\frac{df}{dx} = \lim_{dx \to 0} \frac{f(x + dx) - f(x)}{dx}.$$

Simpson's Rule

$$\int_a^b f(x)\, dx \doteq \frac{b - a}{6} \left[f(a) + 4f\left(\frac{a + b}{2}\right) + f(b) \right].$$

(This relationship is *exact* if $f(x)$ is a polynomial of degree no higher than a cubic.)

Solution of ODEs by Separation of Variables

Consider an ordinary differential equation (ODE) of the form:

$$\frac{dy}{dx} = f(x)g(y),$$

where $f(x)$ and $g(y)$ are functions of only x and y, respectively (and which may possibly be constants). If we know that $y = y_0$ when $x = x_0$, then the solution $y = y(x)$ of the ODE is obtained from:

$$\int_{y_0}^{y} \frac{dy}{g(y)} = \int_{x_0}^{x} f(x)\, dx.$$

Numerical Solution of Differential Equations by Euler's Method

Consider the following differential equation, which occurred (with different notation) in the tank evacuation problem of Example 2.1:

$$\frac{dp}{dt} = f(p) = -cp, \tag{1}$$

in which $c = 0.001$ reciprocal seconds. We can approximate Eqn. (1) by its *finite-difference approximation* (FDA):

$$\frac{p_{i+1} - p_i}{\Delta t} \doteq f(p_i) = -cp_i. \tag{2}$$

Here, p_i and p_{i+1} denote the values of the dependent variable at the beginning and end of the *i*th *time step*, of duration Δt. Note that the derivative function $f(p)$ is approximated at the *beginning* of the time step.

If we know the value p_i at the beginning of the time step, then its value at the end of the step is approximately:

$$p_{i+1} \doteq p_i + f(p_i)\Delta t = p_i - c\Delta t\, p_i. \tag{3}$$

Now at $t = 0$, $p_0 = 1$ (the initial pressure was one bar). Hence, arbitrarily choosing a time step of $\Delta t = 100$ seconds, so that $c\Delta t = 0.1$, the values of the pressure after 100, 200, and 300 seconds are given by the following *approximations*:

$$p_1 \doteq p_0 + f(p_0)\Delta t = p_0 - 0.1p_0 = 1.0 - 0.1 \times 1 = 0.9,$$

$$p_2 \doteq p_1 + f(p_1)\Delta t = p_1 - 0.1p_1 = 0.9 - 0.1 \times 0.9 = 0.81,$$

$$p_3 \doteq p_2 + f(p_2)\Delta t = p_2 - 0.1p_2 = 0.81 - 0.1 \times 0.81 = 0.729.$$

A comparison of these values with those given by the exact solution, $p = e^{-0.001t}$, namely 0.905, 0.819, and 0.741, shows that this simple numerical procedure has generated answers that are surprisingly good. If required, the accuracy could *easily* be improved by taking a smaller time step (and performing more calculations).

This numerical technique can readily be adapted to the solution of more complicated differential equations, and whose analytical solution is either difficult or impossible. Further, because of the repetitive nature of the calculations, Euler's method is readily implemented by spreadsheets, as in Table A.2.

Table A.2 Spreadsheet Implementation of Euler's Method

	A	B	C
1	c (1/s)	p_0 (bar)	
2	0.001	1.0	
3			
4	t (s)	p (bar)	dp/dt (bar/s)
5	0	1.000 (= B2)	0.001000 (= − A2*B5)
6	100	0.900 (= B5 + C5*(A6 − A5))	0.000900 (= − A2*B6)
7	200	0.810 (= B6 + C6*(A7 − A6))	0.000810 (= − A2*B7)
8	300	0.729 (= B7 + C7*(A8 − A7))	0.000729 (= − A2*B8)

Relating to this spreadsheet, observe the following:

1. Digits and letters identify the various rows and columns.
2. In row 5 onwards, the expressions in parentheses are actually entered into the cells and cause the indicated numerical values to appear.
3. Once a basic formula has been entered, such as that in cell B6 or C5, it can be "pasted" into subsequent cells in its column; the row counter will thereby be incremented automatically.
4. Dollar signs indicate an absolute cell address, which will *not* change when a formula is pasted into subsequent cells.
5. The parameters c and p_0 are *not* built into any of the formulas, but are allocated to their own cells. Thus, not only are the values of these parameters immediately apparent, but they can readily be changed in just two cells without having to alter any of the subsequent formulas.
6. The time step is computed by the spreadsheet, as in (A6 - A5), with the advantage that there is no need to maintain constant time intervals throughout the calculations. For example, a small time step could be used if finer detail were needed at some particular point in time.

Curvature

Over an infinitesimally small length, a curve AB may be approximated by the arc of a circle of *radius of curvature* R, subtending an angle $d\theta$ at the center O of the circle, as in Fig. A.1. The angle $d\theta$ is the difference between the angles to the horizontal made by the tangents to the curve at A and B.

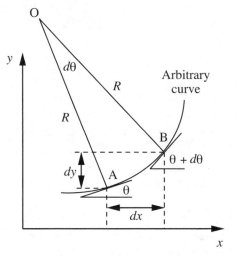

Fig. A.1   Notation for establishing curvature.

Noting that the arc length is $\sqrt{(dx)^2 + (dy)^2}$, using Taylor's expansion, recalling that $d(\tan^{-1} z)/dz = 1/(1 + z^2)$, and observing that the slope of the tangent at A is dy/dx, there results:

$$d\theta = \frac{\sqrt{(dx)^2 + (dy)^2}}{R} = \underbrace{\tan^{-1}\left[\frac{dy}{dx} + \frac{d}{dx}\left(\frac{dy}{dx}\right)dx\right]}_{\theta + d\theta} - \underbrace{\tan^{-1}\left(\frac{dy}{dx}\right)}_{\theta}$$

$$= \tan^{-1}\left(\frac{dy}{dx}\right) + \frac{\dfrac{d^2y}{dx^2}}{1 + \left(\dfrac{dy}{dx}\right)^2}dx - \tan^{-1}\left(\frac{dy}{dx}\right) = \frac{\dfrac{d^2y}{dx^2}}{1 + \left(\dfrac{dy}{dx}\right)^2}dx.$$

Division by dx and rearrangement gives the local *curvature* C, which is the reciprocal of the radius of curvature:

$$C = \frac{1}{R} = \frac{\dfrac{d^2y}{dx^2}}{\left[1 + \left(\dfrac{dy}{dx}\right)^2\right]^{3/2}}.$$

For small slopes dy/dx, $C \doteq d^2y/dx^2$.

Leibnitz's Rule

The following relation—Leibnitz's rule—is useful for differentiating an integral with respect to a variable (x in this case) that appears in any or all of the integrand, the lower limit, and the upper limit:

$$\frac{d}{dx}\int_{a(x)}^{b(x)} f(x,y)\,dy = f(x,b)\frac{db}{dx} - f(x,a)\frac{da}{dx} + \int_{a(x)}^{b(x)} \frac{\partial f(x,y)}{\partial x}\,dy.$$

Successive Substitutions

The method of successive substitutions is occasionally useful for solving equations of the form $f(x) = 0$, if a starting estimate x_1 of the root α is available. First, rearrange the equation so that x appears explicitly on the left-hand side, giving $x = F(x)$, where $F(x)$ is some new function of x. Then compute successive estimates from $x_{i+1} = F(x_i), i = 1, 2, \ldots$. The method will converge to a root provided that the rearranged function is not very sensitive to changes in x—specifically, that $|F(x)| < 1$ for values of x between the starting estimate x_1 and the root α.

For example, if $f(x) = x^3 - 3x + 1 = 0$, a possible rearrangement is $x = (x^3 + 1)/3$. Starting with $x_1 = 0.5$, the method gives $x_2 = (0.5^3 + 1)/3 = 0.375$, $x_3 = (0.375^3 + 1)/3 \doteq 0.3509$. Convergence soon occurs to $x \doteq 0.3473$.

For a given roughness ratio and Reynolds number, the Colebrook and White equation, (3.39), is an ideal candidate for solution for the friction factor f_F by successive substitution, because the right-hand side—which embodies a logarithm—is very insensitive to changes in f_F. A longer illustration is given in Example 3.3.

Appendix B

ANSWERS TO THE TRUE/FALSE ASSERTIONS

Chapter 1

(a)	F	(b)	T	(c)	F	(d)	T	(e)	F
(f)	T	(g)	F	(h)	F	(i)	T	(j)	T
(k)	T	(l)	T	(m)	F	(n)	F	(o)	F
(p)	F	(q)	T	(r)	T	(s)	T	(t)	T
(u)	F	(v)	F	(w)	F	(x)	F	(y)	F
(z)	F	(A)	T						

Chapter 2

(a)	F	(b)	T	(c)	F	(d)	F	(e)	T
(f)	F	(g)	F	(h)	T	(i)	F	(j)	T
(k)	F	(l)	F	(m)	F	(n)	T	(o)	F
(p)	F	(q)	F	(r)	F	(s)	F	(t)	F
(u)	T	(v)	F						

Chapter 3

(a)	T	(b)	F	(c)	F	(d)	F	(e)	F
(f)	F	(g)	T	(h)	F	(i)	T	(j)	T
(k)	F	(l)	F	(m)	T	(n)	T	(o)	T
(p)	F	(q)	F	(r)	F	(s)	F	(t)	T
(u)	T	(v)	F	(w)	F	(x)	T	(y)	F
(z)	F	(A)	T	(B)	T	(C)	F	(D)	F
(E)	F	(F)	T	(G)	T	(H)	F	(I)	F
(J)	T								

Chapter 4

(a)	F	(b)	F	(c)	T	(d)	F	(e)	F
(f)	T	(g)	F	(h)	T	(i)	F	(j)	F
(k)	F	(l)	T	(m)	F	(n)	T	(o)	T
(p)	F	(q)	F	(r)	F	(s)	F	(t)	T
(u)	F	(v)	F	(w)	F	(x)	T	(y)	T
(z)	F	(A)	F	(B)	F	(C)	F	(D)	T
(E)	T	(F)	F	(G)	T	(H)	F	(I)	T
(J)	T	(K)	F	(L)	F				

Chapter 5

(a)	F	(b)	T	(c)	T	(d)	F	(e)	T
(f)	F	(g)	F	(h)	T	(i)	F	(j)	T
(k)	T	(l)	F	(m)	F	(n)	F	(o)	T
(p)	F	(q)	F	(r)	F	(s)	F	(t)	T

Chapter 6

(a)	T	(b)	T	(c)	F	(d)	T	(e)	F
(f)	F	(g)	F	(h)	T	(i)	T	(j)	F
(k)	T	(l)	F	(m)	F	(n)	F	(o)	T

Chapter 7

(a)	F	(b)	F	(c)	T	(d)	F	(e)	F
(f)	F	(g)	T	(h)	F	(i)	F	(j)	F
(k)	F	(l)	F	(m)	T	(n)	F	(o)	F
(p)	T	(q)	T	(r)	F	(s)	F	(t)	F
(u)	F	(v)	F	(w)	T				

Chapter 8

(a)	T	(b)	T	(c)	F	(d)	F	(e)	T
(f)	F	(g)	T	(h)	T	(i)	F	(j)	T
(k)	T	(l)	F	(m)	F	(n)	T	(o)	T
(p)	T	(q)	T	(r)	F	(s)	F			

Chapter 9

(a)	F	(b)	T	(c)	T	(d)	T	(e)	T
(f)	F	(g)	T	(h)	T	(i)	T	(j)	F
(k)	F	(l)	F	(m)	T	(n)	F	(o)	T
(p)	T	(q)	T	(r)	T	(s)	F	(t)	F
(u)	T								

Chapter 10

(a)	T	(b)	F	(c)	T	(d)	T	(e)	F
(f)	F	(g)	F	(h)	T	(i)	T	(j)	T
(k)	T	(l)	F	(m)	T	(n)	F	(o)	T
(p)	F	(q)	F	(r)	T	(s)	F	(t)	F
(u)	F	(v)	F	(w)	T	(x)	T	(y)	F
(z)	T								

Chapter 11

(a)	F	(b)	T	(c)	T	(d)	F	(e)	T
(f)	F	(g)	F	(h)	F	(i)	T	(j)	F
(k)	T	(l)	F	(m)	T	(n)	T	(o)	F
(p)	T	(q)	T	(r)	F	(s)	T	(t)	F
(u)	T	(v)	T	(w)	T	(x)	T	(y)	T
(z)	T	(A)	F	(B)	T	(C)	T		

Chapter 12

(a)	T	(b)	F	(c)	F	(d)	T	(e)	T
(f)	F	(g)	T	(h)	T	(i)	F	(j)	F
(k)	F	(l)	F						

Appendix C

SOME VECTOR AND TENSOR OPERATIONS

T ABLES 5.2–5.5 gave details of the most frequently used vector operations (gradient and Laplacian of a scalar, divergence and curl of a vector) in rectangular, cylindrical, and spherical coordinates. Here, we give a few more definitions for rectangular Cartesian coordinates, sufficient for the needs of this book.[1]

Dyadic Product of Two Vectors

The dyadic product $\mathbf{uv}$ of two vectors $\mathbf{u}$ and $\mathbf{v}$ is a tensor, such that it and its transpose $(\mathbf{uv})^T$ are defined by:

$$\mathbf{uv} = \begin{pmatrix} u_x v_x & u_x v_y & u_x v_z \\ u_y v_x & u_y v_y & u_y v_z \\ u_z v_x & u_z v_y & u_z v_z \end{pmatrix}, \qquad (\mathbf{uv})^T = \begin{pmatrix} u_x v_x & u_y v_x & u_z v_x \\ u_x v_y & u_y v_y & u_z v_y \\ u_x v_z & u_y v_z & u_z v_z \end{pmatrix}. \qquad (C.1)$$

If $\mathbf{u}$ is the nabla operator, we then have the following special dyadic products:

$$\nabla\mathbf{v} = \begin{pmatrix} \dfrac{\partial v_x}{\partial x} & \dfrac{\partial v_y}{\partial x} & \dfrac{\partial v_z}{\partial x} \\[2mm] \dfrac{\partial v_x}{\partial y} & \dfrac{\partial v_y}{\partial y} & \dfrac{\partial v_z}{\partial y} \\[2mm] \dfrac{\partial v_x}{\partial z} & \dfrac{\partial v_y}{\partial z} & \dfrac{\partial v_z}{\partial z} \end{pmatrix}, \qquad (\nabla\mathbf{v})^T = \begin{pmatrix} \dfrac{\partial v_x}{\partial x} & \dfrac{\partial v_x}{\partial y} & \dfrac{\partial v_x}{\partial z} \\[2mm] \dfrac{\partial v_y}{\partial x} & \dfrac{\partial v_y}{\partial y} & \dfrac{\partial v_y}{\partial z} \\[2mm] \dfrac{\partial v_z}{\partial x} & \dfrac{\partial v_z}{\partial y} & \dfrac{\partial v_z}{\partial z} \end{pmatrix}. \qquad (C.2)$$

Such products are very useful in summarizing, for example, the strain-rate tensor that appears in Section 5.7:

$$\dot{\gamma} = \begin{pmatrix} 2\dfrac{\partial v_x}{\partial x} & \left(\dfrac{\partial v_x}{\partial y} + \dfrac{\partial v_y}{\partial x}\right) & \left(\dfrac{\partial v_x}{\partial z} + \dfrac{\partial v_z}{\partial x}\right) \\[3mm] \left(\dfrac{\partial v_y}{\partial x} + \dfrac{\partial v_x}{\partial y}\right) & 2\dfrac{\partial v_y}{\partial y} & \left(\dfrac{\partial v_y}{\partial z} + \dfrac{\partial v_z}{\partial y}\right) \\[3mm] \left(\dfrac{\partial v_z}{\partial x} + \dfrac{\partial v_x}{\partial z}\right) & \left(\dfrac{\partial v_z}{\partial y} + \dfrac{\partial v_y}{\partial z}\right) & 2\dfrac{\partial v_z}{\partial z} \end{pmatrix} = \nabla\mathbf{v} + (\nabla\mathbf{v})^T. \quad (C.3)$$

---

[1] The reader who needs to learn more about vector and tensor operations, including those in cylindrical and spherical coordinate systems, is referred to Appendix A of the excellent text: R.B. Bird, W.E. Stewart, and E.N. Lightfoot, *Transport Phenomena*, 2nd ed., Wiley & Sons, New York (2002).

The "Divergence" of a Tensor

Consider first the tensor:

$$
\boldsymbol{\tau} = \begin{pmatrix} \tau_{xx} & \tau_{xy} & \tau_{xz} \\ \tau_{yx} & \tau_{yy} & \tau_{yz} \\ \tau_{zx} & \tau_{zy} & \tau_{zz} \end{pmatrix}.
\tag{C.4}
$$

The dot product of the nabla operator with a *symmetric* tensor, $\nabla \cdot \boldsymbol{\tau}$ (in which $\tau_{yx} = \tau_{xy}$, etc.), is a vector with the following components in each of the three coordinate directions:

$$
(\nabla \cdot \boldsymbol{\tau})_x = \frac{\partial \tau_{xx}}{\partial x} + \frac{\partial \tau_{xy}}{\partial y} + \frac{\partial \tau_{xz}}{\partial z},
$$

$$
(\nabla \cdot \boldsymbol{\tau})_y = \frac{\partial \tau_{yx}}{\partial x} + \frac{\partial \tau_{yy}}{\partial y} + \frac{\partial \tau_{yz}}{\partial z},
\tag{C.5}
$$

$$
(\nabla \cdot \boldsymbol{\tau})_z = \frac{\partial \tau_{zx}}{\partial x} + \frac{\partial \tau_{zy}}{\partial y} + \frac{\partial \tau_{zz}}{\partial z}.
$$

To obtain the first of these relations, it is as though we had considered the first row of the tensor to be a vector, and we had then taken its divergence, etc. The form $\nabla \cdot \boldsymbol{\tau}$ enables us to summarize the three momentum balances in compact vector form for the case of variable viscosity, as in Example 5.9.

The "Laplacian" of a Vector

The product of the operator ∇^2 with a vector $\mathbf{v}$, as in $\nabla^2 \mathbf{v}$, is a vector with the following components:

$$
(\nabla^2 \mathbf{v})_x = \frac{\partial^2 v_x}{\partial x^2} + \frac{\partial^2 v_x}{\partial y^2} + \frac{\partial^2 v_x}{\partial z^2},
$$

$$
(\nabla^2 \mathbf{v})_y = \frac{\partial^2 v_y}{\partial x^2} + \frac{\partial^2 v_y}{\partial y^2} + \frac{\partial^2 v_y}{\partial z^2},
\tag{C.6}
$$

$$
(\nabla^2 \mathbf{v})_z = \frac{\partial^2 v_z}{\partial x^2} + \frac{\partial^2 v_z}{\partial y^2} + \frac{\partial^2 v_z}{\partial z^2}.
$$

That is, we obtain three Laplacians of the individual components v_x, v_y, and v_z, one for each coordinate direction. As seen in Section 5.7, this enables a compact representation of the constant-viscosity terms in the Navier-Stokes equations.

INDEX

THE AUTHORS

We hope that you have enjoyed reading and learning from this book. If you have any questions or comments, you are most welcome to contact us at the addresses that appear in our website, `http://www.engin.umich.edu/~fmche`. Here is some information about ourselves.

James O. Wilkes

James O. Wilkes was born in Southampton, England, and lived in Shropshire during the Second World War. As a chemical engineering student at Emmanuel College, he obtained his bachelor's degree from the University of Cambridge in 1955. The English-Speaking Union awarded him a King George VI Memorial Fellowship to the University of Michigan, from which he received a master's degree in 1956 and a PhD in 1963. Also in chemical engineering, he was a faculty member at the University of Cambridge from 1956 to 1960, and at the University of Michigan from 1960 to 2000. At Michigan, he was department chairman from 1971 to 1977, and Assistant Dean for Admissions in the College of Engineering from 1989 to 1994. He was named an Arthur F. Thurnau Professor from 1989 to 1992. His research interests are in numerical methods, polymer processing, and computational fluid mechanics.

Professor Wilkes received his organ performance diploma, Associate of the Trinity College of Music (London), in 1951, and his Service-Playing Certificate from the American Guild of Organists in 1981. In addition to music, his hobbies include hiking in North Wales, New Zealand, and the American West; tennis; gardening; reading; and writing. He is author of *Pipe Organs of Ann Arbor* (1995), *A Century of Chemical Engineering at the University of Michigan* (2002), and coauthor of *Applied Numerical Methods* (Wiley, 1969) and *Digital Computing and Numerical Methods* (Wiley, 1973). He is currently editing his grandfather's manuscript, *Place Names of Hampshire and the Isle of Wight*.

Stacy G. Birmingham

Stacy G. Birmingham was born in Long Beach, CA. She received her BS, MS, and PhD degrees from Carnegie Mellon University. She is currently the Dean of the Albert A. Hopeman, Jr., School of Science and Engineering and a Professor of Mechanical Engineering at Grove City College in Grove City, Pennsylvania. Prior to joining Grove City College in 2003, she was on the faculty of the chemical engineering department at the University of Michigan from 1998 to 2003, where she held joint appointments in both chemical engineering and the macromolecular science and engineering program. At Michigan, she received a National Science Foundation Presidential Young Investigator Award, the '38E Award for teaching, research, and service, and was Dow Corning Assistant Professor from 1989 to 1992.

Professor Birmingham's research interests encompass the role of complex fluids in materials processing and biomimetics, including the design of ceramics and drug-delivery systems. She has published numerous articles and made more than 120 invited and contributed technical presentations in this area. Her teaching interests range from fluid mechanics and heat transfer to engineering graphics and design. She greatly enjoys spending time with her husband and three children, and her hobbies include skiing, running, and gardening.

Chi-Yang Cheng

Chi-Yang Cheng was born in Yang Ming Shan, Taiwan, the Republic of China. He received his bachelor's degree from National Taiwan University, and MS and PhD degrees in Mechanical Engineering from the University of Rhode Island. He was a faculty member at Grove City College in western Pennsylvania from 1991 to 2002. Currently he is with Fluent Incorporated in Lebanon, New Hampshire. His research interests include fluid dynamics, computational methods, and object oriented numerics/programming. He is a devoted user of Donald Knuth's TEX and also Leslie Lamport's LATEX implementation and its extensions for document preparation. In 1999, he used the skills to typeset a book of history of the Second Sino-Japanese War (1931–1945) written by his brother. His hobbies include hiking, gardening, reading, and enjoying classical music.

Brian J. Kirby

Brian J. Kirby joined the Sibley School of Mechanical and Aerospace Engineering at Cornell University in August 2004, where he is an assistant professor and currently directs the Micro/Nanofluidics Laboratory. Previously, he was a Senior Member of the Technical Staff in the Microfluidics Department at Sandia National Laboratories in Livermore, California, where he worked from 2001 to 2004 on microfluidic systems, applied primarily to counterbioterrorism. From 1996 to 2001 he worked as a graduate student in the High Temperature Gasdynamics Laboratory at Stanford University, developing laser spectroscopy techniques for imaging gases in flames for applications in combustion and aerothermopropulsion. From 1994 to 1996 he worked as a graduate student in the Variable Gravity Research Laboratory at the University of Michigan, studying multiphase heat-transfer processes; at Hewlett-Packard Laboratories in Palo Alto, CA, studying fluid mechanics processes in hard-drive stacks; and in the Gas Dynamics Research laboratories in the Aerospace Engineering Department at the University of Michigan, studying soot formation processes in low-pressure diffusion flames.

Prof. Kirby has taught in various capacities in the Mechanical and Aerospace Engineering Department at Cornell, the Mechanical Engineering Department at Stanford, and the Chemical Engineering Department at the University of Michigan. His degrees are: PhD 2001, Stanford University, Mechanical Engineering; MSE 1996, University of Michigan, Mechanical Engineering; and BSE 1994, University of Michigan, Aerospace Engineering.

informIT

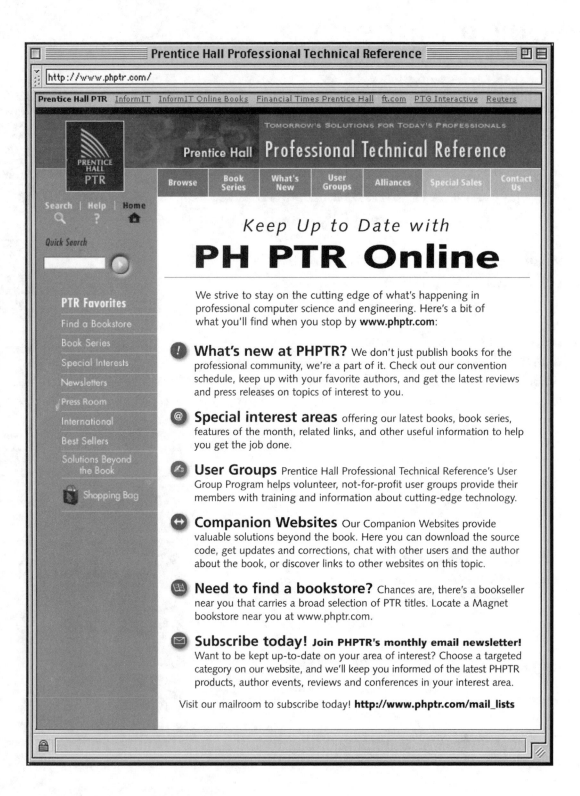